Shooter's Bible

ABOUT OUR COVER

The side-by-side double rifles featured on this year's front cover represent the best of the new and the old produced by Heym of Germany. The rifle at left—the Model 88B Safari—is currently imported exclusively by JägerSport, Ltd. Chambered in .375 H & H, it has elegant scroll engraving on the receiver and an air-dried walnut stock that is hand-checkered and hand-rubbed with an oil finish. The Heym rifle at right was built in the mid-1930s and features sideplates designed by Manton. Chambered for .405 Winchester (10mm), its receiver is color-casehardened with attractive full engraving. This rifle and its gold-plated accessories furnished by Lou Alessandri & Son, Rehoboth, Mass.

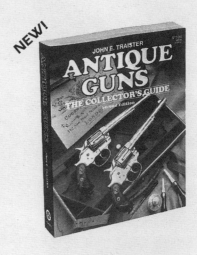
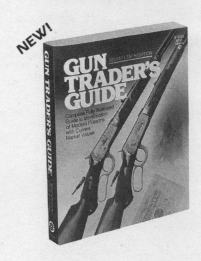

NO. 86
1995 EDITION

EDITOR:
William S. Jarrett

PRODUCTION & DESIGN:
Charlene Cruson Step

FIREARMS CONSULTANTS:
Bill Meade
Vincent A. Pestilli
Paul Rochelle
Robert A. Scanlon

COVER PHOTOGRAPHER:
Ray Wells

PUBLISHER:
Robert E. Weise

PRESIDENT
Brian T. Herrick

Shooter's Bible®

STOEGER PUBLISHING COMPANY

Every effort has been made to record specifications and descriptions of guns, ammunition and accessories accurately, but the Publisher can take no responsibility for errors or omissions. The prices shown for guns, ammunition and accessories are manufacturers' suggested retail prices (unless otherwise noted) and are furnished for information only. These were in effect at press time and are subject to change without notice. Purchasers of this book have complete freedom of choice in pricing for resale.

Published by Stoeger Publishing Company
55 Ruta Court
South Hackensack, New Jersey 07606

Library of Congress Catalog Card No.: 63-6200
International Standard Book No.: 0-88317-177-5

Manufactured in the United States of America

In the United States:
Distributed to the book trade and to the sporting goods trade by Stoeger Industries, 55 Ruta Court
South Hackensack, New Jersey 07606

In Canada:
Distributed to the book trade and to the sporting goods trade by Stoeger Canada Ltd.
1801 Wentworth Street, Unit 16
Whitby, Ontario, L1N 8R6, Canada

Contents

FOREWORD

As noted in the introduction to "50 Years Ago in Shooter's Bible" on page 8, the year 1945 was indeed a momentous one in the history of our nation. Now, half a century later, we are pleased to release this—our 86th—edition in a long line of annual publications whose main purpose is to keep you abreast of new products and updated information in the shooting sports.

The articles that follow "50 Years Ago" also deserve your attention, beginning with Wayne van Zwoll's excellent piece on hunting mule deer in the high country. It's followed by half a dozen or more other articles, including Steve Irwin's look at duck stamps and Jim Casada's absorbing biography of Fred Selous, the great African hunter. Ralph Quinn continues to grace these pages, this time with a full description of the Sportmen's Team Challenge (STC) competition, and veteran writer Don Lewis explains the do's and don'ts for beginning shotgunners. In addition, you'll find articles of lasting interest on the quarter-inch caliber (by Wilf Pyle), the potential hazards of low-power ammo (by C. Rodney James), and Gary Brown's historical coverage of the .300 and .375 Holland & Holland cartridges via Winchester's Model 70. We sincerely hope these articles prove enlightening and entertaining for you.

The article section concludes with "Manufacturers' Showcase," featuring a wide variety of products for shooting sports enthusiasts of all types. The "Showcase" is followed by our annual coverage of gun and gun-related specifications, starting with handguns. Some of the new models of special interest are Bernardelli's PO 18 pistols, Emerging Technologies' Laseraims pistols, Texas Arms' Defender Derringer, Daewoo's High Capacity models, and Brown Precision's High Country pistol. Especially noteworthy is the return of Stoeger's American Eagle Luger®.

The rifle section marks the return also of Kimber and the addition of several other new manufacturers to these pages, including Brown Precision, Cooper Arms and Pedersoli's replica rifles. Among the shotgun producers, we welcome Lanber, Laurona, Maverick of Mossberg and A. H. Fox. Gone is Precision Sports' "600" Series. The Black Powder section adds White Systems, and in the Sights & Scopes department we've added Emerging Technologies and Swift. These firearms sections are followed by individual coverage of ammunition, ballistics, powder, bullets and reloading equipment of all kinds.

Readers are also encouraged to check our Reference Section in the back of the book. There you'll find a listing of the newest books on firearms and related subjects, a summary of Discontinued Models, plus a full lineup of the manufacturers and suppliers (addresses, phone and fax numbers included) that are covered in the Specifications and Manufacturers Showcase sections.

Finally, as in the past, we've included two important indexes: Caliberfinder and Gunfinder. These handy references will help you find every firearm featured in this 86th edition of SHOOTER'S BIBLE—by cartridge size (Caliberfinder) and by type of arm by page (Gunfinder).

We are always interested in improving the content and appearance of "America's favorite gun book," and call your attention to the new graphics approach we've used throughout these pages this year. Should you have any suggestions, recommendations or comments of any kind to make concerning this edition, please contact us—we're always happy to hear from you. Meanwhile, here's to another good season of shooting and outdoor sports!

William S. Jarrett
Editor

Articles

50 YEARS AGO IN SHOOTERS BIBLE

To most Americans over 60, the year 1945 carries special meaning. It marked the end of World War II and promised a return to normalcy and a new sense of prosperity. After the celebrations had ended and ration books were discarded, America's veterans went back to school, found civilian jobs and generally picked up where they left off years before.

SHOOTER'S BIBLE, which had never stopped publication during the war years, was there too, and we've reprinted a dozen pages from that edition of 50 years ago. Included in that issue was this foreword by A. F. Stoeger, the founder and then president of the company that still bears his name:

> With the termination of the war, we all look forward with great anticipation to a great revival in hunting, fishing, camping and other outdoor sports. Many millions who had never before handled firearms have come to know and appreciate their use and care, and have learned the thrill of marksmanship.

You'll also note we've added brief headlines for each month covering the year's major news and sports events, some of which may surprise or enlighten you. As for the firearms available in 1945, we note Mossberg advertised that it owned the industry's only line of "streamlined rifles and bolt action shotguns." Smith & Wesson's best-seller was the .357 Magnum for $60.00. And Stoeger Industries, which re-introduces its American Eagle Luger® in this edition, devoted two pages to that famous pistol ($75.00 for a slightly used Grade I model).

We hope you enjoy this look at what went on during that pivotal year half a century ago.

ITHACA SHOTGUNS

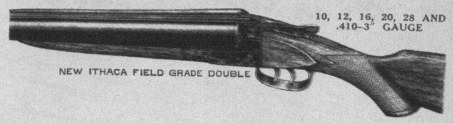

10, 12, 16, 20, 28 AND .410—3" GAUGE

NEW ITHACA FIELD GRADE DOUBLE

FIELD GRADE

The Ithaca Field Gun. For skeet, trap or game shooting, it's the least expensive Ithaca we can build. It will last a lifetime.
BARRELS—Smokeless Powder steel, proof tested with a double powder load.
STOCK AND FOREND—Black walnut, hand checkered, solid where stock joins the frame to prevent splitting, full pistol grip.
LOCKS—Same mechanical construction as in higher grade Ithacas and this lightning lock will improve any man's shooting.
PRICE—Field grade, $59.75. Ejector, if wanted, costs $17.50 extra. Ventilated rib, if wanted, $20.00. Beaver tail forearm, if wanted, $15.00. Ithaca Selective Single

Trigger, if wanted, $24.55. Ithaca soft rubber recoil pad, if wanted, $3.00. Ivory Sights, if wanted, $1.51. Add $16.32 for the Magnum 10 gauge.

ITHACA SKEET MODEL FIELD GRADE

ITHACA SKEET GUNS

Available in all grades, Field, No. 2, No. 4, No. 5, No. 7, and $1,000.00 Grade. Special specifications best suited to skeet and upland game shooting as well: Beavertail forend, Selective Single Trigger, Automatic Ejectors, and Ivory Sights. Skeet Boring (excellent for field shooting, being an open type of boring) 26" barrels are standard. 14"x2½" are standard stock specifications.

Reasonable variations at no extra charge. Furnished in .410, 28, 20, 16, and 12 gauge.
Prices of Skeet guns with extras as described: Field Grade—$118.31, No. 2—$133.06, No. 4—$189.68, No. 5—$257.61, No. 7—$438.17, $1000.00 Grade—$1056.06.
A ventilated rib is sometimes desired on a Skeet gun. If so, add to above prices: $20.00 for Field and No. 2 Grades and $30.00 for No. 4, 5, 7, or $1000.00 Grades.

The Ithaca No. 2, thoroughly well-made and reliable for skeet, game or trap, at a moderate price.
BARRELS—Fluid steel of extra grade.
STOCK AND FOREND—Black walnut with nice color and finish, neatly hand checkered full pistol grip.
LOCKS—Lightning fast locks that will stand use and misuse. You do not have to strain your wrist to cock an Ithaca.
ENGRAVING—Frame, top lever, foreend iron, trigger plate and guard engraved by hand. Bird scene and leaf design on each side of the frame, leaf design on trigger plate and guard.
PRICE—No. 2, $74.50. Ejector, if wanted, $17.50 extra. Ventilated rib, if wanted, $20.00.

ITHACA NO. 2
10, 12, 16, 20, 28 AND .410—3" GAUGE

Beaver tail forearm, if wanted, $15.00. Selective single trigger, if wanted, $24.55. Soft rubber recoil pad, if wanted, $3.00. Ivory sights, if wanted, $1.51. Add $16.32 for the Magnum 10 gauge.

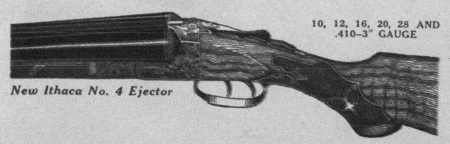

New Ithaca No. 4 Ejector

10, 12, 16, 20, 28 AND .410—3" GAUGE

ITHACA NO. 4 EJECTOR

Best Fluid steel. Stock nicely figured, carefully fitted, handsomely checkered walnut made with full pistol grip unless otherwise ordered. Locks carefully adjusted to get a quick and smooth working lock. Engraving: Frame, top lever, guard and trigger plate hand engraved in an artistic manner. Duck scene on one side of frame; pheasant scene on other side; leaf and flower design on the guard and made in .410, 20, 16, 12 and 10 gauge magnum.

No. 4 Ejector $146.12

QUALITY ARMS BY HARRINGTON & RICHARDSON

The well known firm of Harrington & Richardson has recently brought out and will continue to produce new Rifles and Shotguns, some of which can now be announced and catalogued. These new arms are shown in this edition of the Shooter's Bible and will soon be available from Stoeger Arms Corporation.

Harrington & Richardson, famous for skilled craftsmanship for more than seventy years, converted their entire facilities to the production of war weapons, including the H & R Reising Submachine Gun, the H & R Semiautomatic Rifle, H & R Line Throwing Kits,

H & R Flare Pistols, and the H & R Defender .38—a five shot revolver embodying the best features of their previous line of twenty-six different revolver models.

The broad war-time experience of this company has resulted in the development of new gun types and refinements in older types for postwar sporting arms which will be reflected in the new Harrington & Richardson line.

Being produced at the present for civilian use is the H & R .38 Defender Revolver.

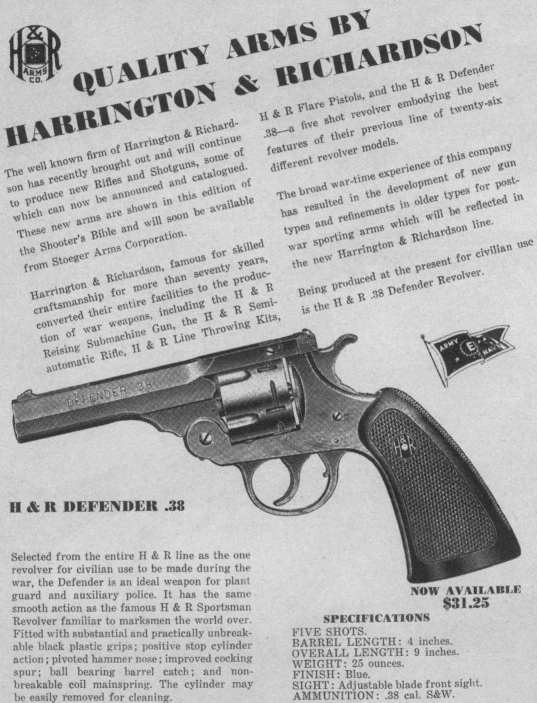

H & R DEFENDER .38

NOW AVAILABLE
$31.25

Selected from the entire H & R line as the one revolver for civilian use to be made during the war, the Defender is an ideal weapon for plant guard and auxiliary police. It has the same smooth action as the famous H & R Sportsman Revolver familiar to marksmen the world over. Fitted with substantial and practically unbreakable black plastic grips; positive stop cylinder action; pivoted hammer nose; improved cocking spur; ball bearing barrel catch; and nonbreakable coil mainspring. The cylinder may be easily removed for cleaning.

SPECIFICATIONS

FIVE SHOTS.
BARREL LENGTH: 4 inches.
OVERALL LENGTH: 9 inches.
WEIGHT: 25 ounces.
FINISH: Blue.
SIGHT: Adjustable blade front sight.
AMMUNITION: .38 cal. S&W.

THE GUN FOR CIVILIAN DEFENSE

PARKER SHOTGUNS

PARKER V. H. E.

Hand-fitted, hand-finished, and made to customer's exact specifications throughout. Stock and fore-end of selected American walnut, hand checkered. German silver name shield inlaid in stock. Stock dimensions, unless otherwise specified, 14 inches long, 2½ inches drop at heel, 1⅝ inches drop at comb. Stocks made to order with lengths from 13½ to 14½ inches, and with drops from 2 to 3¼ inches without extra charge. Full pistol grip with cap, option of straight or half pistol grip. Hard rubber butt plate. Line engraving. Ivory sights if desired. Automatic ejectors. Any boring of barrels.

"V. H. E." Grade with double triggers $140.25
"V. H. E." Grade with selective single trigger 169.11

Raised ventilated rib, extra	27.75
Soft rubber recoil pad, extra	5.55
Skeleton steel butt plate, extra	8.88
Oil finishing stock and fore-end, extra	8.88
Stock outside of prescribed limits	15.14
Extra set of interchangeable barrels	72.65

12, 16, 20, 28 and .410 GAUGE

Beaver tail forearm, Extra $14.98

10, 12, 16, 20, 28, and .410 GAUGE

Beaver tail forearm, Extra $14.85

PARKER G. H. E.

Figured American walnut stock and fore-end, selected for natural beauty. Nicely hand checkered. German silver name shield inlaid in stock. Stock dimensions, unless otherwise specified, 14 inches long, 2½ inches drop at heel, 1⅝ inches drop at comb. Stocks made to order with lengths from 13½ to 14½ inches and with drops from 2 to 3¼ inches without extra charge. Option of straight or half pistol grip. Full pistol grip with cap. Hard rubber butt plate. Game birds and scroll engraving. Ivory sights if desired. Automatic ejectors. Any boring of barrels.

"G. H. E." Grade with double triggers $160.43
"G. H. E." Grade with selective single trigger 189.29

Beavertail fore-end, extra	14.98
Raised ventilated rib, extra	27.75
Soft rubber recoil pad, extra	5.55
Skeleton steel butt plate, extra	8.88
Oil finishing stock and fore-end, extra	8.88
Stock outside of prescribed limits	15.14
Extra set of interchangeable barrels	83.75

PARKER D. H. E.

Stock and fore-end of fancy walnut, finely hand checkered. Sterling silver name plate inlaid in stock. Stock custom-built to any specifications desired without extra charge, including cheek piece, Monte Carlo or cast off, and any style of grip. Rubber recoil pad or skeleton steel butt plate. Engraving is game scenes and scroll. Nickel plated triggers. Ivory sights if desired. Automatic ejectors. Any boring of barrels.

"D. H. E." Grade with double triggers $196.76
"D. H. E." Grade with selective single trigger 229.04

Raised ventilated rib, extra	30.27
Extra set of interchangeable barrels	110.99

10, 12, 16, 20, 28 and .410 GAUGE

Beaver tail forearm, Extra $19.17

10, 12, 16, 20, 28 and .410 GAUGE

Beaver tail forearm, Extra $25.23

PARKER C. H. E.

Selected high grade walnut stock and fore-end, handsomely checkered. Sterling silver name plate inlaid in stock. Stock with any specifications desired including cheek piece, Monte Carlo or cast off, and any style of grip. Rubber recoil pad or skeleton steel butt plate. Engraving is game scenes and scroll. Nickel plated triggers. Ivory sights if desired. Automatic ejectors. Made in 10, 12, 16, 20, 28, and .410 gauges. Any boring of barrels.

"C. H. E." Grade with double triggers $292.61
"C. H. E." Grade with selective single trigger 324.90

Raised ventilated rib, extra	35.22
Extra set of interchangeable barrels	149.33

PARKER B. H. E.

Stock and fore-end of high grade walnut with fine grain and beautiful figure. Handsomely checkered. Mounted with solid gold name plate in stock or in pistol grip cap. Custom-built stock to any measurement desired, including cheek piece, Monte Carlo or cast off, and any style of grip. Rubber recoil pad or engraved skeleton steel butt plate. Scroll engraving and life-like hunting scenes. Nickel plated triggers. Ivory sights if desired. Automatic ejectors. Made in 10, 12, 16, 20, 28, and .410 gauges. Any boring of barrels.

"B. H. E." Grade with double triggers $393.51
"B. H. E." Grade with selective trigger 425.80

Raised ventilated rib, extra	35.32
Extra set of interchangeable barrels	176.58

10, 12, 16, 20, 28, and .410 GAUGE

Beaver tail forearm, Extra $29.26

STOEGER SELECTED USED LUGERS

Special Luger Combination Tool for Loading and Stripping
Sold Out

GRADE 1

Only such pistols are included in this grade as have met our very exacting requirements of perfection. Pistols are to all intents and purposes in original factory condition thruout, but may sometimes show slight holster wear. Usually available only in cal. 7.65 mm with 3⅝" barrel $75.00

GRADE 2

Lugers in this grade are in perfect mechanical condition with very good barrel, and show no abuse. A gun you may depend upon.
Usually available in both calibers with 3⅝" barrel, or in 9 mm only with 7" or 8" barrel $60.00

GRADE 3

A good serviceable weapon for field or defense use. Blueing may show wear and barrel erosion, but still a good and powerful shooter.
Usually available in both calibers with 3⅝" barrel, or in 9 mm only with 7" or 8" barrel $45.00

STOEGER SELECTED USED LUGERS

Because of the impossibility of obtaining new Lugers, we have undertaken to supply selected used genuine Lugers. These pistols are all carefully checked, thus giving the purchaser assurance of a satisfactory weapon.

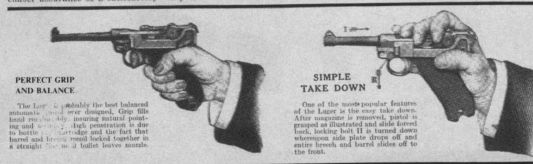

PERFECT GRIP AND BALANCE

The Luger is probably the best balanced automatic pistol ever designed. Grip fills hand comfortably, insuring natural pointing and accuracy. High penetration is due to bottle neck cartridge and the fact that barrel and breech recoil locked together in a straight line until bullet leaves muzzle.

SIMPLE TAKE DOWN

One of the most popular features of the Luger is the easy take down. After magazine is removed, pistol is grasped as illustrated and slide forced back, locking bolt II is turned down whereupon side plate drops off and entire breech and barrel slides off to the front.

GENUINE LUGER BARRELS

Cal. 7.65 mm, 3⅝" long, new $15.00
Cal. 7.65 mm, 3⅝" long, good 10.00
Cal. 9 mm, 7" long, outside shopworn, inside excellent (with Tangent rear sight) 17.50
Cal. 9 mm, 7" long, outside shopworn, inside good (with Tangent rear sight) 10.00

GENUINE D.W.M. LUGER BARREL

In addition we have an odd lot of both 7.65 MM and 9 MM calibers in various lengths. Most are slightly used, some are new; special prices on request.
Note: We have specially imported genuine Luger factory barrel removing and fitting machines, and as this operation is sometimes very difficult otherwise, we recommend that pistol be sent to us for fitting, which we undertake at a charge of $3.50.

With the addition of the 32 shot Magazine Drum shown at right, the Luger instantly becomes a gun which can be fired as rapidly as the user is able to pull the trigger. This drum can be attached as easily as the regular magazine and is so placed that it balances the pistol perfectly. Luger drums were used during the late war but as their manufacture was subsequently forbidden we can offer them only used, but in perfect working order. An important feature is that the number of cartridges in the magazine is always shown by means of an indicator.

LUGER 32-SHOT DRUM MAGAZINE

Caliber 9 MM Only

Loading is facilitated by a clever design thru which the tension of the powerful feeder spring is relieved during loading. To accomplish this the folding arm is extended and given a turn to the right until bottom at lever joint engages in recess shown in illustration. After loading, a pressure on the button releases catch and feeder spring.
Because of the bottle-neck shape of the 7.65 MM cartridge this magazine will function properly only with the 9 MM cartridge.

Luger 32 shot magazine drum.....Sold Out

ZEPHYR SHOTGUNS

The Zephyr shotgun is the result of our many years of experience in development of a high quality, hand finished shotgun at a price within the reach of the average shooter. This gun was first introduced by us a number of years before the war and was an immediate success, combining as it does all the features of the very highest grade hand built English shotguns together with perfect proportion and balance at a price of less than half that of competitive guns of comparable quality.

During the war in Europe, it was of course impossible to supply our Zephyr shotguns, and while at the moment of writing they are still not available, we hope and expect that they will soon be so, and welcome inquiries regarding same. So far as prices are concerned, those shown here are still the pre-war prices, which will undoubtedly have to be increased somewhat, but will be dependent upon new foreign costs and whatever new import tariffs will be in effect.

The Zephyr shotgun is produced to meet the exacting demands of experienced shooters who demand a light weight, well balanced shotgun for skeet or field shooting. The barrels are carefully bored and consistently even patterns are the result.

In construction, the well known Anson & Deeley box lock system has been used in connection with the excellent Purdy style top lock which is not only neat, clean and effective, but also makes insertion and extraction of shells easier. The forend which is of English style is released by means of the Purdy style push button. The ejectors are of the Southgate type. The rib is flat and matted to prevent glare. The safety is automatic. The triggers are checkered to prevent slipping of the finger even when gloves are worn, and is carefully chromium plated to prevent rust.

In addition to the checkering of the pistol grip and forearm, the checks of the stock are carefully checkered in the form of a shield extending out into a diamond shape inlay of genuine black Buffalo horn. The action itself, which is made with integral side clips to prevent any possibility of looseness is supplied with artistic light scroll engraving. The trigger guard is also neatly engraved with scroll work and a flying partridge and extends, in guns with pistol grip, all the way to the horn pistol grip cap with which it cuts off flush. The stock is of selected French Walnut and has an oval silver name plate imbedded. Instead of a butt plate, the butt end is tastefully checkered.

All Zephyr guns have Automatic Ejectors.

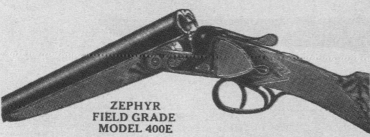

ZEPHYR FIELD GRADE MODEL 400E

The Field Grade is available in 12, 16, 20, 28 and .410 ga. with barrel length of from 25" to 30" in various weights, chokes and stock dimensions. However, this model is regularly made in very light weights; the .410 ga. weighs about 4½ lbs., the 28 ga. about 5 lbs., the 20 ga. 5½ lbs., the 16 ga., 5¾ lbs., and the 12 ga. 6¼ lbs. This gun may also be had on special order with almost any specifications at no extra charge; delivery time about four months.

Zephyr, Field Grade, 12, 16 or 20 gauge..$115.00
Zephyr, Field Grade, 28 or .410 gauge... 125.00
Extra for Selective Single Trigger........ 35.00

The Zephyr Skeet Grade is about the same as the Field Grade except that it is supplied with Beaver Tail forend and on special order with selective or non-selective single trigger. The Zephyr Skeet gun is available in all gauges from .410 to 12 ga. with barrel length varying from 25" to 28". The Beaver Tail forend, itself, is of excellent design and has enjoyed popularity on the Skeet fields. It is specially bored for skeet shooting.

Zephyr, Skeet Grade, 12, 16 or 20 gauge..$137.50

Zephyr, Skeet Grade, 28 or .410 gauge;.... 147.50

Extra for Selective Single Trigger........ 35.00

Extra for Non-Selective Single Trigger.... 27.50

ZEPHYR SKEET GRADE MODEL 401E

ZEPHYR DE LUXE MODEL 402E

The Zephyr De Luxe Grade was built to satisfy the demand of those who wish a gun of even finer appearance and still smoother workmanship. The De Luxe Grade is basically the same as the Field Grade except that the entire action is specially carefully hand honed for the greatest possible smoothness. The engraving through-out is more elaborate. The checkering is of particularly pleasing effect and the walnut of exceptional quality and beauty.

Zephyr, De Luxe Grade, 12, 16 or 20 gauge $150.00
Zephyr, De Luxe Grade, 28 or .410 gauge 160.00
Extra for Selective Single Trigger.... 35.00

MOSSBERG AUTO AND BOLT ACTION .22 CAL. RIFLES

MODEL 51M AUTOMATIC RIFLE

MANNLICHER TYPE

**SHOOTS
.22 LONG RIFLE
STANDARD OR HIGH SPEED**

51M—A high grade automatic rifle handling without any change .22 regular or high speed Long Rifle cartridges, either lubricated or dry, embodying new principles perfected after five years of research. Built up cheek-piece, non-breakable molded trigger guard and a formed steel buttplate, quick detachable swivels, red and green "traffic light" safety indicators, chrome trigger, No. 1A hooded ramp front sight with four permanently attached inserts. No. 2A open sporting rear sight, No. 4 microclick peep sight with No. 4D single aperture disc.

SPECIFICATIONS
Barrel: 20" round tapered with crown muzzle. Stock: Genuine American walnut, oil finished, two piece custom type. Magazine: Holds 15 Long Rifle cartridges. Length: 40". Weight: 7 lbs.

Price.................................. **$23.55**

MODEL 46M TUBULAR REPEATER

MANNLICHER TYPE

**SHOOTS
.22 SHORT, LONG, OR LONG RIFLE
INTERCHANGEABLY**

46M—Genuine walnut, oil finished stock with custom design cheekpiece and grooved comb also new "Safety Cover Plate," safety indicators, chrome-plated trigger, streamlined molded trigger guard with finger grooves, molded buttplate, flush take down screw, detachable swivels, new "Mauser" type bolt handle, hooded ramp front sight with 4 permanently attached inserts, No. 2A rear sight, No. 4 Micro click peep sight, adjustable trigger pull.

SPECIFICATIONS
Barrel: 23" tapered—Crown muzzle chambered for .22 short, long and long rifle regular or high speed ammunition. Stock: Genuine American walnut, oil finished, 2-piece custom design. Magazine: Holds .22 short, 18 long or 15 long rifle cartridges. Length: 40". Weight: 7 lbs.
Price.................................. **$20.40**

MODEL 42M CLIP REPEATER

**SHOOTS
.22 SHORT, LONG, OR LONG RIFLE
INTERCHANGEABLY**

MANNLICHER TYPE
WITH TRAP BUTT PLATE AND SPARE MAGAZINE

42M—This rifle has all the features of the 46M and differs only that it is a clip magazine repeater with detachable clip of Double Duty type as illustrated and described on page 4. Magazine holds 7 shots and one in chamber. An added feature though is the "trap door" buttplate and cut out stock in which we insert an extra clip.

SPECIFICATIONS
Barrel: 23" tapered—Crown muzzle chambered for .22 short, long and long rifle, regular or high speed ammunition. Stock: Genuine American walnut, oil finished, 2-piece custom design. Length 40". Weight: 6¾ lbs.
Price.................................. **$17.70**

MODEL 46 B

Model 46 B **$17.90**

This rifle is the most popular of the Mossberg line. A tubular magazine repeater with 26-inch barrel, equipped with ramp front sight, open rear sight and No. 4 peep sight with selective disc. Trigger pull is adjustable from 2½ to 5 pounds. Drilled and topped for Mossberg side mounting scopes. The rifle has a solid finished Walnut stock with cheek piece with full pistol grip, molded butt plate and detachable swivels. Length overall 43¾ inches. Weight: 7 pounds.

**TARGET
TUBULAR LOADER**

A TRIO OF TARGET FAVORITES

SMITH & WESSON

K-22 MASTERPIECE

$40.00

SHORT ACTION! CLICK SIGHTS! HEAVY FRAME!

Truly a masterpiece of the gunmaker's art, the new K-22 is indeed a worthy successor to the long line of famous Smith & Wesson target revolvers.

Built on the heavy .38 Military and Police Target frame, the K-22 Masterpiece embodies the last word in precision manufacture and craftsmanship and brings target shooters for the first time these three outstanding features in a commercial gun:

Speed Lock! Shorter, faster, easier cocking action. Smooth as velvet. Allows grip to remain unchanged during fast fire.

Micrometer, Click Sights for both windage and elevation. One click moves point of impact 1 inch at 50 yards — ½ inch at 25 yards. Strong construction throughout. Will not shoot loose.

Built-In Anti-Backlash Trigger! No adjustment screws or gadgets. Trigger travel to rear positively stops when hammer starts to fall.

SPECIFICATIONS

AMMUNITION: Any R.F. .22 short, long or long rifle
NUMBER OF SHOTS: 6
BARREL: 6 inches
LENGTH: 11⅜ inches
WEIGHT: 35 ounces
SIGHTS: Front 1/10 or ⅛-inch Plain Patridge, Call Gold or Call Brilliant Bead. Rear: U or square notch, adjustable micrometer click sights
STOCKS: Checkered Circassian walnut with S & W Monograms. Choice of square or Magna type
FINISH: S & W Blue

K-32 TARGET

$38.00

Introduced to the target shooting world after 5 years of exhausting testing in the hands of some of North America's most exacting and discriminating shooters. Companion gun to the K-22 Masterpiece and the .38 Military and Police Target, the K-32 now makes it possible for the .32 S & W Long Cartridge to take its rightful place in popularity on the target range.

SPECIFICATIONS

CALIBER: .32 S & W Long
NUMBER OF SHOTS: 6
BARREL: 6 inches
LENGTH: 11⅜ inches
WEIGHT: 34 ounces
SIGHTS: Front 1/10 or ⅛-inch Plain Patridge, Call Gold or Call Brilliant Bead. Rear: U or Square notch, adjustable for windage and elevation

STOCKS: Checkered Circassian walnut with S & W Monograms. Choice of square or Magna type
FINISH: S & W Blue

AMMUNITION
.32 S & W Long
.32 S & W
.32 S & W Mid-Range
.32 Colt New Police

K-38 TARGET

$38.00

Target version of the great Military and Police Model. This famous revolver has broken literally hundreds of records and is regarded by many as the finest gun of its type ever offered. Beautifully balanced and equipped with grooved trigger and tangs to prevent slipping. S & W's special, case-hardened hammer and trigger assure superbly "crisp," unvarying trigger pull for years of shooting.

SPECIFICATIONS

CALIBER: .38 S & W Special
NUMBER OF SHOTS: 6
BARREL: 6 inches
LENGTH: 11⅜ inches
WEIGHT: 32¼ ounces
SIGHTS: Front, 1/10 or ⅛-inch Plain Patridge, Call Gold or Call Brilliant Bead. Rear: U or square notch, adjustable for windage and elevation

STOCKS: Checkered Circassian walnut with S & W Monograms. Choice of square or Magna type
FINISH: S & W Blue

AMMUNITION
.38 S & W Special
.38 S & W Special Super Police
.38 S & W Special Mid-Range
.38 Short Colt
.38 Colt Special

WINCHESTER
TRADE MARK

MODEL 64

CALIBERS
.25-35, .30-30
AND .32 SPECIAL

MODEL 64 STANDARD

This is an all-round, lightweight game rifle, perfectly adapted for deer, black bear, or any other medium sized game. It is dependable, beautifully made, light, and handles accurately and quickly. This is one of the most popular hunting rifles ever produced by Winchester. It was developed from the famous Model 94, which more than a million shooters have bought. It has improved NRA type stock—shot gun butt of selected walnut, with wide full comb and full pistol grip. The steel butt plate is checkered and has plenty of pitch. The frame is solid and the 24″ barrel is round tapered. The 2/3rds magazine holds five cartridges —one in the chamber, which makes a 6 shot repeater. Mechanism refinements give even smoother operation, and lighter trigger pull. Improved sights: Front, long matted ramp, with hunting bead; rear, new quick-elevating Rocky Mountain type with adjusting slide. Weight approximately 7 lbs.

With the world famous
lever action!

MODEL 64—LEVER ACTION REPEATING RIFLE

Solid frame only—two-thirds magazine

24″ or 20″ round barrel tapered. Shotgun butt, pistol grip stock. Bead front sight on ramp with sight cover. Winchester No. 22H open sporting rear, round top. Flat top furnished if specified. Add letter "F" to symbol. 5-shot magazine. Weight about 7 lbs.

For .25-35 Winchester, .30 (.30-30) Winchester, .32 Winchester Special, with 24″ barrel; For .25-35 Winchester, .30 (.30-30) Winchester, .32 Winchester Special, with 20″ barrel.

Model 64 rifle with 20″ barrel is also furnished with Lyman No. 56 receiver sight in place of Winchester rear sight as follows: For .25-35 Winchester, .30 (.30-30) Winchester, .32 Winchester Special, with 20″ barrel.

PRICE: With 20″ or 24″ barrel......... $62.40

With Lyman No. 56 Receiver Sight
(20″ barrel only) $71.85

CALIBERS
.30-30 AND .32 SPECIAL

MODEL 64 DEER RIFLE

This is a rifle which is designed especially for particular deer hunters. It has a wide, semi-beaver tail, handsomely checkered forearm; full comb stock—well pitched at the heel —with checkered steel butt plate and checkered full pistol grip; an added sling-strap with quick detachable swivels, which is helpful both in shooting and in resting on the stand or trailing.

This gun is of course not just for deer, but equally important in hunting any game which requires use of a powerful rifle. The same basic specification for the Standard Model 64 apply to this model. Stock and forearm of selected walnut, both finely checkered; grip with hard rubber cap. The combination of well-pitched broad butt with checkered steel butt plate, full style well rounded comb, and pistol grip, is valuable in snap shots and rapid fire. Solid frame. Weight (strap included) approximately 7¾ lbs.

The De Luxe Rifle
for
Deer Hunters!

MODEL 64—DEER RIFLE— LEVER ACTION REPEATING

Solid frame only—two-thirds magazine

24″ or 20″ round barrel, tapered. Pistol grip stock with rubber pistol grip cap, shotgun butt; semi-beavertail forearm, stock and forearm checkered; 1″ leather sling strap with quick detachable swivels. Sights as on Standard Model 64.

For .30 (.30-30) Winchester, .32 Winchester Special, with 24″ barrel.

For .30 (.30-30) Winchester, .32 Winchester Special, with 20″ barrel.

PRICE: $75.70

Rifle with 20″ barrel and Lyman 56 Receiver Sight

For .30 (.30-30) Winchester, .32 Winchester Special, with 20″ barrel.

PRICE: $85.20

COLT AUTOMATIC PISTOLS

ACE .22 AUTOMATIC PISTOL CAL. .22 LONG RIFLE

The ACE is designed especially for shooters of the Government Model and Super .38 Automatic Pistols—and has also been in demand by shooters for all around service. Built on the same frame as the Government Model and has the same safety features. Special super-precisioned barrel and hand finished target action. Exceptionally smooth operation and unusually accurate. Rear sight is of target design with adjustments for both elevation and windage. Allows economical target practice for military men, using .22 caliber ammunition in an arm of the same design as the regular military model. For Regular and High Speed Greased Cartridges.

Price $53.75
Extra Magazine. 3.00

SPECIFICATIONS

Ammunition: .22 Long Rifle Greased cartridges. Regular or High Speed.
Magazine Capacity: 10 cartridges.
Length of Barrel: 4¾ inches.
Length Over All: 8¼ inches.

Action: Hand finished.
Weight: 38 ounces.
Sights: Front sight fixed. Rear sight adjustable for both elevation and windage.

Trigger and Hammer Spur: Checked.
Arched Housing: Checked.
Stocks: Checked Walnut.
Finish: Blued.

Price $44.75
Extra Magazine. 1.80

SUPER .38 AUTOMATIC PISTOL CALIBER .38

For the big game hunter, and the lover of the outdoors, the Super .38 offers an arm of unsurpassed power and efficiency. It is built on the same frame as the Government Model and has all of the safety features found in this famous gun. It is especially popular because of the powerful Super .38 cartridges which it handles—having a muzzle velocity of approximately 1300 foot seconds. Will stop any animal on the American continent and is a favorite for use as an auxiliary arm for big game hunting trips. Magazine holds 9 cartridges.

SPECIFICATIONS

Sights: Fixed Patridge type.
Trigger and Hammer Spur: Checked.
Arched Housing: Checked.
Stocks: Checked Walnut.
Finish: Blued. Nickel Finish at extra cost of $5.00.

Ammunition: .38 Automatic cartridges.
Magazine Capacity: 9 cartridges.
Length of Barrel: 5 inches.
Length Over All: 8½ inches.
Weight: 39 ounces.

GOVERNMENT MODEL AUTOMATIC PISTOL CAL. .45

The Colt Arched Housing is illustrated above—used on all heavy frame Colt models. It provides a more secure and more comfortable grip.

The Colt Government Model is the most famous Automatic Pistol in the world. It has for years been the Official side arm of the United States Army, Navy and Marine Corps, as well as the military organizations of many foreign countries. Extremely powerful and absolutely dependable. Magazine holds seven cartridges and magazines can be replaced with great speed. Rugged and simple, it has withstood the most rigorous tests by the United States Government and proved itself unsurpassed in reliability and efficiency.

Price $44.75
Extra Magazine. 1.80

SPECIFICATIONS

Ammunition: .45 Automatic cartridges.
Magazine Capacity: 7 cartridges.
Length of Barrel: 5 inches.
Length Over All: 8½ inches.
Sights: Fixed Patridge type.
Weight: 39 ounces.

Trigger and Hammer Spur: Checked.
Arched Housing: Checked.
Stocks: Checked Walnut.
Finish: Blued. Nickel Finish $5.00 extra.

Price $52.75

COLT NATIONAL MATCH CALIBER .45

The regulation Government Model side arm perfected for match competition. Identical in size and operation, but with velvet-smooth hand-honed target action and a super-precisioned match barrel. Full grip, fine balance, three safety features. Now with adjustable rear sight and ramp type front sight, Colt's National Match brings you accuracy, power and smoothness never before equalled in a caliber .45 automatic pistol.

COLT SUPER MATCH CALIBER .38

With the exception that it is chambered for the high-powered .38 automatic cartridge, the Super Match Automatic Pistol is identical in every way with the National Match Model. It has the same velvet-smooth action, precision match barrel, same dependable safety features, same checked arched housing, same firm non-slipping grip. Accuracy, of course, is further increased by the new sights now available; ramp type front and adjustable rear. The Colt Super Match answers every demand in a caliber .38 automatic for competitive shooting—and possesses tremendous power for the big game hunter.

Prices: National Match and Super Match with adjustable sight $52.75
National Match and Super Match with fixed sights 47.25

SPECIFICATIONS

Ammunition: .38 Automatic cartridges.
Magazine Capacity: 9 cartridges.
Length of Barrel: 5 inches.
Length Over All: 8½ inches.
Weight: 39 ounces.

Action: Hand honed, velvet-smooth.
Stocks: Checked Walnut.
Sights: Adjustable rear, with Adjustments for elevation and windage. Ramp front sight.

Trigger and Hammer Spur: Checked.
Arched Housing: Checked.
Finish: Blued. Can be furnished in nickel finish at extra cost of $5.00.

DOUBLE ADJUSTABLE REAR SIGHT AND A RAMP TYPE FIXED FRONT SIGHT WITH SERRATED FACE

Here is a beautiful and efficient new rear sight for the Colt National Match and Super Match Automatic Pistols. It is designed especially for these two arms, constructed with precision, and adjustable for both windage and elevation. Take a close look at the illustration. Note the simplicity of this new sight, how extremely easy it is to adjust and to set accurately. It's just the finest hand gun sight ever made. And we mean just that. A host of shooters are going to like the new ramp type rugged sight out front, too. All of which means cleaner definition, higher and more consistent scoring.

ADJUSTABLE SIGHTS ON OLDER PISTOLS

You don't have to buy a new gun to enjoy the truly remarkable advantages of this new rear sight. For seven dollars and seventy-five cents, we will equip your Government Model and Super .38, as well as your National Match Model, or your Super Match, with this new sight combination. This includes the cost of the sight, recutting the sight slide cut, labor and targeting. It's a lot of value for $9.30.

STEVENS SHOTGUNS

STEVENS No. 22-410

OVER & UNDER RIFLE & SHOTGUN
.22 CAL. RIFLE BARREL OVER .410 BORE SHOTGUN BARREL
TAKE DOWN—SINGLE TRIGGER

This Over & Under Gun has a .22 Long Rifle caliber barrel over a .410 bore shotgun barrel. An easily operated button on the right side of the frame instantly selects barrel to be fired. Ideal for vermin and small game.

SPECIFICATIONS

Barrels: Length 24". Rifle barrel "precision" rifled for accuracy. Shotgun barrel bored full choke. New combination rifle and shotgun stainless steel bead front sight and adjustable rear sight. Take-down.

Stock: Tenite. Pistol grip. Fluted comb. Hard rubber butt plate.

Fore-end: Tenite. Equipped with special fastening which positively prevents gun from shooting loose.

Action: Single trigger. Low rebounding hammer. Separate extractors. Top lever operates either to the right or left to open action.

Frame: Handsome case-hardened finish. Slide button on right side for instant selection of barrel to be fired.

Weight: About 6 lbs.

Ammunition: Top barrel, .22 short, long, and long rifle, regular or high speed lower barrel, .410 bore, chambered for 3" shells.

Price $20.20
No. 240—410-410 26" barrel, 3" chamber 20.20

STEVENS No. 530M

DOUBLE BARREL SHOTGUN WITH THE
EPOCH MAKING NEW "TENITE" STOCK AND FOREARM
AVAILABLE IN 12, 16, 20, AND .410 GAUGE

"TENITE" STOCK AND FORE-END

TENITE is an ideal material for gunstocks. It is of hornlike hardness, is chip-proof and crack-proof, and has great resistance to breakage. It is absolutely weather-proof and withstands moisture, heat and cold without swelling, shrinking or warping. Its rich, lustrous, permanent buried finish resists scratching. Its uniform weight assures perfect balance.

SPECIFICATIONS

Barrels: Blued forged steel with matted rib, fitted with two white bead Colasta sights. Lengths: 12 ga., 28, and 30"; 16 ga., 28"; 20 ga., 28". Bar-

rels bored right modified and left full choke. Chambered for 2¾ in. shells.

Stock: Tenite. Rich burl walnut appearance. Checkered full pistol grip, capped. Fluted comb; paneled sides. Length 14", drop 2¾".

Fore-end: Tenite. Checkered. With special fastening which positively prevents gun from shooting loose.

Action: Hammerless. Rugged lock-up. Coil springs. Fast hammer fall. Positive extraction. Design of frame prevents powder residue from blowing back into action.

Frame: Handsome case-hardened finish.

Weight: 12 gauge, about 7¼ lbs.; 16 gauge, about 7 lbs.; 20 gauge, about 6½ lbs.

Price $29.95

BOLT ACTION 20 GAUGE. BLACK TIP REPEATING SHOTGUN

Model 258

The gun can be easily taken down. Chambered for 2¾ inch shell. Weight about 6¼ lbs. Length overall 46 inches.

Price $17.10
Extra Magazine 1.25

A new design in a 20 Gauge repeater for those preferring bolt action. These guns have a magazine capacity of 3 shots, 2 in detachable clip magazine, self cocking and independent safety. The stock is one piece American walnut with full pistol grip, black tip forearm and hard rubber butt plate. The gun is fitted with a round 26 inch tapered barrel full choke crowned muzzle.

Model 58 is of the same design as model 258 but made for 410 Gauge and comes with 24 inch barrel full choke. This gun is a 4 shot repeater and takes a 3 shot clip detachable magazine for the 3 inch shell. Weight about 5½ lbs.

Price $15.35
Extra Magazine 1.25

BOLT ACTION 410 GAUGE REPEATING SHOTGUN

Model 59

This model is of the latest design using a tubular magazine. An advantage over the clip loader you will recognize readily,

taking 6 shots, 5 in the magazine and one in the chamber, doing away with the loss of magazine clips. Of bolt action construction with all the features of the models 258 and 58. Handling the 410 Gauge 2½ and 3 inch shell. Weight about 6 lbs. Length overall 44 inches.

Price $18.40

REG. U.S. PAT. OFF.

MODEL 31 TRAP GUNS

MODEL 31TC "TARGET" GRADE
(With Ventilated Rib)

MODEL 31S "TRAP SPECIAL" GRADE

Here are two Remington Model 31 pump guns for the dyed-in-the-wool confirmed trap shooter. Raised ribs, recoil pads, front and rear sights, and special fore-ends are among the features that specially adapt these grades to the needs of top-notch trapshooters.

SPECIFICATIONS and PRICES

Model 31TC "Target" Grade. Chambered for 2¾-inch shells; 3 or 5-shot capacity. Ventilated rib. Barrel 30 or 32 inches. Full choke; modified at extra charge. Front and rear sights. Fore-end as illustrated, but option of long extension beavertail. Hawkins recoil pad. Checkered stock and fore-end of selected high grade walnut. Standard stock dimensions, 14⅜ inches long, 1⅞ inches drop at comb. Full pistol grip with cap; option of straight grip. Made to order stock lengths 13½ to 15 inches without extra charge. Weight, about 8 pounds.

Model 31S "Trap Special" Grade. Similar to 31TC but with solid rib, stock of standard American walnut, dimensions 14⅜ inches long, 1⅞ inches drop at heel, 1½ inches drop at comb. Beavertail fore-end. Half pistol grip.

No. 31TC "Target" Grade
   with ventilated rib... **$113.35**
Extra barrel with ventilated rib **50.55**
No. 31S "Trap Special" Grade **82.30**

Extra barrel with solid rib.. **$35.05**
Special stock dimensions..... **18.85**
Special long range choke
   boring, extra **5.50**

MODEL 31 SKEET GUN

The popular choice among skeet shooters who prefer slide action. Smooth action with short fore-end stroke is lightning fast to operate, and absolutely dependable.

SPECIFICATIONS and PRICES

Model 31 "Skeet" Grade. Chambered for 2¾-inch shells. Solid rib, top of receiver matted. Cross bolt safety. American walnut stock and fore-end, handsomely checkered. Full pistol grip with cap. Bakelite butt plate. Ivory front sight. White metal rear sight on ribbed barrels. Beavertail fore-end. 3 or 5-shot capacity. Weight, 20 gauge, about 7 pounds; 16 gauge, about 7¼ pounds, 12 gauge, about 8 pounds.

The Skeet gun with
"the ball-bearing action"

No. 31 "SKEET" GRADE with raised solid rib **$79.10**
No. 31 "SKEET" GRADE with ventilated rib (12 gauge only) **94.60**
No. 31 "SKEET" GRADE with plain barrel **69.15**

MARLIN'S
OVER AND UNDER SHOTGUNS

PRICE
$52.75

Model 90 stands as a major advance in firearm design by Marlin. Amazingly simple and easy in operation, this fine gun is strongly built for a lifetime of dependable use. Beautifully proportioned, perfectly balanced, it points easily, handles fast, gives even patterns with a far-reaching, hard-hitting range. BARRELS: Special analysis steel barrels chambered for 2¾" shells. Matted top barrel—single sighting plane. 12-gauge barrel lengths—26", 28", and 30"; 16 and 20-gauge barrel lengths—26" and 28" only. BORING: 26" barrel—top barrel modified choke, bottom improved cylinder. 28" and 30" barrels—top full, bottom modified choke. ACTION: Hammerless, cocks on opening. Positive, automatic safety. Take down. Easily accessible for cleaning and oiling. Front trigger fires bottom barrel. FRAME: Direct-line locking, one-piece frame. STOCK AND FOREARM: Genuine American black walnut, specially finished and checkered. Full pistol grip stock—length about 14¼"; drop at comb, about 1⅝"; drop at heel, about 2¼". Forearm large hand-filling and designed to protect the hand from heat of barrels. WEIGHT: 12-gauge about 7½ lbs., 16 and 20-gauge about 6¼ lbs.

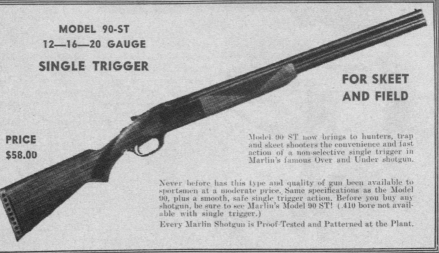

MODEL 90-ST
12—16—20 GAUGE
SINGLE TRIGGER

FOR SKEET
AND FIELD

PRICE
$58.00

Model 90-ST now brings to hunters, trap and skeet shooters the convenience and fast action of a non-selective single trigger in Marlin's famous Over and Under shotgun.

Never before has this type and quality of gun been available to sportsmen at a moderate price. Same specifications as the Model 90, plus a smooth, safe single trigger action. Before you buy any shotgun, be sure to see Marlin's Model 90 ST! (.410 bore not available with single trigger.)

Every Marlin Shotgun is Proof-Tested and Patterned at the Plant.

FINE GUNS—LOW PRICES

MODEL 90—Both Barrels .410 Bore—$52.75

This model is built along the same general specifications as the Model 90 in 12, 16 & 20 gauges but proportionately smaller. Rubber buttplate is supplied instead of recoil pad. The .410 bore barrels are chambered for 3" shells. Length of barrels—26". Boring—both barrels full choke. Weight about 5¾ lbs.

MODEL 90—Top Bbl. .22 Cal.—Lower .410—$52.75

This is an ideal all-around game and vermin gun. Top barrel will shoot .22 caliber short, long and long rifle cartridges. Bottom barrel (.410 bore) will be bored full choke. Other specifications same as Model 90—.410 bore. Weight about 5¾ lbs.

MODEL 90—Standard Two Trigger Gun—$52.75

 Illustration at the left gives you a close up of the fine details of the excellently balanced standard two trigger Marlin Over and Under which is fully described at top of page.

MOUNTAIN MULEYS DON'T ADVERTISE

by Wayne van Zwoll

A really smart 7th-grader once asked me: "If big mule deer bucks live in high places, why aren't they easy to find?" Good question. High places are mostly bare. They cover little acreage because most of a mountain is in its skirts. Trails will take you anywhere on top that bucks are likely to be.

Most of my hunting has been on mountains, partly because that is where you'll find mature mule deer, but mostly because I like hiking the mountains. If mountains don't ring your bell, you're better off hunting easier places, because big bucks can be maddeningly scarce above timber, where they should stick out like beach balls.

One day last fall, for example, I still-hunted a promising alpine ridge from dawn till dusk. My route took me up timbered shoulders, onto a curved spine where bucks like to bed to catch the autumn sun and a fly-chasing breeze. Crisp, clear weather and calm winds helped me that day. I did lots of glassing, and because I knew the country, no time was wasted wandering into dead spots. Still, I saw one deer, period.

You can say that was a failed effort, that I should have gone elsewhere, but the truth is, on this ridge I didn't expect to see a lot of deer. If you don't *think* you'll see deer, you probably won't. In fact, a low level of anticipation can ac-

tually hide deer from you. On the other hand, I knew from previous hunts that I'd not have to dodge hordes of mule deer that day. Still, my expectations were realistic.

I hunted with a positive attitude, though, because I always draw more from a hunt than the mere prospect of a kill. Alpine hiking brings great physical satisfaction in itself, and probing likely pockets with binoculars keeps anticipation alive. Besides, this was a "bucky" ridge, where the odds of seeing a truly big buck were better than in coverts that held more animals. Indeed, the deer I spotted was a crackerjack buck—the best I'd seen in several Oregon seasons.

WHERE TO LOOK
AND WHAT TO LOOK FOR

Mountaintops offer bucks the solitude they like during summer, when their antlers are growing. In small bands and individually, bucks eat, loaf and sleep where they can spot or smell predators at a distance, where people rarely visit and where they can move from shade to sunlight, into and out of the wind and from one drainage to another in just a few steps. Forage is plentiful; and because high country stays cool late, plants there remain succulent after valley vegetation has turned coarse. Slowly melting banks of snow

When you move into heavy cover, go slowly and be ready. Bucks are hard to spot if they're motionless; but you boost your odds if you stop often and look hard for pieces of deer.

trickle water all summer, irrigating forbs that deer like to eat.

In autumn, when antlers harden and bucks grow restive, there's still plenty of forage for deer, which need only about four pounds (dry weight) per day. A short growing season and thin alpine soils deny stockmen the forage production they need for sheep and cattle; but deer don't assemble in herds of hundreds. The quality of late-season alpine forage exceeds that of grasses and forbs down-country where, by August, protein and palatability have been sapped by hot, dry weather.

Hunting on top can be addictive even when

you don't see lots of deer. You'll have less competition than on mountain skirts, where logging roads offer easy access. You'll also find buck-to-doe ratios higher. One of my favorite alpine basins seldom shows me a doe. It's simply a bucky spot. Explaining why one place attracts bucks while another that looks equally promising holds only an occasional deer should be easy—but it isn't. After a season or two hunting the high country, though, you'll quickly spot the most productive places.

These places are, first of all, secure. They give the buck a panoramic view but hide him from intruders. They offer good wind coverage of ma-

Intensive, methodical glassing is a must when hunting mule deer above timberline. This hunter would be less visible sitting just below that ridge, where the rocks break up his silhouette.

jor trails and adjacent patches of cover where a man or predator might make a concealed stalk. They're close to forage and water, but not so close as to attract other animals and compromise security. They are off-trail places, for the most part, but within a jump of several escape routes. A buck likes to have options: Up, down and cross-slope paths give him flexibility, so he can escape wherever danger appears. He'll want to choose the route that gives him the best wind coverage and enables him to slip unexposed into an adjacent drainage.

Big bucks would rather hold tight or sneak away than jump and run when you come close. So good buck coverts are thick enough to hide deer and large enough to accommodate some deer movement inside. There's plenty of exposure to the sun in these pockets, but there's also quick access to shade and shelter from storms. The terrain can be steep but not severe, because bucks prefer easy walking. They also want sure footing should they need to run from danger.

Not long ago, I approached a bucky spot carefully from downwind when suddenly a big mule deer jumped up in front of me and barreled past me into the open. The buck must have been just as startled, because when he erupted from his bed, he cracked his head on a suspended deadfall and fell to his knees. Then he scrambled, broke cover, and in two bounds he dropped out of sight behind a bush, skidding into view again on his nose. He had stepped in a badger hole! By the time I recovered, he was away, dodging expertly through the sage.

This young buck fell at woods-edge low on a mountain. Big deer commonly live as high as forage and cover permit. But they'll move readily into timber, bedding in thickets to elude hunters.

This buck had smelled another hunter on a ridge behind him and had apparently worked his way to the leeward side of his available cover. I suspect he planned to lie tight there, letting the hunter pass. But then I moved in, close enough to force the buck into discarding his first plan. The explosive rush had its desired effect, but this deer was sprinting where he'd not intended to. Twice he'd been brought up short and could have been injured falling. And the delays, though they threw off my timing, held him in jeopardy. Needless to say, most bucks like to have their escape options rehearsed!

Pay attention to the cover that disgorges deer. Ask yourself, "What makes this cover better than other cover? Why did that buck choose to be here rather than somewhere else?" You'll soon form a good idea of what deer look for when they bed, forage and travel. You'll learn why some trails get lots of autumn traffic while others show only old tracks; why bucks that feel hunting

pressure move to certain draws; and why the old deer you want don't live near the young deer that are easier to see.

Mature bucks may act lazy, but they're not sedentary. They'll travel to find forage, escape hunters, seek thermal cover. They'll frequent sunny open slopes after a storm, when the brush is soaked. They like the first rays of sun after a cold night or when hunters have driven them by day into the confines of dense cover. They'd rather stay near ridgetop than in canyon bottoms or even in alpine basins, where surrounding slopes limit escape options and a noisy stream smothers the sound of approaching danger. To avoid hunters, though, they'll dive into spruce bottoms without hesitation.

GLASSING STRATEGICALLY

Knowing what attracts bucks is only a first step in the hunt. You need a strategy as well, one that shows you the deer you want and also en-

This single-shot 7x57 with Leupold 3x scope helped the author take one of his biggest deer.

sures a shot. That means you must spot bucks before they detect you, an assignment best accomplished with binoculars. Seeing deer is mostly a matter of looking in the right places at the right times. But when you look, it's important to look with a passion. I think of the poster in the locker room at our local high school: "Go hard or go home." Looking for a buck isn't like looking for a hamburger stand from your automobile. First, bucks don't advertise. Second, you won't see as many bucks on a hunt as you will hamburger stands at the roadside. So missing one can be costly.

Technique is important too. Some hunters spend a lot of time in one place. This can be productive if you've chosen a good place and the mule deer are undisturbed. Another ploy is to walk slowly, glassing as you move along a ridge. This works best for me late in the season, after the bucks have been schooled. By moving, you change the way you see, each stop offering a new perspective. When deer seem scarce and you have limited time, try hiking briskly between glassing stations. Instead of combing a smaller slice of country in detail, hunt the heads of several drainages cursorily. This is tough work physically; but when the deer have kegged up in heavy cover, it will boost your odds of finding one up and moving.

Successful mule deer hunters use all these tactics in alpine terrain, but, inevitably, circumstances dictate the choice. On opening day, I pick a relatively small chunk of deer range and wring it out, glassing from one spot for dawn's first hour, then poking along just below ridgeline in a cover that most likely will hold bucks. East slopes are preferable in the morning, because the sun hits them first. Late in the season, you want to do more walking and cover more country. That way, you'll be better prepared for future hunts and thus increase your chances for seeing bucks. You'll be less concerned about botching an opportunity because you're aware by now how the deer act under pressure. You're not looking for the perfect shot as much as you are a deer that's making a mistake.

Never glass when you're moving. Take a tip from the deer: When they want to get a good look at something, they stand very still and look a long time. Here's the key to successful glassing: first,

sit or lie prone. Brace your elbows, using both hands to steady the binoculars. Next, in grid-like fashion, move your eyes around the field of view. Don't just look for the shape of the deer, but for colors and reflections that don't quite fit the landscape. When you've finished one piece of the mountain, shift your glasses to the side and examine an adjacent piece, resting your eyes briefly between each session. Look around especially close by, to catch new activity that may not have appeared in the restricted field of the glasses.

When you do move, pay attention to the things bucks think about. Bucks look for people where people have appeared before: on horse trails and logging roads, on the spines of ridges and in meadows. They know people don't like to negotiate thick timber (which is why many mature mule deer bed there). And they know that people make lots of noise when moving on shale or through dry woodlands; so they bed or loaf where they can scan quieter approaches with their noses and eyes, turning their ears toward the noisy routes. Bucks also know that thermals drift uphill on warm, sunny days and down during the night, that prevailing winds cancel thermals, that wind skating across a ridge can spill down the lee side like water over a dam, that canyons funnel wind in new directions while the tops of mountains draw little worthwhile information from the wind.

Mule deer rely most heavily on their noses to detect hunters. Once a deer smells you, it needs no further proof of who or what you are. If it senses you are unaware of it, the deer may stay still; or, if it can do so safely, it will sneak away. Should you indicate by a sudden movement that you've seen the deer, it will likely bound off. One thing is certain, a deer that smells you will not forget about you.

Keeping the wind to your face isn't always possible in broken terrain; but hunting crosswind can be as effective as hunting into the wind. Recently, during the evening thermal shift, I made a successful stalk cross a series of feeder drainages located down-canyon. While the main canyon was funneling air to my back, those feeders gave me side drafts. By moving briskly, I could outpace the air barreling down the primary drainage, although it was interrupted by the feeder drafts.

The 10x50 binoculars (center) are too bulky and heavy for deer hunting. Compact glasses like the 8x20s and 8x24s (right) perform poorly in dim light. The best choice among these Redfield products is the 7x35 (top left). The Electric Zoom binoculars below it are mainly for hunters who would rather play with gadgets than hunt.

A caveat: When you climb to a good vantage point in the dark for your dawn hunt, remember, you're putting yourself at the mercy of the night thermals. They can wash your scent, which behaves a lot like water, into those pockets of cover that the sun will soon illuminate. So it's important to use topography in blocking or redirecting the flow of your scent. Control your scent, and you control the hunt. Pay no attention to the wind, and you might as well stay at home and watch football.

Having the wind in your face won't help you if your silhouette pops up on a ridge. Mule deer can see great distances, and they pick up readily on movement. They're not so good at spotting your motionless form if it's broken in front or back by natural-looking objects. When you move or sit in mule deer country, stay close to trees

High country gets snow early. Be prepared with the right gear and clothing for cold weather. Drink more than you think you need. Snow throttles your pace, but shows where mule deer are moving.

and protruding rocks. They absorb your silhouette and hide your movements. Crawl over ridges, glassing near before you glass far. If you're stalking a buck, bend as low as necessary to expose less of your body; that way, even if the deer does see you, it may mistake you for an animal.

While darkness helps hide your approach on a dawn hunt, you'll have no such luxury when sneaking to a vantage point in the afternoon. Deer that will leave their bed for an evening snack will instead lie tight during daytime, especially if they spot you first. You can't be too careful when closing in on promising cover, whether it's a thicket 100 yards away or a draw cross-canyon beyond rifle shot. I've watched mule deer sneak away from hunters who were so far off they mistakingly took no precautions. By the time they're

handloads. They also come chambered for any factory cartridge one might want to use on mule deer.

Of the myriad cartridges suitable for deer hunting, the .270 Winchester, .280 Remington and .30-06 Springfield excel. Each can deliver 1,200 foot-pounds of energy at 400 yards; each shoots flat enough that, when zeroed at 200 yards, will kill with a top-of-the-shoulder hold to beyond 300. As for the middle-weight bullets best suited to deer—including 140-grain in the .270, 150 in the .280, 165 in the .30-06—recoil is quite manageable.

Among alternative cartridges, a strong case can be made for Remington's .25-06. In short actions, the .284 and .308 Winchesters are excellent choices, along with the 7mm Mauser. Those who

This Kimber Big Game Rifle in .270 has a compact Leupold scope in low rings. It's an ideal outfit for hunting mountain mule deer. The .280 and .30-06 are also excellent cartridges.

old enough to grow big antlers, most bucks have learned that rifles can reach great distances.

GOOD EQUIPMENT, BODY CONDITIONING AND PRACTICE, PRACTICE, PRACTICE

Neither a powerful cartridge nor an accurate rifle can undo a stalk scuttled by your own carelessness. But top-rung equipment can bring more shooting opportunities and help you make the most of them. Hunters after mule deer in the mountains commonly choose bolt rifles because they're strong and relatively simple. They're less prone to malfunction in rain and snow than autoloading rifles, and they have greater chambering and extraction leverage than any other design. Bolt-action rifles are generally considered more accurate, too, and are more forgiving with

favor lighter rifles or less recoil choose the .257 Roberts, 6mm Remington and .243 Winchester. No matter what cartridge you choose, your ability to place a soft-nosed bullet through the forward ribs into the lungs will determine your success as a rifleman afield. There is no such thing as a lucky shot. You decide when to squeeze the trigger, and you are responsible for the result.

Long shooting is never a good idea when there's a reasonable chance to shoot close. In mule deer country, kills at extreme range comprise a small fraction of the total. Sadly, many bucks are crippled and lost by hunters who bang away at outlandish distances but lack the skill to make consistent hits. As range increases, wind and bullet drop markedly increase bullet dispersion. Those tiny muscle quivers that bounce your bullets around the target at 100 yards will cause a

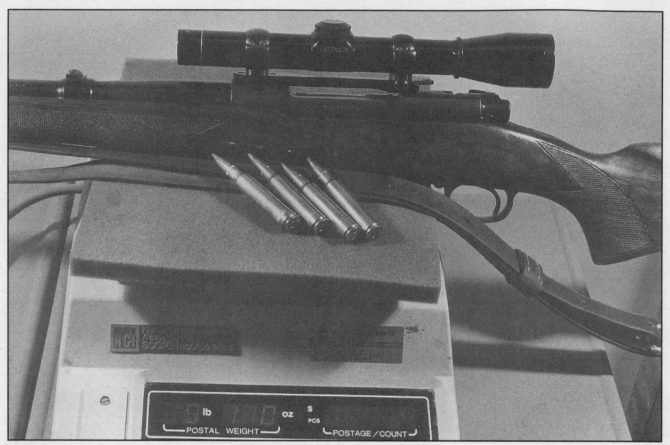

If you're hunting both elk and deer, a belted magnum cartridge makes sense. But short, light barrels aren't practical on magnums. This .338 Model 70 Winchester weighs a hefty 9.7 pounds trailside.

miss or a crippling hit at 400. Some hunters who anticipate long shots use rifles chambered for magnum cartridges, which flatten trajectories somewhat and provide an extra measure of bullet energy down the pike. The best of belted deer cartridges are the .264 Winchester, .270 and 7mm Weatherby, and 7mm Remington Magnums. You don't need heavier bullets than these rounds—unless you're after elk on the same hunt. Mule deer and elk seasons often overlap, in which case the 7mm or .300 magnum with 160- or 180-grain bullets becomes a great deer cartridge.

Some hunters make the mistake of choosing a magnum cartridge for a lightweight rifle, but a short barrel may offer no more velocity than if they'd used a standard cartridge. Barrels on magnum deer rifles should be at least 24 inches long, but 26 is better—excluding muzzle brakes, which don't contribute to velocity. Brakes reduce

recoil, but the increased blast hammers your ears, so be sure to wear ear protection. Actually, I recommend removing the brake when hunting and replacing it with a cap to protect the barrel threads. It could save your hearing.

Big scopes are in vogue these days, even on trim, lightweight "mountain" rifles. Big objective lenses may indeed let in more light, but the eye can only use so much. Much depends upon the quality and coatings of the lenses in your scope, and the exit pupil (a number derived by dividing objective lens diameter in millimeters by the scope's magnification). In the dimmest light at dawn, an exit pupil of 6 is the most your eye can use. A 4x scope with a 40mm objective lens gives you an EP of 10, which is more than one needs. A 6x scope with a 42mm objective (EP=7) will provide all the light your eye can use in total darkness.

Variable scopes with 50mm objectives become practical only when they're cranked up to 8x or higher. At low magnification, you gain nothing for the extra weight and expense; in fact, a 4x scope is all that's needed for mountain-top mule deer hunting (a 6x is better, though, if you anticipate long shooting). If a fixed-power scope seems too primitive, stay with 2-7x or 2.5-8x variable and mount them low, where they belong.

Except for a couple of brief seconds before the shot, you'll be looking through binoculars, not a scope. Buy the best you can afford. It's amazing how hunters will spend a small fortune on a scope but carry discount-store binoculars. Cheap binoculars can transmit distorted or fuzzy images, or they may be out of collimation (i.e., the barrels are misaligned). They may also fog or leak moisture. Because magnification and portability matter, an exit pupil of 4 is acceptable. Choose a 7x35 for all-around hunting; and for long distances, try an 8x40, or even a 10x40.

Scopes and binoculars made in Europe—including Zeiss and Leica, Swarovski, Kahles and Schmidt & Bender—are superb instruments; but Leupold, Burris, Redfield, Nikon and Bausch & Lomb in the U.S. offer top-notch products as well. These optics are so bright and sharp that it would be hard for most riflemen to distinguish them from German optics costing much more.

As for other gear, go light. Carry a cheap plastic compass, a 4-inch folding knife, a dependable flashlight, and a roll of fluorescent marking ribbon to flag a kill or help unravel a blood trail. Other helpful items include a film can stuffed with fire-starting chips, a small case of waterproof matches, and a full-size camera (although the compact automatics make more sense for most deer hunters). Pack a lunch of apples, raisins, cheese and bagels, and don't forget a plastic canteen in a cloth holster to hang from the belt along with an eight-round soft leather cartridge pouch.

As for wearing apparel, I stuff gloves and, in cold weather, a down vest into the big back pocket of my tightly woven Filson wool jacket. All outer clothes should be wool, which insulates well, wet or dry, and is quiet in brush. Avoid hats with wide brims, which sometimes strike the scope or sight, spoiling the shooter's aim. Jogging shoes are good enough in warm weather, but Vibram-soled leather boots are advisable later. Traction is important, too, especially when you must cross scree or ford a stream or hop from one snowy boulder to the next.

Many a well-equipped hunter neglects his most valuable asset: his body. Mountain-top mule deer live at 7,000 to 12,000 feet. Unless you've exercised your legs, lungs and heart regularly in the off-season, that thin air will soon leave you short of breath. Running and bicycling are simple

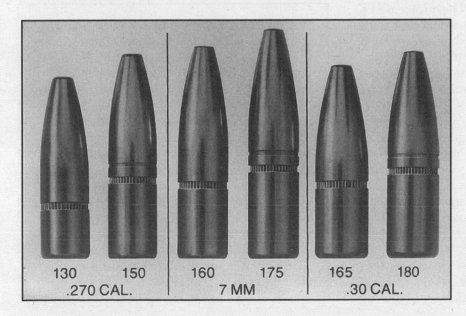

130   150   160   175   165   180
.270 CAL.   7 MM   .30 CAL.

You don't need "controlled-expansion" bullets for deer; but elk make them worth the expense. Of the bullet pairs shown, the light ones are adequate for deer and faster out the muzzle.

A shooting sling like Brownell's Latigo shown here steadies you. Carrying straps, especially the wide "cobra" style, are useless as shooting aids.

exercises, but they rank among the best for hunters. You needn't live at high elevations to enjoy the benefits, either; simply find some hills and quicken your pace to boost the load.

While poor body condition accounts for more failures than poor shooting, the latter is a close second. Most hunters show up with very good equipment; indeed, some have spent much more on their rifles than necessary. The problem is a lack of practice. Some have not even zeroed their rifles; and others have never shot except at a bench. Few know where their bullets will strike beyond a 200-yard zero range. One hunter claimed the bullets from his magnum wouldn't drop more than 2 inches at 300 yards.

One good way to prepare yourself for high-country mule deer is by shooting a pellet rifle or .22 before the season begins. Practice shooting from field positions to learn the value of a sling and how to get into the loop quickly. Without pounding your shoulder or draining your savings, shoot several hundred rounds from your deer rifle. Practice. Practice. And then practice some more.

All this fussing in the off-season can pay big dividends come opening day. Even when the deer win, you won't feel you've lost. Mountains do

something to your sense of perspective. You might even forget the hunt for a moment and strike off on a ridgeline trail just to see where it leads. That's not a bad habit to get into, because there's something about mountain air that does a body good, and your mind and spirit, too.

WAYNE VAN ZWOLL is a writer known for his expertise on big-game hunting and the technical aspects of shooting. He has written for most of the major sporting magazines, including *Field and Stream*, *Rifle* and *Bugle*. Among the books he has authored are: *Mastering Mule Deer* (1988, North American Hunting Club), *America's Great Gunmakers*, (1992, Stoeger Publishing) and *Elk Rifles, Cartridges and Hunting Tactics* (1993). No stranger to the high country, van Zwoll has been a range conservationist with the Bureau of Land Management, a wildlife agent with the Washington State Dept. of Game and a field director for the Rocky Mountain Elk Foundation.

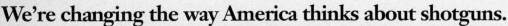

NOW ONE GUN DOES IT ALL.

The new Beretta A390ST.
The most versatile gun you'll ever own.

A semi-automatic shotgun that would reliably handle every size and type of load has been the ambition of shotgunners and gunmakers alike for decades. With the Beretta A390ST, that goal has been achieved. To a Beretta gas system already renowned for its reliability, we added a specially designed self-regulating pressure valve. This unique valve automatically adjusts the gas pressure in the system to perfectly handle any 12 gauge load from factory 2¾" target loads to the heaviest 3" magnums.

Think what this unprecedented versatility means in the field. You can use this gun for any and every type of shooting you do. You can hunt everything from upland birds, to waterfowl, to turkey, to deer, all with one gun—no adjustments of any kind are required to accommodate different loads.

Easy on the shoulder, easy on itself.

The A390ST handles its wide range of loads with remarkably little punishment to shooter or gun. With some of the propellant energy redirected to drive the action, and excess gases vented through its self-compensating gas system, you'll notice far less felt recoil. The A390's regulating valve also relieves needless stress on operating parts when heavy loads are used, minimizing wear and tear.

A new world standard of performance. An old world standard of craftsmanship.

From its machined receiver, to its hand-fitted working parts, to its select American walnut stock, the A390ST shows its 500-year Beretta heritage in many ways. Extensive use of high-strength stainless steels and anodized alloys makes the Beretta A390ST almost impervious to stress and corrosion. Elegant in its simplicity, it will require only minimal maintenance over the years.

The dream autoloader is no longer a dream. Available in standard or matte finish with 24", 26", 28" or 30" Mobilchoke® barrels, the Beretta A390ST is here now. See it at your authorized Beretta dealer, or contact Beretta U.S.A. Corp., 17601 Beretta Dr., Accokeek, MD 20607, (301) 283-2191.

The screw-in Mobilchoke® system expands versatility for hunting everything from upland game to waterfowl. Beretta's "SP" choke tubes are engineered to withstand the stress of non-toxic shot.

New magazine cut-off system lets the shooter slip in a different load without emptying the magazine. Flush mounted cut-off lever can't hang up on clothing. Ambidextrous safety button can be reversed for use from either side.

New Stock Drop System brings the fit of a custom stock within everyone's reach. Cast-On/Cast-Off Spacer allows quick adjustment of both the vertical and horizontal angle of the stock.

Beretta's simple, rugged, one-piece stainless steel piston drives the action. The system is self cleaning, with nothing to break, lubricate, or wear out—no "O" rings to gum up or fail.

The action bar and sleeve is machined and fabricated of high-strength steel to match the hardness of the piston...thus eliminating work hardening of these critical parts. No mild steel stampings or plastic is used in a Beretta action.

The heart of the new Beretta self-compensating gas system is this unique self-regulating valve.

When firing light loads, the valve remains closed, utilizing all expanding gases entering the cylinder to drive the piston and action. A larger gas port to the cylinder assures that there will be enough pressure to drive the action even with target or promotional loads.

With medium-heavy loads, the gas regulation valve opens part way, retaining sufficient gas pressure to drive the piston, while venting excess gas downward, harmlessly away from barrel and shooter.

With the heaviest loads, the valve opens fully, rapidly venting gases not required to work the action. This eliminates needless stress on operating system parts and reduces felt recoil to the shooter.

All gas vents are protected by metal to prevent cauterizing of the wood and future cracking problems.

The barrel is made of high-strength nickel-chrome-moly steel. Exclusive hard chromed bore ensures truer, more consistent patterns and resists pitting, corrosion.

Beretta A390ST
Beretta U.S.A.

AMERICAN EAGLE

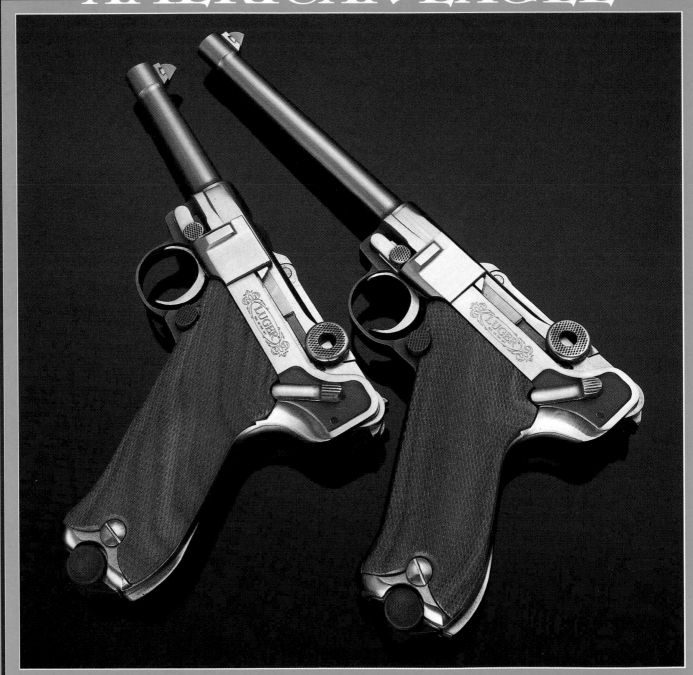

YESTERDAY'S TRADITION - TODAY'S TECHNOLOGY

OUTFITTING THE NEW SHOTGUNNER

by Don Lewis

The shotgun has never enjoyed the popularity of the rifle; in fact, it has always played second fiddle. One reason could be that the shotgun doesn't share the rifle's nostalgic past; it didn't tame the wilderness or make any great contributions toward opening up the West. Worse yet, it has never played a major role in the military. It's fair to point out, however, that beaver trappers on the plains and in the Rockies found some use for the shotgun as far back as the early 1800s. But in general, over the decades, time has embellished the deeds and importance of the rifle. Nostalgia surrounds its accomplishments in defending the frontier, cracking with deadly authority on the Santa Fe and Oregon trails, and putting meat on the frontiersman's table. From the Battle of Bunker Hill to the sands of Saudi Arabia, the rifle spoke for freedom.

The shotgun differs from a rifle in that it is almost always a smoothbore—the one exception being the rifled shotgun barrel designed for slugs used in big-game hunting. The rifle fires a single projectile, while the shotgun features an expanding charge of shot consisting of many lead or steel pellets. Since the velocity of these small pellets falls off quickly, the shotgun is useful for short-range firing only. As the shot charge travels away from the muzzle, it spreads into an ever-widening pattern. Obviously, no shot charge has enough velocity and kinetic energy to be effective on live targets much beyond 50 yards.

No one knows for sure when the shotgun was invented, but there is evidence that some type of shotgun arrived on this hemisphere with Columbus. Compared to the compact, lightweight shotgun of today, those early fowling pieces—which were probably matchlocks—were heavy and had short stocks and barrels. The barrel was little more than a brass or iron tube fitted on a straight stock that was held under the arm or against the chest.

At first, shooters carried a piece of rope saturated with saltpeter. It burned without a flame—something like a punk. When the shooter was ready to fire, he aimed and simultaneously touched the burning rope to the touchhole. A major improvement was the serpentine, an S-shaped piece of metal that held the glowing match. When the shooter pulled back the bottom half of the serpentine, the burning end dropped down into the flashpan. This was an improvement, but it was still no picnic for shotgun shooters.

Down through the decades, firearm ignition systems improved steadily, but the muzzleloader shotgun remained the top choice among small-game hunters. Shortly after the Civil War, self-

The author helps his granddaughter, Jamie, adjust to the 28-gauge H & R Topper Jr. Classic shotgun. Note how her face is pressed against the comb of the stock.

contained shotgun shells and breechloading shotguns arrived on the scene; still, it took another 30 years before the muzzleloading shotgun lost its popularity. In any event, the breechloader with self-contained shotshell was here to stay and has since evolved into a highly efficient short-range firearm.

WHAT KIND OF SHOTGUN SHOULD YOU BUY?

In selecting a shotgun, the beginner is faced with numerous questions, starting with, "What type of shotgun should I buy?" Five distinctive types are available today: (1) break-open single-barrel (single-shot); (2) break open double-barrel (side-by-side or over/under); (3) pump or slide-action; (4) semiautomatic (autoloader); and (5) bolt-action.

The basic shotgun is a *single-shot* with a visible hammer that is cocked manually. This outfit is the least expensive, although it's fair to point out that single-barrel trap guns can cost well over a thousand dollars. Still, the single-shot outfit is probably the wisest and safest choice for the beginner. An excited, inexperienced hunter using one of these guns can't fire a second shot inadvertently and quickly learns the value of making the first shot count as well.

The *two-barrel outfit* is probably the epitome of hunting shotguns, although dedicated pump and autoloader fans will put up strong arguments on behalf of their favorites. As with the single-shot, the two-barrel shotgun exposes the chambers when open, enabling the shooter to see whether or not the gun is loaded and to check for barrel obstructions. Also, when open, the gun cannot fire, which is another important safety factor. Moreover, the double or side-by-side offers the potential for having two separate choke constrictions. In a sense, the two-barrel shotgun is

two separate outfits, with two separate firing mechanisms or locks, two separate barrels and two chokes. An ideal choke setup for small-game hunting is improved cylinder and modified; but dove, pheasant and waterfowl hunters whose shots are often long may opt for modified and full.

The *pump* or *slide-action* shotgun has a long history that dates back to 1839, when Sam Colt developed his 6-shot cylinder-type firearm. Most pump-actions carry extra shells in a tube located below the barrel. During the loading process, a powerful spring is compressed. When the last shell has been inserted, a latch or lock catches its rim, holding it and all previously loaded shells. Then, as the slide is pulled back, the lock is depressed and the spring forces the rearmost shell into a carrier that has dropped down to receive the new round. When the slide is moved forward, the carrier places the shell in front of the chamber. At this point, the bolt slides the shell into the chamber and the carrier returns to its pick-up position.

As for speed, many veteran hunters can work

Jamie Lewis proudly displays her H & R Jr. Classic shotgun. Chambered for 28 gauge and with its stock made shorter for smaller shooters, the H & R is a perfect starter shotgun.

a slide-action shotgun with unbelievable dexterity. But to enjoy the repeating benefits offered by the pump action, compactness is sacrificed. Unlike the double, whose parts are closely integrated, the pump gun requires a long receiver to accommodate the sliding-breech mechanism. Additional room must also be allowed for loading and unloading inside the receiver, making the slide-action shotgun even longer.

The *autoloader* or *semiautomatic* is a modified version of the pump shotgun. Its slide has been removed, and the mechanism is activated by energy derived from the fired shell. The autoloader eliminated the manual work required by a slide-action; a simple pull of the trigger sets the entire working mechanism into action. The sear is released, the hammer falls against the firing pin and sets off the shell in the chamber.

In most of today's autoloaders, some of the gas behind the shot charge is vented through a small hole in the bottom of the barrel into the chamber. The compression of this gas inside the chamber activates a piston that drives the breech mechanism backward against a spring. During this operation, the fired case is extracted and ejected, and the magazine retainer is moved aside, allowing a fresh round to slide onto the carrier. The spring then activates the bolt, which picks up the new shell and guides it completely into the chamber, allowing the bolt to lock at the last instant. The autoloader, with its long receiver, requires lots of attention. For instance, the power piston assembly must be kept clean if the shotgun is to function flawlessly. Moreover, because it reloads instantly and is ready to fire, it is not a wise choice for the beginner. In the excitement of the hunt, an autoloader can easily discharge by accident with possibly tragic results.

THE ABCs OF CHOKES AND GAUGES

Another question the new shotgunner faces is how to choose the correct gauge. There are only five shotgun gauges: 10, 12, 16, 20 and 28. Technically speaking, the .410 (bore) is not a gauge; but in a practical sense, it is included in the shotgun lineup, bringing the total number of gauges to six. Somewhere in the past, it was decided to establish the gauge of a barrel by the number of lead balls that, given the barrel diameter, it takes to make one pound. For instance, a bore that ac-

This young hunter has scored with Remington's Model 870 Express Magnum Youth model. Its 12 1/2 inch-stock and 21-inch barrel contribute to its weight of only 6 pounds.

cepts 16 balls to the pound is a 16 gauge, and so on. This explains why, in gauges, the smaller the number, the larger the diameter.

It's a common but mistaken belief that larger gauges, especially in long barrels, shoot harder or farther than smaller ones. A quick check of a shotgun reloading manual reveals that muzzle velocities for all shotgun gauges, including the .410 bore, average between 1,150 and 1,350 feet per second. Aside from a few exceptions, there are no significant differences in shotgun muzzle velocities.

Likewise, long barrels do not shoot harder or farther than short barrels. Generally speaking, powder charges burn completely and maximum pressure is reached in less than 21 inches of barrel. Little acceleration of the shot charge occurs beyond that point; moreover, a long barrel can do nothing to increase the speed of a shot charge. Long barrels do, however, offer a longer sighting plane for long-range shooting; they swing smoother, and they place the muzzle blast farther from the shooter's ears.

Patterns of the same diameter are made from barrels of the same length and choking. The larger 12-gauge bore does not produce a larger pattern

The 12-gauge Stevens single-shot "Dreadnaught" with 36-inch barrel and "full taper choke" boring was advertised during the gaslight era as a "super range" shotgun. Full choke was believed to mean long range, which indicated power. Such myths of the past still sway thousands of small-game hunters who are obsessed with power, although modern firearms have tended to dispel some of that myth.

than does a 20-gauge bore, assuming both barrels are the same length with identical choke constrictions. It's an established fact that all barrels with the same degree of choke throw similar patterns. The choke stamping on the barrel indicates the degree of choke installed by the manufacturer, but not all manufacturers agree on what degree of choke constitutes full, modified or cylinder. On top of that, shotgun bore diameters of the same gauge and choke constriction vary significantly among shotgun barrel makers. It is the wise hunter who, in patterning his shotgun, knows for certain that the choke stamping on his barrel is correct. In fact, a shotgun should be patterned with each type of shell commonly used for hunting.

The subject of choke is cloaked in mystery. It has a strange fascination for most hunters, and probably more so with beginners. In their minds, full choke means long range, indicating power. The Stevens single-shot "Dreadnaught" with its 36-inch barrel and "full taper choke" boring, was advertised during the gaslight era as a "super range" shotgun. Such myths of the past still sway thousands of small-game hunters who are obsessed with power. Our forefathers thought in terms of large gauges and long barrels; today,

we've evolved from large bores to magnum shells, so we're still on the long-range power track. Paradoxically, power, long-range potential and tight choking are not as important in normal small-game hunting as many people think.

The primary purpose of choke constriction is to control the size of the pattern at normal shooting distances. The ultimate goal for small-game hunters is to attain patterns of sufficient diameter and density to make clean kills in the field. Since the majority of small-game shots are less than 30 yards, tight chokes should be avoided. For example, at 30 yards, full-choke patterns measure roughly 30 inches in diameter. Modified choking increases the pattern diameter to three feet, while improved cylinder expands the diameter to nearly five feet. It's not hard to see why the small-game hunter—especially the beginner—should be wary of too much choke.

FITTING SHOTGUN TO SHOOTER

For any small-game hunter to be successful with a shotgun, proper fit is essential. While the fit of a shotgun *must* meet the physical configurations of the hunter's body, it cannot be determined solely by the shooter's physique. After all, there are short shooters with long arms and

A young enthusiast fires Remington's compact Model 870 Youth shotgun in 20 gauge. This model has the right dimensions for the beginner. Note that his cheek is resting comfortably on the comb of the stock and his shooting eye falls directly over the sighting plane. This position means that the stock is the correct length. Butt pitch and comb drop are also correct.

tall shooters with short arms. Even jowl thickness and neck length cannot always be determined by one's height.

The basic problem in establishing proper fit is stock length. The standard 14-inch stock (as measured from the face of the trigger to the end of the butt pad) is too long for many hunters. A shotgun must come up in one fluid motion and snug firmly into the shoulder pocket without catching in the armpit. Field shooting relies on reflex action—there's no time to adjust to improper fit. The belief that shortening a stock destroys the balance and other dimensions of a shotgun is pure hogwash. When the stock length

is correct for the shooter, the other stock dimensions are usually close enough for good field shooting.

The Achilles' heel for many young hunters, male or female, is bulk and stock length. Young hunters and petite older female hunters find the extra bulk and width of a standard shotgun hard to handle in brush and tiring on an all-day trek. But there's light at the end of the tunnel: H & R's Topper Classic Youth single-shot (.410 or 20 gauge), with its 22-inch modified barrel and 12½-inch stock; and Winchester's Model 1300 Ranger 20-gauge Ladies/Youth outfit with a 22-inch barrel that includes Winchester's Winchoke® setup

in full, modified and improved cylinder. These are excellent starting outfits, and they can become lifetime companions for small (and small-game) hunters.

A used shotgun is less expensive, of course, and its stock can be easily shortened to fit a beginner. But it must be pointed out that removing an inch or so from a stock does nothing to reduce bulk and weight. The argument in favor of buying a used shotgun and reducing its stock and barrel length is that it can take several years before a new hunter really knows what type, gauge and choke setup is best for him or her. Actually, the difference in cost between a used and new shotgun is not all that great, especially if the used

gun is extensively customized. There's not much gain for the beginning hunter in simply cutting off the stock without also reducing barrel length and installing screw-in choke tubes or adjustable choke devices, all of which add considerably to the purchase price of a used shotgun.

SELECTING THE RIGHT AMMO

Choosing the right gauge for the novice shotgunner deserves more thought than automatically selecting the .410 bore. The fact is, it's a poor choice from a ballistics standpoint. True, the .410 has less muzzle blast and recoil, and it is probably lighter than many models; but because of its small shot charge, it really isn't a beginner's shotgun shell. Ballistics clearly indicate the disadvantages of the .410 bore in the hands of inexperienced shooters.

The 2½-inch shell with its half-ounce shot charge should be used on the skeet field, not on game. Even the 3-inch case increases the shot charge to only 11/16 of an ounce. Half an ounce of No. 6 shot contains roughly 112 pellets, with about 250 pellets in the 11/16-ounce load. Another factor that is overlooked with the .410 bore is its long shot column, which causes more pellets to rub the bore and become deformed in the process. As a result, they leave the muzzle at a greater angle of divergence from the aiming point, automatically reducing the number of pellets within the effective pattern (approx. 25 inches). This tiny shell should be patterned at 25 yards—never at 40 yards.

The truth is that the .410 bore, while an efficient hunting shotgun in the hands of an expert shotgunner, is no match for the 20 gauge as a beginner's shotgun. The 20-gauge shell is not only a super small-game hunting round, it's also ideal for beginners. The 2¾-inch case holds up to 1⅛ ounces of shot, while the 3-inch case contains up to 1¼ ounces of shot. The recoil of the 20 gauge is not significantly greater than the .410 bore, and its larger effective pattern and shorter shot column (which produces fewer deformed pellets) give the beginner a decided advantage. After nearly 60 years of small-game hunting with 20-gauge outfits, this writer is convinced it's the top field shell for all hunters—and doubly so for new hunters.

Not all beginning shotgunners are small in stature. Jim Astleford (left) uses a regular 12-gauge Browning over/under. His son, John (center), needs a much shorter stock and has chosen a Remington Youth Model in 20 gauge. Nick Fetterman (right), age 11, can't buy a hunting license until he is 12, but the Remington 870 Youth outfit should fit him perfectly, too.

Above all, the beginner must clearly understand the nomenclature of his or her shotgun, and how that shotgun functions. It's imperative to know how to load and unload, and how the safety or hammer operates. All these things must be learned firsthand in range practice and patterning sessions under the supervision of a competent instructor. Field safety is the paramount goal in hunting, and the new shotgunner who has a thorough knowledge of how his/her shotgun works will be a much safer—and more successful—hunter.

Beginning shotgunners should practice by firing several boxes of ammo under the supervision of an experienced shooter. Note the ear covers to protect the youngster from the blast.

Women hunters generally require shorter stocks. Veteran hunter Helen Lewis (the author's wife and photographer of most of the photos that appear in this article) uses a Remington Model 11-LT 20 Lightweight for rabbits. The choke is improved cylinder.

DON LEWIS is a retired corporate executive who currently tests and evaluates firearms and shooting gear. A regular contributor to SHOOTER'S BIBLE, Lewis writes gun-related articles for other publications, including *Handloader's Digest*, *Deer Trail* and *Central Pennsylvania Afield*. He has written the gun column for *The Pennsylvania Game News* for the past 30 years and is now outdoor writer for *Leader Times*. Lewis also contributes to *Varmint Hunter* magazine and *Pennsylvania Woods & Waters*. His recent book, *The Shooter's Corner*, was published by the Pennsylvania Game Commission.

THE DUCK STAMP: AMERICA'S FAVORITE "ARTISTIC LICENSE"

by R. Stephen Irwin, M.D.

For fowl and man alike, the 1920s and early 1930s were not particularly good years. The Great Crash of 1929 and the Depression that followed created an economic crisis that crushed the human spirit and forever changed an entire generation of Americans. To make things worse, nature went on a rampage as well. Land lying between the Rockies and the Missouri-Mississippi waterway had been over-grazed with cattle and sheep and then put to the plow, having its protective grass matting ruthlessly ripped away. A prolonged drought seared the scalped land and dried the soil to powder. When the winds came, the skies became black with windblown topsoil and the great plains that had once been America's breadbasket now became the Dust Bowl. One old-timer related that he could look out his window and count the Dakota farms blowing by.

The plains of the north-central portion of this disaster-stricken country were once dotted with thousands of small shallow ponds known as potholes, and these were interlaced with millions of acres of grass and marshlands. This habitat, before it was drained for agriculture and urban growth, provided ideal breeding and nesting areas for America's ducks and was fondly referred to by hunters and conservationists as the "duck factory."

Years of land misuse and a withering drought caused havoc among the country's duck population. Fortunately, this did not go unnoticed by duck hunters and some pioneering conservationists, including Ray Holland (former editor of *Field and Stream* magazine), George Lawyer, John Burnham and A. S. Houghton. These men, realizing that seasons and limits alone could not save the ducks, knew what was needed: the preservation of remaining waterfowl habitat and restoration of nesting and breeding areas, as well as protective cover for resting waterfowl among their migration routes.

Some relief arrived with the enactment by the U. S. Congress of the Migratory Bird Conservation Act of 1929. This law expanded the National Wildlife Refuge System that had been established in 1903 and authorized the government to acquire wetlands for waterfowl habitat. This law, which was admittedly a stopgap measure at best, made no provision for the procurement of the funds needed to purchase the land. Even in President Franklin Roosevelt's proconservation administration, the budget was stretched so thin that the welfare of America's duck population rated a low priority. Still, if our ducks were not to become extinct, something clearly had to be done. To this end, Roosevelt ap-

J. N. "Ding" Darling sits at his desk at the Des Moines Register in October 1949. A nationally syndicated political cartoonist, Darling earned his reputation as "the best friend ducks ever had" when, as a staunch conservationist, he served on President Roosevelt's 1934 Duck Committee.

pointed a special committee of three members to address the crisis and recommend a program for its resolution. The President's "Duck Committee," which was established in January 1934, consisted of: Thomas H. Beck, chairman of the Connecticut State Board of Fisheries and Game; Aldo Leopold, a biologist at the University of Wisconsin (and author of the environmental classic, *A Sand County Almanac*); and J. N. "Ding" Darling.

"DING" DARLING— DUCKS' BEST FRIEND

A nationally syndicated political cartoonist for the Des Moines Register, Darling had won a Pulitzer Prize in 1924. As an ardent hunter and strong supporter of conservation, many of his cartoons stressed the importance of good management of wildlife and natural resources. A staunch Republican, he occasionally ridiculed

After the Duck Stamp Act was signed into law in 1934 with conservation of wetlands as its goal, one of "Ding" Darling's unfinished artistic renditions (above) landed on the first stamp issued to waterfowl hunters.

Roosevelt through his cartoons, thus making him a somewhat unlikely choice for this Democratic president's committee. Darling's strong reputation as a conservationist prevailed, however, and his subsequent efforts to save the nation's waterfowl over the next few years earned him the title: "The best friend ducks ever had."

Roosevelt's Duck Committee sent its final report to the President in February 1934. A comprehensive plan for restoring and preserving migratory waterfowl habitat in the U.S., it was designed chiefly to supplement and support the Migratory Bird Conservation Act of 1929. Paramount among the Duck Committee's goals was to raise funds through a national hunting license, with the revenue gained used for the acquisition of wetlands. The idea, which had been tossed

about for some time by a number of conservationists, was to use a revenue stamp as the national hunting license for waterfowl. The concept of a hunting stamp was not difficult to sell to the nation's No. 1 stamp collecting enthusiast, Franklin D. Roosevelt himself. The result, the Migratory Bird Hunting Stamp Act—commonly known as the Duck Stamp Act—was signed into law by the President on March 16, 1934. It required that every migratory waterfowl hunter 16 years or older must purchase a federal stamp validated by the hunter's signature and affixed to his state hunting license. All revenue from the sale of these duck stamps, except for the cost of printing and distribution, was to be used for acquisition of wetlands for waterfowl.

Just six days before the Duck Stamp Act was

From the 1930s into the 1940s, duck stamps cost buyers only one dollar each. The revenue was and is still used today by the federal government to acquire and preserve wetlands. Francis Lee Jaques was the famous artist who rendered this illustration in 1940.

signed, FDR appointed "Ding" Darling as Chief of the Bureau of Biological Survey (which later became the U.S. Fish and Wildlife Service). Darling's experience on the Duck Committee and his background as a cartoonist served him well. He was asked early on to prepare some sketches for the Bureau of Engraving and Printing to be used for the first duck-stamp design. After discussing the matter with Hal Sheldon, Chief of Division of Public Relations at the Department of Interior, it was agreed that the stamp should have a horizontal format and perhaps feature two or more ducks in flight.

Darling, who was a hard worker by reputation, immediately started making some sketches. Because the law was finally passed by Congress and signed by the President in barely enough time to get the stamps out before the start of hunting season, the Postage Stamp Engraving Department was in a rush to get the design. One day, while Ding was out to lunch, someone from the Bureau of Engraving and Printing snatched up an unfinished design from Darling's desk and took it to the engravers. Darling never even had a chance to retouch the drawing. His sketch of two mallards—a hen and a drake landing in a windy marsh—soon became history as the first Federal Duck Stamp. Moreover, it started what was to become the most significant contribution ever made to waterfowl conservation in the U.S.

AMERICA'S FAVORITE TAX

A total of 635,000 of the first duck stamps were issued in 1934 for $1.00 each and started

what remains one of the most successful and most popular taxes ever levied upon the American public. With funds from the sale of the duck stamps, mistakes could be remedied and avoided by restoring drained land to the country's wildife and preserving marshlands not yet destroyed. Many of the 400 national wildlife refuges in the U.S. have been paid for entirely or in part by duck-stamp receipts.

Incidentally, duck stamps aid wildlife other than waterfowl. One-third of the nation's endangered or threatened species find food and shelter in wetland habitats provided by the program, and the enhancement of the marine life food chain provides nourishment for coastal spawning and nursery grounds. The natural filtering action of wetlands purifies our water and diminishes the chance of flooding by preventing rapid runoff.

These Canada geese were drawn by Leslie C. Kouba for the 1958 stamp.

Waterfowl habitats that once sold for as little as a dollar an acre now cost a thousand times that price; commensurate with these ever-rising costs of land acquisition, the face value of the duck stamp has also increased over the years. The 1949 stamp cost $2 and by 1950 the price had risen to $3. Today, the Migratory Bird Hunting Stamp purchased from your local post office or sporting goods dealer costs $15.

The contributions made by the Federal Duck Stamp Program to ducks specifically and conservation in general have become legend, its achievements far exceeding the expectations of the men who conceived and initiated the program. But, beyond that, the duck stamp has attained yet another plateau of success, one that its originators could never have dreamed of: the impact of the stamps themselves on the field of wildlife art. After "Ding" Darling's hastily designed first stamp in 1934, other prominent wildlife artists were commissioned by the Bureau of Biological Survey to contribute their designs for the annual stamp. The list of these early duck-stamp designers consists of nearly every giant in the fledgling arena of wildlife art. For example, Frank W. Benson designed the 1935 stamp, followed by Richard E. Bishop in 1936. Others included Roland H. Clark, Lynn Bogue Hunt, A. Lasell Ripley,

Eye appeal, accuracy and artistic integrity are the prime criteria in selecting art for the duck stamps. For as Hal Sheldon said in 1936, "If you never kill a duck, you will still have acquired something that gives any sportsman a thrill whenever he looks at it." Above are Maynard Reece's cinnamon teal.

Owen J. Gromme, and the incomparable Francis Lee Jaques. If John James Audubon himself had still been alive, he would have undoubtedly been commissioned to design a stamp.

These talented artists helped propagate the popularity of the stamp series and make it a national forum for wildlife art. As early as 1936, Hal Sheldon summed up the aesthetic appeal he and his colleagues planned for the stamps: "It is the plan of the Biological Survey that each issue of the stamp should have a value beyond that of the privileges which its possession conveys. If you never kill a duck, you will still have acquired something that gives any sportsman a thrill whenever he looks at it." Certainly, Sheldon's prophecy has been fulfilled.

A BONANZA FOR WILDLIFE ARTISTS

As the recognition and prestige of Federal Duck Stamp art design grew, other wildlife artists began to submit unsolicited waterfowl illustrations in hopes of joining the list of distinguished artists who had already created a duck stamp. It became clear that a new method of selecting an artist was needed. Fish and Wildlife Service artist Robert Hines, who had produced the 1946 stamp, suggested that an art contest be held to determine each year's design. Service officials agreed with him, and for the next 30 years Hines was in charge of coordinating the annual contest. There being no criteria or set formula for choosing judges in those days, mostly regional wildlife people were brought in and told simply to rate the art by number. Whoever had the most points at the end would be declared the winner.

In 1936, Ed Thomas and Ralph Terrill, art dealers for the venerable New York sporting goods store of Abercrombie and Fitch, asked duck-stamp artist Richard E. Bishop to produce a print of his design and sell it framed with an issue of the stamp. The idea proved an immediate success. The first two duck-stamp artists— Darling and Benson—were then asked to reproduce prints of their stamps to bring the series up to date. The result has been the longest running limited edition print series in history, and one that continues today.

Duck-stamp artists receive no remuneration from the government for contributing the winning design and are given only an album containing a

Selection of the wildlife art for any particular year is conducted in a contest sponsored by the Department of the Interior. Although the winning artist is not paid for the art that appears on the stamp, the sale of art prints that follows is enormously lucrative. Thus, the contest has become very competitive. Claremont G. Pritchard drew these winning hooded mergansers for 1968.

sheet of stamps. They can, however, receive the proceeds from print sales made privately. Over the years, as duck stamps have become more popular, the demand for prints has flourished. As a result, duck-stamp prints are now issued in editions numbering in the tens of thousands, thus generating significant revenue for the winning artists.

Predictably, word soon reached the art world that duck-stamp print sales could make an artist wealthy. Indeed, it has become the world's richest art contest, with more money at stake than the Nobel Prize affords. In 1981, for example, 2,099 entries were received for judging (the number has since leveled off, however, to about 500–600 per year).

Any artist can enter the duck-stamp com-petition by submitting a horizontal 7×10-inch waterfowl design and a $50 entry fee. Obviously, with the large number of entries and the financial windfall to be gained by the winning artist, the stakes are considerably greater than when Roger Pruess submitted his design along with seven other artists back in 1949. The judges, who are unpaid, include ornithologists, philatelists, artists, outdoor writers and biologists. Eye appeal, accuracy and artistic integrity are the prime criteria in selection. To ensure anonymity, each entry is assigned a number—the only identification visible to the judging panel. The contest is held at the Department of Interior, starting on the first Thursday in November.

Not surprisingly, a few flaps have arisen over the years. John A. Ruthven's 1960 stamp depicted

a drake and a hen redhead with four ducklings, a subject considered a breech of biological integrity since the drake leaves the hen once the young are hatched. One of the most popular stamps ever was Maynard Reece's 1959 edition featuring John M. Olin's field-trial champion black labrador, Kingbuck, with a dead mallard in its mouth. Some eyebrows were raised over this use of a dead duck as a central theme on the stamp, but the artist's intent clearly was to encourage the use of retrievers to save downed game, not kill it. James Fisher's 1975 stamp shows a canvasback decoy and anchor weight in the foreground with three live canvasbacks flying in the background. The contest rules were changed after this stamp was chosen, requiring the principal (foreground) subject to be a living waterfowl. A single duck had never been portrayed until Albert E. Gilbert's drake hooded merganser appeared in 1978. And until Edward J. Bierly's 1970 rendition of Ross's geese, all stamps were printed in a single color. Multi-color stamps and prints became the rule after 1970 (except for Alderson Magee's 1976 Canada geese, which were executed in scratch board). The first woman to win the design contest was Nancy Howe, whose king eider was selected for the 1991 stamp.

Subjects chosen have represented the hunter's usual fare, from mallards, pintails, canvasbacks, Canada geese, redheads and teal to species that most wildfowlers have never encountered, such as trumpeter swans, nene geese, old squaws, king eiders and black-bellied whistling ducks. Two species that have not appeared to date are the red-breasted merganser and Barrow's golden eye, causing one to wonder if their time is now due.

Both duck stamps and the stamp prints are eminently collectible. The early edition prints, being of very limited edition, are extremely rare. A few complete sets have been assembled and are valued around $250,000. The task of putting together another complete run would be formidable. Current print editions are issued in great numbers, but the demand for them is so strong that their value continues to appreciate as well. Imagine, if someone had purchased every duck stamp from 1934 to the present, he or she would have expended less than $250 for a collection worth more than $4,000. A mint condition 1934 issue alone is now worth $950. Most duck stamps are signed, pasted onto a state hunting license, maybe run through the wash a time or two, and then disposed of. Perhaps less than 1 in 20 duck stamps in the past have been rescued by collectors; even fewer have been kept in mint, unsigned condition. All stamps not sold are destroyed three years after issue, thus preserving the value of more recent issues. Even those who've never hunted ducks but have an interest in wildlife are potential buyers. For the current cost of $15, not only are funds allocated for wetland preservation, but the owner can enjoy a fine piece of miniature wildlife art.

For decades, the art world relegated the realistic style of wildlife painting to the less prestigious realm of "illustration." Thanks to the duck stamps and the prints that have spun off from them, wildlife art has gained in prestige and acceptance, not just by hunters and wildlife lovers, but by fine art collectors and appreciators as well. Today, the Federal Duck Stamp Program can proudly claim a double legacy. It brought the ducks back, thus fulfilling the fervent hope of all those who dreamed, planned and campaigned on behalf of the program. At the same time, it has helped wildlife and sporting art to gain a lofty niche in the demanding world of fine art.

DR. STEPHEN IRWIN combines his love of big-game hunting and sport fishing with his artistic eye and talent for writing. From his home in Montana, he writes on the history of hunting and fishing as well as such specialized topics as antique fishing lures, duck decoys, Kentucky flintlocks and sporting art. His articles have appeared in most of the major outdoor publications. He has also authored the book, *The Providers: Hunting and Fishing Methods of the Indians and Eskimos*. This is Dr. Irwin's sixth in a series of articles for SHOOTER'S BIBLE.

THE STC: FAST, FURIOUS AND FUN SHOOTING

by Ralph Quinn

The scene is right out of a Wild West Show or a shooting gallery at the local carnival, say first-time observers. Falling metal plates, exploding targets and flurries of clay pigeons are only part of the fast-action shooting and win-or-lose drama that spectators and competitors enjoy when the Sportsman's Team Challenge (STC) comes to town. Often billed as "America's Shooting Sports Showcase," this event places the whole range of shooting sports—rifles, handguns and shotguns—into one wildly exciting format that's open to all, including youngsters. By stressing good individual skills and team strategy rather than individual scores, the STC has broadened the natural appeal of a sport on the move. Add to this the relaxed atmosphere among competitors—not to mention $140,000 in prize money—and it's easy to see why the STC is the fastest-growing shooting sport in the U.S. today.

Originated and produced by the National Shooting Sports Foundation (NSSF) in 1987, the STC was always meant to provide public awareness and understanding of the safety and recreational uses of firearms, and to provide hunters and target shooters the opportunity to increase their participation in shooting events. From its inception, STC has emerged as the country's second highest paying shooting championship (Coor's Schuetzenfest, held annually in Golden, Colorado, is still No. 1).

From the start, the STC differed from other major competitions. Instead of arising out of local or regional events, it appeared as a national championship promoting a coast-to-coast regional network of 13 "Preview" shoots. Out of these familiarization tourneys, a strong following has emerged, with more than 100 teams participating in the STC Championship held annually at the Markham Park Regional Target Range in Ft. Lauderdale, Florida.

HOW THE "CHALLENGE" WORKS

Competitors enter as three-person teams in any one of three classes: Open, Sportsman or Industry. They then participate in six events—handgun, rifle, combo and three shotgun events—with a possible score of 100 points per category. A perfect aggregate score for all events is 600 points.

Each team fires a preliminary round, with the top three teams in each class competing in the championship round. Winners in each class are determined by cumulative points in the championship round. Since scores are based on team performance instead of individual results, strategy becomes an important aspect of winning.

Harold Willis, team captain of the "Carolina Blues" Sportsman Class, squares off on the handgun range. This event is a stern test of a shooter's skill and accuracy.

There's an old saying among competitive shooters: the course of fire dictates the equipment needed. The STC is a prime example of that adage. Since all events are timed, a premium is placed on firearms that are quick and easy to load, lock and fire. From the start, each Team Challenge course of fire is designed to emphasize good shooting skills, not highly modified custom guns.

As the photos on these pages indicate, many competitors head to the line with the kind of handguns, rifles and shotguns most shooters and hunters already own. Subtle modifications to stock, magazine or trigger can, however, make the difference in close competition.

THE HANDGUN EVENT

With a premium place on speed and accuracy in all Team Challenge shooting, the handgun of choice is the semiautomatic. Its firepower—rimfire or centerfire—plus a larger magazine capacity, allow shooters to concentrate on target

strategy, not gun mechanics. Popular rimfire choices include the bull-barreled Ruger Mark I, Smith & Wesson's Model 41, Beretta's Model 89 Target, and the Hi-Standard Citation II and Trophy II models reintroduced by Mitchell Arms.

Even though these handguns all perform well out of the box, most shooters have the triggers honed or replaced with drop-in units. Magazine wells and ports are also opened, the goal being to compete with a quick-loading, slick-shooting pistol that won't jam. That way, shooters can concentrate on target acquisition and follow-through.

Because most centerfire pistols chambered for 9mm, 38 Super, 40 S&W and 45 ACP have significant recoil, controlling muzzle lift is all-important in rapid-fire pistol shooting. Although many semiautomatic fans prefer stock guns like SIG/Sauer's P 229, Ruger's M 89 and 90, Colt's Gold Cup Match or S&W's 6000 Series, there's a growing trend toward drop-in muzzle compen-

sator/barrel kits now available from several sources. Using these devices not only diminishes felt recoil, it also puts added accuracy into each shot. By reducing barrel-to-slide fit and muzzle lift, shooters can more often than not increase their scores from "acceptable" to "outstanding" levels.

If money is no object, several major pistol manufacturers offer factory-direct competition models in a wide variety of calibers. Tanfoglio (European American Armory) offers its "Witness" line of semiautomatic pistols (Gold, Silver and Multi-Class), the top choice of 1990 IPSC World Champion Doug Koenig and two-time Bianchi and STC winner Brian Enos. Other standouts include Springfield Armory's 1911 A1 Factory Comp in 38 Super and 45 ACP, Auto Ordnance's comp-style target pistol in 45 ACP, and Laserarms' Series I in 45 ACP with dual-port compensator.

The easiest way to practice for any of the three sections of the handgun event is by placing six paper plates (4 and 6 inches in diameter) at 20 yards. Those who prefer the realistic sound of lead on metal can use one of Birchwood Casey's "World of Targets" steel swingers. Initially, shooters should try to hit half the targets every six shots, pausing a second or two between rounds. For the action-style event, shooters should set two 10-inch square cardboard targets at 10 and 20 yards and try to place six shots in each alternating square in about 20 seconds. For the long-range event, shooters should place several 4-inch cardboard squares at 35 yards; hitting one square in four or five shots is average.

Using the individual target layout, team members should shoot from each section of the event. More than any other segment of the STC, shooters will perform better or worse at any one of the three courses. Since participants are al-

Women are frequent contestants at STC shoots. Using a semiautomatic handgun enables Judy Woolley of the Smith & Wesson team to compete at all levels.

This Model 1911-1A 9mm pistol has been customized by Clark Custom Guns (Stratford, TX 79084). Note the recoil master compensator, which is used to control recoil and enhance accuracy.

lowed to select which leg of the event they'll compete in, the goal is to have the right shooter in the right section. Accuracy and speed are equally important.

Whether to scope or not is another consideration. For shooters with 20/20 vision and strong techniques, it's possible to clean target banks quickly and efficiently; but if your eyes are suspect, an illuminated dot scope using a C.P.M.I., Gault or World Class Carbon Fiber unit on both rimfire and centerfire pistols is highly recommended, particularly for long-range events.

SHOOTING STRATEGY

As with all other forms of reactive and action-style shooting such as IPSC, USPSA and Bianchi, STC competitors should concentrate on mental preparation for the shot, then execute cleanly, with economy of movement. Keeping things simple and relaxed is all-important. For events that are timed, proficiency is the name of the game. The shooter should first locate the target (brightly colored metallic targets are used in the handgun events), thrust the gun into the field of view, and fire as the sights come into the target scoring area. The eyes, head and upper body are kept still; the knees do the work.

After firing, the shooter must follow through, making a mental note about where the sight was following recoil. By working hard on these basic functions and analyzing each shot, the shooter's technique will soon become second nature—

smooth and fluid. One thing all top-flight shooters have in common is *economy of movement*. They are smooth beyond belief—and they are *fast*.

THE RIFLE EVENT

Picture the classic carnival shooting gallery, stretch it out to 90 × s 45 yards, and you'll have a good mental image of what the Texas Challenge Rifle Event is all about. While it confronts individual skills, this phase of the STC tests each team's ability to formulate a winning strategy. Forty colored metallic targets of various shapes and sizes are set out (in banks of 10) at 45, 50, 75 and 90 yards. Additionally, two exploding targets are placed at 45 and 60 yards. Shooters can take out any target they wish before time expires, with 2 minutes allowed for the Sportsman's and Industry Class and 1 minute 45 seconds for Open Class teams. Each target is assigned points based on distance, color and size. Targets of the same size and shape might be worth 1 point, for example, escalating to 2 or more when the targets are placed farther away. Exploding targets are worth 2 points, with bonus points awarded for shooters who clean a bank of targets.

The Combo event combines handgun and rifle shooting. Three shooters—two using rimfire and the other a .22 rifle—fire simultaneously at 50 knock-down targets: ten 4-inch targets at 25 yards, five 4-inch targets and five 2-inch targets at 45, 50 and 60 yards each. In addition, five 2-inch targets are hidden behind one or more of these targets. In scoring, 2-inchers are worth 3 points at 45 yards, 4 points at 50 yards, and 6 points at 60 yards. Five 3-inch targets set at 80 yards count 3 points each. This event poses a rigorous test of team strategy, forcing members to plan ahead about which targets each shooter must work on.

Accurate offhand shooting at a rate of one shot every 3 seconds is what the STC rifle event is all about. Shooters can easily practice for this event by using paper plates or cutouts of 2 to 6 inches, with four targets set between 50 and 60 yards. Aiming at alternating targets, the shooters see how many they can hit within one minute, using no more than 20 rounds. Those shooters who put 10 or more shots in the various targets can consider themselves well on the road to success.

By using steel swingers and a dueling post, this shooter practices getting on target. Since proficiency is all important here, scoped handguns (Ruger Mark I) help shooters shave seconds in timed events; stance should be balanced and relaxed, with eyes locked on the target. For Combo events, the Ruger Mark I and the 10/22 on the table are extremely popular.

Playing as a team requires setting the same targets out in three separate banks at 50, 60 and 75 yards. Team members take turns shooting at each bank. Each member should concentrate on hitting all the targets in one bank. Some teams practice for this event by firing collectively at each bank, starting with the closest one. As most teams have found, however, there's no single "winning formula."

During a recent STC Preview shoot, a poll was taken among the shooters to determine their favorite rifle designs. As expected, the semi-automatic came out on top, in particular Ruger's 10/22. Some competitors favored out-of-the-box pieces, while others took the exotic route, including laminated thumbhole stocks, air-gauged bull barrels, and target turret scopes. The demand for these full-house rifles has risen so sharply that many manufacturers now offer drop-in parts, i.e., air-gauged barrels, hammers, magazines and stocks for the do-it-yourself shooter. The degree of accuracy obtained with these rifles has been truly astounding, with groups of less than a half-inch at 50 yards a common occurrence. One primary reason for this phenomenal shooting is the newly developed Bentz semiautomatic match reamer made by Clymer. With its reduced body taper of 1 1/2-degree throat angle, this device helps shooters squeeze every last bit of accuracy out of their match-grade ammunition without sacrificing ejection reliability.

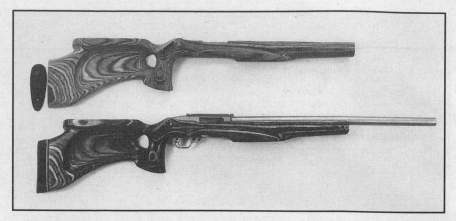

Bull-barreled 10/22s and turret target scopes are preferred by top rifle competitors. A laminated thumbhole stock helps shooters to lock on target and hold steady. This detailed photo shows two views (unfinished and finished) of the customized Ruger 10/22 rifle. It features a bull barrel and laminated, thumbhole silhouette stock of beautifully grained birch.

THE SHOTGUN EVENT

Imagine five computer-controlled traps throwing 50 fluorescent orange clay targets in less than 50 seconds at three shooters, each of whom is limited to two loaded shells! That will give you an idea why competitors describe the STC shotgun event as "fast, furious and mind-boggling."

Three separate events test shotgun-shooting skills under simulated field situations. First, the "Flurry" event pits three gunners at individual stations, firing simultaneously at 50 clay targets thrown in rapid succession. Five traps are set at 35 yards in front of the gunners, and when "pull" is called, five birds rocket over all the stations. Subsequent targets are tossed at 1-second intervals.

The second event is the "Flush," which is similar to the Flurry, in that 50 birds come in quick succession. But here two traps are positioned to the right and two to the left of the shooters, with a fifth trap placed 17 yards in front. The left and right machines toss fixed-angle crossing targets, while the front trap oscillates, producing a variety of angled targets.

In the "Mixed Bag" event, gunners are positioned at individual stations. Again, side traps are set in the Flush position, but the front traps toss rising targets going away. At "pull," the shooting begins from the left station with double targets. After that, selection is random, so gunners can't tell which trap is up next. It's possible to score a total of 100 points, with each broken target worth two points.

Shooters can simulate the experience of taking clay targets tossed in quick and random succession by attending a skeet or trap range. At a combination skeet or trap field, shooters station themselves in front of station 5 and, with their feet positioned comfortably, load two shells, placing their guns in the ready position. Here the puller has the option of tossing singles or doubles from any combination of high-low or trap houses. A score of 18 to 25 is considered respectable. In

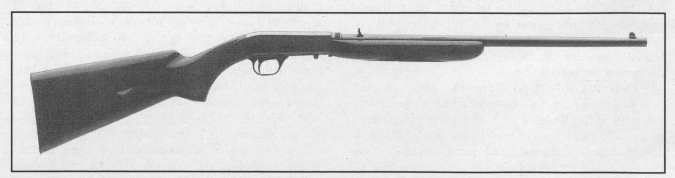

Browning's semiautomatic .22 rifle is also a popular choice among STC shooters.

Reactive rimfire targets come in various geometric shapes, with point values depending on distance and size. Targets are spray-painted after each firing.

the field, a similar situation can be created by using two or more Trius traps set off to the side. A Sporting Clays course is another good way to hone your shotgunning skills for fast-moving targets at various angles and elevations.

Using the same setups, team members can plan strategy by placing themselves in a line about five yards apart and a few yards in front of skeet stations 3 and 5. The puller then throws

12 pairs of targets at 2-second intervals. After a few rounds, the question of whether gunners should shoot in rotation or use a "zone" strategy will become obvious. From then on, it's only a matter of trigger time.

With fast and furious gunning and targets set under 35 yards, the ideal shotgun for STC events is typically semiautomatic and open bored (i.e., skeet, improved cylinder or modified). Gas-

Shooters who seek the realistic sound of lead on steel can practice for STC events using reactive targets. These Birchwood Casey "World of Target" designs are excellent for home-range practice.

During the shotgun event, shooters and spectators alike are treated to some fast and furious action. This STC member is working on "zone" targets during the Flurry segment.

Station	Lead
3 high house	3 feet
3 low house	3 feet
5 high/low house	3 feet
1 high house	3 inches under
1 low house	1 foot ahead
7 high house	1 foot ahead
7 low house	no lead

At trap ranges, work from stations 1 and 5, with birds angling hard left and right. Both require leads of 35 to 36 inches.

One final word: For all those who are interested in competing in the fastest-growing shooting sport in the U.S., start by forming a three-man team and "learn-by-doing" at any one of 13 STC Preview matches held nationwide throughout the year. Practice events are held the day before each Preview competition, so teams who show up early can learn on the job. They may also find themselves rubbing shoulders with many of the country's top-flight shooters, who might just teach them something new!

According to Larry Ference, STC's Preview Shooting Coordinator (National Shooting Sports Foundation, 555 Danbury St., Wilton, CT 06897), "STC is the most fun shooting event three people can have and at the same time, if they're lucky, take home some impressive prize money." By planning ahead, readers can get in on this exciting action during the coming year. Take advantage of it.

operated models from Remington, Beretta and Browning are also popular, because their designs lessen felt recoil over a long day of shooting. To help loading, serious gunners add "speed" paddles to what are otherwise stock guns.

The aiming method in this style of shotgunning is called "pointing out." When the target appears, the gunner centers the bird and in one fluid movement swings, fires and follows through. Again, economy of movement is extremely important to success. The gunner must lock head and eyes on the pigeon, then in one single movement bring the shotgun to the cheek, track the target and pull the trigger.

For those who haven't been on a skeet range in some time, here are a few numbers to remember:

RALPH F. QUINN is an award-winning, full-time freelance outdoor writer and video producer with credits in most major publications, including *Rifle, Wing & Shot, Petersen's Hunting, Field & Stream, Sigarms Quarterly & Handgunning,* among others. As a dedicated shooter, he regularly participates in Skeet, Sporting Clays and IPSC events nationwide. Quinn, who lives what he teaches, bases the advice offered in this article on more than 33 years of competitive shooting.

HOLLAND & HOLLAND VIA WINCHESTER

by Gary M. Brown

The year 1937 began an exciting and eventful period in U.S. firearms history. Not only did Winchester's Model 70—the "Rifleman's Rifle"—make its debut, but also the Great Depression, while still in existence, had finally begun to ease its grip on the American economy. Outdoorsmen who had been unable to enjoy the sport of hunting, even in their own locales, were beginning to plan ambitious ventures that would take them to the far reaches of North America and beyond. To the fortunate few, hunts in Africa once again were possible. Indeed, the 1939 edition of *Game Trails* advised those who were setting out on safari to take their trophies with American-made firearms. That had been the goal of former President Theodore Roosevelt, who, during his African trek in 1909–10, vowed to use his U.S.-built Model 1895 lever-action in .405 Winchester centerfire as much as possible. Roosevelt soon found it advisable, however, to use his English-made side-by-side double rifle against the larger African beasts.

The .405 W.C.F. of that era was, in fact, the most powerful U.S.-made round available. Although it was later offered in bolt action and single shot, it was chambered mostly in the Browning-designed Model 1895 lever gun. Until the introduction of the .375 H & H Magnum cartridge in the Model 70, the .405 remained the most powerful shell chambered in any Winchester-made firearm.

The English had pioneered the development of several huge rounds for hunting in Africa, some dating back to the blackpowder era. Even the more modern smokeless cartridges carried on the British tradition of favoring large-bore diameters. Shells like the .600 and .577 Nitro Express were considered not only proper, but also necessary for the taking of rhino and elephant, using mostly side-by-side double rifles. By the turn of the century, several English riflemakers had begun making the newly developed turnbolt action for their African sporting pieces. Adapted from the Mauser design, these trendsetters rejected the flanged (rimmed) cases of the earlier "Elephant Guns" and the .303 British round of similar design. Some, such as the .404 Jeffery and .416 Rigby (both recently revived in the U.S.) were rimless; i.e., while technically they possessed a rim to be grasped by the extractor, their diameters were identical to the other end of the case. The advantage of this was smoother feeding through the staggered box machines common to most bolt-action designs of that era. Others, such as the .375 Belted Rimless Magnum and Holland & Holland's Super .30 Magnum Rimless (intro-

duced in 1912, and 1925, respectively) featured a belt around the base, just above the rim, on which the case headspaced.

H & H made available nearly identical cartridges in the flanged (rimmed) configuration for those who absolutely demanded the new shells for their side-by-side doubles. The flanged Super .30 and .375 Magnum were later loaded to slightly reduced levels of performance, a modest concession to their theoretically weaker rimmed cases. Soon, widespread use of smokeless powder and its resulting higher energies and velocities dictated a modest reduction of British bore diameters. Westley Richards, Jeffery, Rigby, Holland & Holland and others subsequently built some of the finest, most powerful bolt-action rifles ever offered commercially.

Basically, the belted rimless concept realized the best of both worlds because of its rimmed and rimless case designs. While the belt provided a precise headspacing area—equivalent to the flange of the rimmed case—the system combined the smooth magazine-feeding characteristics of the rimless round. Any added strength afforded to the case's web area was considered a bonus.

The acquisition in late 1931 of the Winchester Repeating Arms Company by the Olin family, owners of the competitive Western Cartridge Company, was not incidental to the eventual chambering of the belted version of the Super .30 and .375 H & H in Winchester-made rifles. John Olin, who had developed the "Super-X" high-velocity shotgun shell, was a "high-performance" advocate within the firearms industry. Western had, in fact, made shells for both H & H belted magnums from about 1925, long before any U.S.-produced gun was available to fire them.

The Model 54—Winchester's first all-commercial centerfire sporting rifle—was introduced in 1925. While it incorporated many advanced features based on the Mauser design, it probably lacked the strength to take the stress of the planned magnum rounds. The Model 54 needed many improvements unrelated to strength. Its Mauser-type rotating safety made it nearly impossible to low-mount a scope directly over the bore. The nonhinged floorplate meant that, to unload the piece, shells had to be worked through the action; moreover, its two-stage trigger pull was nothing short of terrible. Clearly, Winchester/Olin needed a "new" high-performance rifle.

These boxes of .300 H & H and .375 H & H "collector" ammunition include Western .375s (note the "Bear" logo) with 300-grain Silvertips, and Winchester's 270-grain soft-points. A similar box of Winchester H & Hs (lower left) contains scarce 150-grain rounds. The Remington Box (lower right) contains 180-grain boat-tails meant for target use.

THE MODEL 70 FULFILLS A NEED

That dream came true when production of the Model 70 began in 1936. While most of the earliest M-70s were chambered in .30-06 Govt., the .300 Magnum (which was really a redesignated Super .30 belted/rimless) and the .375 H & H Magnum also made their appearance the following year when the rifle itself was introduced to the public. The gun's hinged floorplate allowed rounds to be removed from the magazine without having to cycle them through the action. The M-70's completely independent bolt stop provided that extra measure of safety advantage over the Model 54. Its fully adjustable single-stage trigger is still considered one of the best ever designed for any rifle.

Common to both the Model 70 and the Model 54 were their Mauser-type, nonrotating, claw extractors. This system not only provided positive extraction of both loaded and fired rounds, it also afforded controlled feeding of each shell from the moment one emerged between the magazine guide rails until it actually entered the gun chamber. The frame-mounted, spring-loaded ejector needed only a small notch in the bolt face, allowing the left (top) locking lug to remain totally intact—a great improvement indeed over conventional Mauser actions.

The Model 70's special magnum-length action, which allowed chambering of the super-long H & H rounds, caused much conjecture. In fact, all M-70 actions made before 1964 are identical in length, whether chambered for the small .22 Hornet or the large .375 H & H Magnum. Several modifications were needed to handle these differing shell lengths. Magazine followers of various sorts were used as were sheet steel baffles (to ensure that magazine boxes had the right dimensions for a given cartridge). A bolt-stop extension, attached to the extractor collar, limited overall bolt travel to that of the proper round. A final alteration, required only for the two H & H Magnums, was to relieve the forward receiver ring slightly and the receiver bridge considerably, thus allowing the rounds to fit and feed through the M-70 action.

An early Winchester brochure announced the new Model 70 and its nine original chamberings—.22 Hornet, .220 Swift, .250-3000 Savage, .257 Roberts, .270 W.C.F., 7×57mm Mauser, .30-06, .300 H & H Magnum and .375 H & H Magnum. The new belted cartridges received special attention as follows:

.300 H & H MAGNUM—A new development in .30 caliber modern high intensity ammunition, known for extremely consistent performance and stability, with great power... and fine accuracy up to the longest practical hunting ranges. With 180-gr. bullet muzzle velocity is 3060 f.p.s. With 1800-gr. hollow point or 220-gr. soft-point bullet, [it is] a highly satisfactory choice for long range use on North American and Old World thin-skinned game.

.375 H & H MAGNUM—Here is the maximum of efficiency in abundance of power, with a range of bullet choice of 235, 270 and 300 grains, in a modern cartridge of high intensity. Consult the ballistics table and observe the very high sustained energies at 300 yards, far in excess of the most powerful .30-06 [cartridge] at the muzzle.

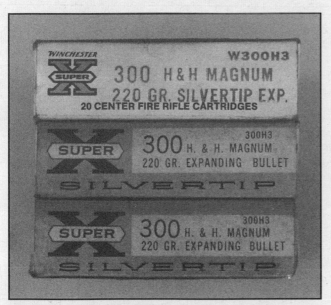

These discontinued Winchester-Western 220-grain Silvertip 300 H & H rounds can be used for large, potentially dangerous North American big game when shooting a Model 70 .300 H & H Magnum.

Both rounds were clearly intended by Winchester for use on the largest and most dangerous North American game, and selectively on all manner of great animals found on the African continent as well.

From their introduction and throughout production, the Model 70 Sporter versions in .300 H & H Magnum (Standard and Super grades only) featured a tapered and relatively lightweight 26-inch barrel. The .300 H & H was also popular as a target round, offered at various times as a Target Model with 26-inch heavy barrel and a super-heavy "Bull" gun with 28-inch barrel.

A MATTER OF BALLISTICS

From early on, U.S. ballistics for the .300 H & H cartridge were formidable. Offered initially in bullet weights of 180 and 220 grains, each load greatly exceeded the output of its primary competitor: the .30-06 Springfield round. For in-

This author's own Model 70, which is chambered in .300 H & H Magnum and bears serial #525312, features a Monte Carlo stock with plastic buttplate. Made in 1961, it represents extremely late production for any .300 H & H. It remains a scarce item that represents one of only 1,297 late H & Hs produced with that type stock. Fitted with the relatively lightweight 26-inch Sporter barrel, it has a Lyman folding rear sight, plus a ramped, hooded front sight. One of the few real advantages of the Model 70 chambered for .375 and .300 H & H rounds, by the way, is that the Holland & Holland magazines hold four rounds (total capacity of five rounds), while Winchester's Short Magnums have a capacity of only three cartridges (total four rounds).

The photograph shown on page 62 is an early post-World War II M-70 standard rifle originally chambered by Winchester for the .300 H & H Magnum. Before a single round was fired through

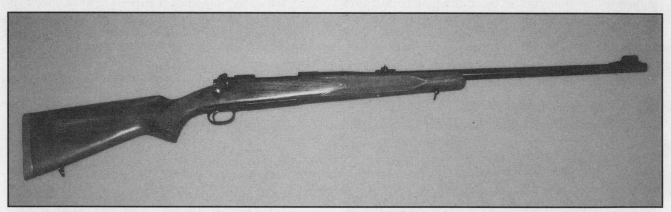

The Winchester Model 70 in .375 H & H Magnum has several desirable features, including a low-comb classic stock and #70B rocking one-piece/two-bladed rear sight. This gun is an excellent choice for close-range shooting of dangerous game.

stance, the .300 H & H 180-grain hollowpoint yielded 3060 fps and 3745 ft-lbs of energy at the muzzle, whereas the "Super Speed" 180-grain .30-06 managed only 2710 fps at 2940 ft-lbs at its spout. Eventually, factory 150-grain bullets were offered for the .300; today, only a 180-grain projectile is commercially loaded for the round. Ironically, as later factory .30-06 fodder was increased in ballistic performance, the .300 H & H was reduced. The modern "Super .30" has only a slight edge over '06s of identical bullet weight.

this big gun following its acquisition in the early 1950s, it was rechambered to the then popular .300 Weatherby Magnum. The result was a super-powerful rifle of proven performance. It has taken countless numbers of big-game animals and remains as tight and accurate as it was nearly 40 years ago, a testimony to the original Winchester design as well as to the gun's rifling.

Current popularity of the .300 H & H has suffered due to competition from other .30-caliber belted magnums later produced by Weatherby,

Winchester and others. Remington includes the .300 H & H in its annual Model 700 classic series, and Winchester recently revived the round in its latest M-70 models. The shell remains adequate for taking all North American game and most African plains animals. Unlike the .300 version, the .375 H & H was not discontinued at the end of pre-1964 Model 70 production.

Barrel lengths and shapes of the various Model 70s chambered for the .375 H & H differ significantly. The first few Model 70s offered in

While standard open rear sights were fitted to .300 H & H Model 70 rifles, the .375s generally featured dual-leaf open sights. The earliest of these was the Lyman 6W dual-folding leaf. The taller of the two blades was sighted-in for long range (approx. 500 yards), while the shorter leaf was zeroed in from point-blank range out to about 200 yards. Beginning in 1953, Winchester's #70B roll-over, one-piece, two-leafed sight was introduced. This improved version rocked back and forth on a large, centrally mounted pivot screw

This group of pre-1964 Winchester Model 70s features the author's Model 70 in .375 H & H Magnum (top), made in 1954; a Model 70 (center) that left the factory in .300 H & H but was later rechambered to .300 Weatherby Magnum; and the author's mint Model 70 in .300 H & H Magnum.

.375 were produced with a standard 24-inch lightweight barrel. Recoil in these Sporters was substandard at best. Almost immediately, a medium-heavy, virtually straight (nontapered) 24-inch barrel replaced the thinner tube. Shortly after World War II, a 25-inch contoured, medium-heavy variation replaced the earlier 24-inch straight barrel. This barrel shape was used throughout the remainder of production, including some very late .375 H & Hs redesignated as "Alaskans," along with M-70s chambered for the .338 Winchester Magnum cartridge.

that also allowed adjustment for windage. Both of these earlier Lyman 6W versions were "drift adjustable only" for windage; hence, the Model 70B was a big improvement.

The author's Standard rifle (#319190 shown left), chambered for .375 H & H Magnum and made in 1954, features Winchester's #70B dual-bladed rear sight and a low-comb stock. Winchester had earlier introduced a Monte Carlo stock (without cheekpiece) for its Standard Grade rifles intended for scope-mounted guns. Since the .375 H & H shell is considered adequate for most

This pre-1964 Winchester Model 70 (originally in .300 H & H) was rechambered to .300 Weatherby Magnum before firing a single round. It has taken numerous big-game animals and is nearly as tight and accurate as it was when new.

large game, including Cape buffalo, rhino and even elephant, and since such beasts are usually taken at relatively short ranges with open sights, this rifle's classic low-comb stock design is a definite plus. Like all Winchester factory-produced .375 H & H Model 70s, this is equipped with a rubber recoil pad. Earlier guns made between 1937 and 1961 featured solid red rubber pads, which later were generally of the webbed or ventilated type. Except for some graying on the bolt knob and a few honest dings in the wood, this rifle still grades out as "near mint."

Because this .375 H & H is fired sparingly, factory ammo is almost always used. Current Winchester and Remington standard production .375 H & H rounds are loaded with a variety of 270- and 300-grain projectiles, including a blunt-tipped 300-grain full-metal-case (FMC) round for use on the largest African species. Federal's Premium Safari custom line also offers factory-loaded .375 H & H ammo in a 250-grain boattail soft-point, a 300-grain Nosler Partition Expanding and a 300-grain solid. Fortunately, today's factory .375s haven't been loaded down as drastically as their .300 H & H brethren. For example, the 270-grain pointed soft-point in the Winchester ballistics table appearing in the 1940 edition of SHOOTER'S BIBLE cites muzzle velocity of 2720 fps and muzzle energy of 4440 ft-lbs. This edition of SHOOTER'S BIBLE lists Remington's .375 H & H 270-grain pointed soft-point with a muzzle velocity of 2690 fps and muzzle energy of 4337 ft-lbs.

NOTE:

Shooters and collectors of the pre-1964 Model 70 should pay careful attention to the ".300 Magnum" chamber marking of the guns first produced in this caliber. This designation *always* indicates that the rifle is chambered for the .300 H & H Magnum. Later, guns bearing the H & H designation were included in the imprint. A problem arises when unscrupulous or unknowing individuals confuse the early marking with the much later .300 Win. Mag. round, which was introduced in mid-1963 and was used for only about six months in the pre-1964 Model 70. While the .300 Win. Mag. shell will not fit in guns marked ".300 Magnum," such pieces can be rechambered to accept the newer shell. The serial number range of actual .300 Win. Mags is always above 500,000; but serial numbers can be altered. So remember, authentic .300 Win. Mags are always marked "MODEL 70—.300 WIN. MAG." Other markings are likely fakes.

Both the .300 and .375 rounds were originally intended by Holland & Holland for medium to large game found in Africa. Adopted by Winchester in 1937, Model 70s in .375 and .300 H & H found fame all across North America as

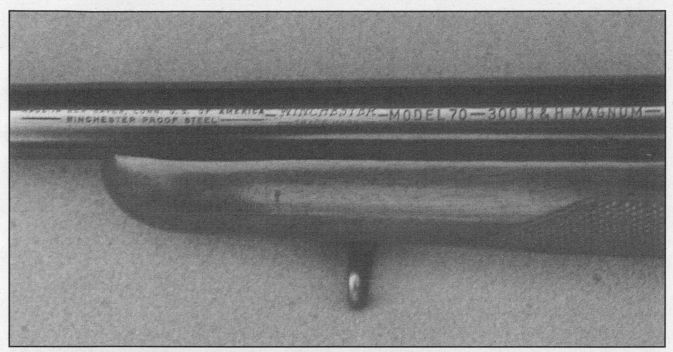

The recent .300 H & H Magnum Model 70s are chamber-marked, as illustrated. Early guns in this caliber were stamped ".300 Magnum." Some unscrupulous collectors have tried to sell such rifles as the much later—and more valuable—.300 Winchester "Short" Magnum.

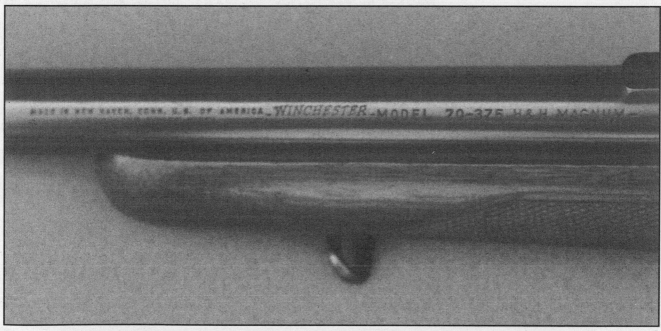

Note the .375 H & H Magnum chamber marking on this Model 70. Although early M-70s in this caliber were marked ".375 Magnum," little confusion is associated with this round, since there was never a .375 Winchester "Short" Magnum cartridge.

In his classic African hunting guidebook, Safari, Elmer Keith, a self-admitted big-bore fanatic, claimed the .375 H & H Magnum inadequate for Africa's big game, albeit acceptable as a "plains rifle." The .300 H & H received little mention in the book at all.

well. Older original .300 H & Hs are still admired and sought after by hunters and target shooters alike.

So after more than 80 years, the .375 H & H cartridge is still regarded as the best all-around shell ever devised. As a true Englishman might say, "Holland & Holland via Winchester has indeed been a smashing success."

GARY M. BROWN is a nationally published freelance writer of firearms. He is also a contributing editor for several gun-related periodicals. A graduate of the University of Florida's College of Journalism and Communications, Brown resides in north Florida.

After 91 Years Browning Strikes Gold.

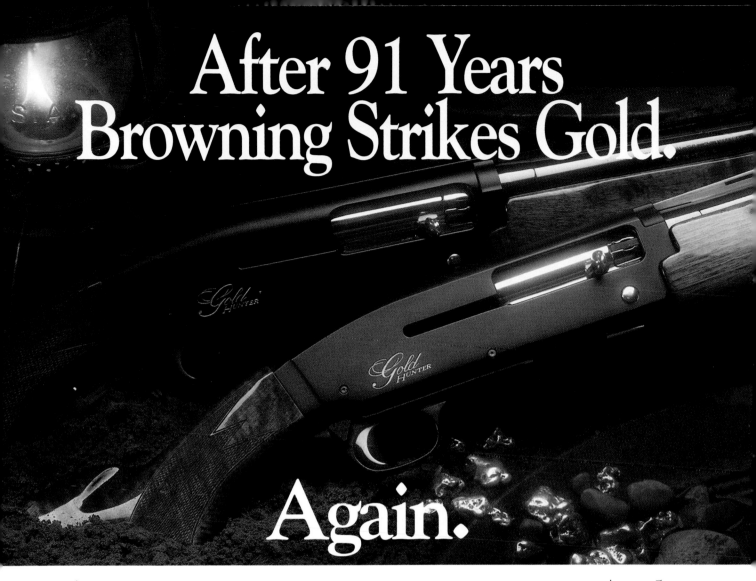

Again.

A NEW GAS-OPERATED SEMI-AUTO WORTHY TO SUCCEED THE LEGENDARY AUTO-5.

Today, nearly 91 years after the introduction of the A-5 — the world's first successful semi-automatic shotgun — Browning has again struck the mother lode with the new Gold shotgun.

Shoots all loads for all game. Shoot all loads in any order, from 2 3/4" field loads to 3" magnums. With light 1 oz. loads most of the energy from expanding gases is used to operate the action. With heavy magnum loads, excess gases are vented away. This results in a higher level of efficiency and reduced recoil to your shoulder. A self-cleaning gas system ensures shot-to-shot reliablity under all conditions.

Speedier operation with speed loading. The Gold is the only gas auto with the advantage of speed loading. Just slide a shell into the magazine

and it's automatically sent to the chamber, ready to fire. The oversized triangular safety allows quick operation while keeping your eye on your target.

Easy take-down with no tools. No auto-loading shotgun is easier to take down than the Gold. There are no loose O-rings or springs that can pop out and get lost. You can completely

Gold

disassemble the barrel, forearm, gas system, bolt and trigger group in minutes — with no tools.

Extraordinary pointability, balance and craftsmanship. The Gold comes to the shoulder swiftly and naturally. The smooth, contoured lines of the forearm and receiver contribute to the Gold's pointing ease. The exceptional between-the-hands balance is more like that of an over & under than a semi-auto. Through every inch of the Gold shotgun, fine Belgian craftsmanship is evident.

All 12 ga. models have back-bored barrels for better patterns. Gold 20 ga. models are lighter weight with trimmer overall dimensions. Both have the advantage of the interchangeable Invector choke tube system.

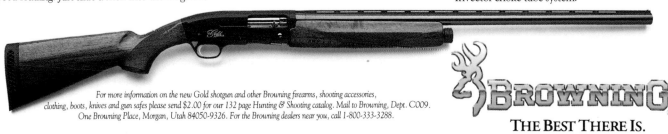

For more information on the new Gold shotgun and other Browning firearms, shooting accessories, clothing, boots, knives and gun safes please send $2.00 for our 132 page Hunting & Shooting catalog. Mail to Browning, Dept. C009. One Browning Place, Morgan, Utah 84050-9326. For the Browning dealers near you, call 1-800-333-3288.

BROWNING
THE BEST THERE IS.

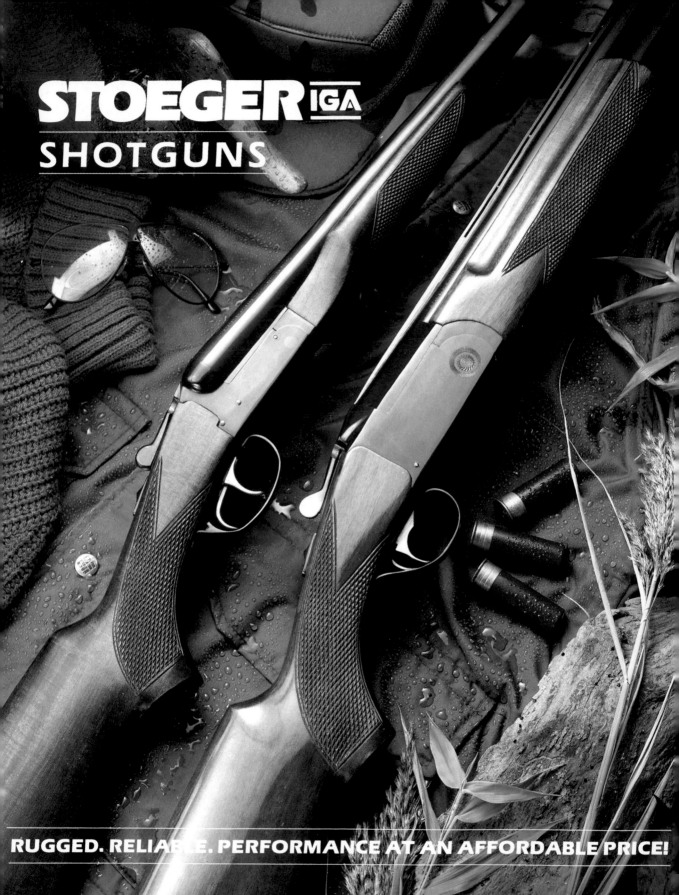

PEERLESS

Living Up To The Legend

TRIM, PERFECTLY BALANCED, NO NONSENSE. THOSE VIRTUES HAVE MADE THE OVER-AND-UNDER THE WINGSHOOTER'S FAVORITE FOR GENERATIONS. NOW REMINGTON PROUDLY INTRODUCES A DOUBLE SHOTGUN TRULY WORTHY OF THE NAME: PEERLESS™. THIS 12-GAUGE MASTERPIECE BLENDS INSTINCTIVE HANDLING WITH TECHNICAL ADVANCES THAT SET IT APART. FEW GUNS CAN MATCH ITS PERFORMANCE, NONE CAN MATCH ITS VALUE. TO FULLY APPRECIATE IT SEE AND SHOULDER ONE TODAY. IT'S WHAT YOU'RE SHOOTING FOR.

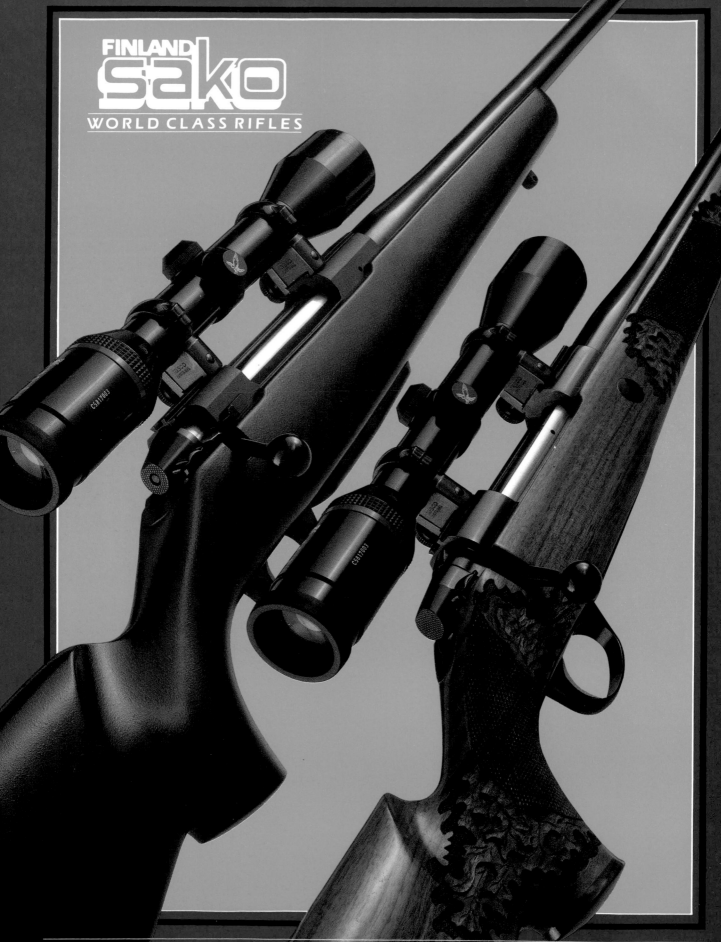

FINLAND **sako**

WORLD CLASS RIFLES

COMMITMENT TO EXCELLENCE—A SAKO TRADITION

F. C. SELOUS: AFRICA'S GREATEST HUNTER

by Jim Casada

Frederick Courteney Selous was one of those rare individuals whose boyhood dreams actually came true. No less an authority than Theodore Roosevelt once called him, "the greatest of the world's big-game hunters." Certainly few knowledgeable sportsmen of the late Victorian and Edwardian eras would have disputed Roosevelt's assessment of Selous' African achievements. Thanks to his prowess afield and the literary grace he brought to the dozens of books and articles he wrote about his exploits as a hunter and explorer, Selous gained a considerable measure of fame and immense popularity among his contemporaries.

There is no doubt that when it came to sporting mastery in Africa, Selous stood in a class by himself. Humble and unassuming, he never thought of himself in that way, but by the same token, he always knew what he wanted to do in life. When his father first introduced the boy with the piercing blue eyes to the headmaster at England's famed Rugby School, the lad immediately declared that he wanted to be like David Livingstone. He may never have been clawed by a lion as that famed African missionary and explorer had been, but in virtually every other way Selous' career in Africa was even more exciting and adventurous than that of his boyhood idol.

Selous' exploits in Africa later served as the model for H. Rider Haggard's renowned fictional character, Allan Quatermain; indeed, Selous' own books rank among the most important ever written on big game hunting. He was, in short, a man who squeezed every drop of goodness from the sponge of life.

THE EARLY YEARS

Selous (pronounced *Sell-oo*) was born in Regent's Park, London, on New Year's Eve, 1851. From the time he could walk, he developed a great love of nature and the outdoors, and on numerous occasions during his youth his preoccupation with sport and adventure got him into trouble. During his days at Bruce Castle School in Tottenham, where he began his formal education, he constantly incurred the wrath of his tutors through mischievous escapades and daydreams of hunting in Africa. He sometimes neglected his lessons to read the African travelogs of such great sportsmen of the time as William Cornwallis Harris and Roualeyn Gordon Cumming.

The lad nonetheless managed to gain entrance to Rugby College—famed for its abilities to shape the character of men who epitomized all that was good and great in Englishmen. Had Selous not been fortunate enough to have the kindly

J.M. Wilson as his housemaster, however, he might not have lasted long there. As Wilson later reminisced:

> He [Selous] breaks every rule; he lets himself out of a dormitory window to go birds'-nesting; he is constantly complained of by neighbours for trespassing; he fastened up an assistant master in a cowshed into which he had chased the young villain early one morning; somehow the youngster scrambled out, and fastened the door on the outside, so that the master missed morning school.

While a student at Rugby, Selous developed the athletic abilities and superb fitness that later

Even as a lad, adventure was his middle name. Note the bow and arrow in Selous' hand in this photograph taken shortly before he entered Rugby.

stood him in such good stead in the field. He played on several varsity sports teams, including swimming, track, rugby, cricket and rifle shooting. He was constantly in and out of trouble at school for taking small game with his "tweaker" (a small-bore rifle he kept hidden at a nearby farmhouse), and at one point he dared to rob some eggs from a nearby farm. Fortunately, Selous' speed as a swimmer and runner enabled him to outdistance the irate gamekeeper.

His derring-do at this point in life ironically gave Selous a foretaste of the dangers that would be his daily lot in Africa. While home for the Christmas holidays, for example, he went ice-skating at a nearby park when suddenly the ice broke, plunging dozens of other skaters to frigid deaths. Selous, however, calmly spread-eagled himself on a floating piece of ice and carefully maneuvered from slab to slab until he reached safety. It was that kind of coolness in times of crisis that served him well in tight spots on his later African hunts.

Along with his daredevil pranks a more serious side began to develop in the young student. He was a founding member of Rugby's Natural History Society and excelled in geography and natural history. All the while, he read incessantly about anything having to do with travel in Africa. After leaving Rugby in 1868, he began taking short hunting trips on the European continent. Finally, while still a teenager, Selous landed at Algoa Bay in South Africa. He had £400 in his pocket and a burning desire to launch a career as a professional hunter. He managed to make his way to the kraal of Lobengula, the powerful and widely feared leader of the warlike Matabele. Selous later recorded their meeting:

> He asked what I had come to do. I said I had come to hunt elephants, upon which he burst out laughing and said, 'Was it not steinbucks (a diminutive deer-like animal) that you came to hunt? Why, you're only a boy.' I replied that, although a boy, I wished to hunt elephants, and asked his permission to do so, upon which he made further disparaging remarks regarding my youthful appearance, and then rose without giving me any answer.

In his early African hunting days, Selous dressed ruggedly in this unique, self-designed outfit. Note the spear and at right the large-bore muzzleloader he used on large game.

Nonetheless, Selous got the necessary permission. Using the cumbersome but inexpensive large-bore muzzleloaders favored by Boer hunters, he quickly developed into a skilled hunter. With a Hottentot companion he nicknamed "Cigar," Selous killed 78 elephants from 1872 to 1874. During one 120-day period in 1873 he and a partner, George Wood, collected some 5,000 pounds of ivory while hunting along the Zambezi River. Other big game—rhinos, lions, Cape buffalo, hippos along with lesser game for the pot— also fell regularly to his guns. Today, Selous' carefully maintained game lists read like a catalog of slaughter. But it must be remembered that he was hunting in an age when the supply of game in Africa seemed virtually inexhaustible, and when his gun provided the food for as many as a dozen men. Also, to his considerable credit, he became in later years an ardent conservationist who spoke out stridently on the need for strict game laws throughout Africa.

For a full quarter century, Selous led a life bursting with danger, romance, excitement and adventure on the African veldt. Time and again, he traveled through country seldom, if ever, seen by white men. His unusual but practical hunting

Selous makes entries to his diary while at a campsite in southern Africa. His daily hunting endeavors yielded much for the camp table (above) as well as for his trophy walls.

outfit consisted of leather shoes, close-fitting shorts, a comfortable short-sleeved shirt tightened about his waist, a belt holding an ammunition pouch, and a bush hat. Thus attired, he was ready for all eventualities. It was, as Teddy Roosevelt wrote many years later, "a wild and dangerous life, and [it] could have been led only by a man with a heart of steel and a frame of iron."

Selous was, indeed, brave almost to the point of foolhardiness, a man who simply did not know what the word "quit" meant. He faced imminent danger so many times that such incidents became almost commonplace. He always maintained that Cape buffalo were the most dangerous of all of Africa's big-game animals, and doubtless this view came about following a brush with death

at the hands of a wounded bull. After Selous' muzzleloader had twice misfired, the angry beast charged just as Selous shot at point-blank range. Seemingly unaffected, the animal fatally gored the hunter's horse and tossed Selous high into the air. A second charge by the buffalo sideswiped Selous, dislocating his shoulder. Fortunately, the bull, dazed by the bullet, turned abruptly and trotted off into the bush.

On another occasion, Selous lost his horse while chasing giraffes and wandered for several days through desert-like terrain, without food or water, until rescue arrived with death only moments away.

Such experiences were in truth routine for old African hunting hands. What set Selous apart from other great sportsmen were his extraordi-

nary powers of observation and his diverse interests and talents; but, most significantly, he was blessed with an incomparable ability to describe his adventures both as a writer and raconteur.

A MAN OF LITERARY TALENT

Selous' writing skills were not obvious at first. Aside from a short break in 1875–76, hunting the Dark Continent was his sole objective during the initial decade of his trips. But, in 1881, he came home determined to write a book on his experiences. The result was *A Hunter's Wanderings in Africa*—possibly the single most popular book ever written on African sport and one that has been reprinted at least a dozen times. The book earned the budding author a fair amount of money and a lot of favorable attention. As a result, whenever he returned to Africa, Selous carried with him a number of commissions from

At age 42, Selous' second major book, Travel and Adventure in South-East Africa *(1893), was published, contributing to his growing reputation as a writer. In an 1893 issue of* Review of Reviews, *this Victorian-style portrait appeared with an article profiling the author.*

museums and private collectors to obtain specimens or trophy heads.

Selous' interests subsequently began to branch out. He developed into a superb natural historian, recognized as an international authority on such widely differing subjects as butterflies and protective coloration in animals. He also became a noted explorer, and his achievements in this regard were enough to earn him the coveted Founder's Medal of the Royal Geographical Society.

Selous continued with his literary efforts, writing regularly for magazines such as *The Field*. His lively, readable articles were just what the prissy Victorian public craved. His second major book—*Travel and Adventure in South-East Africa* (1893)—describes his adventures from 1881 to 1890. At the time this volume appeared, Selous was still a relatively young man of 42, and much of his African career still lay ahead of him.

He had become increasingly concerned with political affairs during the so-called "scramble," which saw the European powers vying for control of African territory. As a result, he was hired in 1891 by the famous British empire-builder, Cecil Rhodes, to lead a "Pioneer Corps" of settlers into the region (this later became Rhodesia, or modern Zimbabwe). He performed these duties well and shortly thereafter played a major role in putting down an uprising known as the "First Matabele War."

During one of his return trips to England, Selous met and fell in love with Gladys Maddy, the strikingly beautiful daughter of a Gloucestershire parson. They married in 1894 and soon had two sons, Fred and Harold. Sadly and ironically, Fred died in combat during World War I precisely one year after his father's death in the East African campaign. For the moment, though, such tragedies lay far in the future, and Selous' wandering inclinations pulled him as strongly as ever.

As was true of so many of the great pioneers, it was almost as if he were addicted to the continent. Having tested, in his own words, its "wild freedom, wonderful vistas, and incomparable sport," he was drawn back to it time and again. He tried his hand at farming, land management, gold mining and other enterprises, but the hunting instinct always remained strong in Selous.

This rough-and-ready-looking group comprised the Pioneer Corps officers who led the first English settlers into Rhodesia in the 1890s. Selous sits in the second row, far right, chewing on a straw.

In 1896, the family returned to England. Selous had recently fought in the Second Matabele War and described his experiences in a book entitled, *Sunshine and Storm in Rhodesia.* He and his family settled in Surrey, where Selous eventually built a home and a splendid big-game museum. He was never there for long, though, for the hunter's horn repeatedly called him to sporting forays in Africa, Europe and North America.

By now Selous had become an accomplished and widely acclaimed author—so much so that his writing became the family's principal source of income. He contributed to a number of the anthologies and multi-volume sporting collections that were so popular at the turn of the century; and he continued to work steadily on books, among which were *Sport and Travel: East and West* (1900) and what some consider his greatest

work: *African Nature Notes and Reminiscences* (1907).

During the decade or so leading to World War I, Selous was constantly in the public eye, thanks to his writings and his well-known friendships with such notables as Charles Sheldon and Theodore Roosevelt. Selous' extended Alaskan hunting trip with Sheldon, in fact, was later described in detail in Sheldon's book, *The Wilderness of Denali.* Selous and Roosevelt corresponded regularly, and eventually Selous visited the then U.S. president in the White House, where he held the President and his dinner guests enthralled evening after evening as he related his African experiences.

Gradually, Selous eased into the mold of a sporting squire. He enjoyed entertaining fellow sportsmen at his home, which was close enough

Spawning an incredible array of African fauna is one wall of Frederick Selous' trophy room at his home in Surrey, England.

BOOKS BY FREDERICK SELOUS

A Hunter's Wanderings in Africa (1881)
Travel and Adventure in South-East Africa (1893)
Sunshine and Storm in Rhodesia (1896)
Sport and Travel: East and West (1900)
Recent Hunting Trips in British North America (1907)
African Nature Notes and Reminiscences (1908)
The Gun at Home and Abroad: The Big Game of Africa and Europe with J.G. Millais and Abel Chapman (1914)

to London to permit regular visits to his clubs and learned societies. There were always invitations for grouse hunting and deer. stalking in Scotland, and the next chase—be it bird-nesting, butterfly gathering or big-game taking abroad— was never far off. In short, Selous lived a sportsman's dream during these prime years of his manhood, all the while making one trip after another back to Africa.

When World War I broke out in 1914, Selous, now 63, eagerly sought active service. His age proved a temporary problem, but through sheer persistence in the face of official objections, plus his obvious excellent health, he secured a commission early in 1915 as Intelligence Officer for

the 25th Royal Fusiliers. Officially known as the "Legion of Frontiersmen," the 25th was nick-named "The Old and Bold." Destined for service in East Africa, its ranks included Selous' long-time hunting friend, the American millionaire W. N. MacMillan, a man whose size matched his wealth (his sword belt measured 64 inches). Other fighting colleagues were Cherry Kearton—who would eventually gain acclaim as a big-game photographer, a servant from Buckingham Palace, a Park Lane millionaire, a professional strong man, half a dozen bartenders, two Texas cowboys, and a former Honduran Army general. Selous felt right at home in such interesting and multi-talented company.

His tracking skills, his knowledge of African terrain and his understanding of the continent's ways proved invaluable once Selous and his comrades arrived in East Africa. He was soon promoted to the rank of captain and was awarded the Distinguished Service Order. His unflagging energy, not to mention the game he was able to obtain for the regimental mess, made him the envy of men a third his age.

On January 4, 1917, while out on a morning reconnaissance, Selous was shot and killed by a sniper. So well loved was he that some of the askaris (native soldiers) accompanying him went wild with rage and routed nearby German forces, despite being outnumbered five to one. Clad in his khaki shorts and slouch hat, with a jaunty bandanna about his neck, Selous was buried beneath a tamarind tree with only a crude wooden cross and a pile of stones to mark his grave. Fittingly, the area eventually became a part of the Selous Game Preserve, thereby perpetuating the keen interest in preservation Selous had espoused.

From a modern perspective, Fred Selous epitomized what African sport was all about during the Victorian heyday of big-game hunting. He was among the fortunate few who enjoyed the opportunity to lead a full, useful life while doing precisely what he enjoyed most. Historically, he is important for a variety of reasons, representing as he did the best of the imperial ethos that made Britain great. He was, according to Cecil Rhodes, "The man above all others to whom we owe Rhodesia to the British crown."

But most of all, Frederick Selous was a peerless hunter. That is how he is remembered most today, and, given his deep love for sport, this is fitting. As Roosevelt wrote in African Game Trails, "Probably no other hunter who has ever lived combined Selous' experience with his skill as a hunter and his power of accurate observation and narration." Books from Selous' prolific pen form a lasting testament to his abilities, and they endure as a memorial to him. Through them, the man who truly deserved the title his contemporaries gave him—"Nimrod of Africa"—still lives.

JIM CASADA, a regular contributor to SHOOTER'S BIBLE, is presently completing work on a book that brings together long-forgotten writings by Fred Selous that have never appeared in book form. Co-editor of *Turkey & Turkey Hunting* magazine, Casada serves on the staffs of several other publications, including *North American Fisherman, Southern Outdoors*, and *Sporting Classics*. He is also the editor of a trilogy of books: *Hunting and Home in the Southern Heartland, Tales of Whitetails*, and *America's Greatest Game Bird: Archibald Rutledge's Turkey Hunting Tales*. Casada is also the author of *Modern Fly Fishing*. He currently serves as secretary-treasurer of the Outdoor Writers Association of America and as president of the Southeastern Outdoor Press Association.

THE FASCINATION WITH QUARTER-INCHERS

by Wilf E. Pyle

Technology dominates our lives today. Computers, lasers, organ transplants, trips to outer space—all tell us that modern technology is well ahead of most people as we race toward the start of the third millennium. For many of us, using the technology that surrounds us has become a worrisome challenge, like programming the VCR.

That feeling wasn't always a part of our lives. Early cartridge designers, for example, eagerly embraced the technology of their time, and from their efforts evolved many modern hunting cartridges. Any change that might increase cartridge velocity and striking energy has, throughout history, been eagerly accepted. The downside of this group effort is that not all cartridges have survived or become successful commercial rounds.

In cartridge evolution, more of the .25 calibers (often referred to as "quarter-inchers") have moved from wildcat status to commercial success than any other calibers in the past 50 years. The quarter-inchers have indeed held a longtime fascination for North American shooters and cartridge aficionados. Most well-known shooters and firearms writers have, in fact, favored the .25 caliber at some point in their careers: Ned Crossman, Charles Newton, Harvey Lovell, Richard Simmons, Parker Ackley, Adolph Nieder, A. E. Mashburn, Lyle Kilbourn, Charles Askins, Elmer Keith, Grove Wotkyns, Col. Townsend Whelen, Jack O'Connor and many others.

The reasons for this high level of interest are readily apparent. The .22 caliber can include bullet weights that run the entire range of shooting applications, from varmint to big-game hunting. For some, it represents the single most versatile group of cartridges ever produced. For young or female shooters, the .257 Roberts and .250 Savage are favorites, mostly because they produce light recoil, mild report and high accuracy. As a result, these cartridges are use more frequently than others, therefore producing better shooters who tend to be pleased with their choice of rifles and cartridges.

Quarter-inch calibers do well because many of them offer high velocity that translates into flatter trajectories and better long-range accuracy. Charles Newton, a designer of many different cartridges, once had a .40-90 Sharps case necked down to .25 caliber and outfitted with a 100-grain bullet. Tests showed that the velocity of this round was indeed high for the time, but not as great as a .30-40 Krag necked to .25 caliber. Shooters are always interested in improving velocity, and the .25 calibers have throughout their history provided a sure route to high velocity.

The .25 calibers are best known for their accuracy, flat shooting, adequate power, broad bullet selection and good commercial cartridge choice. Students of firearms must remember, however, that high-quality rifles like the one pictured are why the various .25s are still so successful in the U.S.

THE .25-20 SUCCESS STORY

The earliest successful commercial .25 caliber was the .25-20 single shot developed in 1882 by J. Francis Rabbeth, a firearms writer/shooter of his day. His cartridge, which produced a muzzle velocity of 1600 fps with an 86-grain bullet, was highly successful, proving its accuracy and popularity especially in rifles built then by Stevens and Maynard. Rabbeth's cartridge marked the beginning of our fascination with small-caliber cartridges and contributed greatly to the idea that small calibers can also be highly accurate.

From the start, these smaller calibers were closely associated with varmint and small-game shooting. Other cartridges, such as the .45-70, which was then in decline, the new .30-40 Krag and the then radical .30-06 were mostly reserved for big-game hunting. Moreover, these cartridges were physically smaller, so they could easily be carried in greater numbers, further emphasizing their reputation as a small cartridge for small game.

The first quarter-inch cartridge to receive widespread use was the .25-20, a shortened version of the .25-20 single shot. Until the arrival of the .30-06 some 12 years later, more rifles were chambered for this round than for any other cartridge. Winchester, Marlin, Remington and Stevens all chambered their rifles for it, and Marlin continues to offer its Model 1894 Classic in .25-

20. Its short overall length allowed generous magazine capacities on the early Winchesters and Marlin lever rifles especially.

The .25-20 also became recognized as a useful tool because its bullets, which traveled at intermediate velocities, did little damage to the meat portions of edible small game like rabbits and squirrels, yet it readily dispatched troublesome pests like skunks and porcupines.

Introduced in 1893, the .25-20 remains immensely successful, with its factory-loaded 60-grain hollow-point bullet offering a mild 2250 fps. Today, any shooter can wander into a shooting sports store and buy a box of .25-20s, a rare tribute indeed to a cartridge that is now over 100 year old. Its continuing popularity also indicates how little the needs of shooters have changed.

The reputation earned by the .25-20s set the stage for the development of the .25-25 Stevens in 1895 and the .25-21 Stevens in 1897. The .25-25 was the first straight-walled .25-caliber cartridge manufactured by Stevens; but its long length caused extraction problems that forced the company to back off about half an inch, thereby creating the .25-21. Both cartridges were a response to the old tenet that black powder would not burn well in bottleneck cases, a notion that persists to this day among some shooters. The result was a demand for performance levels like the .25-20, but in a straight-walled case. Per-

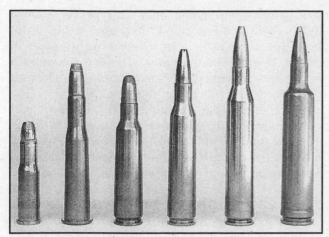

Pictured is the modern family of .25-caliber cartridges (from left): .25-20, .25-35, .250 Savage, .257 Roberts, .25-06 and .257 Weatherby Magnum.

formance was slightly better than the .25-20, but firearm choice was restricted to single-shot, target-type rifles like the Stevens 44 series. As a result, these rounds soon declined in popularity.

Marlin then offered the .25-36, and Winchester followed with its .25-35 in 1895. Little difference existed between the two, although the Marlin version produced just over 1000 foot-pounds of striking energy at 100, while the .25-35 gave slightly less. The .25-35 prospered under Winchester's promotion and went on to gain a mild reputation as a deer and varmint cartridge. Using the 87-grain bullet, it produced slightly less than 2400 fps at the muzzle, which led to the chambering of several single-shot rifles for it. Savage also chambered its Model 99 in .25-35, allowing reloaders to improve down-range performance.

Another .25 Remington, a rimless version of the .25-35, enjoyed some short-lived popularity. Introduced in 1906, it was designed around a series of Remington rifles, most notably the Model 14 pump, the Model 8 semiautomatic and the Model 30 bolt gun. The Model 30 in .25 Remington is an excellent varmint outfit. Since it is extremely rare, if you ever see one, buy it.

NEWTON'S PRINCIPLE: THE .250-3000

The .250-3000, invented by Charles Newton and produced by Savage, was the first real dual-purpose cartridge that proved useful for varmints

and deer-sized animals alike. Newton originally designed the .250-3000 to shoot a 100-grain bullet, but Savage correctly saw the publicity value of having a lighter bullet that could travel at 3000 fps—a new high in commercial cartridge performance. An 87-grain bullet was chosen, and the rest is history.

Flat trajectory, accuracy and excellent killing power firmly established the .250-3000 within the American shooting community, especially when it was teamed up with the Savage Model 99 rifle, offered in .250-3000 from 1916 to 1960. Meanwhile, around 1953 the velocity of the 87-grain bullet was boosted slightly to 3030 from 3000 fps, whereupon the name was changed to the .250 Savage. Stevens brought back the rifle and .250-3000 cartridge in 1971, and for several years its Model 99 was offered in a neat little carbine model with a straight grip and plain

When teamed with this Savage Model 99 rifle, the .250-3000 remains an excellent performer. It was the first real dual-purpose cartridge that fired spire-point bullets in a lever rifle.

stock, called the Model 99E. Unfortunately, it's no longer listed, but it proved well ahead of its time by 20 years or so.

Remington and Winchester still produce the factory 87-grain loads, although recent catalogs don't use them. A 100-grain loading also exists, but the 87 remains the one most wanted by varmint and small-game hunters. With a velocity of 2342 fps at 200 yards, it's a pleasing cartridge to shoot; when sighted-in on targets at 200 yards, it will drop eight inches low at 300. Not a bad performance for an old cartridge with a new name. For many years, the .250 Savage received little attention, suffering under the onslaught of the 6mm and .243. Today, probably more rifles are available in .250 Savage than at any other time.

Originally, the .250-3000 sported a 1:14″ twist, which often failed to stabilize heavier bullet weights (above 100 grains), especially the round-nosed 117-grain bullet commonly used by reloaders for the .25-35. Modern barrels now feature rifling that runs 1:10″, a twist that provides better stability for all bullet weights. For those who hunt small game only and prefer to use lightweight bullets, this rifling isn't a big problem. But in situations in which hunters wish to purchase older, secondhand rifles for use on both small and intermediate game, a quick check of the rifling is advisable.

In the late 1800s, no single .25 cartridge could deliver the accuracy and velocity needed to bring down big game reliably. The .250 Savage worked hard at filling this role, but the death of Charles Newton in the early 1930s stifled development.

THE .257 ROBERTS LEADS THE WAY

At about this time, Dr. Franklin Mann, a well-known firearms writer and experimenter, began working with Adolph Niedner. Applying the technology of their time to experiment exclusively with .25-caliber cartridges, these two men necked down the Krag and Springfield cases to produce a special case they called the "Hamburg." The necked-down Krag produced the best accuracy with the powders available then and led to another series of cartridges: the .25 Niedner, .25 Whelen, .25 Griffin & Howe and the .25 Spring-

Pyle shows off his favorite Model 54 heavy Barrel Winchester rifle in .250-3000. The original one sported a 1:14″ twist that didn't always stabilize heavier bullet weights above 100 grains—especially the round-nosed 117-grain bullet. Modern barrels feature rifling that runs 1:10″, which provides stability for all bullet weights.

The author's wife, Janet, shows the .257 Roberts in a Model 70 XTR Winchester Featherweight. For many women shooters, this caliber has proven the best cartridge for most types of shooting. Modern loads now allow this cartridge to compete with the .243 Winchester.

field Niedner, plus experiments with the .25-36 Marlin, .25-35 Winchester and .250-300 Savage.

Col. Townsend Whelen, Major Ned Roberts and F. J. Sage tested thousands of rounds in different .25 calibers and concluded that the 7×57 cartridge case was best suited for necking down to .25 caliber. As an army officer in charge of ammunition development, Whelen needed no introduction to the capability of the 7×57 Mauser cartridge.

The three men took their cartridge design—the 7×57 necked down to .25 caliber—to Niedner, who immediately began making custom rifles for the .257 Roberts. Originally loaded in bullet weights of 87, 100 and 117, it was introduced by Remington in 1934. As the .257 Remington, its popularity grew until it became a well-balanced, useful and dual-purpose cartridge, one that Niedner later dubbed, "The most useful cartridge ever developed." The 1936 *Western Cartridge Company Rifle and Pistol Ammunition Handbook* lists the .257 Roberts in 100 grain as leaving the muzzle at 2,900 fps, a scant difference of 90 fps over the .250-3000 cartridge carrying the same bullet.

THE 25.06 EMERGES

Today, many U.S. and foreign rifle companies feature the .25-06 as the preferred .25 cartridge for deer and woodchuck shooters. The cartridge

arose from a custom rifle designed by Niedner for a client in the early 1920s. It's really the .30-06 necked down to .25 caliber, with slightly more than 17 degrees of shoulder. In the early years, poor powders prevented the .25-06 from reaching its full potential, especially given the competition from several wildcats, including the .25 Sniper Magnum and two close cousins: the Niedner .25 High Power Special and the Whelen .25 High Power. The .25-06 was said at the time to benefit from something called "Commercial Combustion Conversion," now known as free boring.

The overall accuracy and potential for using heavier bullets prompted Remington to manufacture the .25-06 in 1969 as a commercial round. With an 87-grain bullet, the cartridge produces 3440 fps at the muzzle with two-thirds of its velocity remaining at 300 yards. At this distance, 954 fps of energy remains, which is plenty for long-range hits on all game. Many shooters are impressed with the flat trajectory of the .25-06; for example, when printing dead on at 200 yards and only six inches low at 300 yards. In fact, the .25-06 shoots flatter than the .220 Swift; with heavier bullets of good sectional density, it's much less sensitive to wind.

As with any dual-purpose cartridge, good bullet selection is fundamental to success. When going after varmints, choose only 87-grain

Long-range shooting for any North American game is where the .25-06 excels. With heavier bullets, it offers greater flexibility than any of the other earlier .25 calibers. The trade-off, however, is in greater kick and louder report.

weights available from Remington or 90-grain offerings from Winchester. Heavier bullets for big game include the Winchester 120-grain positive expanding point and a similar bullet from Remington. Remington also lists a 122-grain extended-range version that is worth trying. The 117-grain boat-tail point from Federal is also quite effective. The only complaint about the .25-06 is its magnum-like characteristic. It makes a loud report and kicks more than any other .25 caliber.

THE MODERN MAGNUMS

In this totally modern era of shooting, the .257 Weatherby Magnum is considered the epitome of .25 calibers. Based on the shortened, necked-down British Holland & Holland Magnum, it tops the .25-06 by about 150 fps with any given bullet weight, making it indisputably the hottest commercial .25 caliber on the market. First available in 1944, the .257 Weatherby Magnum delivers outstanding performance at velocities that are more acceptable today than when the cartridge was first introduced. It's a fast, flat-shooting cartridge that comes in an excellent choice of commercially available bullet weights, from the 87-grain that interests small-game hunters to the 120-grain partition bullet that's definitely good for elk. Weatherby uses Nosler bullets known for

their explosive expansion and penetration characteristics on all types of game. Muzzle velocity for the 87-grain bullet is a whopping 3825 fps out of a standard Weatherby 26-inch barrel; and at 300 yards it still produces 2818 fps along with 1535 ft-lbs. of energy. When sighted in at 300 yards, it prints seven inches low at 400 yards, with a point-blank range of 425 yards.

While the cost of these cartridges is a factor the average shooter must consider—and the rifles that shoot them aren't cheap, either—the evolution of the quarter-inchers certainly assures power and accuracy, especially in the hands of the experienced gunner.

WILF PYLE is an avid sportsman who has hunted nearly all game species with a wide variety of firearms. A well-known authority on sporting arms and reloading, he has a passion for sporting rifles and their designs. His books include *Small Game and Varmint Hunting* and *Hunting Predators for Hides and Profit* (Stoeger Publishing). He has also co-authored *The Hunter's Book of the Pronghorn Antelope* (New Century).

BEWARE THE DANGERS OF LOW-POWER AMMO

by C. Rodney James

The blast and recoil of a big-bore rifle, the look of a pointed soft point in a shiny brass cartridge, the heavy slug in a .45 all gain instant respect—but not so the .22. Yet, in a statistical sense, low-powered firearms cause more injuries than the high-powered variety, primarily because those who use them most—the beginners—are often unaware of the dangers involved.

Every box of rimfire ammunition is marked with a warning that indicates the range—usually about one mile—and like most warnings we're exposed to regularly, such as highway speed limits, it is often ignored. In reality, the figures you see on the ammo boxes indicate "extreme range," and they are actually less important than certain other figures that do *not* appear on the box. Range can be defined in four different ways:

1. Effective Range—the distance that a particular gun/cartridge combination can shoot with accuracy and enough power to kill the type of animal for which it was intended;

2. Lethal Range—the distance at which this same combination can deliver a lethal injury to a person;

3. Hazardous Range—the maximum distance at which a bullet can inflict a nonfatal injury, and

4. Extreme Range—the point where the spent bullet has stopped its forward movement and thus poses no threat to human life.

The effective range of a .22 long-rifle cartridge varies according to the type of gun used and the shooter's skill. About 25 yards is the best that can be expected with most handguns of this caliber. An expert shooter armed with a top-quality target rifle fitted with a telescope can extend this range to about 125 yards. Charles S. Landis, the renowned small-bore shooter, once killed a woodchuck at 150–175 yards with a single shot from his Winchester Model 52 target rifle, using a high-speed long-rifle hollow-point bullet. Literally a one-shot feat, Landis did not press his luck. Most of his shots were kept under 125 yards; indeed, for most light sporting .22 rifles, the most effective range is about 50 yards.

The lethal range of the .22 rimfire may actually be five or more times its effective range. One published report told of a boy who was killed by a .22 short "fired from a cheap and rusty gun at a range of 639 yards." That's close to three-eighths of a mile! Another example, recorded in 1929, involved a groundskeeper at a target range in New Jersey. Ignoring the posted warnings, this man proceeded to clean up some debris behind

Bullets capable of penetrating a half-inch sheet of pine plywood, as this photo indicates, are certainly capable of causing fatalities. Here, a hyper-velocity long-rifle hollow-point struck the ground one yard in front of the target, shattering on impact. The vertical line shown on the recording target was used to determine if a regular pattern of left or right ricochet occurred (it did not).

the 600-yard target butts while shooters were firing a 200-yard match at targets placed 400 yards in front of him. A long-rifle bullet plowed deep into a sinus cavity between the man's ear and nose (prompt medical attention fortunately save his life). Calculations later set the velocity of this bullet at about 575 feet per second (fps) with an energy of 29 foot-pounds at the 600-yard point where the bullet actually struck.

The extreme range of a standard velocity long-rifle bullet (1120–1160 fps) was calculated in the 1920s by researchers at the National Rifle Association at about 1500 yards. For high-velocity loading (1285–1375 fps), add about 65 yards. It takes about this distance for the high-velocity load to drop to the muzzle velocity of the standard velocity load. At the far end of the bullet's flight, however, the difference is negligible.

Some years ago, the question was raised: What is the minimum force required for a bullet

to kill a person, assuming it hits a vital spot? One notable expert, James Hamby (editor of the A.F.T.E. Journal), went through more than a thousand documents on the subject and reported no official definition was extant. That situation is still unchanged. In weapons tests designed to determine casualty rates, the U.S. Army has concluded that any bullet capable of passing through a one-inch pine board can inflict a serious or possibly lethal injury. Information from fatal shootings indicate that far less energy is needed.

The smallest standard cartridge that will chamber in a .22 long-rifle gun is the .22 BB cap. With a report no louder than a polite hand clap, it can, with a 15- to 18-grain round or conical bullet, deliver 24 foot-pounds of energy at the muzzle. At 100 yards, there are still 13 foot-pounds left. The .22 BB cap's close relative, the CB (conical bullet) cap, holds a 29-grain bullet and delivers 35 foot-pounds.

Fatalities caused by BB caps are indeed a matter of record. A boy in Athens, Greece, once killed his father accidentally with a Flobert rifle using the 5mm Flobert (.22 BB cap). The bullet struck his father above the right breast, pierced the lung, passed under the aorta and came close to entering the man's heart. And in his book, *Hunting with the Twenty-Two*, Charles Landis made the following comment: "About the year 1922 our local police records showed that a woman shot herself with a .22 CB cap and died almost immediately."

I once conducted my own experiment with a BB cap fired from 25 yards at a target 32/100-inch thick (to simulate the human skull) with a Winchester Model 52 target rifle. The bullet penetrated the target with ease. I repeated the experiment using a sheet of half-inch pine plywood, and again the BB cap penetrated the full thickness. Plywood is, of course, no substitute for flesh; it's merely one means of determining the relative damage these low-power bullets can cause.

Using the rule of thumb—that anything passing through a half-inch of plywood is capable of inflicting a lethal injury—this experiment be-

TABLE I. Range and Penetration of CCI .22 High Velocity Long Rifle Ammo

Distance (Yds)	Penetration-- Plywood (Inches)			Foot Lbs.	Velocity (F.S.)
25--75	2.2		Lethal	147	1285
100	2.1			93	1030
125	2.1			87	991
150	2.0			81	958
175	2.0			76	927
200	1.9			72	898
225	1.9			67	871
250	1.7			63	845
275	1.7			60	821
300	1.6			56	798
325	1.6			54	777
350	1.4			50	755
375	1.4			48	735
400	1.3			45	715
425	1.3			43	696
450	1.1			40	677
475	1.1	Estimated		39	659
500	1.0	"		37	647
525	1.0	"		35	624
550	.9	"		33	607
575	.9	"		31	591
600	.7	"		29	575
625	.7	"		28	559
650	.6	"		26	544
675	.6	"		25	529
700	.5	"		23	515
725	.5	"		22	501
750	.4-----"-----Dangerous			21	487
775	.4	"		20	475
800	.3	"		19	461
825	.3	"		18	449
850	.2	"		17	437
875	.2	"		16	425
900	.2	"		15	414
925	.1	"		14	403
950	.1	"		14	391
975	.1	"		14	381
1000	.1	"		13	371

The above figures were calculated on the basis of Ingall's Tables. Actual tests were concluded at 450 yards.

The small size of BB and CB caps belie their lethality. The short case rounds have the same velocity as the down-loaded short and long cartridges now offered by most U.S. manufacturers. Shown above (left to right) are three European BB caps, a Canadian BB, two obsolete U.S. and British CBs, two down-loaded short and long rounds, and a long rifle (far right) for comparison.

came a kind of litmus test for measuring a variety of low-powered cartridges. Moreover, the results seemed more useful than a collection of kinetic energy numbers and undocumented commentaries about shocking power, stopping power and the magic wrought by high velocities. How could we ignore projectile size, shape and weight? After all, a steel dart or a softball striking a person with the same amount of energy will cause radically different results.

According to the best medical minds on the subject of wound ballistics, lethality generally depends on the depth and size of the permanent cavity, and whether this cavity passes through any vital organs. Continuing these experiments, I selected the Winchester Model 52 as the test rifle. High-velocity ammunition was fired at a solid plywood target at distances to 450 yards. Wind and topography made hitting beyond this range impossible. As the figures in Table I attest, the lethal range of long-rifle cartridges is close to half a mile, with the danger zone calculated to about 1,000 yards.

TABLE II. Barrel length, Velocity and Bullet Energy Winchester High Velocity .22 Rimfire (From Cochrane)

Barrel Length	.22 Short Vel	Ft. Lbs.	.22 Long Vel	Ft. Lbs.	.22 Long Rifle Vel	Ft. Lbs
22	1135	82.8	1251	100.	1213	130.6
21	1121	80.9	1246	99.9	1237	135.9
20	1108	79.1	1233	97.8	1212	130.4
19	1098	77.6	1237	98.5	1199	127.7
18	1119	80.7	1231	97.5	1212	130.4
17	1117	80.3	1224	96.4	1216	131.2
16	1138	83.4	1248	100.	1216	131.2
15	1124	81.3	1230	97.4	1205	129.0
14	1111	79.5	1230	97.4	1195	127.0
13	1115	80.1	1230	97.4	1200	128.0
12	1111	79.5	1230	97.4	1200	128.0
11	1119	80.9	1210	94.3	1200	128.0
10	1111	79.5	1220	95.9	1210	130.0
09	1095	77.0	1200	92.8	1158	119.1
08	1103	78.4	1186	90.6	1136	114.6
07	1075	74.5	1158	88.4	1115	110.4
06	1064	72.9	1124	81.3	1087	104.9
05	1024	67.5	1099	77.8	1049	97.8
04	997	63.9	1042	69.9	1010	90.6
03	946	57.7	958	59.2	940	78.5
02	877	49.6	862	47.9	820	59.7
01	716	33.0	632	25.7	606	32.6

HANDGUNS AND LOW-POWER AMMO

With their shorter barrels, handguns are generally thought to be not as far ranging and therefore less deadly than rifles. Given the same caliber ammunition, however, this notion offers little real comfort. Though the effective range of a handgun may be a dozen yards or so, its lethal range is nearly equal to a rifle of the same caliber.

In 1979, D. W. Cochrane, a firearm examiner for the Royal Canadian Mounted Police, conducted a revealing study. He was looking for a way to determine whether or not a gun used in a crime was a cut-down rifle (a common weapon in Canada) by analyzing the velocity and energy of fired bullets. Using a Cooey bolt-action rifle with a 22-inch barrel, Cochrane chronographed three makes of .22 short, long and long-rifle ammunition, shooting five-round samples of each

type. With each successive test, he cut back the barrel 1 inch, concluding the experiment with bullets shot from a 1-inch barrel. Cochrane anticipated a gradually descending velocity curve, but he was surprised to find modest increases between 22 and 16 inches. Below 16 inches, velocity dropped only slightly, hitting a plateau at 8 inches. Below 8 inches, plateaus were reached at 4 inches and then 2 inches, and finally 1 inch. The figures shown in Table II are for Winchester ammunition only, although no significant differences have been found using other brands.

The clear message here is that pistols like Thompson/Center's Contender with a 14- to 16-inch barrel, Browning's Buck Mark, or any silhouette pistol with more than a 9-inch barrel, all maintain *rifle* velocities. Even a modest 4-inch barrel delivers better than 83 percent of a rifle's velocity with the same long-rifle cartridge.

SHOTGUNS AND LOW-POWER AMMO

The gun of choice for hunting in populous areas is almost always the shotgun. But even here

Buckshot loads will pass through 1.75 inches of hard pine or plywood at 20 yards. Shown here are a 12-gauge OOB (top), a handloaded .410 OOB load at comparable velocity (center), and a .310 muzzleloader ball used as a buckshot substitute in a .410 (bottom).

Long-rifle tracers form a consistent group. Fired from a sitting position, the bullets shown at right struck the ground first at 25 yards, then again at about 200 yards.

the notion that these guns present short-range hazards only should be taken under advisement. The hazardous range of a shotgun can vary from under 200 yards to nearly a mile, depending on what's in it. As with rifles, shotgun report and recoil make a big impression on beginning shooters. The roar of a 12 or 16 gauge is admittedly unforgettable. But the relatively modest report of a .410 or 28-gauge shotgun is deceptive. The charge of shot from the .410 travels at the same velocity and has the same range as a similar loading in 12 gauge—there simply is less shot. While the .410 is really an expert's gun, its mild report and low recoil often make it the beginner's favorite as well.

For most hunting purposes, shot ranging from #8 to #4 is the most common. Because round balls have poor aerodynamic form, the extreme range of most birdshot is under 350 yards. The lethal range, using the plywood test, is about 17 yards for both #4 and #6 shot. According to a formula devised by ballistician Journee, the maximum range of #6 is 242 yards, while the maximum range of larger birdshot should be figured at a minimum of 300 yards—400 yards for those sizes that fall between birdshot and buckshot. The largest buckshot load—00—weighs about 56 grains with a maximum range of 750 yards. Be-

cause of their poor form, these shot slow down rapidly, but they are still considered lethal to about 400 yards and are certainly hazardous to about 600 yards.

When it come to lethality, shotgun slugs are a different matter. The larger, heavier slugs hit with far greater force than the lighter ones. From 20 gauge and up, slugs are considered high-powered ammunition with a range of 1500 yards or more (in sabot loads using a .50-caliber slug). Oddly enough, range warnings that once appeared on slug loads sold 25 to the box are no longer included.

Since no 20-gauge slug is manufactured, that leaves the .410 as the low-power loading on the market. With an 88- to 96-grain slug at a muzzle velocity of 1470 fps and a muzzle energy of 460 foot-pounds, the .410 is woefully inadequate for deer—about like a light, high-velocity .38 pistol load. Maximum range is about 850 yards with a lethal potential to about 425 yards. In actual tests, .410 slugs penetrated .6 inch of plywood at 225 yards and kept going.

LETHAL RICOCHETS: CAUSE FOR CONCERN

Ricochets are the bane of any shooter's existence. Their inherent dangers have long been

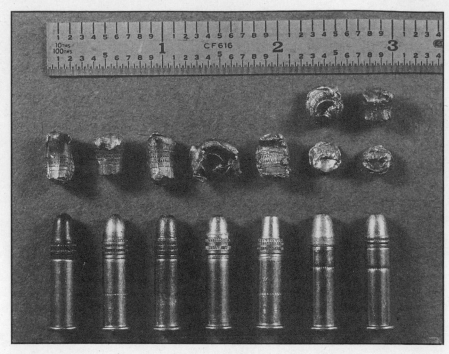

This sampling of long-rifle hunting and target bullets includes (left to right): CCI Standard Velocity, Remington High Power, Western Dynapoint, CCI High Velocity hollow point, Remington Yellow Jacket High Power, CCI Stinger and an obsolete Winchester Xpediter. Above the cartridges are bullets recovered from ricochets at 25 yards. The hyper-velocity Stinger and Xpediter bullets broke up at 25 yards, leaving only the bases intact. The two bullets on the top row ricocheted at 75 yards and stayed together.

preached, but little hard evidence has been available. Extensive tests conducted by this author as well as others have proved that bullets have a critical angle of entry—even in water—below which they will ricochet. That angle is 5.75 degrees and translates roughly to a point on ground level about 24 feet from a standing adult shooter. The angle of exit of a ricochet from the water is about twice as steep as from solid turf, where ricochets for long-rifle loads and .410 slugs can vary from about 2.5 degrees to 8.5 degrees. Some long-rifle hollowpoints will break up on turf out to about 25 yards, but beyond that everything will ricochet. There are cases on record of lethal ricochets traveling nearly a mile. One case, which involved a high-powered .303 British rifle, came off water. Fortunately, birdshot ricochets remain insignificant because of their low weight; generally, they flatten out on hard surfaces such as stone or concrete. But steel shot is exceedingly prone to ricochet damage because it is hard and springy; when fired at hard surfaces such as a steel or concrete wall, it can ricochet almost straight back into the shooter's face.

For target shooting, a safe backstop is a must. The best types are earth and gravel banks steep enough to avoid ricochets. If a target butt must be constructed, make it a double wall with closed ends. The space between the front and rear walls—about a foot or two for medium-power ammo—should be filled with pea gravel, which does a nice job of shredding most lead bullets.

In summary, the .22 long rifle is a fun and safe cartridge indeed, but the only way to keep it that way is to understand its full potential.

C. RODNEY JAMES has published technical and historical articles in several popular and forensic firearms publications, including *Gun Digest, Handloader's Digest, Guns Illustrated, Guns Magazine* and the *A.F.T.E. Journal* (the official publication of the Association of Firearm & Tool Mark Examiners). James, who holds a Ph.D. in mass communications studies from Ohio State University, has also written for *The Atlantic Monthly, Movie & TV Marketing* and several professional journals. His scholarly book on the National Film Board of Canada, called *Film as a National Art*, has been published by The New York Times.

MANUFACTURERS' SHOWCASE

GLASER SAFETY SLUG AMMO

GLASER SAFETY SLUG's state-of-the-art, professional-grade personal defense ammunition is now offered in two bullet styles: BLUE uses a #12 compressed shot core for maximum ricochet protection, and SILVER uses a #6 compressed shot core for maximum penetration. The manufacturing process results in outstanding accuracy, with documented groups of less than an inch at 100 yards! That's why GLASER has been the top choice of professional and private law enforcement agencies worldwide for over 15 years. Currently available in every caliber from 25 ACP through 30-06, including 40 S&W, 10mm, 223 and 7.62 × 39. For a free brochure contact:

GLASER SAFETY SLUG, INC.,
P.O. Box 8223, Foster City, CA 94404

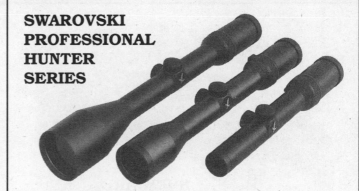

SWAROVSKI PROFESSIONAL HUNTER SERIES

When the engineers at SWAROVSKI OPTIK in Austria designed this radically new line of premium 30mm PH riflescopes, they machined the entire scope tube and central turret housing from one solid bar of steel or aluminum alloy. The "zoom system" tube assembly and reticle are supported and cushioned by a unique "Four Point Coil Spring Suspension" to handle the heaviest recoil shock and to maintain an accurate "zero" year after year. These revolutionary scopes are stronger, lighter and more dependable than comparable scopes—with shock-absorbing eyepieces, large fields of view, fine image quality, the highest light transmission and the best reticle technology. SWAROVSKI's premium optical performance is legendary. For a catalog and more information, contact:

SWAROVSKI OPTIK NORTH AMERICA, LTD.
One Wholesale Way, Cranston, RI 02920

Aimpoint® 2X RED DOT SIGHT

AIMPOINT introduces its new 30mm Aimpoint 500 2 Power, a fixed, low-power electronic sight with a floating red dot. It's the only unit of its kind with built-in magnification. The shooter now has the speed and accuracy of a red dot sight combined with the advantages of a low-power scope. Because the magnification is in the objective lens instead of the ocular lens (as with previous screw-in attachments), the dot covers only 1.5″ at 200 yards. The 5000 2 Power can be used on all types of firearms and comes complete with 30mm rings and all accessories. Suggested retail price is $399.95. For more information write:

AIMPOINT
580 Herndon Pkwy., Suite 500, Dept. SB, Herndon, VA 22070
Phone: 703-471-6828

KLEEN-BORE

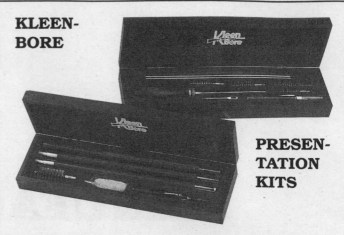

PRESEN-TATION KITS

The ultimate in gun care and cleaning sets, these kits make ideal gifts for the discriminating sportsman. "The Shotgunner," available in 12 or 20 gauge, features KLEEN-BORE'S three-piece, walnut-finished laminated hardwood rod with solid brass hardware, phosphor bronze brush and cotton mop, patch holder and Precision Shooter with Formula 3 gun conditioner. "The Marksman," with patch holder, Precision Shooter and .22, .30, .38/.357/9mm and .45 phosphor bronze brushes, has a four-section, stainless-steel rod with laminated hardwood handle and brass muzzle guard. Presented in select, solid wood cases.

KLEEN-BORE INC.
20 Ladd Avenue, Northampton, MA 01060
Phone: 413-586-7240

MANUFACTURERS' SHOWCASE

B-SQUARE SHOTGUN SADDLE MOUNTS

B-SQUARE Shotgun Saddle Mounts are now available for most popular 12-gauge guns. These newly designed mounts straddle the receiver and fit the top of the gun tightly. All mounts have a standard dovetail base and "see-thru" design allowing continued use of the gun's sight. Standard dovetail rings can be used. B-SQUARE shotgun mounts do not require gunsmithing, have a blued finish, and attach to the gun's side with included hardware. Mounts available for: Remington 870/1100; Mossberg 500, 5500 and 835; Winchester 1400/1300/1200; Ithaca 37/87; and Browning A-5 shotguns. The mounts retail for $49.95 at your local dealer, or call B-SQUARE toll-free for a free catalog.

B-SQUARE CO.
P.O. Box 11281, Fort Worth, TX 76110
Phone: 1-800-433-2909 (toll-free) or 817-923-0964

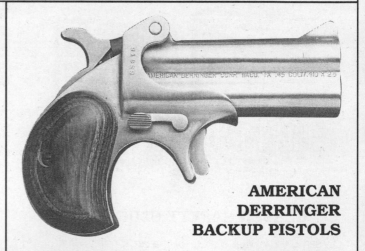

AMERICAN DERRINGER BACKUP PISTOLS

Designed to be the ultimate in short-range backup pistols, this gun has no equal. Over 10 years were spent in developing and refining this pistol to make it the finest derringer ever manufactured. The smallest and most powerful pocket pistol ever made, it is built from the finest high-tensile strength stainless steel—strong enough to handle the 44-Magnum cartridge if you are man enough to shoot it! Over 60 different rifle and pistol calibers are available. Classic styling and smooth lines give these derringers a classic look.

AMERICAN DERRINGER CORPORATION
127 N. Lacy Drive, Waco, Texas 76705
Phone: 817-799-9111 Fax: 817-799-7935

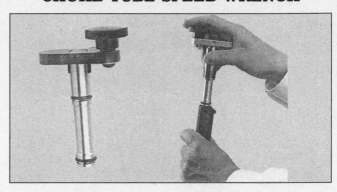
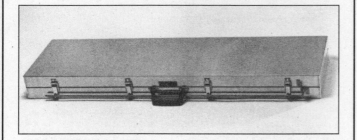

GLASER FEATHERWEIGHT BIPOD

At less than six ounces—half the usual weight—the GLASER bipod offers the discriminating shooter the ultimate in strength and sleek beauty. A frontal area 4¹/₂ times smaller than other bipods greatly reduces snag hazards. Uneven terrain is automatically compensated for up to 33 degrees. Deployment and retraction require only single and silent one-hand movements that take less than a second. The bipod fits all sporter, varmint and most paramilitary firearms. The basic mount permits front or rear mounting to the forearm, rather than the barrel, for target accuracy. Hidden or quick-detachable, customized mounting accessories are available. For a free brochure contact:

GLASER SAFETY SLUG, INC.
P. O. Box 8223, Foster City, CA 94404

SWIFT MODEL 649 RIFLESCOPE
4-12x, 50mm
Waterproof, Multi-Coated

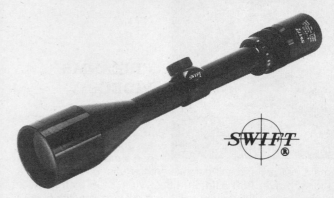

With a maximum effective diameter of 50mm, this wide-field, multi-coated scope is especially effective in poor light conditions—yet its power range of 4 to 12 makes it useful under most hunting conditions. This hard anodized, fog-proof instrument is equipped with self-centering Quadraplex reticle and is available in two finishes: regular (649) and matte (649M). Shock tested. Gift boxed. For more information contact:

SWIFT INSTRUMENTS, INC.
952 Dorchester Ave., Boston, MA 02125
P.O. Box 562, San Jose, CA 95106

MANUFACTURERS' SHOWCASE

KOWA HIGH RESOLUTION SPOTTING SCOPES

KOWA's TSN-1 super high-performance, multi-coated 7mm lens offers a sharper image, wider-than-usual field of view and increased light-gathering capabilities of no less than 60% over conventional spotting scopes. The TSN-1 is also capable of high-quality photo applications. Bayonet mount makes changing the eyepieces quick and easy by a simple insert and turn. Eight interchangeable eyepieces are available, including straight-type design.

KOWA OPTIMED, INC.
20001 S. Vermont Ave., Torrance, CA 90502
Phone: 310-327-1913

PELTOR RANGE PARTNER

PELTOR's combo package contains a Bull's-Eye "6" Shotgunner hearing protector and one pair of Model 2500 Deluxe Amber shooting glasses with high-impact polycarbonate lenses. Metal frame with large one-piece lens provides ultimate vision. Hard-coated for superior scratch resistance. Meets ANSI Z87.1-1989 Standards.

PELTOR, INC.
41 Commercial Way, East Povidence, RI 02914
Phone: 401-438-4800 Fax: 401-434-1708

BELL & CARLSON COMPOSITE GUN STOCKS

All DURALITE, CARBELITE™ and PREMIER gun stocks are composed of hand-placed layers of Kevlar, graphite and fiberglass. The stocks are bound with structural urethane and chopped fiberglass throughout the entire cavity. This creates a dramatic reduction in felt recoil to the shoulder and allows for a secure solid feel when firing. BELL & CARLSON composite stocks are engineered to stand up to hard use, temperature extremes and moisture. All stocks are factory-fit to standard barrel contours as listed in our current model section. To find out more about our complete product line, write or call for a free catalog.

BELL & CARLSON
Dept. SB95, 509 N. 5th Atwood, Kansas 67730
Phone: 913-626-3204

RIZZINI SHOTGUNS

Master craftsmen at RIZZINI are known the world over for building fine over/under sporting shotguns. Although relatively new to America, RIZZINI guns are already available in a wide variety of gauges and specifications for trap, skeet, sporting clays and field shooting. All RIZZINIs have special steel barrels tested at Proof House 1200 BAR. The hand-engraved metal work, premium walnut stocks and wood-to-metal fit are excellent. Interchangeable chokes and extra barrels are also available, depending on model and value range. The new RIZZINI Models 2000 and 2000 SP (shown) are excellent examples of sporting clay guns suitable for field and all-around competition. For more information, call or write:

LOU ALESSANDRI & SON, LTD.
24 French Street, Rehoboth, MA 02769
Phone: 1-800-248-5052 (toll free)

MANUFACTURERS' SHOWCASE

WIDEVIEW
NEW BLACKPOWDER MOUNT

The new WIDEVIEW Black Powder Mount pictured above fits the Thompson/Center Renegade and Hawken rifles with no drilling or tapping. The barrel still comes off for easy cleaning, and the original iron sight is used. With Wideview's Ultra Precision See-Thru mounts included, you may use a scope or iron sights. No hammer modification is needed.

WIDEVIEW SCOPE MOUNT CORP.
26110 Michigan Ave., Inkster, MI 38141 (313) 274-1238

B-SQUARE introduces the little laser that is big on performance. At only 1.1″ x 1.1″ x .6″, the Compact Mini Laser delivers 5mW of power (Class IIIa), while operating on common A76 size batteries (lithium or alkaline). Visibility is 1.0″ at 25 yards. Features an omnidirectional screw-type aiming method with windage and elevation adjustments and is the only laser with an "Air-Lock" feature. Both moisture-proof and shock-resistant, the B-SQUARE Compact Mini Laser carries a lifetime warranty. Service is available simply by calling B-SQUARE toll-free. Mounting systems are available for trigger guard, under barrel, and long guns. The vertical T-Slot design makes them quick-detachable and ensures no change in zero. Contact:

B-SQUARE CO.
P.O. Box 11281, Fort Worth, TX 76110
Phone: 1-800-433-2909 (toll-free) or 817-923-0964

HARRINGTON & RICHARDSON .410 TAMER™ SHOTGUN

NEW!

H&R 1871, Inc., announces its new short-barreled .410 snake gun produced under the HARRINGTON & RICHARDSON trademark. Called the Tamer™, this new shotgun features H&R's safe and reliable single-shot action with its highly respected transfer bar safety. The stock, a modified thumbhole design that actually sports a full pistol grip, features a recessed open side containing a shell holder for storing .410 ammunition. This specialized shotgun has a matte, electroless nickel finish for extra durability, and its unique configuration makes it ideal as a camp gun or hiking companion. It is also an excellent choice for home defense or for shark or halibut fishing. The Tamer's 20-inch barrels are chambered for 3-inch shotshells with Full choke.

H&R 1871, Inc.
60 Industrial Rowe, Gardner, MA 01440
Phone: 508-632-9393

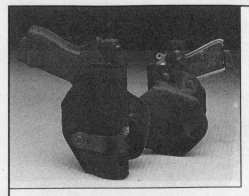

NEW FROM SHOOTING SYSTEMS

BALLISTIC NYLON PADDLE HOLSTER

This exceptional paddle holster is as concealable as the best belt holster, and yet is even more comfortable and convenient to wear. The secret is its unique polymer paddle with twin stabilizer prongs—simply slip the paddle into the waistband. The paddle's unique shape actually transfers the handgun's weight to your hip—ensuring unparalleled comfort and stability. The holster body, made from quilted ballistic nylon, features the patented "PowerBand" device for adjustable drawing tension. The patented Nichols "SightStrip" protector inside accommodates adjustable front and rear sights. Thumb-break straps are adjustable and reversible. AFPH is available in black only and weighs just 7 oz. Price: $39.95.

SHOOTING SYSTEMS GROUP, INC.
1075 Headquarters Park, Fenton, Missouri 63026-2478
Phone: 1-800-325-3049 (toll-free)

MANUFACTURERS' SHOWCASE

PELTOR BULL'S-EYE ULTIMATE 10

This new PELTOR product offers completely new technology for the hearing protection market. The use of twin cups ensures minimized resonance and thus achieves maximum high and low-frequency attenuation. Designed for use with large caliber or magnum rounds. **NRR 29 dB—now the highest noise reduction muff-type hearing protector available worldwide.**

PELTOR, INC.
41 Commercial Way, East Providence, RI 02914
Phone: 401-438-4800 Fax: 401-434-1708

MG-42 KIT

Dress up your Ruger 10/22 with GLASER's MG-42 Stock Kit. With the simplest of tools and in less than 10 minutes, you can assemble this 2/3 replica, WWII MG-42, using your Ruger 10/22 and our stock kit. The MG-44 kit incorporates front and rear sights, with both windage and elevation adjustments. Fully assembled, the new MG-42 weighs no more than the old 10/22. For a final authentic touch, add the Featherweight Bipod to your new MG-42. With the enhancement of accurate prone and bench shooting, the bipod will complete the MG-42 transformation just short of the actual 8mm machine gun belt. For a free brochure contact:

GLASER SAFETY SLUG, INC.
P.O. Box 8223, Foster City, CA 94404

L.A. EXTENSIONS

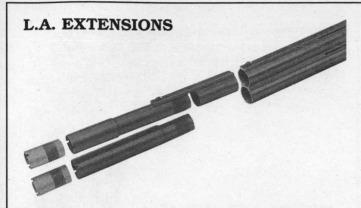

New 2½-inch, 4-inch, and 6-inch L.A. Extensions will transform your favorite double- or single-barrel shotgun into a 30-inch, 32-inch or 34-inch competitor or fowling piece for a fraction of the cost of a new gun or extra set of barrels. The extensions will improve most any shotgun pattern and point of impact. The screw-in-tubes are overbored, made of tool steel and use threads from your present choke system. They will handle the softest target or toughest steel loads. To mount the extension tube system, remove your present screw-in chokes, screw in the L.A. Extensions, which are ordered to match your choke system, and then screw in the new L.A. Choke of your choice. For more information, call or write:

LOU ALESSANDRI & SON, LTD.
24 French Street, Rehoboth, MA 02769 Phone: 1-800-248-5672 (toll-free)

TRIUS TRAPS

TRIUS has set the standard for the industry. From "behind the barn" shooters to upstart Sporting/Hunters' Clay ranges, the easy-cocking, lay-on loading make TRIUS Traps easy to operate. Singles, doubles plus piggy-back doubles offer unparalleled variety. **Birdshooter**—quality at a budget price. **Model 92**—a best seller with high-angle clip and can thrower. **Trapmaster**—sit-down comfort plus pivoting action. **SC-92**—heavy-duty trap for all sporting clay targets except battue and rabbit. **BAT2**—same as SC92 plus will throw single/double battue targets. **Rabbitmaster**—designed to throw "rabbit disc" clay targets along the ground.

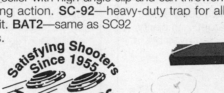

TRIUS PRODUCTS, INC.
P.O. Box 25, Cleves, Ohio 45002

MANUFACTURERS' SHOWCASE

NEW BIPOD MODELS

B-SQUARE bipods are now available in several models. The Rigid Bipod is available with Swivel Stud "Sporter" or Barrel Clamp "Service" attachment. The "Tilt" Bipod provides the same rigid support "canting" from side to side for fine-tuning your aim. "Tilt" bipods are also available with both swivel stud and barrel clamp attachment. The "Roto-Tilt" Bipod offers everything you could want in a bipod: rigid support, side-to-side "canting," swivels in a 30-degree angle, enabling the shooter to follow perfectly aimed shots. Available with swivel stud attachment only. All bipods available in blue or stainless and feature an Unlimited Leg Extension System with 7-inch leg extenders (sold separately). For more information, call or write:

B-SQUARE CO.
P.O. Box 11281, Fort Worth, TX 76110
Phone: 1-800-433-2909 (toll-free) or 817-923-0964

BAIKAL IJ-70 MAKAROV PISTOL

From Russia, the Baikal IJ-70 is the original Russian Makarov pistol. In use since the end of WW II, the Makarov has been widely adopted throughout the former Warsaw Pact countries, the former Soviet Union and China. FEATURES: • Double action, 9mm Makarov (9x18) • Frame-mounted safety with decocking lever • Rear target sights adjustable for windage and elevation • All-blued steel frame and slide • Black composite grips • Two 8-round magazines, cleaning rod, universal tool and leather holster included • Also available in .380 ACP in either blued steel or nickel finish

K.B.I. INC.
P.O. Box 6346, Harrisburg, PA 17112
Phone: 717-540-8518

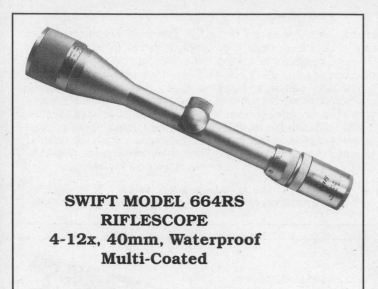

SWIFT MODEL 664RS
RIFLESCOPE
4-12x, 40mm, Waterproof
Multi-Coated

SWIFT's 664 RS 4-12x, 40mm riflescope is totally adaptable for the hunt. Spot your target easily with 4x, then bring it instantly 12 times closer for perfect placement. It is also excellent for gas-powered air rifles. Self-centering Quadraplex reticle and multi-coated optics bring you on target with clarity. Waterproof and fogproof, the SWIFT 664 is also available in black and matte finish. For more information contact:

SWIFT INSTRUMENTS, INC.
952 Dorchester Ave., Boston, MA 02125
P.O. Box 562, San Jose, CA 95106

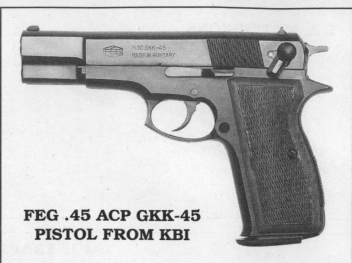

FEG .45 ACP GKK-45
PISTOL FROM KBI

Patterned after the world-famous Browning Hi-Power, and functioning like a Smith & Wesson, the new FEG GKK-45 from Hungary incorporates styling and performance with big-bore .45 ACP power. FEATURES: • Double-action .45 ACP • Slide-mounted safety with decocking lever • High-lustre blued steel frame and slide; Industrial hard-chrome version available • Highly visible three dot sights • Checkered walnut grips • Two 8-round magazines and cleaning rod included

K.B.I. INC.
P.O. Box 6346, Harrisburg, PA 17112
Phone: 717-540-8518

MANUFACTURERS' SHOWCASE

PALSA ADD-A-PAD

ADD-A-PAD is made from shock-absorbent, blended neoprene with a specially formulated adhesive backing.

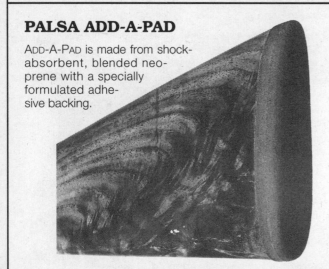

The ADD-A-PAD can be quickly installed simply by pressing a pad on the end of the butt, trimming it with a knife, then sanding it to the exact shape of the stock. The package includes two ¼-inch and one ½-inch pads, thus allowing the use of any pad or combination of pads. The result is an economical "do-it-yourself" pad that looks professionally installed. ADD-A-PAD costs $10.95 and comes with complete instructions. Call or write for more information.

PALSA OUTDOOR PRODUCTS
P.O. Box 81336, Lincoln, Nebraska 68501-1336
Phone: 402-488-5288 Fax: 402-488-2321

FREE!

RELOADING
CATALOG
FROM
MIDWAY

Free three-month subscription to our reloading catalog. If you reload for rifle or pistol, you need our free 64-page catalog containing thousands of products from more than 100 manufacturers. We offer one of the world's largest selections of reloading and shooting products. FREE. Call or write:

MIDWAY
5875-D Van Horn Tavern Rd., Columbia, MO 65203
Phone: 1-800-243-3220 (toll-free)

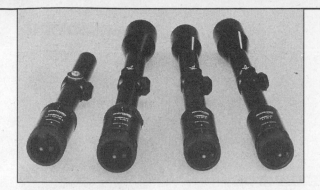

SWAROVSKI'S AMERICAN LINE

SWAROVSKI OPTIK North America's most popular "A" or American line consists of four models in 1-inch tubes. The 4x32, 6x36 and variable models in 1.5-4x20 and 3-9x36 make up what many consider the highest performance, highest value riflescopes in North America. Housed in alloy tubes for light weight and with telescopic eyepieces, these scopes are compact in design for easy mounting and balance. The optical image is brilliantly clear and sharp. That's because SWAROVSKI uses the very best optical glass and guarantees the full multi-coating of all surfaces with its patented Swarotop® optical coating process. The scopes are dry, nitrogen-filled, and waterproof/submersible. For a catalog and more info, contact:

SWAROVSKI OPTIK NORTH AMERICA, LTD.
One Wholesale Way, Cranston, RI 02920

KLEEN-BORE TW25 B LUBRICANT

Unmatched by other lubes, including CLP, TW25 B is available in tubes and aerosol cans with no ozone-depleting chemicals. This lubricant proved itself during the Desert Storm War, when constant exposure to salt water, sand and dust, plus extreme temperatures and heavy load-bearing conditions were the norm. A little goes a long way. It lasts for a long time, and a thin coat in the bore makes cleaning much easier. Recommended by major manufacturers of military weapons, TW25 B prevents corrosion, eliminates galling on stainless steel, reduces friction and is harmless to wood, plastic and rubber.

KLEEN-BORE INC.
20 Ladd Avenue, Northampton, MA 01060
Phone: 413-586-7240

MANUFACTURERS' SHOWCASE

LYMAN'S NEW "E-ZEE CASE GAUGE" FOR RELOADERS

LYMAN PRODUCTS announces a new universal case-length gauge for reloaders, called the "E-Zee Case Gauge." Because of its simple, efficient design, it will quickly and accurately measure case length for both fired and resized cartridge cases. Included on the gauge are the correct case lengths of over 50 of the most popular pistol and rifle calibers as recommended by SAAMI. Simply by aligning the cartridge case against the correct cartridge reading on the gauge scale, the reloader knows immediately whether the case is too long or too short. The handy tool is precision-machined from high-strength aluminum. Both inches and metric scales are provided. For more information, contact:

ED SCHMITT, LYMAN PRODUCTS
Route 147, Dept. 596, Middlefield, CT 06455
Phone: 1-800-22-LYMAN (toll-free)

.45 ACP DOUBLE ACTION ONLY FROM AMT

AMT announces production of the smallest .45 ACP auto pistol ever built. This achievement tacks another "First and Only" to AMT's impressive array of innovative handguns. Designed as a law enforcement backup or personal protection arm, this latest auto pistol features the smooth, safe AMT Double Action Only (Locking Barrel) engineering of its popular predecessor, the compact .380 Backup. Only 3/4 inch longer than the .380, AMT's new .45 ACP provides an added measure of confidence through greater power in an extraordinarily small frame. Stainless steel with checkered black fiberglass grips, this pistol weighs only 23 ounces. 5-round capacity.

AMT
6226 Santos Diaz St., Irwindale, CA 91702
Phone: 818-334-6629

93

Dress up your favorite semiauto sporter stock with BUTLER CREEK'S featherweight, space-age synthetic stocks. These specially formulated polymer stocks are designed for a drop-in fit right at home. All come with a quality-molded recoil pad and quick disconnect swivel studs; a slim, more comfortable pistol-grip area with double palm swells for right- or left-handed shooters; raised, diamond-cut checkering for a lengthened trigger pull to enhance your shooting comfort. We'll send you a free catalog just for mentioning that you saw our ad in the 1995 SHOOTER'S BIBLE. For further information, contact:

BUTLER CREEK CORPORATION
290 Arden Drive, Belgrade, MT 59714
Phone: 1-800-423-8327 (toll-free) or 406-388-1356
Fax: 406-388-7204

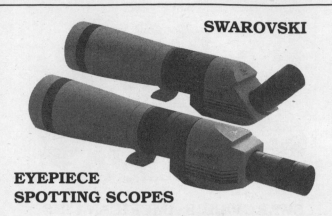

SWAROVSKI

EYEPIECE SPOTTING SCOPES

Two regular models—the AT-80 and ST-80—along with two others with high-density (low-dispersion) glass—AT-80 HD and ST-80 HD—make up this line of fixed tube spotting scopes from SWAROVSKI OPTIK. These AT (angled) and ST (straight) scopes have extra-large 80mm objective lenses for high light transmission and are part of a larger family of spotting scopes used for hunting and wildlife observation. They share a variety of fixed and variable eyepieces from 20x to 60x, as well as 35mm camera adapters. The scopes incorporate the finest optical glasses, with multi-layered lens coatings on all surfaces as well as extra large, phase-corrected prisms. The AT and ST scopes deliver the highest definition, edge-to-edge sharpness and color fidelity. Waterproof/submersible.

SWAROVSKI OPTIK NORTH AMERICA, LTD.
One Wholesale Way, Cranston, RI 02920

MANUFACTURERS' SHOWCASE

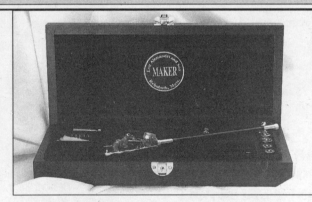

LOU ALESSANDRI PREMIUM GUN-CLEANING KITS

The gun-cleaning kits from LOU ALESSANDRI & SON, LTD. are works of art. Each locking case is hand-crafted in cherry wood with a hand-rubbed finish. The machined, bronze ferrules for the ebony cleaning rods are plated in 24-carat gold, as are the bronze snap caps, cleaning brushes and jags. These fully functional kits are truly heirloom quality. The canoe and retriever motif for the handle is done in bronze and sterling silver along with the gold-plated oil bottle. The kits come with 12-ga. and 20-ga. accessories and are fully felt-lined. Priced at about $1,300. For a catalog and more information, contact:

LOU ALESSANDRI & SON, LTD.
24 French Street, Rehoboth, MA 02769
Phone: 1-800-248-5652 (toll-free)

THE SHOTGUN NEWS

Established in 1946, The SHOTGUN NEWS offers some of the finest gun buys in the United States. Published three times each month, with over 150,000 readers, this national publication has helped thousands of gun enthusiasts locate, buy, trade and sell antique, modern, military and sporting firearms and accessories. As the cover says, SHOTGUN NEWS is "The Trading Post for Anything That Shoots." Call TOLL FREE 1-800-345-6923 and receive a one-year subscription (36 issues) for just $20.00 (use your Master or Visa card). For more information, call 1-402-463-4589 or write:

THE SHOTGUN NEWS
P. O. Box 669, Hastings, NE 68902

VEC 91 ELECTRONIC
CASELESS CARTRIDGE RIFLE

Benchrest accuracy in a hunting rifle. That's what you get from the newest firearms technology to come along in 100 years. Introduced just over a year ago in the U.S., the VEC 91 from Voere in Austria delivers hole-in-hole performance in .223 caliber. The 6-pound rifle has a 5-round magazine and is fired electronically, using a trigger/micro switch so there is zero lock time. The UCC caseless propellant and primer are totally consumed on ignition, pushing a 55-grain .223 bullet out of the barrel at better than 3200 fps. Because the propellant is so stable—it can be submersed in water with no effect—the velocity is the same round after round. That means extreme accuracy round after round as well. Larger calibers, such as 6mm, will also be introduced soon. For a brochure and more information, contact:

JÄGERSPORT
One Wholesale Way, Cranston, RI 02920
Phone: 414-944-9682

AIMPOINT COMP RED DOT SIGHT

The AIMPOINT COMP is the first electronic red dot sight developed for competitive shooting. Its design is based on requirements from competitive shooters all over the world. With a true 30mm field of view, it's the lightest sight in its class, weighing only 4.8 ounces. The patented parallax-free system and durable construction make the AIMPOINT COMP second to none. The AIMPOINT COMP is available in stainless or our **new medium blue** finish.

AIMPOINT INC.
580 Herndon Parkway, Suite 500 Herndon, VA 22070
Phone: 203-471-6828

MANUFACTURERS' SHOWCASE

SWIFT MODEL 667 FIRE-FLY SCOPE
1x, 30mm, Fog-proof, Ruby Coated

This compact ($5^3/_8$-inch) scope with its Variable Intensity FLOATING RED DOT is almost universal in use. Because it has no parallax and unlimited eye relief, it permits shooting with both eyes open. It is, therefore, especially suitable for rapid shooting with handguns, carbines, shotguns, paintball guns and bows. Under any light condition, it is a fast-aligning electronic sight for pinpoint accuracy. It comes complete with the following accessories: a set of 30mm ring mounts fitting Weaver bases, Allen wrench, lenscaps with elastic string, extension tubes, polarizing filter, and lithium battery CR2032, LED Red-Dot with Rheostat, and complete instruction manual. Gift boxed. For more information, contact:

SWIFT INSTRUMENTS, INC.
952 Dorchester Ave., Boston, MA 02125
P.O. Box 562, San Jose, CA 95106

HARRINGTON & RICHARDSON
N.W.T.F.-SPONSORED
YOUTH TURKEY GUN

HARRINGTON & RICHARDSON's new 20-gauge Youth Turkey Gun, sponsored by the National Wild Turkey Federation, is perfect for short-range turkey hunting. This single shot gun with its 22-inch full-choke barrel is designed to deliver excellent patterns for the young turkey hunter without the dramatic recoil of the adult-sized 12-gauge turkey guns. Made in America, this safe and reliable gun features H&R's single-shot action with transfer bar safety and automatic ejection. With the sale of each Youth Turkey gun, H&R 1871, INC., will make a substantial contribution to N.W.T.F.'s programs for habitat restoration so that young hunters may have the opportunity to enjoy the sport of turkey hunting for generations to come.

H&R 1871, INC.
60 Industrial Rowe, Gardner MA 01440
Phone: 508-632-9393

95

FINLAND **SAKO**
FINNFIRE .22 L.R.

THE TERM "PREMIUM .22 RIMFIRE RIFLE" HAS BEEN REDEFINED WITH THE INTRODUCTION OF THE SAKO FINNFIRE .22 L.R. BOLT ACTION RIFLE

Handguns

For addresses and phone numbers of manufacturers and distributors included in this section, turn to *DIRECTORY OF MANUFACTURERS AND SUPPLIERS* at the back of the book.

ACTION ARMS/BRNO PISTOLS

MODEL CZ-75 STANDARD

MODEL CZ-75 COMPACT

SPECIFICATIONS
Calibers: 9mm and 40 S&W (SA/DA or DA only)
Capacity: 15 rounds (12 in 40 S&W)
Barrel length: 4.7″   **Overall length:** 8.1″
Weight: 34.3 oz. (unloaded)
Sights: Fixed blade front; drift adjustable rear
Safety: Thumb safety; cocked and locked or double action
Grip: Polymer
Finish: Mil-spec enamel (matte, black polymer or polish blue)
Prices: $479.00 (black polymer); **$499.00** (matte); **$519.00** (polish blue). Add **$40.00** in 40 S&W for each finish

SPECIFICATIONS
Caliber: 9mm
Capacity: 13 rounds (15 in Semi-Compact)
Barrel length: 3.9″   **Overall length:** 7.3″
Weight: 32.1 oz. (unloaded)
Sights: Fixed blade front (white line front sight pinned in place); drift adjustable rear
Finish: Mil-spec black polymer (matte or polish blue)
Safety: Thumb safety; cocked and locked or double action
Features: Square trigger guard; rounded hammer; non-glare ribbed sighting; walnut grips
Prices: $519.00; $539.00 (matte); **$559.00** (polish blue); same prices for Semi-Compact

MODEL CZ-83

MODEL CZ-85 STANDARD

SPECIFICATIONS
Calibers: 380 ACP and 32 ACP
Capacity: 12 rounds (15 in 32 ACP)
Barrel length: 3.8″   **Overall length:** 6.8″
Weight: 26 oz. (unloaded)
Safety: Thumb safety; cocked and locked or double action
Sights: Fixed blade front; drift adjustable rear
Grip: Polymer
Finish: Blue
Price: $389.00

SPECIFICATIONS
Calibers: 9mm and 9×21mm   **Capacity:** 15 rounds
Barrel length: 4.7″   **Overall length:** 8.1″
Weight: 35.7 oz. (unloaded)
Sights: Fixed blade front; drift adjustable rear
Safety: Thumb safety; cocked and locked or double action
Grip: Walnut
Finish: Mil-spec black polymer (matte or polish blue)
Features: Ambidextrous controls; squared-off, finger-rest trigger guard; non-glare serrated slide top; trigger stop
Price: $515.00 (black polymer); **$537.00** (matte); **$559.00** (polish blue)
MODEL CZ-85 COMBAT w/fully adjustable rear sight, round combat-style hammer, walnut grip; free-dropping magazine.
   Price: $619.00 (polymer); **$645.00** (matte); **$669.00** (blue)

AMERICAN ARMS PISTOLS

MODEL PK-22 DA SEMIAUTO
$209.00

SPECIFICATIONS
Caliber: 22 LR
Capacity: 8-shot clip
Barrel length: 3¹/₃″
Overall length: 6¹/₃″
Weight: 22 oz. (empty)
Sights: Fixed; blade front, "V"-notch rear
Grip: Black polymer

MODEL P-98 CLASSIC SEMIAUTO
$219.00

SPECIFICATIONS
Caliber: 22 LR
Capacity: 8-shot clip
Barrel length: 5″
Overall length: 8¹/₈″
Weight: 26 oz. (empty)
Sights: Fixed blade front; adjustable square-notch rear
Grip: Black polymer

REGULATOR SINGLE ACTION REVOLVER
$305.00
TWO-CYLINDER SET $349.00

SPECIFICATIONS
Calibers: 45 Long Colt, 44-40, 357 Mag.
Barrel lengths: 4³/₄″, 5¹/₂″ and 7¹/₂″
Action: Single Action
Sights: Fixed
Safety: Half cock
Features: Brass trigger guard and back strap; two-cylinder combinations available (45 L.C./45 ACP and 44-40/44 Special)
Also available:
BUCKHORN SA. Same as Regulator but with stronger frame for 44 Mag. **$339.00**
   With target sights 349.00

MODEL CX-22 DA SEMIAUTO
$209.00

SPECIFICATIONS
Caliber: 22 LR
Capacity: 8-shot clip
Barrel length: 3¹/₃″
Overall length: 6¹/₃″
Weight: 22 oz. (empty)
Sights: Fixed; blade front, "V"-notch rear
Grip: Black polymer
Also available:
MODEL PX-22 (7-shot magazine):   **$198.00**

AMERICAN DERRINGER PISTOLS

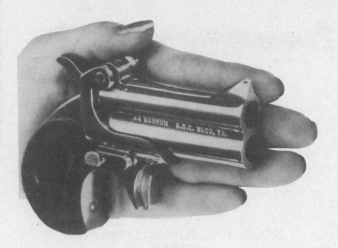

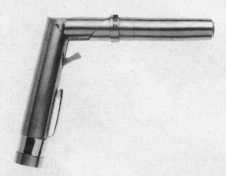

MODEL 2 STEEL "PEN" PISTOL

SPECIFICATIONS
Calibers: 22 LR, 25 Auto, 32 Auto
Barrel length: 2″
Overall length: 5.6″ (4.2″ in pistol format)
Weight: 5 oz.
Price: $199.00-226.50

Also available:
MODEL 7 Ultra Lightweight (7½ oz.) Single Actions
22 LR, 22 Mag. Rimfire, 32 Mag., 38 Special,
   380 Auto **$220.00**
44 Special **500.00**

MODEL 10 (10 oz.)
45 Colt **$320.00**
45 Auto **257.00**

MODEL 11 Lightweight (11 oz.) Double Derringer
38 Special **$205.00**

MODEL 1

SPECIFICATIONS
Overall length: 4.82″
Barrel length: 3″
Weight: 15 oz. (in 45 Auto)
Action: Single action w/automatic barrel selection
Number of shots: 2

Calibers:	Prices
22 Long Rifle w/rosewood grips	$225.00
22 Magnum Rimfire w/Zebra wood grips	225.00
32 Magnum/32 S&W Long	235.00
32-20	235.00
357 Magnum w/rosewood grips	257.00
357 Maximum w/rosewood grips	265.00
38 Special w/rosewood grips	225.00
38 Super w/rosewood grips	243.00
38 Special +P+ (Police)	233.00
38 Special Shot Shells	243.00
380 Auto, 9mm Luger	224.00
10mm Auto, 40 S&W, 45 Auto, 30 M-1 Carbine . . .	257.00
45-70 (single shot)	312.00
45 Colt, 2½″ Snake (45-cal. rifled barrel), 44-40 Win., 44 Special	320.00
45 Win. Mag., 44 Magnum, 41 Magnum	385.00
30-30 Win., 223 Rem. Comm. Ammo dual calibers	375.00
Engraved	855.00

Also available:
LADY DERRINGER (Stainless Steel Double)

38 Special, 32 Magnum, 32 S&W	$235.00
Engraved	750.00

MODEL 3
Stainless Steel Single Shot Derringer
(not shown)

SPECIFICATIONS
Calibers: 32 Magnum, 38 Special
Barrel length: 2.5″
Overall length: 4.9″
Weight: 8.5 oz.
Safety: Manual "Hammer-Block"
Grips: Rosewood
Price: $120.00

38 DOUBLE ACTION DERRINGER (14.5 oz.)

38 Special	$250.00
38 Special Lady Derringer w/syn. ivory grips	265.00
9mm Luger	275.00
357 Magnum, 40 S&W	300.00

High Standard DA Double Derringer
22 LR, 22 Rimfire $169.50

AMERICAN DERRINGER PISTOLS

MODEL 4

MODEL 4
Stainless Steel Double Derringer

SPECIFICATIONS
Calibers: 45 Colt or 3″ .410
Barrel length: 4.1″
Overall length: 6″
Weight: 16.5 oz.
Number of shots: 2
Finish: Satin or high-polish stainless steel
Price: $352.00
   With oversize grips 382.00
Also available:
In 45 Auto, 45 Colt, 357 Mag., 357 Maximum $369.00
In 45-70 w/oversized grips 495.00
Alaskan Survival Model in 45-70 . . . . 387.50
   In 44 Mag. w/oversized grips 422.00

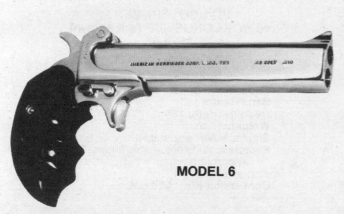

MODEL 6

MODEL 6
Stainless Steel Double Derringer

SPECIFICATIONS
Calibers: .410, 45 Colt
Number of shots: 2
Barrel length: 6″
Overall length: 8.2″
Weight: 21 oz.
Price: Grey matte finish $350.00
   Satin finish 362.50
   High polish finish 387.50
   W/oversized grips, **add** 35.00

COP AND MINI-COP 4-SHOT DA DERRINGERS
(not shown)

22 Magnum Rimfire (Mini-Cop) $312.50
357 Mag. or 38 Special (Cop) 375.00

125th ANNIVERSARY DOUBLE DERRINGER
COMMEMORATIVE (1866–1991)

38 Special $225.00
44-40 or 45 Colt 320.00
Engraved models 750.00

SEMMERLING LM-4 DOUBLE ACTION

**SEMMERLING
LM-4 DOUBLE ACTION**

SPECIFICATIONS
Caliber: 45 ACP or 9mm
Action: Double action
Capacity: 5 rounds
Overall length: 5″
Price: Blued finish (manual repeating) $1750.00
   Stainless steel 1875.00

AMT PISTOLS

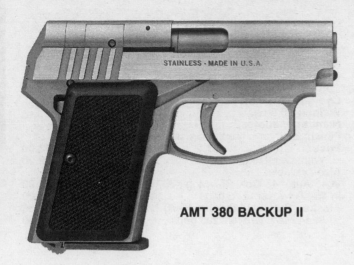

AMT 380 BACKUP II

AMT 380 BACKUP II
$295.99 (Single Action)
$309.99 (Double Action)

SPECIFICATIONS
Caliber: 380 ACP
Capacity: 5 shots
Barrel length: 2 1/2"
Overall length: 5"
Weight: 18 oz.
Width: 11/16"
Sights: Open
Grips: Carbon fiber

ON DUTY
DOUBLE ACTION PISTOL
(not shown)
$469.95 (40 S&W, 9mm)
$529.99 (45 ACP)

SPECIFICATIONS
Caliber: 40 S&W, 45 ACP and 9mm Luger
Capacity: 13 rounds
Barrel length: 4 1/2"
Overall length: 7 3/4"
Weight: 32 oz.
Features: Stainless steel slide and barrel; carbon fiber grips; inertia firing pin; steel recoil shoulder; white 3 dot sighting system; trigger disconnector safety; light let-off double action.
Also available: **Decocker model** (same price).

1911 GOVERNMENT
45 ACP LONGSLIDE (not shown)
$575.95

SPECIFICATIONS
Caliber: 45 ACP
Capacity: 7 shots
Barrel length: 7"
Overall length: 10 1/2"
Weight: 46 oz.
Sights: Millett adjustable
Features: Wide adjustable trigger; Neoprene wraparound grips
Also available:
Conversion Kit: $299.99

1911 GOVERNMENT MODEL
$475.95

SPECIFICATIONS
Caliber: 45 ACP
Capacity: 7 shots
Barrel length: 5"
Overall length: 8 1/2"
Weight: 38 oz.
Width: 1 1/4"
Sights: Fixed
Features: Long grip safety; rubber wraparound Neoprene grips; beveled magazine well; wide adjustable trigger
Also available:
1911 HARDBALLER. Same specifications as Standard Model, but with Millett adjustable sights and matte rib.
Price: $529.95
Conversion Kit 279.00

1911 GOVERNMENT

AMT PISTOLS

22 AUTOMAG II

22 AUTOMAG II RIMFIRE MAGNUM
$385.95

The only production semiautomatic handgun in this caliber, the Automag II is ideal for the small-game hunter or shooting enthusiast who wants more power and accuracy in a light, trim handgun. The pistol features a bold open-slide design and employs a unique gas-channeling system for smooth, trouble-free action.

SPECIFICATIONS
Caliber: 22 Rimfire Magnum
Barrel lengths: $3^3/_8''$, $4^1/_2''$ or $6''$
Magazine capacity: 9 shots ($4^1/_2''$ & $6''$), 7 shots ($3^3/_8''$)
Weight: 32 oz.
Sights: Millett adjustable (white outline rear; red ramp)
Features: Squared trigger guard; grooved carbon fiber grips

AUTOMAG III
$459.95

SPECIFICATIONS
Caliber: 30 M1 Carbine, 9mm Win. Mag.
Capacity: 8 shots
Barrel length: $6^3/_8''$
Overall length: $10^1/_2''$
Weight: 43 oz.
Sights: Millett adjustable
Grips: Carbon fiber
Finish: Stainless steel

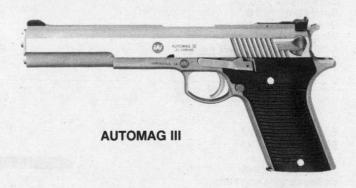

AUTOMAG III

AUTOMAG IV
$679.99

SPECIFICATIONS
Caliber: 45 Win. Mag.
Capacity: 7 shots
Barrel lengths: $6^1/_2''$
Overall length: $10^1/_2''$
Weight: 46 oz.
Sights: Millett adjustable
Grips: Carbon fiber
Finish: Stainless steel

Also available:
Automag V 50 Cal. A.E.:   **$899.99**

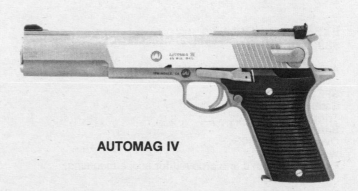

AUTOMAG IV

ANSCHUTZ PISTOLS

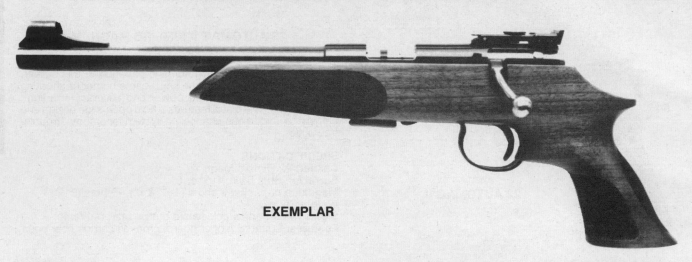

EXEMPLAR

EXEMPLAR
$499.50

SPECIFICATIONS
Calibers: 22 Long Rifle
Capacity: 5-shot clip
Barrel length: 10″
Overall length: 19″
Weight: 3¹/₃ lbs.
Action: Match 64
Trigger pull: 9.85 oz., two-stage adjustable

Safety: Slide
Sights: Hooded ramp post front; open notched rear; adjustable for windage and elevation
Stock: European walnut

Also available:
EXEMPLAR LEFT featuring right-hand operating bolt.
Price: **$499.50**

EXEMPLAR XIV
$522.00

SPECIFICATIONS
Calibers: 22 Long Rifle
Barrel length: 14″
Overall length: 23″
Weight: 4.15 lbs.
Action: Match 64
Trigger pull: 9.85 oz., two-stage
Safety: Slide

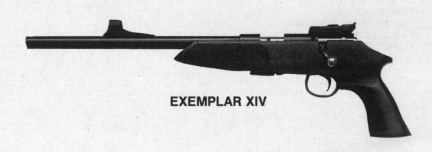

EXEMPLAR XIV

EXEMPLAR HORNET
$899.00

A centerfire version with Match 54 action.

SPECIFICATIONS
Caliber: 22 Hornet
Trigger pull: 19.6 oz.
Barrel: 10″
Overall length: 20″
Weight: 4.35 lbs.
Features: Tapped and grooved for scope mounting; wing safety

EXEMPLAR HORNET

AUTO-ORDNANCE PISTOLS

MODEL 1911A1 THOMPSON

SPECIFICATIONS
Calibers: 45 ACP, 9mm and 38 Super; also 10mm
Capacity: 9 rounds (9mm & 38 Super); 7 rounds (45 ACP)
Barrel length: 5″   **Overall length:** 8 1/2″
Weight: 39 oz.
Sights: Blade front; rear adjustable for windage
Stock: Checkered plastic with medallion
Prices:
In 9mm and 38 Super $415.00
In 10mm 420.95
In 45 ACP 388.95
Satin nickel or Duo-Tone finish (45 cal. only) 405.00
PIT BULL MODEL (45 ACP w/3 1/2″ barrel) 420.95
WW II PARKERIZED PISTOL (45 cal. only) 379.25

**MODEL 1911A1
THOMPSON (9mm)**

MODEL 1911 "THE GENERAL"
(not shown)

SPECIFICATIONS
Caliber: 45 ACP   **Capacity:** 7 rounds
Barrel length: 4 1/2″   **Overall length:** 7 3/4″
Weight: 37 oz.
Stock: Black textured, rubber wraparound with
   medallion
Feature: Full-length recoil guide system
Price: $427.95

**MODEL 1911A1
DUO-TONE**

MODEL 1911A1 COMPETITION

SPECIFICATIONS
Caliber: 45
Weight: 42 oz.
Barrel length: 5″   **Overall length:** 10″
Stock: Rubber wraparound
Sights: 3 white dot sight system
Price: $615.00

MODEL 1927A5 (not shown)

SPECIFICATIONS
Caliber: 45 ACP
Capacity: 30 rounds
Barrel length: 13 1/2″ (finned)   **Overall length:** 26″
Weight: 7 lbs.
Sights: Blade front; adjustable open rear
Stock: Walnut rear grip, vertical forend
Price: $740.00

**MODEL 1911A1
COMPETITION**

BERETTA PISTOLS

COMPACT FRAME PISTOLS

8000/8040 COUGAR SERIES

Beretta's 8000/8040 Cougar Series semiautomatics use a proven locked-breech system with a rotating barrel. This design makes the pistol compact and easy to conceal and operate with today's high-powered 9mm, .41 AE and .40 caliber ammunition. When the pistol is fired, the initial thrust of recoil energy is partially absorbed as it pushes slide and barrel back, with the barrel rotating by cam action against a tooth on the rigid central block. When the barrel has turned about 30 degrees, the locking lugs on the barrel clear the locking recesses, which free the slide to continue rearward. The recoil spring absorbs the remaining recoil energy as the slide extracts and ejects the spent shell casing, rotates the hammer, and then reverses direction to chamber the next round. By channeling part of the recoil energy into barrel rotation and by partially absorbing the barrel and slide recoil shock through the central block before it is transferred to the frame, the Cougar shows an unusually low felt recoil.

MODEL 8000 COUGAR

General specifications are included in the table below. **Prices not set.**

SPECIFICATIONS COUGAR SERIES

Compact Frame Pistols	Caliber				Magazine Capacity	Action		Overall Length	Barrel Length	Overall Width	Overall Height	Sight Radius	Weight Unloaded
	9mm x19	9mm x21IMI	.41 A.E.	.40 S.W.	rounds	Single	Double	mm inches	mm inches	mm inches	mm inches	mm inches	gr ounces
8000 Cougar G	●				15	●	●	180 7	92 3.6	38 1.5	140 5.5	132 5.2	950 33.5
8000 Cougar G Inox			●		10								
8000 Cougar G Combo		●			15	●	●	180 7	92 3.6	38 1.5	140 5.5	132 5.2	950 33.5
8000 Cougar G Inox Combo			●		10								
8040 Cougar G				●	11	●	●	180 7	92 3.6	38 1.5	140 5.5	132 5.2	950 33.5
8040 Cougar G Inox													
8000 Cougar F	●				15	●	●	180 7	92 3.6	38 1.5	140 5.5	132 5.2	950 33.5
8000 Cougar F Inox			●		10								
8000 Cougar F Combo		●			15	●	●	180 7	92 3.6	38 1.5	140 5.5	132 5.2	950 33.5
8000 Cougar F Inox Combo			●		10								
8040 Cougar F				●	11	●	●	180 7	92 3.6	38 1.5	140 5.5	132 5.2	950 33.5
8040 Cougar F Inox													
8000 Cougar D	●				15		●	180 7	92 3.6	38 1.5	140 5.5	132 5.2	940 33.1
8000 Cougar D Inox			●		10								
8000 Cougar D Combo		●			15		●	180 7	92 3.6	38 1.5	140 5.5	132 5.2	940 33.1
8000 Cougar D Inox Combo			●		10								
8040 Cougar D				●	11		●	180 7	92 3.6	38 1.5	140 5.5	132 5.2	940 33.1
8040 Cougar D Inox													

BERETTA PISTOLS

SMALL FRAME PISTOLS

MODEL 21 BOBCAT DA SEMIAUTOMATIC
$235.00 ($247.00 Nickel)

A safe, dependable, accurate small-bore pistol in 22 LR or 25 Auto. Easy to load with its unique barrel tip-up system.

SPECIFICATIONS
Caliber: 22 LR or 25 Auto. **Magazine capacity:** 7 rounds (22 LR); 8 rounds (25 Auto). **Overall length:** 4.9". **Barrel length:** 2.4". **Weight:** 11.5 oz. (25 ACP); 11.8 oz. (22 LR) **Sights:** Blade front; V-notch rear. **Safety:** Thumb operated. **Grips:** Plastic or Walnut. **Frame:** Forged aluminum.
Also available:
Model 21 Engraved $285.00
W/Plastic grips, matte black finish 185.00

MODEL 21 BOBCAT

MODEL 950 BS JETFIRE

MODEL 950 BS JETFIRE
SINGLE ACTION SEMIAUTOMATIC

SPECIFICATIONS
Caliber: 25 ACP. **Barrel length:** 2.4". **Overall length:** 4.7". **Overall height:** 3.4". **Safety:** External, thumb-operated. **Magazine capacity:** 8 rounds. **Sights:** Blade front; V-notch rear. **Weight:** 9.9 oz. **Frame:** Forged aluminum.

Model 950 BS $180.00
Model 950 BS Nickel 210.00
Model 950 EL Engraved 260.00
Model 950 BS Plastic grips, matte black finish . . . 150.00

MEDIUM FRAME PISTOLS

MODEL 84 CHEETAH

This pistol is pocket size with a large magazine capacity. The first shot (with hammer down, chamber loaded) can be fired by a double-action pull on the trigger without cocking the hammer manually.

The pistol also features a favorable grip angle for natural pointing, positive thumb safety (designed for both right- and left-handed operation), quick takedown (by means of special takedown button) and a conveniently located magazine release. Black plastic grips. Wood grips extra.

SPECIFICATIONS
Caliber: 380 Auto (9mm Short). **Weight:** 1 lb. 7 oz. (approx.).
Barrel length: 3³/₄₄". (approx.) **Overall length:** 6¹/₂". (approx.)
Sights: Fixed front and rear. **Magazine capacity:** 13 rounds.
Height overall: 4¹/₄" (approx.).

Model 84 w/Plastic $525.00
Model 84 w/Wood 555.00
Model 84 w/Wood Nickel 600.00

MODEL 84 CHEETAH

BERETTA PISTOLS

MEDIUM FRAME PISTOLS

MODEL 85F CHEETAH

This double-action semiautomatic pistol features walnut or plastic grips, matte black finish on steel slide, barrel and anodized forged aluminum, ambidextrous safety, and a single line 8-round magazine.

SPECIFICATIONS
Caliber: 380 Auto. **Barrel length:** 3.82″. **Weight:** 21.9 oz. (empty). **Overall length:** 6.8″. **Overall height:** 4.8″. **Capacity:** 8 rounds. **Sights:** Blade integral with slide (front); square notched bar, dovetailed to slide (rear).
Prices:
Model 85 w/Plastic (8 rounds)	$485.00
Model 85 w/Wood	510.00
Model 85 w/Wood Nickel	550.00
Model 87 w/Wood (22 LR)	490.00

MODEL 85F CHEETAH

MODEL 86 CHEETAH

MODEL 86 CHEETAH

SPECIFICATIONS
Caliber: 380 Auto. **Barrel length:** 4.33″. **Overall length:** 7.33″. **Capacity:** 8 rounds. **Weight:** 22.6 oz. **Sight radius:** 5.0″. **Overall height:** 4.8″. **Overall width:** 1.4″. **Grip:** Walnut. **Features:** Same as other Medium Frame, straight blow-back models, plus safety and convenience of a tip-up barrel (rounds can be loaded directly into chamber without operating the slide).
Price: $510.00

MODEL 89 SPORT GOLD STANDARD

This sophisticated single-action, target pistol features an eight-round magazine, adjustable target sights, and target-style contoured walnut grips with thumb rest.

SPECIFICATIONS
Caliber: 22 LR. **Barrel length:** 6″. **Overall length:** 9¹/₂″. **Height:** 5.3″. **Weight:** 41 oz.
Price: $735.00

MODEL 89 SPORT GOLD

BERETTA PISTOLS

LARGE FRAME PISTOLS

SPECIFICATIONS MODELS 92, 96 AND 98

Large Frame Pistols	Caliber			Magazine Capacity	Overall Length	Barrel Length	Overall Width	Overall Height	Sight Radius	Weight Unloaded
	9mm x19	9mm x21IMI	.40 S.&W.	rounds	mm inches	mm inches	mm inches	mm inches	mm inches	gr ounces
92FS	●			15	217	125	38	137	155	975
98FS		●			8.5	4.9	1.5	5.4	6.1	34.4
96			●	10						
92FS De Luxe	●			15	217	125	38	137	155	975
98FS De Luxe		●			8.5	4.9	1.5	5.4	6.1	34.4
92FS Inox	●			15	217	125	38	137	155	975
98FS Inox		●			8.5	4.9	1.5	5.4	6.1	34.4
92FS Inox Golden	●			15	217	125	38	137	155	975
98FS Inox Golden		●			8.5	4.9	1.5	5.4	6.1	34.4
92FS Competition Conversion Kit	●			15	277 10.9	185 7.3	46 1.8	147 5.8	230 9	1120 39.5
98FS Target		●		15	242 9.5	150 5.9	46 1.8	137 5.4	176 6.9	1050 37
92FS Centurion	●			15	197	109	38	137	147	940
96 Centurion			●	10	7.8	4.3	1.5	5.4	5.8	33.2
92D	●			15	217	125	38	137	155	960
96D			●	10	8.5	4.9	1.5	5.4	6.1	33.8
92DS	●			15	217	125	38	137	155	970
96DS			●	10	8.5	4.9	1.5	5.4	6.1	34.2
92G	●			15	217	125	38	137	155	975
96G			●	10	8.5	4.9	1.5	5.4	6.1	34.4
92 Brigadier FS	●			15	217	125	38.5	140	161	1000
98 Brigadier FS		●			8.5	4.9	1.5	5.5	6.3	35.3
96 Brigadier			●	10						
92 Brigadier D	●			15	217 8.5	125 4.9	36 1.4	140 5.5	161 6.3	1020 36
92 Stock	●			15	217	125	44.5	140	161	1000
98 Stock		●			8.5	4.9	1.7	5.5	6.3	35.3
96 Stock			●	10						

Prices:
Model 92FS Plastic . . . . $ 625.00
  (Wood grips **$20.00** additional)
Model 92F Stainless w/3-dot
  sights **755.00**
Model 92FS Plastic Centurion
  9mm/40 cal. w/3-dot sights . . **625.00**
  w/tritium sights **680.00**
Model 92D (DA only, bobbed hammer)
  w/3-dot sights **585.00**
  w/tritium sight **680.00**
  Deluxe gold-plated engraved . . **5430.00**
Model 92F-EL Stainless **1240.00**
Model 96 w/3-dot sights **640.00**
  w/tritium sights **730.00**
Model 96D (DA only) **605.00**
  w/tritium sights **685.00**
Model 96 Centurion **640.00**
  w/tritium sights **730.00**
Optional:
3-dot sights (same price) and tritium sights
(**$75.00** additional)

MODEL 96

MODEL 92F (9mm)

MODEL 92D

BERNARDELLI PISTOLS

MODEL PO18 TARGET PISTOL
$499.00 (Black Plastic)
$549.00 (Chrome)

SPECIFICATIONS
Caliber: 9mm
Capacity: 16 rounds
Barrel length: 4.8″   **Overall length:** 8.25″
Weight: 34.2 oz.
Sights: Low micrometric sights adjustable for windage and elevation.   **Sight radius:** 5.4″
Features: Thumb safety decocks hammer; magazine press button release reversible for right- and left-hand shooters; hardened steel barrel; can be carried cocked and locked; squared and serrated trigger guard and grip; frame and barrel forged in steel and milled with CNC machines; manual thumb, half cock, magazine and auto-locking firing-pin block safeties; low-profile three-dot interchangeable combat sights

MODEL PO18 COMPACT TARGET PISTOL
$552.00 (Black Plastic)
$600.00 (Chrome)

SPECIFICATIONS
Calibers: 380 and 9mm   **Capacity:** 14 rounds
Barrel length: 4″   **Overall length:** 7.44″
Weight: 31.7 oz.
Sight radius: 5.4″
Grips: Walnut or plastic
Features: Same as Model PO18

MODEL PO10 TARGET PISTOL
$600.00

SPECIFICATIONS
Caliber: 22 LR   **Capacity:** 5 or 10 rounds
Barrel length: 5.9″   **Weight:** 40 oz.
Sight radius: 7¹/₂″
Sights: Interchangeable front sight; rear sight adjustable for windage and elevation
Features: All steel construction; external hammer with safety notch; external slide catch for hold-open device; inertia safe firing pin; oil-finished walnut grips for right- and left-hand shooters; matte black or chrome finish; pivoted trigger with adjustable weight and take-ups

BERSA AUTOMATIC PISTOLS

THUNDER 9 DOUBLE ACTION
$414.95

SPECIFICATIONS
Caliber: 9mm
Capacity: 15
Action: Double
Barrel length: 4"
Overall length: 7 3/8"
Weight: 30 oz.
Height: 5 1/2"
Sights: Blade front (integral w/slide); fully adjustable rear
Safety: Manual, firing pin and decocking lever
Grips: Black polymer
Finish: Matte blue
Features: Reversible extended magazine release; adjustable trigger release; "Link-Free" design system (ensures positive lockup); instant disassembly; ambidextrous slide release

THUNDER 9 DOUBLE ACTION

MODEL 23 DOUBLE ACTION (not shown)
$287.95 ($321.95 in Nickel)

SPECIFICATIONS
Caliber: 22 LR
Capacity: 10 shots
Action: Blowback
Barrel length: 3 1/2"
Weight: 24 1/2 oz.
Sights: Blade front; notch-bar dovetailed rear, adjustable for windage
Grips: Walnut
Finish: Blue or satin nickel

MODEL 83 DOUBLE ACTION

MODEL 83 DOUBLE ACTION
$287.95 ($321.95 in Nickel)

SPECIFICATIONS
Caliber: 380 Auto
Capacity: 7 shots
Action: Blowback
Barrel length: 3 1/2"
Weight: 25 3/4 oz.
Sights: Front blade sight integral on slide; rear sight square notched, adjustable for windage
Grips: Custom wood

MODEL 85 DOUBLE ACTION
$339.95 ($386.95 in Nickel)

SPECIFICATIONS
Caliber: 380 Auto
Capacity: 13 shots
Barrel length: 3 1/2"
Overall length: 6 5/8"
Weight: 26.45 oz.

Also available:
MODEL 86 CUSTOM UNDERCOVER DA. Same specifications as Model 85, except with military non-glare matte finish and 3-dot sight system. **$374.95 ($403.95 in Nickel)**

MODEL 85 DOUBLE ACTION

BROWNING AUTOMATIC PISTOLS

**9mm HI-POWER
SINGLE ACTION**

9mm HI-POWER SINGLE ACTION

The Browning 9mm Parabellum, also known as the 9mm Browning Hi-Power, is now available in 40 S&W. Both come with either a fixed-blade front sight and a windage-adjustable rear sight or a non-glare rear sight, screw adjustable for both windage and elevation. The front sight is an 1/8-inch-wide blade mounted on a ramp. The rear surface of the blade is serrated to prevent glare. All models have an ambidextrous safety. See table below for specifications.

Prices:

Standard Mark III 9mm and 40 S&W w/matte finish, fixed sights, molded grip	**$506.95**
9mm, polished blue, adj. sights, walnut grip	**585.95**
9mm, polished blue, fixed sights	**537.95**
HP Practical	
w/fixed sights, molded rubber grip	**579.95**
w/adjustable sights	**627.95**
Silver chrome w/adj. sights, molded rubber grip	**596.95**
Capitan w/polished blue finish, tangent sights, walnut grip	**634.95**

9mm SEMIAUTOMATIC PISTOL

	SINGLE ACTION FIXED SIGHTS	SINGLE ACTION ADJUSTABLE SIGHTS
Finish	Polished Blue, Matte, or Nickel	Polished Blue
Capacity of Magazine	13 (10 in 40 S&W)	13 (10 in 40 S&W)
Overall Length	7 3/4″	7 3/4″
Barrel Length	4 21/32″	4 21/32″
Height	5″	5″
Weight (Empty)	32 oz.; 35 oz. in 40 S&W 36 oz. (HP Practical)	32 oz.; 33 oz. w/tangent sight and 35 oz. in 40 S&W
Sight Radius	6 5/16″	6 3/8″
Ammunition	9mm Luger (Parabellum) or 40 S&W	9mm Luger (Parabellum) or 40 S&W
Grips	Checkered Walnut or Contoured Molded	Checkered Walnut or Contoured Molded
Front Sights	1/8″	1/8″ wide on ramp
Rear Sights	Drift adjustable for windage and elevation.	Drift adjustable for windage and elevation. Square Notch.

MODEL BDM 9mm DOUBLE ACTION

Browning's Model BDM (for Browning Double Mode) pistol brings shooters into a new realm of convenience and safety by combining the best advantages of double-action pistols with those of the revolver. In just seconds, the shooter can set the BDM to conventional double-action "pistol" mode or to the all-new double-action "revolver" mode.

SPECIFICATIONS
Caliber: 9mm Luger
Capacity: 10 or 15 rounds
Barrel length: 4.73″
Overall length: 7.85″
Weight: 31 oz. (empty)
Sight radius: 6.26″
Sights: Low-profile front (removable); rear screw adjustable for windage; includes 3-dot sight system
Finish: Matte blue
Features: Dual-purpose ambidextrous decocking lever/safety designed with a short stroke for easy operation (also functions as slide release); contoured grip is checkered on all four sides
Price: **$573.95**

**MODEL BDM
9mm DOUBLE ACTION**

BROWNING AUTOMATIC PISTOLS

MODEL BDA-380

BUCK MARK 5.5 TARGET

MODEL BDA-380

A high-powered, double-action semiautomatic
pistol with fixed sights in 380 caliber.

SPECIFICATIONS
Capacity: 13 shots
Barrel length: 3¹³/₁₆″
Overall length: 6³/₄″
Weight: 32 oz.
Grips: Walnut
Prices:
Nickel Finish **$624.95**
Standard Finish **592.95**

MICRO BUCK MARK

Features light weight,
Pro Target Sight,
bull barrel and
black molded
composite grips.

BUCK MARK SEMIAUTOMATIC PISTOL SPECIFICATIONS

	MICRO BUCK MARK	BUCK MARK STD. BUCK MARK PLUS	BUCK MARK* 5.5 TARGET/FIELD	BUCK MARK* SILHOUETTE	BUCK MARK* VARMINT	BUCK MARK* UNLTD. SILHOUETTE
Capacity	10	10	10	10	10	10
Overall length	8″	9 ¹/₂″	9 ⁵/₈″	14″	14″	18 ¹¹/₁₆″
Barrel length	4″	5 ¹/₂″	5 ¹/₂″	9 ⁷/₈″	9 ⁷/₈″	14″
Height	5³/₈″	5 ³/₈″	5 ⁵/₁₆″	5 ¹⁵/₁₆″	5 ⁵/₁₆″	5 ¹⁵/₁₆″
Weight (empty)	32 oz.	32 oz.	35 oz.	53 oz.	48 oz.	64 oz.
Sight radius	6¹/₂″	8″	8 ¹/₄″	13″	—	15″
Ammunition	22 LR	22 LR	22 LR	22 LR	22 LR	22 LR
Grips	Laminated wood	Black molded; Impregnated hardwood	Contoured walnut	Contoured walnut	Contoured walnut	Contoured walnut
Front Sights	Ramp front	¹/₈″ wide	Inter-changeable Post*	Interchangeable Post*	None*	Interchangeable Post*
Rear Sights	Pro Target Sight	Pro Target Sight	Pro Target Sight	Pro Target Sight	None*	Pro Target Sight
Prices:	$241.95 (Std.) $282.95 (Nickel) $293.95 (Plus)	$241.95 (Std.) $293.95 (Buck Mark Plus) $282.95 (Nickel)	$385.95 $434.95 (Gold Target)	$406.95	$365.95	$499.95

* Buck Mark Target, Silhouette and Varmint models supplied with full-length top rib designed to accept most standard clamp-style scope rings. Additional accessories for Silhouette and Target models available. Finger-groove grip option on all target models (same prices).

CHARTER ARMS REVOLVERS

BULLDOG PUG 44 SPECIAL

SPECIFICATIONS
Caliber: 44 Special. **Type of action:** 5-shot, single- and double-action. **Barrel length:** 3″. **Overall length:** 7³/₄″. **Height:** 5″. **Weight:** 19 oz. **Grips:** Neoprene or American walnut hand-checkered bulldog grips. **Sights:** Patridge-type, ⁹/₆₄″ wide front; square-notched rear. **Finish:** High-luster Service Blue or stainless steel.

Prices:
Blued finish Pug $267.50
Nickel (Electroless) Pug 289.50

**BULLDOG PUG
44 SPECIAL**

OFF-DUTY 38 SPECIAL

SPECIFICATIONS
Calibers: 22 LR and 38 Special. **Type of action:** 5-shot, single and double action. **Barrel length:** 2″. **Overall length:** 6¹/₄″. **Height:** 4¹/₄″. **Weight:** 16 oz. (matte black); 17 oz. (stainless). **Grips:** Select-a-grip (9 colors) or Neoprene. **Sights:** Patridge-type ramp front (with ''red dot'' feature); square-notch rear on stainless.

Prices:
Matte black finish $199.00
Nickel (Electroless) 239.75

OFF-DUTY 38 SPECIAL

POLICE UNDERCOVER

SPECIFICATIONS
Caliber: 32 H&R Magnum and 38 Special. **Type of action:** 6-shot, single and double action. **Barrel length:** 2″. **Height:** 4¹/₂″. **Weight:** 17¹/₂ oz. (2″ barrel) and 19 oz. (4″ barrel.) **Grips:** Checkered walnut panel. **Sights:** Patridge-type ramp front; square-notch rear. **Finish:** Blue.

Prices:
Blued $237.75
Nickel (Electroless) 252.00

POLICE UNDERCOVER

COLT AUTOMATIC PISTOLS

DOUBLE EAGLE MK SERIES 90

SPECIFICATIONS
Caliber: 45 ACP
Capacity: 8 rounds
Barrel length: 5″ (Std.); 4¼″ (D.E. Combat Commander); 3½″ (D.E. Officer's ACP)
Overall length: 8½″ (Std.); 7¾″ (D.E. Combat Commander); 7¼″ (D.E. Officer's ACP)
Weight: 35 to 39 oz. (approx.)
Sights: WDS with sight radius 5¼″ (Officer's ACP) to 6½″ (Std.)
Prices:

DOUBLE EAGLE	$697.00
D.E. COMBAT COMMANDER	697.00
D.E. OFFICER'S ACP	697.00

DOUBLE EAGLE

MODEL M1991A1

MODEL M1991A1 PISTOL

SPECIFICATIONS
Caliber: 45 ACP
Capacity: 7 rounds
Barrel length: 5″
Overall length: 8½″
Sight radius: 6½″
Grips: Black composition
Finish: Parkerized
Features: Custom-molded carry case
Price: $517.00

Also available:

COMPACT M1991A1 with 3½″ barrel	$517.00
COMMANDER M1991A1 with 4¼″ barrel and 7-round capacity	517.00

COMBAT COMMANDER MKIV SERIES 80

The semiautomatic Combat Commander, available in 45 ACP and 38 Super, features an all-steel frame that supplies the pistol with an extra measure of heft and stability. This Colt pistol also offers 3-Dot high-profile sights, lanyard-style hammer and thumb and beavertail grip safety. Also available in lightweight version with alloy frame (45 ACP only).

SPECIFICATIONS

Caliber	Weight (ounces)	Overall Length	Magazine Rounds	Finish	Price
45 ACP	36	7¾″	8	Blue	$707.95
45 ACP	36	7¾″	8	Stainless	759.95
45 ACP LW	27½	7¾″	8	Blue	707.00
38 Super	37	7¾″	9	Stainless	759.00

COMBAT COMMANDER
4¼″ barrel only

COLT AUTOMATIC PISTOLS

MKIV SERIES 80

GOLD CUP NATIONAL MATCH

SPECIFICATIONS
Caliber: 45 ACP
Capacity: 8 rounds
Barrel length: 5″
Weight: 39 oz.
Overall length: 8½″
Sights: Colt Elliason sights; adjustable rear for windage and elevation
Hammer: Serrated rounded hammer
Stock: Rubber combat
Finish: Colt blue, stainless or "Ultimate" bright stainless steel
Prices: $ 899.00 Blue
963.00 Stainless steel
1032.00 Bright stainless
Also available:
COMBAT ELITE in 45 ACP or 38 Super; features Accro Adjustable sights, beavertail grip safety. **Price: $860.00**

**GOLD CUP
NATIONAL MATCH**

GOVERNMENT MODEL

GOVERNMENT MODEL
MKIV SERIES 80 SEMIAUTOMATIC

These full-size automatic pistols with 5-inch barrels are available in 45 ACP and 38 Super. The Government Model's special features include high-profile 3-dot sights, grip and thumb safeties, and rubber combat stocks.

SPECIFICATIONS
Calibers: 38 Super and 45 ACP
Barrel length: 5″
Overall length: 8½″
Capacity: 9 rounds (38 Super); 7 rounds (45 ACP)
Weight: 38 oz.
Prices: $707.00 45 ACP blue
759.00 45 ACP stainless
829.00 45 ACP bright stainless
707.00 38 Super blue
759.00 38 Super stainless
829.00 38 Super bright stainless

GOVERNMENT MODEL 380
MKIV SERIES 80 SEMIAUTOMATIC

This scaled-down version of the 1911A1 Colt Government Model does not include a grip safety. It incorporates the use of a firing pin safety to provide for a safe method to carry a round in the chamber in a "cocked and locked" mode. Available in matte stainless steel finish with black composition stocks.

SPECIFICATIONS
Caliber: 380 ACP
Magazine capacity: 7 rounds
Barrel length: 3.25″
Overall length: 6″
Height: 4.4″
Weight : 21.75 oz. (empty)
Sights: Fixed ramp blade front; fixed square notch rear
Grip: Composition stocks
Prices: $443.00 Blue
504.00 Satin nickel
473.00 Stainless steel
Also available:
POCKETLITE MODEL (14¾ oz.), blue finish only. **$443.00**

**380 GOVERNMENT
POCKETLITE**

COLT AUTOMATIC PISTOLS

MKIV SERIES 80

DELTA ELITE AND DELTA GOLD CUP

The proven design and reliability of Colt's Government Model has been combined with the powerful 10mm auto cartridge to produce a highly effective shooting system for hunting, law enforcement and personal protection. The velocity and energy of the 10mm cartridge make this pistol ideal for the serious handgun hunter and the law enforcement professional who insist on down-range stopping power.

SPECIFICATIONS
Type: 0 Frame, semiautomatic pistol
Caliber: 10mm **Magazine capacity:** 8 rounds
Barrel length: 5″ **Overall length:** 8 1/2″
Weight (empty): 38 oz.
Sights: 3-dot, high-profile front and rear combat sights; Accro rear sight adj. for windage and elevation (on Delta Gold Cup only)
Sight radius: 6 1/2″ (3 dot sight system), 6 3/4″ (adjustable sights)
Grips: Rubber combat stocks with Delta medallion
Safety: Trigger safety lock (thumb safety) is located on left-hand side of receiver; grip safety is located on backstrap; internal firing pin safety
Rifling: 6 groove, left-hand twist, one turn in 16″
Price: $774.00

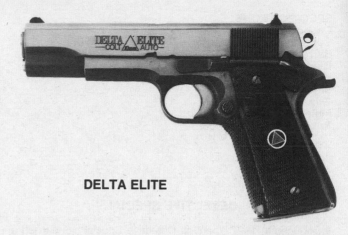

DELTA ELITE

Also available:
DELTA GOLD CUP. Same specifications as Delta Elite, except 39 oz. weight and 6 3/4″ sight radius. Stainless. **$987.00**

COLT MUSTANG .380

This backup automatic has four times the knockdown power of most 25 ACP automatics. It is a smaller version of the 380 Government Model.

SPECIFICATIONS
Caliber: 380 ACP **Capacity:** 6 rounds
Barrel length: 2.75″ **Overall length:** 5.5″
Height: 3.9″ **Weight:** 18.5 oz.
Prices: $443.00 Standard blue
     504.00 Nickel
     473.00 Stainless steel
Also available:
MUSTANG POCKETLITE 380 with aluminum alloy receiver; 1/2″ shorter than standard Govt. 380; weighs only 12.5 oz.
Prices: $443.00; $473.00 in nickel.
MUSTANG PLUS II features full grip length (Govt. 380 model only) with shorter compact barrel and slide (Mustang .380 model only); **weight:** 20 oz. **Price: $443.00** blue; **$473.00** stainless steel.

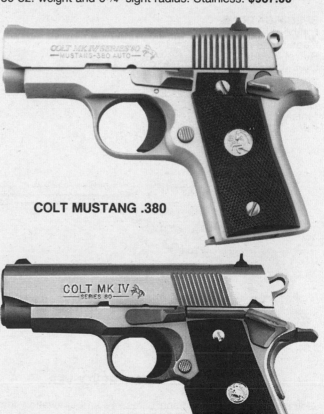

COLT MUSTANG .380

COLT OFFICER'S 45 ACP

SPECIFICATIONS
Caliber: 45 ACP **Capacity:** 6 rounds
Barrel length: 3 1/2″ **Overall length:** 7 1/4″
Weight: 34 oz.
Prices: $759.00 Stainless steel
     707.00 Standard blue
     829.00 Ultimate stainless
Also available:
OFFICER'S LW w/aluminum alloy frame (24 oz.) and blued finish. **Price: $707.00**

COLT OFFICER'S 45 ACP

COLT PISTOLS/REVOLVERS

CADET SEMIAUTOMATIC DA

SPECIFICATIONS
Caliber: 22 LR
Capacity: 11 rounds
Barrel length: 4 1/2"
Overall length: 8 5/8"
Weight: 48 oz.
Sight radius: 5.75"
Grips: Rubber polymer
Finish: Stainless steel
Price: To be determined

CADET SEMIAUTOMATIC DA

DETECTIVE SPECIAL

Introduced in 1927, the Colt Detective Special has earned a legendary reputation in the law enforcement field and for personal protection. Constructed of forged, high-strength alloy steel, the updated Detective Special is based on the Colt "D" frame.

SPECIFICATIONS
Caliber: 38 Special
Capacity: 6 rounds
Barrel length: 2"
Overall length: 7"
Weight: 21 oz.
Sights: Ramp front
Grips: Black composition
Finish: Blue
Price: $383.95

DETECTIVE SPECIAL

SINGLE ACTION ARMY
(Nickel Finish)

SINGLE ACTION ARMY REVOLVER

Colt's maintains the tradition of quality and innovation that Samuel Colt began more than a century and a half ago. Single Action Army revolvers continue to be highly prized collectible arms and are offered in full nickel finish or in Royal Blue with color casehardened frame, without engraving unless otherwise specified by the purchaser. Grips are American walnut.
Price: **$1213.00**

SINGLE ACTION ARMY SPECIFICATIONS

Caliber	Bbl. Length (inches)	Finish	Approx. Weight (ozs.)	O/A Length (inches)	Grips	Medallions
45LC	4 3/4	CC/B	40	10 1/4	Walnut	Gold
45LC	4 3/4	N	40	10 1/4	Walnut	Nickel
45LC	5 1/2	CC/B	42	11	Walnut	Gold
45LC	5 1/2	N	42	11	Walnut	Nickel
45LC	7 1/2	CC/B	43	13	Walnut	Gold
44-40	4 3/4	CC/B	40	10 1/4	Walnut	Gold
44-40	4 3/4	N	40	10 1/4	Walnut	Nickel
44-40	5 1/2	CC/B	42	11	Walnut	Gold
44-40	5 1/2	N	42	11	Walnut	Nickel
38-40	4 3/4	CC/B	40	10 1/4	Walnut	Gold
38-40	4 3/4	N	40	10 1/4	Walnut	Nickel
38-40	5 1/2	CC/B	42	11	Walnut	Gold
38-40	5 1/2	N	42	11	Walnut	Nickel

N—Nickel   CC/B—Colorcase frame/Royal blue cylinder & barrel

COLT REVOLVERS

KING COBRA DOUBLE ACTION

This "snake" revolver features a solid barrel rib, full-length ejector rod housing, red ramp front sight, white outline adjustable rear sight, and "gripper" rubber combat grips.

SPECIFICATIONS
Calibers: 357 Mag./38 Special
Capacity: 6 rounds
Barrel lengths: 4″ and 6″
Overall length: 9″ (4″ bbl.); 11″ (6″ bbl.)
Weight: 42 oz. (4″); 46 oz. (6″)
Finish: Stainless
Price: $437.00

KING COBRA

ANACONDA DOUBLE ACTION

SPECIFICATIONS
Calibers: 44 Magnum, 44 Special and 45 Colt (6″ barrel only)
Capacity: 6 rounds
Barrel lengths: 4″, 6″, 8″
Overall length: 9⅝″, 11⅝″, 13⅝″
Weight: 47 oz. (4″), 53 oz. (6″), 59 oz. (8″)
Sights: Red insert front; adjustable white outline rear
Sight radius: 5¾″ (4″), 7¾″ (6″), 9¾″ (8″)
Grips: Black neoprene combat-style with finger grooves
Finish: Matte stainless steel
Price: $587.00 All barrel lengths

**ANACONDA
(6″ barrel)**

PYTHON (8″ barrel)

PYTHON PREMIUM DOUBLE ACTION

The Colt Python revolver, suitable for hunting, target shooting and police use, is chambered for the powerful 357 Magnum cartridge. Python features include ventilated rib, fast cocking, wide-spur hammer, trigger and rubber grips, adjustable rear and ramp-type front sights, grooved.

SPECIFICATIONS
Calibers: 357 Mag./38 Special
Barrel lengths: 4″, 6″ and 8″
Overall length: 9½″, 11½″, 13½″

Weight: 38 oz. (4″); 43½ oz. (6″); 48 oz. (8″)
Stock: Rubber combat (4″) or rubber target (6″, 8″)
Finish: Colt high-polish royal blue, stainless steel and "Ultimate" bright stainless steel
Prices: $798.00 Royal Blue
          885.00 Stainless steel
          917.00 "Ultimate" Bright Stainless Steel

COONAN ARMS

357 MAGNUM PISTOL
5″ Barrel (top)
6″ Barrel (middle)
Compensated Barrel (bottom)

357 MAGNUM PISTOL

SPECIFICATIONS
Caliber: 357 Magnum
Magazine capacity: 7 rounds + 1
Barrel length: 5″ (6″ or Compensated barrel optional)
Overall length: 8.3″
Weight: 48 oz. (loaded)
Height: 5.6″
Width: 1.3″

Sights: Ramp front; fixed rear, adjustable for windage only
Grips: Smooth black walnut (checkered grips optional)
Finish: Stainless steel and alloy steel
Features: Linkless barrel; recoil-operated; extended slide catch and thumb lock
Prices: With 5″ barrel $720.00
With 6″ barrel 755.00
With Compensated barrel 999.00

"CADET" COMPACT MODEL

SPECIFICATIONS
Caliber: 357 Magnum
Magazine capacity: 6 rounds + 1
Barrel length: 3.9″
Overall length: 7.8″
Weight: 39 oz.
Height: 5.3″
Width: 1.3″
Sights: Ramp front; fixed rear, adjustable for windage only
Grips: Smooth black walnut
Features: Linkless bull barrel; full-length guide rod; recoil-operated (Browning falling-block design); extended slide catch and thumb lock for one-hand operation
Price: $841.00

"CADET" COMPACT

DAEWOO PISTOLS

MODEL DP51/DH40 PISTOL
$390.00 (9mm) $420.00 (40 cal.)

SPECIFICATIONS
Calibers: 9mm Parabellum (Model DP51) and 40 S&W (Model DH40)
Capacity: 13 rounds (9mm); 12 rounds (40 cal.)
Barrel length: 4¼″   **Overall length:** 7″
Weight: 28 oz.
Muzzle velocity: 1150 fps
Sights: Blade front (⅛″); square notch rear, drift adjustable with 3 self-luminous dots
Safety: Ambidextrous manual safety, automatic firing-pin block
Feature: Patented Fastfire action with light 5–6 lb. trigger pull for first-shot accuracy

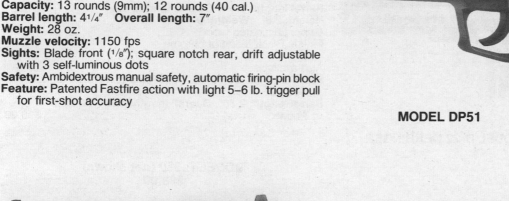

MODEL DP51

MODEL DH40

MODEL DH45 HIGH CAPACITY
DOUBLE ACTION PISTOL (not shown)
$500.00

SPECIFICATIONS
Caliber: 45 ACP
Capacity: 13 rounds
Barrel length: 5″   **Overall length:** 8.1″
Weight: 35 oz.
Sights: ½″ blade front; drift adjustable rear (3-dot sight system)
Action: Short stroke double action; hammerless striker design
Safety: Ambidextrous external safety with internal firing pin

MODEL DP52 PISTOL
$320.00

SPECIFICATIONS
Caliber: 22 LR
Capacity: 10 rounds
Barrel length: 3.8″   **Overall length:** 6.7″
Weight: 23 oz.
Width: 1.18″
Sights: ⅛″ front blade; drift adjustable rear (3 white-dot system)

MODEL DP52

DAVIS PISTOLS

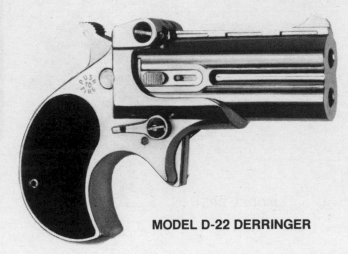

MODEL D-22 DERRINGER

D-SERIES DERRINGERS
$75.00

SPECIFICATIONS
Calibers: 22 LR, 22 Mag., 25 Auto, 32 Auto
Capacity: 2 shot
Barrel length: 2.4″   **Overall length:** 4″
Height: 2.8″   **Weight:** 9.5 oz.
Grips: Laminated wood
Finish: Black teflon or chrome

Also available:
BIG BORE 38 SPECIAL D-SERIES (32 H&R Magnum, 38 Special)
Barrel length: 2.75″. **Overall length:** 4.65″. **Weight:** 11.5 oz.
Price: **$98.00**

MODEL P-380 (not shown)
$98.00

SPECIFICATIONS
Caliber: 380 Auto
Magazine capacity: 5 rounds
Barrel length: 2.8″
Overall length: 5.4″
Height: 4″
Weight: 22 oz. (empty)

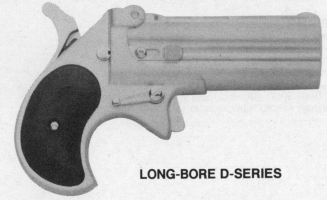

LONG-BORE D-SERIES

LONG-BORE D-SERIES

SPECIFICATIONS
Calibers: 22 LR, 22 WMR, 32 Auto, 32 H&R Mag., 380 Auto, 9mm, 38 Special
Capacity: 2 rounds
Barrel length: 3.75″   **Overall length:** 5.65″
Overall height: 3.31″   **Weight:** 13 oz.
Price: NA

MODEL P-32
$87.50

SPECIFICATIONS
Caliber: 32 Auto
Magazine capacity: 6 rounds
Barrel length: 2.8″   **Overall length:** 5.4″
Height: 4″   **Weight** (empty): 22 oz.
Grips: Laminated wood
Finish: Black teflon or chrome

MODEL P-32

EMERGING TECHNOLOGIES

LASERAIM ARMS

LASERAIM SERIES I PISTOLS

SPECIFICATIONS
Calibers: 45 ACP, 10mm
Capacity: 7+1 (45 ACP) and 8+1 (10mm)
Barrel lengths: $3^7/_8''$ and $5^1/_2''$
Overall length: $8^3/_4''$ ($3^7/_8''$) and $10^1/_2''$
Weight: 46 oz. ($3^7/_8''$) and 52 oz.
Features: Adjustable Millet sights, ambidextrous safety, beavertail tang, non-glare slide serration, beveled magazine well, extended slide release
Price: **$552.95**
w/Adjustable Sights) **579.95**

Also available:
AUTO ILLUSION w/red dot and fixed sights **$649.95**
DREAM TEAM w/Laseraim Laser, fixed sights,
HOTDOT **694.95**

The Series I pistol features a dual port compensated barrel and vented slide to reduce recoil and improve control. Other features include stainless-steel construction, ramped barrel, accurized barrel bushing and fixed sights (Laseraim's "HOT-DOT" sight or Auto Illusion electronic red dot sight available as options).

LASERAIM SERIES III PISTOLS

SPECIFICATIONS
Calibers: 45 ACP and 10mm
Capacity: 8+1 (10mm) and 7+1 (45 ACP)
Barrel length: 5"   **Overall length:** $7^5/_8''$
Weight: 43 oz.
Overall height: $5^5/_8''$
Features: Same as Series I
Price: **$533.95**
w/Adjustable Sights **559.95**

Also available:
SERIES II w/$3^3/_8''$ or 5" barrel, fixed sights **$399.95**
w/Adjustable sights, add **30.00**

The Series III pistol features a ported barrel and slide to reduce recoil and improve control. Includes stainless-steel construction, ramped barrel, accurized barrel bushing and comes with adjustable and laser sights (optional).

EMF/DAKOTA

SINGLE ACTION REVOLVERS

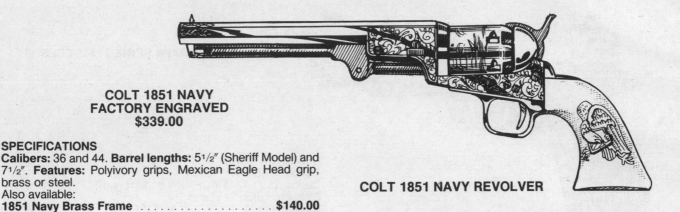

COLT 1851 NAVY FACTORY ENGRAVED
$339.00

SPECIFICATIONS
Calibers: 36 and 44. **Barrel lengths:** 5¹/₂″ (Sheriff Model) and 7¹/₂″. **Features:** Polyivory grips, Mexican Eagle Head grip, brass or steel.
Also available:

1851 Navy Brass Frame	**$140.00**
Engraved	160.00
With nickel trim	172.00
With steel frame	172.00
Engraved	285.00

COLT 1851 NAVY REVOLVER

HARTFORD SCROLL ENGRAVED

HARTFORD SCROLL ENGRAVED SINGLE ACTION REVOLVER
$840.00   ($1000.00 in Nickel)

SPECIFICATIONS
Calibers: 22, 45 Long Colt, 357 Magnum, 44-40. **Barrel lengths:** 4⁵/₈″, 5¹/₂″ and 7¹/₂″. **Features:** Classic original-type scroll engraving.

HARTFORD MODEL

HARTFORD MODELS
$600.00   ($680.00 in Nickel)

EMF's Hartford Single Action revolvers are available in the following calibers: 32-20, 38-40, 44-40, 44 Special and 45 Long Colt. **Barrel lengths:** 4³/₄″, 5¹/₂″ and 7¹/₂″. All models feature steel back straps, trigger guards and forged frame. Identical to the original Colts.

HARTFORD MODELS "CAVALRY COLT" AND "ARTILLERY"
$655.00

The Model 1873 Government Model Cavalry revolver is an exact reproduction of the original Colt made for the U.S. Cavalry in caliber 45 Long Colt with barrel length of 7¹/₂″. The Artillery Model has 5¹/₂″ barrel.
Also available:
Sheriff's Model (3¹/₂″ barrel) $655.00

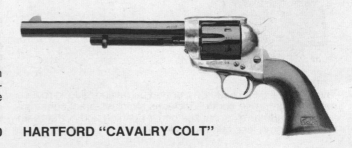

HARTFORD "CAVALRY COLT"

EMF/DAKOTA

MODEL 1873 $460.00
($560.00 w/Extra Cylinder)

DAKOTA TARGET
$500.00

SPECIFICATIONS
Calibers: 44-40, 357 Mag., 45 Long Colt. **Barrel lengths:** 4³/₄″,
5¹/₂″, 7¹/₂″. **Finish:** Engraved models, blue or nickel. **Special
feature:** Each gun is fitted with second caliber.

SPECIFICATIONS
Calibers: 45 Long Colt, 357 Magnum, 22 LR. **Barrel lengths:**
5¹/₂″ and 7¹/₂″. **Finish:** Polished blue. **Special features:** Casehardened frame, one-piece walnut grips, brass back strap,
ramp front blade target sight and adjustable rear sight.

1873 DAKOTA SINGLE ACTION
REVOLVER $460.00 ($585.00 in Nickel)

PINKERTON DETECTIVE SA
$680.00

SPECIFICATIONS
Calibers: 357 Mag., 44-40, 45 Long Colt. **Barrel lengths:** 4³/₄″,
5¹/₂″ and 7¹/₂″. **Finish:** Blued, casehardened frame. **Grips:** One-
piece walnut. **Features:** Set screw for cylinder pin release;
parts are interchangeable with early Colts.

SPECIFICATIONS
Caliber: 45 Long Colt. **Barrel length:** 4″. **Grip:** Bird's-head.

MODEL 1875 "OUTLAW"
$465.00 w/Brass Trigger Guard
$550.00 in Nickel

MODEL 1890 REMINGTON POLICE
$470.00 w/Brass Trigger Guard
$560.00 in Nickel

SPECIFICATIONS
Calibers: 44-40, 45 Long Colt, 357. **Barrel length:** 5¹/₂″ and
7¹/₂″. **Finish:** Blue or nickel. **Special features:** Casehardened
frame, walnut grips; an exact replica of the Remington No. 3
revolver produced from 1875 to 1889.
Factory Engraved Model (7¹/₂″ bbl.) **$600.00**
  In nickel 710.00
With steel trigger guard 475.00

SPECIFICATIONS
Calibers: 44-40, 45 Long Colt and 357 Magnum. **Barrel length:**
5³/₄″. **Finish:** Blue or nickel. **Features:** Original design (1891–
1894) with lanyard ring in buttstock; casehardened frame;
walnut grips.
Engraved Model **$620.00**
In nickel 725.00
With steel trigger guard 480.00

ERMA TARGET ARMS

MODEL 777 SPORTING REVOLVER

SPECIFICATIONS
Caliber: 357 Magnum
Capacity: 6 cartridges
Barrel length: 4″ and 5¹/₂″
Overall length: 9.7″ and 11.3″
Weight: 43.7 oz. w/5¹/₂″ barrel
Sight radius: 6.4″ and 8″
Grip: Checkered walnut
Price: $1420.00

Also available:
MODEL 773 MATCH (32 S&W Wadcutter).
   Same specifications as Model 777 (5¹/₂″ barrel), but with adjustable match grip and 6″ barrel. **Weight:** 47.3 oz. **Price: $1371.00**
MODEL 772 MATCH (22 Long Rifle).
   Same specifications as Model 773, except weight is 47¹/₄ ounces. **Price: $1371.00**
Both Match revolvers feature micrometer rear sights (adjustable for windage and elevation), interchangeable front and rear sight blades, adjustable triggers, polished blue finish and 6-shot capacity.

MODEL 777 STANDARD

MODEL ESP 85A GOLDEN TARGET

Caliber 22 Long Rifle with case. Same specifications as Model ESP 85A below, but with fully interchangeable barrels for 22 LR or 32 S&W. Adjustable trigger. **Price: 2100.00**

**MODEL ESP 85A
GOLDEN TARGET**

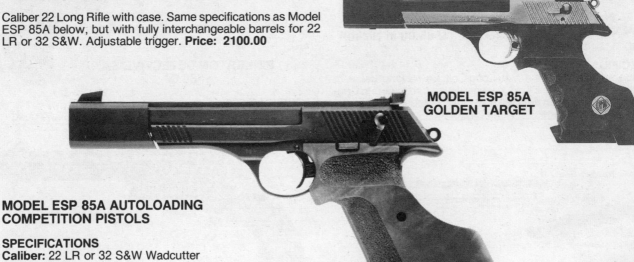

MODEL ESP 85A AUTOLOADING COMPETITION PISTOLS

SPECIFICATIONS
Caliber: 22 LR or 32 S&W Wadcutter
Action: Semiautomatic
Capacity: 5 cartridges (8 in 22 LR optional)
Barrel length: 6″
Overall length: 10″
Weight: 40 oz.
Sight radius: 7.8″
Sights: Micrometer rear sight; fully adjustable interchangeable front and rear sight blade (3.5mm/4.0mm)
Grip: Checkered walnut grip with thumbrest

MODEL ESP 85A

Prices:

ESP 85A MATCH 32 S&W	$1714.00
Left Hand	1744.00
Target adjustable grip, **add**	169.50
Left hand, **add**	199.00
ESP 85A MATCH 22 LR	1645.00
Left hand	1675.00
ESP 85A CHROME MATCH 22 LR	1753.00
Left hand	1783.00
In 32 S&W	1822.00
Left hand	1852.00
Conversion Units 22 LR	823.00
In 32 S&W	890.00

EUROPEAN AMERICAN ARMORY

ASTRA MODEL A-70
$340.00 (Blue)
$370.00 (Nickel or Stainless Steel)

SPECIFICATIONS
Calibers: 9mm and 40 S&W
Capacity: 8 rounds
Barrel length: 3½″
Overall length: 6½″
Weight: 23.3 oz.
Finish: Blue, nickel or stainless steel

ASTRA MODEL A-70
Blued

ASTRA MODEL A-75
Blued

ASTRA MODEL A-75
$395.00 (Blue)
$425.00 (Nickel or Stainless Steel)

SPECIFICATIONS
Calibers: 9mm, 40 S&W and 45 ACP
Capacity: 8 rounds (7 in 40 S&W)
Barrel length: 3½″
Overall length: 6½″
Weight: 31 oz. (23½ oz. in Airweight model)
Finish: Blue, nickel or stainless steel
Also available: **AIRWEIGHT 9mm**

ASTRA MODEL A-100
$425.00 (Blue)
$450.00 (Nickel or Stainless Steel)

SPECIFICATIONS
Calibers: 9mm, 40 S&W and 45 ACP
Capacity: 15 rounds (9 in 45 ACP)
Barrel length: 3.75″
Overall length: 7″
Weight: 34 oz.
Finish: Blue, nickel or stainless steel

ASTRA MODEL A-100
Blued

EUROPEAN AMERICAN ARMORY

WITNESS DOUBLE-ACTION PISTOLS

WITNESS

SPECIFICATIONS
Calibers: 9mm, 40 S&W and 45 ACP
Capacity: 16 rounds (9mm); 12 rounds (40 S&W); 10 rounds (45 ACP)
Barrel length: 4¹/₂″
Overall length: 8.1″
Weight: 33 oz.
Finish: Blue, chrome, blue/chrome, stainless steel
Prices:
9mm Blue $390.00
Chrome or blue/chrome. 425.00
Stainless steel 460.00
40 S&W Blue 415.00
Chrome or blue/chrome. 445.00
Stainless steel 485.00
45 ACP Blue 498.00
Chrome or blue/chrome. 525.00
Stainless steel 565.00

WITNESS

WITNESS FAB

This all-steel semiautomatic pistol features a special ambidextrous hammer drop safety/decocker system now required by many law enforcement agencies.

SPECIFICATIONS
Calibers: 9mm, 40 S&W, 45 ACP (full size only)
Capacity: 16 rounds (9mm); 13 rounds (9mm compact); 12 rounds (40 S&W); 10 rounds (45 ACP); 9 rounds (40 S&W Compact)
Barrel length: 4¹/₂″ (3⁵/₈″ Compact)
Overall length: 8.1″ (4¹/₂″ Compact)
Weight: 33 oz. (30 oz. Compact)
Finish: Blue, hard chrome or Duo-Tone
Optional: Tritium night sights; extended magazine release; rubber grips
Prices:
9mm Blue (full and Compact) $365.00
40 S&W Blue (full and Compact) 395.00
45 ACP 395.00

WITNESS FAB

WITNESS GOLD TEAM

SPECIFICATIONS
Calibers: 9mm, 40 S&W, 38 Super, 9×21mm, 10mm, 45 ACP
Capacity: 20 rounds (38 Super); 19 rounds (9mm and 9×21mm); 17 rounds (10mm); 14 rounds (40 S&W); 11 rounds (45 ACP)
Barrel length: 5¹/₄″
Overall length: 10¹/₂″
Weight: 38 oz.
Finish: Hard chrome
Features: Triple chamber comp, S/A trigger, extended safety, competition hammer, checkered front strap and backstrap, low-profile competition grips, square trigger guard
Price: $2100.00

WITNESS GOLD TEAM

EUROPEAN AMERICAN ARMORY

EUROPEAN SINGLE ACTION/ DOUBLE ACTION COMPACTS

SPECIFICATIONS
Calibers: 380 ACP and 32 ACP (Single action only)
Capacity: 7 rounds
Barrel length: 3⁷/₈″
Overall length: 6¹/₂″
Weight: 26 oz.
Finish: Blue, chrome or blue/chrome Duo-Tone
Prices:
32 ACP & 380 ACP Blue.................... $153.00
Chrome finish 165.00
380 ACP DA Blue 185.00
Chrome 200.00
EUROPEAN LADY 380 ACP Blue/gold 217.00

EUROPEAN SA/DA COMPACT

WITNESS SUBCOMPACT (not shown)

SPECIFICATIONS
Calibers: 9mm, 38 Super, 40 S&W, 45 ACP. **Capacity:** 13 rounds (9mm); 9 rounds (40 S&W). **Barrel length:** 3.66″ **Overall length:** 7.24″. **Weight:** 30 oz. **Finish:** Blue, chrome or blue/chrome.

Prices:
9mm Blue $390.00
Chrome or blue/chrome 425.00
40 S&W Blue 415.00
Chrome or blue/chrome 445.00
45 ACP/38 Super Blue 498.00
Chrome or blue/chrome..................... 525.00

WINDICATOR DOUBLE ACTION (STANDARD GRADE)

SPECIFICATIONS
Calibers: 22 LR, 22 LR/22 WMR, 38 Special, 357 Mag.
Capacity: 6 rounds (38 Special only); 8 rounds (22 LR and 22 LR/22 WMR)
Barrel lengths: 2″, 4″ and 6″
Finish: Blue only
Features: Swing-out cylinder; black rubber grips; hammer block safety
Prices:
22 LR w/4″ barrel $180.00
22 LR w/6″ barrel 190.00
22 LR/22 WMR w/4″ barrel 233.00
22 LR/22 WMR w/6″ barrel 243.00
38 Special w/2″ barrel 179.00
38 Special w/4″ barrel 190.00

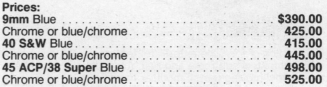

WINDICATOR REVOLVER

BIG BORE BOUNTY HUNTER SA

SPECIFICATIONS
Calibers: 357 Mag., 45 Long Colt and 44 Mag.
Capacity: 6 rounds
Barrel lengths: 4¹/₂″ and 7¹/₂″
Sights: Fixed
Finish: Blue, blue/gold, color casehardened, chrome
Features: Transfer bar safety, 3-position hammer; hammer-forged barrel; walnut grips
Prices:
4¹/₂″ barrel, blued $285.00
7¹/₂″ barrel, blued 288.00
4¹/₂″ barrel, color casehardened receiver 295.00
7¹/₂″ barrel, color casehardened receiver 299.00
4¹/₂″ barrel, blue/gold 299.00
7¹/₂″ barrel, blue/gold 299.00
4¹/₂″ barrel, chrome 325.00
7¹/₂″ barrel, chrome 330.00

BIG BORE BOUNTY HUNTER SINGLE ACTION

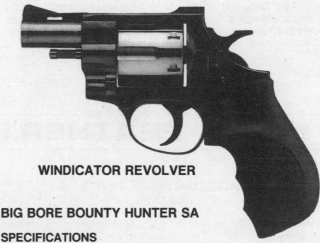

EUROPEAN AMERICAN ARMORY

BENELLI MODEL MP90S TARGET PISTOL
$1296.70 (22LR) $1501.50 (32 WC)

This match-grade silhouette pistol features a chromium/molybdenum steel subframe and bolt, fully adjustable grip (with no internal moving parts), modular firing system and a barrel axis that maintains alignment with the shooter's forearm and shoulder. Trigger is adjustable for take-up and breaking point, longitudinally or axially, without altering other trigger adjustments. An extra-large ejection port dissipates gas and reduces recoil.

SPECIFICATIONS
Calibers: 22 LR or 32 Wadcutter
Capacity: 5 rounds (detachable magazine)
Barrel length: 4.33″   **Overall length:** 11.7″
Weight: 41.3 oz.
Finish: Blued barrel and frame
Grip: Adjustable, contoured wooden target grip

FEATHER HANDGUNS

GUARDIAN ANGEL PISTOL
$119.95

The Guardian Angel is a compact, lightweight handgun built to last with virtually all stainless-steel parts. It features an interchangeable dropping block that slips easily in and out of the top of the frame with the push of a button. The double-action firing mechanism automatically chooses the next barrel for instant response on the second shot. Safety features include a full trigger guard, a loading-block safety that automatically disengages for firing only when the block is firmly in place; and a passive internal firing-pin-block safety that is always ''on'' and disengages only when the trigger is pulled almost fully forward. The gun cannot fire because of an accidental fall or blow.

SPECIFICATIONS
Calibers: 22 LR and 22 WMR
Capacity: 2 rounds (removable block)
Barrel length: 2″   **Overall length:** 5″
Weight: 12 oz.

GUARDIAN ANGEL PISTOL

FREEDOM ARMS

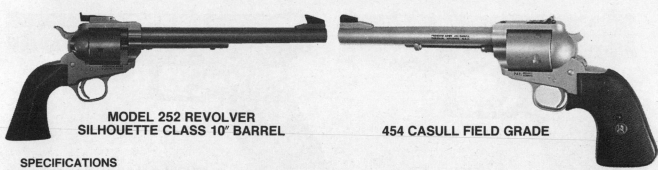

MODEL 252 REVOLVER
SILHOUETTE CLASS 10″ BARREL

454 CASULL FIELD GRADE

SPECIFICATIONS
Caliber: 22 LR (optional 22 Magnum cylinder)
Barrel lengths: 5¹/₈″, 7¹/₂″ (Varmint Class) and 10″ (Silhouette Class)
Sights: Silhouette competition sights (Silhouette Class); adjustable rear express sight; removable front express blade
Grips: Black micarta (Silhouette Class); black and green laminated hardwood (Varmint Class)
Finish: Stainless steel
Features: Dual firing pin; lightened hammer; pre-set trigger stop; accepts all sights and/or scope mounts
Prices:
Silhouette Class **$1350.00**
Varmint Class 1295.00
22 Mag. Cylinder 233.00

SILHOUETTE/COMPETITION MODELS
(not shown)

SPECIFICATIONS
Calibers: 357 Magnum and 44 Rem. Mag.
Barrel lengths: 9″ (357 Mag.) and 10″ (44 Rem. Mag.)
Sights: Silhouette competition   **Grips:** Pachmayr
Trigger: Pre-set stop; trigger over travel screw
Finish: Field Grade
Price: **$1267.00**

454 CASULL PREMIER & FIELD GRADES

SPECIFICATIONS
Calibers: 454 Casull, 44 Rem. Mag.
Action: Single action   **Capacity:** 5 rounds
Barrel lengths: 4³/₄″, 6″, 7¹/₂″, 10″
Overall length: 14″ (w/7¹/₂″ barrel)
Weight: 3 lbs. 2 oz. (w/7¹/₂″ barrel)
Safety: Patented sliding bar
Sights: Notched rear; blade front (optional adjustable rear and replaceable front blade)
Grips: Impregnated hardwood or rubber Pachmayr
Finish: Brushed stainless
Features: Patented interchangeable forcing cone bushing (optional); ISGW silhouette, Millett competition and express sights are optional; SSK T'SOB 3-ring scope mount optional; optional cylinder in 454 Casull, 45 ACP, 45 Win. Mag. ($233.00)
Prices:
MODEL FA-454AS (Premier Grade)
W/adjustable sights **$1480.00**
W/fixed sights 1396.00
44 Remington w/adjustable sights 1480.00
MODEL 555 50 Action Express 1480.00
MODEL FA-454FGAS (Field Grade)
With stainless steel matte finish, adj. sight,
  Pachmayr presentation grips 1175.00
W/fixed sights 1119.00
44 Remington w/adjustable sights 1175.00

MODEL 353 REVOLVER
FIELD GRADE 7¹/₂″ BARREL

SPECIFICATIONS
Caliber: 357 Magnum
Action: Single action   **Capacity:** 5 shots
Barrel lengths: 4³/₄″, 6″, 7¹/₂″, 9″
Sights: Removable front blade; adjustable rear
Grips: Pachmayr Presentation grips (Premier Grade has impregnated hardwood grips
Finish: Non-glare Field Grade (standard model); Premier Grade brushed finish (all stainless steel)
Prices:
Field Grade **$1175.00**
Premier Grade 1480.00

MODEL 555 PREMIER GRADE (50 AE)
$1175.00

GLOCK PISTOLS

MODEL 17
$608.95

MODEL 17L COMPETITION
$806.75

SPECIFICATIONS
Caliber: 9mm Parabellum
Magazine capacity: 17 rounds (19 rounds optional)
Barrel length: 4¹/₂″ (hexagonal profile with right-hand twist
Overall length: 7.28″
Weight: 22 oz. (without magazine)
Sights: Fixed or adjustable rear sights
Also available:
MODEL 19 COMPACT 15 rounds. **Barrel length:** 4″. **Overall
length:** 6.85″. **Weight:** 21 oz. **Price:** $608.95
MODEL 21 in 45 ACP (13 round capacity). **Price:** $670.50

SPECIFICATIONS
Caliber: 9mm Parabellum
Magazine capacity: 17 rounds
Barrel length: 6.02″
Overall length: 8.85″
Weight: 23.35 oz. (without magazine)
Sights: Fixed or adjustable rear sights

Also available:
MODEL 22 (Sport and Service models) in 40 S&W. **Overall
length:** 7.28″. **Capacity:** 15 rounds. **Price:** $608.95
MODEL 23 (Compact Sport and Service models) in 40 S&W.
Overall length: 6.85″. **Capacity:** 13 rounds. **Price:** $608.95

MODEL 20
$670.50

MODEL 24 COMPETITION
$800.00

SPECIFICATIONS
Caliber: 10mm
Magazine capacity: 15 rounds
Action: Double action
Barrel length: 4.6″
Overall length: 7.59″
Height: 5.47″ (w/sights)
Weight: 26.35 oz. (empty)
Sights: Fixed or adjustable
Features: 3 safeties, "safe-action" system, polymer frame

SPECIFICATIONS
Caliber: 40 S&W
Capacity: 15 rounds
Barrel length: 6.02″
Overall length: 8.85″
Weight: 26.5 oz. (empty)
Safety: Manual trigger safety; passive firing block and drop
safety
Finish: Matte (Tenifer process); nonglare

GRENDEL PISTOLS

MODEL P-30
$225.00

SPECIFICATIONS
Caliber: 22 WRF Magnum
Capacity: 30 rounds
Barrel length: 5"
Overall length: 8½"
Weight: 21 oz. (empty)
Sight radius: 7.2"
Features: Fixed barrel, flat trajectory, low recoil, ambidextrous safety levers, front sight adjustable for windage.
Also available:
Model P-30M (w/removable muzzle brake); **barrel length:** 5.6"; **overall length:** 10"; **weight:** 22 oz. **Price: $235.00**
Model P-31 with 11" barrel; **weight:** 48 oz. **Price: $345.00**

MODEL P-30

MODEL P-12
$175.00   ($195.00 in Nickel)

SPECIFICATIONS
Caliber: 380 ACP
Capacity: 12 rounds
Barrel length: 3"
Overall length: 5.3"
Weight: 13 oz. (empty)
Sight radius: 4½"
Features: Low inertia safety hammer system; glass reinforced Zytel magazine; solid steel slide w/firing pin and extractor; polymer DuPont ST-800 grip

MODEL P-12

HÄMMERLI U.S.A. PISTOLS

MODEL 280 TARGET PISTOL

MODEL 280 TARGET PISTOL
$1465.00 ($1650.00 in 32 S&W)

SPECIFICATIONS
Calibers: 22 LR and 32 S&W
Capacity: 6 rounds (22 LR); 5 rounds (32 S&W)
Barrel length: 4.58"
Weight: (excluding counterweights) 34.92 oz. (22 LR); 38.8 oz. (32 S&W)
Sight radius: 8.66"
Also available:
MODEL 280 TARGET PISTOL COMBO
   With carrying case $2490.00

HÄMMERLI U.S.A. PISTOLS

MODEL 160 FREE PISTOL
$1910.00 ($2247.00 Left Hand)

SPECIFICATIONS
Caliber: 22 LR
Overall length: 17.2″
Weight: 45.6 oz.
Trigger action: Infinitely variable set trigger weight; cocking lever located on left of receiver; trigger length variable along weapon axis
Sights: Sight radius 14.8″; micrometer rear sight adj. for windage and elevation
Locking action: Martini-type locking action w/side-mounted locking lever
Barrel: Free floating, cold swaged precision barrel w/low axis relative to the hand
Ignition: Horizontal firing pin (hammerless) in line w/barrel axis; firing pin travel 0.15″
Grips: Selected walnut w/adj. hand rest for direct arm to barrel extension

MODEL 162 ELECTRONIC PISTOL
$2095.00 ($2419.00 Left Hand)

SPECIFICATIONS:
Same as **Model 160** except trigger action is electronic.
Features: Short lock time (1.7 milliseconds between trigger actuation and firing pin impact), light trigger pull, and extended battery life.

MODEL 208S
STANDARD PISTOL
$1695.00

SPECIFICATIONS:
Caliber: 22 LR
Barrel length: 6″
Overall length: 10.2″
Weight: 37.3 oz. (w/accessories)
Capacity: 9 rounds
Sight radius: 8.3″
Sights: Micrometer rear sight w/notch width; standard front blade

MODEL 212 HUNTER'S PISTOL
$1395.00

SPECIFICATIONS
Caliber: 22 LR
Barrel length: 5″
Overall length: 8.6″
Weight: 31 oz.
Capacity: 8 rounds
Sights: Blade front; notched rear

HARRINGTON & RICHARDSON

SPORTSMAN 999 REVOLVER
$279.95

SPECIFICATIONS
Calibers: 22 Short, Long, Long Rifle
Action: Single and double action
Capacity: 9 rounds
Barrel lengths: 4″ and 6″ (both fluted)
Weight: 30 oz. (w/4″ barrel); 34 oz. (w/6″ barrel)
Sights: Windage adjustable rear; elevation adjustable front
Grips: Walnut-finished hardwood
Finish: Blue
Features: Top-break loading with auto shell ejection

SPORTSMAN 999 REVOLVER

**MODEL 949
CLASSIC WESTERN**

MODEL 949 CLASSIC WESTERN REVOLVER
$174.95

The Model 949 is a 9-shot revolver featuring a double-action mechanism, windage adjustable rear sight and true Western styling. Includes loading gate, shrouded ejector rod, case-colored frame and backstrap, two-piece walnut-stained hardwood grips and transfer bar safety system. **Calibers:** 22 Short, Long and Long Rifle. **Barrel length:** 5¹/₂″.

HECKLER & KOCH PISTOLS

**MODEL HK USP $635.00
($655.00 w/Control Lever
on Right)**

**MODEL P7M8
SELF-LOADING PISTOL
$1100.00 (Blue or Nickel)**

SPECIFICATIONS
Calibers: 9mm and 40 S&W
Capacity: 13 rounds (40 S&W); 15 rounds (9mm)
Operating system: Short recoil, modified Browning action
Barrel length: 4.25″   **Overall length:** 7.64″
Weight: 1.74 lbs. (40 S&W); 1.66 lbs. (9mm)
Height: 5.35″
Sights: Adjustable 3-Dot
Grips/stock: Polymer receiver and integral grips

SPECIFICATIONS
Caliber: 9mm×19 (Luger)   **Capacity:** 8 rounds
Barrel length: 4.13″   **Overall length:** 6.73″
Weight: 1.75 lbs. (empty)
Sight radius: 5.83″   **Sights:** Adjustable rear
Finish: Blue or nickel
Also available: **MODEL P7M13** with same barrel length, but slightly longer overall (6.9″), heavier (1.87 lbs.) and 13-round capacity **$1330.00** (Blue or nickel)

HECKLER & KOCH PISTOLS

MODEL P7K3

SPECIFICATIONS
Calibers: 22 LR, 380   **Capacity:** 8 rounds
Barrel length: 3.8″   **Overall length:** 6.3″
Weight: 1.65 lbs. (empty)
Sight radius: 5.5″   **Sights:** Adjustable rear
Also available:
22 LR Conversion Kit **$525.00**; Tritium Sights (orange, yellow or green rear with green front) **$85.00**

MODEL P7K3
$1100.00

MODEL P7M10
$1315.00 (Blue or Nickel)

MODEL P7M10

SPECIFICATIONS
Caliber: 40 S&W   **Capacity:** 10 rounds
Operating system: Recoil operated; retarded inertia bolt
Barrel length: 4.13″   **Overall length:** 6.9″
Weight: 2.69 lbs. (empty)
Sights: Adjustable rear
Finish: Blue or nickel

HELWAN PISTOLS

THE BRIGADIER
$260.00

Originally designed and produced by Beretta, the classic styling of the "Brigadier" will be immediately recognized as the model that preceded the current U.S. Military Service Pistol. The Helwan Brigadier was one of the pistols that served with distinction in combat during the Persian Gulf Desert Storm operations. Employing a breechblock mechanism similar to the Walther P 38, the Brigadier carries an 8-round magazine (with button release) that is fitted with a contoured finger-rest extension for added stability and control. It also features a muted, non-reflective black matte finish on both slide and frame, plus all-steel construction, single-action trigger and cross-bolt safety.

SPECIFICATIONS
Caliber: 9mm Parabellum
Capacity: 8 rounds
Barrel length: 4.5″
Overall length: 8″
Weight: 32.6 oz.
Finish: Matte black

THE BRIGADIER

HERITAGE MANUFACTURING

ROUGH RIDER SINGLE ACTION

SPECIFICATIONS
Caliber: 22 LR or 22 LR/22 WMR
Capacity: 6 rounds
Barrel lengths: 3", 4³/₄", 6¹/₂", 9"
Weight: 31 to 38 oz.
Sights: Blade front, fixed rear
Grips: Goncalo Alves
Finish: Blue or nickel
Features: Rotating hammer block safety; brass accent screws
Prices:
22 LR (4³/₄", 6¹/₂" barrel) . . . . $ 99.95
  With 9" barrel 119.95
22 LR/22 WRM Combo (4³/₄", 6¹/₂" barrel) 119.95
  With 9" barrel only 129.95
  With nickel, **add** 10.00

**ROUGH RIDER
SINGLE ACTION**

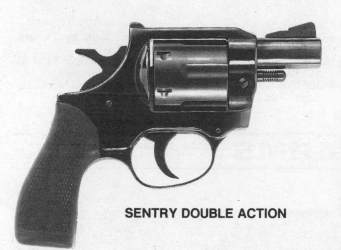

SENTRY DOUBLE ACTION

SENTRY DOUBLE ACTION
$109.95 – $119.95

SPECIFICATIONS
Calibers: 38 Special or 9mm
Capacity: 6 rounds
Barrel length: 2" or 4"
Weight: 21 oz.
Sights: Ramped front, fixed rear
Grips: Black polymer
Finish: Blue or nickel
Features: Internal hammer block; additional safety plug in cylinder

MODEL H25B/H25G SEMIAUTOMATIC

SPECIFICATIONS
Caliber: 25 ACP
Capacity: 6 rounds
Barrel length: 2¹/₄"
Weight: 15 oz.
Grips: Smooth hardwood (H25B); checkered walnut (H25G)
Sights: Fixed
Features: Exposed hammer w/half-cock position; gold trigger, safety and hammer (H25G)
Prices:
Model H25B $64.95
Model H25G 79.95

HERITAGE MODEL H25B

HI-POINT FIREARMS

JS-SERIES

**MODEL JS-9mm
$139.95**

**MODEL JS-9mm
COMPACT**

SPECIFICATIONS
Caliber: 9mm Parabellum
Capacity: 8 shots
Barrel length: 4.75″
Overall length: 7.72″
Weight: 48 oz.
Sights: Fixed low-profile
Features: Quick on-off thumb safety; nonglare
military black finish

Also available:
MODEL JS45 in 45 ACP. **Weight:** 44 oz. **Capacity:** 7 shots.
   Price: $148.95
MODEL JS-9mm COMPACT w/3¹/₂″ barrel. **Price: $124.95**
   ($132.95 black polymer)
MODEL JS-40 in 40 S&W. Same specifications as Model JS-
   9mm. **Price $148.95**

KAHR ARMS

**MODEL K9 PISTOL
$595.00**

All key components of the Kahr K9—frame, slide, barrel, etc.—
are made from 4140 steel, allowing the pistol to chamber reliably and fire virtually any commercial 9mm ammo, including
+P rounds. The frame and sighting surfaces are matte blued,
and the sides of the slide carry a polished blue finish. The grips
are crafted from exotic wood.

   The unique trigger system holds the striker in a partially
cocked state; then, a pull of the trigger completes the cocking
cycle and releases the striker. Recoil on firing partially cocks
the striker for the next trigger pull. This design allows a lighter-
than-normal "DA" pull that is consistent from shot to shot. A
trigger-activated firing-pin block prevents accidental discharge.
Like a double-action revolver, no other safeties are needed.

MODEL K9 PISTOL

SPECIFICATIONS
Caliber: 9mm and 9mm+P
Capacity: 7 rounds
Barrel length: 3¹/₂″
Overall length: 6″
Weight: 24 oz. (empty)

Sights: Drift-adjustable, low-profile white bar-dot combat
   sights
Grips: Wraparound exotic wood
Finish: Blued; polished slide; matte frame & sighting surface

KBI PISTOLS

MODEL PSP-25
$249.00 ($299.00 Hard Chrome)

SPECIFICATIONS
Caliber: 25 ACP
Capacity: 6 rounds
Barrel length: 2¹/₈″
Overall length: 4¹/₈″
Weight: 9.5 oz. (empty)
Height: 2⁷/₈″
Finish: Blue or chrome
Features: Dual safety system; all-steel construction

MODEL PSP-25

FEG MODEL PJK-9HP
$349.00 ($429.00 Chrome)

SPECIFICATIONS
Caliber: 9mm Para.
Magazine capacity: 13 rounds
Action: Single
Barrel length: 4³/₄″
Overall length: 8″
Weight: 21 oz.
Grips: Hand-checkered walnut
Safety: Thumb safety
Sights: Three-dot system
Finish: Blue or chrome
Features: Two 13-round magazines, cleaning rod

FEG MODEL PJK-9HP

FEG MODEL SMC-380
$279.00

SPECIFICATIONS
Caliber: 380 ACP
Capacity: 6 rounds
Action: Double
Barrel length: 3¹/₂″
Overall length: 6.1″
Weight: 18¹/₂ oz.
Safety: Thumb safety
Grips: Black composite
Features: High-luster blued steel slide; blue anodized aluminum alloy frame

FEG MODEL SMC-380

KBI HANDGUNS

FEG GKK-45 AUTO PISTOL
$449.00 ($499.00 Chrome)

SPECIFICATIONS
Caliber: 45 ACP
Capacity: 8 rounds
Barrel length: 4²/₃″
Overall length: 8¹/₁₆″
Weight: 34 oz.
Sights: 3 dot, blade front; rear adjustable for windage
Stock: Hand-checkered walnut
Features: Double action; polished blue or hard chrome finish; combat trigger guard; two magazines and cleaning rod standard

FEG SMC-22 AUTO PISTOL (not shown)
$299.00

SPECIFICATIONS
Caliber: 22 LR
Capacity: 8 rounds
Barrel length: 3¹/₂″
Overall length: 6¹/₈″
Weight: 18¹/₂ oz.
Stock: Checkered composition w/thumbrest
Sights: Blade front; rear adjustable for windage
Features: Alloy frame; steel slide; double action; blue finish; two magazines and cleaning rod standard

BAIKAL IJ-70 AUTO PISTOL
$199.00 (9×18 Blue)
$219.00 (380 Blue)
$234.00 (380 Nickel)

SPECIFICATIONS
Caliber: 9×18 Makarov, 380 ACP
Capacity: 8 rounds
Barrel length: 4″
Overall length: 6¹/₄″
Weight: 25 oz.
Stock: Checkered composition
Sights: Blade front; rear adjustable for windage and elevation
Features: Double action; all steel; frame-mounted safety with decocker; two magazines, cleaning rod, universal tool and leather holster standard

L.A.R. GRIZZLY

MARK I
GRIZZLY 45 WIN MAG $920.00
(357 Magnum $933.00)

This semiautomatic pistol is a direct descendant of the tried and trusted 1911-type .45 automatic, but with the added advantage of increased caliber capacity.

SPECIFICATIONS
Calibers: 45 Win. Mag., 45 ACP, 357 Mag., 10mm
Barrel lengths: 5.4″ and 6½″
Overall length: 10½″
Weight (empty): 48 oz.
Height: 5¾″
Sights: Fixed, ramped blade (front); fully adjustable for elevation and windage (rear)

Magazine capacity: 7 rounds
Grips: Checkered rubber, nonslip, combat-type
Safeties: Grip depressor, manual thumb, slide-out-of-battery disconnect
Materials: Mil spec 4140 steel slide and receiver with non-corrosive, heat-treated, special alloy steels for other parts

Also available:
Grizzly 44 Magnum Mark 4 w/adj. sights $933.00
Grizzly Win Mag Conversion Units 214.00
  In 357 Magnum 228.00
Win Mag Compensator 110.00
  in 50 caliber 120.00

MARK 4 GRIZZLY 44 MAG.

GRIZZLY WIN MAG
6½″ BARREL

GRIZZLY WIN MAG
8″ BARREL

GRIZZLY 50 MARK 5

Also available:
GRIZZLY WIN MAG (8″ and 10″)
Model G-WM8 (8″ barrel in 45 Win. Mag., 45 ACP
  or 357/45 Grizzly Win. Mag. $1313.00
Model G357M8 (8″ barrel in 357 Magnum) 1337.50
Model G-WM10 (10″ barrel in 45 Win. Mag., 45 ACP
  or 357/45 Grizzly Win. Mag. 1375.00
Model G357M10 (10″ barrel in 357 Magnum) 1400.00

GRIZZLY 50 MARK 5
$1060.00

SPECIFICATIONS
Caliber: 50 AE
Capacity: 6 rounds
Barrel lengths: 5.4″ and 6.5″
Overall length: 10⅝″
Sights: Fixed front; fully adjustable rear

LLAMA AUTOMATIC PISTOLS

**SMALL-FRAME
AUTOMATIC PISTOL
(Satin Chrome Finish)
Calibers 22 and 380
$341.95**

**SMALL-FRAME AUTOMATIC
(Deep Blue Finish)
Calibers 22 and 380
$248.95**

LLAMA AUTOMATIC PISTOL SPECIFICATIONS

Type:	Small-Frame Auto Pistols		Compact-Frame Auto Pistols			Large-Frame Auto Pistols
Calibers:	22 LR	380 Auto	45 Auto			45 Auto
Frame:	Precision machined from high-strength steel; serrated front strap, checkered (curved) backstrap		Precision machined from high-strength steel; serrated front strap, checkered (curved) backstrap			Precision machined from high-strength steel; plain front strap, checkered (curved) backstrap
Trigger:	Serrated		Serrated			Serrated
Hammer:	External; wide spur, serrated		External; wide spur, serrated			External; wide spur, serrated
Operation:	Straight blow-back		Locked breech			Locked breech
Loaded Chamber Indicator:	No	Yes	No	No	No	Yes
Safeties:	Side lever thumb safety, grip safety		Side lever thumb safety, grip safety			Side lever thumb safety, grip safety
Grips:	Modified thumbrest black plastic grips		Matte black polymer			Matte black polymer
Sights:	Patridge-type front; square-notch rear adjustable for windage		Patridge-type front; square-notch rear adjustable for windage			Patridge-type front; square-notch rear adjustable for windage
Sight Radius:	4¹/₄″		6¹/₄″			6¹/₄″
Magazine Capacity:	8-shot	7-shot	7-shot			7-shot
Weight:	23 oz.		37 oz.	37 oz.	34 oz.	36 oz.
Barrel Length:	3¹¹/₁₆″		4¹/₄″			5¹/₈″
Overall Length:	6¹/₂″		7⁷/₈″			8¹/₂″
Height:	4³/₈″		5⁷/₁₆″			5⁵/₁₆″
Finish:	Std. models: High-polished, deep blue. Deluxe models: Satin chrome		Std. models: High-polished, deep blue. Deluxe models: Satin chrome			Std. models: High-polished, blue. Deluxe models: Satin chrome

LLAMA AUTOMATIC PISTOLS

COMPACT 45
$299.95

45 STANDARD BLUE
$299.95

LARGE-FRAME 45 AUTOMATIC
(Deep Blue Finish)
$299.95

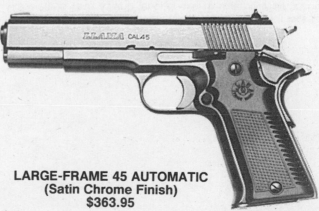

LARGE-FRAME 45 AUTOMATIC
(Satin Chrome Finish)
$363.95

AMERICAN EAGLE LUGER®

**AMERICAN EAGLE
LUGER®**

P-08 MODEL 9mm AMERICAN EAGLE LUGER®
STAINLESS STEEL

It is doubtful that there ever was a pistol created that evokes the nostalgia or mystique of the Luger® pistol. Since its beginnings at the turn of the 20th century, the name Luger® conjures memories of the past. Stoeger Industries is indeed proud to have owned the name Luger® since the late 1920s and is equally proud of the stainless-steel version that graces this page.

The "American Eagle" name was introduced around 1900 to capture the American marketplace. It served its purpose well, the name having become legendary along with the Luger® name. The "American Eagle" inscribed on a Luger® also distinguishes a firearm of exceptional quality over some inexpensive models that have been manufactured in the past.

Constructed entirely of stainless steel, the gun is available in 9mm Parabellum only, with either a 4″ or 6″ barrel, each with deeply checkered American walnut grips.

The name Luger®, combined with Stoeger's reputation of selling only quality merchandise since 1918, assures the owner of complete satisfaction.

SPECIFICATIONS
Caliber: 9mm Parabellum
Barrel lengths: 4″ and 6″
Weight: 31 oz.
Grips: Deeply checkered American walnut
Features: All stainless steel construction
Price: $695.00

MAGNUM RESEARCH

DESERT EAGLE PISTOLS

SPECIFICATIONS	357 MAGNUM	41/44 MAGNUM
Length, with 6-inch barrel	10.6 inches	10.6 inches
Height	5.6 inches	5.7 inches
Width	1.25 inches	1.25 inches
Trigger reach	2.75 inches	2.75 inches
Sight radius (wh 6-inch barrel)	8.5 inches	8.5 inches
Additional available barrels	14 inch	14 inch & 10 inch
Weight	See below	See below
Bore rifling — Six rib	Polygonal: 1 turn in 14 inches	Polygonal: 1 turn in 18 inches
Method of operation	Gas operated	Gas operated
Method of locking	Rotating bolt	Rotating bolt
Magazine capacity	9 rounds (plus one in chamber)	8 rounds (plus one in chamber)

**DESERT EAGLE MARK I
(10″ Barrel)**

DESERT EAGLE — WEIGHT TABLES
357 Magnum

Frame	Without Magazine		With Empty Magazine	
	6″ Barrel	14″ Barrel	6″ Barrel	14″ Barrel
	ounces	ounces	ounces	ounces
Aluminum	47.8	55.0	51.9	59.1
Steel	58.3	65.5	62.4	96.6
Stainless	58.3	65.5	62.4	69.6

41/44 Magnum

Frame	Without Magazine		With Empty Magazine	
	6″ Barrel	14″ Barrel	6″ Barrel	14″ Barrel
	ounces	ounces	ounces	ounces
Aluminum	52.3	61.0	56.4	65.1
Steel	62.8	71.5	66.9	75.6
Stainless	62.8	71.5	66.9	75.6

DESERT EAGLE PISTOLS

357 MAGNUM
$789.00 Standard Parkerized
    Finish (6″ Barrel)
 839.00 Stainless Steel (6″)
 939.00 Standard (10″ barrel)
 989.00 Stainless (10″ barrel)
 949.00 Standard (14″ barrel)
 999.00 Stainless (14″ barrel)

41 MAGNUM (6″ & 14″ Barrels)
$799.00 Std. Parkerized Finish
 849.00 Stainless Steel

44 MAGNUM (6″, 10″, 14″ Barrels)
$ 899.00 Standard Parkerized Finish
  949.00 Stainless Steel (6″ Barrel)
1099.00 Standard (10″ Barrel)
1159.00 Stainless (10″ Barrel)
1119.00 Standard (14″ Barrel)
1169.00 Stainless (14″ Barrel)

**DESERT EAGLE 50 MAGNUM
$1249.00**

Features: 10½″ barrel; weight: 72.4 oz.;
height: 5.9″; sight radius: 8.3″

MAGNUM RESEARCH

MOUNTAIN EAGLE

BABY EAGLE

LONE EAGLE
With Barreled Action

MOUNTAIN EAGLE
$239.00

This affordable, lightweight triangular barreled pistol with minimum recoil is ideal for plinkers, target shooters and varmint hunters. The barrel is made of hybrid injection-molded polymer and steel, and the standard 15-round magazine is made of high-grade, semi-transparent polycarbonate resin. It uses a constant force spring to load all 15 rounds easily. The receiver is made of machined T-6 alloy.

SPECIFICATIONS
Caliber: 22 LR
Barrel length: 6 1/2"
Overall length: 10.6"
Weight: 21 oz.
Sights: Standard orange in front, black in rear (adjustable for windage and elevation
Grip: One-piece injection-molded, checkered conventional contour side panels, horizontally textured front and back panels
Also available:
TARGET EDITION with 8" barrel, 2-stroke target trigger, jeweled bolt, adj. sights, range case. **$279.00**

BABY EAGLE PISTOL
$569.00

SPECIFICATIONS
Calibers: 9mm, 40 S&W, 41 AE
Capacity: 10 rounds (40 S&W), 11 rounds (41 AE), 16 rounds (9mm)
Barrel length: 4.72"
Overall length: 8.14"
Weight: 35.37 oz. (empty)
Sights: Combat
Features: Extra-long slide rail; combat-style trigger guard; decocking safety; polygonal rifling; ambidextrous thumb safety; all-steel construction; double action
Also available:
Conversion Kit (9mm to 41 AE). **$239.00**
GT Model with green tea finish. **$169.00**
Tritium Sights (fixed or adjustable)

LONE EAGLE SINGLE SHOT PISTOL
$344.00

This specialty pistol is designed for hunters, silhouette enthusiasts, long-range target shooters and other marksmen. The pistol can fire 15 different calibers of ammunition. Available with interchangeable 14-inch barreled actions, the gun is accurate and engineered to handle the high internal pressures of centerfire rifle ammunition. The design of the grip and its placement toward the center of the barrel provide the Lone Eagle with balance, reduced actual and felt recoil, and a high level of comfort for the shooter. **Calibers:** 22-250, 223 Rem., 22 Hornet, 22 Win. Mag., 243 Rem., 30-30, 30-06, 308 Win., 357 Maximum, 358 Win., 35 Rem., 44 Mag., 444 Marlin, 7mm-08, 7mm Bench Rest.
Also available:
LONE EAGLE with barreled action in calibers 7mm-08, 308 or 30-06. Features integral muzzle brake, silver matte Ecoloy II satin finish, satin scope mount and travel case. **$489.00**

McMILLAN

WOLVERINE PISTOL
$1700.00

This competition-ready pistol is built and designed around the 1911 Colt auto pistol frame. Two models—the **Combat** and **Competition Match**—are available with metal finish options (electroless nickel, NP³, black Teflon and deep blue), plus wood or Pachmayr grips. Other features include: a compensator integral to the barrel; round burr-style hammer; oversized button-style catch for quick clip release; high grip system to reduce muzzle rise and felt recoil; knurled backstrap and low-profile adjustable sights; skeletonized aluminum trigger with hand-tuned medium weight pull.

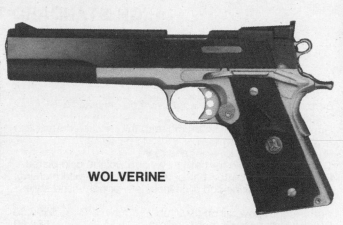

WOLVERINE

SPECIFICATIONS
Calibers: 45 ACP, 45 Italian, 38 Super, 38 Wadcutter, 9mm, 10mm
Action: Single
Barrel length: 6″ (5″ in Officer's Model)

MITCHELL ARMS

SINGLE ACTION ARMY REVOLVERS

The Mitchell Arms Single Action Army Model Revolver is a modern version of the original "gun that won the West," adopted by the U.S. Army in 1873. Faithful to the original design, these revolvers—Bat Masterson, Cowboy, U.S. Army and U.S. Cavalry—are made of modern materials and use up-to-date technology.

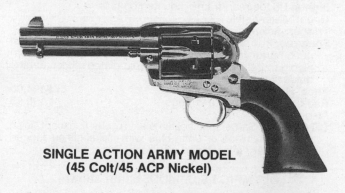

SINGLE ACTION ARMY MODEL
(45 Colt/45 ACP Nickel)

SPECIFICATIONS
Calibers: 357 Mag., 44 Mag., 45 Colt/45 ACP dual cylinder
Barrel lengths: 4³/4″, 5¹/2″ and 7¹/2″
Frame: Forged steel, fully machined with traditional color casehardening; two-piece style backstrap made of solid brass
Action: Traditional single action with safety position and half-cock position for loading and unloading
Sights: Rear sight is fully adjustable for windage and elevation. Front sight is two-step ramp style with non-glare serrations. Fixed sight models feature deep notch with fixed blade front sight.
Grip: One-piece solid walnut grip built in the style of the old blackpowder revolvers
Accuracy: High-grade steel barrel honed for accuracy with smooth lands and grooves; precise alignment between cylinder and barrel. Fully qualified for big game hunting or silhouette shooting.

HIGH STANDARD SIGNATURE SERIES COMPETITION PISTOLS

Prices:
Single Action Army

Blued finish (all calibers)	$399.00
Nickel finish	439.00
Stainless steel backstrap/trigger guard	494.00
Dual Cylinders	
45 Colt, 4³/4″ barrel	549.00
Same as above in nickel	588.00

Caliber: 45 ACP. **Capacity:** 8 and 13 rounds. **Barrel length:** 5″. **Sights:** Fully interchangeable front sight; drift adjustable rear combat sight. **Grips:** American walnut. **Finish:** Royal blue or stainless steel. **Price:** $529.00–$749.00

MITCHELL ARMS

HIGH STANDARD TARGET PISTOLS

VICTOR II

SPECIFICATIONS
Caliber: 22 LR
Barrel lengths: 4½" (full-length vent rib only) and 5½"
Sights: Rib-mounted
Grips: Military-style with thumb rest; full checkered American walnut
Frame: Stippled grip, front and rear
Trigger: Adjustable for both travel and weight; gold-plated
Features: Push-button barrel takedown; gold-filled marking; safety, magazine and slide lock; left- or right-hand styles
Prices:
W/full-length dovetail rib (5½" bbl.) **$599.00**
W/full-length vent rib (both barrels) **569.00**
W/full-length vent rib/Weaver-style base **648.00**
Detachable barrel weights (2 or 3 oz.) **add** **20.00**

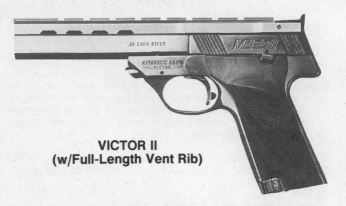

**VICTOR II
(w/Full-Length Vent Rib)**

TROPHY II

SPECIFICATIONS
Caliber: 22 LR
Barrel length: Interchangeable 5½" (Bull) or 7½" (Fluted)
Sights: Bridge rear; frame mounted on rails
Grips: Military-style
Trigger: Adjustable for travel and weight; gold-plated
Features: Full checkered American walnut thumbrest grips; push-button barrel takedown; all roll marks gold filled; stippled grip frame (front and rear); gold-plated trigger, safety, magazine release and slide lock
Price: Royal blue or stainless **$494.00**
Red dot scope ultralite w/rings **199.95**
Also available:
CITATION II. Same specifications as Trophy II, except finish is satin stainless or satin blue with nickel-plated trigger, safety, mag. release and slide lock. **Price: $468.00**

TROPHY II

OLYMPIC I.S.U. MODEL (not shown)

SPECIFICATIONS
Caliber: 22 Short, Long
Barrel length: 6¾" w/integral stabilizer
Sights: Bridge rear sight; frame-mounted on rails
Trigger: Adjustable for travel and weight
Grips: Full checkered with thumb rest
Finish: Royal blue or stainless steel
Features: Push-button barrel takedown; stippled grip frame (front and rear); military-style grip; adjustable barrel weights
Price: Blue or stainless **$599.00**

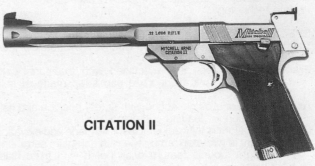

CITATION II

SHARPSHOOTER II

SPECIFICATIONS
Caliber: 22 Rimfire
Barrel length: 5½" bull barrel
Sights: Slide-mounted adjustable rear
Features: Push-button takedown; special barrel weights (optional); smooth grip frame; satin finish
Price: $379.00
Also available:
SPORT KING II 22 LR w/4½" and 5½" barrels. **Price: $312.00**

SHARPSHOOTER II

MOA MAXIMUM PISTOL

MAXIMUM

This single-shot pistol with its unique falling-block action performs like a finely tuned rifle. The single piece receiver of stainless steel is mated to a Douglas barrel for optimum accuracy and strength.

SPECIFICATIONS
Calibers: 22 Hornet to 358 Win.
Barrel lengths: 8½″, 10½″ and 14″
Weight: 3 lbs. 8 oz. (8¾″ bbl.); 3 lbs. 13 oz. (10½″ bbl.); 4 lbs. 3 oz. (14″ bbl.)
Prices:
Stainless receiver, blued barrel **$622.00**
Stainless receiver and barrel **677.00**
Extra barrels (blue) **164.00**
For stainless, **add** **58.00**
Muzzle brake **125.00**

NAVY ARMS REPLICAS

1873 SINGLE ACTION

1895 U.S. ARTILLERY MODEL (not shown)

Same specifications as the U.S. Cavalry Model, but with a 5½″ barrel as issued to Artillery units. **Caliber:** 45 Long Colt.
Price: **$480.00**

1873 U.S. CAVALRY MODEL (not shown)

An exact replica of the original U.S. Government issue Colt Single Action Army, complete with Arsenal stampings and inspector's cartouche. **Caliber:** 45 Long Colt. **Barrel length:** 7½″. **Overall length:** 13¼″. **Weight:** 2 lbs. 7 oz. **Sights:** Blade front; notch rear. **Grips:** Walnut.
Price: **$480.00**

TT-OLYMPIA PISTOL

The TT-Olympia is a faithful reproduction of the famous Walther .22 target pistol that won the gold medal at the 1936 Olympics in Berlin. **Caliber:** 22 LR. **Barrel length:** 4⅝″. **Overall length:** 8″. **Weight:** 1 lb. 11 oz. **Sights:** Blade front; adjustable rear.
Price: **$290.00**

1873 COLT-STYLE SINGLE ACTION REVOLVERS

The classic 1873 Single Action is the most famous of all the "six shooters." From its adoption by the U.S. Army in 1873 to the present, it still retains its place as America's most popular revolver. **Calibers:** 44-40 or 45 Long Colt. **Barrel lengths:** 3″, 4¾″, 5½″ or 7½″. **Overall length:** 10¾″ (5½″ barrel). **Weight:** 2¼ lbs. **Sights:** Blade front; notch rear. **Grips:** Walnut.
Prices: . . . . **$340.00 to $455.00**

1875 SCHOFIELD REVOLVER (not shown)

A favorite side arm of Jesse James and General George Armstrong, the 1875 Schofield revolver was one of the legendary handguns of the Old West. **Caliber:** .44-40, 45 LC. **Barrel length:** 5″ (Wells Fargo Model) or 7″ (U.S. Cavalry Model). **Overall length:** 10¾″ or 12¾″. **Weight:** 2 lbs. 7 oz. **Sights:** Blade front; notch rear. **Features:** Top-break, automatic ejector single action.
Price: **$795.00**

TT-OLYMPIA PISTOL

NEW ENGLAND FIREARMS

STANDARD REVOLVER
$124.95 ($134.95 in Nickel)

SPECIFICATIONS
Calibers: 22 S, L or LR
Capacity: 9 shots
Barrel lengths: 2¹/₂″ and 4″
Overall length: 7″ (2¹/₂″ barrel) and 8¹/₂″ (4″ barrel)
Weight: 25 oz. (2¹/₂″ bbl.) and 28 (4″ bbl.)
Sights: Blade front; fixed rear
Grips: American hardwood, walnut finish
Finish: Blue or nickel
Also available in 5-shot 32 H&R Mag. Blue **$114.95**

**STANDARD MODEL
(22 LR, 2¹/₂″ Barrel)**

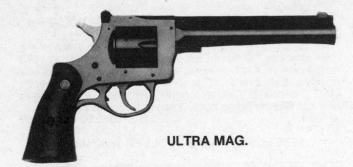

ULTRA MAG.

ULTRA MODEL (6″ Barrel)

ULTRA AND ULTRA MAG. REVOLVERS
$149.95

SPECIFICATIONS
Calibers: 22 Short, Long, Long Rifle (Ultra); 22 Win. Mag.
   (Ultra Mag.)
Capacity: 9 shots (22 LR); 6 shots (22 Win. Mag.)
Barrel lengths: 4″ and 6″
Overall length: 10⁵/₈″
Weight: 36 oz. (6″ barrel)
Sights: Blade on rib front; fully adjustable rear
Grips: American hardwood, walnut finish
Also available:
LADY ULTRA in 5-shot 32 H&R Magnum. **Barrel length:** 3″.
   Overall length: 7¹/₂″. **Weight:** 31 oz. **Price:** $149.95

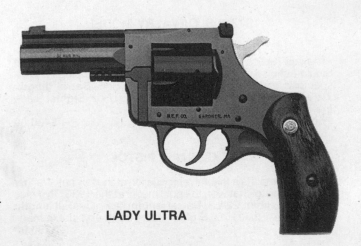

LADY ULTRA

NORTH AMERICAN ARMS

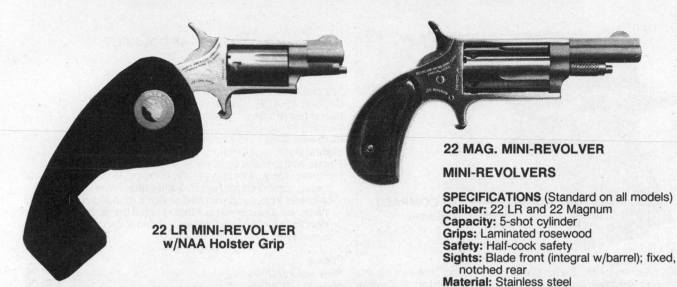

22 LR MINI-REVOLVER
w/NAA Holster Grip

22 MAG. MINI-REVOLVER

MINI-REVOLVERS

SPECIFICATIONS (Standard on all models)
Caliber: 22 LR and 22 Magnum
Capacity: 5-shot cylinder
Grips: Laminated rosewood
Safety: Half-cock safety
Sights: Blade front (integral w/barrel); fixed, notched rear
Material: Stainless steel
Finish: Matte with brushed sides

SPECIFICATIONS: MINI-REVOLVERS & MINI-MASTER SERIES

Model	Weight	Barrel Length	Overall Length	Overall Height	Overall Width	Prices
NAA-MMT-M	10.7 oz.	4″	7³/₄″	3⁷/₈″	⁷/₈″	$279.00
NAA-MMT-L	10.7 oz.	4″	7³/₄″	3⁷/₈″	⁷/₈″	279.00
*NAA-BW-M	8.8 oz.	2″	5⁷/₈″	3⁷/₈″	⁷/₈″	235.00
*NAA-BW-L	8.8 oz.	2″	5⁷/₈″	3⁷/₈″	⁷/₈″	235.00
NAA-22LR**	4.5 oz.	1¹/₈″	4¹/₄″	2³/₈″	13/16″	157.00
NAA-22LLR**	4.6 oz.	1⁵/₈″	4³/₄″	2³/₈	13/16″	157.00
*NAA-22MS	5.9 oz.	1¹/₈″	5″	2⁷/₈″	⁷/₈″	178.00
*NAA-22M	6.2 oz.	1⁵/₈″	5³/₈″	2⁷/₈″	⁷/₈″	178.00

* Available with Conversion Cylinder chambered for 22 Long Rifle   ** Available with holster grip (**$188.00**)

MINI-MASTER SERIES

SPECIFICATIONS (Standard on all models)
Calibers: 22 LR (NAA-MMT-L, NAA-BW-L) and 22 Magnum (NAA-MMT-M, NAA-BW-M)
Barrel: Heavy vent

Rifling: 8 land and grooves, 1/12 R.H. button broach twist
Grips: Oversized black rubber
Cylinder: Bull
Sights: Front integral with barrel; rear Millett adjustable white outlined (elevation only) or low-profile fixed

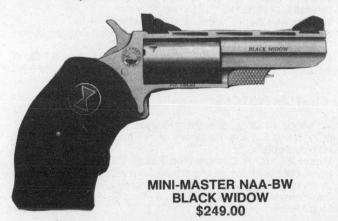

MINI-MASTER NAA-BW
BLACK WIDOW
$249.00

MINI-MASTER NAA-MMT-M
(22 Mag. 4″ Barrel)
$279.00

PARA-ORDNANCE

MODEL P12 • 45 COMPACT

MODEL P12 • 45 COMPACT
With 3¹/₂″ Barrel

SPECIFICATIONS
Caliber: 45 ACP   **Capacity:** 11 + 1
Barrel length: 3¹/₂″   **Overall length:** 7″
Weight: 24 oz. (alloy); 33 oz. (steel)
Receiver: Alloy, steel or stainless steel
Sights: 3-dot sight system
Finish: Matte black or silver
Features: Manual thumb safety; firing-pin lock safety; trigger guard contoured for higher, tighter grip; flared ejection port; grooved front strap rounded smooth; rounded combat-style hammer; solid barrel bushing; ramped barrel; high-capacity double-column magazine with bumper pad; beveled magazine well

Also available:
MODEL P14 • 45. Same specifications and features as the P12 with the following exceptions: **Barrel length:** 5″. **Overall length:** 8¹/₂″. **Weight:** 28 oz. (alloy) and 38 oz. (steel). **Capacity:** 13 + 1. **Prices:** $716.25 (Steel); $650.00 (Alloy)

Prices:
Alloy w/13 rounds	**$650.00**
Alloy w/12 rounds	650.00
Steel w/13 rounds	712.50
Steel w/12 rounds	708.75

REMINGTON LONG-RANGE PISTOLS

MODEL XP-100
BOLT-ACTION SILHOUETTE

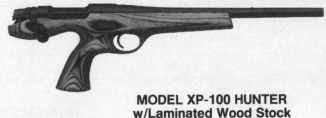

MODEL XP-100 HUNTER
w/Laminated Wood Stock

These unique single-shot centerfire, bolt-action pistols have become legends for their strength, precision, balance and accuracy. Chambered for 223 Rem., 35 Rem. and 7mm-08 Rem. with a 14¹/₂″ barrel, they are also available in 7mm BR, which many feel is the ideal factory-made metallic silhouette handgun for "unlimited" events. All XP-100 handguns have one-piece Du Pont "Zytel" nylon stocks with universal grips, two-position thumb safety switches, receivers drilled and tapped for scope mounts or receiver sights, and match-type grooved triggers.

SPECIFICATIONS
Calibers: 223 Rem., 35 Rem., 7mm BR Rem. and 7mm-08 Rem.
Barrel length: 14¹/₂″ (10¹/₂″ also in 7mm BR Rem.)
Overall length: 21¹/₄″ (17¹/₄″ w/10¹/₂″ barrel)
Weight: 4³/₈ lbs. (3⁷/₈ lbs. w/10¹/₂″ barrel)
Stock: American walnut (10¹/₂″ barrel model) and laminated wood
Model XP-100 (Silhouette & Hunter) **$548.00**
w/10¹/₂″ barrel 625.00

XP-100 LONG-RANGE CUSTOM PISTOLS

Remington's Model XP-100 Custom pistol is chambered for the 22-250 Rem., 6mm BR Rem., 223 Rem., 250 Savage, 7mm BR, 7mm-08 Rem., 308 Win. and 35 Rem. All XP-100 Custom pistols are hand-crafted from select English walnut in right- and left-hand versions. All chamberings except the 35 Rem. (standard barrel only) are offered in a choice of standard 14¹/₂″ barrels with adjustable rear leaf and front bead sights or 15¹/₂″ barrels without sights. Receivers are drilled and tapped

for scope mounts. **Weight** averages 4¹/₂ lbs. for standard models and 5¹/₂ lbs. for heavy barrel models. **$945.00**
Also available:
Model XP-100 R Custom Bolt Action Centerfire Repeater. Same specifications as the XP-100, but with Kevlar-reinforced stock and sling swivel studs. **Calibers:** 22-250, 223 Rem., 250 Savage, 7mm-08 Rem., 308 Win., 35 Rem., 350 Rem. Mag. **Price:** **$840.00**

ROSSI REVOLVERS

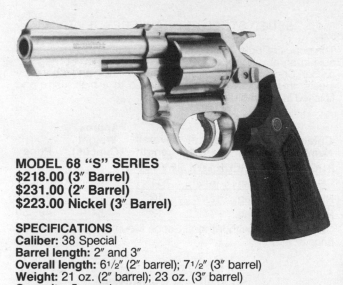

MODEL 68 "S" SERIES
$218.00 (3″ Barrel)
$231.00 (2″ Barrel)
$223.00 Nickel (3″ Barrel)

SPECIFICATIONS
Caliber: 38 Special
Barrel length: 2″ and 3″
Overall length: 6¹/₂″ (2″ barrel); 7¹/₂″ (3″ barrel)
Weight: 21 oz. (2″ barrel); 23 oz. (3″ barrel)
Capacity: 5 rounds
Finish: Blue or nickel

MODEL 720 (not shown)
$312.00

SPECIFICATIONS
Caliber: 44 S&W Special
Capacity: 5 shots
Barrel length: 3″
Overall length: 8″
Weight: 27¹/₂ oz.
Sights: Adjustable rear; red insert front
Finish: Stainless steel
Features: Rubber combat grips; full ejector rod shroud
Also available:
MODEL 720 "COVERT SPECIAL." Double action only, hammerless, notched sight channel in frame. **Price: $312.00**

MODEL M88 "S" SERIES
$249.00 (3″ Barrel)
$265.00 (2″ Barrel)

SPECIFICATIONS
Caliber: 38 Special
Capacity: 5 rounds, swing-out cylinder
Barrel lengths: 2″ and 3″
Overall length: 6¹/₂″ (2″ barrel); 7¹/₂″ (3″ barrel)
Weight: 21 oz. (2″); 23 oz. (3″)
Sights: Ramp front, square notch rear adjustable for windage
Finish: Stainless steel

MODEL 515 & 518 (not shown)
$275.00 (22 LR)
$290.00 (22 Mag.)

SPECIFICATIONS
Calibers: 22 LR (Model 518) and 22 Mag. (Model 515)
Capacity: 6 rounds
Barrel length: 4″
Overall length: 9″
Weight: 30 oz.

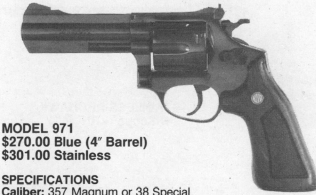

MODEL 971
$270.00 Blue (4″ Barrel)
$301.00 Stainless

SPECIFICATIONS
Caliber: 357 Magnum or 38 Special
Capacity: 6 rounds
Barrel lengths: 2¹/₂″, 4″ and 6″
Overall length: 8⁵/₁₆″ w/2¹/₂″ bbl.; 9³/₁₆″ w/4″ bbl.; 11³/₁₆″ w/6″ bbl.
Weight: 22 oz. (2¹/₂″ bbl.); 35.4 oz. (4″ bbl.); 40.5 oz. (6″ bbl.)
Finish: Blue or stainless steel
Also available:
MODEL 971 COMPACT GUN 357 Mag. only w/3¹/₄″ barrel (32 oz.). **Price: $301.00**

MODEL 851
$270.00

SPECIFICATIONS
Caliber: 38 Special
Capacity: 6 rounds
Barrel length: 4″
Overall length: 7¹/₂″
Weight: 30 oz.
Frame: Medium
Finish: Stainless

RUGER REVOLVERS

BLUED REDHAWK REVOLVER

**STAINLESS REDHAWK
REVOLVER**

**STAINLESS REDHAWK
w/Scope (KRH-44R)**

**SUPER REDHAWK STAINLESS
DOUBLE-ACTION REVOLVER**

BLUED STEEL REDHAWK REVOLVER

The popular Ruger Redhawk® double-action revolver is available in an alloy steel model with blued finish in 44 Magnum caliber. Constructed of hardened chrome-moly and other alloy steels, this Redhawk is satin polished to a high lustre and finished in a rich blue.

Catalog Number	Caliber	Barrel Length	Overall Length	Approx. Weight (Ounces)	Price
RUGER BLUED REDHAWK REVOLVER					
RH-445	44 Mag.	5$1/2$″	11″	49	**$458.50**
RH-44	44 Mag.	7$1/2$″	13″	54	**458.50**
RH-44R*	44 Mag.	7$1/2$″	13″	54	**496.50**

* Scope model, with Integral Scope Mounts, 1″ Ruger Scope rings.

STAINLESS REDHAWK DOUBLE-ACTION REVOLVER

There is no other revolver like the Ruger Redhawk. Knowledgeable sportsmen reaching for perfection in a big bore revolver will find that the Redhawk demonstrates its superiority at the target, whether silhouette shooting or hunting. The scope sight model incorporates the patented Ruger integral Scope Mounting System with 1″ stainless steel Ruger scope rings.

Catalog Number	Caliber	Barrel Length	Overall Length	Approx. Weight (Ounces)	Price
RUGER STAINLESS REDHAWK REVOLVER					
KRH-445	44 Mag.	5$1/2$″	11″	49	**$516.75**
KRH-44	44 Mag.	7$1/2$″	13″	54	**516.75**
KRH-44R*	44 Mag.	7$1/2$″	13″	54	**557.25**

* Scope model, with Integral Scope Mounts, 1″ Stainless Steel Ruger Scope rings.

SUPER REDHAWK STAINLESS DOUBLE-ACTION REVOLVER

The **Super Redhawk** double-action revolver in stainless steel features a heavy extended frame with 7$1/2$″ and 9$1/2$″ barrels. Cushioned grip panels contain Goncalo Alves wood grip panel inserts to provide comfortable, nonslip hold. Comes with integral scope mounts and 1″ stainless steel Ruger scope rings.

SPECIFICATIONS
Caliber: 44 Magnum
Barrel length: 7$1/2$″ and 9$1/2$″
Overall length: 13″ w/7$1/2$″ bbl.; 15″ w/9$1/2$″ bbl.
Weight (empty): 53 oz. (7$1/2$″ bbl.); 58 oz. (9$1/2$″ bbl.)
Sight radius: 9$1/2$″ (7$1/2$″ bbl.); 11$1/4$″ (9$1/2$″ bbl.)
Finish: Stainless steel; satin polished

KSRH-7 (7$1/2$″ barrel) **$557.25**
KSRH-9 (9$1/2$″ barrel) **557.25**

RUGER SINGLE ACTION REVOLVERS

SPECIFICATIONS: VAQUERO SINGLE-ACTION REVOLVER

**VAQUERO SINGLE ACTION
$394.00 (Model BNV455)**

Catalog Number	Caliber	Barrel Length	Overall Length	Approx. Weight (Ounces)	Finish*
BNV40	44-40	4⅝"	10¼"	39	CB
KBNV40	44-40	4⅝"	10¼"	39	SSG
BNV405	44-40	5½"	11½"	40	CB
KBNV405	44-40	5½"	11½"	40	SSG
BNV407	44-40	7½"	13⅛"	41	CB
KBNV407	44-40	7½"	13⅛"	41	SSG
BNV475	44 Mag.	5½"	11½"	40	CB
KBNV475	44 Mag.	5½"	11½"	40	SSG
BNV477	44 Mag.	7½"	13⅛"	41	CB
KBNV477	44 Mag.	7½"	13⅛"	41	SSG
BNV44	45 LC	4⅝"	10¼"	39	CB
KBNV44	45 LC	4⅝"	10¼"	39	SSG
BNV455	45 LC	5½"	11½"	40	CB
KBNV455	45 LC	5½".	11½"	40	SSG
BNV45	45 LC	7½"	13⅛"	41	CB
KBNV45	45 LC	7½"	13⅛"	41	SSG

* Finish: high-gloss stainless steel (SSG); color case finish on steel cylinder frame w/blued steel grip, barrel and cylinder (CB). LC = Long Colt.

SPECIFICATIONS: NEW MODEL BLACKHAWK AND BLACKHAWK CONVERTIBLE*

Catalog Number	Caliber	Finish**	Barrel Length	Overall Length	Approx. Weight (Oz.)	Price
BN31	.30 Carbine	B	7½"	13⅛"	44	**$328.00**
BN34	.357 Mag. + +	B	4⅝"	10⅜"	40	**328.00**
KBN34	.357 Mag. + +	SS	4⅝"	10⅜"	40	**404.00**
BN36	.357 Mag. + +	B	6½"	12¼"	42	**328.00**
KBN36	.357 Mag. + +	SS	6½"	12½"	42	**404.00**
BN34X*	.357 Mag. + +	B	4⅝"	10⅜"	40	**343.50**
BN36X*	.357 Mag. + +	B	6½"	12¼"	42	**343.50**
BN41	.41 Mag.	B	4⅝"	10¼"	38	**328.00**
BN42	.41 Mag.	B	6½"	12⅛"	40	**328.00**
BN44	.45 Long Colt	B	4⅝"	10¼"	39	**328.00**
KBN44	.45 Long Colt	SS	4⅝"	10¼"	39	**424.00**
BN455	.45 Long Colt	B	5½"	11⅛"	39	**328.00**
BN45	.45 Long Colt	B	7½"	13⅛"	41	**328.00**
KBN45	.45 Long Colt	SS	7½"	13⅛"	41	**424.00**

* Convertible: this model is designated by an X in the Catalog Number, and comes with an extra cylinder. The extra cylinder is 9mm Parabellum and can be instantly interchanged without the use of tools. Price of the Convertible model includes the extra cylinder.
** Finish: blued (B); stainless steel (SS); color case finish on the steel cylinder frame with blued steel grip, barrel, and cylinder (CB). Models KBN34, KBN36, KBN44 and KBN45 are available in high-gloss finish (same price).
+ + Revolvers chambered for the .357 Magnum cartridge also accept factory-loaded .38 Special cartridges.

RUGER REVOLVERS

NEW MODEL SUPER BLACKHAWK SINGLE ACTION REVOLVER

SPECIFICATIONS
Caliber: 44 Magnum; interchangeable with 44 Special
Barrel lengths: 5½″, 7½″, 10½″
Overall length: 13⅜″ (7½″ barrel)
Weight: 48 oz. (7½″ bbl.) and 51 oz. (10½″ bbl.)
Frame: Chrome molybdenum steel or stainless steel
Springs: Music wire springs throughout
Sights: Patridge style, ramp front matted blade ⅛″ wide; rear sight click and adjustable for windage and elevation

NEW MODEL SUPER BLACKHAWK SINGLE ACTION REVOLVER

Grip frame: Chrome molybdenum or stainless steel, enlarged and contoured to minimize recoil effect
Trigger: Wide spur, low contour, sharply serrated for convenient cocking with minimum disturbance of grip
Finish: Polished and blued or brushed satin stainless steel
Prices:

KS45N	5½″ bbl., brushed or high-gloss stainless	**$413.75**
KS458N	4⅝″ bbl., brushed or high-gloss stainless	413.75
KS47N	7½″ bbl., brushed or high-gloss stainless	413.75
KS411N	10½″ bull bbl., stainless steel	413.75
KSH7NH	7½″ bbl., scope rings, stainless	479.50
S45N	5½″ bbl., blued	378.50
S458N	4⅝″ bbl., blued	378.50
S47N	7½″ bbl., blued	378.50
S411N	10½″ bull bbl., blued	378.50

NEW MODEL SUPER SINGLE-SIX REVOLVER

SPECIFICATIONS
Caliber: 22 LR (fitted with WMR cylinder)
Barrel lengths: 4⅝″, 5½″, 6½″, 9½″; stainless steel model in 5½″ and 6½″ lengths only
Weight (approx.): 33 oz. (with 5½″ barrel); 38 oz. (with 9½″ barrel)
Sights: Patridge-type ramp front sight; rear sight click adjustable for elevation and windage; protected by integral frame ribs. Fixed sight model available with 5½″ or 6½″ barrel (same prices as adj. sight models).
Finish: Blue or stainless steel
Prices:

In blue	**$281.00**
In brushed or high-gloss stainless steel (convertible 5½″ and 6½″ barrels only)	354.00

**FIXED SIGHT
NEW MODEL SINGLE-SIX
(W/Extra Cylinder)**

Also available:
BEARCAT SINGLE ACTION with 4″ barrel, walnut grips

Blue	**$298.00**
Stainless	325.00

NEW MODEL SINGLE-SIX SSM™ REVOLVER

SPECIFICATIONS
Caliber: 32 Magnum; also handles 32 S&W and 32 S&W Long
Barrel lengths: 4⅝″, 5½″, 6½″, 9½″
Weight (approx.): 34 oz. (with 6½″ barrel)
Price: **$281.00**

**NEW MODEL
SINGLE-SIX SSM™**

RUGER REVOLVERS

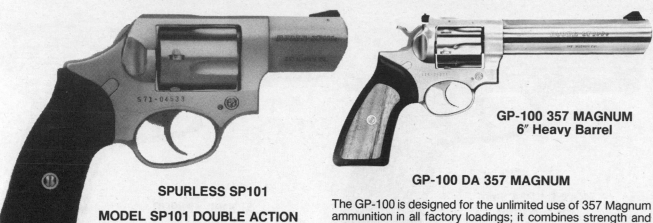

GP-100 357 MAGNUM
6″ Heavy Barrel

SPURLESS SP101

MODEL SP101 DOUBLE ACTION
$408.00

This critically acclaimed small frame SP101 model has added 6-shot capacity in 32 Magnum and 5-shot capacity in 9mm and 357 Magnum.

SPECIFICATIONS

Catalog Number	Caliber	Capacity	Sights	Barrel Length	Approx. Wt. (Oz.)
KSP-221	22 LR	6	Adj.	2¼″	32
KSP-240	22 LR	6	Adj.	4″	33
KSP-241	22 LR	6	Adj.	4″	34
KSP-3231	32 Mag.	6	Adj.	3¹⁄₁₆″	30
KSP-3241	32 Mag.	6	Adj.	4″	33
KSP-921	9×19mm	5	Fixed	2¼″	25
KSP-931	9×19mm	5	Fixed	3¹⁄₁₆″	27
KSP-821	38 + P	5	Fixed	2¼″	25
KSP-821L	38 + P	5	Fixed	2¼″	26
KSP-831	38 + P	5	Fixed	3¹⁄₁₆″	27
KSP-321X	357 Mag.*	5	Fixed	2¼″	25
KSP-321XL	357 Mag.*	5	Fixed	2¼″	25
KSP-331X	357 Mag.*	5	Fixed	3¹⁄₁₆″	27

* Revolvers chambered for .357 Magnum also accept 38 Special cartridges. Model KSP-240 has short shroud; all others have full. Spurless hammer models are designated by "L" in catalog no.

BISLEY SINGLE ACTION
TARGET GUN (not shown)

The Bisley single action was originally used at the British National Rifle Association matches held in Bisley, England, in the 1890s. Today's Ruger Bisleys are offered in two frame sizes, chambered from 22 LR to 45 Long Colt. These revolvers are the target-model versions of the Ruger single-action line.

Special features: Unfluted cylinder rollmarked with classic foliate engraving pattern (or fluted cylinder without engraving); hammer is low with smoothly curved, deeply checkered wide spur positioned for easy cocking.

Prices:
22 LR or 32 Magnum **$328.75**
357 Mag., 41 Mag., 44 Mag., 45 Long Colt **391.00**

GP-100 DA 357 MAGNUM

The GP-100 is designed for the unlimited use of 357 Magnum ammunition in all factory loadings; it combines strength and reliability with accuracy and shooting comfort. (Revolvers chambered for the 357 Magnum cartridge also accept the 38 Special cartridge.)

SPECIFICATIONS

Catalog Number	Finish*	Sights†	Shroud††	Barrel Length	Wt. (Oz.)	Prices
GP-141	B	A	F	4″	41	$413.50
GP-160	B	A	S	6″	43	413.50
GP-161	B	A	F	6″	46	413.50
GPF-331	B	F	F	3″	36	397.00
GPF-340	B	F	S	4″	37	397.00
GPF-341	B	F	F	4″	38	397.00
KGP-141	S	A	F	4″	41	446.50
KGP-160	S	A	S	6″	43	446.50
KGP-161	S	A	F	6″	46	446.50
KGPF-330	S	F	S	3″	35	430.00
KGPF-331	S	F	F	3″	36	430.00
KGPF-340	S	F	S	4″	37	430.00
KGPF-341	S	F	F	4″	38	430.00
KGPF-840*	S	F	S	4″	37	430.00
KGPF-841*	S	F	F	4″	38	430.00

* B = blued; S = stainless. † A = adjustable; F = fixed. †† F = full; S = short. * 38 Special only.

BISLEY SPECIFICATIONS

Catalog Number	Caliber	Barrel Length	Overall Length	Sights	Approx. Wt. (Oz.)
RB22AW	.22 LR	6½″	11½″	Adj.	41
RB32W	.32 Mag.	6½″	11½″	Fixed*	41
RB32AW	.32 Mag.	6½″	11½″	Adj.	41
RB35W	.357 Mag.	7½″	13″	Adj.	48
RB41W	.41 Mag.	7½″	13″	Adj.	48
RB44W	.44 Mag.	7½″	13″	Adj.	48
RB45W	.45 Long Colt	7½″	13″	Adj.	48

* Dovetail rear sight adjustable for windage only.

RUGER P-SERIES PISTOLS

**MODEL KP94 9mm
(5³/₄" Barrel)**

MODEL KP89DC

P-SERIES PISTOLS

Ruger's P-Series pistols are now available in two Compact models—the P93 Double-Action-Only and the P93 Decock-Only. These Ruger Compacts feature a one-piece Zytel grip frame that decreases their weight to 31 ounces with a loaded 15-round magazine. The mechanism is recoil-operated, double action, and autoloading with a tilting barrel, link-actuated (as in the M1911A1).

GENERAL SPECIFICATIONS (see also table below for additional specifications and prices)

Barrel length: 3⁹/₁₀"
Overall length: 7³/₁₀"
Weight: 24 oz. (empty magazine)
Height: 5¹/₂"
Width: 1¹/₂"
Sight radius: 5"
Sights: Square notch rear, drift adjustable for windage; square post front (both sights have white dots for rapid target acquisition)

SPECIFICATIONS: P-SERIES PISTOLS

Catalog Number	Model	Finish	Caliber	Mag. Cap.	Price
KP89 (C)*	Safety-Convertible*	Stainless	9mm & 30 Luger	15	$497.00
P89	Safety	Blued	9mm	15	410.00
KP89	Safety	Stainless	9mm	15	452.00
P89DC	Decock-Only	Blued	9mm	15	410.00
KP89DC	Decock-Only	Stainless	9mm	15	452.00
KP89DAO	Dbl.-Action-Only	Stainless	9mm	15	452.00
KP90	Safety	Stainless	45 ACP	7	488.65
KP90DC	Decock-Only	Stainless	45 ACP	7	488.65
KP91DC	Decock-Only	Stainless	40 Auto	11	488.65
KP91DAO	Dbl.-Action-Only	Stainless	40 Auto	11	488.65
KP93DC	Decock-Only Compact	Stainless	9mm	15	520.00
KP93DAO	Dbl.-Action-Only Compact	Stainless	9mm	15	520.00
KP94	Manual Safety	Stainless	9mm	15	520.00
KP94DC	Decock-Only	Stainless	9mm	15	520.00
KP94DAO	Dbl.-Action-Only	Stainless	9mm	15	520.00
KP944	Manual Safety	Stainless	40 Auto	11	520.00
KP944DC	Decock-Only	Stainless	40 Auto	11	520.00
KP944DAO	Dbl.-Action-Only	Stainless	40 Auto	11	520.00

* Safety-Convertible Model. This model comes with two interchangeable barrels: one in 9mm Luger and one in .30 Luger. The barrels can be interchanged without the use of tools. Convertible model is designated by an "X" in its Catalog Number; price includes both barrels.

RUGER 22 AUTOMATIC PISTOLS

MARK II STANDARD MODEL STAINLESS

MARK II STANDARD MODEL

The Ruger Mark II models represent continuing refinements of the original Ruger Standard and Mark I Target Model pistols. More than two million of this series of autoloading rimfire pistol have been produced since 1949.

The bolts on all Ruger Mark II pistols lock open automatically when the last cartridge is fired, if the magazine is in the pistol. The bolt can be operated manually with the safety in the "on" position for added security while loading and unloading. A bolt stop can be activated manually to lock the bolt open.

The Ruger Mark II pistol uses 22 Long Rifle ammunition in a detachable, 10-shot magazine (standard on all Mark II models except Model 22/45, whose 10-shot magazine is not interchangeable with other Mark II magazines). Designed for easy insertion and removal, the Mark II magazine is equipped with a magazine follower button for convenience in reloading.

For additional specifications, please see the chart on the next page.

MARK II GOVERNMENT MODEL

MARK II TARGET MODEL

RUGER 22 AUTOMATIC PISTOLS

MARK II BULL BARREL MODEL

MARK II 22/45 w/Zytel Frame

SPECIFICATIONS: RUGER 22 MARK II PISTOLS

Catalog Number	Model*	Finish**	Barrel Length	Overall Length	Approx. Wt. (Oz.)	Price
MK-4	Std.	B	4³/₄″	8⁵/₁₆″	35	$252.00
KMK-4	Std.	SS	4³/₄″	8⁵/₁₆″	35	330.25
KP-4	Std.	SS	4³/₄″	8¹³/₁₆″	28	280.00
MK-6	Std.	B	6″	10⁵/₁₆″	37	252.00
KMK-6	Std.	SS	6″	10⁵/₁₆″	37	330.25
MK-514	Target	B	5¹/₄″	9⁷/₁₆″	38	310.50
KMK-514	Target	SS	5¹/₄″	9⁷/₁₆″	38	389.00
KP-514	Target	SS	5¹/₄″	9⁵/₁₆″	32	330.00
MK-678	Target	B	6⁷/₈″	11¹/₈″	42	310.50
KMK-678	Target	SS	6⁷/₈″	11¹/₈″	42	389.00
MK-512	Bull	B	5¹/₂″	9³/₄″	42	310.50
KMK-512	Bull	SS	5¹/₂″	9³/₄″	42	389.00
KP-512	Bull	SS	5¹/₂″	9⁹/₁₆″	35	330.00
MK-10	Bull	B	10″	14⁵/₁₆″	51	294.50
KMK-10	Bull	SS	10″	14⁵/₁₆″	51	373.00
MK-678G	Bull	B	6⁷/₈″	11¹/₈″	46	356.50
KMK-678G	Bull	SS	6⁷/₈″	11¹/₈″	46	427.25
KMK-678GC	Bull	SS	6⁷/₈″	11¹/₈″	45	441.00

* Model: standard (Std.).   ** Finish: blued (B); stainless steel (SS).

SIG SAUER PISTOLS

MODEL P220 "AMERICAN"

MODEL P220 "AMERICAN"

SPECIFICATIONS
Calibers: 38 Super, 45 ACP
Capacity: 9 rounds; 7 rounds in 45 ACP
Barrel length: 4.4" **Overall length:** 7.79"
Weight (empty): 26½ oz.; 25.7 oz. in 45 ACP
Finish: Blue or K-Kote
Prices:
45 ACP Blued $780.00
  W/"Siglite" night sights 880.00
  W/Nickel slide 805.00
  W/Nickel slide & "Siglite" night sights 905.00
  W/K-Kote 850.00
  W/K-Kote and "Siglite" night sights 950.00

MODEL P225

MODEL P225

SPECIFICATIONS
Caliber: 9mm Parabellum **Capacity:** 8 rounds
Barrel length: 3.85" **Overall length:** 7"
Weight (empty): 26.1 oz.
Finish: Blue or K-Kote
Prices:
Blued finish $775.00
Blued w/"Siglite" night sights 875.00
W/K-Kote 845.00
W/K-Kote and "Siglite" night sights 945.00
W/Nickel slide 800.00
W/Nickel slide and "Siglite" night sights 900.00

MODEL P226

MODEL P226

SPECIFICATIONS
Caliber: 9mm Parabellum
Capacity: 15 rounds (20 rounds optional)
Barrel length: 4.4" **Overall length:** 7¾"
Weight (empty): 26.5 oz.
Finish: Blue or K-Kote
Prices:
Blued finish $805.00
Blued finish, nickel slide 830.00
Blued w/"Siglite" night sights 905.00
W/K-Kote 875.00
K-Kote w/"Siglite" night sights 975.00
W/Nickel slide 830.00
W/Nickel slide & "Siglite" night sights 930.00
Blued finish, double action only 805.00
  W/"Siglite" night sight, DAO 905.00
  W/Nickel slide, DAO 830.00
  W/K-Kote, DAO 875.00
  W/K-Kote, "Siglite" night sight, DAO 975.00

SIG SAUER PISTOLS

MODEL P228

MODEL P228

SPECIFICATIONS
Caliber: 9mm **Capacity:** 13 rounds
Barrel length: 3.86″ **Overall length:** 7.08″
Weight (empty): 26.1 oz.
Finish: Blue or K-Kote
Prices:
Blued finish **$805.00**
Blued w/"Siglite" night sights 905.00
W/Nickel slide 830.00
W/Nickel slide & "Siglite" night sights 930.00
W/K-Kote 875.00
W/K-Kote and "Siglite" night sights 975.00
Double Action Only
  Blued finish 805.00
  Blued finish, "Siglite" night sights 905.00
  Nickel slide 830.00
  Nickel slide, "Siglite" night sights 930.00

MODEL P229

MODEL P229

SPECIFICATIONS
Calibers: 9mm, 40 S&W **Capacity:** 12 rounds
Barrel length: 3.86″ **Overall length:** 7.08″
Weight (empty): 27.54 oz.
Finish: Stainless steel, black frame in aluminum alloy
Features: Stainless steel slide; DA/SA or DA only; automatic
  firing pin lock
Prices:
Model P229 . . . . **$875.00**
  W/"Siglite night sight **975.00**

MODEL P230

MODEL P230

SPECIFICATIONS
Caliber: 9mm Short (380 ACP)
Capacity: 7 rounds
Barrel length: 3.6″ **Overall length:** 6.6″
Weight (empty): 16¼ oz.; 20.8 oz. in stainless steel
Finish: Blue or stainless steel
Prices:
Blued finish **$510.00**
Stainless steel 595.00

SMITH & WESSON PISTOLS

COMPACT SERIES

MODEL 3900 COMPACT SERIES

SPECIFICATIONS
Caliber: 9mm Parabellum DA Autoloading Luger
Capacity: 8 rounds
Barrel length: 3¹/₂″
Overall length: 6⁷/₈″
Weight (empty): 25 oz.
Sights: Post w/white dot front; fixed rear adj. for windage only w/2 white dots. Adjustable sight models include micrometer click, adj. for windage and elevation w/2 white dots. Deduct **$25** for fixed sights.
Finish: Blue (Model 3914); satin stainless (Model 3913)
Prices:
MODEL 3913 . . . . **$622.00**
MODEL 3914 . . . . **562.00**
MODEL 3913 LADYSMITH (stainless) **640.00**
MODEL 3913NL (DA, stainless slide, alloy frame) . . **622.00**
MODEL 3953 (Double action only, stainless) **622.00**

MODEL 3913 DA
Fixed Sight

MODEL 4000 COMPACT SERIES (not shown)

SPECIFICATIONS
Caliber: 40 S&W
Capacity: 8 rounds
Barrel length: 3¹/₂″
Overall length: 7″
Weight: 27 oz.
Sights: White dot front; fixed w/two-dot rear
Price: **$722.00**

MODEL 4500 COMPACT SERIES

SPECIFICATIONS
Caliber: 45 ACP
Capacity: 7 rounds
Barrel length: 3³/₄″
Overall length: 7¹/₄″
Weight: 34 oz.
Sights: White dot front; fixed two-dot rear
Price: **$774.00**

MODEL 4516 COMPACT

MODEL 6900 COMPACT SERIES

SPECIFICATIONS
Caliber: 9mm Parabellum DA Autoloading Luger
Capacity: 12 rounds
Barrel length: 3¹/₂″
Overall length: 6⁷/₈″
Weight (empty): 26¹/₂ oz.
Sights: Post w/white dot front; fixed rear, adj. for windage only w/2 white dots
Stocks: Delrin one-piece wraparound, arched backstrap, textured surface
Finish: Blue (Model 6904); clear anodized/satin stainless (Model 6906)
Prices:
MODEL 6904 . . . . **$614.00**
MODEL 6906 . . . . **677.00**
MODEL 6906 Fixed Novak night sight **788.00**
MODEL 6946 DA only, fixed sights **677.00**

MODEL 6904 DA
Fixed Sight

SMITH & WESSON PISTOLS

FULL-SIZE DOUBLE-ACTION PISTOLS

Smith & Wesson's double-action semiautomatic Third Generation line includes the following features: fixed barrel bushing for greater accuracy • smoother trigger pull plus a slimmer, contoured grip and lateral relief cut where trigger guard meets frame • three-dot sights • wraparound grips • beveled magazine well for easier reloading • ambidextrous safety lever secured by spring-loaded plunger • low-glare bead-blasted finish.

MODEL 4006
With Fixed Sight

MODEL 4046

MODEL 4000 SERIES

SPECIFICATIONS
Caliber: 40 S&W
Capacity: 11 rounds
Barrel length: 4″
Overall length: 7⁷/₈″
Weight: 38¹/₂ oz. (with fixed sights)
Sights: Post w/white dot front; fixed or adjustable w/white two-dot rear
Stocks: Straight backstrap, Xenoy wraparound
Finish: Stainless steel
Prices:
MODEL 4006 w/fixed sights $745.00
   Same as above w/adj. sights 775.00
   w/fixed night sight 855.00
MODEL 4043 Double action only 727.00
MODEL 4046 w/fixed sights, DA only 745.00
   Double action only, fixed Tritium night sight 855.00

MODEL 4500 SERIES

SPECIFICATIONS
Caliber: 45 ACP Autoloading DA
Capacity: 8 rounds (Model 4506); 7 rounds (Model 4516)
Barrel length: 5″ (Model 4506); 3³/₄″ (Model 4516)
Overall length: 8⁵/₈″ (Model 4506); 7¹/₈″ (Model 4516)
Weight (empty): 38¹/₂ oz. (Model 4506); 34¹/₂ oz. (Model 4516)
Sights: Post w/white dot front; fixed rear, adj. for windage only. Adj. sight incl. micrometer click, adj. for windage and elevation w/2 white dots. Add **$29.00** for adj. sights.
Stocks: Delrin one-piece wraparound, arched backstrap, textured surface
Finish: Satin stainless
Prices:
MODEL 4506 w/adj. sights, 5″ bbl. $806.00
   With fixed sights 774.00
MODEL 4566 w/4¹/₄″ bbl., fixed sights 774.00
MODEL 4586 DA only, 4¹/₄″ bbl., fixed sights 774.00

MODEL 4506 DA
Fixed Sight

SMITH & WESSON PISTOLS

FULL-SIZE DOUBLE ACTION PISTOLS

MODEL 5900 SERIES

SPECIFICATIONS
Caliber: 9mm Parabellum DA Autoloading Luger
Capacity: 15 rounds
Barrel length: 4″
Overall length: 7 1/2″
Weight (empty): 28 1/2 oz. (Models 5903, 5904, 5946); 37 1/2 oz. (Model 5906); 38 oz. (Model 5906 w/adj. sight)
Sights: Post w/white dot front; fixed rear, adj. for windage only w/2 white dots. Adjustable sight models include micrometer click, adj. for windage and elevation w/2 white dots.
Finish: Blue (Model 5904); satin stainless (Models 5903 and 5906)
Prices:

MODEL 5903 Satin stainless	$690.00
MODEL 5904 Blue	642.00
MODEL 5906 Satin stainless	742.00
With fixed sights	707.00
With Tritium night sight	817.00
MODEL 5946 Double action only	707.00

**MODEL 5904 DA
Fixed Sight**

MODEL 411

SPECIFICATIONS
Caliber: 40 S&W
Capacity: 11 rounds + 1
Barrel length: 4″
Overall length: 7 1/2″
Weight: 29.4 oz.
Sights: Post w/white dot front; fixed rear
Grips: One-piece Xenoy straight backstrap
Features: Right-hand slide-mounted manual safety; decocking lever; aluminum alloy frame; blue carbon steel slide; non-reflective finish
Price: $525.00

MODEL 411

MODEL 915 9mm

SPECIFICATIONS
Caliber: 9mm
Capacity: 15 rounds
Barrel length: 4″
Overall length: 7 1/2″
Weight: 28 1/2 oz.
Sights: Post w/white dot front; fixed rear
Grips: One-piece Xenoy straight backstrap
Features: Right-hand slide-mounted manual safety; decocking lever; aluminum alloy frame; blue carbon steel slide; non-reflective finish
Price: $467.00

MODEL 915 9mm

SMITH & WESSON TARGET PISTOLS

MODEL NO. 41 RIMFIRE
$753.00 (Blue Only)

SPECIFICATIONS
Caliber: 22 Long Rifle
Magazine capacity: 12 rounds
Barrel lengths: 5$\frac{1}{2}$″ and 7″
Weight: 44 oz. (5$\frac{1}{2}$″ barrel)
Sights: Front, $\frac{1}{8}$″ Patridge undercut; rear, S&W micrometer click sight adjustable for windage and elevation
Stocks: Checkered walnut with modified thumb rest, equally adaptable to right- or left-handed shooters
Finish: S&W Bright blue
Trigger: .365″ width; with S&W grooving and an adjustable trigger stop

MODEL NO. 41

MODEL 422 RIMFIRE 22 SA
$235.00 (Fixed Sight)
$290.00 (Adjustable Sight)

SPECIFICATIONS
Caliber: 22 LR
Capacity: 12 rounds (magazine furnished)
Barrel lengths: 4$\frac{1}{2}$″ and 6″
Overall length: 7$\frac{1}{2}$″ (4$\frac{1}{2}$″ barrel) and 9″ (6″ barrel)
Weight: 22 oz. (4$\frac{1}{2}$″ barrel) and 23 oz. (6″ barrel)
Stock: Plastic (field version) and checkered walnut w/S&W monogram (target version)
Front sight: Serrated ramp w/.125″ blade (field version); Patridge w/.125″ blade (target version)
Rear sight: Fixed sight w/.125″ blade (field version): adjustable sight w/.125″ blade (target version)
Hammer: .250″ internal
Trigger: .312″ serrated
Also available:
MODEL 622. Same specifications as Model 422 in stainless steel. **Price: $284.00.** (Add **$53.00** for adj. sights).

MODEL 422

MODEL 2213/2214 RIMFIRE
"SPORTSMAN" (not shown)
$269.00 (Blue) $314.00 (Stainless)

SPECIFICATIONS
Caliber: 22 LR. **Capacity:** 8 rounds. **Barrel length:** 3″. **Overall length:** 6$\frac{1}{8}$″. **Weight:** 18 oz. **Finish:** Blue carbon steel slide and alloy frame (**Model 2213** has stainless steel slide w/alloy frame)

MODEL 2206
$327.00 (Fixed Sights)
$385.00 (Adj. Sights)

SPECIFICATIONS
Caliber: 22 LR. **Capacity:** 12 rounds. **Barrel length:** 6″. **Overall length:** 9″. **Weight:** 39 oz. **Finish:** Stainless steel.
Also available:
Model 2206 TARGET w/adj. target sight, drilled and tapped. **Price: $433.00**

MODEL 2206 TARGET

SMITH & WESSON REVOLVERS

SMALL FRAME

LADYSMITH HANDGUNS
MODEL 36-LS $404.00 (Blue)
MODEL 60-LS $456.00 (Stainless)

SPECIFICATIONS
Caliber: 38
Capacity: 5 shots
Barrel length: 2″
Overall length: 6⁵/₁₆″
Weight: 20 oz.
Sights: Serrated ramp front; fixed notch rear
Grips: Contoured laminated rosewood
Finish: Glossy deep blue or stainless
Features: Both models come with soft-side LadySmith carry case

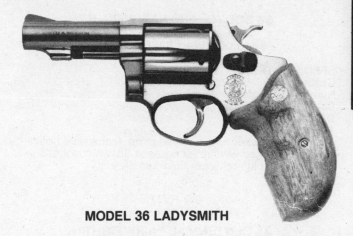

MODEL 36 LADYSMITH

MODEL 36
38 CHIEFS SPECIAL
$374.00

SPECIFICATIONS
Caliber: 38 S&W Special
Number of shots: 5
Barrel length: 2″ or 3″
Overall length: 6¹/₂″ with 2″ barrel
Weight: 19¹/₂ oz. (2″ barrel); 21¹/₂ oz. (3″ barrel)
Sights: Serrated ramp front; fixed square notch rear
Grips: Checkered walnut Service
Finish: S&W blue carbon steel or nickel
Features: .312″ smooth combat-style trigger; .240″ service hammer
MODEL 37 CHIEFS SPECIAL AIRWEIGHT: Same as Model 36, except finish is blue or nickel aluminum alloy; wt. 13¹/₂ oz.; 2″ barrel only
Blue **$408.00**
With nickel finish **424.00**

MODEL 36
38 CHIEFS SPECIAL

MODEL 60
38 CHIEFS SPECIAL STAINLESS
$427.00 (2″ Barrel) $453.00 (3″ Barrel)

SPECIFICATIONS
Caliber: 38 S&W Special
Number of shots: 5
Barrel lengths: 2″ and 3″
Overall length: 6⁵/₁₆″ (2″ barrel); 7¹/₂″ (3″ barrel)
Weight: 19¹/₂ oz. (2″ barrel); 21¹/₂ oz. (3″ barrel); 24¹/₂ oz. (3″ full lug barrel)
Sights: Micrometer click rear, adj. for windage and elevation; pinned black front (3″ full lug model only); standard sights as on Model 36
Grips: Checked walnut Service with S&W monograms; Santoprene combat-style on 3″ full lug model
Finish: Stainless steel
Features: .312″ smooth combat-style trigger (.347″ serrated trigger on 3″ full lug model); .240″ service hammer (.375″ semi-target hammer on 3″ full lug model)

MODEL 60
38 CHIEFS SPECIAL

SMITH & WESSON REVOLVERS

SMALL FRAME

38 BODYGUARD "AIRWEIGHT"
MODEL 38
$440.00 Blue $455.00 Nickel

SPECIFICATIONS
Caliber: 38 S&W Special
Number of shots: 5
Barrel length: 2″
Overall length: 6³/₈″
Weight: 14 oz.
Sights: Front, fixed ¹/₁₀″ serrated ramp; rear square notch
Stocks: Checked walnut Service with S&W monograms
Finish: S&W blue or nickel aluminum alloy

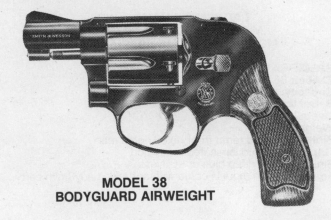

**MODEL 38
BODYGUARD AIRWEIGHT**

38 CENTENNIAL "AIRWEIGHT"
MODEL 442 (not shown)
$423.00 (Blue) $438.00 (Nickel)

SPECIFICATIONS
Caliber: 38 Special
Capacity: 5 rounds
Barrel length: 2″
Overall length: 6⁷/₁₆″
Weight: 15.8 oz.
Sights: Serrated ramp front; fixed square notch rear
Finish: Matte blue or satin nickel

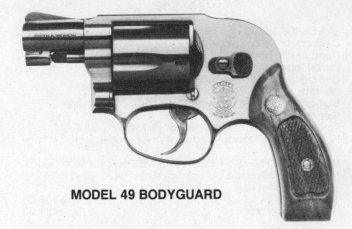

MODEL 49 BODYGUARD

38 BODYGUARD MODEL 49
$405.00

SPECIFICATIONS
Caliber: 38 S&W Special
Number of shots: 5
Barrel length: 2″
Overall length: 6¹/₄″
Weight (empty): 20 oz.
Sights: Serrated ramp (front); fixed square notch (rear)
Finish: S&W blue

MODEL 649 BODYGUARD
$464.00

SPECIFICATIONS
Caliber: 38 Special
Capacity: 5 shots
Barrel length: 2″
Overall length: 6¹/₄″
Weight: 20 oz.
Sights: Serrated ramp front, fixed square notch rear
Grips: Round butt; checkered walnut Service
Finish: Stainless steel

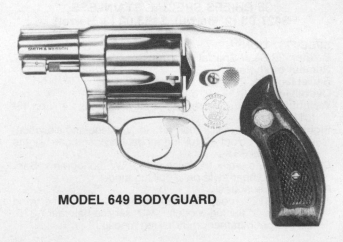

MODEL 649 BODYGUARD

SMITH & WESSON REVOLVERS

SMALL FRAME

MODEL 63 22/32 KIT GUN
$453.00 (2″ Barrel)
$458.00 (4″ Barrel)

SPECIFICATIONS
Caliber: 22 Long Rifle
Number of shots: 6
Barrel lengths: 2″ and 4″
Weight: 22 oz. (2″ barrel); 24½ oz. (4″ barrel)
Sights: ⅛″ red ramp front sight; rear sight is black stainless steel S&W micrometer click square-notch, adjustable for windage and elevation
Stocks: Square butt
Finish: Satin stainless

**MODEL 63
22/32 KIT GUN**

MODEL 640 CENTENNIAL (not shown)
$464.00

SPECIFICATIONS
Caliber: 38 S&W Special (+P))
Capacity: 5 rounds
Barrel length: 2″
Overall length: 6⁵/₁₆″
Weight: 21 oz.
Sights: Serrated ramp front; fixed square notch rear
Features: Fully concealed hammer; smooth hardwood service stock

MODEL 940 CENTENNIAL
$470.00

SPECIFICATIONS
Caliber: 9mm Parabellum
Capacity: 5 rounds
Barrel length: 2″
Overall length: 6⁷/₁₆″
Weight: 23 oz.
Sights: Serrated ramp front; fixed square notch rear
Grips: Santoprene combat grips
Feature: Fully concealed hammer

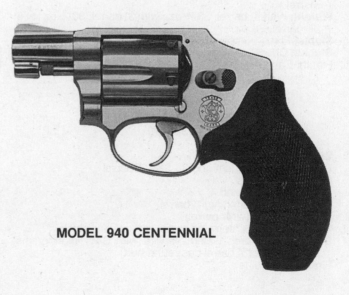

MODEL 940 CENTENNIAL

22 MAGNUM KIT GUN
MODEL 651 (not shown)
$455.00

SPECIFICATIONS
Caliber: 22 Magnum
Number of shots: 6
Barrel length: 4″
Overall length: 8¹¹/₁₆″
Weight: 24½ oz.
Sights: Red ramp front; micrometer click rear, adjustable for windage and elevation
Grips: Checkered premium hardwood Service
Finish: Stainless steel
Features: .375″ hammer; .312″ smooth combat trigger

HANDGUNS

SMITH & WESSON REVOLVERS

MEDIUM FRAME

38 MILITARY & POLICE
MODEL 10
$368.00 (2″ Bbl.) $375.00 (4″ Bbl.)

SPECIFICATIONS
Caliber: 38 S&W Special
Capacity: 6 shots
Barrel length: 2″, 4″ (also 4″ heavy barrel)
Weight: 33½ oz. with 4″ barrel
Sights: Front, fixed ⅛″ serrated ramp; rear square notch
Stocks: Checkered walnut Service with S&W monograms, round or square butt
Finish: S&W blue

MODEL 10

38 MILITARY & POLICE STAINLESS
MODEL 64
$411.00 (2″ Bbl.) $419.00 (3″ & 4″ Bbl.)

SPECIFICATIONS
Calibers: 38 S&W Special, 38 S&W Special Mid Range
Capacity: 6 shots
Barrel length: 4″ heavy barrel, square butt; 3″ heavy barrel, round butt; 2″ regular barrel, round butt
Overall length: 9¼″ w/4″ barrel; 7⅞″ w/3″ barrel; 6⅞″ w/2″ barrel
Weight: With 4″ barrel, 33½ oz.; with 3″ barrel, 30½ oz.; with 2″ barrel, 28 oz.
Sights: Fixed, ⅛″ serrated ramp front; square notch rear
Stocks: Checkered walnut Service with S&W monograms
Finish: Satin stainless

MODEL 64

357 MILITARY & POLICE
MODEL 13 (HEAVY BARREL)
$386.00

SPECIFICATIONS
Caliber: 357 Magnum and 38 S&W Special
Capacity: 6 shots
Barrel length: 3″ and 4″
Overall length: 9¼″ (w/4″ barrel)
Weight: 34 oz. (w/4″ barrel)
Sights: Front, ⅛″ serrated ramp; rear square notch
Stocks: Checkered walnut service with S&W monograms, square butt (3″ barrel has round butt)
Finish: S&W blue

MODEL 13

357 MILITARY & POLICE
MODEL 65 (HEAVY BARREL)
$423.00

SPECIFICATIONS
Same specifications as Model 13, except Model 65 is stainless steel. Available with matte finish.
Also available:
MODEL 65 LADYSMITH. Same specifications as model 65 but with 3″ barrel only. Also features rosewood laminate stock, glass beaded. **Price $456.00**

MODEL 65

SMITH & WESSON REVOLVERS

MEDIUM FRAME

K-38 MASTERPIECE
MODEL 14
$461.00

MODEL 14

SPECIFICATIONS
Caliber: 38 S&W Special
Barrel length: 6″ full lug barrel
Overall length: 11 1/8″
Weight: 47 oz.
Sights: Micrometer click rear, adjustable for windage and elevation; pinned black Patridge-style front
Grips: Combat-style premium hardwood
Finish: Blue carbon steel
Features: .500 target hammer; .312″ smooth combat trigger

38 COMBAT MASTERPIECE
MODEL 15
$407.00

MODEL 15

SPECIFICATIONS
Caliber: 38 S&W Special
Number of shots: 6
Barrel length: 4″
Overall length: 9 5/16″
Weight (loaded): 32 oz.
Sights: Serrated ramp front; S&W micrometer click sight adjustable for windage and elevation
Stocks: Checkered walnut service with S&W monograms
Finish: S&W blue
Features: .375″ semi-target hammer; .312″ smooth combat-style trigger
Also available:
MODEL 67. Same specifications as above but w/red ramp front sight and adj. rear sight; stainless steel. **Price: $462.00**

K-22 MASTERPIECE
MODEL 617
$455.00 (4″ barrel) $460.00 (6″ barrel)
$485.00 (6″ full lug barrel w/target trigger, hammer)
$496.00 (8 3/4″ barrel)

MODEL 17

SPECIFICATIONS
Caliber: 22 Long Rifle
Number of shots: 6
Barrel length: 4″, 6″ or 8 3/8″
Overall length: 9 1/8″ (4″ barrel); 11 1/8″ (6″ barrel); 13 1/2″ (8 3/8″ barrel)
Weight (loaded): 42 oz. with 4″ barrel; 48 oz. with 6″ barrel; 54 oz. with 8 3/8″ barrel
Sights: Front, 1/8″ plain Patridge; rear, S&W micrometer click sight adjustable for windage and elevation
Stocks: Checkered walnut Service with S&W monograms
Finish: S&W blue
Also available:
MODEL 648. Same as Model 617 but in 22 Magnum (6″ barrel; 47 oz.) $464.00

SMITH & WESSON REVOLVERS

MEDIUM FRAME

357 COMBAT MAGNUM
MODEL 19
$408.00 (2½″ Bbl.) $417.00 (4″ Bbl.)
$421.00 (6″ Barrel)

MODEL 19

SPECIFICATIONS
Caliber: 357 S&W Magnum (actual bullet dia. 38 S&W Spec.)
Number of shots: 6
Barrel length: 2½″, 4″ and 6″
Overall length: 9½″ with 4″ barrel; 7½″ with 2½″ barrel; 11⅜″ with 6″ barrel
Weight: 30½ oz. (2½″ barrel); 36 oz. (4″ barrel); 39 oz. (6″ barrel)
Sights: Front, ⅛″ Baughman Quick Draw on 2½″ or 4″ barrel, ⅛″ Patridge on 6″ barrel; rear, S&W micrometer click sight adjustable for windage and elevation
Stocks: Checkered Goncalo Alves Target with S&W monograms
Finish: S&W bright blue
Features: 2½″ barrel has round butt and synthetic grip
Also available with red ramp front sight, insert and white outline rear. **$443.00** (4″ barrel only)

MODEL 66

357 COMBAT MAGNUM
MODEL 66
$461.00 (2½″ Bbl.) $467.00 (4″ & 6″ Bbl.)

SPECIFICATIONS
Caliber: 357 Magnum (actual bullet dia. 38 S&W Spec.)
Number of shots: 6
Barrel length: 4″ or 6″ with square butt; 2½″ with round butt
Overall length: 7½″ with 2½″ barrel; 9½″ with 4″ barrel; 11⅜″ with 6″ barrel
Weight: 30½ oz. with 2½″ barrel; 36 oz. with 4″ barrel; 39 oz. with 6″ barrel
Sights: Front: ⅛″. Rear: S&W Red Ramp on ramp base, S&W Micrometer Click sight, adjustable for windage and elevation; for white outline rear sight, add **$5.00**
Stocks: Checked Goncalo Alves target with square butt with S&W monograms
Finish: Satin stainless
Trigger: S&W grooving with an adjustable trigger stop
Ammunition: 357 S&W Magnum, 38 S&W Special Hi-Speed, 38 S&W Special, 38 S&W Special Mid Range

DISTINGUISHED COMBAT MAGNUM
MODEL 586
$457.00 (4″ Bbl.) $461.00 (6″ Bbl.)

SPECIFICATIONS
Caliber: 357 Magnum
Capacity: 6 shots
Barrel lengths: 4″ and 6″
Overall length: 9 9/16″ with 4″ barrel; 11 5/16″ with 6″ barrel
Weight: 41 oz. with 4″ barrel; 46 oz. with 6″ barrel
Sights: Front is S&W Red Ramp; rear is S&W Micrometer Click adjustable for windage and elevation; White outline notch. Option with 6″ barrel only—plain Patridge front with black outline notch; for white outline rear sight, add **$4.00**
Stocks: Checkered Goncalo Alves with speedloader cutaway
Finish: S&W Blue
MODEL 686: Same as Model 586, except also available with 2½″ barrel (35¾ oz.) and 8⅜″ barrel (53 oz.). All models have stainless steel finish, combat or target stock and/or trigger; adj. sights optional. **Price: $476.00 to $510.00**

MODEL 586

SMITH & WESSON REVOLVERS

LARGE FRAME

357 MAGNUM MODEL 27
$486.00

SPECIFICATIONS
Caliber: 357 Magnum (actual bullet dia. 38 S&W Spec.)
Number of shots: 6
Barrel length: 6″
Overall length: 11⁵/₁₆″
Weight: 45¹/₂ oz.
Sights: Front, Patridge on ramp base; rear, S&W Micrometer Click sight adjustable for windage and elevation
Stocks: Checkered hardwood target
Frame: Finely checked top strap and barrel rib
Finish: Blue carbon steel

MODEL 27

44 MAGNUM MODEL 29
$549.00 (6″ Bbl.) $560.00 (8³/₈″ Bbl.)

SPECIFICATIONS
Caliber: 44 Magnum
Number of shots: 6
Barrel lengths: 6″ and 8³/₈″
Overall length: 11⁷/₈″ with 6″ barrel; 13⁷/₈″ with 8³/₈″ barrel
Weight: 47 oz. with 6″ barrel; 51¹/₂ oz. with 8³/₈″ barrel
Sights: Front, Red Ramp on ramp base; rear, S&W Micrometer Click sight adjustable for windage and elevation; white outline notch
Stocks: Checkered, highly grained hardwood target type
Hammer: Checkered target type
Trigger: Grooved target type
Finish: Blue carbon steel
Also available:
MODEL 29 CLASSIC in blue carbon steel w/full lug barrel, interchangeable front sights, Hogue combat rubber grips, round butt.
　With 5″ or 6¹/₂″ barrel $591.00
　With 8³/₈″ barrel 603.00

MODEL 29

MODEL 629
$581.00 (4″ Bbl.) $586.00 (6″ Bbl.)
$600.00 (8³/₈″ Barrel)

SPECIFICATIONS
Calibers: 44 Magnum, 44 S&W Special
Capacity: 6 shots
Barrel lengths: 4″, 6″, 8³/₈″
Overall length: 9⁵/₈″, 11³/₈″, 13⁷/₈″
Weight (empty): 44 oz. (4″); 47 oz. (6″ barrel); 51¹/₂ oz. (8³/₈″)
Sights: S&W Red Ramp front; plain blade rear w/S&W Micrometer Click; adj. for windage and elevation; scope mount
Stock: Checkered hardwood target or synthetic
Finish: Stainless steel
Also available:
MODEL 629 BACKPACKER. With 3″ barrel (40 oz. weight), round butt and rubber grip. **Price:** $581.00.

MODEL 629 BACKPACKER

SMITH & WESSON REVOLVERS

LARGE FRAME

MODEL 629 CLASSIC
$623.00 (5″ & 6¹/₂″ Bbl.)
$643.00 (8³/₈″ Bbl.)

SPECIFICATIONS
Calibers: 44 Magnum, 44 S&W Special
Capacity: 6 rounds
Barrel length: 5″, 6¹/₂″, 8³/₈″
Overall Length: 10¹/₂″, 12″, 13⁷/₈″
Weight: 51 oz.(5″ barrel); 52 oz.(6¹/₂″ barrel); 54 oz. (8³/₈″ barrel)
Also available:
MODEL 629 CLASSIC DX. Same features as Model 629 Classic above, plus two sets of grips and five interchangeable front sights and proof target.
With 6¹/₂″ barrel $803.00
With 8³/₈″ barrel 829.00

MODEL 629 CLASSIC DX

MODEL 657 STAINLESS
$523.00

SPECIFICATIONS
Caliber: 41 Magnum
Capacity: 6 shots
Barrel length: 6″
Overall length: 11³/₈″
Weight (empty): 48 oz.
Sights: Serrated ramp on ramp base (front); Blue S&W Micrometer Click sight adj. for windage and elevation (rear)
Finish: Satin stainless steel

MODEL 657

MODEL 625
$591.00

SPECIFICATIONS
Caliber: 45 ACP
Capacity: 6 shots
Barrel length: 5″ full lug barrel
Overall length: 10³/₈″
Weight (empty): 45 oz.
Sights: Front, Patridge on ramp base; S&W micrometer click rear, adjustable for windage and elevation
Stock: Pachmayr SK/GR gripper, round butt
Finish: Stainless steel

MODEL 625

SPRINGFIELD PISTOLS

**MODEL 1911-A1
STANDARD**

**MODEL 1911-A1
MIL-SPEC COMPACT 45 ACP**

**MODEL 1911-A1
TROPHY MATCH 45 ACP**

**MODEL 1911-A1
CHAMPION**

MODEL 1911-A1 STANDARD

An exact duplicate of the M1911-A1 pistol that served the U.S. Armed Forces for more than 70 years, this model has been precision manufactured from forged parts, including a forged frame, then hand-assembled.

SPECIFICATIONS
Calibers: 9mm Parabellum, 38 Super and 45 ACP
Capacity: 9 + 1 in chamber (9mm/38 Super); 8 + 1 in chamber (45 ACP)
Barrel length: 5.04″   **Overall length:** 8.59″
Weight: 35.62 oz.
Trigger pull: 5 to 6.5 lbs.
Sight radius: 6.281″
Rifling: 1 turn in 16; right-hand, 4-groove (9mm); left-hand, 6-groove (45 ACP)

MODEL 1911-A1	Prices
45 ACP/9mm, Blued	$489.00
45 ACP, Parkerized finish	449.00
45 ACP, Stainless finish	532.00
45 ACP/9mm, Bi-Tone finish	829.00
9mm, Blued	529.00
9mm Stainless finish	569.00

1911-A1 TROPHY MATCH 45 ACP	
Blued	**899.00**
Stainless Steel	**936.00**
1911-A1 DEFENDER w/fixed combat sights, bobbed hammer, walnut grips, beveled magazine well, extended thumb safety, 45 ACP, Bi-Tone	**959.00**
1911-A1 CHAMPION with 1/2″ shortened slide and barrel, 45 ACP, stainless finish	**558.00**
With Blued finish	**529.00**
With Compensator	**829.00**
Mil-Spec 45 ACP Parkerized	**449.00**
1911-A1 38 SUPER	
Blued finish	**529.00**
Parkerized	**489.00**
1911-A1 COMPACT 45 ACP	
Blued finish	**529.00**
Stainless finish	**558.00**
Lightweight 45 ACP w/blued finish	**499.00**
Mil-Spec 45 ACP Parkerized	**449.00**
1911-A1 FACTORY COMP Blued	
In 38 Super	**929.00**
In 45 ACP	**869.00**

SPRINGFIELD PISTOLS

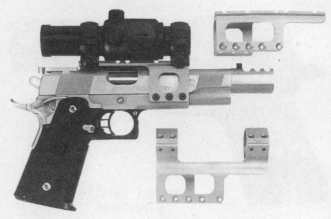

**MODEL XM4™ HIGH-CAPACITY
1911-A1 PISTOL**

MODEL XM4™ HIGH-CAPACITY 1911-A1

A special joint venture between a leading Israeli arms maker and Springfield, Inc., has resulted in a significant advance in the field of high-capacity semiautomatic combat pistol design. Known as the XM4™, this hybrid pistol consists of Sringfield's 1911-A1 stainless-steel slide combined with an Israeli 1911-A1 high-capacity frame made of a high-strength, lightweight and durable polymer material, called Xanex™, which is impervious to solvents and chemicals commonly used in gun cleaning—and can therefore never rust.

The XM4™ stainless-steel slide assembly is fit precisely to the stainless-steel frame rails, the same as all Springfield 1911 pistols. It is available in a Standard Model with a 5″ barrel or in a shortened Champion Model with a Commander-length slide. The XM4™ also uses Springfield's proprietary custom modifications for elevating the shooting hand as high as possible on the pistol, known as the High Hand™ grip. The pistol's frame contains a double-column magazine. The trigger reach (distance from rear of the frame's grip to the trigger) is the same as a standard 1911-A1, as is the length of the grip (distance from front strap to mainspring housing). The top of the stainless-steel slide is finely serrated to reduce glare and enhance appearance. The sights are rugged combat sights with three dots for low-light shooting conditions. The ejection port has been lowered and flared for reliable ejection. Other features and specifications include:

SPECIFICATIONS
Calibers: 45 ACP and 9mm
Capacity: 13+1 (45 ACP); 18+1 (9mm)
Barrel length: 4¼″ (Champion); 5″ (Standard)
Overall length: 7¾″ (Champion); 8¼″ (Standard)
Grip width: 1¼″
Weight: 25⅞ oz. (Champion); 29¼ oz. (Standard)
Sight radius: 5¾″

Prices:
Standard $689.00
Champion 699.00

HIGH-CAPACITY FULL-HOUSE RACEGUN
$2830.00

SPECIFICATIONS
Calibers: 45 ACP, 9×25 Dillon, 38 Super (custom calibers available on request)
Capacity: 15+1 rounds (45 ACP); 17+1 rounds (9×25 Dillon); 17+1 (38 Super)
Barrel: Custom match-grade barrel, match-grade chamber and leade
Sights: Bo-Mar rear, dovetail front; Springfield scope mount (optional extra); C-More "heads-up display" optical sight (optional extra)
Safeties: Extended ambidextrous thumb safety, recessed hammer beavertail grip safety
Compensator: Progressive triple port expansion chamber design, tapered cone system
Trigger pull: 3½ lbs.
Checkering: 20 or 30 lines per inch (custom on request) on frontstrap, trigger guard, mainspring housing

Other custom Raceguns available include:
Compact Comp 1911-A1 and LIGHTWEIGHT COMPACT COMP (45 ACP)
Trophy Master Distinguished (45 ACP & 38 Super)
Trophy Master Distinguished "Limited" (45 ACP)
Trophy Master Expert (45 ACP, 9mm, 38 Super)
Springfield Custom "Carry Gun" (45 ACP)
Basic Competition Pistol (45 ACP)
National Match Hardball Pistol (45 ACP)
Bull's-Eye Wadcutter Pistol (45 ACP)
Prices on request. For address and phone number, see Directory of Manufacturers & Suppliers.

STAR AUTOMATIC PISTOLS

FIRESTAR PLUS

MODELS M40, M43 & M45 FIRESTAR
9mm Parabellum, 40 S&W or 45 ACP

This pocket-sized Firestar pistol features all-steel construction, a triple-dot sight system (fully adjustable rear), and ambidextrous safety. The Acculine barrel design reseats and locks the barrel after each shot. Checkered rubber grips.

SPECIFICATIONS
Barrel length: 3.39″ (3.6″ 45 ACP)
Overall length: 6¹/₂″ (6.85″ 45 ACP)
Weight: 30.35 oz. (35 oz. 45 ACP)
Capacity: 7 rounds (6 rounds in 40 S&W and 45 ACP)
Prices:
Firestar M40 Blue finish, 40 S&W **$471.00**
   Starvel finish **497.00**
Firestar M43 Blue finish. 9mm Para. **453.00**
   Starvel finish, 9mm Para. **480.00**
Firestar M45 Blue finish, 45 ACP **494.00**
   Starvel finish, 45 ACP **521.00**

MEGASTAR

MEGASTAR

The Megastar is the first production pistol designed especially to tame the potent 10mm magnum cartridge. Recoil is mild and durability enhanced. The fast button release magazine delivers 14 rounds in 10mm (12 in 45 ACP). All-steel Megastars are made from precision-machined castings and feature Star's reverse taper "Acculine" barrel contour for out-of-box accuracy. Available in blue or rust-resistant Starvel finish with decocking lever. Polymer grips provide non-slip surface for positive control.

SPECIFICATIONS
Caliber: 45 ACP and 10 mm magnum
Capacity: 12 rounds (45 ACP); 14 rounds (10mm)
Barrel length: 4.6″   **Overall length:** 8¹/₂″
Weight: 47.6 oz.
Prices:
Blued finish **$653.00**
Starvel finish **682.00**

ULTRASTAR

The new Ultrastar features a slim profile, light weight and first shot double-action speed. The use of polymers makes this pistol exceptionally strong and durable. Other features include a triple-dot sight system, ambidextrous two-position manual safety (safe and safe decock), all-steel internal mechanism and barrel, slide-mounted on rails inside frame.

SPECIFICATIONS
Caliber: 9mm Parabellum   **Capacity:** 9 rounds
Barrel length: 3.57″   **Overall length:** 7″
Weight: 26 oz.
Price: Not set

ULTRASTAR

TAURUS PISTOLS

MODEL PT-58

SPECIFICATIONS
Caliber: 380 ACP
Action: Semiautomatic double action
Capacity: Staggered 13 shot
Barrel length: 4″
Overall length: 7.2″
Weight: 30 oz.
Hammer: Exposed
Sights: Front, drift adjustable; rear, notched bar
  dovetailed to slide, 3-dot combat
Grips: Smooth Brazilian walnut
Finish: Blue, satin nickel or stainless steel
Features: tri-position safety system

MODEL PT-58
$445.00 (Blue)
$477.00 (Nickel)
$506.00 (Stainless)

MODEL PT-92 AF

Caliber: 9mm Parabellum
Action: Semiautomatic double action
Capacity: Staggered 15-shot magazine
Hammer: Exposed
Barrel length: 5″
Overall length: 8½″
Height: 5.39″
Width: 1.45″
Weight: 34 oz. (empty)
Rifling: R.H., 6 grooves
Sights: Front, fixed; rear, drift adjustable, 3-dot combat
Safeties: (a) Ambidextrous manual safety locking trigger
  mechanism and slide in locked position; (b) half-cock po-
  sition; (c) inertia operated firing pin; (d) chamber loaded
  indicator
Slide: Hold open upon firing last cartridge
Grips: Smooth Brazilian walnut
Finish: Blue, satin nickel or stainless steel

Also available:
MODEL PT 99. Same specifications as Model PT 92, but has
  micrometer click-adjustable rear sight. **$532.00** (Blue);
  $577.00 (Nickel); **$606.00** (Stainless).

MODEL PT-92 AF
$492.00 (Blue)
$532.00 (Nickel)
$559.00 (Stainless)

MODEL PT-92 AFC

SPECIFICATIONS
Caliber: 9mm Parabellum
Capacity: 13 rounds
Barrel length: 4.25″
Overall length: 7.5″
Weight: 31 oz.
Sights: Fixed front; drift-adjustable rear, 3-dot combat
Stocks: Brazilian hardwood
Slide: Last shot held open
Safety: Manual, ambidextrous hammer drop; inertia firing pin;
  chamber load indicator
Finish: Blue, satin nickel or stainless steel

MODEL PT-92 AFC
$492.00 (Blue)
$559.00 (Stainless)

TAURUS PISTOLS/REVOLVERS

MODEL PT-908
$492.00 (Blue)
$559.00 (Stainless)

SPECIFICATIONS
Caliber: 9mm Parabellum
Action: Double action
Capacity: 8 shot
Barrel length: 3.8″
Overall length: 7.05″
Weight: 30 oz.
Sights: Drift adjustable front and rear, 3-dot combat
Stock: Rubber
Finish: Blue, satin nickel or stainless steel

MODEL PT 101
$542.00 (Blue)
$587.00 (Nickel)
$619.00 (Stainless)

SPECIFICATIONS
Caliber: 40 S&W
Capacity: 15 shots
Barrel length: 5″
Weight: 34 oz. (empty)
Sights: Micrometer click-adjustable rear sight
Grips: Brazilian hardwood
Finish: Blue, satin nickel or stainless steel

Also available:
MODEL PT 100. Same specifications as Model PT 101, but has fixed sights. **Prices: $502.00** (blue); **$542.00** (nickel); **$569.00** (stainless)

MODEL PT 22
$193.00

SPECIFICATIONS
Caliber: 22 LR
Action: Semiautomatic
Capacity: 9 shots
Barrel length: 2³/₄″
Overall length: 5¹/₄″
Weight: 12.3 oz.
Sights: Fixed
Grips: Brazilian hardwood
Finish: Blue

Also available:
MODEL PT 25. Same price and specifications as Model PT 22, except magazine holds 8 rounds in 25 ACP.

MODEL 44

SPECIFICATIONS
Caliber: 44 Mag.   **Capacity:** 6 rounds
Barrel length: 4″ (heavy, solid); 6¹/₂″ and 8³/₈″ (vent. rib)
Weight: 44³/₄ oz. (4″); 52¹/₂ oz. (6¹/₂″); 57¹/₄ oz. (8³/₈″)
Sights: Serrated ramp front; rear micrometer click, adjustable for windage and elevation
Finish: Blue or stainless steel
Stock: Brazilian hardwood
Features: Compensated barrel; transfer bar safety
Prices:

4″ barrel blue	$418.00
stainless steel	480.00
6¹/₂″ and 8³/₈″ blue	435.00
stainless steel	500.00

TAURUS REVOLVERS

MODEL 66
$313.00 (2¹/₂″ Blue)
$323.00 (4″ and 6″ Blue)
$393.00 (Stainless)

SPECIFICATIONS
Caliber: 357 Magnum/38 Special
Action: Double
Capacity: 6 shot
Barrel lengths: 2¹/₂″, 4″, 6″
Weight: 35 oz. (4″ barrel)
Sights: Serrated ramp front; rear, micrometer click
 adjustable for windage and elevation
Stock: Brazilian hardwood
Finish: Royal blue or stainless steel
Prices (add'l.): $323.00 (4″ and 6″ Blue w/recoil compensator);
 $403.00 (4″ and 6″ Stainless w/recoil compensator)

MODEL 83
$260.00 (Blue)
$309.00 (Stainless)

SPECIFICATIONS
Caliber: 38 Special
Action: Double
Number of shots: 6
Barrel length: 4″
Weight: 34 oz.
Sights: Patridge-type front; rear, micrometer click
 adjustable for windage and elevation
Stock: Brazilian hardwood
Finish: Blue

MODEL 85
$276.00 (Blue)
$337.00 (Stainless Steel)

SPECIFICATIONS
Caliber: 38 Special
Capacity: 5 shot
Action: Double
Barrel length: 2″ and 3″
Weight: 21 oz. (2″ barrel)
Sights: Notch rear sight, fixed sight
Stock: Brazilian hardwood
Finish: Blue or stainless steel
Also available:
MODEL 85CH. Same specifications as Model 85, except has
 concealed hammer and 2″ barrel only. **Prices: $276.00**
 (Blue); **$337.00** (Stainless Steel)

MODEL 86 TARGET MASTER
$352.00

SPECIFICATIONS
Caliber: 38 Special
Capacity: 6 shot
Action: Double
Barrel length: 6″
Weight: 34 oz.
Sights: Patridge-type front; rear, micrometer click
 adjustable for windage and elevation
Stock: Brazilian hardwood
Finish: Bright royal blue

TAURUS REVOLVERS

MODEL 80
$248.00 (Blue)
$299.00 (Stainless)

SPECIFICATIONS
Caliber: 38 Special
Capacity: 6 shot
Action: Double
Barrel lengths: 3″, 4″
Weight: 30 oz. (4″ barrel)
Sights: Notched rear; serrated ramp front
Stock: Brazilian hardwood
Finish: Blue or stainless

MODEL 80

MODEL 82
$248.00 (Blue)
$299.00 (Stainless)

SPECIFICATIONS
Caliber: 38 Special
Capacity: 6 shot
Action: Double
Barrel lengths: 3″, 4″
Weight: 34 oz. (4″ barrel)
Sights: Notched rear; serrated ramp front
Stock: Brazilian hardwood
Finish: Blue or stainless

MODEL 82

MODEL 94
$288.00 (Blue)
$339.00 (Stainless)

SPECIFICATIONS
Caliber: 22 LR
Number of shots: 9
Action: Double
Barrel lengths: 3″ and 4″
Weight: 25 oz.
Sights: Serrated ramp front; rear micrometer click
  adjustable for windage and elevation
Stock: Brazilian hardwood
Finish: Blue or stainless steel

Also available:
MODEL 941 in 22 Magnum, 8-shot capacity; ejector shroud.
  In blue **$310.00**
  In stainless steel 367.00

MODEL 941

TAURUS REVOLVERS

MODEL 65 (2¹/₂" Barrel)
$285.00 (Blue)
$358.00 (Stainless)

SPECIFICATIONS
Caliber: 357 Magnum
Capacity: 6 shot
Barrel lengths: 2¹/₂", 4"
Weight: 34 oz.
Sights: Rear square notch; serrated front ramp
Action: Double
Stock: Brazilian hardwood
Finish: Royal blue or stainless

MODEL 96
$352.00

MODEL 96

SPECIFICATIONS
Caliber: 22 LR
Number of shots: 6
Action: Double
Barrel length: 6"
Weight: 34 oz.
Sights: Patridge-type front; rear, micrometer click adjustable for windage and elevation
Stock: Brazilian hardwood
Finish: Blue only

MODEL 431
$281.00 (Blue)
$351.00 (Stainless)

MODEL 431

SPECIFICATIONS
Caliber: 44 Special
Capacity: 5 shots
Action: Double
Barrel length: 3" or 4" w/ejector shroud; heavy, solid rib barrel
Weight: 35 oz. (4" barrel)
Sights: Notched rear; serrated ramp front
Safety: Transfer bar
Stock: Brazilian hardwood
Finish: Blue or stainless steel

MODEL 441
$307.00 (Blue)
$386.00 (Stainless)

MODEL 441

SPECIFICATIONS
Caliber: 44 Special
Capacity: 5 shots
Action: Double
Barrel lengths: 3", 4" or 6" w/ejector shroud; heavy, solid rib barrel
Weight: 40¹/₄ oz. (6" barrel)
Sights: Serrated ramp front; rear, micrometer click adjustable for windage and elevation
Safety: Transfer bar
Stock: Brazilian hardwood
Finish: Blue or stainless steel

TAURUS REVOLVERS

MODEL 669
$322.00 (4″ and 6″ Blue)
$402.00 (4″ and 6″ Stainless)

SPECIFICATIONS
Caliber: 357 Magnum
Capacity: 6 shots
Action: Double
Barrel lengths: 4″ and 6″
Weight: 37 oz. (4″ barrel)
Sights: Serrated ramp front; rear, micrometer click adjustable for windage and elevation
Stock: Brazilian hardwood
Finish: Royal blue or stainless
Features: Recoil compensator (optional) **$19.00** additional

MODEL 669

MODEL 689
$335.00 (Blue)
$416.00 (Stainless)

The Model 689 has the same specifications as the Model 669, except vent rib is featured.

MODEL 689 STAINLESS

MODEL 741
$254.00 (Blue)
$342.00 (Stainless)

SPECIFICATIONS
Caliber: 32 H&R Magnum
Capacity: 6 shot
Action: Double
Barrel length: 3″ or 4″
Weight: 20 oz. (w/3″ barrel)
Sights: Serrated ramp front; rear, micrometer click adjustable for windage and elevation
Stock: Brazilian hardwood
Finish: Blue or stainless steel

MODEL 741

MODEL 761
$326.00 (Blue)

Similar to Model 741, except has 6″ barrel, weight of 34 oz. and blue finish only.

MODEL 761

TEXAS ARMS

DEFENDER DERRINGER
$299.00

This single-action derringer is the first to feature a rebounding hammer and retracting firing pins. This system, along with the positive crossbolt safety, makes this a safer derringer. Its spring-loaded cammed locking lever provides a tighter barrel/frame fit, assuring faster and easier loading and unloading. An interchangeable barrel system allows different caliber barrels to be used on the same gun frame.

SPECIFICATIONS
Calibers: 9mm Luger, 357 Magnum, 38 Special, 44 Magnum, 45 Auto, 45 LC/.410 shot shell
Barrel length: 3″ (octagonal)
Overall length: 5″
Weight: 16 to 21 oz.
Height: 3.75″
Width: 1.25″ across grips
Finish: Bead-blasted, gun-metal grey
Features: Removable trigger guard; automatic shell extractor; improved grip angle distributes recoil

DEFENDER DERRINGER

THOMPSON/CENTER

CONTENDER HUNTER

CONTENDER HUNTER

Chambered in 7-30 Waters, 223 Rem., 30-30 Win., 35 Rem., 45-70 Government, 375 Win. and. 44 Rem. Mag., the most popular commercially loaded cartridges available to handgunners. **Barrel length:** 14″. **Overall length:** 16″. **Weight:** 4 lbs. (approx.). **Features:** T/C Muzzle Tamer (to reduce recoil); a mounted T/C Recoil Proof 2.5X scope with lighted reticle, QD sling swivels and nylon sling, plus suede leather carrying case. **Price:** $740.00

THOMPSON/CENTER

BULL BARREL

CONTENDER
BULL BARREL MODELS

These pistols with 10-inch barrel feature fully adjustable Patridge-style iron sights. All stainless steel models (including the Super "14" and Super "16" below) are equipped with Rynite finger-groove grip with rubber recoil cushion and matching Rynite forend, plus Cougar etching on the steel frame.

Standard and Custom calibers available:
22 LR, 22 Hornet, 22 Win. Mag., 223 Rem., 300 Whisper, 30-30 Win., 32-20 Win., 7mm T.C.U., 357 Mag., 357 Rem. Max., 44 Mag. and 45 Colt/.410
Bull Barrel (less internal choke) **$435.00–445.00**
Bull Barrel Stainless **465.00–475.00**
Standard calibers w/internal choke 45 Colt/.410 . . . **440.00**
Vent Rib Model **455.00**
With **Match Grade Barrel** (22 LR only) **455.00**

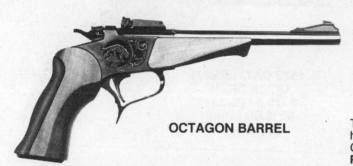

OCTAGON BARREL

CONTENDER
OCTAGON BARREL MODELS

This standard barrel is interchangeable with any model listed here. Available in 10-inch length, it is supplied with iron sights. Octagon barrel is available in 22 LR. No external choke.
Price: **$435.00**

**CONTENDER SUPER "14"
STAINLESS STEEL**

CONTENDER SUPER "14" MODELS

Chambered in 10 calibers (22 LR, 22 LR Match Grade Chamber, 22 Hornet, 223 Rem., 7-30 Waters, 300 Whisper, 30-30 Win., 35 Rem., 375 Win. and 44 Mag.), this gun is equipped with a 14-inch bull barrel, fully adjustable target rear sight and Patridge-style ramped front sight with 13½-inch sight radius.
Overall length: 18¼". **Weight:** 3½ lbs.
Prices: Blued **$445.00–475.00**
        Stainless **475.00–485.00**
14" Vent Rib Model in 45 Colt/.410 . . . . **475.00**

CONTENDER SUPER "16"
VENTILATED RIB/INTERNAL CHOKE MODELS
(not shown)

Featuring a raised ventilated (7/16-inch wide) rib, this Contender model is available in 45 Colt/.410 caliber. A patented detachable choke (1⅞ inches long) screws into the muzzle internally.

Barrel length: 16¼" inches.
Prices: Blued **$450.00–455.00**
        Stainless **480.00–500.00**
10" Vent Rib Model w/internal choke **485.00**

A. UBERTI REPLICAS

1871 ROLLING BLOCK TARGET PISTOL

SPECIFICATIONS
Calibers: 22 LR, 22 Magnum, 357 Magnum, 45 L.C.
Capacity: Single shot
Barrel length: 9¹/₂″ (half-octagon/half-round or full round Navy Style)
Overall length: 14″
Weight: 2.75 lbs.
Sights: Fully adjustable rear; ramp front or open sight on Navy Style barrel
Grip and forend: Walnut
Trigger guard: Brass
Frame: Color casehardened steel

**1871 ROLLING BLOCK
TARGET PISTOL
$380.00**

1873 CATTLEMAN QUICK DRAW

SPECIFICATIONS
Calibers: 38-40, 357 Magnum, 44 Special, 44-40, 45 L.C., 45 ACP
Capacity: 6 shots
Barrel lengths: 4³/₄″, 5¹/₂″, 7¹/₂″ round tapered; 10″, 12″, 18″ (Buntline)
Overall length: 10³/₄″ w/5¹/₂″ barrel
Weight: 2.42 lbs.
Grip: One-piece walnut
Frame: Color casehardened steel; also available in charcoal blue (**$40.00** extra) or nickel (**$50.00** extra)
Also available:
45 L.C./45 ACP Convertible $455.00
In brass 425.00

**1873 CATTLEMAN
QUICK DRAW
$375.00 (Brass)
$415.00 (Steel)**

1875 REMINGTON ARMY S.A. "OUTLAW"

SPECIFICATIONS
Calibers: 357 Magnum, 45 Long Colt, 44-40, 45 ACP
Capacity: 6 shots
Barrel length: 7¹/₂ ″ round tapered
Overall length: 13³/₄″
Weight: 2.75 lbs.
Grips: Two-piece walnut
Frame: Color casehardened steel
Also available:
In nickel plate $450.00
45 L.C./45 ACP Convertible 450.00

**1875 REMINGTON ARMY
S.A. "OUTLAW"
$405.00**

BUCKHORN SINGLE ACTION LARGE FRAME

SPECIFICATIONS
Calibers: 44-40, 44 Magnum
Capacity: 6 shots
Barrel length: 4³/₄″ or 7¹/₂″
Overall length: 11³/₄″
Weight: 2.5 lbs.
Sights: Open or target
Finish: Black
Also available:
In nickel plate $455.00
44-40 Mag. Convertible 460.00
BUCKHORN TARGET (7¹/₂″ barrel) 448.00
Convertible (44-40/44 Mag.) 499.00
BUCKHORN BUNTLINE 448.00–475.00
Convertible 499.00–530.00

**BUCKHORN SINGLE ACTION
LARGE FRAME
$410.00 (4³/₄″ Barrel)
$415.00 (7¹/₂″ Barrel)**

UNIQUE PISTOLS

MODEL DES 69U
$1195.00

SPECIFICATIONS
Caliber: 22 LR
Capacity: 5- or 6-shot magazine
Barrel length: 5.9″
Overall length: 11.2″
Weight: 40.2 oz. (empty)
Height: 5.5″
Width: 1.97″
Sights: Micrometric rear; lateral and vertical correction by clicks
Safety: Manual
Features: Orthopedic French walnut grip with adjustable hand rest; external hammer
Also available:
Model DES 32U in 32 S&W Long Wadcutter. Designed for centerfire U.I.T. and military rapid fire. Other specifications same as Model DES 69U. **Price: $1295.00**

MODEL I.S. INTERNATIONAL SILHOUETTE
$1095.00

SPECIFICATIONS
Calibers: 22 LR, 22 Magnum, 7mm TCU, 357 Magnum, 44 Magnum
Barrel length: 10″
Overall length: 14.5″
Weight: 38 oz.
Height: 6.5″
Width: 1.5″
Sights: Micrometric rear; lateral and vertical correction by clicks; interchangeable front sight; dovetailed grooves for scope
Features: French walnut grip; interchangeable shroud/barrel assembly; external hammer; firing adjustments
Also available:
International Sport w/light alloy frame in 22 LR and 22 Mag.
    Price: **$795.00**

MODEL DES 2000U
$1350.00

SPECIFICATIONS
Caliber: 22 Short
Barrel length: 5.9″
Overall length: 11.4″
Weight: 43.4 oz. (empty)
Height: 5.3″
Width: 1.97″
Sights: Micrometric rear; lateral and vertical correction by clicks
Features: French walnut grips with adjustable hand rest; left-hand grips available; external hammer; dry firing device; slide stop catch; anti-recoil device

WALTHER PISTOLS

The Walther double-action system combines the principles of the double-action revolver with the advantages of the modern pistol without the disadvantages inherent in either design.

Models PPK and PPK/S differ only in the overall length of the barrel and slide. Both models offer the same features, including compact form, light weight, easy handling and absolute safety. Both models can be carried with a loaded chamber and closed hammer, but ready to fire either single- or dou-

ble-action. Both models are provided with a live round indicator pin to signal a loaded chamber. An automatic internal safety blocks the hammer to prevent accidental striking of the firing pin, except with a deliberate pull of the trigger. Sights are provided with white markings for high visibility in poor light. Rich Walther blue/black finish is standard and each pistol is complete with an extra magazine with finger rest extension.

MODEL PPK & PPK/S

MODEL PPK & PPK/S

SPECIFICATIONS
Caliber: 380 ACP
Barrel length: 3.2″
Overall length: 6.1″
Height: 4.28″
Weight: 21 oz. (PPK); 23 oz. (PPK/S)
Finish: Walther blue or stainless steel
Price: **$610.00**

MODEL PP

MODEL PP DOUBLE ACTION

SPECIFICATIONS
Caliber: 22 LR, 32 ACP or 380 ACP
Barrel length: 3.8″ (32 ACP)
Overall length: 6.7″ (32 ACP)
Weight: 23.5 oz. (32 ACP)
Prices:
In 22 LR . . . . $ 783.00
In 32 ACP 1206.00
In 380 ACP 1206.00

MODEL TPH DOUBLE ACTION

Walther's Model TPH is considered by government agents and professional lawmen to be one of the top undercover/backup guns available. A scaled-down version of Walther's PP-PPK series.

SPECIFICATIONS
Calibers: 22 LR and 25 ACP
Capacity: 6 rounds
Barrel length: 2.3″
Overall length: 5³/₈″
Weight: 14 oz.
Finish: Walther blue or stainless steel
Price: (All models) **$458.00**

MODEL TPH

WALTHER PISTOLS

MODEL P-38 DOUBLE ACTION
With Presentation Case

The Walther P-38 is a double-action, locked-breech, semiautomatic pistol with an external hammer. Its compact form, light weight and easy handling are combined with the superb performance of the 9mm Luger Parabellum cartridge. The P-38 is equipped with both a manual and automatic safety, which allows it to be carried safely while the chamber is loaded.

SPECIFICATIONS
Caliber: 9mm Parabellum
Capacity: 8 rounds
Barrel length: 5″
Overall length: 8¹/₂″
Weight: 28 oz.
Finish: Blue
Price: $824.00

MODEL P-38

MODEL P-5 DA

SPECIFICATIONS
Caliber: 9mm Parabellum
Capacity: 8 rounds
Barrel length: 3¹/₂″
Overall length: 7″
Weight: 28 oz.
Finish: Blue
Features: Four automatic built-in safety functions; lightweight alloy frame; supplied with two magazines
Price: $1096.00
MODEL P-5 COMPACT (3.1″ barrel) 1096.00

MODEL P-5 DA

MODEL P-88 DA

SPECIFICATIONS
Caliber: 9mm Parabellum
Capacity: 15 rounds
Barrel length: 4″
Overall length: 7³/₈″
Weight: 31¹/₂ oz.
Finish: Blue
Sights: Rear adjustable for windage and elevation
Features: Internal safeties; ambidextrous de-cocking lever and magazine release button; lightweight alloy frame; loaded chamber indicator
Price: $1129.00
Also available:
MODEL P-88 COMPACT w/3.8″ barrel. **Weight:** 29 oz. **Overall length:** 7.1″. **Capacity:** 14 rounds. **Price:** $1725.00

MODEL P-88 DA

DAN WESSON REVOLVERS

357 MAGNUM REVOLVERS

Introduced in 1935, the 357 Magnum is still the top-selling handgun caliber. It makes an excellent hunting sidearm, and many law enforcement agencies have adopted it as a duty caliber. Take your pick of Dan Wesson 357s; then, add to its versatility with an additional barrel assembly option to alter it to your other needs.

SPECIFICATIONS
Action: Six-shot double and single action. **Ammunition:** 357 Magnum, 38 Special Hi-speed, 38 Special Mid-range. **Typical dimension:** 4″ barrel revolver, 9¼″×5¾″. **Trigger:** Smooth, wide tang (³/₈″) with overtravel adjustment. **Hammer:** Wide spur (³/₈″) with short double-action travel. **Sights: Models 14 and 714,** ¹/₈″ fixed serrated front; fixed rear integral with frame.

Models 15 and 715, ¹/₈″ serrated interchangeable front blade; red insert standard, yellow and white available; rear notch (.125, .080, or white outline) adjustable for windage and elevation; graduated click. 10″ barrel assemblies have special front sights and instructions. **Rifling:** Six lands and grooves, right-hand twist, 1 turn in 18.75 inches (2½″ thru 8″ lengths); six lands & grooves, right-hand twist, 1 turn in 14 inches (10″ bbl.). **Note:** All 2½″ guns shipped with undercover grips. 4″ guns are shipped with service grips and the balance have oversized target grips.

Price:
Pistol Pac Models 357 Magnum . . . . **$615.00** (Blue) to
$888.00 (Stainless)

357 MAGNUMS

Model	Type	Barrel Lengths/ Weight in Ounces					Finish
		2½″	4″	6″	8″	10″	
14-2	Service	30	34	38	NA	NA	BB
15-2	Target	32	36	40	44	50	BB
15-2V	Target	32	35	39	43	49	BB
15-2VH	Target	32	37	42	47	55	BB
714	Service	30	34	38	NA	NA	SSS
715	Target	32	36	40	45	50	SSS
715-V	Target	32	35	40	43	49	SSS
715-VH	Target	32	37	42	49	55	SSS

* BB=Brite Blue.   SSS=Satin Stainless Steel.

357 MAGNUM

38 SPECIAL REVOLVERS

For decades a favorite of security and law enforcement agencies, the 38 special still maintains it's reputation as a fine caliber for sportsmen and target shooters. Dan Wesson offers a choice of barrel lengths in either service or target configuration.

SPECIFICATIONS
Action: Six-shot double and single action. **Ammunition:** 38 Special Hi-speed, 38 Special Mid-range. **Typical dimension:** 4″ barrel revolver, 9¼″ × 5¾″. **Trigger:** Smooth, wide tang

(³/₈″) with overtravel adjustment. **Hammer:** Wide spur (³/₈″) with short double travel. **Sights:** Models 8 and 708, ¹/₈″ fixed serrated front; fixed rear integral with frame. Models 9 and 709, ¹/₈″ serrated interchangeable front blade; red insert standard, yellow and white available; rear, standard notch (.125, .080, or white outline) adjustable for windage and elevation; graduated click. **Rifling:** Six lands and grooves, right-hand twist, 1 turn in 18.75 inches. **Note:** All 2½″ guns shipped with undercover grips. 4″ guns are shipped with service grips and the balance have oversized target grips.

Price:
Pistol Pac Models 38 Special . . . . **$615.00** (Blue) to
$888.00 (Stainless)

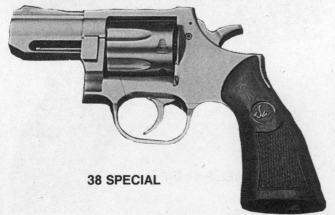

38 SPECIAL

38 SPECIALS

Model	Type	Barrel Lengths/ Weight in Ounces				Finish
		2½″	4″	6″	8″	
8-2	Service	30	34	38	NA	BB
9-2	Target	32	36	40	44	BB
9-2V	Target	32	35	39	43	BB
9-2VH	Target	32	37	42	47	BB
708	Service	30	34	38	NA	SSS
709	Target	32	36	40	44	SSS
709-V	Target	32	35	39	43	SSS
709-VH	Target	32	37	42	47	SSS

* BB=Brite Blue.   SSS=Satin Stainless Steel.

DAN WESSON REVOLVERS

22 SILHOUETTE REVOLVER

SPECIFICATIONS

This six-shot, single-action-only revolver is available with 10″ vented or vent heavy barrel assembly; incorporates a new cylinder manufacturing process that enhances the inherent accuracy of the Wesson revolver. Shipped with a combat-style grip, .080 narrow notch rear sight blade, a Patridge-style front sight blade to match. **Caliber:** 22 rimfire. **Type:** Target. **Weight:** 55 to 62 oz. **Finish:** Bright blue or stainless steel.
Price: . . . . **$459.75 to $516.40**

22 SILHOUETTE REVOLVER

FIXED BARREL REVOLVERS
$249.00–$274.00 (Service Model)
$259.00–$298.00 (Target Model)

These revolvers retain the advantages of a barrel in tension (minimal barrel whip) without the interchangeable features of earlier Wesson models.

SPECIFICATIONS

Calibers: 357 Magnum and 38 Special. **Capacity:** 6 shots. **Barrel lengths:** 3″, 4″, 5″ and 6″ (Target); 2¹⁄₂″ and 4″ (Service). **Overall length:** 8³⁄₄″ w/3″ barrel; 11³⁄₄″ w/6″ barrel. **Height:** 5³⁄₄″. **Weight:** 37 oz. to 45 oz. **Sight radius:** 5¹⁄₈″ (3″); 8¹⁄₈″ (6″). **Sights:** Adjustable rear target (Target model); fixed service sight (Service model). **Finish:** Brushed stainless steel or High Brite Blue (Target); brushed stainless steel or satin blue (Service).
    Also available in 44 Mag./44 Special. **Barrel lengths:** 4″, 5″, 6″ and 8″. **Price: $400.00** (Blue); **$442.00** (Stainless).

357 MAG. FIXED BARREL REVOLVER

MODEL 738P

SPECIFICATIONS

Caliber: 38 + P. **Capacity:** 5 shots. **Barrel length:** 6¹⁄₂″. **Weight:** 24.6 oz. **Sights:** Fixed. **Finish:** Stainless steel or blue. **Grip:** Pauferro wood or rubber.
Price: Stainless **$285.00**
(Blue model price not set)

MODEL 738P

.45 PIN GUN
$654.00–$762.00

SPECIFICATIONS

Caliber: 45 Auto. **Barrel length:** 5″ (5.28″ compensated shroud). **Overall length:** 12¹⁄₂″. **Weight:** 54 oz. **Sight radius:** 8.375″. **Finish:** Brushed stainless steel or High Brite Blue steel.

45 PIN GUN

DAN WESSON REVOLVERS

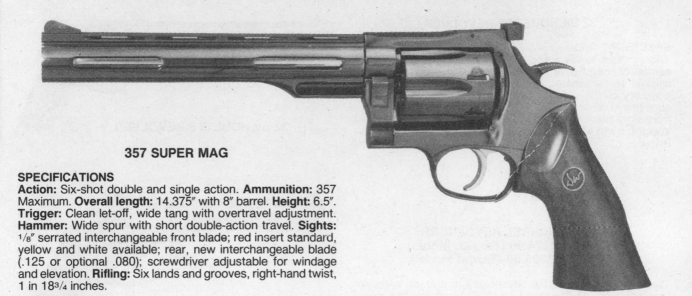

357 SUPER MAG

SPECIFICATIONS
Action: Six-shot double and single action. **Ammunition:** 357 Maximum. **Overall length:** 14.375″ with 8″ barrel. **Height:** 6.5″. **Trigger:** Clean let-off, wide tang with overtravel adjustment. **Hammer:** Wide spur with short double-action travel. **Sights:** 1/8″ serrated interchangeable front blade; red insert standard, yellow and white available; rear, new interchangeable blade (.125 or optional .080); screwdriver adjustable for windage and elevation. **Rifling:** Six lands and grooves, right-hand twist, 1 in 18³/₄ inches.

SPECIFICATIONS

Model	Caliber	Type	Barrel Lengths/ Weight in Ounces			Finish	Prices*
			6″	8″	10″		
740-V	357 Max	Target	59.5	64	65	Stainless	$550.00–$683.00
740-VH	357 Max	Target	62	72	76	Stainless	
740-V8S	357 Max	Target		62		Stainless	

* Model 40 (Blue): **$488.00–613.00**

32/32-20 MAGNUM SIX SHOT

This target and small-game gun offers a high muzzle velocity and a flat trajectory for better accuracy. **Action:** Six-shot double and single. **Calibers:** 32 H&R Magnum, 32 S&W Long, 32 Colt new police cartridges interchangeable. **Barrel length:** 4″. **Overall length:** 9¹/₄″ (w/4″ barrel). **Trigger:** Smooth, wide tang (³/₈″) w/overtravel adjustment. **Hammer:** Wide spur (³/₈″) w/short double-action travel. **Sights:** Front—1/8″ serrated, interchangeable blade, red insert standard (yellow and white available); rear—interchangeable blade for wide or narrow notch sight picture (wide notch standard, narrow notch available), adj. for windage and elevation, graduated click. **Rifling:** Six lands and grooves, right-hand twist 1:18³/₄″. **Finish:** Blue or stainless steel.

SPECIFICATIONS

Model	Caliber	Type	Barrel Lengths/ Weight in Ounces				Finish	Pistol Pac Prices*
			2¹/₂″	4″	6″	8″		
32	.32 Magnum	Target	35	39	43	48	BB	$615.00–$818.00
32V	.32 Magnum	Target	35	39	43	48	BB	
32VH	.32 Magnum	Target	35	40	46	53	BB	
732	.32 Magnum	Target	35	39	43	48	SSS	$689.00–$893.00
732V	.32 Magnum	Target	35	39	43	48	SSS	
732VH	.32 Magnum	Target	35	40	46	53	SSS	

DAN WESSON REVOLVERS

41, 44 MAGNUM AND 45 COLT REVOLVERS

The Dan Wesson 41, 44 Magnum and 45 Colt revolvers are available with a patented "Power Control" to reduce muzzle flip. Each has a one-piece frame and patented gain bolt for maximum strength.

SPECIFICATIONS
Action: Six-shot double- and single-action. **Typical dimension:** 6″ barrel revolver, 12″×6.″ **Trigger:** Smooth, wide tang (³/₈″) with overtravel adjustment. **Hammer:** Wide checkered spur with short double-action travel. **Sights:** Front, ¹/₈″ serrated in-terchangeable blade; red insert standard, yellow and white available; rear, standard notch (.125, .080, or white outline) adjustable for windage and elevation; click graduated. **Rifling:** Eight lands and grooves, right-hand twist, 1 turn in 18.75 inches.

Prices:
Pistol Pac Models (All calibers)
Blue $657.00–710.00
Stainless Steel 726.00–815.00

Model	Caliber	Type	Barrel Lengths/ Weight in Ounces				Finish
			4″	6″	8″	10″*	
41-V	41 Magnum	Target	48	53	58	64	BB
41-VH	41 Magnum	Target	49	56	64	69	BB
44-V	44 Magnum	Target	48	53	58	64	BB
44-VH	44 Magnum	Target	49	56	64	69	BB
741-V	41 Magnum	Target	48	53	58	64	SSS
741-VH	41 Magnum	Target	49	56	64	69	SSS
744-V	44 Magnum	Target	48	53	58	64	SSS
744-VH	44 Magnum	Target	49	56	64	69	SSS
745-V	45 Colt	Target	48	53	58	64	SSS
745-VH	45 Colt	Target	49	56	64	69	SSS

* BB=Brite Blue.   SSS=Satin Stainless Steel.

41/44 MAGNUM AND 45 COLT REVOLVERS

445 SUPERMAG REVOLVERS

With muzzle velocities in the 1650 fps range, and chamber pressures and recoil comparable to the 44 Magnum, the 445 Supermag has already won considerable renown in silhouette competition. As a hunting cartridge, it is more than adequate for any species of game on the American continent. **Action:** Six-shot double and single. **Type:** Target. **Caliber:** 445 Su-permag. **Overall length:** 14.375″ w/8″ barrel. **Trigger:** Clean let-off, widg tane with overtravel adjustment. **Hammer:** Wide spur with short double-action travel. **Sights:** ¹/₈″ serrated, in-terchangeable front blade, red insert standard (yellow and white available); rear—interchangeable blade for wide or narrow notch sight picture, wide notch standard (narrow notch avail-able), adj. for windage and elevation. **Rifling:** Six lands and grooves, right-hand twist 1:18³/₄″.

Model	Barrel Lengths/ Weight in Ounces			Finish	Prices
	6″	8″	10″		
445-V	59.5	62	65	BB	
445-VH	62	72	76	BB	$516.00–$615.00
445-VHS	—	64	—	BB	
445-VS	—	60	64	BB	
7445-V	59.5	62	65	SS	
7445-VH	62	72	76	SS	$592.00–$683.00
7445-VHS	—	64	—	SS	
7445-VS	—	60	64	SS	

* BB=Brite Blue.   SS=Stainless Steel.

445 SUPERMAG

DAN WESSON REVOLVERS

22 RIMFIRE and 22 WIN. MAGNUM REVOLVERS

Built on the same frames as the Dan Wesson 357 Magnum, these 22 rimfires offer the heft and balance of fine target revolvers. Affordable fun for the beginner or the expert.

SPECIFICATIONS
Action: Six-shot double and single action. **Ammunition:** Models 22 & 722, 22 Long Rifle; Models 22M & 722M, 22 Win. Mag. **Typical dimension:** 4″ barrel revolver, 9¼″×5¾″. **Trigger:** Smooth, wide tang (³/₈″) with overtravel adjustment. **Hammer:** Wide spur (³/₈″) with short double-action travel. **Sights:** Front, ⅛″ serrated, interchangeable blade; red insert standard, yellow and white available; rear, standard wide notch (.125, .080, or white outline) adjustable for windage and elevation; graduated click. **Rifling:** Models 22 and 722, six lands and grooves, right-hand twist, 1 turn in 12 inches; Models 22M and 722M, six lands and grooves, right-hand twist, 1 turn in 16 inches. **Note:** All 2½″ guns are shipped with undercover grips. 4″ guns are shipped with service grips and the balance have oversized target grips.
Prices:
Pistol Pac Models 22 thru 722M . . . . . . $637.00 (Blue)—
$923.00 (Stainless)

Model	Caliber	Type	Barrel Lengths/ Weight in Ounces				Finish
			2¼″	4″	6″	8″	
22	22 LR	Target	36	40	44	49	BB
22-V	22 LR	Target	36	40	44	49	BB
22-VH	22 LR	Target	36	41	47	54	BB
22-M	22 Win Mag	Target	36	40	44	49	BB
22M-V	22 Win Mag	Target	36	40	44	49	BB
22M-VH	22 Win Mag	Target	36	41	47	54	BB
722	22 LR	Target	36	40	44	49	SSS
722-V	22 LR	Target	36	40	44	49	SSS
722-VH	22 LR	Target	36	41	47	54	SSS
722M	22 Win Mag	Target	36	40	44	49	SSS
722M-V	22 Win Mag	Target	36	40	44	49	SSS
722M-VH	22 Win Mag	Target	36	41	47	54	SSS

* BB=Brite Blue.   SSS=Satin Stainless Steel.

22 RIMFIRE/22 WIN. MAG.
(Blue)

22 RIMFIRE/22 WIN. MAG.
(Stainless)

HUNTER PACS

Offered in all magnum calibers and include the following:
1. Gun with vent heavy 8″ shroud.
2. A vent 8″ shroud only, equipped with Burris scope mounts and Burris scope in either 1½x4X variable or fixed 2X.
3. Barrel changing tool and Wesson emblem packed in a custom-fitted carrying case. Available in either bright blue or stainless steel.

Prices: $691.00 (32 Mag. w/blue finish, mounts only) to **$1145.00** (445 Supermag, stainless steel w/1½X-4X scope)

HUNTER PAC

DAN WESSON REVOLVERS

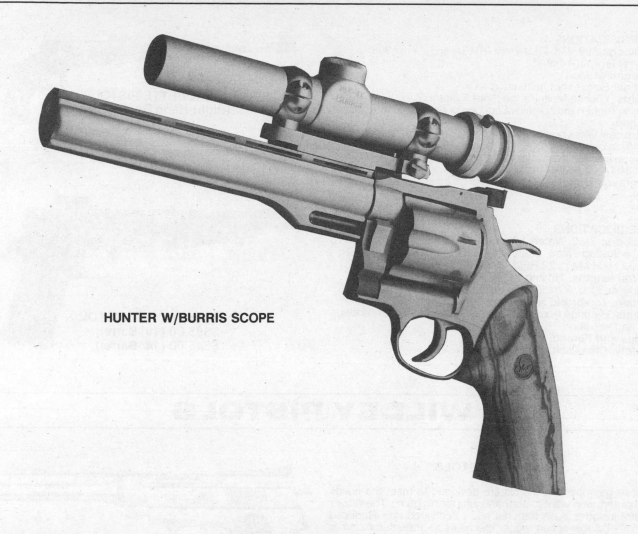

HUNTER W/BURRIS SCOPE

HUNTER SERIES

Open Hunter: Features a dovetailed Iron Sight Gunworks rear sight for quick target acquisition. The Alan Taylor throated 7½ inch barrel provides improved accuracy.

Compensated Open Hunter: For those who prefer open sights and less muzzle flip for quick follow-up shots. A shorter barrel (6″ with 7″ shroud) retains sight radius while allowing compensator system.

Scoped Hunter: No open sights and a 7½-inch barrel. Features a Burris base and rings mounted on shroud. Drilled, tapped and factory mounted.

Compensated Scoped Hunter: Same specifications as above, but with 6-inch barrel (7-inch shroud).

GENERAL SPECIFICATIONS

Calibers: 44 Magnum (Buck Series); 445 Super Mag. (Grizzly Series); 357 Super Mag. (Varmint Series); 41 Mag. (Boar Series)

Barrel lengths: See above
Shrouds: Vent heavy on Open and Scoped Hunter; full round (Lightweight) on Compensated models (approx. 1″ longer than inner barrel); non-fluted cylinder
Overall length: Approx. 14″
Weight: 4 lbs. (approx.)
Action: Double and single (creep-free)
Grips: Hogue rubber, finger-grooved
Finish: Bright blue steel or satin stainless steel
Prices:

Open Hunter	**$805.00**
w/Stainless Steel	849.00
Open Hunter Compensated	837.00
w/Stainless Steel	880.00
Scoped Hunter	838.00
w/Stainless Steel	881.00
Scoped Hunter Compensated	870.00
w/Stainless Steel	914.00

WICHITA ARMS PISTOLS

SPECIFICATIONS
Calibers: 308 Win. F.L., 7mm IHMSA and 7mm×308
Barrel length: 14¹⁵/₁₆″
Weight: 4¹/₂ lbs.
Action: Single-shot bolt action
Sights: Wichita Multi-Range Sight System
Grips: Right-hand center walnut grip or right-hand rear walnut grip
Features: Glass bedded; bolt ground to precision fit; adjustable Wichita trigger
Also available:
WICHITA CLASSIC PISTOL **$2950.00**
Engraved Model **4850.00**

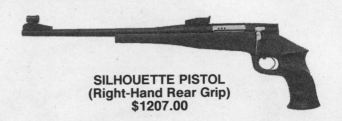

SILHOUETTE PISTOL
(Right-Hand Rear Grip)
$1207.00

SPECIFICATIONS
Calibers: 7-30 Waters, 7mm Super Mag., 7R (30-30 Win. necked to 7mm), 30-30 Win., 357 Mag., 357 Super Mag., 32 H&H Mag., 22 RFM, 22 LR
Barrel lengths: 10″ and 14″ (10¹/₂″ for centerfire calibers)
Weight: 3 lbs. 2 oz. (10″ barrel); 4 lbs. 7 oz. (14″ barrel)
Action: Top-break, single-shot, single action only
Sights: Patridge front sight; rear sight adjustable for windage and elevation
Grips and Forend: Walnut
Safety: Cross bolt

INTERNATIONAL PISTOL
$595.00 (10″ Barrel)
$645.00 (14″ Barrel)

WILDEY PISTOLS

WILDEY PISTOLS

These gas-operated pistols are designed to meet the needs of hunters who want to use handguns for big game. The Wildey pistol includes such features as: • Ventilated rib • Reduced recoil • Double-action trigger mechanism • Patented hammer and trigger blocks and rebounding fire pin • Sights adjustable for windage and elevation • Stainless construction • Fixed barrel for increased accuracy • Increased action strength (with 3-lug and exposed face rotary bolt) • Selective single or autoloading capability • Ability to handle high-pressure loads

SPECIFICATIONS
Calibers: 45 Win. Mag., 475 Wildey Mag.
Capacity: 7 shots
Barrel lengths: 5″, 6″, 7″, 8″, 10″, 12″
Overall length: 11″ with 7″ barrel
Weight: 64 oz. with 5″ barrel
Height: 6″

SURVIVOR MODEL in 45 Win. Mag.	**Prices**
5″, 6″ or 7″ models	**$1295.00**
Same model w/8″ or 10″ barrels	1295.00
With square trigger guard, **add**	35.00
With new vent rib, **add**	30.00
SURVIVOR MODEL in 475 Wildey Mag.	
8″ or 10″ barrels	1316.00
With square trigger guard, **add**	35.00
With new vent rib, **add**	30.00

HUNTER MODEL in 45 Win. Mag.	
8″, 10″ or 12″ barrels	**$1400.00**
With square trigger guard, **add**	35.00
HUNTER MODEL in 475 Wildey Mag.	
8″ or 10″ barrels	1400.00
With 12″ barrel	1449.00
W/square trigger guard (8″, 10″, 12″ barrels)	
add	35.00

Also available:
Interchangeable barrel extension assemblies **$523.00** (5″ barrel) to **$648.95** (12″ barrel).

Rifles

For addresses and phone numbers of manufacturers and distributors included in this section, turn to *DIRECTORY OF MANUFACTURERS AND SUPPLIERS* at the back of the book.

ACTION ARMS

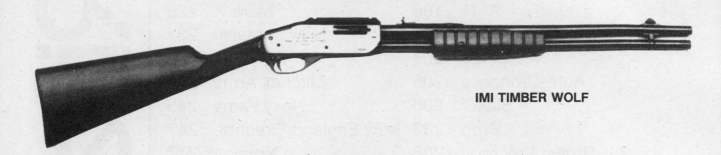

IMI TIMBER WOLF

IMI TIMBER WOLF
$299.00 ($385.00 Chrome)

SPECIFICATIONS
Caliber: 357 Magnum
Capacity: 10 rounds (8 rounds in 44 Mag.)
Barrel length: 18½"
Overall length: 36½"
Weight: 5½ lbs. (empty)

Sights: Fixed blade front; adjustable rear; integral scope mount on receiver
Stock: Walnut
Operation: Locked breech
Safety system: Push button on trigger guard

AMT RIFLES

SMALL GAME HUNTER II RIFLE

SPECIFICATIONS
Caliber: 22 LR Semiautomatic
Capacity: 10 rounds
Barrel length: 22"
Overall length: 40½"
Weight: 6 lbs.
Features: Uncle Mike's swivel studs for strap attachment; removable recoil pad provides storage for ammo, cleaning rod and survival knife; grooves for scope mount; synthetic stock is black matte fiberglass filled nylon, checkered at forearm and grip

Also available:
22 MAGNUM RIMFIRE (MATCH GRADE)
   w/free-floating 20" barrel $459.99

SMALL GAME HUNTER II RIFLE
$299.99
(Scope not included)

ANSCHUTZ SPORTER RIFLES

MODEL 1700D CUSTOM $1267.00 (22 LR)
$1426.00 (22 Hornet & 222 Rem.)
$1163.00 (22 LR Featherweight)
$1392.00 (22 LR Featherweight Deluxe)

MODEL 1700D CLASSIC
$1236.00 (22 LR) $1395.00 (22 Hornet & 222 Rem.)

BAVARIAN 1700
$1267.00 (22 LR) $1426.00 (22 Hornet & 222 Rem.)

MODEL 1700D GRAPHITE CUSTOM
$1233.00 (22 LR)

MANNLICHER 1733D
$1539.00

SPECIFICATIONS

MODEL	Custom	Bavarian	Classic	Mannlicher 1733D	Graphite Custom	1700 FWT
Calibers	22 Long Rifle, 22 Hornet, 222 Remington, 22 Hornet				22 LR	22 LR
Length—Overall	43″	43″	43″	39″	41″	41″
Barrel	24″	24″	24″	19³/₄″	22″	22″
Pull	14″	14″	14″	14″	14″	14″
Drop at—Comb	1¹/₄″	1¹/₄″	1¹/₄″	1¹/₄″	1¹/₄″	1¹/₄″
Monte Carlo	1″	—	—	1³/₄″	1″	1″
Heel	1¹/₂″	1¹/₂″	1¹/₂″	2³/₈″	1¹/₂″	1¹/₂″
Average Weight (lbs.)	7¹/₂	7¹/₂	6³/₄	6¹/₄	7¹/₄	6¹/₄
Trigger—Single Stage 5096 (.222 Rem., 5095)	•	•	•	•	•	•
Swivel Studs	•	•	•	³/₄″ swivels	•	•
Rate of Twist	Right Hand—one turn in 16.5″ for .22 LR; 1-16″ for .22 Hornet; 1-14″ for .222 Rem.					

ANSCHUTZ SPORTER RIFLES

MATCH 64 SPORTER MODELS

MODEL 1416DCL (22 LR) AND 1516DCL (22 Magnum) CLASSIC
$695.00 (1416DCL) $704.00 (1516DCL)
$729.00 (22 LR Left Hand)

MODEL 1416D (22 LR) AND 1516D (22 Magnum) CUSTOM
$708.00 (22 LR) $716.00 (22 Magnum)
(not shown)

1418D MANNLICHER
$1537.00 (22 LR)

MODEL 525 SPORTER
$508.00   ($199.00 Add'l. for Meistergrade)

SPECIFICATIONS

MODEL	Classic 1416D** 1516D	Custom 1416D 1516D	Mannlicher* 1418D 1518D	Model 525 Sporter
Length—Overall	41″	41″	38″	43″
Barrel	22 1/2″	22 1/2″	19 3/4″	24″
Pull	14″	14″	14″	14″
Drop at—Comb	1 1/4″	1 1/4″	1 1/4″	1 1/8″
Monte Carlo	1 1/2″	1 1/2″	1 1/2″	1 3/4″
Heel	1 1/2″	2 1/2″	2 1/2″	2 5/8″
Average Weight	5 1/2 lbs.	6 lbs.	5 1/2 lbs.	6 1/2 lbs.
Rate of Twist Right Hand—one turn in 16.5″ for .22 LR; 1–16″ for .22 Mag				
Takedown Bolt Action With Removable Firing Pin	•	•		
3/4″ Swivel			•	•
Swivel Studs	•	•		

ANSCHUTZ MATCH RIFLES

MODEL BR-50 BENCH REST RIFLE
$3075.00

SPECIFICATIONS
Caliber: 22 LR.
Barrel length: 19³/₄″ without muzzle weight; 23″ with muzzle weight.

Overall length: 37³/₄″–42¹/₂″.
Weight: 11 lbs.
Trigger: No. 5015 two-stage adjustable target trigger (3.9 oz. weight).
Stock: European hardwood with adjustable spacers.

BIATHLON RIFLES SPECIFICATIONS

Model	1827B	1827BT
Barrel	21¹/₂″ (³/₄″ dia.)	21¹/₂″ (³/₄″ dia.)
Action	Super Match 54	T. Bolt
Trigger	1 lb. 3.5 oz. 2 stage #5020 adjustable* from 3¹/₂ oz. to 2 lbs.	1 lb. 3.5 oz., 2 stage #5020 adjustable* from 3¹/₂ oz. to 2 lbs.
Safety	Slide safety	Slide safety
Stock	European Walnut, cheek piece, stippled pistol grip and front stock	European Walnut, cheek piece, stippled pistol grip and front stock
Sights	6827 Sight Set with snow caps furnished with rifle and 10 click adjustment.	6827 Sight Set with snow caps furnished with rifle and 10 click adjustment.
Overall length	42¹/₂″	42¹/₂″
Weight	8¹/₂ lbs. with sights	8¹/₂ lbs. with sights
Price	$2233.00	$3449.00 ($3794.00 L.H.)

MODEL 64MS

MODEL 54.18MS-REP DELUXE

METALLIC SILHOUETTE RIFLES

Prices:
64MS	**$ 916.00**
Left Hand	960.00
30054.18MS	1488.00
Left Hand	1670.00
54.18MS-REP	1896.00
54.18MS-REP DELUXE	2280.00
1700 FWT	1163.00
1700 FWT DELUXE	1392.00

SPECIFICATIONS AND FEATURES (22 LR)

Model	64MS	54.18MS	54.18MS-REP DELUXE*
Grooved for scope	●	●	●
Tapped for scope mount	●	●	●
Overall length	39.5″	41″	41–49″
Barrel length	21¹/₂″	22″	22–30″
Length of pull	13¹/₂″	13³/₄″	13³/₄″
High cheekpiece with Monte Carlo effect	●	●	●
Drop at Comb	1¹/₂″	1¹/₂″	1¹/₂″
Average weight	8 lbs.	8 lbs. 6 oz.	7 lbs. 12 oz.
Trigger: Stage Factory adjusted weight Adjustable weight	#5091 Two 5.3 oz. 4.9–7 oz.	#5018 Two 3.9 oz. 2.1–8.6 oz.	#5018 Two 3.9 oz. 2.1–8.6 oz.
Safety	Slide	Slide	Slide
True Left-hand Model	●	●	

ANSCHUTZ INT'L. TARGET RIFLES

MODEL 1913
SUPER MATCH
$3212.00 ($3182.00 Left Hand)

MODEL 1910 SUPER MATCH II
(not shown)
$2709.00 ($2843.00 Left Hand)

MODEL 1911 PRONE MATCH
$2134.00

MODEL 1808DRT (not shown)
SUPER RUNNING $1759.00

MODEL 1903D (not shown)
$1039.00 ($1093.00 Left Hand)

MODEL 1907
ISU STANDARD
$1827.00 ($1916.00 Left Hand)
$1897.00 (Walnut) $1990.00 (Walnut L.H.)

INTERNATIONAL MATCH RIFLES: SPECIFICATIONS AND FEATURES

MODEL	1913	1911	1910	1907	DRT-Super
Barrel Length **O/D**	27¼" ⁷/₈" (23.4 mm)	27¼" ⁷/₈" (23.4 mm)	27¼" ⁷/₈" (23.4 mm)	26" ⁷/₈" (22 mm)	32½" ⁷/₈" (22 mm)
Stock	Int'l.- Thumb Hole Adj. Palm Rest Adj. Hand Rest	Prone	Int'l.- Thumb Hole	Standard	Thumb Hole
Cheek Piece **Butt Plate**	Adj. Adj. Hook 10 Way Hook	Adj. Adj. 4 Way	Adj. Adj. Hook 10 Way Hook	Removable Adj. 4 Way	Adj. Adj. 4 Way
Recommended **Sights**	6820, 6823 *6820 Left	6820, 6823 *6820 Left	6820, 6823 *6820 Left	6820, 6823 *6820 Left	Grooved for Scope Mounts
Overall Length	45"-46"	45"-46"	45"-46"	43¾"-44½"	50½"
Overall Length **to Hook**	49.6"-51.2"		49.6"-51.2"		
Weight (approx) **without sights**	15.2 lbs.	11.7 lbs.	13.7 lbs.	11.2 lbs.	9.4 lbs.
True Left- **Hand Version**	1913 Left	1911 Left	1910 Left	1907 Left	1808 Left

Trigger
  Stage
  Factory Set Wt.
  Adjust. Wt.
	#5018
	Two
	3.9 oz.
	2.1-8.6 oz.
		#5018
		Two
		3.9 oz.
		2.1-8.6 oz.
			#5018
			Two
			3.9 oz.
			2.1-8.6 oz.
				#5018
				Two
				3.9 oz.
				2.1-8.6 oz.
					5020D
					Single
					1.2 lbs.
					14 oz.-2.4 lbs.

ANSCHUTZ TARGET RIFLES

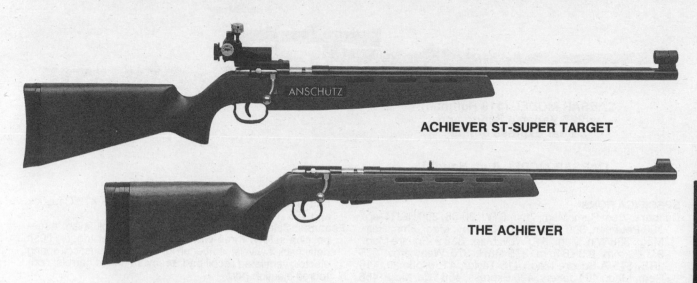

ACHIEVER ST-SUPER TARGET

THE ACHIEVER

THE ACHIEVER
$395.00

This rifle has been designed especially for young shooters and is equally at home on range or field. It meets all NRA recommendations as an ideal training rifle.

SPECIFICATIONS
Caliber: 22 LR **Capacity:** 5- or 10-shot clips available
Action: Mark 2000-type repeating
Barrel length: 19 1/2" **Overall length:** 35 1/2"–36 2/3"
Weight: 5 lbs.
Trigger: #5066-two stage (2.6 lbs.)

Safety: Slide
Sights: Hooded ramp front; Lyman folding-leaf rear, adjustable for elevation
Stock pull: 11 7/8"–13"
Stock: European hardwood
Also available:
ACHIEVER ST-SUPER TARGET.
Barrel length: 22". **Overall length:** 38 3/4".
Weight: 6 1/2 lbs.
Price: $485.00

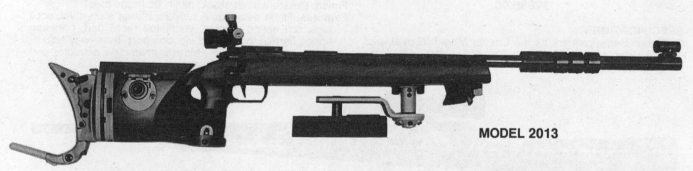

MODEL 2013

MODEL 2007 SUPER MATCH 54 ISU STANDARD
$2675.00 ($2807.00 Left Hand)
$2750.00 (Walnut)
(not shown)

MODEL 2013 SUPER MATCH 54
$3812.00 ($4004.00 Left Hand)

SPECIFICATIONS
Caliber: 22 LR
Barrel length: 19 3/4" **Overall length:** 43" to 45 1/2"
Weight: 12.5 lbs. (without sights)
Trigger: #5018, two-stage, 3.9 oz. (factory set wt.)
Sights: #6802 Set
Stock: International with adjustable palm rest and hand rest
Feature: Adjustable cheekpiece

SPECIFICATIONS
Caliber: 22 LR
Barrel length: 19 3/4" **Overall length:** 43 1/2" to 44 1/2"
Weight: 10.8 lbs. (without sights)
Trigger: #5018, two-stage, 3.9 oz. (factory set wt.)
Sights: #6802 Set
Stock: Standard ISU
Feature: Adjustable cheekpiece

A-SQUARE RIFLES

CAESAR MODEL (416 Hoffman)
w/2x7 Variable Scope and
3-Leaf Express Sights

CAESAR MODEL (Left Hand)
$2950.00

SPECIFICATIONS
Calibers: 7mm Rem. Mag., 7mm STW, 30-06, 300 H&H Mag., 300 Petersen, 300 Win. Mag., 300 Wby. Mag., 8mm Rem. Mag., 338 Win. Mag., 340 Wby. Mag., 338 A-Square Mag., 9.3×62mm, 9.3×64mm, 375 H&H, 375 Weatherby, 375 JRS, 375 A-Square Mag., 416 Taylor, 416 Hoffman, 416 Rem. Mag., 404 Jeffery, 425 Express, 458 Win. Mag., 458

Lott, 450 Ackley Mag., 460 Short A-Square, 470 Capstick and 495 A-Square Mag.
Features: Selected Claro walnut stock with oil finish; three-position safety; three-way adjustable target trigger; flush detachable swivels; leather sling; dual recoil lugs; coil spring ejector; ventilated recoil pad; premium honed barrels; contoured ejection port

HANNIBAL MODEL
$2895.00

SPECIFICATIONS
Calibers: Same calibers as the Caesar Model (above), plus 378 Wby. Mag., 416 Rigby, 416 Wby. Mag., 460 Wby. Mag. and 500 A-Square

HANNIBAL MODEL (416 Rigby)
w/2xLER Scope and 3-Leaf
Express Sights

Barrel lengths: 20″ to 26″
Length of pull: 12″ to 15¼″
Finish: Deluxe walnut stock; oil finish; matte blue
Features: Flush detachable swivels, leather sling, dual recoil lugs, coil spring ejector, ventilated recoil pad, premium honed barrels, contoured ejection port, three-way adjustable target-style trigger, Mauser-style claw extractor and controlled feed, positive safety

HAMILCAR
$2795.00

HAMILCAR

SPECIFICATIONS
Calibers: 25-06, 257 Wby., 6.5×55 Swedish, 264 Win., 270 Win., 270 Wby., 7×57 Mauser, 280 Rem., 7mm Rem., 7mm Wby., 7mm STW, 30-06, 300 Win., 300 Wby., 338-06, 9.3×62
Barrel lengths: 20″ to 26″

Length of pull: 12″ to 15¼″
Finish: Deluxe walnut stock; oil finish; matte blue
Features: Flush detachable swivels; leather sling, coil spring ejector, vent. recoil pad; honed barrels; contoured ejection port; target-style adjustable trigger; Mauser-style claw extractor; controlled feed; positive safety

AUTO-ORDNANCE

THOMPSON MODEL M1 CARBINE
$747.50

SPECIFICATIONS
Caliber: 45 ACP
Barrel length: 16½″
Overall length: 38″
Weight: 11½ lbs.
Sights: Blade front; fixed rear
Stock: Walnut stock and forend
Finish: Military black
Features: Side cocking lever; frame and receiver milled from solid steel

THOMPSON DELUXE MODEL 1927 A1
$770.00 (45 Cal.)

SPECIFICATIONS
Caliber: 45 ACP
Barrel length: 16″
Overall length: 42″
Weight: 11½ lbs.
Sights: Blade front; open rear adjustable
Stock: Walnut stock; vertical forend

Also available:
THOMPSON 1927A-1C LIGHTWEIGHT (45 cal.). Same as the 1927A1 model, but 20% lighter. **Price: $742.00**
THOMPSON MODEL 1927A3. Same as above, but in caliber 22 LR. **Weight:** 7 lbs. **Price: $510.50**

VIOLIN CARRYING CASE (for gun, drum & extra magazines) $105.00

BLASER RIFLES

MODEL R 84 BOLT ACTION

SPECIFICATIONS
Calibers: (interchangeable)
   Standard: 22-250, 243 Win., 6mm Rem., 25-06, 270 Win.,
      280 Rem., 30-06
   Magnum: 257 Weatherby Mag., 264 Win. Mag., 7mm Rem.
      Mag., 300 Win. Mag., 300 Weatherby Mag., 338
      Win. Mag., 375 H&H
Barrel lengths: 23″ (Standard) and 24″ (Magnum)

Overall length: 41″ (Standard) and 42″ (Magnum)
Weight: (w/scope mounts) 7 lbs. (Standard) and 7¼ lbs.
   (Magnum)
Safety: Locks firing pin and bolt handle
Stock: Two-piece Turkish walnut stock and forend; solid black
   recoil pad, handcut checkering (18 lines/inch, borderless)
Length of pull: 13¾″
Prices:
Std. & Mag. calibers w/scope mounts $2300.00
L.H. Std. and Mag. cal. w/scope mounts 2350.00
Interchangeable barrels 600.00
Deluxe Model Right Hand 2600.00
   Left Hand 2650.00
Super Deluxe Model Right Hand 2950.00
   Left Hand 3000.00

BRNO RIFLES

MODEL 600

MODEL 602

Prices:
Model 600 Standard walnut stock $609.00
Model 601 Short action walnut 609.00
Model 602 Magnum action walnut stock 835.00
   Synthetic stock 745.00

SPECIFICATIONS: MODEL 600 BOLT-ACTION SERIES

| Model | Action Type | Caliber | Barrel Specifications | | | Overall Length | Weight | Magazine Capacity | Sighted In |
			Length	Rifling Twist	#Lands				
600	Bolt (Std.)	7mm Mauser 270 Win. 30-06 Spfld.	23.6 in	1 in 10″ 1 in 10″ 1 in 9″	4 4 4	43.7 in.	7.2 lbs.	5	110 yd.
601	Bolt (Short)	243 Win. 308 Win.	23.6 in.	1 in 12″ 1 in 10″	4 4	43.1 in.	8.2 lbs.	5	110 yd.
602	Bolt (Magnum)	300 Win. Mag. 375 H&H 458 Win. Mag.	25.2 in.	1 in 10″ 1 in 12″ 1 in 12″	4 4 4	45.3 in.	9.4 lbs.	5 4 4	110 yd.

BRNO RIFLES

MODEL 452 DELUXE

MODEL 452

SPECIFICATIONS
Caliber: 22 LR
Capacity: 5 rounds
Barrel length: 23.6"
Overall length: 42.6"
Weight: 6.1 lbs. (unloaded)

Safety: Thumb safety
Sights: Hooded front; tangent adjustable rear
Stock: European hardwood or checkered walnut
Finish: Lacquer
Features: Firing-pin blocking safety; 5- and 10-round detachable magazines; receiver grooved for .22 scope mounts; hammer-forged barrel threaded to receiver
Prices:
    w/Hardwood stock $305.00
    w/Deluxe walnut stock 349.00

MODEL 527 MINI-MAUSER

MODEL 527 MINI-MAUSER

SPECIFICATIONS
Calibers: 22 Hornet, 222 Rem., 223 Rem.
Capacity: 5 rounds
Barrel length: 23.6" (standard or heavy)
Overall length: 42.4"

Weight: 6.2 lbs. (unloaded)
Safety: Thumb safety
Sights: Hooded front; drift adjustable rear
Stock: Checkered walnut
Finish: Lacquer
Features: Mauser-type bolt and non-rotating claw extractor; hammer-forged barrel; integral dovetail scope bases w/recoil stop; detachable 5-round magazine
Price: $665.00

MODEL 537

MODEL 537

SPECIFICATIONS
Calibers: 270 Win., 308 Win., 30-06
Capacity: 5 rounds
Barrel length: 23.6"
Overall length: 44.7"
Weight: 7.9 lbs. (unloaded)
Safety: Thumb safety (locks trigger and bolt)

Sights: Hooded front; adjustable rear
Stock: Checkered walnut or synthetic
Finish: Lacquer
Features: Streamlined bolt shroud w/cocking indicator; integral dovetail scope bases w/recoil stop; controlled-feed claw extractor; forged, one-piece bolt
Price: $669.00

Also available:
MODEL 537 MOUNTAIN CARBINE. Barrel length: 19".
   Weight: 7.1 lbs. **Price: $699.00** (243 Win. only)

BROWN PRECISION RIFLES

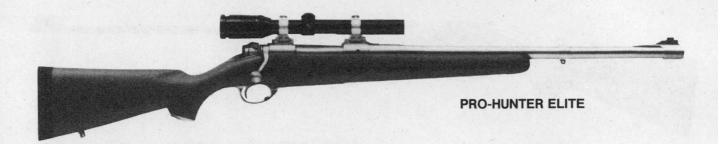

PRO-HUNTER ELITE

PRO-HUNTER ELITE

Designed for the serious game hunter or guide, this version of Brown Precision's Pro-Hunter rifle is custom built, utilizing standard features and equipment. The rifle begins as a Winchester Model 70 Super Grade action with controlled feed claw extractor. The action is finely tuned for smooth operation, and a Speedlock firing-pin spring is installed. The trigger is tuned to crisp let-off at each customer's specified weight. A Shilen Match Grade stainless-steel barrel is custom crowned at the desired barrel length and hand-fitted to the action. A barrel band front QD sling swivel is attached permanently to the barrel.

The Pro-Hunter Elite features the customer's choice of express rear sight or custom Dave Talley removable peep sight and banded front ramp sight with European dovetail and replaceable brass bead. An optional flip-up white night sight is also available, as is a set of Dave Talley detachable T.N.T.

scope mount rings and bases installed with Brown's Magnum Duty 8×40 screws.

All metal parts are finished in either matte electroless nickel or black Teflon. The barreled action is glass bedded to a custom Brown Precision fiberglass stock. Brown's Alaskan configuration stock features a Dave Talley trapdoor grip cap designed to carry the rear peep sight and/or replacement front sight beads. A premium 1″ recoil-reducing buttpad is installed, and the stock painted according to customer choice. Weight ranges from 7 to 15 lbs., depending on barrel length, contour and options.

Optional equipment includes custom steel drop box magazine, KDF or Answer System muzzle brake, Mag-Na-Port, Zeiss, Swarovski or Leupold scope and Americase aluminum hard case. **Price** **$3170.00**

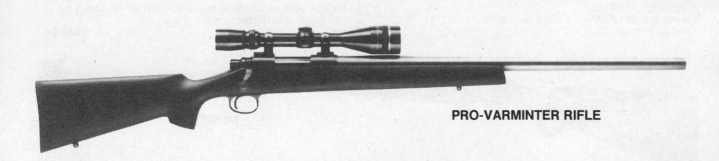

PRO-VARMINTER RIFLE

PRO-VARMINTER RIFLE

The standard Pro-Varminter is built on the Remington 700 or Remington 40X action (right or left hand) and features a hand-fitted Shilen Match Grade Heavy Benchrest stainless-steel barrel in bright or bead-blasted finish. The barreled action is custom-bedded in Brown Precision's Varmint Special Hunter Bench or 40X Benchrest-style custom fiberglass, Kevlar or graphite stock.

Other standard features include custom barrel length and

contour, trigger tuned for crisp pull to customer's specified weight, custom length of pull, and choice of recoil pad. Additional options include metal finishes, muzzle brakes, Leupold target or varmint scopes and others.

Prices:

Right-hand Model 700 Action	**$1625.00**
For Left-hand Model, **add**	110.00
40X Action	2599.00

BROWN PRECISION RIFLES

HIGH COUNTRY YOUTH RIFLE

This custom rifle has all the same features as the standard High Country rifle, but scaled-down to fit the younger or smaller shooter. Based on the Remington Model 7 barreled action, it is available in calibers .223, .243, 7mm-08, 6mm and .308.

The rifle features a fiberglass, Kevlar or graphite stock shortened to fit. As the shooter grows, the stock can be lengthened, a new recoil pad installed and the stock refinished. Custom features and options include choice of actions, custom barrels, chamberings, muzzle brakes, metal finishes, scopes and accessories.

All Youth Rifles include a deluxe package of shooting, reloading and hunting accessories and information to increase a young shooter's interest. **Price:** starts at **$1165.00**

TACTICAL ELITE RIFLE

Brown Precision's Tactical Elite is built on a Remington 700 action and features a bead-blasted Shilen Select Match Grade Heavy Benchrest Stainless Steel barrel custom-chambered for .223 Remington, .308 Winchester, .300 Winchester Magnum (or any standard or wildcat caliber). A nonreflective custom black Teflon metal finish on all metal surfaces ensures smooth bolt operation and 100 percent weatherproofing. The barreled action is bedded in a target-style stock with high rollover comb/cheekpiece, vertical pistol grip and palmswell. The stock is an advanced, custom fiberglass/Kevlar/graphite composite for maximum durability and rigidity. QD sling swivel studs and swivels are included on the stock. Standard stock paint color is flat black (camouflage patterns are also available).

Other standard features include: three-way adjustable buttplate/recoil pad assembly with length of pull, vertical and cant angle adjustments, custom barrel length and contour, and trigger tuned for a crisp pull to customer's specifications. Options include muzzle brakes, Leupold or Kahles police scopes, and others, and are priced accordingly.
Price: **$2395.00**

CUSTOM TEAM CHALLENGER

This custom rifle was designed for use in the Chevy Trucks Sportsman's Team Challenge shooting event. It's also used in metallic silhouette competition as well as in the field for small game and varmints. Custom built on the Ruger 10/22 semiautomatic rimfire action, which features an extended magazine release, a simplified bolt release and finely tuned trigger, this rifle is fitted with either a Brown Precision fiberglass or Kevlar stock with custom length of pull up to 15". The stock can be shortened at the butt and later relengthened and repainted to accommodate growing youth shooters. Stock color is also optional. To facilitate shooting with scopes, including large objective target or varmint scopes, the lightweight stock is of classic styling with a high comb. The absence of a cheekpiece accommodates either right- or left-handed shooters, while the stock's flat-bottom, 1³/₄"-wide forearm ensures maximum comfort in both offhand and rest shooting. Barrels are custom-length Shilen Match Grade .920" diameter straight or lightweight tapered.

Prices:

With blued action/barrel	**$690.00**
With blued action/stainless barrel	740.00
With stainless action/stainless barrel	789.00

RIFLES

BROWNING RIFLES

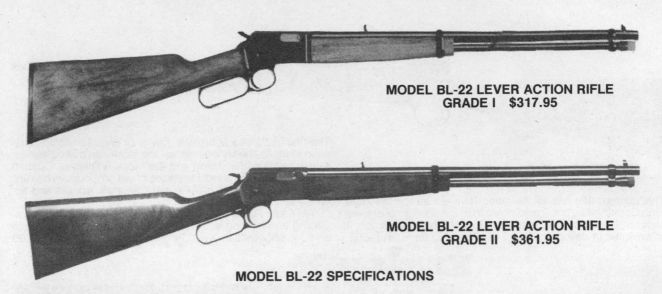

**MODEL BL-22 LEVER ACTION RIFLE
GRADE I $317.95**

**MODEL BL-22 LEVER ACTION RIFLE
GRADE II $361.95**

MODEL BL-22 SPECIFICATIONS

Action: Short throw lever action. Lever travels through an arc of only 33 degrees and carries the trigger with it, preventing finger pinch between lever and trigger on the upward swing. The lever cycle ejects the fired shell, cocks the hammer and feeds a fresh round into the chamber. **Magazine:** Rifle is designed to handle 22 caliber ammunition *in any combination* from tubular magazine. Magazine capacity is 15 Long Rifles, 17 Longs and 22 Shorts. The positive magazine latch opens and closes easily from any position. **Safety:** A unique disconnect system prevents firing until the lever and breech are fully closed and pressure is released from and reapplied to the trigger. An inertia firing pin and an exposed hammer with a half-cock position are other safety features. **Receiver:** Forged and milled steel. Grooved. All parts are machine-finished and hand-fitted. **Trigger:** Clean and crisp without creep. Average

pull 5 pounds. Trigger gold-plated on Grade II model. **Stock and forearm:** Forearm and straight grip butt stock are shaped from select, polished walnut. Hand checkered on Grade II model. Stock dimensions:

Length of Pull . . . . 13½″
Drop at Comb . . . . 1⅝″
Drop at Heel . . . . 2¼″

Sights: Precision, adjustable folding leaf rear sight. Raised bead front sight. **Scopes:** Grooved receiver will accept the Browning 22 riflescope (Model 1217) and two-piece ring mount (Model 9417) as well as most other groove or tip-off type mounts or receiver sights. **Engraving:** Grade II receiver and trigger guard are engraved with tasteful scroll designs. **Barrel length:** 20″; recessed muzzle. **Overall length:** 36¾″. **Weight:** 5 pounds.

**MODEL 81 BLR HIGH POWER
$529.95 (Short Action)
$561.95 (Long Action)**

Short Action Calibers: 22-250 Rem., 223 Rem., 243 Win., 7mm-08 Rem., 284 Win. and 308 Win. **Action:** Lever action with rotating head, multiple lug breech bolt with recessed bolt face. Side ejection. **Barrel length:** 20″. Individually machined from forged, heat treated chrome-moly steel; crowned muzzle. **Rifling:** 243 Win., one turn in 10″; 308 and 358 Win., one turn in 12″. **Magazine:** Detachable, 4-round capacity. **Overall length:** 39¾″. **Approximate Weight:** 6 lbs. 15 oz. **Trigger:** Wide, grooved finger piece. Short crisp pull of 4½ pounds. Travels with lever. **Receiver:** Non-glare top. Drilled and tapped to accept most top scope mounts. Forged and milled steel. All parts are machine-finished and hand-fitted. Surface deeply

polished. **Sights:** Low profile, square notch, screw adjustable rear sight. Gold bead on a hooded raised ramp front sight. Sight radius: 17¾″. **Safety:** Exposed, 3-position hammer. Trigger disconnect system. Inertia firing pin. **Stock and forearm:** Select walnut with tough oil finish and sure-grip checkering, contoured for use with either open sights or scope. Straight grip stock. Deluxe recoil pad installed.

Long Action Calibers: 270, 30-06, 7mm Magnum. **Barrel length:** 22″ (30-06, 270) and 24″ (7mm Mag.). **Overall length:** 42½″ (30-06, 270) and 44½″ (7mm Mag.). **Weight:** 8 lbs. 8 oz. (30-06, 270) and 8 lbs. 13 oz. (7mm Mag.). Other features/specifications same as Short Action.

BROWNING RIFLES

MODEL 1885 SINGLE SHOT
$853.95

Calibers: 22-250, 223, 30-06, 270, 7mm Rem. Mag., 45-70 Govt. **Bolt system:** Falling block. **Barrel length:** 28″ (recessed muzzle). **Overall length:** 43¹⁄₂″. **Weight:** 8 lbs. 11 oz. (7mm Rem. Mag.) to 8 lbs. 14 oz. (45-70 Govt.). **Action:** High-wall type, single shot, lever action. **Sights:** Drilled and tapped for scope mounts; two-piece scope base available. Open sights on 45-70 Govt. only. **Hammer:** Exposed, serrated, three-position with inertia sear. **Stock and Forearm:** Select Walnut, straight grip stock and Schnabel forearm with cut checkering. Recoil pad standard.

MODEL A-BOLT 22 BOLT ACTION
$393.95
$403.95 with Sights

Caliber: 22 LR. **Barrel length:** 22″. **Overall length:** 40¹⁄₄″. **Average weight:** 5 lbs. 9 oz. **Action:** Short throw bolt. Bolt cycles a round with 60° of bolt rotation. Firing pin acts as secondary extractor and ejector, snapping out fired rounds at prescribed speed. **Magazine:** Five and 15-shot magazine standard. Magazine/clip ejects with a push on magazine latch button. **Trigger:** Gold colored, screw adjustable. Pre-set at approx. 4 lbs. **Stock:** Laminated walnut, classic style with pistol grip. **Length of pull:** 13³⁄₄″. **Drop at comb:** ³⁄₄″. **Drop at heel:** 1¹⁄₂″. **Sights:** Available with or without sights (add **$10** for sights). Ramp front and adjustable folding leaf rear on open sight model. **Scopes:** Grooved receiver for 22 mount. Drilled and tapped for full-size scope mounts.

Also available:
Grade I (22 Magnum, open sights) **$464.95**
Grade I (22 Magnum without sights) 454.95
Gold Medallion Model (22 LR, no sights) 521.95

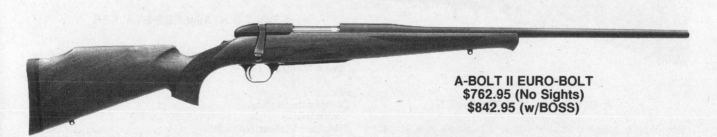

A-BOLT II EURO-BOLT
$762.95 (No Sights)
$842.95 (w/BOSS)

This new A-Bolt II rifle features a schnabel-style forearm and a rounded bolt shroud plus a continental-style cheekpiece that provides improved handling and shooting comfort. The finish is low-luster blueing and satin-finished walnut. See specifications above for A-Bolt II models.

BROWNING RIFLES

A-BOLT II BOLT-ACTION RIFLE

A-BOLT II BOLT-ACTION CENTERFIRE RIFLES

Calibers: Long Action 25-06 Rem., 270 Win., 280 Rem., 30-06 Sprg.; **Long Action Magnum** 375 H&H, 7mm Rem. Mag., 300 Win. Mag., 338 Win. Mag. **Action:** Short throw bolt of 60 degrees. Plunger-type ejector. **Magazine:** Detachable. Depress the magazine latch and the hinged floorplate swings down. The magazine can be removed from the floorplate for reloading or safety reasons. **Trigger:** Adjustable within the average range of 3 to 6 pounds. Also grooved to provide sure finger control. **Stock and forearm:** Stock is select grade American walnut cut to the lines of a classic sporter with a full pistol grip.

BOSS (Ballistic Optimizing Shooting System) is now optional on all A-Bolt II models (except standard on Varmint). BOSS adjusts barrel vibrations to allow a bullet to leave the rifle muzzle at the most advantageous point in the barrel oscillation, thereby fine-tuning accuracy with any brand of ammunition regardless of caliber.

Scopes: Closed. Clean tapered barrel. Receiver is drilled and tapped for a scope mount; or select **Hunter** model has open sights. **Barrel lengths:** 20″ (Micro Medallion); 22″ (Short

and Long Action); 26″ (Long Action Magnum). Hammer-forged rifling where a precision machined mandrel is inserted into the bore. The mandrel is a reproduction of the rifling in reverse. As hammer forces are applied to the exterior of the barrel, the barrel is actually molded around the mandrel to produce flawless rifling and to guarantee a straight bore. Free-floated. **Overall length:** 44¼″. **Weight:** 7 lbs. 8 oz. in Magnum; 6 lbs. 8 oz. in Short Action; 7 lbs. in Standard (Long Action).

Short Action A-Bolt available in 22 Hornet, 22-250 Rem., 223 Rem., 243 Win., 257 Roberts, 284 Win., 7mm-08 Rem. and 308 Win.

Prices:

Hunter	**$559.95**
with BOSS	639.95
with open sights	629.95
Medallion No sights	652.95
with BOSS	732.95
Medallion Left-Hand	679.95
with BOSS	759.95
Medallion 375 H&H w/Open sights	756.95
Medallion 375 H&H L.H., open sights	782.95
Micro Medallion No sights	652.95
Gold Medallion	876.95
with BOSS	956.95

A-BOLT II COMPOSITE STALKER

A-BOLT II STAINLESS STALKER

A-BOLT II STALKER SERIES

Browning's A-Bolt II Stalker series is available in a Stainless version or Composite stock version. The graphite-fiberglass composite stock resists the nicks and scrapes of hard hunting and is resistant to weather and humidity. Its recoil-absorbing properties also make shooting a more pleasant experience. The stock is checkered for a good grip and has a non-glare textured finish. The A-Bolt Composite or Stainless Stalker share the same features of Browning's A-Bolt Hunter rifle

(above). All exposed metal surfaces of the Composite Stalker have a non-glare matte blued finish.

Prices:

Composite Stalker	**$576.95**
with BOSS	656.95
Stainless Stalker No sights	726.95
with BOSS	803.95
Stainless Stalker Left Hand	749.95
with BOSS	829.95
Stainless Stalker 375 H&H w/Open sights	827.95
Same as above in Left Hand	853.95
Varmint 22-250 Rem. and 223 Rem.	
w/heavy barrel and BOSS	856.95

BROWNING RIFLES

22 SEMIAUTOMATIC RIMFIRE RIFLES
GRADES I AND VI

SPECIFICATIONS
Caliber: 22 LR **Overall length:** 37". **Barrel length** 19¼".
Weight: 4 lbs. 4 oz. **Safety:** Cross-bolt type. **Capacity:** 11
cartridges in magazine, 1 chamber. **Trigger:** Grade I is
blued; Grade VI is gold colored. **Sights:** Gold bead front,
adjustable folding leaf ear; drilled and tapped for Browning
scope mounts. **Length of pull:** 13¾". **Drop at comb:** 1³/₁₆".
Drop at heel: 2⅝". **Stock & Forearm:** Grade I, select walnut
with checkering (18 lines/inch); Grade VI, high-grade walnut
with checkering (22/lines inch).
Prices:
Grade I $363.95
Grade VI 747.95

GRADE VI ENGRAVED
(24-Karat Gold Plated)

BAR MARK II SAFARI

BAR MARK II SAFARI
SEMIAUTOMATIC RIFLES

The BAR has been upgraded to include an engraved receiver,
a redesigned bolt release, new gas and buffeting systems, and
a removable trigger assembly. Additional features include:
cross-bolt safety with enlarged head; hinged floorplate, 4-shot
capacity in Standard models (1 in chamber); gold trigger; select
walnut stock and forearm with cut-checkering and swivel
studs; 13¾" length of pull; 2" drop at heel; 1⅝" drop at comb;
and a recoil pad (magnum calibers only).

Prices:
Standard Calibers Open sights $681.95
  with BOSS (no sights) 744.95
Magnum Calibers open sights 731.95
  with BOSS (no sights) 794.95
(Deduct $17 without sights)

SPECIFICATIONS: BAR MARK II SAFARI

Model	Calibers	Barrel Length	Sight Radius*	Overall Length	Average Weight	Rate of Twist (Right Hand)
Magnum	270 Weatherby	24"	19½"	45"	8 lbs. 6 oz.	1 in 10"
Magnum	338 Win. Mag.	24"	19½"	45"	8 lbs. 6 oz.	1 in 12"
Magnum	300 Win. Mag.	24"	19½"	45"	8 lbs. 6 oz.	1 in 10"
Magnum	7mm Rem. Mag.	24"	19½"	45"	8 lbs. 6 oz.	1 in 10"
Standard	243 Win.	22"	17½"	43"	7 lbs. 10 oz.	1 in 10"
Standard	270 Win.	22"	17½"	43"	7 lbs. 9 oz.	1 in 10"
Standard	30-06 Sprgfld.	22"	17½"	43"	7 lbs. 6 oz.	1 in 10"
Standard	308 Win.	22"	17½"	43"	7 lbs. 9 oz.	1 in 12"

* All models are available with or without open sights. All models drilled and tapped for scope mounts.

CLIFTON ARMS

SCOUT RIFLE
$2750.00

Several years ago, in response to Colonel Jeff Cooper's concept of an all-purpose rifle, which he calls the "Scout Rifle," Clifton Arms developed the integral, retractable bipod and its accompanying state-of-the-art composite stock. Further development resulted in an integral butt magazine well for storage of cartridges inside the buttstock. These and other components make up the Clifton Scout Rifle.

SPECIFICATIONS
Calibers: 243, 7mm-08, 30-06, 308, 350 Rem. Mag.
Barrel length: 19″ to 19½″ (longer or shorter lengths available; made with Shilen stainless premium match-grade steel)
Weight: 7 to 8 lbs.
Sights: Forward-mounted Burris 2¾X Scout Scope attached to integral scope base pedestals machined in the barrel; Warne rings; reserve iron sight is square post dovetailed into a ramp integral to the barrel, plus a large aperture "ghost ring" mounted on the receiver bridge.
Features: Standard action is Ruger 77 MKII stainless; metal finish options include Polymax, NP3 and chrome sulphide; left-hand rifles available.

COLT RIFLES

COLT SPORTER LIGHTWEIGHTS
$877.95

The Colt Sporter semiautomatic rifle fires from a closed bolt, is easy to load and unload, and has a buttstock and pistol grip made of tough nylon. A round, ribbed handguard is fiberglass-reinforced to ensure better grip control. **Calibers:** 223 Rem., 7.62×39mm and 9mm. **Barrel length:** 16″. **Overall length:** 34½″ (35½″ in 7.62×39mm) **Weight:** 6.7 lbs. (223 Rem.); 7.1 lbs. (9mm); 7 lbs. 3 oz. (7.62×39mm) **Capacity:** 5 rounds.

SPORTER RIFLES

The Colt sporter is range-selected for top accuracy. It has a 3-9x rubber armored variable power scope mount, carry handle with iron sight, Cordura nylon case and other accessories. **Caliber:** 223 Rem. **Barrel length:** 20″. **Overall length:** 39″ **Weight:** 8 lbs. (Match H-Bar); 7½ lbs. (Target); 8½ lbs. (Competition H-Bar). **Capacity:** 5 rounds.

Prices:
MATCH H-BAR	$ 944.00
TARGET	897.95
COMPETITION H-BAR	989.95
W/Scope, accessories, range-selected	1497.00

COOPER ARMS RIFLES

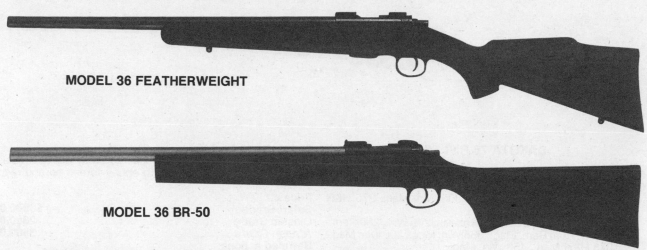

MODEL 36 FEATHERWEIGHT

MODEL 36 BR-50

MODEL 36RF MARKSMAN SERIES

SPECIFICATIONS
Caliber: 22 LR   **Capacity:** 5-shot magazine
Action: bolt-action repeater
Barrel length: 23″ (chrome moly); free-floated barrel w/competition step crown
Stock: AA Claro walnut (AAA Claro walnut in Custom and Custom Classic models; Custom Classic has ebony for-end tip)
Features: Glass-beaded adjustable trigger; bases and rings optional; 3 mid-locking lugs

Prices:
STANDARD MODEL $1125.00
CUSTOM MODEL 1250.00
CUSTOM CLASSIC 1350.00

Also available:
MODEL 36 CF: Same specifications as above, but in centerfire calibers 17 CCM, 22 CCM and 22 Hornet. Prices are same.
MODEL 36 RF/CF FEATHERWEIGHT: Same specifications as Model 36 RF, but with 22″ barrel $1195.00
MODEL 36 RF BR-50: Same specifications as Model 36 RF, but with stainless steel 22″ barrel 1295.00

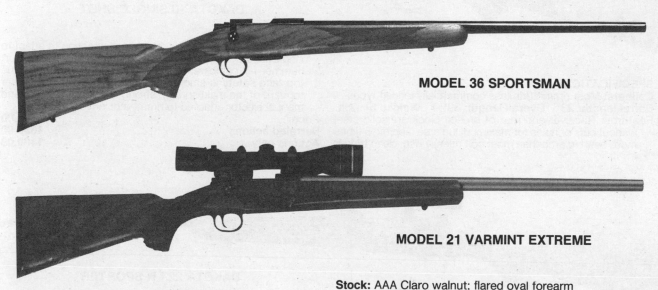

MODEL 36 SPORTSMAN

MODEL 21 VARMINT EXTREME

Stock: AAA Claro walnut; flared oval forearm
Other specifications same as Model 36 RF.
Price $1225.00

Also available:
MODEL 21 CUSTOM CLASSIC. Same as above, but with chrome moly barrel.
Price $1350.00

MODEL 21 VARMINT EXTREME

SPECIFICATIONS
Calibers: 221 Fireball, 222, 223, 17 Rem.
Barrel length: 24″

DAKOTA ARMS

DAKOTA 76 AFRICAN GRADE

DAKOTA 76 RIFLES

SPECIFICATIONS
Calibers:
    Safari Grade: 338 Win. Mag., 300 Win. Mag., 375 H&H Mag., 458 Win. Mag.
    Classic Grade: 22-250, 257 Roberts, 270 Win., 280 Rem., 30-06, 7mm Rem. Mag., 338 Win. Mag., 300 Wln. Mag., 375 H&H Mag., 458 Win. Mag.
    African Grade: 404 Jeffery, 416 Dakota, 416 Rigby, 450 Dakota
Barrel lengths: 21″ or 23″ (Clas.); 23″ only (Saf.); 24″ (Afr.)
Weight: 7¹/₂ lbs. (Classic); 9¹/₂ lbs. (African); 8¹/₂ lbs. (Safari)
Safety: Three-position striker-blocking safety allows bolt operation with safety on

Sights: Ramp front sight; standing leaf rear sight
Stock: Medium fancy walnut stock fitted with recoil pad (Classic); fancy walnut stock with ebony forend tip and recoil pad (Safari)
Prices:
Safari Grade	**$3000.00**
Classic Grade	2300.00
African Grade	3500.00
Barreled actions:	
Safari Grade	1850.00
Classic Grade	1650.00
African Grade	2500.00
Actions:	
Safari Grade	1500.00
Classic Grade	1400.00

DAKOTA 10 SINGLE SHOT

SPECIFICATIONS
Calibers: Most rimmed/rimless commercially loaded types
Barrel length: 23″   **Overall length:** 39½″   **Weight:** 5¹/₂ lbs.
Features: Receiver and rear of breech block are solid steel without cuts or holes for maximum lug area (approx. 8 times more bearing area than most bolt rifles); crisp, clean trigger pull (trigger plate is removable, allowing action to adapt to single set triggers); straight-line coil spring action and short hammer fall combine for extremely fast lock time; unique top tang safety is smooth and quiet (it blocks the striker forward of the main spring); strong, positive extractor and manual ejector adapted to rimmed or rimless cases.

Price:	**$2500.00**
Barreled actions	1650.00
Actions only	1400.00

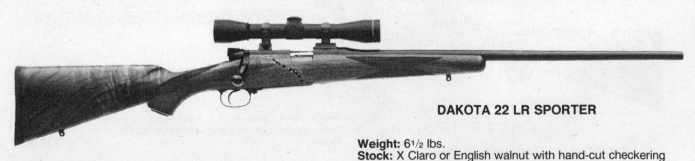

DAKOTA 22 LR SPORTER

SPECIFICATIONS
Calibers: 22 LR   **Capacity:** 5-round clip
Barrel length: 22″ (chrome-moly, 1 turn in 16″)

Weight: 6¹/₂ lbs.
Stock: X Claro or English walnut with hand-cut checkering
Features: Plain bolt handle; swivels and single screw studs; ¹/₂-inch black pad; 13⁵/₈″ length of pull

Price:	**$1500.00**
Barreled actions	1200.00

EAGLE ARMS EA-15 RIFLES

All EA-15 rifles include these standard features: upper and lower receivers precision machined from 7075 T6 aluminum forgings • receivers equipped with push-type pivot pin for easy disassembly • EZ-style forward assist mechanism that clears jammed shells from the chamber • trapdoor-style buttstock • full-size 30-round magazine

MODEL EA-15 MATCH RIFLE
With Action Master
or Match Grade Accessories $1075.00
With Dovetailed Sight
or Aluminum Gas Housing $1115.00

aluminum handguard tube (allows for a free-floating barrel); compensator for reducing muzzle climb; NM Match Trigger Group & Bolt Carrier Group; fixed stock

SPECIFICATIONS
Caliber: 223
Barrel length: 20″ Douglas Premium fluted barrel
Weight: 8 lbs. 5 oz.; 8 lbs. 6 oz. w/Match Grade accessories
Features: One-piece international-style upper receiver; solid

MODEL EA-15 GOLDEN EAGLE MATCH RIFLE
$1075.00

Also available:
EA-15 STANDARD (Wt. 7 lbs.)	**$ 800.00**
EA-15 w/E-2 Accessories (Wt. 8 lbs. 14 oz.)	895.00
Same as above with NM sights (Wt. 9 lbs.) . . .	945.00
EA-15 CARBINE w/E-2 Accessories 16″ barrel,	
Sliding buttstock (Wt. 7 lbs.)	895.00
With 5″ muzzle break	915.00
EA-15 EAGLE EYE with National Match	
High Power accessories (Wt. 14 lbs.)	1495.00
EA-15 EAGLE SPIRIT (Wt. 8 lbs.)	1075.00

SPECIFICATIONS
Caliber: 223
Barrel length: 20″ Douglas Premium extra heavy w/1:9″ twist
Weight: 12 lbs. 12 oz.
Sights: Elevation adjustable NM front sight w/set screw; E-2 style NM rear sight assembly with ½MOA adjustments for windage and elevation
Stock: Fixed

EMF RIFLES

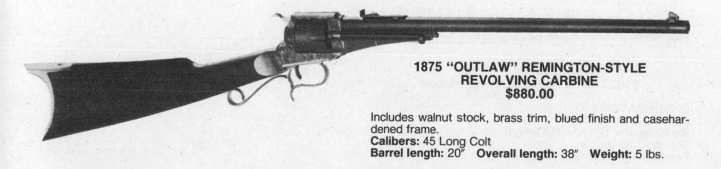

1875 "OUTLAW" REMINGTON-STYLE REVOLVING CARBINE
$880.00

Includes walnut stock, brass trim, blued finish and casehardened frame.
Calibers: 45 Long Colt
Barrel length: 20″ **Overall length:** 38″ **Weight:** 5 lbs.

EMF REPLICA RIFLES

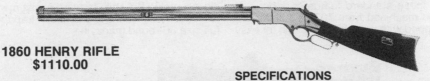

1860 HENRY RIFLE
$1110.00

This lever-action rifle was patented by B. Tyler Henry and produced by the New Haven Arms Company, where Oliver Winchester was then president (he later gave his name to future models and the company itself). Production was developed between 1860 and 1865, with serial numbers 1 to 12000 (plus 2000 additional units in 1866, when the Winchester gun first appeared).

SPECIFICATIONS
Calibers: 44-40 and 45 LC
Barrel length: 24¼"; upper half-octagonal w/magazine tube in one-piece steel
Overall length: 43¾"
Weight: 9¼ lbs.
Stock: Varnished American walnut wood
Features: Polished brass frame; brass buttplate

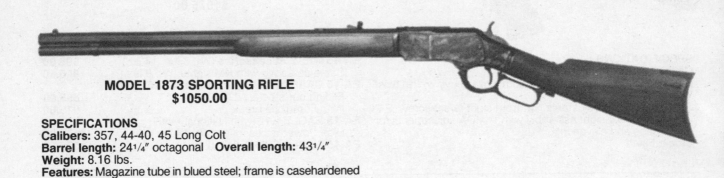

MODEL 1866 YELLOW BOY RIFLE & CARBINE
$848.00 (Rifle) $825.00 (Carbine)

These exact reproductions of guns used over 100 years ago are available in 45 Long Colt, 38 Special and 44-40. Both carbine and rifle are offered with blued finish, walnut stock and brass frame.

MODEL 1873 SPORTING RIFLE
$1050.00

SPECIFICATIONS
Calibers: 357, 44-40, 45 Long Colt
Barrel length: 24¼" octagonal   **Overall length:** 43¼"
Weight: 8.16 lbs.
Features: Magazine tube in blued steel; frame is casehardened steel; stock and forend are walnut wood

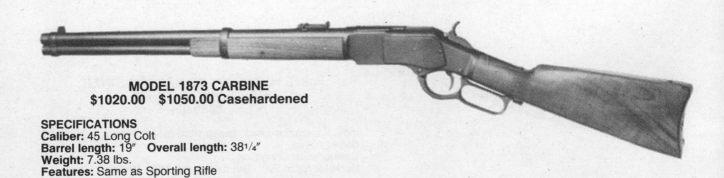

MODEL 1873 CARBINE
$1020.00 $1050.00 Casehardened

SPECIFICATIONS
Caliber: 45 Long Colt
Barrel length: 19"   **Overall length:** 38¼"
Weight: 7.38 lbs.
Features: Same as Sporting Rifle

EUROPEAN AMERICAN ARMORY

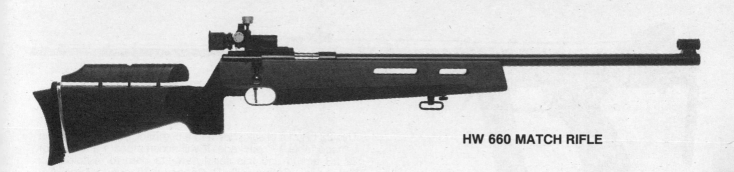

HW 660 MATCH RIFLE

HW 660 MATCH RIFLE (SINGLE SHOT)
$795.00

SPECIFICATIONS
Caliber: 22 LR
Barrel length: 26.8"
Overall length: 45.7"
Weight: 10.8 lbs.
Sights: Match-type aperture rear; hooded ramp front
Finish: Blue

Stock: Stippled walnut
Features: Adjustable match trigger; push-button safety; walnut adjustable buttplate; left-hand stock available

Also available:
HW 60 TARGET RIFLE. Same specifications as Model HW 660, except with target stock. **Price: $705.00**

HW 60 TARGET RIFLE

FEATHER RIFLES

MODEL AT-22
$249.95

This 22 LR rifle breaks down to a compact, easy-to-stow and transportable 17″ package. It will accommodate any kind of 22 LR ammo and has set a new standard for autoloading rimfire rifles. **Caliber:** 22 LR. **Capacity:** 20 rounds. **Type:** Autoloader. **Operation:** Blowback. **Barrel length:** 17″. **Overall length:** 35″ (26″ w/stock folded). **Weight:** 3.25 lbs.

MODEL F2 AT-22

Also available:
MODEL F2 AT-22. Same as Model AT-22, but with fixed buttstock made of high-impact polymer for a more traditional look. **Price:** **$279.95**

MODEL F9 AT-9

MODEL AT-9 (not shown)
$390.00

Feather's AT-9 offers 3″ groups or less at 50 yards using standard, fully adjustable sights. Ideal for competitive use in rapid-fire events like pin-shooting, weekend plinking or personal security. **Caliber:** 9mm. **Capacity:** 25 rounds. **Type:** Autoloader. **Operation:** Blowback. **Barrel length:** 17″. **Overall length:** 35″ (26½″ w/stock folded). **Weight:** 5 lbs.

Also available:
MODEL F9 AT-9. Same as Model AT-9, but with fixed buttstock made of high-impact polymer for a more traditional look. **Price:** **$420.00**

FRANCOTTE RIFLES

August Francotte rifles are available in all calibers for which barrels and chambers are made. All guns are custom made to the customer's specifications; there are no standard models. Most bolt-action rifles use commercial Mauser actions; however, the magnum action is produced by Francotte exclusively for its own production. Side-by-side and mountain rifles use either boxlock or sidelock action. Francotte system sidelocks are back-action type. Options include gold and silver inlay, special engraving and exhibition and museum grade wood. Francotte rifles are distributed in the U.S. by Armes de Chasse (see Directory of Manufacturers and Distributors for details).

BOLT-ACTION RIFLE

SPECIFICATIONS
Calibers: 17 Bee, 7×64, 30-06, 270, 222R, 243W, 308W, 375 H&H, 416 Rigby, 460 Weatherby, 505 Gibbs
Barrel length: To customer's specifications
Weight: 8 to 12 lbs., or to customer's specifications
Stock: A wide selection of wood in all possible styles according to customer preferences
Engraving: Per customer specifications
Sights: All types of sights and scopes
Prices:
BOLT-ACTION RIFLES
Standard Bolt Action (30-06, 270, 7×64, etc.) . . $ 5,572.00
Short Bolt Action (222R, 243W, etc.) 6,930.00
African Action (416 Rigby, 460 Wby., etc.) 9,960.00
Magnum Action (300 WM, 338 WM, 375 H&H, 458 WM 6,180.00

BOXLOCK SIDE-BY-SIDE DOUBLE RIFLES
Std. boxlock double rifle (9.3×74R, 8×57JRS, 7×65R, etc.) $11,513.00
Std. boxlock double (Magnum calibers). 15,293.00
Optional sideplates, **add** 1,700.00

SIDELOCK S/S DOUBLE RIFLES
Std. sidelock double rifle (9.3×74R, 8×57JRS, 7×65R, etc.) $30,000.00
Std. sidelock double (Magnum calibers) 27,761.00
Special safari sidelock . . . . **Price on request**

MOUNTAIN RIFLES
Standard boxlock . . . . $ 9,186.00
Std. boxlock in magnum & rimless calibers **Price on request**
Optional sideplates, **add** 1,158.00
Standard sidelock 20,538.00

CARL GUSTAF RIFLES

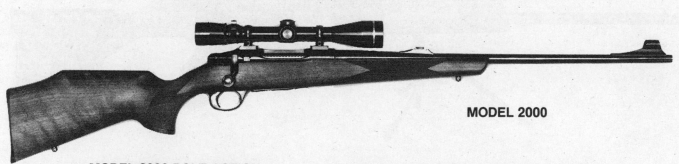

MODEL 2000

MODEL 2000 BOLT ACTION
$1985.00 (w/Sights)
$1875.00 (without Sights)

The Carl Gustaf 2000 rifle, created by one of the oldest riflemakers in the world, uses the latest Swedish hi-tech engineering, design and manufacturing methods to produce this fine hunting piece, while retaining the balance and sleek lines of its famous predecessor, the *Husqvarna*. Each rifle bears the "Crowned C" stamp, the exclusive Royal Swedish symbol of quality and durability.

SPECIFICATIONS
Calibers: 30-06, 308 Win., 6.5×55, 7×64, 9.3×63, 243, 270, 7mm Rem. Mag., 300 Win. Mag.
Capacity: 3-round clip (4-round clip optional)
Barrel length: 24″ **Overall length:** 44″
Weight: 7½ lbs.
Sights: Hooded ramp front; open rear (drilled and tapped for scope mounting)
Stock: Cheekpiece with Monte Carlo; Wundhammer palm swell on pistol grip **Length of pull:** 13¾″

HARRINGTON & RICHARDSON

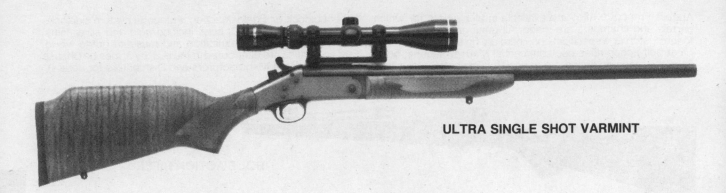

ULTRA SINGLE SHOT VARMINT

ULTRA SINGLE SHOT VARMINT RIFLE
$249.95

SPECIFICATIONS
Calibers: 223 Rem. and 22-250 Rem.
Action: Break-open; side lever release; positive ejection
Barrel length: 22″ heavy varmint profile
Weight: 7 to 8 lbs.
Sights: None (scope mount included)

Length of pull: 14¹/₄″
Drop at comb: 1¹/₄″
Drop at heel: 1¹/₈″
Forend: Semi-beavertail
Stock: Monte Carlo; hand-checkered curly maple
Features: Sling swivels on stock and forend; patented transfer bar safety; automatic ejection; hammer extension; rebated muzzle; Uncle Mike recoil pad

HECKLER & KOCH RIFLES

MODEL HK PSG-1 HIGH PRECISION
MARKSMAN'S RIFLE
$9750.00

SPECIFICATIONS
Caliber: 308 (7.62mm)
Capacity: 5 rounds and 20 rounds
Barrel length: 25.6″
Rifling: 4 groove, polygonal
Twist: 12″, right hand

Overall length: 47.5″
Weight: 17.8 lbs.
Sights: Hensoldt 6×42 telescopic
Stock: Matte black, high-impact plastic
Finish: Matte black, phosphated

HEYM RIFLES

MODEL EXPRESS
$6550.00

SPECIFICATIONS
Calibers: 338 Lapua Magnum, 375 H&H, 416 Rigby, 378 Weatherby Magnum, 460Weatherby Magnum, 450 Ackley, 500 A-Square, 500 Nitro Express

Also available:
MODEL EXPRESS 600 in 600 Nitro Express . . . **$11,350.00**
  For Left-hand Model, **add** **595.00**
MODEL 55B OVER/UNDER (308 Win., 30-06,
  375 H&H, 458 Win. Mag., 470 N.E.) **10,800.00**

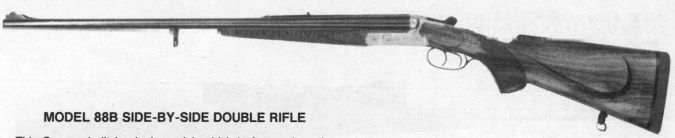

MODEL 88B SIDE-BY-SIDE DOUBLE RIFLE

This German-built boxlock model, which is featured on the front cover of this edition of SHOOTER'S BIBLE, has a modified Anson & Deeley action with standing sears, plus Purdey-type double underlocking lugs and Greener extension with cross-bolt. Actions are furnished with sliding safeties and cocking indicators on the top tang, nonbreakable coil springs, front single set triggers and steel trigger guards.

SPECIFICATIONS
Calibers: 8×57 JRS, 9.3×74R and 30-06
Weight: 8 lbs.
Features: Ejectors, small action

Price: **$12,500.00**
In 375 H&H Mag. **14,450.00**

Also available:
MODEL 88B/SS. Sidelock version of above **$16,600.00**
MODEL 88B "SAFARI": Same general specifications as above, except **calibers:** 375 H&H, 458 Win. Mag., 470 Nitro Express, 500 Nitro Express. **Weight:** 10 lbs., large frame.
Price: **$16,400.00**
Interchangeable barrels for Safari model: **5,870.00**

CALIBER/CARTRIDGE AVAILABILITY

Model	.600 N.E.	.500 A-Square	.500 N.E.	.470 N.E.	.460 Weath. Mag.	.458 Win. Mag.	.416 Rigby	.378 Weath. Mag.	.375 H & H Mag.	.338 Lap. Mag.	.338 Win. Mag.	.300 Win. Mag.	7 mm Rem. Mag.	9,5 x 74 R	9,3 x 64	8 x 75 RS	8 x 68 S	6,5 x 68	6,5 x 65 (R)	9,3 x 62	8 x 57 I (R) S	7 x 65 R	7 x 64	7 x 57 (R)	6,5 x 64 Bre.	6,5 x 57 (R)	6,5 x 55 SM	6 x 62 Freres (R)	6 x 57 (R)	5,6 x 57 (R)	5,6 x 52 R	5,6 x 50 (R) Mag.	.30 R Blaser	.30-06	.308 Win.
Mod. »Express«	●	●	●		●		●	●	●	●																									
Mod. 55 B					●	●			●			●						●	●														●	●	●
Mod. 88 B/BSS									●							●		●	●															●	
Mod. 88 B »Safari«			●	●		●			●																										

Kugelkaliber Calibers

JARRETT CUSTOM RIFLES

PRIVATE COLLECTION SERIES

All Jarrett Private Collection rifles include the following features: Metal finish • Top-mounted bolt release • Break-in and load development with 20 rounds • Setup for switch barrel • Serial number

MODELS 2 AND 3
$3495.00

MODEL 2 features: McMillan Mountain Rifle stock • Remington 700 Magnum receiver • Jarrett #4 tapered barrel (.308 bore w/1 in 12″ twist) • Remington conversion trigger • Olive drab metal finish • Forest camo stock • 3.5x10x50 Leupold scope with A.O. matte finish • Pachmayr decelerator pad and Jarrett muzzle brake kit

MODEL 3 offers: Caliber 7mm STW • Jarrett .284 bore barrel w/1 in 10″ twist (25½″ at crown) • Black textured stock • Leupold base w/30mm Redfield rings. Other specifications same as Model 1. Also available in 300 Win. Mag. with black metal and Old English Pachmayr 1″ pad.

STANDARD HUNTING RIFLE
$2850.00

Features: Caliber 280 IMP • McMillan Classic stock • Remington 700 Long Action • Jarrett barrel with .284 bore and 1 in 10″ twist • Leupold mount system • Blued receiver • Forest camo stock finish • 1″ decelerator pad

LIGHTWEIGHT VARMINT RIFLE
$2850.00

Features: Caliber 223 • Remington XP-100 action • Jarrett barrel w/.224 bore and 1 in 14″ twist (#4 tapered to 23″ at crown) • Leupold mount system • Black Teflon metal finish • Gray textured-finish stock • 6.5x20 scope (optional) • Remington conversion trigger

KDF RIFLES

MODEL K15
$1950.00 ($2000.00 in Magnum)

Trigger: Competition-quality single stage; adjustable for travel, pull and sear engagement

Safety: Located on right-hand side

Stocks: Kevlar composite or laminate stock; Pachmayr decelerator pad; quick detachable swivels; AAA grade walnut stocks with 22-line hand-checkering; ebony forend; pistol-grip cap and crossbolts (walnut stocks may be ordered in classic, schnabel or thumbhole style)

Features and options: Iron sights; recoil arrestor; choice of metal finishes; 3 locking lugs w/large contact area; 60-degree lift for fast loading; box-style magazine system; easily accessible bolt release catch; fully machined, hinged bottom metal

SPECIFICATIONS
Calibers:
> **Standard**—25-06, 270 Win., 280 Rem., 30-06
> **Magnum**—7mm Rem., 300 Win., 300 Wby., 338 Win., 340 Wby., 375 H&H, 411 KDF, 416 Rem., 458 Win.

Capacity: 4 rounds (3 rounds in Magnum)

Barrel lengths: 24″ (26″ in Magnum)

Overall length: 44¹/₂″ (46¹/₂″ in Magnum)

Weight: 8 lbs. (approx.)

Receiver: Drilled and tapped for scope mounts (KDF bases available to take 1″ or 30mm rings)

MODEL K-15 LIGHTWEIGHT
$1990.00   ($2400.00 in Magnum)

KIMBER RIFLES

MODEL 82C CLASSIC

MODEL 82C 22 LR CLASSIC
$696.00

The Kimber 22 Long Rifle Model 82, first introduced in 1980, became a leader in the resurgence of interest in premium-grade rimfire rifles. Not since the demise of the original American-made Winchester 52 Sporter 30 years earlier had there been a .22 sporting rifle suitable for serious rimfire rifle enthusiasts. The latest version of the Kimber Model 82 includes the following features: a large diameter receiver, fully machined from solid steel and threaded at the front • a rear locking bolt with twin horizontally opposed lugs • single-stage trigger adjustable for pressure and overtravel • independent coil spring-operated bolt stop • coil spring-loaded bolt detent in rear of receiver • rocker-style safety • and much more.

SPECIFICATIONS
Caliber: 22 LR

Capacity: 4-shot clip (10-shot optional)

Barrel length: 21″ (1 turn in 16″)

Overall length: 40¹/₂″   **Weight:** 6¹/₂ lbs.

Stock: Standard grade claro walnut

Length of pull: 13¹/₂″

Features: Red rubber buttpad w/Kimber insignia; plain buttstock (no cheekpiece); hand checkering (20 lines per inch); polished steel pistol-grip cap

Also available:

MODEL 82C SUPERAMERICA with AAA fancy-grade claro walnut; genuine ebony forend tip; beaded cheekpiece; black rubber buttstock; hand-checkering (20 lines per inch).
Price: **$1175.00**

MODEL 82C CUSTOM SHOP SUPERAM. Same as above with Neidner-style steel buttplate and numerous options.
Price: **$1250.00**

KRIEGHOFF DOUBLE RIFLES

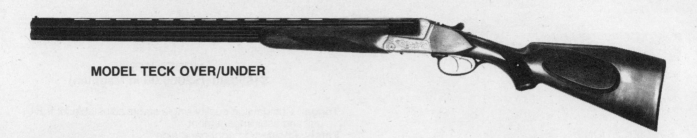

MODEL TECK OVER/UNDER

SPECIFICATIONS
Calibers: 308, 30-06, 300 Win. Mag., 9.3×74R, 8×57JRS, 7×65R, 458 Win. Mag.
Barrel length: 25″   **Weight:** 7½ lbs.
Action: Boxlock; double greener-type crossbolt and double-barrel lug locking, steel receiver
Triggers: Double triggers; single trigger optional
Safety: Located on top tang
Sights: Open sight with right-angle front sight
Stock: German-styled with pistol grip and cheekpiece; oil-finished (14³/₈″ long)
Finish: Nickel-plated steel receiver with satin grey finish
Prices:
Model Teck (Boxlock) . . . . **$ 9,200.00**
  In 458 Win. Mag. **9,900.00**
Teck-Handspanner (16 ga. receiver only;
  7×65R, 30-06, 308 Win.) **10,970.00**
Also available:
TRUMPF SBS (Side-by-side boxlock) **16,150.00**

MODEL ULM OVER/UNDER
(not shown)

SPECIFICATIONS
Calibers: 308 Win., 30-06, 300 Win. Mag., 375 H&H, 458 Win. Mag.
Barrel length: 25″   **Weight:** 7.8 lbs.
Triggers: Double triggers (front trigger=bottom; rear trigger=upper
Safety: Located on top tang
Sights: Open sight w/right-angle front sight
Stock: German-styled with pistol grip and cheekpiece; oil-finished (14³/₈″ long); semi-beavertail forend
Prices:
Model ULM (Sidelock) **$14,580.00**
ULM PRIMUS (Deluxe Sidelock) **20,500.00**
Also available:
NEPTUN DRILLING **14,950.00**

LAURONA RIFLES

MODEL 2000X EXPRESS RIFLE

MODEL 2000X EXPRESS
$2919.00   ($3279.00 Magnum)

SPECIFICATIONS
Calibers: 9.3×74R, 30-06 Spfd., 8×57 JRS, 8×75 RS, 375 H&H Mag.
Barrel length: 23.8″ (25″ in 375 H&H Mag.)
Weight: 8½lbs. (8.15 lbs. in 9.3×74R)

Triggers: Single or double
Stock: Monte Carlo w/rubber recoil pad
Drop at comb: 14″   **Drop at heel:** 20.25″
Features: Tulip forend; automatic ejectors (or extractors); high alloy full-length hinge pin; anti-rust matte black chrome; hand-checkered select-grade walnut with non-glare finish; engravings on receivers; convergency adjustment; drilled and tapped for Leupold mounts

LAKEFIELD SPORTING RIFLES

SPECIFICATIONS

MODEL 90B BIATHLON
$558.95   ($614.95 Left Hand)

MODEL 92S SILHOUETTE
$379.95   ($417.95 Left Hand)

Model:	90B	92S
Caliber:	.22 Long Rifle Only	.22 Long Rifle
Capacity:	5-shot metal magazine	5-shot metal magazine
Action:	Self-cocking bolt action, thumb-operated rotary safety	Self-cocking bolt action, thumb-operated rotary safety
Stock:	One-piece target-type stock with natural finish hardwood; comes with clip holder, carrying & shooting rails, butt hook and hand stop	One-piece high comb, target-type with walnut finish hardwood
Barrel Length:	21″ w/snow cover	21″
Sights:	Receiver peep sights with 1/4 min. click micrometer adjustments; target front sight with inserts	None (receiver drilled and tapped for scope base)
Overall Length:	39⅝″	39⅝″
Approx. Weight:	8¼ lbs.	8 lbs.

MARK I SINGLE SHOT
$124.95   ($137.95 Left Hand)

MODEL 64B SEMIAUTO
$138.95

Also available:
MARK I "SMOOTH BORE" (20¾″ barrel) $124.95
  Left Hand 137.95
MARK I YOUTH (19″ barrel) 124.95
  Left Hand 137.95
MARK II & MARK II YOUTH (19″ barrel) 133.95
MARK II LEFT HAND (20½″ barrel) 146.95

SPECIFICATIONS

Model:	MARK I	MARK II	64B
Caliber:	.22 Short, Long or Long Rifle	.22 Long Rifle only	.22 Long Rifle only
Capacity:	Single shot	10-shot clip magazine	10-shot clip magazine
Action:	Self-cocking bolt action, Thumb-operated rotary safety	Self-cocking bolt action Thumb-operated rotary safety	Semiautomatic side ejection, bolt hold-open device, thumb-operated rotary safety
Stock:	One Piece, Walnut Finish Hardwood, Monte Carlo Type with Full Pistol Grip. Checkering on Pistol Grip and Forend.		
Barrel Length:	20¾″	20¾″	20¼″
Sights:	Open bead front sight, adjustable rear sight, receiver grooved for scope mounting.		
Overall Length:	39½″	39½″	40″
Approx. Weight:	5½ lbs.	5½ lbs.	5½ lbs.

LAKEFIELD SPORTING RIFLES

MODEL 91TR TARGET REPEATER
$474.95 ($521.95 Left Hand)

MODEL 91T SINGLE SHOT TARGET
$445.95 ($489.95 Left Hand)

Model	91TR	91T
Caliber:	.22 Long Rifle	.22 Short, Long or Long Rifle
Capacity	5-shot clip magazine	Single shot
Action:	Self-cocking bolt action, thumb-operated rotary safety	Self-cocking bolt action, thumb-operated rotary safety
Stock:	One-piece, target-type with walnut finish hardwood (also available in natural finish); comes with shooting rail and hand stop	One-piece, target-type walnut finish hardwood (also available in natural finish); comes with shootng rail and hand stop
Sights:	Receiver peep sights with 1/4 min. click micrometer adjustments, target front sight with inserts	Receiver peep sights with 1/4 min. click micrometer adjustments, target front sight with inserts
Overall Length:	43⁵/₈″	43⁵/₈″
Approx. Weight:	8 lbs.	8 lbs.

L.A.R. GRIZZLY RIFLE

BIG BOAR COMPETITOR
$2400.00

SPECIFICATIONS
Caliber: 50 BMG
Capacity: Single shot
Action: Bolt action, bull pup, breechloading
Barrel length: 36″
Overall length: 45¹/₂″
Weight: 28.4 lbs.
Safety: Thumb safety
Features: All-steel construction; receiver made of 4140 alloy steel, heat-treated to 42 R/C; bolt made of 4340 alloy steel; low recoil (like 12 ga. shotgun)

MAGTECH RIFLES

MODEL MT-122.2S

MODEL MT-122.2

SPECIFICATIONS
Calibers: 22 Short, Long, Long Rifle, rimfire
Capacity: 6- or 10-shot clip magazine
Barrel length: 24″ w/6-grooved rifling; free floating
Overall length: 43″
Weight: 6½ lbs.
Sights: Fully adjustable micrometer rear; blade front (Model 122.2T)

Features: Sling swivel studs; ³⁄₈″ grooved receiver; cured Brazilian hardwood stock; double locking bolt safety; red cocking indicator and safety lever
Prices:
MODEL 122.2S w/o sights $150.00
MODEL 122.2R w/adj. rear sight 160.00
MODEL 122.2T w/adj. micrometer rear sight 180.00

MARK X RIFLES

ACTIONS & BARRELED ACTIONS

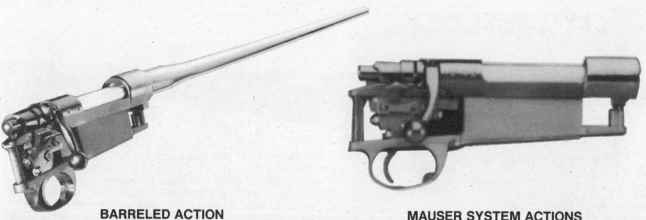

BARRELED ACTION
$331.00

Hand-fitted with premium hammer-forged barrels created from corrosion resistant chrome vanadium steel. Each barreled action is carefully proofed and marked under close government control, ready to drop into the stock of your choice. **Caliber:** 223. **Barrel length:** 24″. **Weight:** 5½ lbs. **Rifling twist:** 10.

MAUSER SYSTEM ACTIONS

Type A: 7×57mm to 30-06. Standard magazine (3³⁄₈″) and bolt face (.470″) $268.00
Type B: 22-250 to 308. Short magazine (2⁷⁄₈″); standard bolt face 268.00
Type C: 7mm Rem. Mag. to 458 Win. Mag. Standard magazine and Magnum bolt face (.532″) 273.00
Mini-Mark X (.17 to .223) 268.00

MARK X RIFLES

MINI-MARK X
$455.00 (Without sights)

SPECIFICATIONS
Calibers: 223, 7.62×39   **Capacity:** 5 rounds
Barrel length: 20″   **Twist:** I turn in 10″
Overall length: 39³/₄″
Weight: 6.35 lbs.
Trigger: Adjustable

MARK X VISCOUNT SPORTER
$471.00   ($486.00 Magnum calibers)

SPECIFICATIONS
Calibers: 30-06, 308 Win.; 300 Win. Mag., 7mm Rem. Mag.
Capacity: 5 rounds; 3 in 7mm Rem. Mag.,
   300 Win. Mag.
Barrel length: 24″   **Overall length:** 45″
Twist: 1 turn in 10″
Weight: 7½ lbs.
Stock: Carbolite with checkered pistol grip and forend, de-
tachable sling swivels

MARK X WHITWORTH
$565.00   ($584.00 in 300 Win. Mag.)

Features forged and machined Mauser System actions . . .
Hammer-forged, chrome, vanadium steel barrels . . . Drilled
and tapped for scope mounts and receiver sights . . . Hooded
ramp front and fully adjustable rear sight . . . All-steel button
release magazine floor plate . . . Detachable sling swivels . . .
Silent sliding thumb safety . . . Prime European walnut stocks
. . . Sculpted, low-profile cheekpiece . . . Rubber recoil butt
plate . . . Steel grip cap.

Calibers: 270 Win., 30-06, 300 Win. Mag. **Barrel length:** 24″.
Overall length: 44″. **Weight:** 7 lbs. **Capacity:** 5 rounds.
Also available:
WHITWORTH EXPRESS RIFLE in 458 Win. Mag. with ad-
justable trigger, express sights. **Price: $703.00**

MARLIN 22 RIFLES

MODEL 60
$153.00

Barrel length: 22″
Overall length: 40 1/2″
Weight: 5 1/2 lbs.
Sights: Ramp front sight; adjustable open rear, receiver grooved for scope mount
Action: Self-loading; side ejection; manual and automatic "last-shot" hold-

SPECIFICATIONS
Caliber: 22 Long Rifle
Capacity: 14-shot tubular magazine with patented closure system

open devices; receiver top has serrated, non-glare finish; cross-bolt safety
Stock: One-piece walnut-finished hardwood Monte Carlo stock with full pistol grip; Mar-Shield® finish

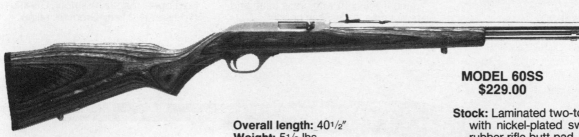

MODEL 60SS
$229.00

Overall length: 40 1/2″
Weight: 5 1/2 lbs.
Sights: Adjustable folding semi-buckhorn rear; ramp front sight with high-visibility post and removable Wide-Scan™ hood

SPECIFICATIONS
Caliber: 22 LR
Capacity: 14 rounds
Barrel length: 22″

Stock: Laminated two-tone Maine birch with nickel-plated swivel studs and rubber rifle butt pad
Features: Micro-Groove® rifling; side ejection; manual bolt hold-open; automatic last-shot bolt hold-open; cross-bolt safety

MODEL 70HC
$161.75

Overall length: 36 3/4″
Weight: 5 1/2 lbs.
Action: Self-loading; side ejection; manual bolt hold-open; receiver top has serrated, non-glare finish; cross-bolt safety

SPECIFICATIONS
Caliber: 22 LR
Capacity: Two 7-shot clip magazines
Barrel length: 18″

Sights: Adjustable open rear, ramp front; receiver grooved for scope mount
Stock: Monte Carlo walnut-finished hardwood with full pistol grip and Mar-Shield® finish

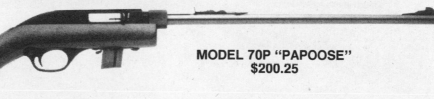

MODEL 70P "PAPOOSE"
$200.25

Overall length: 35 1/4″
Weight: 3 1/4 lbs.
Action: Self-loading; side ejection; manual and "last-shot" bolt hold-open; receiver top has serrated non-glare finish; cross bolt safety

SPECIFICATIONS
Caliber: 22 LR
Capacity: 7-shot clip
Barrel length: 16 1/4″

Sights: Adjustable open rear; ramp front; receiver grooved for scope mount
Stock: Walnut-finished hardwood with full pistol grip and Mar-Shield® finish
Features: Zippered carrying case included

RIFLES

MARLIN 22 RIFLES

MODEL 922 MAGNUM
$377.95

SPECIFICATIONS
Caliber: 22 Win. Mag. Rimfire
Capacity: 7-shot clip magazine
Barrel length: 20 1/2"
Overall length: 39 3/4"

Weight: 6 1/2 lbs.
Sights: Adjustable semi-buckhorn rear; ramp front sight with brass bead and removable Wide-Scan hood™

Stock: Monte Carlo checkered American black walnut with rubber rifle butt pad and swivel studs
Features: Side ejection; manual bolt hold-open; automatic last-shot bolt hold-open; magazine safety; Garand-type safety; Micro-Groove® rifling

MODEL 990L
$223.00

SPECIFICATIONS
Caliber: 22 LR (self-loading)
Capacity: 14 rounds
Barrel length: 22" Micro-Groove®
Overall length: 40 1/2"

Weight: 5.75 lbs.
Sight: Folding semi-buckhorn rear
Stock: Laminated hardwood Monte Carlo
Features: Cross-bolt safety; manual and automatic last-shot bolt hold-open;

solid locking, spring-loaded magazine with patented closure system; swivel studs; rubber rifle butt pad; rustproof receiver grooved for scope mount; gold-plated steel trigger

MODEL 995
$205.75

SPECIFICATIONS
Caliber: 22 Long Rifle
Action: Self-loading
Capacity: 7-shot clip magazine
Barrel: 18" with Micro-Groove® rifling (16 grooves)

Overall length: 36 3/4"
Stock: Monte Carlo genuine American black walnut with full pistol grip; checkering on pistol grip and forend
Sights: Adjustable folding semi-buck-horn rear; ramp front sight with brass bead, Wide-Scan™ hood

Weight: 5 lbs.
Features: Receiver grooved for tip-off scope mount; bolt hold-open device; cross-bolt safety

MARLIN BOLT-ACTION RIFLES

MARLIN 15YN "LITTLE BUCKAROO™"
Single Shot 22 Beginner's Rifle
$156.50

SPECIFICATIONS
Caliber: 22 Short, Long or Long Rifle
Capacity: Single shot
Action: Bolt action; easy-load feed throat; thumb safety; red cocking indicator

Barrel length: 16¼" (16 grooves)
Overall length: 33¼"
Weight: 4¼ lbs.
Sights: Adjustable open rear; ramp front sight

Stock: One-piece walnut finish hardwood Monte Carlo with full pistol grip; tough Mar-Shield® finish

RIFLES

MODEL 25MN
$185.90

SPECIFICATIONS
Caliber: 22 Win. Mag Rimfire (not interchangeable with any other 22 cartridge)

Capacity: 7-shot clip magazine
Barrel length: 22" with Micro-Groove® rifling
Overall length: 41"
Weight: 6 lbs.

Sights: Adjustable open rear, ramp front sight; receiver grooved for scope mount
Stock: One-piece walnut-finished hardwood Monte Carlo with full pistol grip; Mar-Shield® finish

MODEL 25N
$162.60

Same specifications as Model 25MN, except **caliber** 22 LR and **weight** 5½ pounds.

MARLIN 880
$225.35

SPECIFICATIONS
Caliber: 22 Long Rifle
Capacity: 7-shot clip magazine
Action: Bolt action; positive thumb safety; red cocking indicator
Barrel: 22" with Micro-Groove® rifling (16 grooves)

Sights: Adjustable folding semi-buckhorn rear; ramp front with Wide-Scan™ with hood; receiver grooved for scope mount
Overall length: 41"   **Weight:** 5½ lbs.
Stock: Checkered Monte Carlo Ameri-

can black walnut with full pistol grip; tough Mar-Shield® finish; rubber butt pad; swivel studs
Also available: **MODEL 880SS.** Same specifications except with stainless steel (weight 6 lbs.). **$240.80**

MARLIN BOLT-ACTION RIFLES

MODEL 881
$234.75

Specifications same as Marlin 880, except with tubular magazine that holds 17 22 LR cartridges. **Weight:** 6 lbs.

MODEL 882L
$263.45

SPECIFICATIONS
Caliber: 22 WMR (not interchangeable with other 22 cartridges)
Capacity: 7-shot clip magazine
Barrel length: 22″ Micro-Groove®

Overall length: 41″ **Weight:** 6¼ lbs.
Sights: Ramp front w/brass bead and removable Wide-Scan hood; adj. folding semi-buckhorn rear
Stock: Laminated hardwood Monte Carlo w/Mar-Shield® finish

Features: Swivel studs; rubber rifle butt pad; receiver grooved for scope mount; positive thumb safety; red cocking indicator

MODEL 883 MAGNUM
$257.50

SPECIFICATIONS
Caliber: 22 WMR (not interchangeable with other 22 cartridges)
Capacity: 12-shot tubular magazine with patented closure system
Action: Bolt action; positive thumb safety; red cocking indicator

Barrel length: 22″ with Micro-Groove® rifling (20 grooves)
Overall length: 41″ **Weight:** 6 lbs.
Sights: Adjustable folding semi-buckhorn rear; ramp front with Wide-Scan™ hood; receiver grooved for scope mount

Stock: Checkered Monte Carlo American black walnut with full pistol grip; rubber butt pad; swivel studs; tough Mar-Shield® finish
Also available: **MODEL 882 MAGNUM.** Same as Model 883 Magnum, except with 7-shot clip magazine. **$248.50**

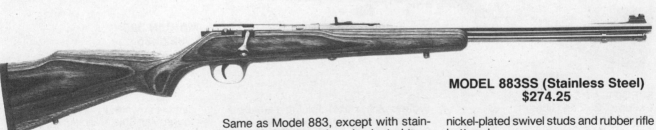

MODEL 883SS (Stainless Steel)
$274.25

Same as Model 883, except with stainless barrel and receiver, laminated two-tone brown Maine birch stock with

nickel-plated swivel studs and rubber rifle butt pad.

MARLIN RIFLES

MODEL 2000A TARGET
$625.00

SPECIFICATIONS
Caliber: 22 Long Rifle
Capacity: Single-shot; 5-shot adapter kit available
Action: Bolt action, 2-stage target trigger, red cocking indicator

Barrel length: 22" Micro-Groove with match chamber, recessed muzzle
Overall length: 41"   **Weight:** 8½ lbs.
Sights: Hooded Lyman front sight with 10 aperture inserts; fully adjustable Lyman target rear peep sight

Stock: Fiberglass + Kevlar with adjustable comb; ambidextrous pistol grip
Also available: **MODEL 2000.** Same specifications as Model 2000A, except without adjustable comb and ambidextrous pistol grip: **$581.90**

MODEL 9 CAMP CARBINE
$384.95

SPECIFICATIONS
Caliber: 9mm
Capacity: 4-shot clip (12-shot avail.)
Action: Self-loading. Manual bolt hold-open. Garand-type safety, magazine safety, loaded chamber indicator. Solid-top, machined steel receiver is

sandblasted to prevent glare, and is drilled/tapped for scope mounting.
Barrel length: 16½" with Micro-Groove® rifling
Overall length: 35½"   **Weight:** 6¾ lbs.
Sights: Adjustable folding rear, ramp front sight with high visibility, orange front sight post; Wide-Scan™ hood.

Stock: Walnut finished hardwood with pistol grip; tough Mar-Shield™ finish; rubber rifle butt pad; swivel studs
Also available: **MODEL 9N** (electroless nickel-plated): **$421.90**

MODEL 45
$384.95

SPECIFICATIONS
Caliber: 45 Auto
Capacity: 7-shot clip
Barrel length: 16½"

Overall length: 35½"
Weight 6.75 lbs.
Sights: Adjustable folding rear; ramp front sight with high visibility, orange front sight post; Wide-Scan™ hood

Stock: Walnut finished hardwood with pistol grip; rubber rifle butt pad; swivel studs

MARLIN LEVER-ACTION CARBINES

MODEL 30AS
$354.25

SPECIFICATIONS
Caliber : 30-30
Capacity: 6-shot tubular magazine
Action: Lever action w/hammer block

safety; solid top receiver w/side ejection
Barrel length: 20″ w/Micro-Groove®
Overall length: 38¼″   **Weight:** 7 lbs.

Sights: Tapped for scope mount and receiver sight; also available in combination w/4x, 32mm, 1″ scope
Stock: Walnut-finish hardwood stock w/ pistol grip; Mar-Shield® finish

MARLIN GOLDEN 39AS
$417.25

The Marlin lever-action 22 is the oldest (since 1891) shoulder gun still being manufactured.

Solid Receiver Top. You can easily mount a scope on your Marlin 39 by screwing on the machined scope adapter base provided. The screw-on base is a neater, more versatile method of mounting a scope on a 22 sporting rifle. The solid top receiver and scope adapter base provide a maximum in eye relief adjustment. If you prefer iron sights, you'll find the 39 receiver clean, flat and sandblasted to prevent glare. Exclusive brass magazine tube.

Micro-Groove® Barrel. Marlin's famous rifling system of multi-grooving has consistently produced fine accuracy because the system grips the bullet more securely, minimizes distortion, and provides a better gas seal.

And the Model 39 maximizes accuracy with the heaviest barrels available on any lever-action 22.

SPECIFICATIONS
Caliber: 22 Short, Long and Long Rifle
Capacity: Tubular magazine holds 26 Short, 21 Long and 19 Long Rifle cartridges
Action: Lever action; solid top receiver; side ejection; one-step takedown; deeply blued metal surfaces; receiver top sandblasted to prevent glare;

hammer block safety; rebounding hammer
Barrel: 24″ with Micro-Groove® rifling (16 grooves)
Overall length: 40″   **Weight:** 6½ lbs.
Sights: Adjustable folding semi-buckhorn rear, ramp front sight with new Wide-Scan™ hood; solid top receiver tapped for scope mount or receiver sight; scope adapter base; offset hammer spur for scope use—works right or left
Stock: Two-piece checkered American black walnut w/fluted comb; full pistol grip and forend; blued-steel forend cap; swivel studs; grip cap; white butt and pistol-grip spacers; Mar-Shield® finish; rubber butt pad

MODEL 39TDS TAKEDOWN
$430.25 (Incl. Carrying Case)

SPECIFICATIONS
Caliber: 22 Short, Long or Long Rifle
Capacity: Tubular mag. holds 16 Short, 12 Long or 11 LR cartridges
Action: Lever action; solid top receiver; side ejection; rebounding hammer; one-step take-down; deep blued metal surfaces; gold-plated trigger

Barrel length: 16½″ lightweight barrel (16 grooves)
Overall length: 32⅝″
Weight: 5¼ lbs.
Safety Hammer block safety
Sights: Adjustable semi-buckhorn folding rear, ramp front with brass bead and Wide-Scan™ hood; top receiver

tapped for scope mount and receiver sight; scope adapter base; offset hammer spur (right or left hand) for scope use
Stock: Two-piece straight-grip American black walnut with scaled-down forearm and blued steel forend cap; Mar-Shield® finish

MARLIN LEVER-ACTION CARBINES

MODEL 1894 CENTURY LIMITED
$1087.00

SPECIFICATIONS
Caliber: 44/40
Capacity: 12-shot tubular magazine
Action: Lever w/side ejection; engraved and case-colored solid-top receiver; hammer block safety

Barrel length: 24″ tapered octagon with conventional rifling
Overall length: 40¾″ **Weight:** 6½ lbs.
Sights: Adjustable semi-buckhorn folding rear, carbine front sight

Stock: Straight-grip checkered semi-fancy American black walnut with crescent brass buttplate; Mar-Shield® finish

MARLIN 1894S
$454.80

SPECIFICATIONS
Calibers: 44 Rem. Mag./44 Special, 45 Colt
Capacity: 10-shot tubular magazine
Action: Lever action w/square finger lever; hammer block safety

Barrel length: 20″ w/ Micro-Groove®
Sights: Ramp front sight w/brass bead; adjustable semi-buckhorn folding rear and Wide-Scan™ hood; solid-top receiver tapped for scope mount or receiver sight

Overall length: 37½″
Weight: 6 lbs.
Stock: Checkered American black walnut stock w/Mar-Shield™ finish; blued steel forend cap; swivel studs

MARLIN 1894CS 357 MAGNUM
$454.80

SPECIFICATIONS
Calibers: 357 Magnum, 38 Special
Capacity: 9-shot tubular magazine
Action: Lever action w/square finger lever; hammer block safety; side ejection; solid top receiver; deeply blued metal surfaces; receiver top sandblasted to prevent glare

Barrel length: 18½″ with Micro-Groove® rifling (12 grooves)
Sights: Adjustable semi-buckhorn folding rear, bead front; solid top receiver tapped for scope mount or receiver sight; offset hammer spur for scope use—adjustable for right or left hand

Overall length: 36″
Weight: 6 lbs.
Stock: Checkered straight-grip two-piece American black walnut with white butt plate spacer; tough Mar-Shield® finish; swivel studs

RIFLES

MARLIN LEVER-ACTION CARBINES

MARLIN 1895SS
$490.25

SPECIFICATIONS
Caliber: 45/70 Government
Capacity: 4-shot tubular magazine
Action: Lever action; hammer block safety; receiver top sandblasted to prevent glare

Barrel: 22″ Micro-Groove® barrel
Sights: Ramp front sight w/brass bead; adjustable semi-buckhorn folding rear and Wide-Scan™ hood; receiver tapped for scope mount or receiver sight

Overall length: 40½″
Weight: 7½ lbs.
Stock: Checkered American black walnut pistol-grip stock w/rubber rifle butt pad and Mar-Shield® finish; white pistol grip and butt spacers

MARLIN 336CS
$416.00 (Without Scope)

SPECIFICATIONS
Calibers: 30-30 Win., and 35 Rem.
Capacity: 6-shot tubular magazine
Action: Lever action w/hammer block safety; deeply blued metal surfaces; receiver top sandblasted to prevent glare

Barrel: 20″ Micro-Groove® barrel
Sights: Adjustable folding semi-buckhorn rear; ramp front sight w/brass bead and Wide-Scan™ hood; tapped for receiver sight and scope mount; offset hammer spur for scope use (works right or left)

Overall length: 38½″
Weight: 7 lbs.
Stock: Checkered American black walnut pistol-grip stock w/fluted comb and Mar-Shield® finish; rubber rifle butt pad; swivel studs

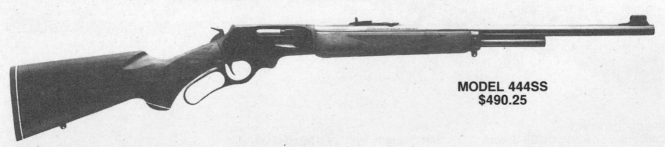

MODEL 444SS
$490.25

SPECIFICATIONS
Caliber: 444 Marlin
Capacity: 5-shot tubular magazine
Barrel: 22″ Micro-Groove®

Overall length: 40½″
Weight: 7½ lbs.
Stock: Checkered American black walnut pistol grip stock with rubber rifle butt pad; swivel studs

Sights: Ramp front sight with brass bead and Wide-Scan® hood; adjustable semi-buckhorn folding rear; receiver tapped for scope mount or receiver sight

McMILLAN SIGNATURE RIFLES

CLASSIC SPORTER
$2400.00

SPECIFICATIONS
Calibers:
   Model SA: 22-250, 243, 6mm Rem., 6mm BR, 7mm BR, 7mm-08, 284, 308, 350 Rem. Mag.
   Model LA: 25-06, 270, 280 Rem., 30-06
   Model MA: 7mm STW, 7mm Rem. Mag., 300 Win. Mag., 300 Weatherby, 300 H&H, 338 Win. Mag., 340 Weatherby, 375 H&H, 416 Rem.

Capacity: 4 rounds; 3 rounds in magnum calibers
Weight: 7 lbs; 7 lbs. 9 oz. in long action
Barrel lengths: 22″, 24″, 26″
Options: Fibergrain; wooden stock, optics, 30mm rings, muzzle brakes, steel floor plates, iron sights

STAINLESS SPORTER
$2550.00

Same basic specifications as the Classic and Standard Sporters, but with stainless steel action and barrel. It is designed to withstand the most adverse weather conditions. Accuracy is guaranteed (3 shot in ½″ at 100 yards). Choice of wood, laminate or McMillan fiberglass stock.

ALASKAN
$3300.00

SPECIFICATIONS
Calibers:
   Model LA: 270, 280, 30-06
   Model MA: 7mm Rem. Mag., 300 Win. Mag., 300 H&H, 300 Weatherby, 358 Win., 340 Weatherby, 375 H&H, 416 Rem.

Other specifications same as the Classic Sporter, except McMillan action is fitted to a match-grade barrel, complete with single-leaf rear sight, barrel band front sight, 1″ detachable rings and mounts, steel floorplate, electroless nickel finish. Monte Carlo stock features cheekpiece, palm swell and special recoil pad.
Also available: Stainless Steel Receiver, **add** **$150.00**

McMILLAN SIGNATURE RIFLES

TALON SPORTER
$2600.00

The all-new action of this model is designed and engineered specifically for the hunting of dangerous (African-type) game animals. Patterned after the renowned pre-64 Model 70, the Talon features a cone breech, controlled feed, claw extractor, positive ejection and three-position safety. Action is available in chromolybdenum and stainless steel. Drilled and tapped for scope mounting in long, short or magnum, left or right hand.

Same basic specifications as McMillan's Signature series, but offered in the following **calibers:**
Standard Action: 22-250, 243, 6mm Rem., 6mm BR, 7mm BR, 7mm-08, 284, 308, 350 Rem. Mag.
Long Action: 25-06, 270, 280 Rem., 30-06
Magnum Action: 7mm STW, 7mm Rem. Mag., 300 Win. Mag., 300 Weatherby, 300 H&H, 338 Win. Mag., 340 Weatherby, 375 H&H, 416 Rem.

VARMINTER
$2400.00

SPECIFICATIONS
Calibers: 223, 22-250, 220 Swift, 243, 6mm Rem., 25-06, 7mm-08, 308, 350 Rem. Mag.
Other specifications same as the Classic Sporter, except the Super Varminter comes with heavy contoured barrel, adjustable trigger, field bipod and hand-bedded fiberglass stock.

TITANIUM MOUNTAIN RIFLE
$3000.00
$3600.00 w/Titanium Barrel

SPECIFICATIONS
Calibers:
   Model LA: 270, 280 Rem., 30-06
   Model MA: 7mm Rem. Mag., 300 Win. Mag.
Weight: 5½ lbs.
Other specifications same as the Classic Sporter, except barrel is made of chrome-moly (titanium alloy light contour match-grade barrel is available at additional cost of **$500.00**).

.300 PHOENIX
$2995.00

Caliber: 300 Phoenix. **Barrel length:** 27½″. **Weight:** 12½ lbs. **Stock:** Fiberglass with adjustable cheekpiece. **Feature:** Available in left-hand action.

McMILLAN SIGNATURE RIFLES

SAFARI
$3600.00 (Magnum)
TALON SAFARI $4200.00 (Super Magnum)

Super Magnum: 300 Phoenix, 338 Lapua, 378 Wby., 416 Rigby, 416 Wby., 460 Wby.
Other specifications same as the Classic Sporter, except for match-grade barrel, positive extraction McMillan Safari action, quick detachable 1″ scope mounts, positive locking steel floorplate, multi-leaf express sights, barrel band ramp front sight, barrel band swivels, and McMillan's Safari stock.

SPECIFICATIONS
Calibers:
  Magnum: 300 Win. Mag., 300 Weatherby, 300 H&H, 338 Win. Mag., 340 Weatherby, 375 H&H, 404 Jeffrey, 416 Rem., 458 Win.

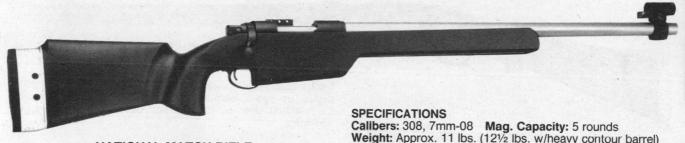

NATIONAL MATCH RIFLE
$2600.00

SPECIFICATIONS
Calibers: 308, 7mm-08   **Mag. Capacity:** 5 rounds
Weight: Approx. 11 lbs. (12½ lbs. w/heavy contour barrel)
Available for right-hand shooters only. Features modified ISU fiberglass stock with adjustable butt plate, stainless steel match barrel with barrel band and Tompkins front sight; McMillan repeating bolt action with clip shot and Canjar trigger. Barrel twist is 1:12″.

LONG RANGE RIFLE
$2600.00

SPECIFICATIONS
Calibers: 300 Win. Mag., 300 Phoenix, 7mm Mag., 338 Lapua
Barrel length: 26″   **Weight:** 14 lbs.
Available in right-hand only. Features a fiberglass stock with adjustable butt plate and cheekpiece. Stainless steel match barrel comes with barrel band and Tompkins front sight. McMillan solid bottom single-shot action and Canjar trigger. Barrel twist is 1:12″.

McMILLAN BENCHREST RIFLE
$2800.00 (not shown)

SPECIFICATIONS
Calibers: 6mm PPC, 243, 6mm BR, 6mm Rem., 308
Built to individual specifications to be competitive in hunter, light varmint and heavy varmint classes. Features solid bottom or repeating bolt action, Canjar trigger, fiberglass stock with recoil pad, stainless steel match-grade barrel and reloading dies. Right- or left-hand models.

RIFLES

MITCHELL ARMS

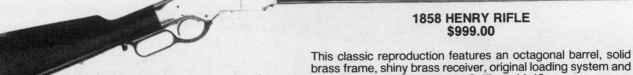

1858 HENRY RIFLE
$999.00

This classic reproduction features an octagonal barrel, solid brass frame, shiny brass receiver, original loading system and solid European walnut stock. **Caliber:** 44-40.

1866 WINCHESTER RIFLE
$829.00

This lever-action Winchester with octagonal barrel has a solid brass frame, original loading system and solid European walnut stock. **Caliber:** 44-40.

1873 WINCHESTER RIFLE
$950.00

Features steel side plates, color casehardened frame and side plates, octagonal barrel, solid walnut buttstock and forend. Lever action. Uses centerfire ammo. **Calibers:** 44-40, 45 Colt, 357 Mag.

HIGH STANDARD
MODEL 9301 BOLT ACTION

front; adjustable rear. **Stock:** American walnut (rosewood grip and forend caps on Models 9301 and 9302).

HIGH STANDARD
BOLT-ACTION RIMFIRE RIFLES

Calibers: 22 LR, 22 Magnum. **Capacity:** 5 and 10 rounds. **Barrel length:** 22½″. **Weight:** 6.5–6.7 lbs. **Sights:** Ramp bead

Prices:

Model 9301	$312.50
Model 9302	317.50
Model 9303	275.00
Model 9304	280.00

HIGH STANDARD
MODEL 15/22 SEMIAUTO

HIGH STANDARD
SEMIAUTOMATIC RIMFIRE RIFLES

Caliber: 22 LR. **Capacity:** 15 and 30 rounds. **Barrel length:** 20½″. **Weight:** 6.4 lbs. **Sights:** Ramp front; adjustable rear.

Stock: American walnut (checkered rosewood grip and forend caps on Model 15/22D).

Prices:	$159.95
Model 15/22D	199.95

NAVY ARMS REPLICA RIFLES

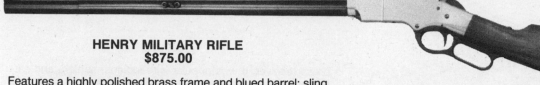

HENRY MILITARY RIFLE
$875.00

Features a highly polished brass frame and blued barrel; sling swivels to the original specifications are located on the left side.

SPECIFICATIONS
Calibers: 44-40 and 44 rimfire
Barrel length: 24″   **Overall length:** 43″
Weight: 9¼ lbs.
Stock: Walnut

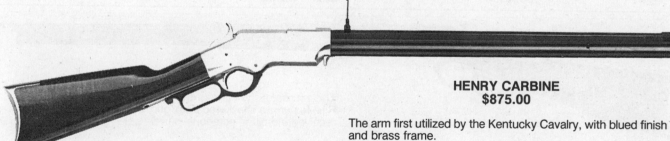

IRON FRAME HENRY
$895.00

Same specifications as the Henry Rifle, except with iron frame. Available with blued or casehardened receiver.

HENRY CARBINE
$875.00

The arm first utilized by the Kentucky Cavalry, with blued finish and brass frame.

SPECIFICATIONS
Caliber: 44-40
Barrel length: 23⅝″   **Overall length:** 45″
Weight: 8¾ lbs.

HENRY TRAPPER MODEL
$875.00

This short, lightweight lever-action arm is ideal for the hunter.

SPECIFICATIONS
Caliber: 44-40
Barrel length: 16½″   **Overall length:** 34½″
Weight: 7¼ lbs.

NAVY ARMS REPLICA RIFLES

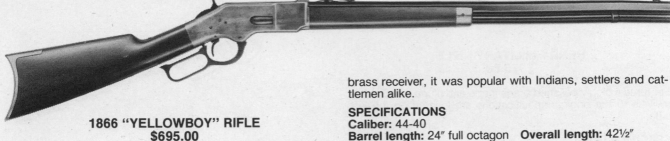

1866 "YELLOWBOY" RIFLE
$695.00

The 1866 model was Oliver Winchester's improved version of the Henry rifle. Called the "Yellowboy" because of its polished

brass receiver, it was popular with Indians, settlers and cattlemen alike.

SPECIFICATIONS
Caliber: 44-40
Barrel length: 24″ full octagon   **Overall length:** 42½″
Weight: 8½ lbs.
Sights: Blade front; open ladder rear
Stock: Walnut

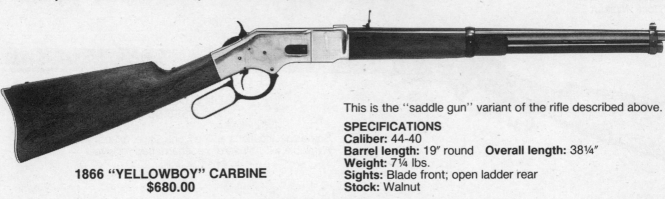

1866 "YELLOWBOY" CARBINE
$680.00

This is the "saddle gun" variant of the rifle described above.

SPECIFICATIONS
Caliber: 44-40
Barrel length: 19″ round   **Overall length:** 38¼″
Weight: 7¼ lbs.
Sights: Blade front; open ladder rear
Stock: Walnut

1873 WINCHESTER SPORTING RIFLE
$895.00

This replica of the state-of-the-art Winchester 1873 Sporting Rifle features a checkered pistol grip, buttstock, casehardened receiver and blued octagonal barrel.

SPECIFICATIONS
Caliber: 44-40 or 45 LC
Barrel length: 24″ or 30″
Overall length: 48¾″ (w/30″ barrel)
Weight: 8 lbs. 14 oz.
Sights: Blade front; buckhorn rear

1873 WINCHESTER-STYLE RIFLE
$840.00

Known as "The Gun That Won the West," the 1873 was the most popular lever-action rifle of its time. This fine replica features a casehardened receiver.

SPECIFICATIONS
Caliber: 44-40 or 45 Long Colt
Barrel length: 24″   **Overall length:** 43″
Weight: 8¼ lbs.
Sights: Blade front; open ladder rear
Stock: Walnut
Also available: **1873 WINCHESTER-STYLE CARBINE**
   (19″ barrel) $820.00

NAVY ARMS REPLICA RIFLES

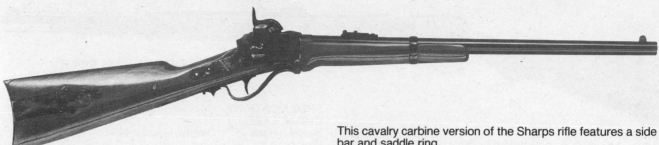

1874 SHARPS CAVALRY CARBINE
$775.00

This cavalry carbine version of the Sharps rifle features a side bar and saddle ring.

SPECIFICATIONS
Caliber: 45-70 percussion
Barrel length: 22″   **Overall length:** 39″
Weight: 7¾ lbs.
Sights: Blade front; military ladder rear
Stock: Walnut

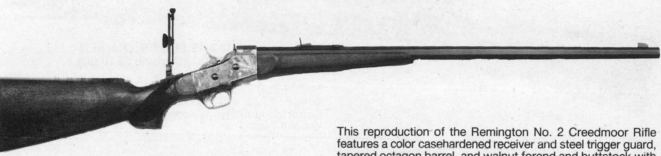

No. 2 CREEDMOOR TARGET RIFLE
$750.00

This reproduction of the Remington No. 2 Creedmoor Rifle features a color casehardened receiver and steel trigger guard, tapered octagon barrel, and walnut forend and buttstock with checkered pistol grip.

SPECIFICATIONS
Caliber: 45-70
Barrel length: 30″, tapered   **Overall length:** 46″
Weight: 9 lbs.
Sights: Globe front, adjustable Creedmoor rear
Stock: Checkered walnut stock and forend

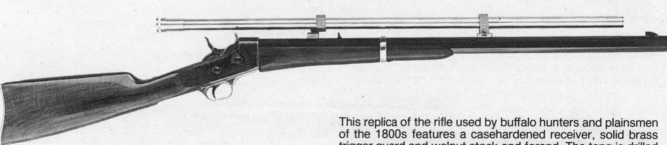

REMINGTON-STYLE ROLLING BLOCK
BUFFALO RIFLE
$530.00

This replica of the rifle used by buffalo hunters and plainsmen of the 1800s features a casehardened receiver, solid brass trigger guard and walnut stock and forend. The tang is drilled and tapped to accept the optional Creedmoor sight.

SPECIFICATIONS
Caliber: 45-70
Barrel length: 26″ or 30″; full octagon or half-round
Sights: Blade front, open notch rear
Stock: Walnut stock and forend
Feature: Shown with optional 32½″ Model 1860 brass telescopic sight **$125.00**; Compact Model (18″) is **$130.00**

NAVY ARMS REPLICA RIFLES

MODEL TU-KKW TRAINING RIFLE
$210.00

The TU-KKW is a replica of the "Kleine Kaliber Wehrsport Gewehr" 22-caliber training rifle used by the Germans in World War II. A full-size, full-weight 98K, it is complete with Mauser-style military sights, bayonet lug and cleaning rod. Unlike the original, this replica features a 5-round detachable magazine.

SPECIFICATIONS
Caliber: 22 LR
Barrel length: 26″   **Overall length:** 44″
Weight: 8 lbs.
Sights: Open military style
Also available:
MODEL TU-KKW SNIPER TRAINER w/26″ barrel
  and 2.75 power Type 89 scope **$310.00**

TU-33/40 SNIPER TRAINER
$310.00   (Carbine $210.00)

Based on the WW II German Mauser G. 33/40 Mountain Carbine, this TU-33/40 is a short, lightweight bolt-action military replica. It comes complete with a 2.75 power Type 89 scope with quick-detachable mount.

SPECIFICATIONS
Caliber: 22 LR w/5-round detachable magazine
Barrel length: 20¾″   **Overall length:** 38″
Weight: 7 lbs. 7 oz.
Sights: Telescopic; Mauser-style tangent rear; barley-corn front open sights
Stock: 98 Mauser style
Features: Bayonet lug, leather military sling

GREENER LIGHT MODEL
HARPOON GUN
$995.00

Designed for large game fish, the Greener Harpoon gun utilizes the time-proven Martini action. The complete outfit consists of gun, harpoons, harpoon lines, line release frames, blank cartridges and cleaning kit—all housed in a carrying case.

SPECIFICATIONS
Caliber: 38 Special (blank)
Barrel length: 20″   **Overall length:** 36″
Weight: 6 lbs. 5 oz.
Stock: Walnut

MODEL EM-331 SPORTING RIFLE
$350.00

Weight: 7 lbs. 8 oz.
Sights: Adjustable open rear; hooded ramp front
Stock: Monte Carlo style with sling swivels and recoil pad
Features: Detachable 5-round magazine (spare magazine included); receiver dovetailed for scope rings

SPECIFICATIONS
Caliber: 7.62×39   **Capacity:** 5 rounds
Barrel length: 21″   **Overall length:** 42″

NEW ENGLAND FIREARMS RIFLES

HANDI-RIFLE

SPECIFICATIONS
Calibers: 22 Hornet, 22-250 Rem., 223 Rem., 243 Win., 270 Win., 30-30 Win., 45-70 Govt.
Action: Break-open; side lever release; positive ejection
Barrel length: 22"
Weight: 7 lbs.
Sights: Ramp front; fully adjustable rear; tapped for scope mounts (22 Hornet, 30-30 Win. and 45-70 Govt. only)
Length of pull: 14¼"
Drop at comb: 1½" (1¼" in Monte Carlo)

Drop at heel: 2⅛" (1⅛" in Monte Carlo)
Stock: American hardwood, walnut finish; full pistol grip
Features: Semi-beavertail forend; patented transfer bar safety; automatic ejection; rebated muzzle; hammer extension; sling swivel studs on stock and forend

Prices:
In 22 Hornet, 223 Rem., 45-70 Govt. **$199.95**
In 22-250 Rem., 243 Win., 270 Win. and 30-30
Win. **199.95**

NORINCO RIFLES

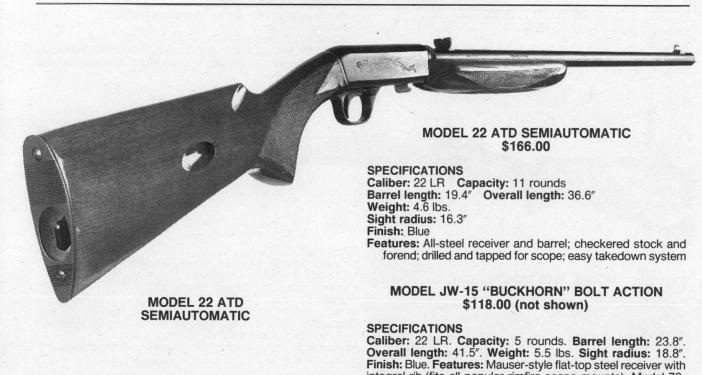

**MODEL 22 ATD
SEMIAUTOMATIC**

MODEL 22 ATD SEMIAUTOMATIC
$166.00

SPECIFICATIONS
Caliber: 22 LR   **Capacity:** 11 rounds
Barrel length: 19.4"   **Overall length:** 36.6"
Weight: 4.6 lbs.
Sight radius: 16.3"
Finish: Blue
Features: All-steel receiver and barrel; checkered stock and forend; drilled and tapped for scope; easy takedown system

MODEL JW-15 "BUCKHORN" BOLT ACTION
$118.00 (not shown)

SPECIFICATIONS
Caliber: 22 LR. **Capacity:** 5 rounds. **Barrel length:** 23.8".
Overall length: 41.5". **Weight:** 5.5 lbs. **Sight radius:** 18.8".
Finish: Blue. **Features:** Mauser-style flat-top steel receiver with integral rib (fits all popular rimfire scope mounts); Model 70-style safety (locks both firing pin and bolt); detachable 5-shot magazine

PARKER-HALE RIFLES

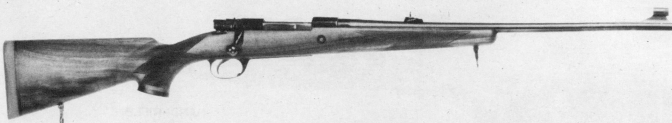

MODEL M81 CLASSIC
$900.00

SPECIFICATIONS
Calibers: 22-250, 243 Win., 6mm Rem., 270 Win., 6.5×55, 7×57, 7×64, 308 Win., 30-06
Capacity: 4 rounds
Barrel length: 24″   **Overall length:** 44¹/₂″
Weight: 7.75 lbs.
Stock: Checkered walnut   **Length of pull:** 13¹/₂″
Features: All-steel trigger guard; adjustable trigger

MODEL M81 AFRICAN
$1050.00

SPECIFICATIONS
Calibers: 375 H&H, 9.3 × 62mm   **Capacity:** 3 rounds
Barrel length: 24″   **Overall length:** 44¹/₂″
Weight: 7.75 lbs.
Stock: Checkered walnut
Features: All-steel trigger guard, adjustable trigger, barrel band front swivel, African express rear sight, hand-engraved receiver

MODEL 1100 LIGHTWEIGHT
$595.00

SPECIFICATIONS
Calibers: 22-250, 243 Win., 6mm Rem., 270 Win., 308 Win., 30-06   **Capacity:** 4 rounds
Barrel length: 22″   **Overall length:** 43″
Weight: 6¹/₂ lbs.
Length of pull: 13¹/₂″
Stock: Monte Carlo style, satin finished walnut with schnabel forend and wraparound checkering
Features: Slim profile barrel, alloy trigger guard and anodized bolt handle

PARKER HALE RIFLES ARE MANUFACTURED BY GIBBS RIFLE COMPANY

PARKER-HALE RIFLES

MODEL 2100 MIDLAND
$375.00

SPECIFICATIONS
Calibers: 22-250, 243 Win., 6mm Rem., 270 Win., 308 Win., 30-06   **Capacity:** 4 rounds
Barrel length: 22″ (24″ in cal. 22-250)   **Overall length:** 43″
Weight: 7 lbs.
Stock: Checkered walnut   **Length of pull:** 13½″
Sights: Hooded ramp front; adjustable flip-up rear

Also available:
MODEL 2600 MIDLAND SPECIAL in 22-250, 243 Win., 270, 6mm Rem., 6.5×55, 7×57, 7×64, 30-06 and 308. **Price: $360.00**
MODEL 2700 LIGHTWEIGHT in 22-250, 243 Win., 6mm Rem., 270 Win., 6.5×55, 7×57, 7×64, 308 Win. and 30-06 **Weight:** 6½ lbs. **Price: $415.00**

MODEL 1200 SUPER
$595.00

SPECIFICATIONS
Calibers: 22-250, 243 Win., 6mm Rem., 270 Win., 308 Win., 30-06   **Capacity:** 4 rounds
Barrel length: 24″   **Overall length:** 44½″

Weight: 7½ lbs.   **Length of pull:** 13½″
Also available:
MODEL 1200 SUPER CLIP in 22-250, 243 Win., 6mm Rem., 270 Win., 30-06 and 308 Win. Same specifications as Model 1200 Super, but weighs 7½ lbs. **Price: $640.00**

MODEL 1000 CLIP
$535.00

SPECIFICATIONS
Calibers: 22-250, 243 Win., 6mm Rem., 270 Win., 6.5×55, 7×57, 7×64, 308 Win. and 30-06   **Capacity:** 4 rounds
Barrel length: 22″   **Overall length:** 43″
Weight: 7¼ lbs.

Stock: Checkered walnut, Monte Carlo style
Features: Detachable magazine, Mauser-style 98 action
Also available:
MODEL 1000 STANDARD. Same specifications as the Model 1000 Clip, but with fixed 4-round magazine. **Price: $495.00**

PARKER HALE RIFLES ARE MANUFACTURED BY GIBBS RIFLE COMPANY

RIFLES

PARKER-HALE RIFLES

MODEL 1100M AFRICAN MAGNUM
$930.00

SPECIFICATIONS
Calibers: 375 H&H Magnum, 458 Win. Mag.
Capacity: 4 rounds
Barrel length: 24″   **Overall length:** 46″

Weight: 9½ lbs.
Stock: Checkered walnut, weighted, with two recoil lugs
Features: Vented recoil pad, shallow ''V'' rear sight, steel magazine with hinged floorplate

MODEL 1300C "SCOUT"
$595.00

SPECIFICATIONS
Calibers: 243 Win., 308 Win.   **Capacity:** 10 rounds
Barrel length: 20″   **Overall length:** 41″

Weight: 8½ lbs.
Stock: Checkered laminated birch wood
Features: Detachable magazine, muzzle brake

M-85 SNIPER RIFLE
$2150.00

SPECIFICATIONS
Caliber: 308 Win. (7.62 NATO)   **Capacity:** 10 or 20 rounds
Barrel length: 24¼″   **Overall length:** 45″
Weight: 12 lbs. 6 oz. (with scope)

Sights: Blade front (adjustable for windage); folding aperture rear (adjustable for elevation)
Stock: Fiberglass McMillan, adjustable for length of pull
Features: M-14 type detachable magazine; adjustable trigger; ''quick-detach'' bipod

PARKER HALE RIFLES ARE MANUFACTURED BY GIBBS RIFLE COMPANY

PEDERSOLI REPLICA RIFLES

ROLLING BLOCK TARGET RIFLE
$620.00   ($690.00 w/Creedmore Sight)

SPECIFICATIONS
Caliber: 45-70 or 357
Barrel length: 30″ octagonal (blued)
Weight: 9¹/₂ lbs. (45-70); 10 lbs. (357)
Sights: Adjustable rear sight; tunnel modified front (all models designed for fitting of Creedmore sight)

Also available:
Cavalry, Infantry, Long Range Creedmoor $550.00

SHARPS CARBINE MODEL 766
$750.00   ($795.00 with Patchbox)

SPECIFICATIONS
Caliber: 54
Barrel length: 22″ round (6 grooves)
Overall length: 39″
Weight: 7¹/₂ lbs.
Sights: Fully adjustable rear; fixed front

Also available:
Sharps 1859 Military Rifle (set trigger, 30″ barrel, 8.4 lbs.).
  Price: $895.00

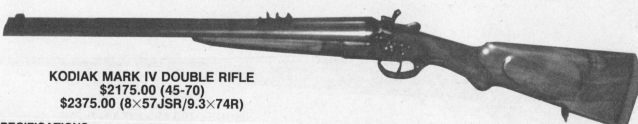

KODIAK MARK IV DOUBLE RIFLE
$2175.00 (45-70)
$2375.00 (8×57JSR/9.3×74R)

SPECIFICATIONS
Calibers: 45-70, 9.3×74R, 8×57JSR
Barrel length: 22″ (24″ 45-70)
Overall length: 39″ (40¹/₂″ 45-70)
Weight: 8.24 lbs. (9.7 lbs. 45/70)

Also available:
Kodiak Mark IV w/interchangeable 20-ga. barrel . $4100.00
  In 8×57JSR or 9.3×74R 4325.00

REMINGTON BOLT-ACTION RIFLES

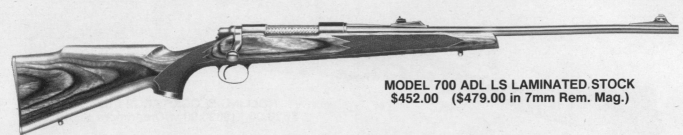

MODEL 700 ADL LS LAMINATED STOCK
$452.00   ($479.00 in 7mm Rem. Mag.)

The Model 700 ADL LS features a traditional wood stock made by laminating alternate strips of light and dark wood with waterproof adhesive and impregnating it with a phenolic resin for greater stability. Other features include low-gloss satin finish, cut checkering, sling swivel studs and open factory sights.

Calibers: 243 Win., 270 Win., 30-06, 308 Win., and 7mm Rem. Mag.
Capacity: 5 (4 in 7mm Rem. Mag.)
Barrel length: 22″ (24″ in 7mm Rem. Mag.)
Weight: 7¼ lbs.
Stock dimensions: Drop at heel 1⅜″; Drop at comb ½″; Length of pull 13⅜″.

MODEL 700 CS (Camo Synthetic)
$581.00   ($608.00 in 7mm Rem.)

The Model 700™ is bedded to a synthetic stock fully camouflaged in Mossy Oak® Bottomland™. Stronger than wood and unaffected by weather, this stock will not warp or swell in rain, snow or heat. The Model 700 Camo Synthetic comes with a non-reflective matte finish on the bolt and is available in nine calibers. See Model 700 Specifications table for additional information.

MODEL 700 MOUNTAIN RIFLE
$532.00

The lean barrel contour on this Model 700 enhances the accuracy of this light (6.75 lbs.) rifle. The pistol grip is pitched lower to position the wrist for better grip. The smooth cheekpiece positions the eye for accurate sighting. Features include semi-finished walnut stock, 20-line deep-cut checkering, hinged magazine floorplate, sling swivel studs, and trim buttpad. Stainless synthetic model also available. For additional specifications, see Model 700 Specifications table.

REMINGTON BOLT-ACTION RIFLES

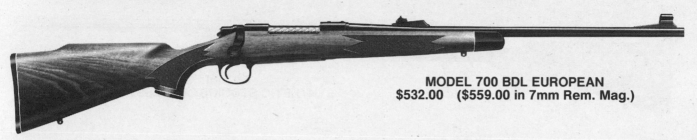

MODEL 700 BDL EUROPEAN
$532.00   ($559.00 in 7mm Rem. Mag.)

The stock on this Model 700 BDL features Monte Carlo comb and raised cheekpiece as well as fine-cut skipline checkering on the pistol grip and forend. Offered in seven calibers from 243 Win. to the 7mm Rem. Mag. Features include hinged floorplate, sling swivel studs, hooded ramp front sight and adjustable rear sight. For additional specifications, see the Model 700 Specifications table.

MODEL 700 BDL

MODEL 700 BDL SS
(Stainless/synthetic)

MODEL 700 BDL

This Model 700 features the Monte Carlo American walnut stock finished to a high gloss with fine-cut skipline checkering. Also includes a hinged floorplate, sling swivels studs, hooded ramp front sight, and adjustable rear sight. Also available in stainless synthetic version (Model 700 BDL SS) with stainless-steel barrel, receiver and bolt plus synthetic stock for maximum weather resistance. For additional specifications, see Model 700 table on the following page.

Prices:
Model 700 BDL

In 17 Rem., 7mm Rem. Mag., 300 Win. Mag., 35 Whelen, 338 Win. Mag.	$559.00
In 222 Rem., 22-250 Rem., 223 Rem., 6mm Rem., 243 Win., 25-06 Rem., 270 Win., 280 Rem., 7mm-08 Rem., 30-06, 308 Win.	532.00
Left Hand in 22-250 Rem., 243 Win., 270 Win., 30-06, and 308 Win.	559.00
Left Hand in 7mm Rem. Mag. and 338 Win. Mag.	585.00
Model BDL SS (Stainless Synthetic)	585.00
In Magnum calibers	612.00

MODEL 700 SENDERO (not shown)
$652.00   ($679.00 Magnum)

Remington's new Sendero rifle combines the accuracy features of the Model 700 Varmint Special with long action and magnum calibers for long-range hunting. The 26" barrel has a heavy varmint profile and features a spherical concave crown. For additional specifications, see Model 700 table on the following page.

REMINGTON BOLT-ACTION RIFLES

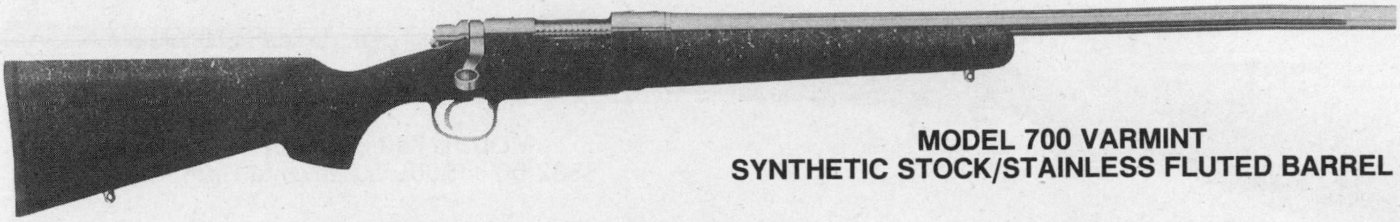

MODEL 700 VARMINT
SYNTHETIC STOCK/STAINLESS FLUTED BARREL

MODEL 700 VS VARMINT SYNTHETIC BOLT-ACTION CENTERFIRE RIFLE

With heavy barrel, synthetic stock and aluminum bedding block **$652.00**
With stainless fluted barrel and synthetic stock in 220 Swift, 22-250 Rem. 223 Rem., 308 Win. . . **799.00**

Also available:
MODEL 700 VS with wood stock in calibers 222 Rem., 22-250 Rem., 223 Rem., 6mm Rem., 243 Win., 7mm-08 Rem. and 308 Win. **$565.00**

SPECIFICATIONS MODEL 700

			Overall Length/Avg. Wt. (lbs.)									Twist
Calibers	Magazine Capacity	Barrel Length	Mtn Rifle	Sendero™[1] (26" Barrel)	Camo Synthetic	BDL Stainless Synthetic	ADL & BDL	BDL European	Varmint Special Wood Stock	Varmint Synthetic[1] (26" Heavy BBL)	Varmint Synthetic[1] (26" Fluted BBL)	R-H, 1 turn in
17 Rem.	5	24"	—	—	—	—	43 5/8"/ 7 1/4	—	—	—	—	9"
220 Swift	4	24"	—	—	—	—	—	—	—	45 1/2"/ 8 7/8★	45 1/2"/ 8 3/8★	14"
222 Rem.	5	24"	—	—	—	—	43 5/8"/ 7 1/4	—	43 1/2"/ 9	—	—	14"
22-250 Rem.	4	24"	—	—	—	—	43 5/8"/ 7 1/4	—	43 1/2"/ 9	45 1/2"/ 8 7/8★	45 1/2"/ 8 3/8★	14"
	4	24"	—	—	—	—	43 5/8"/ 7 1/4	—	—	—	—	14"
223 Rem.	5	24"	—	—	—	43 5/8"/ 6 7/8	43 5/8"/ 7 1/4	—	43 1/2"/ 9	45 1/2"/ 8 7/8★	45 1/2"/ 8 3/8★	12"
6mm Rem.	4	22"	—	—	—	—	41 5/8"/ 7 1/4	—	—	—	—	9 1/8"
	4	24"	—	—	—	43 5/8"/ 6 7/8	—	—	43 1/2"/ 9	—	—	9 1/8"
243 Win.	4	22"	41 5/8"/6 3/4	—	41 5/8"/ 6 3/4	—	41 5/8"/ 7 1/4	41 5/8"/ 7 1/4	—	—	—	9 1/8"
	4	22"	—	—	—	—	41 5/8"/ 7 1/4	—	—	—	—	9 1/8"
	4	24"	—	—	—	43 5/8"/ 6 7/8	—	—	43 1/2"/ 9	—	—	9 1/8"
25-06 Rem.	4	24"	—	46 3/8"/ 9★	—	44 1/2"/ 6 7/8	44 1/2"/ 7 1/4	—	—	—	—	10"
	4	22"	42 1/2"/6 3/4	—	—	—	—	—	—	—	—	10"
257 Roberts	4	22"	41 5/8"/6 3/4	—	—	—	—	—	—	—	—	10"
270 Win.	4	22"	42 1/2"/6 3/4	—	—	—	42 1/2"/ 7 1/4	42 1/2"/ 7 1/4	—	—	—	10"
	4	22"	—	—	—	—	42 1/2"/ 7 1/4	—	—	—	—	10"
	4	24"	—	46 3/8"/ 9★	—	44 1/2"/6 7/8	—	—	—	—	—	10"
280 Rem.	4	22"	42 1/2"/6 3/4	—	—	—	42 1/2"/ 7 1/4	42 1/2"/ 7 1/4	—	—	—	9 1/4"
	4	24"	—	—	—	44 1/2"/6 7/8	—	—	—	—	—	9 1/4"
7mm-08 Rem.	4	22"	41 5/8"/6 3/4	—	—	—	41 5/8"/ 7 1/4	41 5/8"/ 7 1/4	—	—	—	9 1/4"
	4	24"	—	—	—	43 5/8"/ 6 7/8	—	—	43 1/2"/ 9	—	—	9 1/4"
7mm Mauser (7x57)	4	22"	41 5/8"/6 3/4	—	—	—	—	—	—	—	—	9 1/4"
7mm Rem. Mag.[2]	3	24"	—	46 3/8"/ 9★	44 1/2"/ 7	44 1/2"/ 7	44 1/2"/ 7 1/2	44 1/2"/ 7 1/2	—	—	—	9 1/4"
	3	24"	—	—	—	—	44 1/2"/ 7 1/2	—	—	—	—	9 1/4"
7mm Wby. Mag.	3	24"	—	—	—	44 1/2"/ 7	—	—	—	—	—	9 1/4"
30-06	4	22"	42 1/2"/6 3/4	—	42 1/2"/ 6 3/4	—	42 1/2"/ 7 1/4	42 1/2"/ 7 1/4	—	—	—	10"
	4	22"	—	—	—	—	42 1/2"/ 7 1/4	—	—	—	—	10"
	4	24"	—	—	—	44 1/2"/6 7/8	—	—	—	—	—	10"
308 Win.	4	22"	41 5/8"/6 3/4	—	—	—	41 5/8"/ 7 1/4	41 5/8"/ 7 1/4	—	—	—	10"
	4	24"	—	—	—	43 5/8"/ 6 7/8[2]	—	—	43 1/2"/ 9	45 1/2"/ 8 7/8★	45 1/2"/ 8 3/8★	12"[2]
	4	22"	—	—	—	—	41 5/8"/ 7 1/4	—	—	—	—	10"
300 Win. Mag.2	3	24"	—	46 3/8"/ 9★	—	44 1/2"/ 7	44 1/2"/ 7 1/2	—	—	—	—	10"
300 Wby. Mag.2	3	24"	—	—	—	44 1/2"/ 7	—	—	—	—	—	12"
35 Whelen	4	22"	—	—	—	—	42 1/2"/ 7 1/4	—	—	—	—	16"
338 Win. Mag.2	3	24"	—	—	—	44 1/2"/ 7	44 1/2"/ 7 1/2	—	—	—	—	10"
	3	24"	—	—	—	—	44 1/2"/ 7 1/2	—	—	—	—	10"
Stock Dimensions: Length of Pull			13 3/8"	13 3/8"	13 3/8"	13 3/8"	13 3/8"	13 3/8"	13 1/2"	13 3/8"	13 3/8"	
Drop at Comb (from centerline of bore)			3/8"	5/8"	1/2"	1/2"	1/2"	1/2"	15/16"	5/8"	5/8"	
Drop at Heel (from centerline of bore)			3/8"	5/8"	3/8"	3/8"	1 3/8"	1 3/8"	13/16"	5/8"	5/8"	

[1]Sendero™, Varmint Synthetic Stainless Fluted and Varmint Synthetic equipped with a 26" barrel only. LH = Left Hand. All Model 700™ and Model Seven™ rifles come with sling swivel studs. The BDL, BDL European, Camo Synthetic, ADL, and Seven™ (except Seven™ chambered for 17 Remington) are furnished with sights. The BDL Stainless Synthetic, Mountain Rifle, Classic, and Varmint guns have clean barrels. All Remington centerfire rifles are drilled and tapped for scope mounts. [2]Twist for BDL Stainless Synthetic in 308 Win. is R-H, one turn in 10".

REMINGTON BOLT-ACTION RIFLES

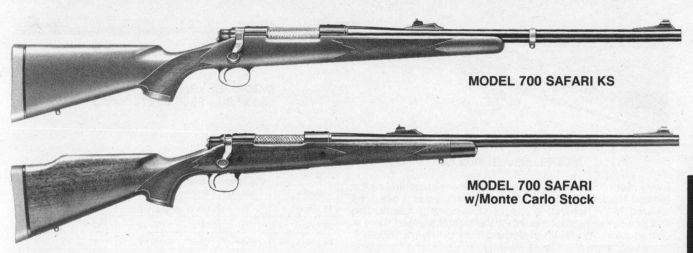

MODEL 700 SAFARI KS

MODEL 700 SAFARI
w/Monte Carlo Stock

Model 700 Safari Grade bolt-action rifles provide big-game hunters with a choice of either wood or synthetic stocks. Model 700 Safari Monte Carlo (with Monte Carlo comb and cheekpiece) and Model 700 Safari Classic (with straight-line classic comb and no cheekpiece) are the satin-finished wood-stock models. Both are decorated with hand-cut checkering 18 lines to the inch and fitted with two reinforcing cross bolts covered with rosewood plugs. The Monte Carlo model also has rosewood pistol-grip and forend caps. All models are fitted with sling swivel studs and 24″ barrels. Synthetic stock has simulated wood-grain finish, reinforced with Kevlar® (KS). **Calibers:** 8mm Rem. Mag., 375 H&H Magnum, 416 Rem. Mag. and 458 Win. Mag. **Capacity:** 3 rounds. **Average weight:** 9 lbs. **Overall**

length: 44 1/2″. **Rate of twist:** 10″ (8mm Rem. Mag.); 12″ (375 H&H Mag.); 14″ (416 Rem. Mag., 458 Win. Mag.). Prices on request.
Also available:
Model 700 APR (AFRICAN PLAINS RIFLE) in 7mm Rem. Mag., 300 Win. Mag., 300 Wby. Mag., 338 Win. Mag. and 375 H&H Mag. **Barrel length:** 26″ (46 1/2″ overall). **Weight:** 7 3/4 lbs. **Price** **$1475.00**
MODEL 700 AWR (Alaskan Wilderness Rifles). Same calibers as Model 700 APR with 24″ barrel (44 1/2″ overall). **Weight:** 6 3/4 lbs. **Price** **$1232.00**
MODEL 700 CLASSIC in 6.5×55 Swedish (22″ barrel). **Weight:** 7 1/4 lbs. **Twist:** 8″ **Price** **$532.00**

MODEL 40-XR KS SPORTER
Target Rimfire Position Rifle w/Kevlar Stock
$1325.00

Action: Bolt action, single shot
Caliber: 22 Long Rifle rimfire   **Capacity:** Single loading
Barrel: 24″ medium weight target barrel countersunk at muzzle. Drilled and tapped for target scope blocks. Fitted with front sight base
Bolt: Artillery style with lock-up at rear; 6 locking lugs, double extractors
Overall length: 43 1/2″   **Average weight:** 10 1/2 lbs.
Sights: Optional at extra cost; Williams Receiver No. FPTK and Redfield Globe front match sight
Safety: Positive serrated thumb safety
Receiver: Drilled and tapped for receiver sight
Trigger: Adjustable from 2 to 4 lbs.

Stock: Position style with Monte Carlo, cheekpiece and thumb groove; five-way adjustable buttplate and full-length guide rail
Also available:
MODEL 40-XR BR with 22″ stainless-steel barrel (heavy contour), 22 LR match chamber and bore dimensioms. Receiver and barrel drilled and tapped for scope mounts (mounted in green, Du Pont Kevlar reinforced fiberglass benchrest stock. Fully adjustable trigger (2 oz. trigger optional).
Price **$1345.00**
(Additional target rifles are available through Remington's Custom Shop.)

REMINGTON BOLT-ACTION RIFLES

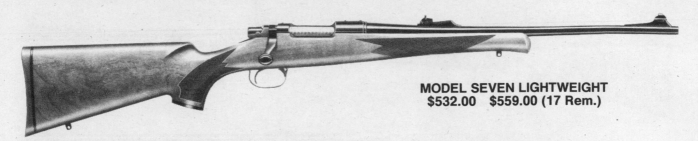

MODEL SEVEN LIGHTWEIGHT
$532.00 $559.00 (17 Rem.)

MODEL SEVEN RIFLES

Every **Model Seven** is built to the accuracy standards of the famous Model 700 and is individually test fired to prove it. Its tapered 18¹/₂″ Remington special steel barrel is free-floating out to a single pressure point at the forend tip. And there is ordnance-quality steel in everything from its fully enclosed bolt and extractor system to its steel trigger guard and floorplate. Ramp front and fully adjustable rear sights, sling swivel studs are standard. The Youth Model features a hardwood stock that is 1 inch shorter for easy control. Chambered in 6mm Rem., 243 Win. and 7mm-08 for less recoil. See table for additional specifications.

SPECIFICATIONS MODEL SEVEN™

Calibers	Clip Mag. Capacity	Barrel Length	Overall Length	Twist R-H 1 turn in	Avg. Wt. (lbs.)
17 Rem.	5	18¹/₂″	37³/₄″	9″	6¹/₄
223 Rem.	5	18¹/₂″	37³/₄″	12″	6¹/₄
243 Win.	4	18¹/₂″	37³/₄″	9¹/₈″	6¹/₄
	4	18¹/₂″	36³/₄″ (Youth)	9¹/₈″	6
	4	20″	39¹/₄″	9¹/₈″	6¹/₄
6mm Rem.	4	18¹/₂″	37³/₄″	9¹/₈″	6¹/₄
	4	18¹/₂″	36³/₄″ (Youth)	9¹/₈″	6
7mm-08 Rem.	4	18¹/₂″	37³/₄″	9¹/₄″	6¹/₄
	4	18¹/₂″	36³/₄″ (Youth)	9¹/₄″	6
	4	20″	39¹/₄″	9¹/₄″	6¹/₄
308 Win.	4	18¹/₂″	37³/₄″	10″	6¹/₄
	4	20″	39¹/₄″	10″	6¹/₄

Stock Dimensions: 13³/₁₆″ length of pull, 9/₁₆″ drop at comb, 5/₁₆″ drop at heel. Youth gun has 12¹/₂″ length of pull. 17. Rem. provided without sights.
Note: New Model Seven Mannlicher and Model Seven KS versions are available from the Remington Custom Shop through your local dealer.

MODEL SEVEN YOUTH
$439.00

MODEL SEVEN STAINLESS SYNTHETIC
$585.00 (243 Win., 7mm-08 Rem., 308 Win.)

REMINGTON REPEATING RIFLES

MODEL 7400 (High Gloss Stock)
$524.00

Calibers: 243 Win., 270 Win., 280 Rem., 30-06, 308 Win., 35 Whelen and 30-06 Carbine (see below)
Capacity: 5 centerfire cartridges (4 in the magazine, 1 in the chamber); extra 4-shot magazine available
Action: Gas-operated; receiver drilled and tapped for scope mounts
Barrel lengths: 22″ (18½″ in 30-06 Carbine)
Weight: 7½ lbs. (7¼ lbs. in 30-06 Carbine and 35 Whelen)

Overall length: 42″ (39⅛″ in 30-06 Carbine)
Sights: Standard blade ramp front; sliding ramp rear
Stock: Satin or high-gloss (270 Win. and 30-06 only) walnut stock and forend; curved pistol grip; also available with Special Purpose non-reflective finish (270 Win. and 30-06 only)
Length of pull: 13⅜″
Drop at heel: 2¼″   **Drop at comb:** 1¹³/₁₆″

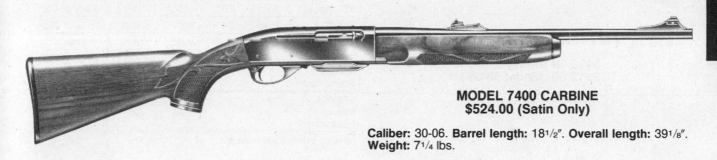

MODEL 7400 CARBINE
$524.00 (Satin Only)

Caliber: 30-06. **Barrel length:** 18½″. **Overall length:** 39⅛″.
Weight: 7¼ lbs.

MODEL 7600 (High Gloss Stock)
$496.00

Calibers: 243 Win., 270 Win., 280 Rem., 30-06, 308 Win., 35 Whelen, and 30-06 Carbine
Capacity: 5-shot capacity in all six calibers (4 in the removable magazine, 1 in the chamber)
Action: Pump action
Barrel length: 22″ (18½″ in 30-06 Carbine)
Overall length: 42″ (39⅛″ in 30-06 Carbine)
Weight: 7½ lbs. (7¼ lbs. in 30-06 Carbine and 35 Whelen)
Sights: Standard blade ramp front sight; sliding ramp rear, both removable

Stock: Satin or high-gloss walnut; also available with Special Purpose non-reflective finish
Length of pull: 13⅜″
Drop at heel: ¹⁵/₁₆″   **Drop at comb:** ⁹/₁₆″
Also available:
MODEL 7600 CARBINE with 18½″ barrel; chambered for 30-06 cartridge. **Price** **$496.00**

RIFLES

REMINGTON RIMFIRE RIFLES

MODEL 522 VIPER (22 LR)
$165.00

Remington's autoloading rimfire rifle utilizes a strong light-weight stock of PET resin that is impervious to changing temperatures and humidity. The receiver is made of a Du Pont high-tech synthetic. All exposed metal work, including barrel,

breech bolt and trigger guard, have a non-glare, black matte finish. Stock shape with slim pistol grip and semi-beavertail forend is proportioned to fit the size and stature of younger or smaller shooters. Other features include: factory-installed centerfire-type iron sights, detachable clip magazine, safety features, primary and secondary sears in trigger mechanism; and a protective ejection port shield.

MODEL 541-T BOLT ACTION
$399.00
$425.00 (Heavy Barrel)
$212.00 (Model 581-S)

RIMFIRE RIFLE SPECIFICATIONS

Model	Action	Barrel Length	Overall Length	Average Wt. (lbs.)	Magazine Capacity
522 Viper	Auto	20″	40″	4⅝	10-Shot Clip
541-T	Bolt	24″	42½″	5⅞*	5-Shot Clip
581-S	Bolt	24″	42½″	5⅞	5-Shot Clip
552 BDL Deluxe Speedmaster	Auto	21″	40″	5¾	15 Long Rifle
572 BDL Deluxe Fieldmaster	Pump	21″	40″	5½	15 Long Rifle

* 6½ lbs. in Heavy Barrel model

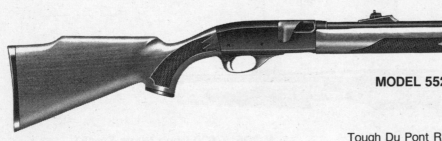

MODEL 552 BDL DELUXE SPEEDMASTER
$265.00

The rimfire semiautomatic 552 BDL Deluxe sports Remington custom-impressed checkering on both stock and forend.

Tough Du Pont RK-W lifetime finish brings out the lustrous beauty of the walnut while protecting it. Sights are ramp-style in front and rugged big-game type fully adjustable in rear.

MODEL 572 BDL DELUXE FIELDMASTER
$279.00

Features of this rifle with big-game feel and appearance are: Du Pont's tough RK-W finish; centerfire-rifle-type rear sight fully adjustable for both vertical and horizontal sight alignment; big-game style ramp front sight; Remington impressed checkering on both stock and forend.

ROSSI RIFLES

PUMP-ACTION GALLERY GUNS

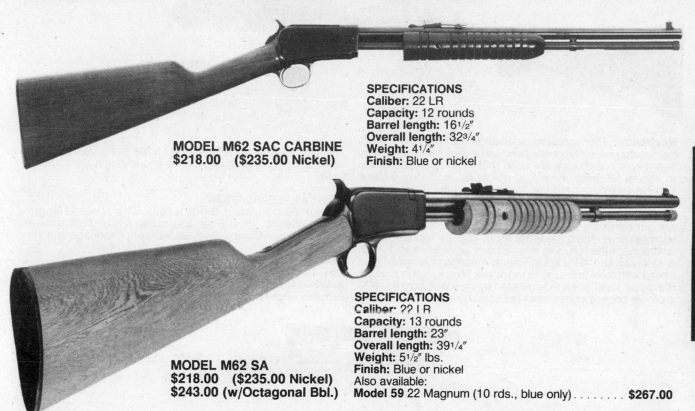

MODEL M62 SAC CARBINE
$218.00   ($235.00 Nickel)

SPECIFICATIONS
Caliber: 22 LR
Capacity: 12 rounds
Barrel length: 16 1/2″
Overall length: 32 3/4″
Weight: 4 1/4″
Finish: Blue or nickel

MODEL M62 SA
$218.00   ($235.00 Nickel)
$243.00 (w/Octagonal Bbl.)

SPECIFICATIONS
Caliber: 22 LR
Capacity: 13 rounds
Barrel length: 23″
Overall length: 39 1/4″
Weight: 5 1/2″ lbs.
Finish: Blue or nickel
Also available:
Model 59 22 Magnum (10 rds., blue only) **$267.00**

LEVER-ACTION OLD WEST CARBINES

MODEL M92 SRC
$334.00

MODEL M92 SRS
$334.00 (not shown)

SPECIFICATIONS
Caliber: 38 Special or 357 Magnum
Capacity: 8 rounds
Barrel length: 16″
Overall length: 33″
Weight: 5 3/4 lbs.
Finish: Blue

SPECIFICATIONS
Caliber: 38 Special or 357 Magnum
Capacity: 10 rounds
Barrel length: 20″
Overall length: 37″
Weight: 6 lbs.
Finish: Blue
Also available:
Model M65SRC (44 Mag., 5 3/4 lbs.) **$350.00**

RIFLES

RUGER CARBINES

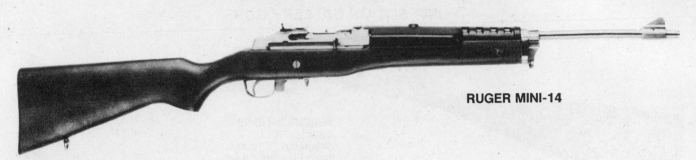

RUGER MINI-14

Mechanism: Gas-operated, semiautomatic. **Materials:** Heat-treated chrome molybdenum and other alloy steels as well as music wire coil springs are used throughout the mechanism to ensure reliability under field-operating conditions. **Safety:** The safety blocks both the hammer and sear. The slide can be cycled when the safety is on. The safety is mounted in the front of the trigger guard so that it may be set to Fire position without removing finger from trigger guard. **Firing pin:** The firing pin is retracted mechanically during the first part of the unlocking of the bolt. The rifle can only be fired when the bolt is safely locked. **Stock:** One-piece American hardwood reinforced with steel liner at stressed areas. Sling swivels standard. Handguard and forearm separated by air space from barrel to promote cooling under rapid-fire conditions. **Field stripping:**

The Carbine can be field-stripped to its eight (8) basic sub-assemblies in a matter of seconds and without use of special tools.

MINI-14 SPECIFICATIONS
Caliber: 223 (5.56mm). **Barrel length:** 18¹/₂″. **Overall length:** 37¹/₄″. **Weight:** 6 lbs. 8 oz. **Magazine:** 5-round, detachable box magazine. **Sights:** Rear adjustable for windage and elevation.
Prices:
Mini-14/5 Blued **$491.50**
K-Mini-14/5 Stainless Steel 542.00
(Scopes rings not included)

MINI-14 RANCH RIFLE

SPECIFICATIONS
Caliber: 223 (5.56mm). **Barrel length:** 18¹/₂″. **Overall length:** 37¹/₄″. **Weight:** 6 lbs. 8 oz. **Magazine:** 5-round detachable

box magazine. **Sights:** Fold-down rear sight; 1″ scope rings
Prices:
Mini-14/5R Blued **$530.00**
K-Mini-14/5R Stainless Steel 580.00

MINI THIRTY

This modified version of the Ruger Ranch rifle is chambered for the 7.62 × 39mm Soviet service cartridge (used in the SKS carbine and AKM rifle). Designed for use with telescopic sights, it features a low, compact scope mounting for greater accuracy and carrying ease, and a buffer in the receiver. A metal peep sight is installed for emergencies. Sling swivels are standard.

SPECIFICATIONS
Caliber: 7.62×39mm. **Barrel length:** 18¹/₂″. **Overall length:** 37¹/₈″. **Weight:** 6 lbs. 14 oz. (empty). **Magazine capacity:** 5 shots. **Rifling:** 6 grooves, right-hand twist, one turn in 10″. **Finish:** Blued or stainless.
Prices:
In Blue **$530.00**
In Stainless steel **580.00**

RUGER CARBINES

STANDARD 10/22 CARBINE

DELUXE 10/22 SPORTER

**MODEL K10/22RBI INTERNATIONAL CARBINE
STAINLESS**

MODEL 10/22 CARBINE 22 LONG RIFLE

Construction of the 10/22 Carbine is rugged and follows the Ruger design practice of building a firearm from integrated sub-assemblies. For example, the trigger housing assembly contains the entire ignition system, which employs a high-speed, swinging hammer to ensure the shortest possible lock time. The barrel is assembled to the receiver by a unique dual-screw dovetail system that provides unusual rigidity and strength—and accounts, in part, for the exceptional accuracy of the 10/22.

SPECIFICATIONS
Mechanism: Blow-back, semiautomatic. **Caliber:** 22 Long Rifle, high-speed or standard-velocity loads. **Barrel:** 18½" long; barrel is assembled to the receiver by unique dual-screw dovetail mounting for added strength and rigidity. **Weight:** 5 lbs. **Overall length:** 37¼". **Sights:** 1/16" brass bead front sight; single folding leaf rear sight, adjustable for elevation; receiver

drilled and tapped for scope blocks or tip-off mount adapter (included).
Magazine: 10-shot capacity, exclusive Ruger rotary design; fits flush into stock. **Trigger:** Curved finger surface, 3/8" wide. **Safety:** Sliding cross-button type; safety locks both sear and hammer and cannot be put in safe position unless gun is cocked. **Stocks:** 10/22 RB is birch; 10/22 SP Deluxe Sporter is American walnut. **Finish:** Polished all over and blued or anodized or brushed satin bright metal.

Model 10/22 RB Standard (Birch stock)	**$201.50**
Model 10/22 DSP Deluxe (Hand-checkered	
American walnut)	**254.50**
Model K10/22 RB Stainless	**236.00**
Model K10/22 RBI International w/full-	
length hardwood stock	**249.50**
Same as above in Stainless Steel	**269.00**

RUGER SINGLE-SHOT RIFLES

The following illustrations show the variations currently offered in the Ruger No. 1 Single-Shot Rifle Series. Ruger No. 1 rifles have a Farquharson-type falling-block action and selected American walnut stocks. Pistol grip and forearm are hand-checkered to a borderless design. Price for any listed model is **$634.00** (except the No. 1 RSI International Model: **$656.00**). Barreled Actions (blued only): **$429.50**

NO. 1A LIGHT SPORTER

Calibers: 243 Win., 270 Win., 30-06, 7×57mm. **Barrel length:** 22″. **Sights:** Adjustable folding-leaf rear sight mounted on quarter rib with ramp front sight base and dovetail-type gold bead front sight; open. **Weight:** 7¼ lbs.

NO. 1S MEDIUM SPORTER

Calibers: 218 Bee, 7mm Rem. Mag., 300 Win. Mag., 338 Win. Mag., 45-70. **Barrel length:** 26″ (22″ in 45-70). **Sights:** (same as above). **Weight:** 8 lbs. (7¼ lbs. in 45-70).

NO. 1B STANDARD RIFLE

Calibers: 218 Bee, 22 Hornet, 22-250, 220 Swift, 223, 243 Win., 6mm Rem., 25-06, 257 Roberts, 270 Win., 270 Wby. Mag., 7mm Rem. Mag., 280, 30-06, 300 Win. Mag., 300 Wby. Mag., 338 Win. Mag. **Barrel:** 26″. **Sights:** Ruger 1″ steel tip-off scope rings. **Weight:** 8 lbs.

NO. 1V SPECIAL VARMINTER

Calibers: 22 PPC, 22-250, 220 Swift, 223, 25-06, 6mm, 6mm PPC. **Barrel length:** 24″ (26″ in 220 Swift). **Sights:** Ruger target scope blocks, heavy barrel and 1″ tip-off scope rings. **Weight:** 9 lbs.

Also available:
NO. 1H TROPICAL RIFLE (24″ heavy barrel) in 375 H&H Mag., 404 Jeffery, 458 Win. Mag., 416 Rigby and 416 Rem. Mag.
NO 1. RSI INTERNATIONAL (20″ lightweight barrel) in 243 Win., 270 Win., 30-06 and 7×57mm

RUGER BOLT-ACTION RIFLES

MODEL 77/22RH HORNET

MODEL 77/22RH HORNET
$452.00   ($469.00 w/Sights)

The Model 77/22RH is Ruger's first truly compact centerfire bolt-action rifle. It features a 77/22 action crafted from heat-treated alloy steel. Exterior surfaces are blued to match the hammer-forged barrel. The action features a right-hand turning bolt with a 90-degree bolt throw, cocking on opening. Fast lock time (2.7 milliseconds) adds to accuracy. A three-position swing-back safety locks the bolt; in its center position firing is blocked, but bolt operation and safe loading and unloading are permitted. When fully forward, the rifle is ready to fire. The American walnut stock has recoil pad, grip cap and sling swivels installed. One-inch diameter scope rings fit integral bases.

SPECIFICATIONS
Caliber: 22 Hornet
Capacity: 6 rounds (detachable rotary magazine)
Barrel length: 20″   **Overall length:** 40″
Weight: 6 lbs. (unloaded)
Sights: Single folding-leaf rear; gold bead front
Length of pull: 13³/₄″
Drop at heel: 2³/₈″   **Drop at comb:** 2″
Finish: Polished and blued, matte, non-glare receiver top

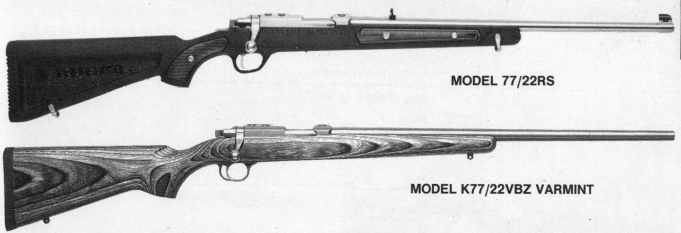

MODEL 77/22RS

MODEL K77/22VBZ VARMINT

MODEL 77/22 RIMFIRE RIFLE

The Ruger 22-caliber rimfire 77/22 bolt-action rifle has been built especially to function with the patented Ruger 10-Shot Rotary Magazine concept. The magazine throat, retaining lips, and ramps that guide the cartridge into the chamber are solid alloy steel that resists bending or deforming.

The 77/22 weighs just under six pounds. Its heavy-duty receiver incorporates the integral scope bases of the patented Ruger Scope Mounting System, with 1-inch Ruger scope rings. With the 3-position safety in its "lock" position, a dead bolt is cammed forward, locking the bolt handle down. In this position the action is locked closed and the handle cannot be raised.

All metal surfaces are finished in non-glare deep blue or satin stainless. Stock is select straight-grain American walnut, hand checkered and finished with durable polyurethane.

An All-Weather, all-stainless steel **MODEL K77/22RS** features a stock made of glass-fiber reinforced Zytel. **Weight:** Approx. 6 lbs.

SPECIFICATIONS
Calibers: 22 LR and 22 Magnum. **Barrel length:** 20″. **Overall length:** 39¹/₄″. **Weight:** 6 lbs. (w/o scope, magazine empty). **Feed:** Detachable 10-Shot Ruger Rotary Magazine.
Prices:

77/22R Blue, w/o sights, 1″ Ruger rings	**$402.00**
77/22RM Blue, walnut stock, plain barrel, no sights, 1″ Ruger rings, 22 Mag.	402.00
77/22RS Blue, sights included, 1″ Ruger rings	424.00
77/22RSM Blue, American walnut, iron sights	424.00
K77/22-RP Synthetic stock, stainless steel, plain barrel with 1″ Ruger rings	419.00
K77/22-RMP Synthetic stock, stainless steel, plain barrel, 1″ Ruger rings	419.00
K77/22-RSP Synthetic stock, stainless steel, gold bead front sight, folding-leaf rear, Ruger 1″ rings	445.00
K77/22RSMP Synthetic stock, metal sights, stainless	445.20
K77/22VBZ Varmint Laminated stock, scope rings, heavy barrel, stainless	485.00

RUGER BOLT-ACTION RIFLES

MARK II SERIES

MODEL M-77R MKII

Integral Base Receiver, 1″ scope rings. No sights. **Calibers:** (Long action) 6mm Rem., 6.5×55mm, 7×57mm, 257 Roberts, 270, 280 Rem., 30-06 (all with 22″ barrels); 7mm Rem. Mag., 300 Win. Mag., 338 Win. Mag. (all with 24″ barrels); and (Short Stroke action) 223, 243, 308 (22″ barrels). **Weight:** Approx. 7 lbs.

Price: **$558.00**
Also available: **M-77LR MKII** (Left Hand).
Calibers: 270, 30-06, 7mm Rem. Mag., 300 Mag. **558.00**

MODEL M-77RS MKII

Integral Base Receiver, Ruger steel 1″ rings, open sights. **Calibers:** 243, 25-06, 270, 7mm Rem. Mag., 30-06, 300 Win. Mag., 308, 338 Win. Mag., 458 Win. Mag. **Weight:** Approx. 7 lbs.

Price: **$617.00**

MODEL M-77RL MKII ULTRA LIGHT

This big-game, bolt-action rifle encompasses the traditional features that have made the Ruger M-77 one of the most popular centerfire rifles in the world. It includes a sliding top tang safety, a one-piece bolt with Mauser-type extractor and diagonal front mounting system. American walnut stock is hand-checkered in a sharp diamond pattern. A rubber recoil pad, pistol-grip cap and studs for mounting quick detachable sling swivels are standard. Available in both long- and short-action versions, with Integral Base Receiver and 1″ Ruger scope rings. **Calibers:** 223, 243, 257, 270, 30-06, 308. **Barrel length:** 20″. **Weight:** Approx. 6 lbs.

Price: **$592.50**

RUGER BOLT-ACTION RIFLES

MARK II SERIES

MODEL M-77VI MK II HEAVY BARREL TARGET

Features Mark II stainless steel bolt action, gray matte finish, two-stage adjustable trigger. No sights. **Calibers:** 22 PPC, 22-250, 220 Swift, 6mm PPC, 223, 243, 25-06 and 308. **Barrel length:** 26″, hammer-forged, free-floating stainless steel. **Weight:** 9³/₄ lbs. **Stock:** Laminated American hardwood with flat forend.

Price: KM77VTMKII **$665.00**

M-77 II MARK II ALL-WEATHER

KM-77RP MKII ALL-WEATHER Receiver w/integral dovetails to accommodate Ruger 1″ rings, no sights, stainless steel, synthetic stock. **Calibers:** 223, 243, 270, 280, 30-06, 7mm Rem. Mag., 300 Win. Mag., 308, 338 Win. Mag. **$558.00**
KM-77RSP MKII ALL-WEATHER Receiver w/integral dovetails to accommodate Ruger 1″ rings, metal sights, stainless steel, synthetic stock. **Calibers:** 243, 270, 7mm Rem. Mag., 30-06, 300 Win. Mag., 338 Win. Mag. **617.00**

RUGER 77RSM MK II MAGNUM RIFLE

This "Bond Street" quality African safari hunting rifle features a sighting rib machined from a single bar of steel; Circassian walnut stock with black forend tip; steel floorplate and latch; a new Ruger Magnum trigger guard with floorplate latch designed flush with the contours of the trigger guard (to eliminate accidental dumping of cartridges); a three-position safety mechanism (*see* illustrations); Express rear sight; and front sight ramp with gold bead sight.
  Calibers: 375 H&H, 404 Jeffery, 416 Rigby. **Capacity:** 4 rounds (375 H&H, 404 Jeffery) and 3 rounds (416 Rigby). **Barrel thread diameter:** 1¹/₈″. **Weight:** 9¹/₄ lbs. (375 H&H, 404 Jeffery); 10¹/₄ lbs. (416 Rigby)

Price: **$1550.00**

RUGER BOLT-ACTION RIFLES

MODEL M-77EXP MARK II EXPRESS RIFLE

For shooters who prefer a sporting rifle of premium quality and price, Ruger offers the M-77 Mark II Express rifle. Its stock is precision machined from a single blank of French walnut. Forend and pistol grip are hand-checkered in a diamond pattern with 18 lines per inch. Hardened alloy steel grip cap, trigger guard and floorplate plus a black buttpad of live rubber, black forend cap and steel studs for quick detachable sling swivels are standard.

Action is standard short or long length in blued chrome-moly steel with stainless steel bolt. A fixed blade-type ejector working through a slot under the left locking lug replaces the plunger-type ejector used in the earlier M-77 models. A three-position wing safety allows the shooter to unload the rifle with safety on. The trigger guard houses the patented floorplate latch, which holds the floorplate closed securely to prevent accidental dumping of cartridges into the magazine.

An integral, solid sight rib extends from the front of the receiver ring. Machined from a solid chrome-moly steel barrel blank, the rib has cross serrations on the upper surface to reduce light reflections. Each rifle is equipped with open metal sights. A blade front sight of blued steel is mounted on a steel ramp with curved rear surface serrated to reduce glare. Rear sights (adjustable for windage and non-folding) are mounted on the sighting rib. The forward rear sight is folding and adjustable for windage. A set of Ruger 1″ scope rings with integral bases is standard.

Calibers: 270, 30-06, 7mm Rem. Mag., 300 Win. Mag., 338 Win. Mag. **Barrel length:** 22″. **Capacity:** 4 rounds (3 rounds in Magnum). **Overall length:** 42$\frac{1}{8}$″. **Weight:** 7$\frac{3}{4}$ lbs. (avg., loaded); 7$\frac{1}{2}$ lbs. (unloaded). **Length of pull:** 13$\frac{1}{2}$″.

Price: **$1550.00**

MODEL M-77RSI INTERNATIONAL MANNLICHER

Mannlicher-type stock, Integral Base Receiver, open sights, Ruger 1″ steel rings. **Calibers:** 243, 270, 30-06, 308. **Barrel length:** 18$\frac{1}{2}$″. **Weight:** Approx. 6 lbs.

Price **$623.50**

RUKO-ARMSCOR RIFLES

MODEL M14P
$119.00

SPECIFICATIONS
Type: Bolt action
Caliber: 22 LR
Capacity: 10 rounds
Barrel length: 23″
Overall length: 41½″

Weight: 7 lbs.
Trigger pull: 14″
Stock: Mahogany
Finish: Blued
Also available:
MODEL M14D Bolt Action Deluxe $139.00

MODEL M1500
$199.00

SPECIFICATIONS
Type: Bolt action deluxe
Caliber: 22 Win. Mag.
Capacity: 5 rounds
Barrel length: 21½″

Overall length: 41¼″
Weight: 6.8 lbs.
Trigger pull: 14″
Stock: Mahogany
Finish: Blued

MODEL M20P
$129.00

SPECIFICATIONS
Type: Semiautomatic
Caliber: 22 LR
Capacity: 15 or 30 rounds
Barrel length: 21″
Overall length: 39¾″

Weight: 6½ lbs.
Trigger pull: 13″
Stock: Mahogany
Finish: Blued
Also available:
MODEL M2000 with deluxe stock $139.00
MODEL M20C (16½″ barrel; 5¼″ lbs.) 159.00

SAKO RIFLES

**FINNFIRE
22 LONG RIFLE**

FINNFIRE 22 LR BOLT-ACTION RIFLE
$685.00

SAKO of Finland, acclaimed as the premier manufacturer of bolt-action centerfire rifles, announces the development of its new .22 Long Rifle Finnfire. Designed by engineers who use only state-of-the-art technology to achieve both form and function and produced by craftsmen to exacting specifications, this premium grade bolt-action rifle exceeds the requirements of even the most demanding firearm enthusiast.

The basic concept in the design of the Finnfire was to make it as similar to its "big brothers" as possible—just scaled down. For example, the single-stage adjustable trigger is a carbon copy of the trigger found on any other big-bore hunting model. The 22-inch barrel is cold-hammered to ensure superior accuracy. **Overall length:** 39½". **Weight:** 5¼ lbs. **Rate of twist:** 16½".

Other outstanding features include:

- European walnut stock
- Luxurious matte lacquer finish
- 50° bolt lift
- Free-floating barrel
- Integral 11mm dovetail for scope mounting
- Two-position safety that locks the bolt
- Cocking indicator
- Five-shot detachable magazine
- Ten-shot magazine available

COMMITMENT TO EXCELLENCE — A SAKO TRADITION

SAKO RIFLES

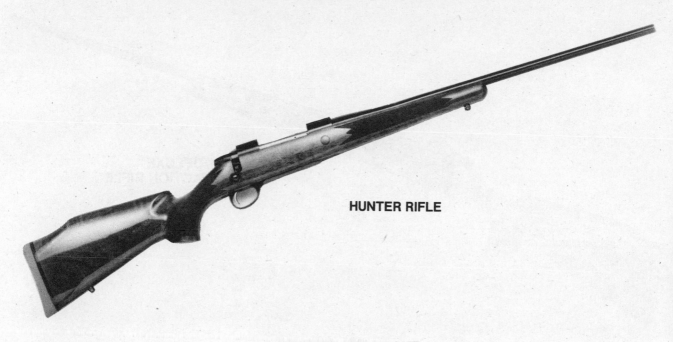

HUNTER RIFLE

HUNTER LIGHTWEIGHT

Here's one case of less being more. SAKO has taken its famed bolt-action, centerfire rifle, redesigned the stock and trimmed the barrel contour. In fact, in any of the short action (S 491-1) calibers—17 Rem., 222 or 223 Rem.—the Hunter weighs in at less than 7 pounds, making it one of the lightest wood stock production rifles in the world.

The same cosmetic upgrading and weight reduction have been applied to the entire Hunter line in all calibers and action lengths, standard and magnum. All the precision, quality and accuracy for which this Finnish rifle has been so justly famous are still here. Now it just weighs less.

The SAKO trigger is a rifleman's delight—smooth, crisp and fully adjustable. If these were the only SAKO features, it would still be the best rifle available. But the real test that sets SAKO apart from all others is its truly outstanding accuracy.

While many factors can affect a rifle's accuracy, 90 percent of any rifle's accuracy potential lies in its barrel. And the creation of superbly accurate barrels is where SAKO excels.

The care that SAKO takes in the cold-hammering processing of each barrel is unparalleled in the industry. For example, after each barrel blank is drilled, it is diamond-lapped and then optically checked for microscopic flaws. This extra care affords the SAKO owner lasting accuracy and a finish that will stay "new" season after season.

You can't buy an unfired SAKO. Every gun is test fired using special overloaded proof cartridges. This test ensures the SAKO owner total safety and uncompromising accuracy. Every barrel must group within SAKO specifications or it's scrapped. Not recycled. Not adjusted. Scrapped. Either a SAKO barrel delivers SAKO accuracy, or it never leaves the factory.

And hand-in-hand with SAKO accuracy is SAKO beauty. Pistol-grip stocks are of genuine European walnut, flawlessly finished and checkered by hand. Also available with a matte lacquer finish.

Prices:
Short Action (S 491-1)
In 17 Rem., 222 Rem., 223 Rem. $1000.00
Medium Action (M 591)
In 22-250 Rem., 7mm-08, 243 Win. &
  308 Win. 1000.00
Long Action (L 691)
In 25-06 Rem., 270 Win., 280 Rem., 30-06 . . . . . 1025.00
In 7mm Rem. Mag., 300 Win. Mag.,
  338 Win. Mag. 1055.00
In 300 Wby. Mag., 375 H&H Mag.,
  416 Rem. Mag. 1070.00

LEFT-HANDED MODELS (Matte Lacquer Finish)
Medium Action (M 591)
In 22-250 Rem., 7mm-08, 243 Win. &
  308 Win. $1085.00
Long Action (L 691)
In 25-06 Rem., 270 Win., 280 Rem., 30-06 . . . . . 1120.00
In 7mm Rem. Mag., 300 Win. Mag.,
  338 Win. Mag. 1135.00
In 300 Wby. Mag., 375 H&H Mag.,
  416 Rem. Mag. 1150.00

SAKO RIFLES

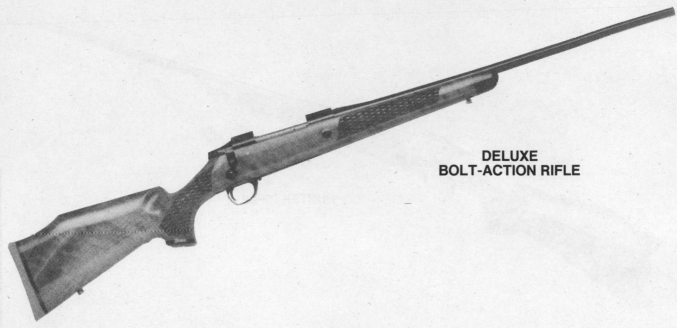

**DELUXE
BOLT-ACTION RIFLE**

DELUXE

All the fine-touch features you expect of the deluxe grade SAKO are here—beautifully grained French walnut, superbly done high-gloss finish, hand-cut checkering, deep rich bluing and rosewood forend tip and grip cap. And of course the accuracy, reliability and superior field performance for which SAKO is so justly famous are still here too. It's all here—it just weighs less than it used to. Think of it as more for less.

In addition, the scope mounting system on these SAKOs is among the strongest in the world. Instead of using separate bases, a tapered dovetail is milled into the receiver, to which the scope rings are mounted. A beautiful system that's been proven by over 20 years of use. SAKO Original Scope Mounts and SAKO scope rings are available in short/medium, and high in one-inch and 30mm.

Prices:
Short Action (S 491)
In 17 Rem., 222 Rem. & 223 Rem. **$1325.00**
Medium Action (M 591)
In 22-250 Rem., 243 Win., 7mm-08
  and 308 Win. **1325.00**
Long Action (L 691)
In 25-06 Rem., 270 Win., 280 Rem., 30-06 . . . . . **1410.00**
In 7mm Rem. Mag., 300 Win. Mag. &
  338 Win. Mag. **1425.00**
In 300 Wby. Mag., 375 H&H Mag.,
  416 Rem. Mag. **1440.00**

SUPER DELUXE

SAKO offers the Super Deluxe for the most discriminating gun buyer. This one-of-a-kind beauty is available on special order.

SUPER DELUXE **$2895.00**

CLASSIC

Classic elegance best describes one of SAKO'S latest models—the CLASSIC—designed for discriminating shooters who demand quality and the traditional clean, graceful lines of the classic style. Available in two action lengths and the most popular calibers (see below). Also available in a left-handed model.

SPECIFICATIONS
Calibers: 243 Win., 270 Win., 30-06, 7mm Rem. Mag.
Barrel length: 22″ and 24″ (Magnum action only)
Capacity: 5 rounds (Medium and Long action); 3 rounds (Magnum action)
Overall length: 42″ and 44″ (Magnum action only)
Weight: 6⅞ lbs. (243 Win.); 7 lbs. (270 Win. and 30-06); 7¼ lbs. (7mm Rem. Mag.)
Finish: Matte lacquer

Prices:
Medium Action (M 591)
In 243 Win. **$1000.00**
Long Action (L 691)
In 270 Win., 30-06 . . . . **1025.00**
In 7mm Rem. Mag. **1055.00**

LEFT-HANDED MODELS (Long Action Only)
In 270 Win. **$1120.00**
In 7mm Rem. Mag. **1135.00**

RIFLES

SAKO RIFLES/CARBINES

SAFARI GRADE

Crafted in the tradition of the classic British express rifles, Safari Grade is truly a professional's rifle. Every feature has been carefully thought out and executed with one goal in mind: performance. The magazine allows four belted magnums to be stored inside (instead of the usual three). The steel floorplate straddles the front of the trigger guard bow for added strength and security.

An express-style quarter rib provides a rigid, non-glare base for the rear sight, which consists of a fixed blade. The front swivel is carried by a contoured barrel band to keep the stud away from the off-hand under the recoil of big calibers. The front sight assembly is also a barrel-band type for maximum strength. The blade sits on a non-glare ramp and is protected by a steel hood.

The Safari's barreled action carries a subtle semi-matte blue, which lends an understated elegance to this eminently practical rifle. The functional, classic-style stock is of European walnut selected especially for its strength of grain orientation as well as for its color and figure. A rosewood forend tip, rosewood pistol grip cap with metal insert suitable for engraving, an elegant, beaded cheekpiece and presentation-style recoil pad complete the stock embellishments.

In **Calibers:** 338 Win. Mag., 375 H&H Mag. and 416 Rem. Mag. See also Specifications Table on page 276.

Price: **$2715.00**

MANNLICHER-STYLE CARBINE

SAKO's Mannlicher-style Carbine combines the handiness and carrying qualities of the traditional, lever-action "deer rifle" with the power of modern, high-performance cartridges. An abbreviated 18½-inch barrel trims the overall length of the Carbine to just over 40 inches in the Long Action (L 691) calibers, and 38 inches in the Medium Action (M 591) calibers. Weight is a highly portable 7 pounds and 6½ pounds, respectively (except in the 338 and 375 H&H calibers, which measure 7½ pounds).

As is appropriate for a rifle of this type, the Carbine is furnished with an excellent set of open sights; the rear is fully adjustable for windage, while the front is a non-glare, serrated ramp with protective hood.

The Mannlicher Carbine is available in the traditional wood stock of European walnut done in a contemporary Monte Carlo style with hand-rubbed oil finish. Hand-finished checkering is standard. The Mannlicher-style full stock Carbine wears SAKO's exclusive two-piece forearm, which joins beneath the barrel band and also features an oil finish. This independent forward section of the forearm eliminates the bedding problems normally associated with the full forestock. A blued steel muzzle cap puts the finishing touches on this European-styled Carbine.

Prices:
Medium Action (M 591)
In 243 Win. and 308 Win. **$1190.00**
Long Action (L 691)
In 270 Win. and 30-06 . . . . **1225.00**
In 338 Win. Mag. **1245.00**
In 375 H&H Mag. **1260.00**

SAKO RIFLES

MODEL TRG-21
$4000.00

SAKO, known for manufacturing the finest and most accurate production sporting rifles available today, presents the ultimate in sharpshooting systems: the **TRG-21 Target Rifle.** Designed for use when nothing less than total precision is demanded, this new SAKO rifle features a cold-hammer forged receiver, "resistance-free" bolt, stainless-steel barrel and a fully adjustable polyurethane stock. Chambered in .308 Win. A wide selection of optional accessories is also available. Designed, crafted and manufactured in Finland. For additional specifications, see the table on page 276.

- Cold-hammer forged receiver
- "Resistance-free" bolt
- Cold-hammer forged, stainless steel barrel
- Three massive locking lugs
- 60° bolt lift
- Free-floating barrel
- Detachable 10-round magazine

- Fully adjustable cheekpiece
- Infinitely adjustable buttplate
- Adjustable two-stage trigger pull
- Trigger adjustable for both length and pull
- Trigger also adjustable for horizontal or vertical pitch
- Safety lever inside the trigger guard
- Reinforced polyurethane stock

Optional features:
- Muzzle brake
- Quick-detachable one-piece scope mount base
- Available with 1″ or 30mm rings
- Collapsible and removable bipod rest
- Quick-detachable sling swivels
- Wide military-type nylon sling

MODEL TRG-S
$759.00   ($799.00 in Magnum)

The new TRG-S has been crafted and designed around SAKO's highly sophisticated and extremely accurate TRG-21 Target Rifle (above). The "resistance-free" bolt and precise balance of the TRG-S, plus its three massive locking lugs and short 60-degree bolt lift, are among the features that attract the shooter's attention. Also of critical importance is the cold-hammer forged receiver—

unparalleled for strength and durability. The detachable 5-round magazine fits securely into the polyurethane stock. The stock, in turn, is molded around a synthetic skeleton that provides additional support and maximum rigidity. For additional specifications, see the table on page 276.

SAKO RIFLES

LAMINATED STOCK MODELS

In response to the growing number of hunters and shooters who seek the strength and stability that a fiberglass stock provides, coupled with the warmth and feel of real wood, Sako features its Laminated Stock models.

Machined from blanks comprised of 36 individual layers of 1/16-inch hardwood veneers that are resin-bonded under extreme pressure, these stocks are virtually inert. Each layer of hardwood has been vacuum-impregnated with a permanent brown dye. The bisecting of various layers of veneers in the shaping of the stock results in a contour-line appearance similar to a piece of slab-sawed walnut. Because all Sako Laminated Stocks are of real wood, each one is unique, with its own shading, color and grain.

These stocks satisfy those whose sensibilities de-mand a rifle of wood and steel, but who also want state-of-the-art performance and practicality. Sako's Laminated Stock provides both, further establishing it among the most progressive manufacturers of sporting rifles—and the *only* one to offer hunters and shooters their choice of walnut, fiberglass or laminated stocks in a wide range of calibers.

Laminated Stock Model Prices:
Medium Action (M 591)
In 22-250, 243 Win., 308 Win. and 7mm-08 .... **$1145.00**
Long Action (L 691)
In 25-06 Rem., 270 Win., 280 Rem. & 30-06 ... **1195.00**
In 7mm Rem. Mag., 300 Win. Mag. and
    338 Win. Mag. **1210.00**
In 375 H&H Mag., 416 Rem. Mag. **1230.00**

FIBERCLASS MODEL

In answer to the increased demand for Sako quality and accuracy in a true "all-weather" rifle, this fiberglass-stock version of the renowned Sako barreled action has been created. Long since proven on the bench rest circuit to be the most stable material for cradling a rifle, fiberglass is extremely strong, light in weight, and un-affected by changes in weather. Because fiberglass is inert, it does not absorb or expel moisture; hence, it cannot swell, shrink or warp. It is impervious to the high humidity of equatorial jungles, the searing heat of arid deserts or the rain and snow of the high mountains. Not only is this rifle lighter than its wood counterpart, it also appeals to the performance-oriented hunter who seeks results over appearance.

Prices:
Long Action (L 691)
In 25-06 Rem., 270 Win., 280 Rem., 30-06 .... **$1355.00**
In 7mm Rem. Mag., 300 Win. Mag. and
    338 Win. Mag. **1370.00**
In 375 H&H Mag., 416 Rem. Mag. **1390.00**

SAKO RIFLES

LEFT-HANDED MODELS

SAKO's Left-Handed models are based on mirror images of the right-handed models SAKO owners have enjoyed for years; the handle, extractor and ejection port all are located on the port side. Naturally, the stock is also reversed, with the cheekpiece on the opposite side and the palm swell on the port side of the grip.

Otherwise, these guns are identical to the right-handed models. That means hammer-forged barrels, one-piece bolts with integral locking lugs and handles, integral scope mount rails, adjustable triggers and Mauser-type inertia ejectors.

SAKO's Left-Handed rifles are available in all Long Action models, while the Hunter grade is available in both Medium and Long Action. The Hunter Grade carries a durable matte lacquer finish with generous-size panels of hand-cut checkering, a presentation-style recoil pad and sling swivel studs installed. The Deluxe model is distinguished by its rosewood forend tip and grip cap, its skip-line checkering and gloss lacquer finish atop a select-grade of highly figured European walnut. The metal work carries a deep, mirror-like blue that looks more like black chrome.

Prices:

Hunter Lightweight (Medium Action)
In 22-250 Rem. 243 Win., 308 Win., 7mm-08 . . . **$1085.00**

Hunter Lightweight (Long Action)
In 25-06, 270 Win., 280 Rem. & 30-06 . . . . **1120.00**
In 7mm Rem. Mag., 300 Win. Mag. and
   338 Win. Mag. **1135.00**
In 300 Wby. Mag., 375 H&H Mag. and
   416 Rem. Mag. **1150.00**

Deluxe (Long Action)
In 25-06, 270 Win. Mag., 280 Rem. & 30-06 . . **$1475.00**
In 7mm Rem. Mag., 300 Win. Mag. and
   338 Win. Mag. **1490.00**
In 300 Wby. Mag., 375 H&H Mag. and
   416 Rem. Mag. **1500.00**

Classic Left-Handed (Long Action, Matte Lacquer Finish)
In 270 Win. **$1120.00**
In 7mm Rem. Mag. **1135.00**

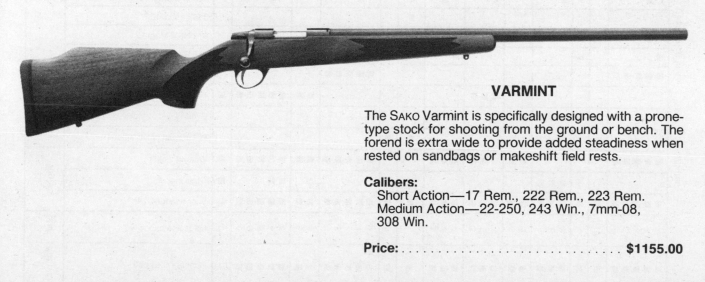

VARMINT

The SAKO Varmint is specifically designed with a prone-type stock for shooting from the ground or bench. The forend is extra wide to provide added steadiness when rested on sandbags or makeshift field rests.

Calibers:
   Short Action—17 Rem., 222 Rem., 223 Rem.
   Medium Action—22-250, 243 Win., 7mm-08,
   308 Win.

Price: **$1155.00**

RIFLES

SAKO RIFLES

	Finnfire	Hunter	Classic	Deluxe	TRG-S	TRG-S Magnum	Fiberclass	Laminated	Varmint	Carbine Mannlicher Style	Safari	TRG-21	Super Deluxe	
Action	L691	S491 M591 L691 L691 L691	M591 L691	S491 M591 L691 L691 L691	*	*	L691 L691 L691	M591 L691 L691	S491 M591 L691 L691	M591 L691 L691	L691	*	S491 M591 L691 L691 L691	Model
Left-handed		■ ■ ■ ■	■	■ ■ ■										
Total length (inches)	39½	41½ 42½ 44 46 46	42½ 44	41½ 42½ 44 46 46	45½	47½	44 46 46	42½ 44 46	43½ 43½ 46	39½ 40½ 40½	44	46½	41½ 42½ 44 46 46	Dimensions
Barrel length (inches)	22	21¼ 21¾ 22 24 24	21¾ 24	21¼ 21¾ 22 24 24	22	24	22 24 24	21¾ 22 24	23 23 24	18½ 18½ 18½	23¾	25¾	21¼ 21¾ 22 24 24	
Weight (lbs)	5¼	6¼ 6½ 7¾ 7¾ 8¼	6 7½	6¼ 6½ 7¾ 7¾ 8¼	7¾	7¼	8	6½ 7¼ 7¾	8¼ 8½	6 7¼ 7¾	8¼	10½	6¼ 6½ 7¾ 7¾ 8¼	
17 Rem/10″		■	■						■				■	Caliber/Rate of Twist
22 LR/16.5″ (Finnfire)	■													
222 Rem/14″		■	■						■					
223 Rem/12″		■	■						■					
22-250 Rem/14″		■					■		■				■	
243 Win/10″		■	■	■					■	■			■	
308 Win/12″		■		■					■	■	■		■	
7mm-08/9½″		■		■					■				■	
25-06 Rem/10″		■		■			■	■					■	
270 Win/10″		■	■	■						■			■	
280 Rem/10″		■		■			■	■					■	
30-06/10″		■	■	■	■		■	■		■			■	
7mm/Rem Mag/9½″		■		■	■		■	■					■	
300 Win Mag/10″		■		■	■		■	■					■	
300 Wby Mag/10″				■			■	■					■	
338 Win Mag/10″				■			■	■			■		■	
338 Lapua/12″						■								
375 H&H Mag/12″		■		■				■		■	■		■	
416 Rem Mag/14″		■		■			■				■		■	
Lacquered				■ ■ ■ ■ ■									■ ■ ■ ■ ■	Stock Finish
Matte Lacquered	■	■ ■ ■ ■ ■	■ ■					■ ■						
Oiled									■ ■	■ ■ ■ ■	■			
Reinforced polyurethane					■	■						■		
Without sights	■	■ ■ ■ ■	■ ■ ■	■ ■ ■ ■ ■			■ ■	■ ■	■ ■				■ ■ ■ ■ ■	Sights
Open sights	■	■ ■								■ ■ ■				
Scope mount rails	■	■ ■ ■ ■	■ ■	■ ■ ■ ■ ■			■ ■	■ ■	■	■	■		■ ■ ■ ■ ■	
Magazine capacity	5***	6 5 5 6 4	5 4	4 4 5 5 4	5	**	5 4	5 5 4	5 4	5 5 3	4	10	4 4 5 5 4	Mag.
Rubber recoil pad		■ ■ ■ ■ ■	■ ■	■ ■ ■ ■ ■	■	■	■ ■ ■	■ ■ ■	■ ■	■ ■ ■	■	■	■ ■ ■ ■ ■	Buttplate

S491 formerly A1; M591 formerly A11; L691 formerly AV

*Cold-Hammer Forged Receiver   **5 except for the .375 H&H which is 4   ***10 round optional

SAKO ACTIONS

Only by building a rifle around a SAKO action do shooters enjoy the choice of three different lengths, each scaled to a specific family of cartridges. The S 491-1 (Short) action is miniaturized in every respect to match the 222 family, which includes everything from 17 Remington to 223 Remington. The M 591 (Medium) action is scaled down to the medium-length cartridges of standard bolt face—22-250, 243, 308, 7mm-08 or similar length cartridges. The L 691 (Long) action is offered in either standard or Magnum bolt face and accommodates cartridges of up to 3.65 inches in overall length, including rounds like the 300 Weatherby and 375 H&H Magnum. **For left-handers, the Medium and Long actions are offered in either standard or Magnum bolt face.** All actions are furnished in-the-white only.

S 491 SHORT ACTION (formerly AI-1)
$495.00
CALIBERS:
17 Rem., 222 Rem.
222 Rem. Mag.
223 Rem.

M 591 MEDIUM ACTION (formerly AII-1)
$500.00
CALIBERS:
22-250 Rem. (M 591-3)
243 Win.
308 Win.
7mm-08

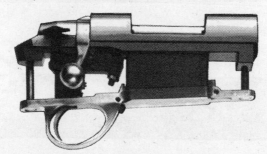

L 691 LONG ACTION (formerly AV-4)
$495.00
CALIBERS:
25-06 Rem. (L 691-1)
270 Win. (L 691-1)
280 Rem. (L 691-1)
30-06 (L 691-1)
7mm Rem. Mag. (L 691-4)
300 Win. Mag. (L 691-4)
300 Wby. Mag. (L 691-4)
338 Win. Mag. (L 691-4)
375 H&H Mag. (L 691-4)
416 Rem. Mag. (L 691-4)

Also available:
LEFT-HANDED ACTIONS
Medium and Long: $545.00

SAUER RIFLES

MODEL 90 LUX

MODEL 90
$1190.00

SPECIFICATIONS
Calibers: 25-06, 270, 30-06, 7mm Rem. Mag., 300 Win. Mag., 300 Wby. Mag., 338 Win., 375 H&H and 458 Win. Mag. (see below)
Barrel length: 24″ or 26″
Overall length: 44″ (standard calibers and 375 H&H, 458 Win. Mag.); 46¹/₈″ (for other calibers)
Weight: 7 lbs. 12 oz. to 10 lbs. 6 oz. (458 Win. Mag.)
Sights: None furnished; drilled and tapped for scope mount
Stock: Monte Carlo cut with sculptured cheekpiece, hand-checkered pistol grip and forend, rosewood pistol grip cap

and forend tip, black rubber recoil pad, and fully inletted sling swivel studs.
Features: Rear bolt cam-activated locking lug action; jeweled bolt with an operating angle of 65°; fully adjustable gold-plated trigger; chamber loaded signal pin; cocking indicator; tang-mounted slide safety with button release; bolt release button (to operate bolt while slide safety is engaged); detachable 3 or 4-round box magazine; sling side scope mounts; leather sling (extra)
Engravings: Four distinctive hand-cut patterns on gray nitride receiver, trigger housing plate, magazine plate and bolt handle (extra). **Prices on request.**

MODEL 90 ENGRAVING #2

MODEL 90 ENGRAVING #4

SAUER .458 SAFARI

SAUER .458 SAFARI

The Sauer .458 Safari features a rear bolt cam-activated locking-lug action with a low operating angle of 65°. It has a gold plated trigger, jeweled bolt, oil finished bubinga stock and deep luster bluing. Safety features include a press bottom slide safety that engages the trigger sear, toggle joint and bolt. The bolt release feature allows the sportsman to unload the rifle while the safety remains engaged to the trigger sear and toggle joint. The Sauer Safari is equipped with a chamber loaded signal pin for positive identification. Specifications include: **Barrel Length:** 24″ (heavy barrel contour). **Overall length:** 44″. **Weight:** 10 lb. 6 oz. **Sights:** Williams open sights (sling swivels included). **Price:** $1650.00

SAVAGE RIFLES

MODEL 110 BOLT-ACTION CENTERFIRE RIFLES

MODEL 110FP TACTICAL

MODEL 110 FP TACTICAL
$409.00

SPECIFICATIONS
Calibers: 223 Rem., 308 Win.
Capacity: 5 rounds (1 in chamber)
Barrel length: 24″ (w/recessed target-style muzzle)
Overall length: 45½″

Weight: 8½ lbs.
Sights: None; drilled and tapped for scope mount
Features: Black matte nonreflective finish on metal parts; bolt coated with titanium nitride; stock made of black graphite/fiberglass filled composite with positive checkering

MODEL 110CY BOLT ACTION

MODEL 110CY
$362.00

SPECIFICATIONS
Calibers: 223 Rem., 243 Win., 270 Win., 300 Savage, 308 Win.
Capacity: 5 rounds (1 in chamber); top-loading internal magazine
Barrel length: 22″ blued

Overall length: 42½″
Weight: 6½ lbs.
Sights: Adjustable: drilled and tapped for scope mounts
Stock: High comb, walnut-stained hardwood w/cut checkering and short pull

SAVAGE RIFLES

MODEL 111GC CLASSIC HUNTER
$407.00

SPECIFICATIONS
Calibers: 270 Win., 30-06 Springfield, 7mm Rem. Mag., 300 Win. Mag.
Capacity: 5 rounds (4 rounds in Magnum calibers)
Overall length: 43½″ (45½″ Magnum calibers)
Weight: 6⅜ lbs.

Sights: Adjustable
Stock: American-style walnut-finished hardwood; cut checkering
Features: Detachable staggered box-type magazine; left-hand model available

MODEL 111FC CLASSIC HUNTER
$418.00

Same specifications as Classic Hunter above, except stock is lightweight graphite/fiberglass-filled composite w/positive checkering. **Calibers:** 270 Win. and 30-06 Springfield only.

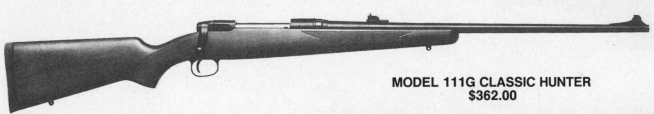

MODEL 111G CLASSIC HUNTER
$362.00

Same specifications as Model 111GC Classic Hunter, except available also in calibers 223 Rem., 22-250 Rem., 243 Win., 300 Savage, 308, 7mm-08. Stock is American-style walnut-finished hardwood with cut-checkering.

MODEL 111F CLASSIC HUNTER
$376.00

Same specifications as Model 111G Classic Hunter, except stock is black non-glare graphite/fiberglass-filled polymer w/ positive checkering.

SAVAGE CENTERFIRE RIFLES

MODEL 112FV
$392.00

MODEL 112FVSS STAINLESS
$495.00

The Model 112FV was designed to meet the demand for a 26″ heavy barrel varmint rifle, incorporating the best features of the Savage Model 110—famous for its accuracy and dependability. Available with black Graphite Fiberglass Polymer stock and recoil pad, the Model 112FV is drilled and tapped for scope mounting. Shipped from the factory with gunlock, ear puffs, target and shooting glasses.

SPECIFICATIONS
Calibers: 22-250 and 223. **Capacity:** 5 rounds. **Barrel length:** 26″. **Overall length:** 47½″. **Weight:** 8⅞ lbs. **Stock:** Graphite/Fiberglass-filled composite.
Also available:
MODEL BVSS VARMINT RIFLE. Same specifications as Model 112FV, except designed for long-range hunting and target-style shooting. **Weight:** 10½ lbs. **Price:** $600.00

MODEL 112 BT COMPETITION GRADE
$1000.00

Calibers: 223 Rem. and 308 Win. **Barrel length:** 26″. **Weight:** 10⅞ lbs.

MODEL 114CU CLASSIC ULTRA
$501.00

Savage's Model 114CU is a high-grade sporting firearm designed with a straight classic American black walnut stock. Cut-checkering, fitted grip cap and recoil pad are complemented by a high-gloss luster metal polish finish. The staggered box-type magazine is removed with a push of a button, making loading and reloading quick and easy.

SPECIFICATIONS
Calibers: 270, 30-06, 7mm Rem Mag., 300 Win. Mag.
Capacity: 4 rounds (Magnum); 5 rounds (270 and 30-06)
Barrel length: 22″ (270 and 30-06) and 24″ (Magnum)
Overall length: 43½″ (270 and 30-06) and 45½″ (Magnum)
Weight: 7½ lbs. (270 and 30-06) and 7¾ lbs. (Magnum)
Sights: Deluxe adjustable; receivers drilled and tapped for scope mounts
Length of pull: 13½″

SAVAGE CENTERFIRE RIFLES

MODEL 116FSS "WEATHER WARRIOR"
$489.00

Savage Arms combines the strength of a black graphite fiberglass polymer stock and the durability of a stainless-steel barrel and receiver in this bolt-action rifle. Major components are made from stainless steel, honed to a low reflective satin finish. Drilled and tapped for scope mounts, the 116FSS is offered in popular long-action calibers. Packed with gunlock, ear puffs and target.

SPECIFICATIONS
Calibers: 223, 243, 270, 30-06, 7mm Rem Mag., 300 Win. Mag., 338 Win. Mag.
Capacity: 4 (7mm Rem. Mag., 300 Win. Mag., 338 Win. Mag.); 5 (223, 243, 270, 30-06)
Barrel length: 22″ (223, 243, 270, 30-06); 24″ (7mm Rem. Mag., 300 Win. Mag., 338 Win. Mag.)
Overall length: 43½″–45½″
Weight: 6½ lbs.

MODEL 116FCS "WEATHER WARRIOR"
$552.00

This bolt-action rifle has the same quality features as the Model 116FSS plus a removable box magazine with recessed push-button release for ease in loading and unloading.

MODEL 116SE SAFARI EXPRESS
$893.00

SPECIFICATIONS
Calibers: 300 Win. Mag., 338, 458,
Capacity: 4 rounds (1 in chamber)

Barrel length: 24″   **Overall length:** 45½″
Weight: 8½ lbs.
Sights: 3-leaf express
Stock: Classic-style select-grade walnut

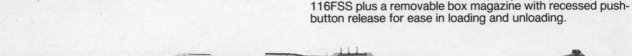

MODEL 116FSK "WEATHER WARRIOR"
$552.00

Features a compact barrel with "shock suppressor" that reduces average linear recoil by more than 30% without loss of Magnum stopping power. Available in left hand.

SPECIFICATIONS
Calibers: 270 Win., 30-06 Sprg., 7mm Rem. Mag., 300 Win. Mag., 338 Win. Mag.

Capacity: 5 rounds (4 in Magnum)
Barrel length: 22″   **Overall length:** 43½″
Weight: 6½ lbs.
Also available:
MODEL 116FSAK. Same specifications as above except includes adj. muzzle brake. **Price: $581.00**

SAVAGE CENTERFIRE RIFLES

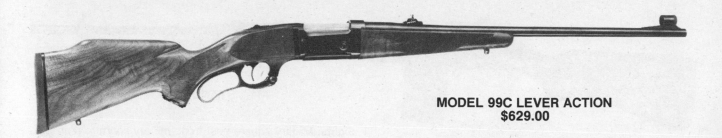

MODEL 99C LEVER ACTION
$629.00

Clip magazine allows for the chambering of pointed, high-velocity big-bore cartridges. **Calibers:** 243 Win., 308 Win. **Action:** Hammerless, lever action, top tang safety. **Magazine:** Detachable clip; holds 4 rounds plus 1 in the chamber. **Stock:** Select walnut with high Monte Carlo and deep-fluted comb.

Cut-checkered stock and forend with swivel studs. Recoil pad and pistol grip cap. **Sights:** Detachable hooded ramp front sight, bead front sight on removable ramp adjustable rear sight. Tapped for top mount scopes. **Barrel length:** 22″. **Overall length:** 42³/₄″. **Weight:** 7³/₄ lbs.

SAVAGE MODEL 24F-12T TURKEY
with Camo Graphite Fiberglass Stock
$414.00

SAVAGE MODEL 24F COMBINATION
RIFLE/SHOTGUN
$400.00

Match a 12- or 20-gauge shotgun with any of four popular centerfire calibers. Frame is color casehardened and barrel is a deep, lustrous blue and tapped, ready for scope mounting. Two-way top opening lever. All models are stocked with tough

Graphite Fiberglass Polymer plus hammerblock safeties that limit hammer travel in the safe position. Other features include interchangeable chokes (extra full tube supplied) and factory swivel studs.

SPECIFICATIONS MODEL 24F COMBINATION RIFLE/SHOTGUN

O/U Comb. Model	Gauge/ Caliber	Choke	Chamber	Barrel Length	O.A. Length	Twist R.H.	Stock
24F-20	12 or 20/22 LR	Mod. Barrel	3″	24″	40¹/₂″	1 in 14″	Black Graphite Fiberglass Polymer
	12 or 20/22 Hor.					1 in 14″	
	12 or 20/223					1 in 14″	
	12 or 20/30-30					1 in 12″	
24F-12T Turkey	12/22 Hor.	Full Mod., IC Choke Tubes	3″	24″	40¹/₂″	1 in 14″	Camo Graphite Fiberglass Polymer
	12/223					1 in 14″	

SPRINGFIELD RIFLES

M1A STANDARD

Sights: Military square post front; military aperture rear, adjustable for windage and elevation
Sight radius: 26³/₄″
Rifling: 6 groove, RH twist, 1 turn in 11″
Finish: Walnut
Price: **$1269.00**

Also available:
BASIC M1A RIFLE w/painted black fiberglass stock, caliber 308 only. **Price:** **$1129.00**

SPECIFICATIONS
Caliber: 308 Win./7.62mm NATO (243 or 7mm-08 optional)
Capacity: 5, 10 or 20-round box magazine
Barrel length: 22″
Overall length: 44¹/₃″
Weight: 9 lbs.

M1A NATIONAL MATCH

Features: Comes with National Match barrel, flash suppressor, gas cylinder, special glass-bedded walnut stock and match-tuned trigger assembly.
Price: **$1598.00**

Also available:
M1A SUPER MATCH. Features heavy match barrel and permanently attached figure-8-style operating rod guide, plus special heavy walnut match stock, longer pistol grip and contoured area behind rear sight for better grip.
Price: **$1929.00**

SPECIFICATIONS
Caliber: 243 Win., 308 Win., 7mm-08 Win.
Barrel length: 22″
Overall length: 44.375″
Trigger pull: 4¹/₂ lbs.
Weight: 10 lbs. (11 lbs. in Super Match)

M1A-A1 BUSH RIFLE

SPECIFICATIONS
Caliber: 308 Win./7.62mm, 243 or 7mm-08 Win.
Barrel length: 18″ (w/o flash suppressor)
Overall length: 40.5″

Weight: 8.75 lbs.   **Sight radius:** 26³/₄″
Prices:
With walnut stock **$1289.00**
With folding stock 1489.00

STEYR-MANNLICHER RIFLES

SPORTER SERIES

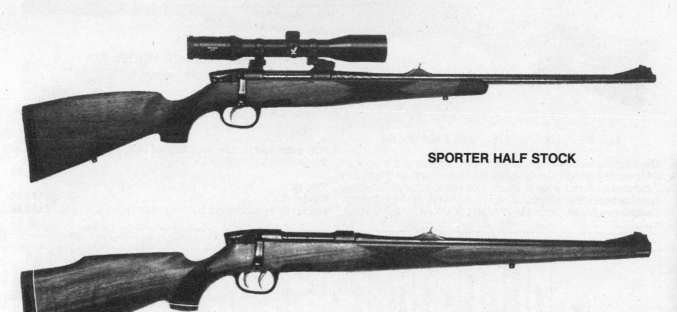

SPORTER HALF STOCK

SPORTER FULL STOCK

SPORTER SERIES

All Sporter models feature hand-checkered wood stocks, a five-round detachable rotary magazine, and a choice of single or double-set triggers. M actions are available in left-hand models. S (Magnum) action are available in half stock only.

SPECIFICATIONS
Calibers: See table on the following page
Barrel length: 20″ (Full Stock); 23.6″ (Half Stock)
Overall length: 39″ (Full)
Weight:
    Model SL—6.16 lbs. (Full) and 6.27 lbs. (Half Stock)
    Model L—6.27 lbs. (Full) and 6.38 lbs. (Half)
    Model M—6.82 lbs. (Full) and 7 lbs. (Half).
Features: SL and L Models have rifle-type rubber butt pad

Prices:
Models SL, L, M Full Stock	$2179.00
Models SL, L, M Half Stock	2023.00
Model M Left Hand Full Stock	2335.00
Model M Left Hand Half Stock	2179.00
Model M Professional (w/black synthetic half stock, 23.6″ barrel and no sights)	995.00
Same as above w/stippled checkered European wood stock (270 Win. and 30-06 calibers)	1125.00
Varmint Rifle Half stock, 26″ heavy barrel	2179.00

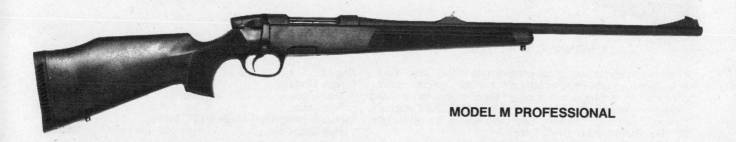

MODEL M PROFESSIONAL

STEYR-MANNLICHER RIFLES

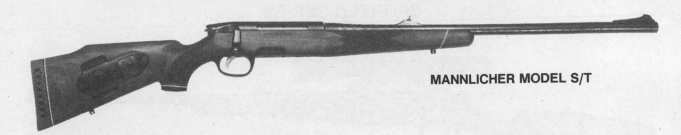

MANNLICHER MODEL S/T

MANNLICHER MODEL S-S/T MAGNUM

The Mannlicher S/T is a heavy-barreled version of the Sporter S Model designed specifically for big game hunting. It features a hand-checkered walnut stock, five-round rotary magazine, optional butt stock magazine, and double-set or single trigger. **Calibers:** 6.5×68, 7mm Rem. Mag., 300 Win. Mag., 8×685, 375 H&H Mag., 338 Win. **Barrel length:** 26″. **Weight:** 8.36 lbs. (Model S); 9 lbs. (Model S/T).

Prices:
Model S **$2179.00**
Model S/T (w/optional butt magazine) **2335.00**

MODELS:	CALIBERS																							
	222 Rem.	222 Rem. Mag.	223 Rem.	5.6 × 50 Mag.	5.6 × 57	243 Win.	308 Win.	6.5 × 57	270 Win.	7 × 64	30-06 Spr.	9.3 × 62	6.5 × 68	7mm Rem. Mag.	300 Win. Mag.	8 × 685	22-250 Rem.	6mm Rem.	6.5 × 55	7.5 Swiss	7 × 57	8 × 57 JS	375 H&H Mag.	458 Win. Mag.
Sporter (SL)	•	•	•	•																				
(L)					•	•	•										•	•						
(M)								•	•	•	•								•	•	•	•		
S and S/T*													•	•	•	•							•	•
Professional (M)								•	•	•	•	•							•	•	•	•		
Luxus (L)					•	•	•																	
(M)								•	•	•	•	•							•		•			
(S)													•	•	•	•								
Varmint	•		•		•	•	•										•							
Match UIT							•																	
SSG						•	•																	

* Also available in 9.3 × 64

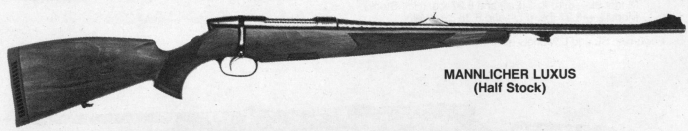

MANNLICHER LUXUS
(Half Stock)

MANNLICHER LUXUS

The Mannlicher Luxus is the premier rifle in the Steyr lineup. It features a hand-checkered walnut stock, smooth action, combination shotgun set trigger, steel in-line three-round magazine (detachable), rear tang slide safety, and European-designed receiver. **Calibers:** See table above. **Barrel length:** 20″ (Full Stock); 23.6″ (Half Stock).

Prices:
Luxus Models
  Half Stock **$2648.00**
  Full Stock **2804.00**
Luxus S (Magnum) Models (26″ barrel,
  Half Stock only) **2804.00**

STEYR-MANNLICHER RIFLES

STEYR SSG

The Steyr SSG features a black synthetic Cycolac stock (walnut optional), heavy Parkerized barrel, five-round standard (and optional 10-round) staggered magazine, heavy-duty milled receiver. **Calibers:** 243 Win. and 308 Win. **Barrel length:** 26″. **Overall length:** 44.5″. **Weight:** 8.5 lbs. **Sights:** Iron sights; hooded ramp front with blade adjustable for elevation; rear standard V-notch adjustable for windage. **Features:** Sliding safety; 1″ swivels.

Prices:
Model SSG-PI Cycolac half-stock (26″ bbl.) **$1995.00**
Model SSG-PII (26″ heavy bbl.) 1995.00
Model SSG Marksman Scope Mount 244.00
Model SSG P-IV Urban In 308 Win.
   w/16³⁄₄″ heavy barrel 2660.00

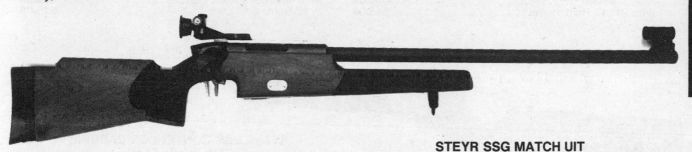

STEYR SSG MATCH UIT

Designed especially for target competition, the Steyr Match UIT features a walnut competition stock, stipple-textured pistol grip, adjustable straight and wide trigger, adjustable first-stage trigger pull, enlarged bolt handle for rapid fire, cold hammer-forged barrel, and non-glare band for sighting. **Caliber:** 308 Win. **Overall length:** 44″. **Weight:** 10 lbs.

Prices:
Steyr Match SPG-UIT **$3995.00**
   10-shot magazine 143.00
Model SPG-CISM 4295.00
Model SPG-T 3695.00

AUG S.A. SEMIAUTOMATIC RIFLE

SPECIFICATIONS
Caliber: 223   **Capacity:** 30-round magazine
Barrel lengths: 16″ or 20″
Price: **$1575.00**
Also available: **AUG** Special Receiver w/Stanag
   Scope Mount 695.00

THOMPSON/CENTER RIFLES

THE CONTENDER CARBINE
$460.00

Available in **9 calibers:** 17 Rem., 22 LR, 22 LR Match, 22 Hornet, 223 Rem., 7×30 Waters, 30-30 Win., 35 Rem. and 375 Win. **Barrels** are 21 inches long and are interchangeable, with adjustable iron sights and tapped and drilled for scope mounts. **Weight:** Only 5 lbs. 3 oz.
Also available:
Contender Vent Rib Carbine
  With standard walnut stock **$505.00**
  With 21″ 17 Rem. barrel **515.00**

Contender Youth Model Carbine
  W/16¼″ bbl., walnut Youth stock **$450.00**
  With 16¼″ 45 Colt/.410 barrel **480.00**
Contender Carbine
  w/Match Grade 22LR barrel **495.00**
  w/Rynite Stock & 21″ barrel **480.00**
  In 17 Rem. **515.00**

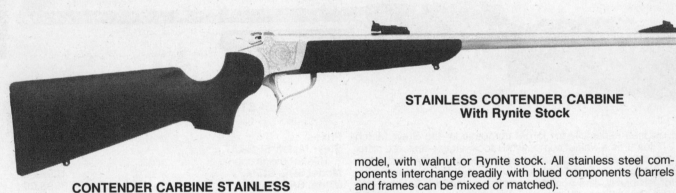

STAINLESS CONTENDER CARBINE
With Rynite Stock

CONTENDER CARBINE STAINLESS

Available in 22LR, 22 Hornet, 223 Rem., 7-30 Waters, 30-30 Win., 375 Win., and .410 bore. Same specifications as standard

model, with walnut or Rynite stock. All stainless steel components interchange readily with blued components (barrels and frames can be mixed or matched).
Prices:
Stainless Carbine Standard **$480.00**
Stainless Carbine w/vent rib **505.00**

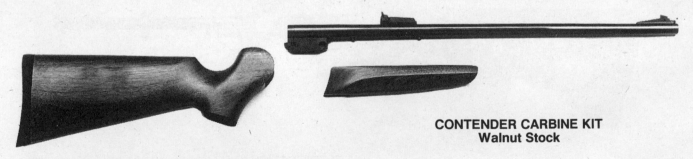

CONTENDER CARBINE KIT
Walnut Stock

CONTENDER CONVERSION KIT

Available in 22LR, 22 Hornet, 223 Rem., 30-30 Win. and .410 smoothbore. Each kit contains a buttstock, blued 21″ barrel, forend and sights.

Prices:
With Walnut stock **$290.00**
With Rynite stock **310.00**

TIKKA RIFLES

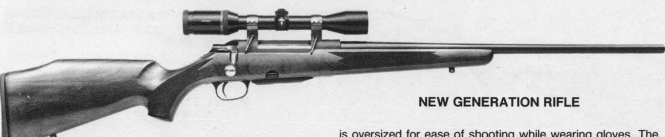

NEW GENERATION RIFLE

With the consolidation of three renowned Finnish firearms manufacturers—Tikka, Sako and Valmet—a "new generation" of Tikka rifles becomes a reality. These new rifles feature a "smooth as silk" bolt action made possible by a sleeve constructed of a space-age synthetic Polyarylamide material reinforced with fiberglass. The overall look of the rifle is enhanced by a walnut stock with matte lacquer finish and diamond point checkering. A short bolt throw allows for rapid firing, and a free-floating barrel increases accuracy. Barrel quality itself is ensured through Tikka's cold-hammered forging process. The trigger guard, made of synthetic materials for added strength,

is oversized for ease of shooting while wearing gloves. The recessed magazine release is located conveniently for quick and safe release. Tikka's wood-to-metal fit reflects the high standards of Finnish craftsmanship throughout. **Calibers:** 223 Rem., 22-250 Rem., 243 Win., 308 Win., 270 Win., 30-06, 7mm Rem. Mag., 300 Win. Mag., and 338 Win. Mag. **Barrel length:** 22½" (24½" in Magnum). **Weight:** 7⅛ lbs.
Prices:
Calibers 223 Rem., 22-250 Rem., 243 Win., 308 Win., 270 Win. and 30-06 . . . . $675.00
Calibers 7mm Rem. Mag., 300 Win. Mag., 338 Win. Mag. 700.00
Magazines (5 rounds) 53.00
(3 rounds) 45.00

PREMIUM GRADE RIFLE

The Tikka Premium Grade rifle is designed and crafted by Sako of Finland for the discriminating shooter. This superb firearm features a detachable magazine, along with a "smooth as silk" bolt that is encased in a polymer sleeve. The luxurious matte lacquer stock incorporates a rollover cheekpiece, rosewood pistol grip cap and forend tip and hand-checkered throughout. The cold-hammered barrel is deeply blued and free floated for maximum accuracy. The two action lengths

eliminate unnecessary weight and each trigger is designed and built to be crisp, clean and travel-free. Available in a wide assortment of calibers.
Prices:
Calibers 223 Rem., 22-250 Rem., 243 Win., 308 Win., 270 Win., 30-06 . . . . $825.00
Calibers 7mm Rem. Mag., 300 Win. Mag., 338 Win. Mag. 850.00
Magazines (5 rounds) 53.00
(3 rounds) 45.00

TIKKA CONTINENTAL
$935.00

The Tikka Continental is designed specifically with a prone-type stock for shooting from ground or bench. The forend is extra wide to provide added steadiness when rested on sandbags or makeshift field rests. The heavy barrel is ideal for varmint or target shooting. **Calibers:** 223 Rem., 22-250 Rem., 243 Win., 308 Win. **Overall length:** 43¾". **Weight:** 8½ lbs.

TIKKA RIFLES

MODEL 512S DOUBLE RIFLE
$1525.00

The renowned Valmet 512S line of fine firearms is now being produced under the Tikka brand name and is being manufactured to the same specifications as the former Valmet. As a result of a joint venture entered into by Sako Ltd., the production facilities for these firearms are now located in Italy. The manufacture of the 512S series is controlled under the rigid quality standards of Sako Ltd., with complete interchangeability of parts between firearms produced in Italy and Finland. Tikka's double rifle offers features and qualities no other action can match: rapid handling and pointing qualities and the silent, immediate availability of a second shot. As such,

this model overcomes the two major drawbacks usually associated with this type of firearm: price and accuracy. Automatic ejectors included.

SPECIFICATIONS
Calibers: 308 Win., 30-06, 9.3×74R
Barrel length: 24″
Overall length: 40″
Weight: 8 1/2 lbs.
Stock: European walnut
Barrel sets: $825.00

TIKKA WHITETAIL/BATTUE

Originally designed by Tikka for wild boar shooting in the French marketplace, this unique rifle is now being introduced to the North American audience because of its proven success. The primary purpose of the rifle is for snap-shooting when quickness is a requirement in the field. The raised quarter-rib, coupled with the wide ''V''-shaped rear sight, allow the shooter a wide field of view. This enables him to zero in on a moving target swiftly. Also features a hooded front sight. A 3-round detachable magazine is available as an option.

The 20½″ barrel (overall length: 40½″) is perfectly balanced and honed to ensure the accuracy for which Tikka is famous. The stock is finished in soft matte lacquer, enhancing its beauty and durability. **Weight:** 7 pounds.

Prices:
In 308 Win., 270 Win., 30-06 .... $715.00
In 7mm Mag., 300 Win. Mag., 338 Win. Mag. **740.00**

A. UBERTI REPLICAS

ALL UBERTI FIREARMS AVAILABLE IN SUPER GRADE, PRESTIGE AND ENGRAVED FINISHES

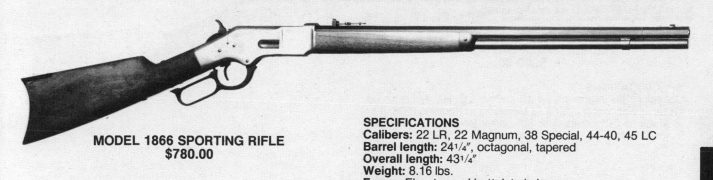

MODEL 1866 SPORTING RIFLE
$780.00

SPECIFICATIONS
Calibers: 22 LR, 22 Magnum, 38 Special, 44-40, 45 LC
Barrel length: 24¼", octagonal, tapered
Overall length: 43¼"
Weight: 8.16 lbs.
Frame: Elevator and buttplate in brass
Stock: Walnut
Sights: Vertically adjustable rear; horizontally adjustable front

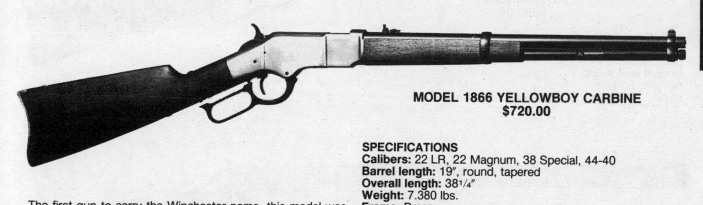

MODEL 1866 YELLOWBOY CARBINE
$720.00

SPECIFICATIONS
Calibers: 22 LR, 22 Magnum, 38 Special, 44-40
Barrel length: 19", round, tapered
Overall length: 38¼"
Weight: 7.380 lbs.
Frame: Brass
Stock and forend: Walnut
Sights: Vertically adjustable rear; horizontally adjustable front

The first gun to carry the Winchester name, this model was born as the 44-caliber rimfire cartridge Henry and is now chambered for 22 LR and 44-40.

MODEL 1871 ROLLING BLOCK
BABY CARBINE
$460.00

SPECIFICATIONS
Calibers: 22 LR, 22 Magnum, 357 Magnum
Barrel length: 22"
Overall length: 35½"
Weight: 4.85 lbs.
Stock & forend: Walnut

Trigger guard: Brass
Sights: Fully adjustable rear; ramp front
Frame: Color-casehardened steel

A. UBERTI REPLICAS

MODEL 1873 SPORTING RIFLE
$900.00

SPECIFICATIONS
Calibers: 357 Magnum, 44-40 and 45 LC. Hand-checkered.
Other specifications same as Model 1866 Sporting Rifle. Also
available in 20″ octagon or 30″ barrel length with pistol-grip
stock (extra).

1873 CARBINE
$890.00

SPECIFICATIONS
Calibers: 357 Mag., 44-40, 45 LC
Barrel length: 19″ round, tapered
Overall length: 38¼″
Weight: 7.38 lbs.
Sights: Fixed front; vertically adjustable rear

HENRY RIFLE
$895.00 (44-40 Cal.)
$900.00 (45 LC)

SPECIFICATIONS
Calibers: 44-40, 45 LC
Barrel length: 24¼″ (half-octagon, with tubular
  magazine)
Overall length: 43¾″
Weight: 9.26 lbs.
Frame: Brass
Stock: Varnished American walnut

HENRY CARBINE (not shown)
$895.00

SPECIFICATIONS
Caliber: 44-40   **Capacity:** 12 shots
Barrel length: 22¼″   **Weight:** 9.04 lbs.

Also available:
HENRY TRAPPER. Barrel length: 16¼″ or 18″. **Overall length:**
  35¾″ or 37¾″. **Weight:** 7.383 lbs. or 7.934 lbs. **Capacity:**
  8 or 9 shots. **Price: $900.00**
HENRY RIFLE w/Steel Frame (24¼″ barrel; 44/40 cal.). **Price:**
  $950.00

ULTRA LIGHT ARMS

MODEL 28
(7mm Rem. Mag.)

MODEL 20
MOUNTAIN RIFLE

MODEL 20 SERIES
$2400.00 ($2500.00 Left Hand)

SPECIFICATIONS
Calibers (Short Action): 6mm Rem., 17 Rem., 22 Hornet, 222 Rem., 222 Rem. Mag., 22-250 Rem., 223 Rem., 243 Win., 250-3000 Savage, 257 Roberts, 257 Ackley, 7×57 Mauser, 7×57 Ackley, 7mm-08 Rem., 284 Win., 300 Savage, 308 Win., 358 Win.
Barrel length: 22″
Weight: 4.75 lbs.
Safety: Two-position safety allows bolt to open or lock with sear blocked
Stock: Kevlar/Graphite composite; choice of 7 or more colors

Also available:
MODEL 24 SERIES (Long Action) in 270 Win., 30-06, 25-06, 7mm Express **Weight:** 5¼ lbs.
  Barrel length: 22″ **$2500.00**
  Same as above in Left-Hand Model **2600.00**
MODEL 28 SERIES (Magnum Action) in 264 Win., 7mm Rem., 300 Win., 338 **Weight:** 5¾ lbs.
  Barrel length: 24″ **2900.00**
  Same as above in Left-Hand Model **3000.00**
MODEL 40 SERIES (Magnum Action) in 300 Wby. and 416 Rigby. **Weight:** 7½ lbs.
  Barrel length: 26″ **2900.00**
  Same as above in Left-Hand Model **3000.00**

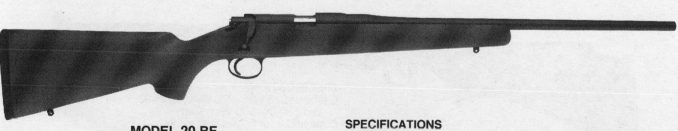

MODEL 20 RF
(Single Shot or Repeater)
$800.00

SPECIFICATIONS
Caliber: 22 LR
Barrel length: 22″ (Douglas Premium #1 Contour)
Weight: 5½ lbs.
Sights: None (drilled and tapped for scope)
Stock: Composite

UNIQUE RIFLES

MODEL T DIOPTRA SPORTER
$700.00

SPECIFICATIONS
Caliber: 22 LR or 22 Magnum bolt action
Capacity: 5 or 10 shots (5 shots only in 22 Mag.)
Barrel length: 23.6"   **Overall length:** 41.1"
Weight: 6.4 lbs.
Sights: Adjustable rear; lateral and vertical correction; dove-tailed grooves for scope or Micro-Match target sight
Features: French walnut Monte Carlo stock; firing adjustment safety (working in firing pin)

MODEL T UIT STANDARD RIFLE
$1250.00

SPECIFICATIONS
Caliber: 22 LR
Barrel length: 25.6"   **Overall length:** 44.1"
Weight: 10.4 lbs.
Sights: Micro-Match target sight
Stock: French walnut
Features: Adjustable buttplate and cheek rest; fully adjustable firing; left-hand stock and action available

MODEL T/SM SILHOUETTE
$825.00

SPECIFICATIONS
Caliber: 22 LR or 22 Magnum
Capacity: 5- or 10-shot magazine (5-shot only in 22 Mag.)
Barrel length: 20.5"   **Overall length:** 38.4"
Weight: 6.6 lbs.
Sights: Dovetailed grooves on receiver for scope or Micro-Match target sight
Stock: French walnut Monte Carlo stock (left-hand stock available)

MODEL TGC CENTERFIRE
$1190.00

SPECIFICATIONS
Calibers: 243 Win., 270 Win., 7mm-08, 7mm Rem. Mag., 308 Win., 30-06, 300 Win. Mag.
Capacity: 3- or 5-shot magazine
Barrel length: 24" bolt action (interchangeable barrel)
Overall length: 44.8"   **Weight:** 8.4 lbs.
Sights: Dovetailed grooves on receiver for scope
Stock: French walnut Monte Carlo stock (left-hand stock available)

VOERE RIFLES

MODEL VEC 91 LIGHTNING

MODEL VEC 91 LIGHTNING BOLT
$2895.00

This first factory-made high-power caseless ammunition sporting and hunting rifle features two small batteries capable of delivering 5,000 shots. The rifle will not fire unless the bolt is fully closed with the sliding safety in forward fire position. The trigger let-off is adjustable for 5 ounces to 7 pounds by an adjustment screw in the trigger guard. A free-floating barrel ensures a high level of accuracy. The rifle features a bolt action with twin forward locking lugs, a double protector from gas leaks, a two-stage fully adjustable electrical trigger, and electronic ignition.

SPECIFICATIONS
Action: Electronic caseless cartridge, bolt action
Caliber: 223 UCC caseless ammo   **Capacity:** 5-shot
Barrel length: 20"   **Overall length:** 39"
Weight: 6 lbs.
Safety: Top tang sliding safety
Sights: Fixed post ramp front; open rear adjustable for windage and elevation
Stock: Select walnut, hand-cut checkered pistol-grip stock with schnabel forend (modified Bavarian buttstock); hand-rubbed oil finish; vented recoil pad; detachable sling swivels

MODEL 2115 SEMIAUTOMATIC

MODEL 2115 SEMIAUTOMATIC
$645.00

SPECIFICATIONS
Action: Semiauto rimfire
Caliber: 22 LR

Capacity: 5-, 10- or 15-round detachable magazine
Barrel length: 18"   **Overall length:** 37½"
Weight: 5.75 lbs.
Safety: Wing-type safety locks firing pin
Sights: Fixed post ramp front; open rear adjustable for elevation and windage
Stock: Hardwood walnut-finished checkered pistol-grip stock; European-style with buttplate and grip caps

MODEL 2150 (not shown)
$1795.00

SPECIFICATIONS
Action: American custom classic K-98 Mauser bolt action
Safety: Three-position safety on bolt shroud
Sights: No sights
Stock: "Old English" recoil pad; shadow cheekpiece; inletted swivel stud; ebony forend cap; steel grip cap
Features: Other specifications same as Model 2165M

Also available:
MODEL 2150M: Same specifications as above with 24" barrel.
Overall length: 46½". **Weight:** 7½ lbs.
Price: $1845.00

RIFLES

VOERE RIFLES

MODEL 2165 BOLT ACTION

MODEL 2165 BOLT ACTION
$1495.00

SPECIFICATIONS
Action: K-98 Mauser-style bolt action w/claw extractor
Calibers: 22-250, 243, 270, 7×57, 7×64, 30-06, 308
Capacity: 5 rounds
Barrel length: 22″   **Overall length:** 44½″
Weight: 7 lbs.
Sights: Fixed post ramp front; open rear adjustable for windage and elevation
Safety: Top tang sliding safety
Stock: Vented recoil pad; cut checkering; select walnut with hand-rubbed oil finish; modified Bavarian buttstock; detachable sling swivels; rosewood forend, grip cap

Also available:
MODEL 2165M: Similar specifications as above, except in magnum calibers 7mm Rem. Mag., 300 Win. Mag. **Capacity:** 3 or 5 rounds. **Barrel length:** 24″. **Overall length:** 46½″. **Weight:** 7½ lbs.
Price: **$1545.00**
MODEL 2155: Similar specifications as above, except in calibers 243, 270 and 30-06 only. **Capacity:** 5 rounds (hinged steel floorplate). **Barrel length:** 20″. No sights
Price: **$995.00**
MODEL 2155M: Same specifications as Model 2165M, except with 22″ barrel.
Price: **$1075.00**

MODEL 2185MR MATCH RIFLE

MODEL 2185/2FS

Safety: Manual safety in trigger guard (push button for bolt hold- open)
Sights: Metallic match sights
Stock: Laminated, stipple checkered, glass-bedded stock; adjustable buttplate

Also available:
MODEL 2185SR. Same specifications as above, except with competition stock. **Weight:** 10½ lbs.
Price: **$3995.00**
MODEL 2185/2FS. Same specifications as above, except with full Mannlicher-style stock. **Weight:** 7¼ lbs.
Price: **$2045.00**
MODEL 2185/1FH. Similar specifications as above, except with modified Bavarian stock. **Calibers:** 30-06, 308 Win. **Weight:** 7¾ lbs.
Price: **$1995.00**

MODEL 2185MR MATCH RIFLE
$4995.00

SPECIFICATIONS
Action Three-locking lug action; gas-operated semiauto match rifle
Calibers: 30-06, 308 Win., 7×64
Capacity: 2-, 3- or 5-round detachable magazine
Barrel length: 20″ free-floating match-grade barrel
Overall length: 45¼″   **Weight:** 11 lbs.

WEATHERBY MARK V RIFLES

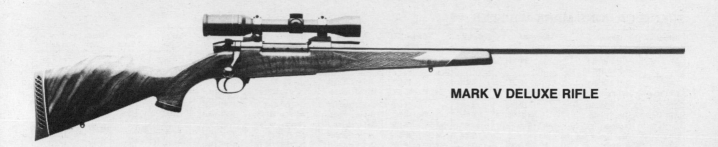

MARK V DELUXE RIFLE

MARK V DELUXE

The Mark V Deluxe stock is made of hand-selected American walnut with skipline checkering, traditional diamond-shaped inlay, rosewood pistol grip cap and forend tip. Monte Carlo design with raised cheekpiece properly positions the shooter while reducing felt recoil. The action and hammer-forged barrel are hand-bedded for accuracy, then deep blued to a high luster finish.

	Prices
MARK V DELUXE (24″ barrel)	
257, 270, 7mm, 300 Wby. Mag., 30-06	$1275.00
MARK V DELUXE (26″ barrel)	
240, 257, 7mm Wby. Mag., 30-06	1289.00
300 Wby. Mag. and 340 Wby. Mag.	1322.00
378 Wby. Mag.	1358.00
416 Wby. Mag. (24″ or 26″ bbl.)	1401.00
460 Wby. Mag. (26″) or Left Hand (24″)	1727.00

MARK V SPORTER

	Prices
MARK V SPORTER (22″ barrel)	
In cal. 270, 30-06	$799.00
MARK V SPORTER (24″ barrel)	
In cal. 257, 270, 7mm, 300 Wby. Mag., 7mm Rem. Mag., 300, 338 Win.	799.00
In cal. 375 H&H	909.00
MARK V SPORTER (26″ barrel)	
In cal. 257, 270, 7mm Wby. Mag.	N/A
In cal. 300, 340 Wby. Mag.	852.00

WEATHERBY MARK V RIFLES

SPECIFICATIONS MARK V RIFLES

Caliber	Model	Barrelled Action	Weight *	Overall Length	Magazine Capacity	Barrel Length/ Contour	Rifling	Length of Pull	Drop at Comb	Monte Carlo	Drop at Heel
22 - .250	Mark V Deluxe Varmintmaster	RH 26"	6 1/2 lbs.	45 3/8"	3+1 in chamber	26" #2	1 - 14" twist	13 1/2"	5/8"	1/4"	1 1/8"
.240 WBY Mag.	Mark V Deluxe	RH 26"	8 lbs.	46 5/8"	4+1 in chamber	26" #2	1-10" twist	13 5/8"	5/8"	1/4"	1 1/8"
	Lazermark	RH 24/26"	8 lbs.	44 5/8" or 46 5/8"	4+1 in chamber	24" #1 or 26" #2	1-10" twist	13 5/8"	5/8"	1/4"	1 1/8"
.257 WBY Mag.	Mark V Sporter	RH 24"	8 lbs.	44 5/8"	3+1 in chamber	24" #1	1-10" twist	13 5/8"	1"	5/8"	1 5/8"
	Mark V Deluxe	RH 24/26"	8 lbs.	44 5/8" or 46 5/8"	3+1 in chamber	24" #1 or 26" #2	1-10" twist	13 5/8"	5/8"	1/4"	1 1/8"
	Lazermark	RH 24" or 26"	8 lbs.	44 5/8" or 46 5/8"	3+1 in chamber	24" #1 or 26" #2	1-10" twist	13 5/8"	5/8"	1/4"	1 1/8"
	Weathermark	RH 24" or 26"	7 1/2 lbs.	44 5/8" or 46 5/8"	3+1 in chamber	24" #1 or 26" #2	1-10" twist	13 1/2"	7/8"	–	1 1/8"
	Alaskan	RH 24" or 26"	7 1/2 lbs.	44 5/8" or 46 5/8"	3+1 in chamber	24" #1 or 26" #2	1-10" twist	13 1/2"	7/8"	–	1 1/8"
.270 WIN.	Mark V Sporter	RH 22"	8 lbs.	42 5/8"	4+1 in chamber	22" #1	1-10" twist	13 5/8"	1"	5/8"	1 5/8"
	Weathermark	RH 22"	7 1/2 lbs.	42 5/8"	4+1 in chamber	22" #1	1-10" twist	13 1/2"	7/8"	–	1 1/8"
	Alaskan	RH 22"	7 1/2 lbs.	42 5/8"	4+1 in chamber	22" #1	1-10" twist	13 1/2"	7/8"	–	1 1/8"
.270 WBY Mag.	Mark V Sporter	RH 24"	8 lbs.	44 5/8"	3+1 in chamber	24" #1	1-10" twist	13 5/8"	1"	5/8"	1 5/8"
	Mark V Deluxe	RH 24/26", LH 24"	8 lbs.	44 5/8" or 46 5/8"	3+1 in chamber	24" #1 or 26" #2	1-10" twist	13 5/8"	5/8"	1/4"	1 1/8"
	Lazermark	RH 24/26", LH 24"	8 lbs.	44 5/8" or 46 5/8"	3+1 in chamber	24" #1 or 26" #2	1-10" twist	13 5/8"	5/8"	1/4"	1 1/8"
	Weathermark	RH 24" or 26"	7 1/2 lbs.	44 5/8" or 46 5/8"	3+1 in chamber	24" #1 or 26" #2	1-10" twist	13 1/2"	7/8"	–	1 1/8"
	Alaskan	RH 24" or 26"	7 1/2 lbs.	44 5/8" or 46 5/8"	3+1 in chamber	24" #1 or 26" #2	1-10" twist	13 1/2"	7/8"	–	1 1/8"
7mm Rem Mag.	Mark V Sporter	RH 24"	8 lbs.	44 5/8"	3+1 in chamber	24" #1	1-10" twist	13 5/8"	1"	5/8"	1 5/8"
	Weathermark	RH 24"	7 1/2 lbs.	44 5/8"	3+1 in chamber	24" #1	1-10" twist	13 1/2"	7/8"	–	1 1/8"
	Alaskan	RH 24"	7 1/2 lbs.	44 5/8"	3+1 in chamber	24" #1	1-10" twist	13 1/2"	7/8"	–	1 1/8"
7mm WBY Mag.	Mark V Sporter	RH 24"	8 lbs.	44 5/8"	3+1 in chamber	24" #1	1-10" twist	13 5/8"	1"	5/8"	1 5/8"
	Mark V Deluxe	RH 24/26", LH 24"	8 lbs.	44 5/8" or 46 5/8"	3+1 in chamber	24" #1 or 26" #2	1-10" twist	13 5/8"	5/8"	1/4"	1 1/8"
	Lazermark	RH 24/26", LH 24"	8 lbs.	44 5/8" or 46 5/8"	3+1 in chamber	24" #1 or 26" #2	1-10" twist	13 5/8"	5/8"	1/4"	1 1/8"
	Weathermark	RH 24" or 26"	7 1/2 lbs.	44 5/8" or 46 5/8"	3+1 in chamber	24" #1 or 26" #2	1-10" twist	13 1/2"	7/8"	–	1 1/8"
	Alaskan	RH 24" or 26"	7 1/2 lbs.	44 5/8" or 46 5/8"	3+1 in chamber	24" #1 or 26" #2	1-10" twist	13 1/2"	7/8"	–	1 1/8"
30-06 Springfield	Mark V Sporter	RH 22"	8 lbs.	42 5/8"	4+1 in chamber	22" #1	1-10" twist	13 5/8"	1"	5/8"	1 5/8"
	Mark V Deluxe	RH 24/26", LH 24"	8 lbs.	44 5/8" or 46 5/8"	4+1 in chamber	24" #1 or 26" #2	1-10" twist	13 5/8"	5/8"	1/4"	1 1/8"
	Lazermark	RH/LH24"	8 lbs.	44 5/8"	4+1 in chamber	24" #1	1-10" twist	13 5/8"	5/8"	1/4"	1 1/8"
	Weathermark	RH 22"	7 1/2 lbs.	42 5/8"	4+1 in chamber	22" #1	1-10" twist	13 1/2"	7/8"	–	1 1/8"
	Alaskan	RH 22"	7 1/2 lbs.	42 5/8"	4+1 in chamber	22" #1	1-10" twist	13 1/2"	7/8"	–	1 1/8"
.300 Win Mag.	Mark V Sporter	RH 24"	8 lbs.	44 1/2"	3+1 in chamber	24" #1	1-10" twist	13 5/8"	1"	5/8"	1 5/8"
	Weathermark	RH 24"	7 1/2 lbs.	44 1/2"	3+1 in chamber	24" #1	1-10" twist	13 1/2"	7/8"	–	1 1/8"
	Alaskan	RH 24"	7 1/2 lbs.	44 1/2"	3+1 in chamber	24" #1	1-10" twist	13 1/2"	7/8"	–	1 1/8"
.300 WBY Mag.	Mark V Sporter	RH 26"	8 lbs.	46 5/8"	3+1 in chamber	26" #2	1-10" twist	13 5/8"	1"	5/8"	1 5/8"
	Mark V Deluxe	RH 24/26", LH 24/26"	8 lbs.	44 5/8" or 46 5/8"	3+1 in chamber	24" #1 or 26" #2	1-10" twist	13 5/8"	5/8"	1/4"	1 1/8"
	Lazermark	RH 24/26", LH 24/26"	8 lbs.	44 5/8" or 46 5/8"	3+1 in chamber	24" #1 or 26" #2	1-10" twist	13 5/8"	5/8"	1/4"	1 1/8"
	Weathermark	RH 24" or 26"	7 1/2 lbs.	44 5/8" or 46 5/8"	3+1 in chamber	24" #1 or 26" #2	1-10" twist	13 1/2"	7/8"	–	1 1/8"
	Alaskan	RH 24" or 26"	7 1/2 lbs.	44 5/8" or 46 5/8"	3+1 in chamber	24" #1 or 26" #2	1-10" twist	13 1/2"	7/8"	–	1 1/8"
.338 Win Mag.	Mark V Sporter	RH 24"	8 lbs.	44 1/2"	3+1 in chamber	24" #2	1-10" twist	13 5/8"	1"	5/8"	1 5/8"
	Weathermark	RH 24"	7 1/2 lbs.	44 1/2"	3+1 in chamber	24" #2	1-10" twist	13 1/2"	7/8"	–	1 1/8"
	Alaskan	RH 24"	7 1/2 lbs.	44 1/2"	3+1 in chamber	24" #2	1-10" twist	13 1/2"	7/8"	–	1 1/8"
.340 WBY Mag.	Mark V Sporter	RH 26"	8 lbs.	46 5/8"	3+1 in chamber	26" #2	1-10" twist	13 5/8"	1"	5/8"	1 5/8"
	Mark V Deluxe	RH/LH 26"	8 lbs.	46 5/8"	3+1 in chamber	26" #2	1-10" twist	13 5/8"	5/8"	1/4"	1 1/8"
	Lazermark	RH/LH 26"	8 lbs.	46 5/8"	3+1 in chamber	26" #2	1-10" twist	13 5/8"	5/8"	1/4"	1 1/8"
	Weathermark	RH 26"	8 lbs.	46 5/8"	3+1 in chamber	26" #2	1-10" twist	13 5/8"	7/8"	1/4"	1 1/8"
	Alaskan	RH 26"	8 lbs.	46 5/8"	3+1 in chamber	26" #2	1-10" twist	13 5/8"	7/8"	1/4"	1 1/8"
.375 H&H Mag.	Mark V Sporter	RH 24"	8 1/2 lbs.	44 5/8"	3+1 in chamber	24" #3	1-12" twist	13 5/8"	1"	5/8"	1 5/8"
	Weathermark	RH 24"	8 lbs.	44 5/8"	3+1 in chamber	24" #3	1-12" twist	13 1/2"	7/8"	–	1 1/8"
	Alaskan	RH 24"	8 lbs.	44 5/8"	3+1 in chamber	24" #3	1-12" twist	13 1/2"	7/8"	–	1 1/8"
.378 WBY Mag.	Mark V Deluxe	RH 26"	8 1/2 lbs.	46 5/8"	2+1 in chamber	26" #3	1-12" twist	13 7/8"	5/8"	1/4"	1 1/8"
	Lazermark	RH 26"	8 1/2 lbs.	46 5/8"	2+1 in chamber	26" #3	1-12" twist	13 7/8"	5/8"	1/4"	1 1/8"
**.416 WBY Mag.	Mark V Deluxe	RH/LH 26"	9 1/2 lbs.	46 3/4"	2+1 in chamber	26" #3.5	1-14" twist	13 7/8"	5/8"	1/4"	1 1/8"
	Lazermark	RH/LH 26"	9 1/2 lbs.	46 3/4"	2+1 in chamber	26" #3.5	1-14" twist	13 7/8"	5/8"	1/4"	1 1/8"
**.460 WBY Mag.	Mark V Deluxe	RH/LH 26"	10 1/2 lbs.	46 3/4"	2+1 in chamber	26" #4	1-16" twist	14"	5/8"	1/4"	1 1/8"
	Lazermark	RH/LH 26"	10 1/2 lbs.	46 3/4"	2+1 in chamber	26" #4	1-16" twist	14"	5/8"	1/4"	1 1/8"

* Weight approximate due to stock density and bore diameter. ** Available with Weatherby Accubrake only. Safari Grade Custom and Crown Custom rifles are also available. Consult your Weatherby dealer or the Weatherby Custom Shop for specifications.

WEATHERBY MARK V RIFLES

LAZERMARK

LAZERMARK (24″ barrel)	Prices
In 240, 257, 270, 7mm, 300 Wby. Mag., 30-06 . . .	**$1395.00**
Same as above w/26″ barrel	**1409.00**
In 300 and 340 Wby. Mag. (26″ bbl. only)	**1445.00**
In 378 Wby. Mag. (26″ barrel only)	**1486.00**
In 416 Wby. Mag. (24″ or 26″ bbl.)	**1533.00**
In 460 Wby. Mag. (24″ or 26″ bbl.)	**1899.00**

ALASKAN

The Alaskan model rifle with Mark V bolt action features Weatherby's Weathermark® composite stock with special non-glare finish and impregnated color for scratch-resistance. The satin, electroless nickel finish is stainless and impervious to rust.

ALASKAN (22″ barrel)	Prices
In cal. 270, 30-06	**$ 875.00**
ALASKAN (24″ barrel)	
In cal. 257, 270, 7mm, 300 Wby. Mag.,	
7mm Rem. Mag., 300, 338 Win.	**875.00**
In cal. 375 H&H	**1039.00**
ALASKAN (26″ barrel)	
In cal. 257, 270, 7mm Wby. Mag.	**912.00**
In cal. 300, 340 Wby. Mag.	**912.00**

WEATHERMARK

The Weathermark® rifle features the Mark V bolt action and a special composite stock guaranteed not to warp, preserving accuracy.

WEATHERMARK (22″ barrel)	Prices
In cal. 270, 30-06	**$ 675.00**
WEATHERMARK (24″ barrel)	
In cal. 257, 270, 7mm, 300 Wby. Mag.,	
7mm Rem. Mag., 300, 338 Win.	**675.00**
In cal. 375 H&H	**802.00**
WEATHERMARK (26″ barrel)	
In cal. 257, 270, 7mm Wby. Mag.	**705.00**
In cal. 300, 340 Wby. Mag.	**705.00**
Also available:	
VARMINTMASTER (26″ barrel)	
In 224 Wby. Mag. and 22-250	**1245.00**

FOR COMPLETE SPECIFICATIONS ON THE ABOVE RIFLES, PLEASE SEE THE TABLE ON THE PREVIOUS PAGE.

WINCHESTER BOLT-ACTION RIFLES

MODEL 70 FEATHERWEIGHT
$562.00 (Walnut)

MODEL 70 FEATHERWEIGHT CLASSIC
$585.00

SPECIFICATIONS MODEL 70 FEATHERWEIGHT

Model	Caliber	Magazine Capacity*	Barrel Length	Overall Length	Nominal Length Of Pull	Nominal Drop At Comb	Nominal Drop At Heel	Nominal Weight (Lbs.)	Rate of Twist 1 Turn In
70 WALNUT FEATHERWEIGHT	22-250 Rem.	5	22″	42″	13½″	9/16″	7/8″	7	14″
Standard Grade Walnut	223 Rem.	6	22	42	13½	9/16	7/8	7	12
	243 Win.	5	22	42	13½	9/16	7/8	7	10
	6.5×55mm Swedish	5	22	42½	13½	9/16	7/8	7	7.87
	270 Win.	5	22	42½	13½	9/16	7/8	7¼	10
	280 Rem.	5	22	42½	13½	9/16	7/8	7¼	10
	7mm-08 Rem.	5	22	42	13½	9/16	7/8	7	9.5
	30-06 Spfld.	5	22	42½	13½	9/16	7/8	7¼	10
	308 Win.	5	22	42	13½	9/16	7/8	7	12
70 WALNUT FEATHERWEIGHT CLASSIC	223 Rem.	6	22″	42″	13½″	9/16″	7/8″	7	12″
	22-250 Rem.	5	22	42	13½	9/16	7/8	7	14
	243 Win.	5	22	42	13½	9/16	7/8	7	10
	308 Win.	5	22	42	13½	9/16	7/8	7	12
	7mm-08 Rem.	5	22	42	13½	9/16	7/8	7	10
	270 Win.	5	22	43	14	9/16	7/8	7¼	10
Standard Grade Walnut	280 Rem.	5	22	43	14	9/16	7/8	7¼	10
Controlled Round Feeding	30-06 Spfld.	5	22	43	14	9/16	7/8	7¼	10

* For additional capacity, add one round in chamber when ready to fire. Drops are measured from center line of bore. Rate of twist is right-hand.

WINCHESTER BOLT-ACTION RIFLES

MODEL 70 LIGHTWEIGHT RIFLE
$498.00

SPECIFICATIONS: MODEL 70 LIGHTWEIGHT

Model	Caliber	Magazine Capacity (A)	Barrel Length	Overall Length	Nominal Length Of Pull	Nominal Drop At Comb	Nominal Drop At Heel	Nominal Weight (Lbs.)	Rate of Twist 1 Turn In
70 WALNUT	223 Rem.	6	22	42	13½	$9/16$	$7/8$	6⅞	12
	243 Win.	5	22	42	13½	$9/16$	$7/8$	6⅞	10
Checkered,	270 Win.	5	22	42½	13½	$9/16$	$7/8$	7	10
No Sights	30-06 Spgfld.	5	22	42½	13½	$9/16$	$7/8$	7	10
	308 Win.	5	22	42	13½	$9/16$	$7/8$	6⅞	12

(A) For additional capacity, add one round in chamber when ready to fire. Drops are measured from center line of bore. Rate of twist is right-hand. No sights.

MODEL 70 CLASSIC SM
(Synthetic Composite Stock, Matte)
$585.00
$634.00 (375 H&H MAGNUM)

SPECIFICATIONS MODEL 70 CLASSIC SM (SYNTHETIC STOCK; CONTROLLED ROUND FEED)

Caliber	Magazine Capacity	Barrel Length	Overall Length	Nominal Length Of Pull	Nominal Drop At Comb	Nominal Drop At Heel	Nominal Weight (Lbs.)	Rate of Twist 1 Turn In	Bases Rings or Sights
270 Win.	5	24	45	14	$9/16$	$7/8$	7⅜	10	B+R
30-06 Spfld.	5	24	45	14	$9/16$	$7/8$	7⅜	10	B+R
7mm Rem. Mag.	3	24	45	14	$9/16$	$7/8$	7⅝	9½	B+R
300 Win. Mag.	3	24	45	14	$9/16$	$7/8$	7⅝	10	B+R
338 Win. Mag.	3	24	45	14	$9/16$	$7/8$	7⅝	10	B+R
375 H&H Mag.	3	24	45	14	$9/16$	$7/8$	8	12	Sights

WINCHESTER BOLT-ACTION RIFLES

MODEL 70 CLASSIC SUPER GRADE

The Winchester Model 70 Classic Super Grade features a bolt with true claw-controlled feeding of belted magnums. The stainless steel claw extractor on the bolt grasps the round from the magazine and delivers it to the chamber and later extracts the spent cartridge. A gas block doubles as bolt stop and the bolt guard rail assures smooth action. Winchester's 3-position safety and field-strippable firing pin are standard equipment. Other features include a satin finish select walnut stock with sculptured cheekpiece designed to direct recoil forces rearward and away from the shooter's cheek; an extra-thick honeycomb recoil; all-steel bottom metal; and chrome molybdenum barrel with cold hammer-forged rifling for optimum accuracy. Specifications are listed in the tables below.

SPECIFICATIONS WINCHESTER MODEL 70 CLASSIC SUPER GRADE RIFLE Price: $792.00

Caliber	Magazine Capacity*	Barrel Length	Overall Length	Nominal Length of Pull	Nominal Drop at Comb	Heel	MC	Nominal Weight (Lbs.)	Rate of Twist 1 Turn in	Bases & Rings or Sights
270 Win.	5	24"	44³/₄"	13³/₄"	9/16"	13/16"	—	7³/₄	10"	B + R
30-06 Spfld.	5	24	44³/₄	13³/₄	9/16	13/16	—	7³/₄	10	B + R
7mm Rem. Mag.	3	26	46³/₄	13³/₄	9/16	13/16	—	8	9¹/₂	B + R
300 Win. Mag.	3	26	46³/₄	13³/₄	9/16	13/16	—	8	10	B + R
338 Win. Mag.	3	26	46³/₄	13³/₄	9/16	13/16	—	8	10	B + R

* For additional capacity, add one round in chamber when ready to fire. Drops are measured from center line of bore. Rate of twist is right-hand.

MODEL 70 CLASSIC STAINLESS (SYNTHETIC COMPOSITE STOCK) Price: $634.00

Caliber	Magazine Capacity	Barrel Length	Overall Length	Nominal Length Of Pull	Nominal Drop At Comb	Heel	Nominal Weight (Lbs.)	Rate of Twist 1 Turn In	Bases & Rings or Sights
223 Rem.	6"	22"	42¹/₄"	13³/₄"	9/16"	7/8"	7	12"	—
22-250 Rem.	5	22	42¹/₄	13³/₄	9/16	7/8	7	14	—
243 Win.	5	22	42¹/₄	13³/₄	9/16	7/8	7	10	—
308 Win.	5	22	42¹/₄	13³/₄	9/16	7/8	7	12	—
270 Win.	5	24	44³/₄	13³/₄	9/16	7/8	7¹/₄	10	—
30-06 Spfld.	5	24	44³/₄	13³/₄	9/16	7/8	7¹/₄	10	—
7mm Rem. Mag.	3	26	46³/₄	13³/₄	9/16	7/8	7¹/₂	9¹/₂	—
300 Win. Mag.	3	26	46³/₄	13³/₄	9/16	7/8	7¹/₂	10	—
300 Weath. Mag.	3	26	46³/₄	13³/₄	9/16	7/8	7¹/₂	10	—
338 Win. Mag.	3	26	46³/₄	13³/₄	9/16	7/8	7¹/₂	10	—

WINCHESTER BOLT-ACTION RIFLES

MODEL 70 SPORTER

**MODEL 70 SPORTER $556.00 ($590.00 in 270 Win.,
30-06, 7mm Rem. Mag. 300 Win. Mag.)
CLASSIC SPORTER $577.00
($613.00 in 270 Win., 30-06, 7mm Rem. Mag.,
300 Win. Mag., 338 Win. Mag.)**

SPECIFICATIONS MODEL 70 SPORTER & CLASSIC SPORTER

Model	Caliber	Magazine Capacity A	Barrel Length	Overall Length	Nominal Length Of Pull	Nominal Drop At Comb	Heel	Nominal Weight (Lbs.)	Rate of Twist 1 Turn In	Sights
70 SPORTER	25-06 Rem.	5	24	44¾"	13¾	9/16	7/8	7½	10	
	264-Win. Mag.	3	26	46¾"	13¾	9/16	7/8	7¾	9	
	270 Win.	5	24	44¾"	13¾	9/16	7/8	7½	10	Sights
	270 Win.	5	24	44¾	13¾	9/16	7/8	7½	10	
	270 Weath. Mag.	3	26	46¾"	13¾	9/16	7/8	7¾	10	
	30-06 Spfld.	5	24	44¾"	13¾	9/16	7/8	7½	10	Sights
	30-06 Spfld.	5	24	44¾"	13¾	9/16	7/8	7½	10	
	7mm Rem. Mag.	3	26	46¾"	13¾	9/16	7/8	7¾	9½	Sights
	7mm Rem. Mag.	3	26	46¾"	13¾	9/16	7/8	7¾	9½	
	300 Win. Mag.	3	26	46¾"	13¾	9/16	7/8	7¾	10	Sights
	300 Win. Mag.	3	26	46¾"	13¾	9/16	7/8	7¾	10	
	300 Weath.	3	26	46¾	13¾	9/16	7/8	7¾	10	
	338 Win. Mag.	3	26	46¾	13¾	9/16	7/8	7¾	10	

MODEL 70 HEAVY BARREL VARMINT RIFLE SPECIFICATIONS Price: $720.00

Model	Caliber	Magazine Capacity (A)	Barrel Length	Overall Length	Nominal Length Of Pull	Nominal Drop At Comb	Heel	Nominal Weight (Lbs.)	Rate of Twist 1 Turn In	Sights
70 VARMINT	220 Swift	5	26"	46"	13½"	¾"	½"	10¾	14"	—
	22-250 Rem.	5	26	46	13½	¾	½	10¾	14	—
	223 Rem.	6	26	46	13½	¾	½	10¾	9	—
	243 Win.	5	26	46	13½	¾	½	10¾	10	—
	308 Win.	5	26	46	13½	¾	½	10¾	12	—

(A) For additional capacity, add one round in chamber when ready to fire. Drops are measured from center line of bore. Rate of twist is right-hand.

WINCHESTER BOLT-ACTION RIFLES

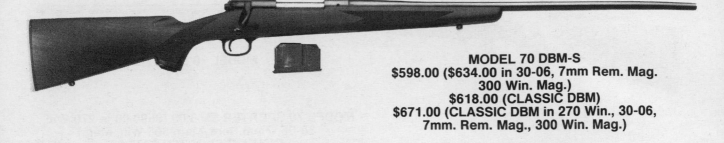

MODEL 70 DBM-S
$598.00 ($634.00 in 30-06, 7mm Rem. Mag.
300 Win. Mag.)
$618.00 (CLASSIC DBM)
$671.00 (CLASSIC DBM in 270 Win., 30-06,
7mm. Rem. Mag., 300 Win. Mag.)

SPECIFICATIONS MODEL 70 DBM (DETACHABLE BOX MAGAZINE)

Caliber	Magazine Capacity*	Barrel Length	Overall Length	Nominal Length Of Pull	Nominal Drop At Comb	Nominal Drop At Heel	Nominal Weight (Lbs.)	Rate of Twist 1 Turn In	Sights
243 Win.	4	24″	45″	14″	9/16″	7/8″	7	10″	
270 Win.	4	24	45	14	9/16	7/8	7 3/8	10	
30-06 Spfld.	4	24	45	14	9/16	7/8	7 3/8	10	
30-06 Spfld.	4	24	45	14	9/16	7/8	7 3/8	10	Sights
7mm Rem. Mag.	3	24	45	14	9/16	7/8	7 3/8	9 1/2	
7mm Rem. Mag.	3	24	45	14	9/16	7/8	7 3/8	9 1/2	Sights
300 Win. Mag.	3	24	45	14	9/16	7/8	7 3/8	10	
300 Win. Mag.	3	24	45	14	9/16	7/8	7 3/8	10	Sights

MODEL 70 CLASSIC DBM (DETACHABLE BOX MAGAZINE)

Caliber	Magazine Capacity*	Barrel Length	Overall Length	Nominal Length Of Pull	Nominal Drop At Comb	Nominal Drop At Heel	Nominal Weight (Lbs.)	Rate of Twist 1 Turn In	Sights
22-250 Rem.	4	24	44 1/4	13 3/4	9/16	13/16	7	14″	
243 Win.	4	24	44 1/4	13 3/4	9/16	13/16	7	10	
270 Win.	4	24	44 3/4	13 3/4	9/16	13/16	7 3/8	10	
270 Win.	4	24	44 3/4	13 3/4	9/16	13/16	7 3/8	10	Sights
284 Win.	3	24	44 1/4	13 3/4	9/16	13/16	7	10	
308 Win.	4	24	44 1/4	13 3/4	9/16	13/16	7	12	
30-06 Spfld.	4	24	44 3/4	13 3/4	9/16	13/16	7 3/8	10	
30-06 Spfld.	4	24	44 3/4	13 3/4	9/16	13/16	7 3/8	10	Sights
7mm Rem. Mag.	3	26	46 3/4	13 3/4	9/16	13/16	7 3/4	9 1/2″	
7mm Rem. Mag.	3	26	46 3/4	13 3/4	9/16	13/16	7 3/4	9 1/2	Sights
300 Win. Mag.	3	26	46 3/4	13 3/4	9/16	13/16	7 3/4	10	
300 Win. Mag.	3	26	46 3/4	13 3/4	9/16	13/16	7 3/4	10	Sights

WINCHESTER BOLT-ACTION RIFLES

**WINCHESTER RANGER®
BOLT ACTION CENTERFIRE RIFLE
$455.00**

The Ranger Bolt Action Rifle comes with an American hardwood stock, a wear-resistant satin walnut finish, ramp bead-post front sight, steel barrel, hinged steel magazine floorplate, three-position safety and engine-turned, anti-bind bolt. The receiver is drilled and tapped for scope mounting; accuracy is enhanced by thermoplastic bedding of the receiver. Barrel and receiver are brushed and blued.

**WINCHESTER RANGER®
YOUTH BOLT ACTION CARBINE
$455.00**

This carbine offers dependable bolt action performance combined with a scaled-down design to fit the younger, smaller shooter. It features anti-bind bolt design, jeweled bolt, three-position safety, contoured recoil pad, ramped bead front sight, semi-buckhorn folding leaf rear sight, hinged steel magazine floorplate, and sling swivels. Receiver is drilled and tapped for scope mounting. Stock is of American hardwood with protective satin walnut finish. Pistol grip, length of pull, overall length, and comb are all tailored to youth dimensions (see table).

SPECIFICATIONS RANGER & YOUTH RIFLE

Model	Caliber	Magazine Capacity	Barrel Length	Overall Length	Nominal Length Of Pull	Nominal Drop At Comb	Nominal Drop At Heel	Nominal Weight (Lbs.)	Rate of Twist 1 Turn in	Bases & Rings Sights
70 RANGER	223 Rem.	6	22″	42″	13$\frac{1}{2}$″	$\frac{9}{16}$″	$\frac{7}{8}$″	6$\frac{3}{4}$	12″	Sights
	243 Win.	5	22	42	13$\frac{1}{2}$	$\frac{9}{16}$	$\frac{7}{8}$	6$\frac{3}{4}$	10	Sights
	270 Win.	5	22	42$\frac{1}{2}$	13$\frac{1}{2}$	$\frac{9}{16}$	$\frac{7}{8}$	7	10	Sights
	30-06 Spfld.	5	22	41$\frac{1}{2}$	13$\frac{1}{2}$	$\frac{9}{16}$	$\frac{7}{8}$	7	10	Sights
70 RANGER LADIES/ YOUTH	243 Win.	5	22	41	12$\frac{1}{2}$	$\frac{3}{4}$	1	6$\frac{1}{2}$	10	Sights
	308 Win.	5	22	41	12$\frac{1}{2}$	$\frac{3}{4}$	1	6$\frac{1}{2}$	12	Sights

For additional capacity, add one round in chamber when ready to fire. Drops are measured from center line of bore. Rate of twist is right-hand.

RIFLES

WINCHESTER RIFLES

LEVER-ACTION CARBINES/RIFLES

MODEL 94 STANDARD WALNUT RIFLE

and forearm have a protective stain finish with precise-cut wraparound checkering. It has a 20-inch barrel with hooded blade front sight and semi-buckhorn rear sight.

The top choice for lever-action styling and craftsmanship. Metal surfaces are highly polished and blued. American walnut stock

Prices:
30-30 Win., checkered **$370.00**
  w/o checkering **342.00**

MODEL 94 WALNUT TRAPPER CARBINE

magazine capacity of five shots (9 in 45 Colt or 44 Rem. Mag./44 S&W Special). **Calibers:** 30-30 Winchester, 357 Mag., 45 Colt, and 44 Rem. Mag./44 S&W Special.

With 16-inch short-barrel lever action and straight forward styling. Compact and fast-handling in dense cover, it has a

Prices:
30-30 Winchester **$342.00**
45 Colt, 44 Rem. Mag./44 S&W Special **361.00**

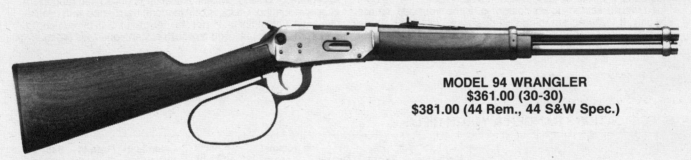

MODEL 94 WRANGLER
$361.00 (30-30)
$381.00 (44 Rem., 44 S&W Spec.)

MODEL 94 SPECIFICATIONS (with 100th Anniversary Receiver Inscription)

Model	Caliber	Magazine Capacity (A)	Barrel Length	Overall Length	Nominal Length Of Pull	Nominal Drop At Comb	Heel	Nominal Weight (Lbs.)	Rate of Twist 1 Turn in	Rings Sights
94 CHECKERED WALNUT	30-30 Win.	6	20″	37³/₄″	13¹/₂″	1¹/₈″	1⁷/₈″	6¹/₂	12″	Rifle
94 STANDARD	30-30 Win.	6	20	37³/₄	13¹/₂	1¹/₈	1⁷/₈	6¹/₂	12	Rifle
94 TRAPPER CARBINE	30-30 Win.	5	16	33³/₄	13¹/₂	1¹/₈	1⁷/₈	6¹/₈	12	Rifle
	44 Rem. Mag.	9	16	33³/₄	13¹/₂	1¹/₈	1⁷/₈	6	38	Rifle
	44 S&W Spec.	9	16	33³/₄	13¹/₂	1¹/₈	1⁷/₈	6	38	Rifle
	45 Colt	9	16	33³/₄	13¹/₂	1¹/₈	1⁷/₈	6	38	Rifle
94 WRANGLER	30-30 Win.	5	16	33³/₄	13¹/₂	1¹/₈	1⁷/₈	6¹/₈	12	Rifle
	44 Rem. Mag.	9	16	33³/₄	13¹/₂	1¹/₈	1⁷/₈	6	12	Rifle
	44 S&W Spec.	9	16	33³/₄	13¹/₂	1¹/₈	1⁷/₈	6	12	Rifle

(A) for additional capacity, add one round in chamber when ready to fire. Drops are measured from center line of bore. Rate of twist is right-hand.

WINCHESTER RIFLES

LEVER ACTION

MODEL 94 RANGER
$302.00 ($355.00 with Scope)

Model 94 Ranger is an economical version of the Model 94. Lever action is smooth and reliable. In 30-30 Winchester, the rapid-firing six-shot magazine capacity provides two more shots than most centerfire hunting rifles.

MODEL 94 BIG BORE WALNUT
$381.00

Winchester's powerful 307 and 356 hunting calibers combined with maximum lever-action power and angled ejection provide hunters with improved performance and economy.

MODEL 94 WIN-TUFF RIFLE
$381.00

Includes all features and specifications of standard Model 94 plus tough laminated hardwood styled for the brush-gunning hunter who wants good concealment and a carbine that can stand up to all kinds of weather.

MODEL 94 SPECIFICATIONS

Model	Caliber	Magazine Capacity (A)	Barrel Length	Overall Length	Nominal Length Of Pull	Nominal Drop At Comb	Nominal Drop At Heel	Nominal Weight (Lbs.)	Rate of Twist 1 Turn in	Sights
94 WIN-TUFF	30-30 Win.	6	20"	37³/4"	13¹/2"	1¹/8"	1⁷/8"	6¹/2	12"	Rifle
94 BIG BORE WALNUT	307 Win.	6	20	37¹/4	13¹/2	1¹/8	1⁷/8	6¹/2	12	Rifle
	356 Win.	6	20	37³/4	13¹/2	1¹/8	1⁷/8	6¹/2	12	Rifle
RANGER	30-30 Win.	6	20	37³/4	13¹/2	1¹/8	1⁷/8	6¹/2	12	Rifle
Scope 4X32 and see-through mounts	30-30 Win.	6	20	37³/4	13¹/2	1¹/8	1⁷/8	6¹/2	12	R/S

(A) For additional capacity, add one round in chamber when ready to fire. Drops are measured from center line of bore. R/S-Rifle sights and Bushnell® Sportview™ scope with mounts. Rate of twist is right-hand.

WINCHESTER RIFLES

MODEL 9422 LEVER-ACTION RIMFIRE RIFLES

These Model 9422 rimfire rifles combine classic 94 styling and handling in ultra-modern lever action 22s of superb craftsmanship. Handling and shooting characteristics are superior because of their carbine-like size.

Positive lever action and bolt design ensure feeding and chambering from any shooting position. The bolt face is T-slotted to guide the cartridge with complete control from magazine to chamber. A color-coded magazine follower shows when the brass magazine tube is empty. Receivers are grooved for scope mounting. Other functional features include exposed hammer with half-cock safety, hooded bead front sight, semi-buckhorn rear sight and side ejection of spent cartridges.

Stock and forearm are American walnut with checkering, high-luster finish, and straight-grip design. Internal parts are carefully finished for smoothness of action.

MODEL 9422 WALNUT

Considered one of the world's finest production sporting arms, this lever action rimfire (shown above) holds 21 Short, 17 Long or 15 Long Rifle cartridges.

Model 9422 Walnut Magnum gives exceptional accuracy at longer ranges than conventional 22 rifles. It is designed specifically for the 22 Winchester Magnum Rimfire cartridge and holds 11 cartridges.

Model 9422 Win-Cam Magnum features laminated non-glare, green-shaded stock and forearm. American hardwood stock is bonded to withstand all weather and climates. **Model 9422 Win-Tuff** is also availale to ensure resistance to changes in weather conditions, or exposure to water and hard knocks.

SPECIFICATIONS MODEL 9422

Model	Caliber	Magazine Capacity	Barrel Length	Overall Length	Nominal Length Of Pull	Nominal Drop At Comb	Nominal Drop At Heel	Nominal Weight (Lbs.)	Rate of Twist 1 Turn in	Sights	Prices
9422 WALNUT	22 S, L, LR	21S,17L,15LR	20½"	37⅛"	13½"	1⅛"	1⅞"	6¼	16"	Rifle	$384.00
	22WMR Mag.	11	20½	37⅛	13½	1⅛	1⅞	6¼	16	Rifle	400.00
9422 WIN-TUFF	22 S, L, LR	21S,17L,15LR	20½"	37⅛"	13½"	1⅛"	1⅞"	6¼	16"	Rifle	384.00
	22WMR Mag.	11	20½	37⅛	13½	1⅛	1⅞	6¼	16	Rifle	400.00
9422 WIN-CAM	22WMR Mag.	11	20½"	37⅛"	13½"	1⅛"	1⅞"	6¼	16"	Rifle	400.00

WMR-Winchester Magnum Rimfire. S-Short, L-Long, LR-Long Rifle. Drops are measured from center line of bore.

MODEL 52B RIMFIRE BOLT ACTION (Limited)
$576.00

SPECIFICATIONS
Caliber: 22 LR
Capacity: 5 rounds
Barrel length: 24"

Overall length: 42⅛"
Length of pull: 13⅝"
Drop at comb: 1⅜"
Drop at heel: 2 5/16"
Weight: 7 lbs.

WINSLOW RIFLES

SPECIFICATIONS
Stock: Choice of two stock models. **The Plainsmaster** offers pinpoint accuracy in open country with full curl pistol grip and flat forearm. **The Bushmaster** offers lighter weight for bush country; slender pistol with palm swell; beavertail forend for light hand comfort. Both styles are of hand-rubbed black walnut. Length of pull—13½ inches; plainsmaster ⅜ inch castoff; Bushmaster ³/₁₆ inch castoff; all rifles are drilled and tapped to incorporate the use of telescopic sights; rifles with receiver or open sights are available on special order; all rifles are equipped with quick detachable sling swivel studs and white-line recoil pad. All Winslow stocks incorporate a slight castoff to deflect recoil, minimizing flinch and muzzle jump. **Magazine:** Staggered box type, four shot. (Blind in the stock has no floorplate). **Action:** Mauser Mark X Action. **Overall length:** 43″ (Standard Model); 45″ (Magnum); all Winslow rifles have company name and serial number and grade engraved on the action

and caliber engraved on barrel. **Barrel:** Douglas barrel premium grade, chrome moly-type steel; all barrels, 20 caliber through 35 caliber, have six lands and grooves; barrels larger than 35 caliber have eight lands and grooves. All barrels are finished to (.2 to .4) micro inches inside the lands and grooves. **Total weight** (without scope): 7 to 7½ lbs. with 24″ barrel in standard calibers 243, 308, 270, etc; 8 to 9 lbs. with 26″ barrel in Magnum calibers 264 Win., 300 Wby., 458 Win., etc. Winslow rifles are made in the following calibers:

Standard cartridges: 22-250, 243 Win., 244 Rem., 257 Roberts, 308 Win., 30-06, 280 Rem., 270 Win., 25-06, 284 Win., 358 Win., and 7mm (7×57).

Magnum cartridges: 300 Weatherby, 300 Win., 338 Win., 358 Norma, 375 H.H., 458 Win., 257 Weatherby, 264 Win., 270 Weatherby, 7mm Weatherby, 7mm Rem., 300 H.H., 308 Norma.

Left-handed models available in most calibers.

WINSLOW BASIC RIFLE

The Basic Rifle, available in the Bushmaster stock, features one ivory diamond inlay in a rose-wood grip cap and ivory trademark in bottom of forearm. Grade 'A' walnut jeweled bolt and follower. **Price: $1750.00 and up.** With **Plainsmaster stock: $100.00** extra. **Left-hand model: $1850.00 and up.**

WINSLOW VARMINT

This 17-caliber rifle is available with Bushmaster stock or Plainsmaster stock, which is a miniature of the original with high roll-over cheekpiece and a round leading edge on the forearm, modified spoon billed pistol grip. Available in 17/222, 17/222 Mag. 17/233, 222 Rem. and 223. Regent grade shown. With **Bushmaster stock: $1750.00 and up.** With **Plainsmaster stock: $100.00** extra. **Left-hand model: $1850.00 and up.**

RIFLES

shotguns

For addresses and phone numbers of manufacturers and distributors included in this section, turn to *DIRECTORY OF MANUFACTURERS AND SUPPLIERS* at the back of the book.

AMERICAN ARMS SHOTGUNS

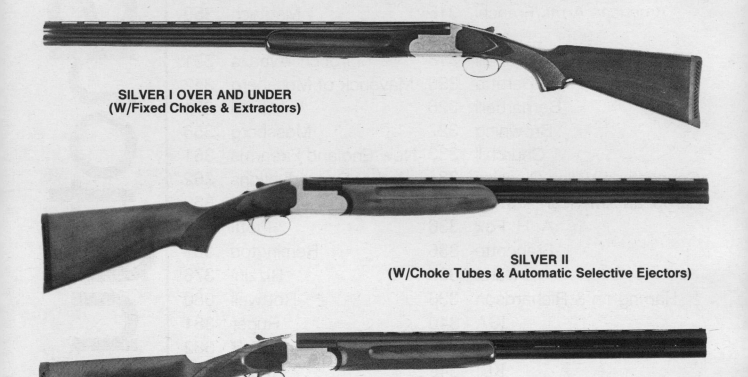

SILVER I OVER AND UNDER
(W/Fixed Chokes & Extractors)

SILVER II
(W/Choke Tubes & Automatic Selective Ejectors)

SILVER SPORTING

SPECIFICATIONS

Model	Gauge	Bbl. Length	Chamber	Chokes	Avg. Weight	Prices
Silver I	12	26″–28″	3″	IC/M-M/F	6 lbs. 15 oz.	$ 549.00
	20	26″–28″	3″	IC/M-M/F	6 lbs. 12 oz.	
	28	26″	2³/₄″	IC/M	5 lbs. 14 oz.	609.00
	.410	26″	3″	IC/M	6 lbs. 6 oz.	
Silver II	12	26″–28″	3″	CT-3	6 lbs. 15 oz.	
	16	26″	2³/₄″	IC/M	6 lbs. 13 oz.	699.00
	20	26″	3″	CT-3	6 lbs. 12 oz.	
	28	26″	2³/₄″	IC/M	5 lbs. 14 oz.	
	.410	26″	3″	IC/M	6 lbs. 6 oz.	1129.00
Silver Lite (not shown)	12	26″	3″	CT-3	6 lbs. 4 oz.	959.00
	20	26″	3″	CT-3	5 lbs. 12 oz.	
	28	26″	2³/₄″	IC/M	6 lbs.	
Sporting	12	28″–30″	2³/₄″	CTS	7 lbs. 6 oz.	899.00

CT-3 Choke Tubes IC/M/F   Cast Off = ³/₈″   CTS = SK/SK/IC/M
Silver I and II: Pull = 14 ¹/₈″; Drop at Comb = 1 ³/₈″; Drop at Heel = 2 ³/₈″
Silver Sporting: Pull = 14 ³/₈″; Drop at Comb = 1 ¹/₂″; Drop at Heel = 2 ³/₈″

AMERICAN ARMS SHOTGUNS

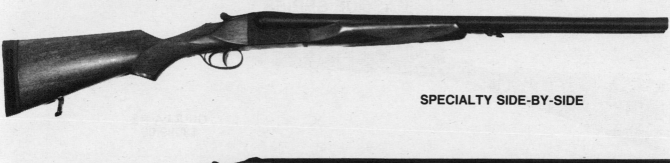

SPECIALTY SIDE-BY-SIDE

SPECIALTY OVER/UNDER

SPECIFICATIONS

Model	Gauge	Bbl. Length	Chamber	Chokes	Avg. Wgt.	Prices
WT/OU	10	26″	3½″	CT-2	9 lbs. 10 oz.	**$989.00**
WS/OU	12	24″, 28″	3½″	CT-3	7 lbs. 2 oz.	739.00
TS/SS	10	26″	3½″	CT-2	10 lbs. 13 oz.	669.00
TS/SS	12	26″	3½″	CT-3	7 lbs. 6 oz.	669.00

CT-3 = Choke tubes IC/M/F.   CT-2 = Choke tubes F/F.   SF = Steel Full Choke.   Drop at Comb = 1⅛″.   Drop at Heel = 2 ⅜″.

BASQUE SERIES

BRITTANY
$809.00

SPECIFICATIONS
Gauges: 12, 20
Chamber: 3″   **Chokes:** CT-3
Barrel length: 26″
Weight: 6 lbs. 7 ozs. (20 ga.); 6 lbs. 15 oz. (12 ga.)

Features: Engraved case-colored frame; single selective trigger with top tang selector; automatic selective ejectors; manual safety; hard chrome-lined barrels; walnut English-style straight stock and semi-beavertail forearm w/cut checkering and oil-rubbed finish; ventilated rubber recoil pad; and choke tubes with key

AMERICAN ARMS SHOTGUNS

BASQUE SERIES

GRULLA #2
$3099.00

Chokes: IC/M (M/F also in 12 ga.)
Features: Hand-fitted and finished high-grade classic double; double triggers; automatic selective ejectors; fixed chokes; concave rib; case-colored sidelock action w/engraving; English-style straight stock; splinter forearm and checkered butt of oil rubbed walnut
Also available in sets (20 & 28 or 28 & .410): **$4219.00**

SPECIFICATIONS
Gauges: 12, 20, 28, .410
Chambers: 2³/₄″ (28 ga.); 3″ (12, 20 & .410 ga.)
Barrel length: 26″ (28″ also in 12 ga.)
Weight: 6 lbs. 4 oz. (12 ga.); 5 lbs. 11 oz. (20 & 28 ga.); 5 lbs. 13 oz. (.410)

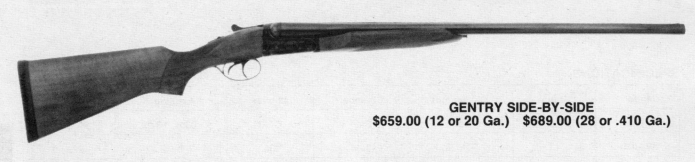

GENTRY SIDE-BY-SIDE
$659.00 (12 or 20 Ga.) $689.00 (28 or .410 Ga.)

Features boxlocks with engraved English-style scrollwork on side plates; one-piece, steel-forged receiver; chrome barrels; manual thumb safety; independent floating firing pin.

SPECIFICATIONS
Gauges: 12, 20, 28, .410
Chambers: 3″ (except 28 gauge, 2³/₄″)

Barrel lengths: 26″, choked IC/M (all gauges); 28″, choked M/F (12 and 20 gauges)
Weight: 6 lbs. 15 oz. (12 ga.); 6 lbs. 7 oz. (20 and .410 ga.); 6 lbs. 5 oz. (28 ga.)
Drop at comb: 1³/₈″   **Drop at heel:** 2³/₈″
Other features: Fitted recoil pad; flat matted rib; walnut pistol-grip stock and beavertail forend with hand-checkering; gold front sight bead

DERBY SIDE-BY-SIDE
$1039.50

Features functioning side locks with English-style hand engraving on side plates; one-piece, steel-forged receiver; chrome barrels; automatic safety

SPECIFICATIONS
Gauges: 12 and 20   **Chambers:** 3″
Barrel lengths: 26″, choked IC/M; 28″, choked M/F (12 gauge)
Weight: 6 lbs. 12 oz. (12 ga.); 6 lbs. 6 oz. (20 ga.)

Sights: Gold bead front sight
Stock: Walnut and splinter forend with hand-checkering
Length of pull: 14¹/₈″
Drop at comb: 1³/₈″   **Drop at heel:** 2³/₈″
Features: Walnut straight stock and splinter forearm; auto selective ejectors; fixed chokes; single non-selective trigger; frame and sidelocks finished with antique silver and machine engraving.

AMERICAN ARMS/FRANCHI

MODEL 48AL (Recoil)
$609.00 $640.00 (Slug Barrel)

SPECIFICATIONS
Gauge: 12/20   **Chamber:** 2³/₄″
Action: Single Action
Choke: Full
Barrel lengths: 24″, 26″, 28″
Weight: 5¹/₂ to 6¹/₂ lbs.
Length of pull: 14¹/₄″
Drop at comb: 1¹/₂″   **Drop at heel:** 2³/₈″

FRANCHI SPAS-12
$769.00

This Franchi 12-gauge shotgun is designed for law enforcement and personal defense. It operates as a pump action and/or semiautomatic action (a manual action selector switch is located at the bottom of the forearm). The barrel features a muzzle protector and is made of chrome-moly steel with a hard-chromed bore and matte finish. A rifle-type aperture rear sight and blade front sight are standard, along with cylinder bore choking (modified and full choke tubes are available as accessories). The muzzle is threaded externally for mounting the SPAS-12 choke tubes (SPAS-12 and standard Franchi tubes are *not* interchangeable; 2³/₄″ only). The receiver/frame is lightweight alloy with non-reflective anodized finish. The fire control system is easily removable for maintenance with two push pins; a magazine cut-off button is located on the right side of the receiver, and the primary crossbolt safety button is in front of the trigger guard. A secondary ''tactical'' safety lever is located on the left side of the trigger guard, and the bolt/carrier latch release button is on the left side of the receiver.

The SPAS-12 is designed to shoot 2³/₄″ shells only. Two safety systems are standard.

SPECIFICATIONS
Gauge: 12; 2³/₄″ chamber
Capacity: 7 rounds   **Choke:** Cylinder bore
Barrel length: 21¹/₂″   **Overall length:** 41″
Weight: 8 lbs. 12 oz.
Stock: Nylon full-length stock with full pistol grip and serrated nylon forearm

LAW-12
$719.00

Same general specifications as the **SPAS-12,** except this high-power 12-gauge shotgun has gas-operated action, plus ambidextrous safety, decocking lever and adjustable sights.

AYA SHOTGUNS

SIDELOCK SHOTGUNS

AYA sidelock shotguns are fitted with London Holland & Holland system sidelocks, double triggers with articulated front trigger, automatic safety and ejectors, cocking indicators, bushed firing pins, replaceable hinge pins and chopper lump barrels. Stocks are of figured walnut with hand-cut checkering and oil finish, complete with a metal oval on the buttstock for engraving of initials.

Exhibition grade wood is available as are many special options, including a true left-hand version and self-opener.

Barrel lengths: 26″, 27″, 28″ and 29″. **Weight:** 5 to 7 pounds, depending on gauge.

Model	Prices
MODEL 1: Sidelock in 12 and 20 gauge with special engraving and exhibition quality wood	**$6181.00**
MODEL 2: Sidelock in 12, 16, 20, 28 gauge and .410 bore	**3219.00**
MODEL 53: Sidelock in 12, 16 and 20 gauge with 3 locking lugs and side clips	**4636.00**
MODEL 56: Sidelock in 12 gauge only with 3 locking lugs and side clips	**7209.00**
MODEL XXV/SL: Sidelock in 12 and 20 gauge only with Churchill-type rib	**3824.00**

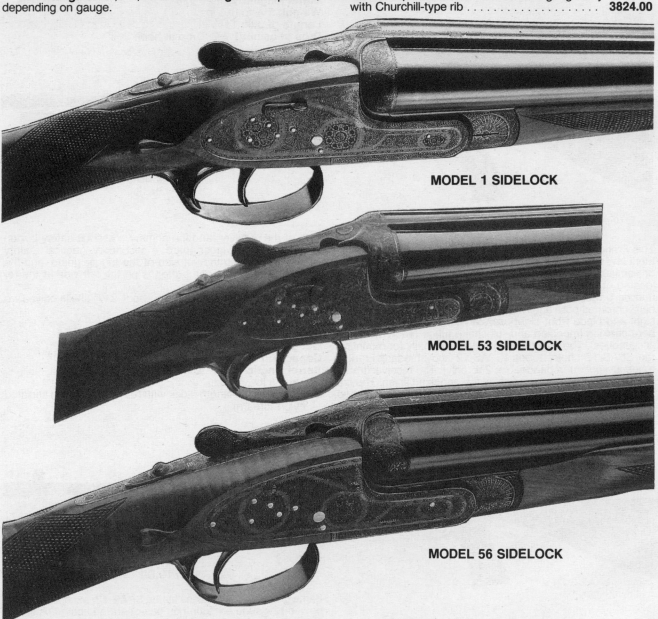

MODEL 1 SIDELOCK

MODEL 53 SIDELOCK

MODEL 56 SIDELOCK

AYA SHOTGUNS

BOXLOCK SHOTGUNS

AYA boxlocks use the Anson & Deeley system with double locking lugs, incorporating detachable cross pin and separate plate to allow easy access to the firing mechanism. Barrels are chopper lump, firing pins are bushed, plus automatic safety and ejectors and metal oval for engraving of initials.

Barrel lengths: 26″, 27″ and 28″. **Weight:** 5 to 7 pounds, depending on gauge.

Model	Price
MODEL XXV BOXLOCK: 12 and 20 gauge only	$ 2,867.00
MODEL 4 BOXLOCK: 12, 16, 20, 28, .410 ga.	1,684.00
MODEL 4 DELUXE BOXLOCK: Same gauges as above	2,966.00
MODEL 37 SUPER A	13,842.00
MODEL AUGUSTA	24,899.00

**MODEL XXV BOXLOCK
(Close-up)**

MODEL XXV BOXLOCK

MODEL 4 BOXLOCK

BENELLI SHOTGUNS
MODEL M1 SUPER 90 SERIES

See table on the following page for specifications.

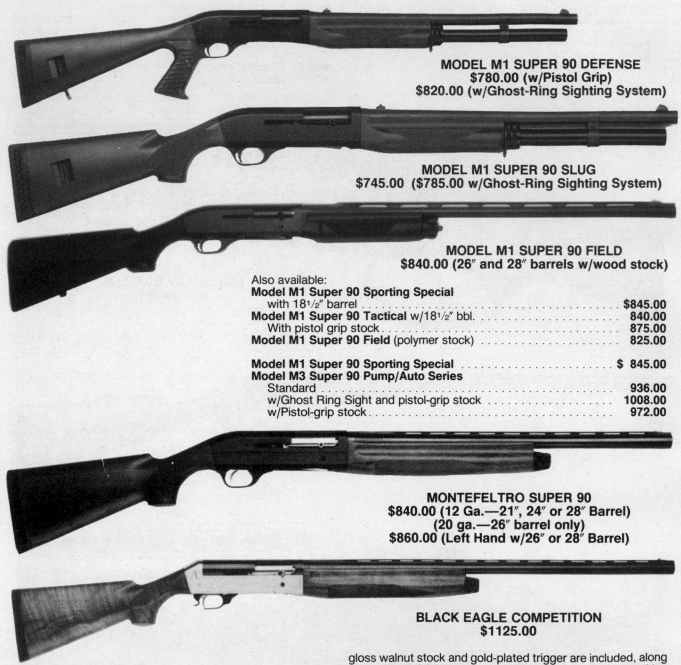

MODEL M1 SUPER 90 DEFENSE
$780.00 (w/Pistol Grip)
$820.00 (w/Ghost-Ring Sighting System)

MODEL M1 SUPER 90 SLUG
$745.00 ($785.00 w/Ghost-Ring Sighting System)

MODEL M1 SUPER 90 FIELD
$840.00 (26″ and 28″ barrels w/wood stock)

Also available:
Model M1 Super 90 Sporting Special
   with 18½″ barrel **$845.00**
Model M1 Super 90 Tactical w/18½″ bbl. 840.00
   With pistol grip stock 875.00
Model M1 Super 90 Field (polymer stock) 825.00

Model M1 Super 90 Sporting Special . . . . **$ 845.00**
Model M3 Super 90 Pump/Auto Series
   Standard 936.00
   w/Ghost Ring Sight and pistol-grip stock 1008.00
   w/Pistol-grip stock 972.00

MONTEFELTRO SUPER 90
$840.00 (12 Ga.—21″, 24″ or 28″ Barrel)
(20 ga.—26″ barrel only)
$860.00 (Left Hand w/26″ or 28″ Barrel)

BLACK EAGLE COMPETITION
$1125.00

Benelli's Black Eagle Competition shotgun combines the best technical features of the Montefeltro Super 90 and the classic design of the old SL 80 Series. It comes standard with a specially designed two-piece receiver of steel and aluminum, adding to its reliability and resistance to wear. A premium high-gloss walnut stock and gold-plated trigger are included, along with a Montefeltro rotating bolt. The Black Eagle Competition has no complex cylinders and pistons to maintain. Features include etched receiver, competition stock and mid-rib bead. Also available:
Black Eagle Limited Edition. One of 1,000 with 26″ vent-rib barrel and satin walnut stock. **Price: $2,000.00**

BENELLI SHOTGUNS

SUPER BLACK EAGLE
$1100.00 (24″ or 26″ barrel)
$1115.00 (28″ barrel

Benelli's Super Black Eagle shotgun offers the advantage of owning one 12-gauge auto that fires every type of 12-gauge currently available. It has the same balance, sighting plane and fast-swinging characteristics whether practicing on the sporting clays course with light target loads or touching off a 3½″ magnum steel load at a high-flying goose.

The Super Black Eagle also features a specially strengthened steel upper receiver mated to the barrel to endure the toughest shotgunning. The alloy lower receiver keeps the overall weight low, making this model as well balanced and point-able as possible. Distinctive high-gloss or satin walnut stocks and a choice of dull finish or blued metal add up to a universal gun for all shotgun hunting and sports.

Stock: Satin walnut (28″) with drop adjustment kit; high-gloss walnut (26″) with drop adjustment kit; or synthetic stock
Finish: Matte black finish on receiver, barrel and bolt (28″); blued finish on receiver and barrel (26″) with bolt mirror polished
Features: Montefeltro rotating bolt with dual locking lugs
For additional specifications, see table below.

Also available:
Custom Slug Gun with 24″ E.R. Shaw rifled barrel for sabot-type slugs and polymer stock. **Price: $1125.00 ($1140.00 with satin wood stock)**

SPECIFICATIONS BENELLI SHOTGUNS

	Gauge (Chamber)	Operation	Magazine Capacity*	Barrel Length	Overall Length	Weight (in lbs.)	Choke	Metal Finish	Stock	Sights
Super Black Eagle (28)	12 (3½ in.)	semi-auto inertia recoil	3	28 in.	49⅝ in.	7.3	S, IC, M, IM, F**	matte	satin walnut or polymer	front & mid rib bead
Super Black Eagle (26)	12 (3½ in.)	semi-auto inertia recoil	3	26 in.	47⅝ in.	7.1	S, IC, M, IM, F**	matte or blued	satin walnut or polymer	front & mid rib bead
Super Black Eagle (24)	12 (3½ in.)	semi-auto inertia recoil	3	24 in.	45⅝ in.	7.0	S, IC, M, IM, F**	matte	polymer	front & mid rib bead
Super Black Eagle Custom Slug	12 (3 in.)	semi-auto inertia recoil	3	24 in.	45½ in.	7.6	rifled barrel	matte	satin walnut or polymer	scope mount base
Black Eagle Competition Gun	12 (3 in.)	semi-auto inertia recoil	4	28 or 26 in.	49⅝ or 47⅝ in.	7.3/7	S, IC, M, IM, F**	blued with etched receiver	satin walnut	front & mid rib bead
Black Eagle Limited Edition	12 (3 in.)	semi-auto inertia recoil	4	26 in.	47⅝ in	7.3/7	S, IC, M, IM, F**	blued with etched & inlaid rec.	satin walnut	front & mid rib bead
Montefeltro Super 90 (28/26)	12 (3 in.)	semi-auto inertia recoil	4	28 or 26 in.	49½ or 47½ in.	7.4/7	S, IC, M, IM, F**	blued	satin walnut	bead
Montefeltro Super 90 (24/21)	12 (3 in.)	semi-auto inertia recoil	4	24 or 21 in.	45½ or 42½ in.	6.9/6.7	S, IC, M, IM, F**	blued	satin walnut	bead
Montefeltro Left Hand	12 (3 in.)	semi-auto inertia recoil	4	28 or 26 in.	49½ or 47½ in.	7.4/7	S, IC, M, IM, F**	blued	satin walnut	bead
Montefeltro 20 Gauge	20 (3 in.)	semi-auto inertia recoil	4	26 in.	47½ in.	5.75	S, IC, M, IM, F**	blued	satin walnut	front & mid rib bead
M1 Super 90 Field (28)	12 (3 in.)	semi-auto inertia recoil	3	28 in.	49½ in.	7.4	S, IC, M, IM, F**	matte	polymer standard or satin walnut	bead
M1 Super 90 Field (26)	12 (3 in.)	semi-auto inertia recoil	3	26 in.	47½ in.	7.3	S, IC, M, IM, F**	matte	polymer standard or satin walnut	bead
M1 Super 90 Field (24)	12 (3 in.)	semi-auto inertia recoil	3	24 in.	45½ in.	7.2	S, IC, M, IM, F**	matte	polymer standard	bead
M1 Super 90 Field (21)	12 (3 in.)	semi-auto inertia recoil	3	21 in.	42½ in.	7	S, IC, M, IM, F**	matte	polymer standard	bead
M1 Super 90 Sporting Special	12 (3 in.)	semi-auto inertia recoil	3	18½ in	39¾ in.	6.5	IC, M, F**	matte	polymer standard	ghost ring
M1 Super 90 Tactical	12 (3 in.)	semi-auto inertia recoil	7	18½ in.	39¾ in.	6.5	IC, M, F**	matte	polymer pistol grip or polymer standard	rifle or ghost ring
M1 Super 90 Slug	12 (3 in.)	semi-auto inertia recoil	7	19¾ in.	41 in.	6.7	Cylinder	matte	polymer standard	rifle or ghost ring
M1 Super 90 Defense	12 (3 in.)	semi-auto inertia recoil	7	19¾ in.	41 in.	7.1	Cylinder	matte	polymer pistol grip	rifle or ghost ring
M1 Super 90 Entry	12 (3 in.)	semi-auto inertia recoil	5	14 in.	35½ in.	6.7	Cylinder	matte	polymer pistol grip or polymer standard	rifle or ghost ring
M3 Super 90	12 (3 in.)	semi-auto/pump inertia recoil	7	19¾ in.	41 in.	7.9	Cylinder	matte	polymer pistol grip or polymer standard	rifle or ghost ring

*Magazine capacity given for 2¾ inch shells   **Skeet, Improved Cylinder, Modified, Improved Modified, Full

BERETTA SHOTGUNS

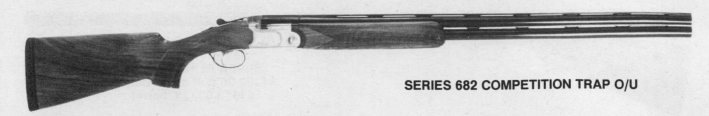

SERIES 682 COMPETITION TRAP O/U

Available in Competition Mono, Over/Under or Mono Trap Over/Under Combo Set, the 12-gauge 682X trap guns boast hand-checkered walnut stock and forend with International or Monte Carlo left- or right-hand stock.

Features: Adjustable gold-plated, single selective sliding trigger for precise length of pull fit; fluorescent competition front sight; step-up top rib; non-reflective black matte finish; low profile improved boxlock action; manual safety with barrel selector; 2³/₄″ chambers; auto ejector; competition recoil pad buttplate; stock with silver oval for initials; silver inscription inlaid on trigger guard; handsome fitted case. **Weight:** Approx. 8 lbs.

Barrel length/Choke	Prices
30″ Imp. Mod./Full (Black or Silver)	$2495.00
30″ or 32″ Mobilchoke® (Black or Silver)	2570.00
Top Single 32″ or 34″ Mobilchoke®	2650.00
Pigeon (Silver)	2760.00
Unsingle.	2650.00
Combo.: 30″ or 32″ Mobilchoke® (Top)	3400.00
30″ IM/F (Top)	3340.00
32″ Mobilchoke® (Mono)	3400.00

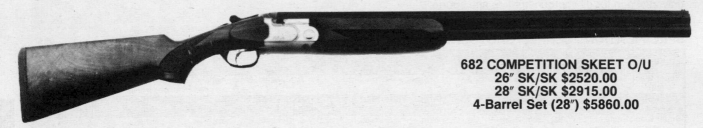

682 COMPETITION SKEET O/U
26″ SK/SK $2520.00
28″ SK/SK $2915.00
4-Barrel Set (28″) $5860.00

This skeet gun sports hand-checkered premium walnut stock, forged and hardened receiver, manual safety with trigger selector, auto ejector, stock with silver oval for initials, silver inlaid on trigger guard. Price includes fitted case.
Gauges: 12; 4-barrel sets in 12, 20, 28 and .410
Action: Low-profile hard chrome-plated boxlock

Trigger: Single adjustable sliding trigger
Barrels: 26″ or 28″ blued barrels with 2³/₄″ chambers
Stock dimensions: Length of pull 14³/₄″; drop at comb 1³/₈″; drop at heel 2¹/₄″
Sights: Fluorescent front and metal middle bead
Weight: Approx. 7¹/₂ lbs.

MODEL 682 SUPER TRAP

MODEL 682 SUPER TRAP

Beretta's 12 gauge over/under shotgun features a revolutionary adjustable stock. Comb is adjustable with interchangeable comb height adjustments. Length of pull is also adjustable with interchangeable butt pad spacers. Also features ported barrels and tapered step rib, satin-finished receiver and adjustable trigger.

Barrel lengths: 30″, 32″, 34″; chamber 2³/₄″
Chokes: Choice of Mobilchoke®, IM/F, Full

Prices:
Model 682 Super Trap O/U (Mobilchoke®)	$2885.00
Same as above with IM/F choke	2820.00
Model 682 Top Single Super Trap	
32″ barrel and Mobilchoke	3060.00
Same as above with Full choke	2990.00
Model 682 Top Combo Super Trap	3790.00
Same as above w/o Mobilchoke®	3865.00

BERETTA SHOTGUNS

12 AND 20 GAUGE SPORTING CLAY SHOTGUNS

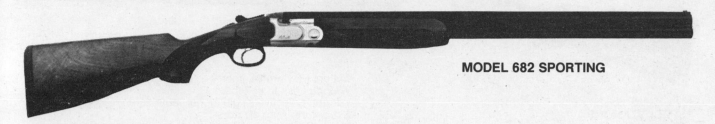

MODEL 682 SPORTING

MODELS 682/686/687 SPORTING CLAYS

This competition-style shotgun for sporting clays features 28″ or 30″ barrels with four flush-mounted screw-in choke tubes (Full, Modified, Improved Cylinder and Skeet), plus hand-checkered stock and forend of fine walnut, 2³/₄″ or 3″ chambers and adjustable trigger. **Model 682 Continental Course Sporting** has tapered rib and schnabel forend. **Model 682 Super Sporting** has ported barrels, adjustable comb height inserts and length of pull. **Model 686 Onyx Sporting** has black matte receiver and **686 Silver Perdiz Sporting** has coin silver receiver with scroll engraving. **Model 687EL** has sideplates with scroll engraving.

Prices:

682 Sporting	$2605.00
682 Continental Course Sporting	2715.00
682 Super Sporting	2925.00
682 Sporting Combo	3470.00
686 Onyx Sporting	1385.00
686 Silver Perdiz Sporting	1425.00
686 Silver Perdiz Sporting Combo	2600.00
687 Silver Perdiz Sporting	2285.00
687 Silver Pigeon Sporting Combo	3405.00
687EL	3225.00
687EELL Diamond Pigeon Sporting	4700.00

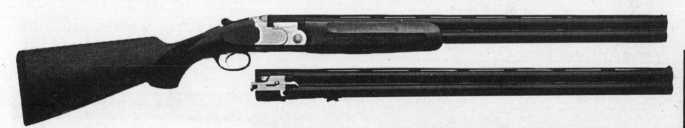

MODEL 686 SILVER PERDIZ SPORTING COMBO

MODELS 686/687 SILVER PERDIZ/PIGEON SPORTING COMBO

This 12-gauge Beretta over/under features interchangeable 28″ and 30″ barrels for versatility in competition at different courses with short and long passing shots. **Chamber:** 3″. Mobilchoke® screw-in tube system. Prices listed above.

MODEL 686 ESSENTIAL
$1215.00

SPECIFICATIONS
Gauge: 12 (3″ chamber)
Choke: MC3 Mobilchoke® (F, M, IC)
Barrel length: 26″ or 28″   **Overall length:** 45.7″
Weight: 6.7 lbs.

Stock: American walnut
Drop at comb: 1.4″   **Drop at heel:** 2.2″
Length of pull: 14¹/₂″
Features: Matte black receiver

BERETTA SHOTGUNS

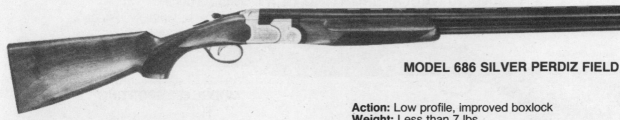

MODEL 686 SILVER PERDIZ FIELD

SPECIFICATIONS
Gauges: 12, 20
Barrels/chokes: 26″ and 28″ with Mobilchoke® screw-in choke tubes

Action: Low profile, improved boxlock
Weight: Less than 7 lbs.
Trigger: Selective single trigger, auto safety
Extractors: Auto ejectors
Stock: Choice walnut, hand-checkered and hand-finished with a tough gloss finish
Price: $1355.00

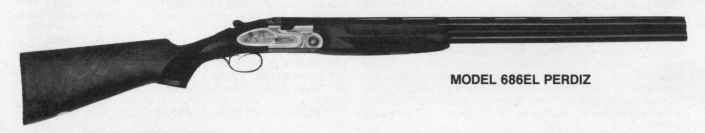

MODEL 686EL PERDIZ

This 12- or 20-gauge over/under field gun features scroll engraving on sideplates, European walnut stock and forend, hard-chromed bores, Mobilchoke® system of interchangeable choke tubes, and gun case. **Price:** $2200.00

Also available:
Model 686L Silver Perdiz (28 ga.) with highly polished silver receiver, traditional blued finish 26″ barrels, and rubber recoil pad, plus Mobilchoke® **Price:** $1355.00

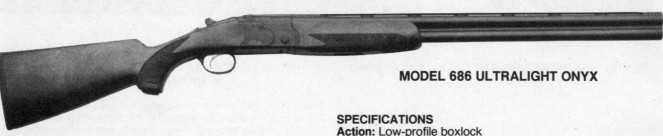

MODEL 686 ULTRALIGHT ONYX

This new 12-gauge over/under field gun features the payload of a 12-gauge gun in a 20-gauge weight (approx. 5 lbs. 13 oz.). Chambered for 2³/₄″, the Ultralight has a matte black finish on its receiver with a gold inlay of the P. Beretta signature.

SPECIFICATIONS
Action: Low-profile boxlock
Barrel length: 26″ or 28″
Trigger: Single, selective gold trigger
Safety: Automatic
Stock: Walnut, hand-checkered
Price: $1525.00

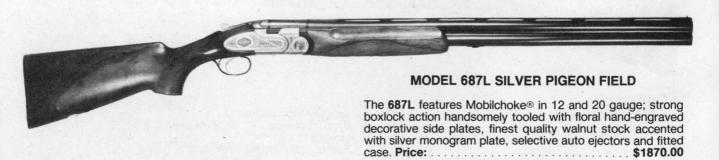

MODEL 687L SILVER PIGEON FIELD

The **687L** features Mobilchoke® in 12 and 20 gauge; strong boxlock action handsomely tooled with floral hand-engraved decorative side plates, finest quality walnut stock accented with silver monogram plate, selective auto ejectors and fitted case. **Price:** $1870.00

BERETTA SHOTGUNS

MODEL 687EL GOLD PIGEON FIELD
(not shown)

Features game scene engraving on receiver with gold highlights. Available in 12, 20 gauge (28 ga. and .410 in small frame).

SPECIFICATIONS
Barrels/chokes: 26″ and 28″ with Mobilchoke®
Action: Low-profile improved boxlock
Weight: 6.8 lbs. (12 ga.)
Trigger: Single selective with manual safety
Extractors: Auto ejectors
Prices:
Model 687EL (12, 20, 28 ga.; 26″ or 28″ bbl.) **$3180.00**
Model 687EL Small Frame (.410) **3320.00**

MODEL 687EELL DIAMOND PIGEON
$4625.00 (not shown)
Model 687EELL Combo (20 and 28 ga.) $5130.00

In 12, 20 or 28 ga., this model features the Mobilchoke® choke system, a special premium walnut stock and exquisitely engraved sideplate with game-scene motifs.

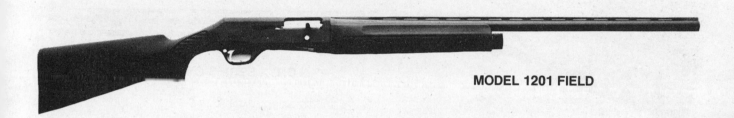

MODEL 1201 FIELD

This All-Weather semiautomatic shotgun features an adjustable space-age technopolymer stock and forend with recoil pad. Lightweight, it sports a unique weather-resistant matte black finish to reduce glare, resist corrosion and aid in heat dispersion; short recoil action for light and heavy loads. **Gauge:** 12. **Chamber:** 3″. **Barrel lengths:** 24″, 26″ and 28″. **Choke:** Mobilchoke® (Full, Modified). **Weight:** 7.4 lbs.
Price: **$625.00**

Also available:
Model 1201 Riot (Law Enforcement) with 20″ barrel
(2¾″ or 3″ shells), IC choke (7 rounds
w/2¾″ chamber; 6 rounds w/3″) **$660.00**
Model 1201 Riot w/pistol grip **705.00**

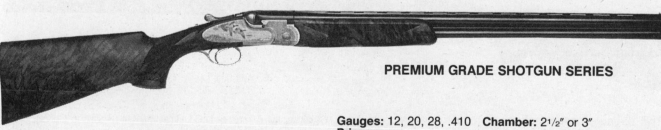

PREMIUM GRADE SHOTGUN SERIES

These hand-crafted over/under and side-by-side shotguns feature custom engraved or game scenes, casehardened, gold-inlay, scroll or floral patterns, all available on receivers. Sidelock action. Stocks are of select European walnut, hand-finished and hand-checkered. Also available in Competition Skeet, Trap, Sporting Clays and Custom Sidelock Side-by-Side models. Barrels are constructed of Boehler high-nickel antinit steel.

Gauges: 12, 20, 28, .410   **Chamber:** 2½″ or 3″
Prices:
SO5 Competition (Sporting Clays, Skeet,
Trap) **$12,900.00**
SO6 O/U Competition (Sporting Clays, Skeet,
Trap) **15,500.00**
SO6 EELL Custom Sidelock (12 gauge only) . . . **26,500.00**
SO9 Custom Sidelock (12, 20, 28, .410 ga.) . . . **28,500.00**
452 EELL Custom Sidelock Side/Side
(12 ga.) **31,000.00**

BERETTA SHOTGUNS

PINTAIL

PINTAIL

This new 12-gauge semiautomatic shotgun with short-recoil operation is available with 24″ or 26″ barrels and Mobilchoke®. A slug version is also available with rifled sights and choke tube. Finish is non-reflective matte on all exposed wood and metal surfaces. Checkered walnut stock and forend; sling swivels.

SPECIFICATIONS
Barrel lengths: 24″, 26″; 24″ Slug
Weight: 7.3 lbs.
Stock: Walnut
Sights: Bead front on vent rib
Price: **$700.00**

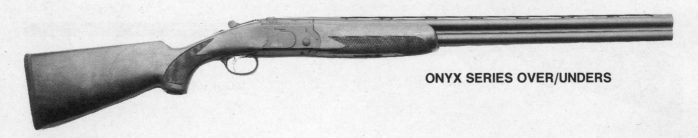

ONYX SERIES OVER/UNDERS

SPECIFICATIONS
Gauges: 12, 20 **Chamber:** 3″
Barrel lengths: 26″ and 28″
Chokes: Mobilchoke® screw-in system
Weight: 6 lbs. 12 oz. (12 ga.); 6.2 lbs. (20 ga.)
Stock: American walnut with recoil pad (English stock available)

Features: Automatic ejectors; matte black finish on barrels and receiver to reduce glare
Price: **$1355.00**
2-Barrel Set (20 & 28 ga.) **2085.00**
Ultralight **1525.00**

MODEL ASE 90

This 12-gauge beauty features drop-out trigger group assembly for ease in cleaning, inspection or in-the-field replacement. Also has wide ventilating top and side rib, hard-chromed bores, and a strong competition-style receiver in coin-silver finish and gold etching with P. Beretta initials.

SPECIFICATIONS
Barrel lengths: 28″ (Pigeon, Skeet, Sporting Clays); 30″ (Trap and Sporting Clays); 30″ and 32″ Combo (Top Combo Trap); 30″ and 34″ Combo (Top Combo Trap)

Chokes: IM/F Trap or MCT (Trap); MC4 (Sporting Clays); SK/SK (Skeet); IM/F (Pigeon)
Prices:
Model ASE 90 (Pigeon, Trap, Skeet) **$8070.00**
Model ASE 90 (Sporting Clays) **8140.00**
Model ASE 90 (Trap Combo) **9480.00**

BERETTA SHOTGUNS

MODEL 390 SILVER MALLARD

This gas-operated semiautomatic has an innovative gas system that handles a variety of loads. A self-regulating valve automatically adjusts gas pressure to handle anything from 2¹/₂″ target loads to heavy 3″ magnums. Matte finish models for turkey/waterfowl, slug and Deluxe models with gold-engraved receiver and deluxe wood are available. Also offered are **Model 390 Super Trap** and **390 Super Skeet** with ported barrels and adjustable comb height and length of pull.

SPECIFICATIONS
Gauge: 12   **Action:** Locked breech, gas-operated
Barrel lengths: 24″, 26″, 28″, 30″; 22″ slug

Weight: 7 lbs.
Sights: Ventilated rib with front bead
Safety: Crossbolt (reversible)
Prices:

390 Silver Mallard Field	$ 775.00
390 Waterfowl/Turkey Matte Finish (Silver Mallard)	775.00
390 Gold Mallard	935.00
390 Super Trap	1210.00
390 Super Skeet	1160.00

MODEL 303 YOUTH GUN

MODEL 303 AUTOLOADER

SPECIFICATIONS
Gauges: 12 and 20
Barrel lengths: 24″, 26″, 28″, 30″, 32″
Weight: 7 lbs. (12 gauge) and 6 lbs. (20 gauge)
Safety: Crossbolt
Action: Locked breech, gas-operated
Sights: Vent rib with front metal bead
Length of pull: 14⁷/₈″
Capacity: Plugged to 2 rounds
Prices:

Upland	$755.00
Field (English)	735.00
Youth	735.00
Skeet	735.00
Trap	735.00
With Mobilchoke® F, IM, M	775.00
Sporting Clays	835.00

MODEL 303 COMPETITION TRAP & SKEET
(not shown)
$735.00   ($775.00 w/Mobilchoke®)

SPECIFICATIONS
Gauges: 12 and 20   **Chamber:** 2³/₄″
Action: Semiautomatic, locked breech, gas-operated
Barrel lengths: 26″ (Skeet); 30″ and 32″ (Trap); 28″ and 30″ (Sporting); wide vent rib
Sights: Ventilated rib with fluorescent front bead, metal middle bead
Stock: Hand-checkered pistol-grip stock and forend of select European walnut

BERNARDELLI SHOTGUNS

Bernardelli shotguns are the creation of the Italian firm of Vincenzo Bernardelli, known for its fine quality firearms and commitment to excellence for more than a century. Most of the long arms featured below can be built with a variety of options, customized for the discriminating sportsman. With the exceptions indicated for each gun respectively, options include choice of barrel lengths and chokes; pistol or straight English grip stock; single selective or non-selective trigger; long tang trigger guard; checkered butt; beavertail forend; hand-cut rib; automatic safety; custom stock dimensions; standard or English recoil pad; extra set of barrels; choice of luggage gun case.

MODEL 112 12 GAUGE
$1435.00 (Single Trigger)
$1375.00 (Double Trigger)

Features extractors or automatic ejectors, English or half pistol-grip stock and splinter forend. **Barrel length:** 26³/₄″ (3″ chamber). **Choke:** Improved Cylinder and Improved Modified. **Safety:** Manual. **Weight:** 6¹/₂ lbs. **Price** (with ejector and multi-choke): **$1600.00**

ROMA S/S SERIES
$1470.00 (ROMA 3) — $3850.00 (ROMA 9)

Features include Anson & Deeley action, Purdey triple lock, concave rib, engraved sideplates, double trigger, ejectors.

HOLLAND & HOLLAND TYPE SIDELOCK SIDE-BY-SIDE

These 12-gauge Holland & Holland-style sidelock side-by-sides feature sidelocks with double safety levers, reinforced breech, three round Purdey locks, automatic ejectors, right trigger folding, striker retaining plates, best-quality walnut stock and finely chiselled high-grade engravings. The eight shotguns in this series differ only in the amount and intricacy of engravings.

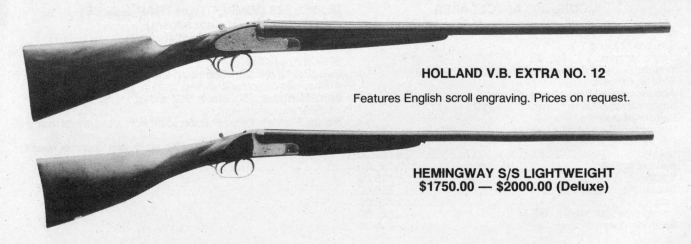

HOLLAND V.B. EXTRA NO. 12

Features English scroll engraving. Prices on request.

HEMINGWAY S/S LIGHTWEIGHT
$1750.00 — $2000.00 (Deluxe)

BROWNING AUTO SHOTGUNS

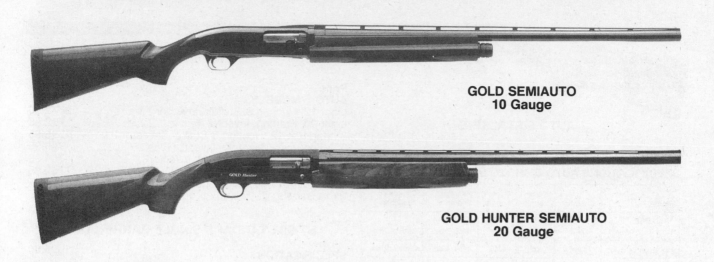

**GOLD SEMIAUTO
10 Gauge**

**GOLD HUNTER SEMIAUTO
20 Gauge**

GOLD HUNTER & STALKER
SEMIAUTOMATIC SHOTGUNS
$687.95 (12 and 20 Gauge)
$931.95 (10 Gauge)

The extremely short piston stroke of the new Gold semi-automatic shotgun limits the amount of gases that can get into the receiver. To operate the action, Browning's gas system utilizes a piston rod, which is not connected to the bolt assembly. This further inhibits gases from entering the receiver. The Gold's self-regulating gas system allows hunters to shoot all loads interchangeably. With light 1 oz. loads, the self-regulating gas system uses most of the energy of the expanding gases to operate the action. With 3" magnums, excess gases are vented out the lateral vents in the piston. A gas valve on the inside of the piston and the piston itself operate independently of each other. This design reduces pressure on the piston, resulting in more efficient operation.

A lightweight composite buffer and lightweight aluminum gas piston help reduce recoil. Further, a lightweight component requires less gas pressure to function and delivers less weight to the shoulder upon recoil. The shorter length of tilting-type bolt design reduces the overall length of the receiver, providing

improved between-the-hands balance for quicker reflexes when hunting fast-flying game. There's no need to fumble around for a bolt release button.

The 10-gauge shotgun in the Gold series permits goose and duck hunters to shoot larger, more powerful steel shot without having to sacrifice pellet count. To get maximum performance from steel shot, the general rule is to shoot steel shot that is two sizes larger than the old lead load. Jumping up to a larger shot size greatly reduces the number of pellets in a 12-gauge shell, providing less dense patterns. A 10-gauge steel load, however, offers dense patterns along with the benefit of a full pattern of large shot.

With most of its weight slightly forward, the Gold controls easily and swings smoothly—ideal for high-flying waterfowl. Wood-to-metal fit is tight. The stock and forearm are of select walnut with a durable gloss finish. Barrel and receiver have a deep, high-polished blued finish. Extra barrels are available for **$239.95** (10 ga.) and **$254.95** (12 and 20 ga.).

SPECIFICATIONS GOLD 10

Chamber	Barrel Length	Overall Length	Average Weight	Chokes
3½"	30"	52"	10 lbs. 13 oz.	Standard Invector
3½"	28"	50"	10 lbs. 10 oz.	Standard Invector
3½"	26"	48"	10 lbs. 7 oz.	Standard Invector

SPECIFICATIONS GOLD 12 AND 20

Gauge	Model	Barrel Length	Overall Length	Average Weight	Chokes Available
12	Hunting	30"	50½"	7 lbs. 9 oz.	Invector-Plus
12	Hunting	28"	48½"	7 lbs. 6 oz.	Invector-Plus
12	Hunting	26"	46½"	7 lbs. 3 oz.	Invector-Plus
20	Hunting	28"	48¼"	6 lbs. 14 oz.	Invector
20	Hunting	26"	46¼"	6 lbs. 12 oz.	Invector

Total capacity for 3" loads is 3 shells in magazine, 1 in chamber; with 2½" loads, 4 in magazine, 1 in chamber. Gold 12 ga. models have vent recoil pad; 20 ga. models have solid pad.

BROWNING AUTOMATIC SHOTGUNS

AUTO 5 STALKER

SPECIFICATIONS AUTO-5 SHOTGUNS

Model	Chamber	Barrel Length	Overall Length	Average Weight	Chokes Available
12 GAUGE Light	2³/₄″	30″	49¹/₂″	8 lbs. 7 oz.	Invector-Plus
Light	2³/₄″	28″	47¹/₂″	8 lbs. 4 oz.	Invector-Plus
Light	2³/₄″	26″	45¹/₂″	8 lbs. 1 oz.	Invector-Plus
Lt. Buck Special	2³/₄″	24″	43¹/₂″	8 lbs.	Slug/buckshot
Light	2³/₄″	22″	41¹/₂″	7 lbs. 13 oz.	Invector-Plus
Magnum	3″	32″	51¹/₄″	9 lbs. 2 oz.	Invector-Plus
Magnum	3″	30″	49¹/₄″	8 lbs. 13 oz.	Invector-Plus
Magnum	3″	28″	47¹/₄″	8 lbs. 11 oz.	Invector-Plus
Magnum	3″	26″	45¹/₄″	8 lbs. 9 oz.	Invector-Plus
Mag. Buck Special	3″	24″	43¹/₄″	8 lbs. 8 oz.	Slug/buckshot
Light Stalker	2³/₄″	28″	47¹/₂″	8 lbs. 4 oz.	Invector-Plus
Light Stalker	2³/₄″	26″	45¹/₂″	8 lbs. 1 oz.	Invector-Plus
Magnum Stalker	3″	30″	49¹/₄″	8 lbs. 13 oz.	Invector-Plus
Magnum Stalker	3″	28″	47¹/₄″	8 lbs. 11 oz.	Invector-Plus
20 GAUGE Light	2³/₄″	28″	47¹/₈″	6 lbs. 10 oz.	Invector
Light	2³/₄″	26″	45¹/₄″	6 lbs. 8 oz.	Invector
Magnum	3″	28″	47¹/₄″	7 lbs. 3 oz.	Invector
Magnum	3″	26″	45¹/₄″	7 lbs. 1 oz.	Invector

Prices:
AUTO-5 MODELS
Light 12, Hunting & Stalker, Invector Plus $772.95
Light 20, Hunting, Invector Plus 772.95
3″ Magnum 12, Hunting & Stalker, Invector Plus . . . 795.95
3″ Magnum 12, Hunting, Invector Plus 795.95
3″ Magnum 20, Hunting, Std. Invector 704.95
Light 12, Buck Special 762.95
3″ Magnum 12 ga. Buck Special 785.95
Extra Barrels $194.95–282.95

BT-99 STANDARD SINGLE BARREL TRAP

SPECIFICATIONS
Gauge: 12 ga. w/2³/₄″ chamber   **Choke:** Invector-Plus
Barrel lengths: 32″ and 34″; ported barrel
Overall length: 48¹/₂″ w/32″ barrel; 50¹/₂″ w/34″ barrel
Weight: 8 lbs. 6 oz. to 8 lbs. 10 oz.
Stock: Conventional or Monte Carlo
Prices: (w/Conventional or Monte Carlo Stock)
Grade I Competition $1288.00
Stainless 1738.00
Pigeon Grade 1505.00
Signature Painted 1323.00
Golden Clays, Invector Plus, ported bbl. 2800.00

SPECIFICATIONS BT-99 PLUS/BT-99 MICRO PLUS

Model	Barrel Length	Overall Length	Average Weight
Micro BT-99 Plus	34″	50¹/₂″	8 lbs. 10 oz.
Micro BT-99 Plus	32″	48¹/₂″	8 lbs. 8 oz.
Micro BT-99 Plus	30″	46¹/₂″	8 lbs. 6 oz.
Micro BT-99 Plus	28″	44¹/₂″	8 lbs. 4 oz.
BT-99 Plus	34″	51¹/₄″	8 lbs. 12 oz.
BT-99 Plus	32″	49¹/₄″	8 lbs. 10 oz.
Citori Plus	32″	49¹/₄″	9 lbs. 7 oz.
Citori Plus	30″	47¹/₄″	9 lbs. 5 oz.

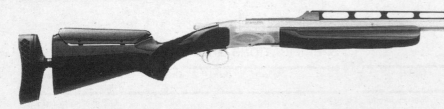

BT-99 PLUS & BT 99 MICRO PLUS

These handsome 12-gauge trap guns come with 28″, 30″, 32″ or 34″ ported barrels (unless specified otherwise) and Invector-Plus choke (2³/₄″ chamber). See specifications chart for additional information.

BT-99 PLUS TRAP STAINLESS STEEL

Prices:
Grade I $1855.00
Grade I without ported barrels 1835.00
Stainless 2240.00
Pigeon Grade 2065.00
Signature Painted 1890.00
Golden Clays, Invector Plus, ported bbl. 3205.00

BROWNING CITORI SHOTGUNS

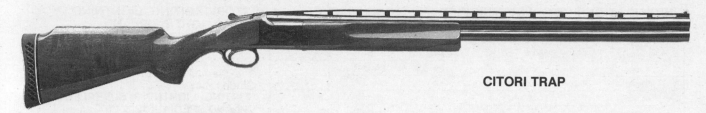

CITORI TRAP

Models	Prices
TRAP MODELS (High Post Target Rib)	
12 Ga. Monte Carlo/Conventional Stock	
Grade I Invector-Plus	$1450.00
Grade III Invector-Plus	1995.00
Grade VI Invector-Plus	2555.00
Golden Clays	2965.00
SKEET MODELS (High Post Target Rib)	
Standard 12 and 20 Gauge	
Grade I Invector-Plus	1450.00
Grade III Invector-Plus	1995.00
Grade VI Invector-Plus	
(12 ga.)	2550.00
Golden Clays	2965.00
Standard 20, 28 Ga., .410 Bore	
Grade I Std. Invector	1420.00
Grade I 20 ga.	1250.00
Grade III Std. Invector	2000.00
Grade III Std. Invector 20 ga. or	
28 ga. and .410 SK/SK	1767.00
Grade VI SK/SK or	
Std. Invector	2518.00
Golden Clays 20 ga.,	
Std. Invector	2518.00
Golden Clays 28 ga., .410	2900.00
Golden Clays SK/SK	2518.00
CITORI PLUS TRAP 12 Ga.	
With Ported Barrels	2030.00
Without Porting	2005.00
Pigeon Grade	2225.00
Signature Painted	2065.00
Golden Clays	3435.00

SPECIFICATIONS CITORI TRAP & SKEET

Gauge	Model	Chamber	Barrel Length	Overall Length	Average Weight	Chokes	Grades Available
CITORI TRAP							
12	Conventional	2¾"	32"	49"	8 lbs. 11 oz.	Invector-Plus	I, III, Golden Clays
12	Monte Carlo	2¾"	32"	49"	8 lbs. 10 oz.	Invector-Plus	I, III, Golden Clays
12	Conventional	2¾"	30"	47"	8 lbs. 7 oz.	Invector-Plus	I, III, Golden Clays
12	Monte Carlo	2¾"	30"	47"	8 lbs. 6 oz.	Invector-Plus	I, III, Golden Clays
CITORI SKEET							
12	Standard	2¾"	28"	45"	8 lbs.	Invector-Plus	I, III, Golden Clays
12	Standard	2¾"	26"	43"	7 lbs. 15 oz.	Invector-Plus	I, III, Golden Clays
20	Standard	2¾"	28"	45"	7 lbs. 4 oz.	Invector-Plus	I, III, Golden Clays
20	Standard	2¾"	26"	43"	7 lbs. 1 oz.	Invector-Plus	I, III, Golden Clays
28	Standard	2¾"	28"	45"	6 lbs. 15 oz.	Invector	I, III, Golden Clays
28	Standard	2¾"	26"	43"	6 lbs. 10 oz.	Invector	I, III, Golden Clays
.410	Standard	2¾"	28"	45"	7 lbs. 6 oz.	Invector	I, III, Golden Clays
.410	Standard	2¾"	26"	43"	7 lbs. 3 oz.	Invector	I, III, Golden Clays
20,28,.410	3 Gauge Set	2¾"	28"	45"	7 lbs. 4 oz.	Invector-Plus, Invector	I, III, Golden Clays
12,20,28,.410	4 Gauge Set	2¾"	28"	45"	8 lbs. 5 oz.	Invector-Plus, Invector	I, III, Golden Clays
12,20,28,.410	4 Gauge Set	2¾"	26"	43"	8 lbs. 3 oz.	Invector-Plus, Invector	I, III, Golden Clays

F=Full, M=Modified, IM=Improved Modified, S=Skeet, Invector=Invector Choke System — Invector & Invector-Plus Trap models: Full, Improved Modified, Modified, and wrench included. Skeet models: Skeet and Skeet tubes only, wrench included. Sets; 12 gauge Invector-Plus, Ported.

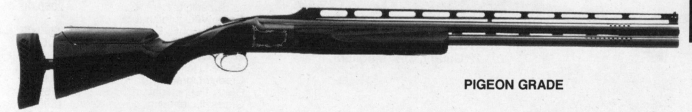

PIGEON GRADE

CITORI PLUS COMBO TRAP O/U

Browning's Citori Plus Combo is two trap guns in one—a single barrel trap gun for singles and an over/under trap gun for doubles. It features Browning's recoil reducer system, fully adjustable stock dimension, back-bored barrels, optional barrel porting, ventilated side ribs, and the Invector-Plus choke tube system.

Available with set of 32" o/u barrels and a 34" single trap barrel or a model with 30" o/u barrels and 32" or 34" single trap barrel. Features include back-bored barrel, blued steel receiver with engraved rosette scroll design, automatic ejectors, barrel selector in top tang safety; gold-colored trigger, manual top tang safety

Price:
Plus Combo Grade I Invector-Plus with ported barrels (furnished w/single barrel plus standard o/u barrels) and fitted luggage case **$3435.00**
Same as above in **Golden Clays** 5200.00

BROWNING CITORI O/U SHOTGUNS

SPECIFICATIONS CITORI OVER/UNDER SHOTGUNS

Gauge	Model	Chamber	Barrel Length	Overall Length	Average Weight	Chokes Available	Grades Available
12	Hunting	3½" Mag.	30"	47"	8 lbs. 10 oz.	Invector-Plus	I
12	Hunting	3½" Mag.	28"	45"	8 lbs. 9 oz.	Invector-Plus	I
12	Hunting	3"	30"	47"	8 lbs. 4 oz.	Invector-Plus	I
12	Hunting	3"	28"	45"	8 lbs. 1 oz.	Invector-Plus	I, III, VI
12	Hunting	3"	26"	43"	7 lbs. 15 oz.	Invector-Plus	I, III, VI
12	Lightning	3"	28"	45"	8 lbs. 1 oz.	Invector-Plus	I, GL, III, VI
12	Lightning	3"	26"	43"	7 lbs. 15 oz.	Invector-Plus	I, GL III, VI
12	Superlight	2¾"	28"	45"	6 lbs. 12 oz.	Invector-Plus	I, III, VI
12	Superlight	2¾"	26"	43"	6 lbs. 10 oz.	Invector-Plus	I, III, VI
12	Upland Special	2¾"	24"	41"	6 lbs. 11 oz.	Invector-Plus	I
20	Hunting	3"	28"	45"	6 lbs. 12 oz.	Invector	I, III, VI
20	Hunting	3"	26"	43"	6 lbs. 10 oz.	Invector	I, III, VI
20	Lightning	3"	28"	45"	6 lbs. 14 oz.	Invector	I, GL, III, VI
20	Lightning	3"	26"	43"	6 lbs. 9 oz.	Invector	I, GL, III, VI
20	Lightning	3"	24"	41"	6 lbs. 6 oz.	Invector	I
20	Micro Lightning	2¾"	24"	41"	6 lbs. 3 oz.	Invector	I, III, VI
20	Superlight	2¾"	26"	43"	6 lbs.	Invector	I, III, VI
20	Upland Special	2¾"	24"	41"	6 lbs.	Invector	I
28	Hunting	2¾"	28"	45"	6 lbs. 11 oz.	M-F	I
28	Hunting	2¾"	26"	43"	6 lbs. 10 oz.	IC-M	I
28	Lightning	2¾"	28"	45"	6 lbs. 11 oz.	M-F	I
28	Lightning	2¾"	26"	43"	6 lbs. 10 oz.	IC-M	I, III, VI
28	Superlight	2¾"	26"	43"	6 lbs. 10 oz.	IC-M	I, III, VI
.410	Hunting	3"	28"	45"	7 lbs.	M-F	I
.410	Hunting	3"	26"	43"	6 lbs. 14 oz.	IC-M	I
.410	Lightning	3"	28"	45"	7 lbs.	M-F	I
.410	Lightning	3"	26"	43"	6 lbs. 14 oz.	IC-M	I, III, VI
.410	Superlight	3"	28"	45"	6 lbs. 14 oz.	M-F	I
.410	Superlight	3"	26"	43"	6 lbs. 13 oz.	IC-M	I, III, VI

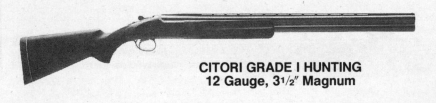

**CITORI GRADE I HUNTING
12 Gauge, 3½" Magnum**

SKEET SETS
4-BARREL SKEET SET
12 gauge with one removable forearm and four sets of barrels, 12, 20, 28 and .410 gauges, high post target rib. (Furnished with fitted luggage case for gun and extra barrels).
Grade 1, Invector System **$4275.00**
Grade I, Skeet & Skeet **4038.00**
Grade III, Invector System **5250.00**
Grade III, Skeet & Skeet **4617.00**
Grade VI, Skeet & Skeet **5225.00**
Golden Clays, Invector System **6480.00**

CITORI HUNTING, LIGHTNING, SUPERLIGHT & UPLAND SPECIAL MODELS

**HUNTING 12 Gauge
3½" Magnum, Invector Plus**
Grade I Hunting **$1303.00**
HUNTING, LIGHTNING, SUPERLIGHT & UPLAND SPECIAL: 12 & 20 Ga., Invector-Plus
Grade I Hunting **$1228.00**
Grade I Lightning 1263.00
Grade I Superlight & Upland
Special 1278.00
Grade I 20 ga. Micro
Lightning 1315.00
Gran Lightning 1712.00
Grade III Hunting 1802.00
Grade III Lightning & Superlight . 1838.00
Grade III 20 ga. Micro
Lightning 1850.00
Grade VI Lightning &
Superlight 2672.00
Grade VI 20 ga. Micro
Lightning 2680.00
HUNTING, LIGHTNING, SUPERLIGHT & UPLAND SPECIAL: 20 Ga., Standard Invector
Grade I Hunting **$1107.00**
Grade I Lightning 1138.00
Grade I Micro Lightning 1167.00
Grade I Superlight & Upland
Special 1154.00
Gran Lightning 1549.00
Grade III Hunting 1630.00
Grade III Lightning 1658.00
Grade III Superlight 1686.00
Grade VI Micro Lightning 1663.00
LIGHTNING AND SUPERLIGHT: 28 Gauge and .410 Bore, Standard Invector
Grade I Lightning **$1305.00**
Grade I Superlight 1325.00
Gran Lightning 1800.00
Grade III Lightning &
Superlight 2050.00
Grade VI Lightning &
Superlight 2880.00
HUNTING, LIGHTNING AND SUPERLIGHT: 28 Gauge and .410 Bore
Grade I Hunting **$1097.00**
Grade I Lightning 1140.00
Grade I Superlight 1159.00
Grade III Lightning 1805.00
Grade III Superlight 1824.00
Grade VI Lightning and
Superlight 2560.00

3-BARREL SKEET SET
20 gauge with one removable forearm and three sets of barrels, 20, 28 and .410 bore.
Grade 1, Invector System **$2975.00**
Grade III, Invector System 3750.00
Grade III, Skeet & Skeet 3382.00
Grade VI, Skeet & Skeet 3990.00
Golden Clays, Invector System 4875.00

BROWNING CITORI O/U SHOTGUNS

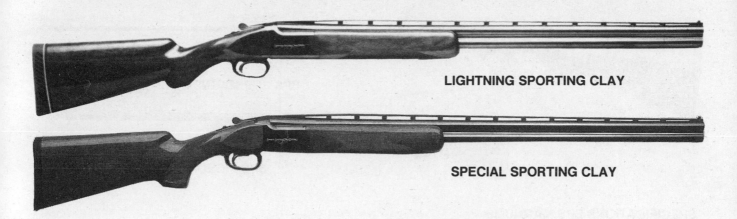

LIGHTNING SPORTING CLAY

SPECIAL SPORTING CLAY

Prices: (w/Invector-Plus Chokes)

Special Sporting (Grade I)* **$1430.00**
Special Sporting, Signature Painted 1465.00
Special Sporting, Pigeon Grade 1630.00
Special Sporting, Grade I, Adj. Comb 1550.00
Special Sporting Golden Clays 2930.00
Lightning Sporting Grade I 1370.00
Lightning Sporting, Gr. I, 3" chamber 1430.00
Lightning Sporting Signature Painted 1405.00
Lightning Sporting, Signature Painted,
    3" chamber 1465.00
    Same as above with low rib 1330.00
Lightning Sporting, Pigeon Gr., 3" chamber 1630.00
    Same as above with low rib 1566.00
Lightning Sporting Golden Clays 2930.00

SPECIFICATIONS CITORI O/U SPORTING CLAYS

Model	Chamber	Barrel Length	Overall Length	Average Weight
Special Sporting	2¾"	32"	49"	8 lbs. 5 oz.
Special Sporting	2¾"	30"	47"	8 lbs. 3 oz.
Special Sporting	2¾"	28"	45"	8 lbs. 1 oz.
Lightning Sporting	3"	30"	47"	8 lbs. 8 oz.
Lightning Sporting	3"	28"	45"	8 lbs. 6 oz.

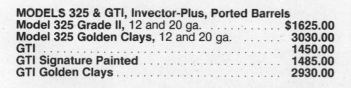

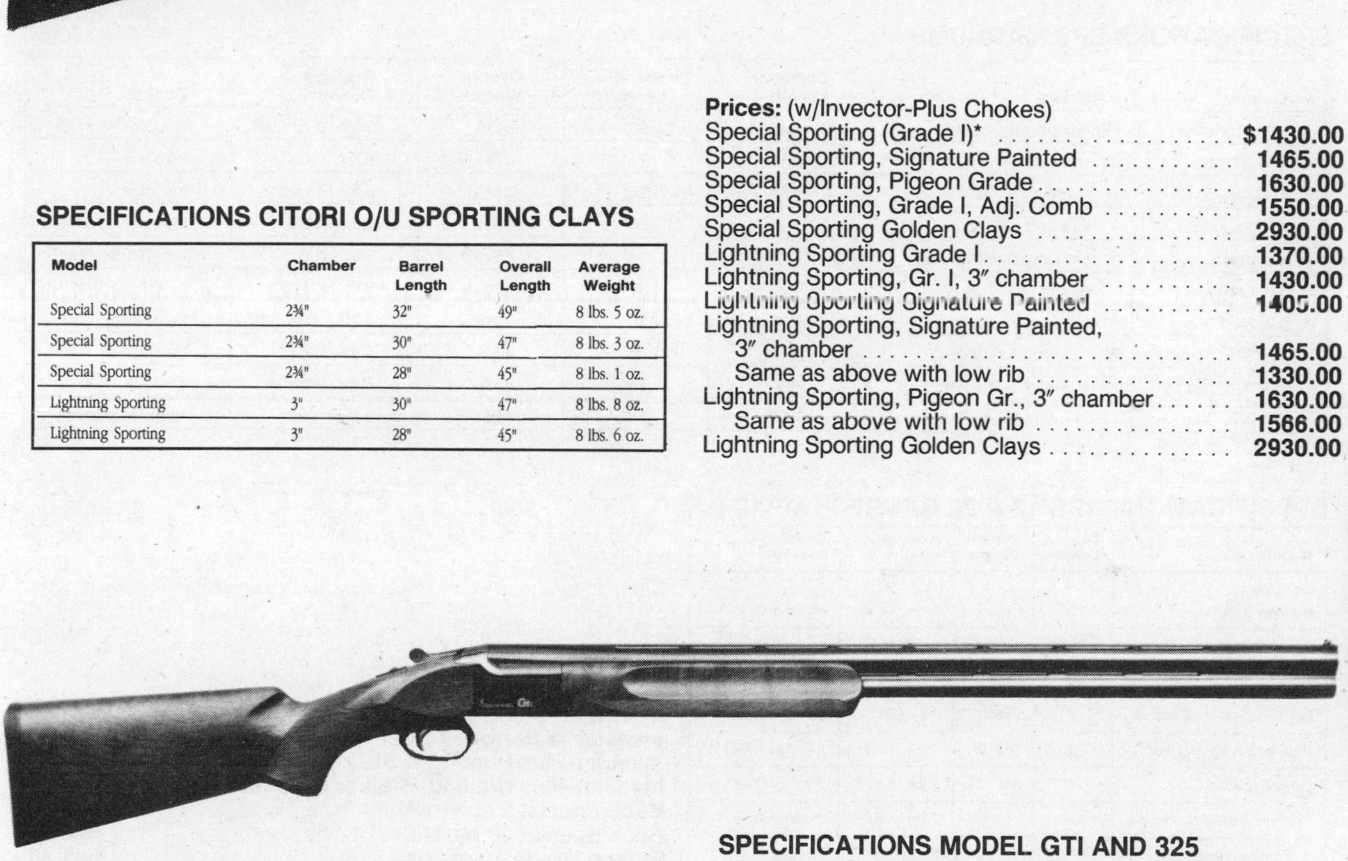

GTI SPORTING CLAY

MODELS 325 & GTI, Invector-Plus, Ported Barrels
Model 325 Grade II, 12 and 20 ga. **$1625.00**
Model 325 Golden Clays, 12 and 20 ga. 3030.00
GTI 1450.00
GTI Signature Painted 1485.00
GTI Golden Clays 2930.00

SPECIFICATIONS MODEL GTI AND 325

Model	Chamber	Barrel Length	Overall Length	Average Weight
GTI 12 ga.	2¾"	30"	47"	8 lbs. 2 oz.
GTI 12 ga.	2¾"	28"	45"	8 lbs.
325 12 ga.	2¾"	32"	49½"	7 lbs. 15 oz.
325 12 ga.	2¾"	30"	47½"	7 lbs. 14 oz.
325 12 ga.	2¾"	28"	45½"	7 lbs. 13 oz.
325 20 ga.	2¾"	30"	47½"	6 lbs. 13 oz.
325 20 ga.	2¾"	28"	45½"	6 lbs. 12 oz.

Invector-Plus=back-bored barrel, interchangeable choke system. 1 Mod. 1 Imp. Cyl. and 1 Skeet tube supplied.

BROWNING SHOTGUNS

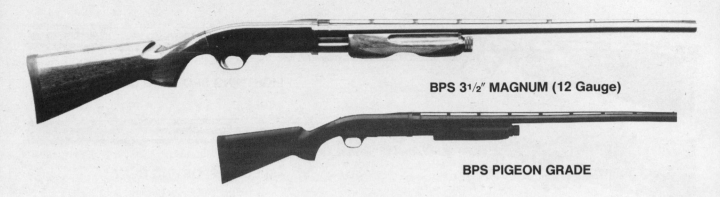

BPS 3½" MAGNUM (12 Gauge)

BPS PIGEON GRADE

SPECIFICATIONS BPS MAGNUMS

Gauge	Model	Chamber Length	Barrel Length	Overall Weight	Average Available	Chokes
10 Magnum	Hunting & Stalker	3½"	30"	51¾"	9 lbs. 8 oz.	Invector
10 Magnum	Hunting & Stalker	3½"	28"	49¾"	9 lbs. 6 oz.	Invector
10 Magnum	Hunting & Stalker	3½"	26"	47¾"	9 lbs. 4 oz.	Invector
10 Magnum	Hunting & Stalker	3½"	24"	45¾"	9 lbs. 4 oz.	Invector
10 Magnum	Hunting Buck Special	3½"	24"	45¾"	9 lbs. 2 oz.	Slug/Buckshot
12, 3½" Mag	Hunting & Stalker	3½"	30"	51¾"	8 lbs. 12 oz.	Invector-Plus
12, 3½" Mag	Hunting & Stalker	3½"	28"	49¾"	8 lbs. 9 oz.	Invector-Plus
12, 3½" Mag	Hunting & Stalker	3½"	26"	47¾"	8 lbs. 6 oz.	Invector-Plus
12, 3½" Mag	Hunting & Stalker	3½"	24"	45¾"	8 lbs. 3 oz.	Invector-Plus
12, 3½" Mag	Hunting Buck Special	3½"	24"	45¾"	8 lbs. 7 oz.	Slug/Buckshot

SPECIFICATIONS BPS 12 & 20 GAUGE PUMP (3")

Model	Barrel Length	Overall Length	Average Weight	Chokes Available
12 GAUGE Hunting & Pigeon	32"	52½"	7 lbs. 14 oz.	Invector-Plus
Hunting, Stalker & Pigeon	30"	50¾"	7 lbs. 12 oz.	Invector-Plus
Hunting, Stalker & Pigeon	28"	48¾"	7 lbs. 11 oz.	Invector-Plus
Hunting, Stalker & Pigeon	26"	46¾"	7 lbs. 10 oz.	Invector-Plus
Standard Buck Special	24"	44¾"	7 lbs. 10 oz.	Slug/Buckshot
Upland Special	22"	42½"	7 lbs. 8 oz.	Invector-Plus
Hunting, Stalker & Pigeon	22"	42½"	7 lbs. 7 oz.	Invector-Plus
Game Gun Turkey Special	20½"	40⅞"	7 lbs. 7 oz.	Invector
Game Gun Deer Special	20½"	40⅞"	7 lbs. 7 oz.	Special Inv./Rifled
20 GAUGE Hunting	28"	48¾"	7 lbs. 1 oz.	Invector-Plus
Hunting	26"	46¾"	7 lbs.	Invector-Plus
Youth/Ladies	22"	41¾"	6 lbs. 11 oz.	Invector-Plus
Upland Special	22"	42½"	6 lbs. 12 oz.	Invector-Plus
28 GAUGE Hunting*	28"	48¾"	7 lbs. 1 oz.	Invector
Hunting*	26"	46¾"	7 lbs.	Invector

* 2¾" chamber

Prices:

Invector Hunting & Stalker 10 ga.	**$615.95**
Invector Waterfowl 10 ga.	787.95
Invector-Plus Hunting & Stalker 12 ga.	615.95
Invector-Plus Hunting, Stalker & Upland Spec.	489.95
Buck Special 12 ga. only	474.95
Buck Special 10 ga. & 3½" 12 ga.	620.95
Pigeon Grade 12 ga. only	651.95
Deer Special Game Gun	554.95
Turkey Special Game Gun	524.95
BPS Youth, Ladies & Upland Special 20 ga. only	398.95
Hunting, Youth, Ladies, Upland Spec. 20 ga.	489.95
Hunting (Standard Invector), 20 ga.	489.95

BROWNING SHOTGUNS

MODEL B-125 CUSTOM
A STYLE

BROWNING CUSTOM GUNS

The original **Superposed B-25,** which has been made since 1931 entirely in Herstal, Belgium, was originally designed by John M. Browning in 1925. It was meant to be, in his own words, "The best gun I ever invented." Consequently, the Browning Superposed design benefits from a culmination of decades of ingenuity and firearms experience.

When Browning died in Belgium shortly before the design was completed on the Superposed, his son, Val Browning, completed the gun. Its design reflects Browning's knowledge of American hunters. He knew they preferred a single sighting plane, a single selective trigger and perfect balance. The Superposed B-125 has the same qualities and features of the original Superposed, including single selective trigger, auto-

matic ejectors, non-detachable sliding forearm and full-width hinge pin. But its parts are sourced worldwide to take advantage of low costs. The B-125 is assembled and finished at Herstal's Custom Gun Shop and is stamped with the legendary, "Made in Belgium."

Prices: B-125 Over/Under

12 Gauge Trap, F-1 Style	$ 4,350.00
12 Gauge Sporting A Style	4,250.00
B Style	4,550.00
C Style	4,900.00
12 & 20 Gauge Hunting, A Style	4,150.00
B Style	4,450.00
C Style	4,850.00
Pigeon Grade	7,100.00
Pointer Grade	8,500.00
Diana Grade	9,100.00
Midas Grade	12,500.00

RECOILLESS TRAP
$1850.00
($1900.00 w/Browning Signature)

SPECIFICATIONS
Gauge: 12, Standard or Micro
Chamber: $2^3/_4$"
Barrel length: 27" or 30"
Overall length: $47^5/_8$" (27" barrel); $51^5/_8$" (30" barrel)
Weight: 8 lbs. 10 oz.

CHURCHILL SHOTGUNS

TURKEY AUTOMATIC SHOTGUNS
$569.95

SPECIFICATIONS
Gauge: 12   **Chokes:** ICT choke system
Barrel lengths: 25" (Standard & Turkey); 26" and 28" (Standard only)

Stock: Hand-checkered walnut w/satin finish
Features: Magazine cut-off (shoots all loads interchangeably w/o alterations); non-glare finish; vent rib w/mid-bead; gold trigger; receiver has engraved turkey scene

CONNECTICUT VALLEY CLASSICS

CLASSIC SPORTER

CVC CLASSIC 101 SPORTER

The designers of the new CVC "Classic Sporter" and "Classic Waterfowler" trace their lineage back to the well-known "Classic Doubles" and Winchester over/under shotguns. They have used the proven strength and durability of the old M-101 design and integrated these qualities with advanced engineering and manufacturing techniques. In addition to the basic specifications listed below, the CVC Classic models feature the following: Frame, monoblock and key integral parts are machined from solid steel bar stock. . . Tang spacer is an integral part of the frame to ensure rigid alignment for solid lockup of buttstock to frame. . . Chrome molybdenum steel barrels; chrome-lined bores and chambers (suitable for steel shot use). . . Barrels have elongated forcing cones for reduced recoil; interchangeable screw-in chokes are included. . . Stock and forend are full, fancy-grade American black walnut, hand-checkered with a low-luster satin finish and fine-line engraving. . . Waterfowler model has non-reflective surface.

Prices:

Classic Sporter Stainless	2750.00
Waterfowler	2195.00

CONNECTICUT VALLEY CLASSICS SPECIFICATIONS

Model	Symbol	Gauge	Barrel Length	Overall Length	Length of Pull	Drop at Comb	Drop at Heel	Nominal Weight (lbs.)
Classic Sporter-Stainless	CV-SS28	12	28″	44⁷⁄₈″	14¹⁄₂″	1¹⁄₂″	2¹⁄₈″	7³⁄₄
	CV-SS30	12	30″	46⁷⁄₈″	14¹⁄₂″	1¹⁄₂″	2¹⁄₈″	7³⁄₄
	CV-SS32	12	32″	48⁷⁄₈″	14¹⁄₂″	1¹⁄₂″	2¹⁄₈″	7³⁄₄
Classic Field-Waterfowler	CV-W30	12	30″	46⁷⁄₈″	14¹⁄₂″	1¹⁄₂″	2¹⁄₈″	7³⁄₄
	CV-W32	12	32″	48⁷⁄₈″	14¹⁄₂″	1¹⁄₂″	2¹⁄₈″	7³⁄₄

Chambers: All Connecticut Valley Classics have 3″ chambers. **Interchangeable Chokes:** All Connecticut Valley Classics have the internal CV Choke System, and each gun includes the following four chokes: Full, Modified, Improved Cylinder and Skeet. CV Choke Systems are fully compatible with previously manufactured Classic Doubles In-Choke and Winchoke interchangeable choke tubes.

Interchangeable Choke System Options: Two options are available at additional cost to the standard CV Choke System: CV Plus Choke System, a 2³⁄₈″ choke tube system for an extra ⁷⁄₈″ choke length using the standard CV choke tube barrel threading; and the Briley Competition Choke System, a factory-installed 2³⁄₄″ flush-mounted system.

EUROPEAN AMERICAN ARMORY

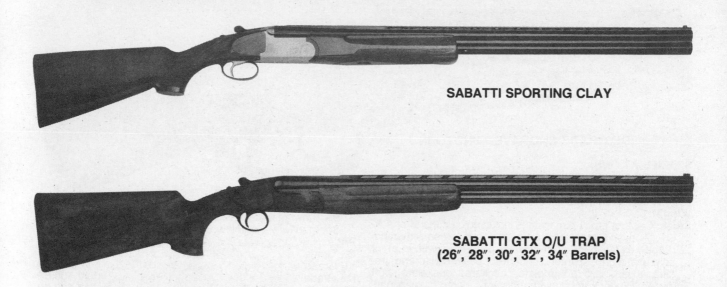

SABATTI SPORTING CLAY

SABATTI GTX O/U TRAP
(26″, 28″, 30″, 32″, 34″ Barrels)

SABATTI SPORTING CLAY PRO
$977.72   ($1014.00 Gold Model)

Barrels are hammer-forged from solid chromium/molybdenum steel and chrome-lined. Actions and monobloc receivers are nickel and nickel-steel, respectively. Stocks are made from European hardwoods. Gold model has gold-filled engravings on both sides of receiver.

SPECIFICATIONS
Gauge: 12 (2 3/4″ chamber)
Choke: Screw-in choke tubes
Barrel length: 28″ or 30″
Features: Hand-checkered walnut pistol grip with wide ejectors; wide vent rib with flourescent front sight; tang-mounted safety; rubber recoil pad; padded security case; full set of choke tubes and wrench included

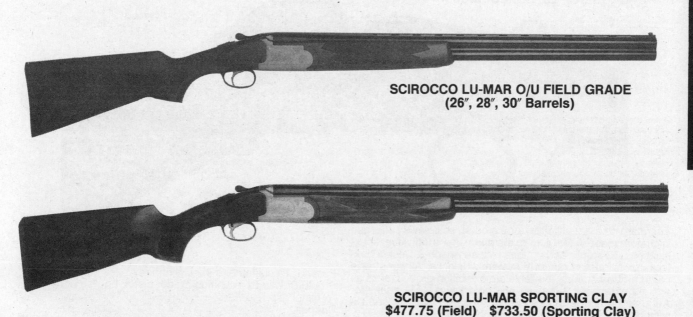

SCIROCCO LU-MAR O/U FIELD GRADE
(26″, 28″, 30″ Barrels)

SCIROCCO LU-MAR SPORTING CLAY

SCIROCCO LU-MAR SPORTING CLAY
$477.75 (Field) $733.50 (Sporting Clay)

Same specifications as Sabatti Shotgun, but with nickel-plated receiver and lightweight forend. Basic model (Field) has fixed chokes; Sporting Clay Model includes five screw-in chokes with single selective trigger and Raybar-style front sight.

A.H. FOX SHOTGUNS

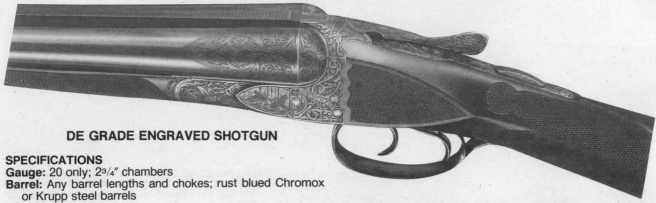

DE GRADE ENGRAVED SHOTGUN

SPECIFICATIONS

Gauge: 20 only; 2³/₄″ chambers
Barrel: Any barrel lengths and chokes; rust blued Chromox or Krupp steel barrels
Weight: 5¹/₂ to 7 lbs.
Stock: Custom stock dimensions including cast; hand-checkered Turkish Circassian walnut stock and forend with hand-rubbed oil finish; straight grip, full pistol grip (with cap), or semi-pistol grip; splinter, schnabel or beavertail forend; traditional pad, hard rubber plate, checkered, or skeleton butt
Features: Boxlock action with automatic ejectors; scalloped, rebated and color casehardened receiver; double or Fox single selective trigger; hand-finished and hand-engraved. This is the same gun that was manufactured between 1905 and 1930 by the A.H. Fox Gun Company of Philadelphia, PA, now manufactured in the U.S. by the Connecticut Shotgun Mfg. Co. (New Britain, CT).

Prices:

CE Grade*	$ 5,650.00
XE Grade	8,500.00
DE Grade	12,500.00
FE Grade	17,500.00
Exhibition Grade	25,000.00

* Grades differ in engraving and inlay, grade of wood and amount of hand finishing needed.

FRANCOTTE SHOTGUNS

CLOSE-UP OF BOXLOCK S6

"CUSTOM" BOXLOCKS/SIDELOCKS

There are no standard Francotte models, since every shotgun is custom made in Belgium to the purchaser's individual specifications. Features and options include Anson & Deeley boxlocks or Auguste Francotte system sidelocks. All guns have custom-fitted stocks. Available are exhibition-grade stocks as well as extensive engraving and gold inlays. U.S. agent for Auguste Francotte of Belgium is Armes de Chasse (see Directory of Manufacturers and Distributors).

SPECIFICATIONS

Gauges: 12, 16, 20, 28, .410; also 24 and 32
Chambers: 2¹/₂″, 2³/₄″ and 3″
Barrel length: To customer's specifications
Forend: To customer's specifications
Stock: Deluxe to exhibition grade; pistol, English or half--pistol grip

Prices:

Basic Boxlock	$ 9,710
Basic Boxlock (28 & .410 ga.)	10,710
Optional sideplates, add	1,158
Basic Sidelock	22,427
Basic Sidelock (28 & .410 ga.)	24,665

GARBI SIDELOCK SHOTGUNS

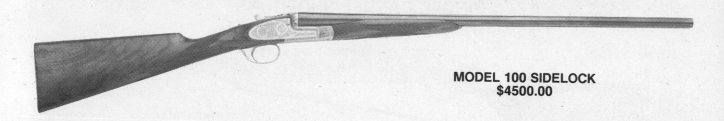

MODEL 100 SIDELOCK
$4500.00

MODEL 100 SIDELOCK

Like this Model 100 shotgun, all Spanish-made Garbi models featured here are Holland & Holland pattern sidelock ejector guns with chopper lump (demibloc) barrels. They are built to English gun standards with regard to design, weight, balance and proportions, and all have the characteristic "feel" associated with the best London guns. All of the models offer fine 24-line hand-checkering, with outstanding quality wood-to-metal and metal-to-metal fit. The Model 100 is available in 12, 16, 20 and 28 gauge and sports Purdey-style fine scroll and rosette engraving, partly done by machine.

MODELS 101, 103A and 103B (not shown)

Available in 12, 16, 20, and 28 gauge, the sidelocks are hand-crafted with hand-engraved receiver and select walnut straight grip stock.

SPECIFICATIONS
Barrels: 25" to 30" in 12 ga.; 25" to 28" in 16, 20 and 28 ga.; high-luster blued finish; smooth concave rib (optional Churchill or level, file-cut rib)
Action: Holland & Holland pattern sidelock; automatic ejectors; double triggers with front trigger hinged; case-hardened

Stock/forend: Straight grip stock with checkered butt (optional pistol grip); hand-rubbed oil finish; classic (splinter) forend (optional beavertail)
Weight: 12 ga. game, 6 lbs. 8 oz. to 6 lbs. 12 oz.; 12 ga. pigeon or wildfowl, 7 lbs.—7lbs. 8 oz.; 16 ga., 6 lbs. 4 oz. to 6 lbs. 10 oz.; 20 ga., 5 lbs. 15 oz.—6 lbs. 4 oz.; 28 ga., 5 lbs. 6 oz.—5 lbs. 10 oz.

Prices:

Model 101	$5750.00
Model 103A	7100.00
Model 103B	9900.00

MODEL 200
$8500.00

Also available:
MODEL 200 in 12, 16, 20 or 28 gauge; features Holland pattern stock ejector double, heavy-duty locks, Continental-style floral and scroll engraving, walnut stock.

HARRINGTON & RICHARDSON

SINGLE BARREL SHOTGUNS

TURKEY MAG
$169.95

SPECIFICATIONS
Gauge: 12 (3½" chamber); Turkey Full screw-in choke
Barrel length: 24"   **Overall length:** 40"
Weight: 6 lbs.
Sights: Bead sights
Stock & forearm: American hardwood with recoil pad; swivel and studs
Finish: Mossy Oak camo coverage and sling

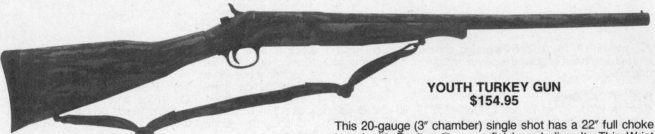

YOUTH TURKEY GUN
$154.95

This 20-gauge (3" chamber) single shot has a 22" full choke barrel with Realtree™ camo finish and sling. Its Thin Wrist stock is equipped with a recoil pad and semi-beavertail forend. Other features include a transfer bar safety and automatic ejection. The weight is only 6 lbs., making it ideal for young shooters to hunt all types of game, including upland and waterfowl.

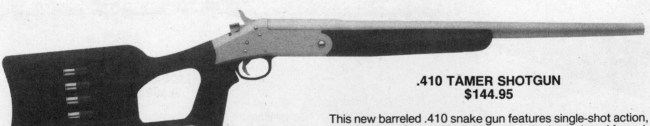

.410 TAMER SHOTGUN
$144.95

This new barreled .410 snake gun features single-shot action, transfer bar safety and high-impact synthetic stock and forend. Stock has a thumbhole design that sports a full pistol grip and a recessed open side, containing a holder for storing ammo. Forend is modified beavertail configuration. Other features include a matte, electroless nickel finish. **Weight:** 5–6 lbs. **Barrel length:** 20" (3" chamber). **Choke:** Full.

HARRINGTON & RICHARDSON

SINGLE BARREL SHOTGUNS

TOPPER MODEL 098
$109.95

SPECIFICATIONS
Gauges: 12, 20 and .410 (3″ chamber)
Chokes: Modified (12 and 20 ga.); Full (.410 ga.)
Barrel lengths: 26″ and 28″ **Weight:** 5 to 6 lbs.

Action: Break-open; side lever release; automatic ejection
Stock: Full pistol grip; American hardwood; black finish with white buttplate spacer
Length of pull: 14 1/2″

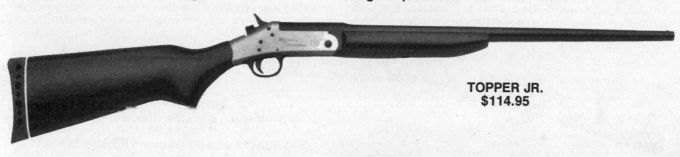

TOPPER JR.
$114.95

SPECIFICATIONS
Gauges: 20 and .410 (3″ chamber)
Chokes: Modified (20 ga.); Full (.410 ga.)

Barrel length: 22″ **Weight:** 5 to 6 lbs.
Stock: Full pistol grip; American hardwood; black finish; white line spacer; recoil pad
Finish: Satin nickel frame; blued barrel

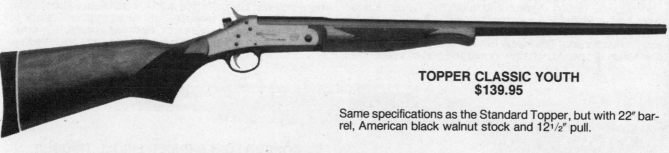

TOPPER CLASSIC YOUTH
$139.95

Same specifications as the Standard Topper, but with 22″ barrel, American black walnut stock and 12 1/2″ pull.

TOPPER DELUXE MODEL 098
$124.95

SPECIFICATIONS
Gauge: 12 (3 1/2″ chamber)
Chokes: Screw-in Modified (Full, Extra Full Turkey and Steel Shot also available)

Action: Break-open; side lever release; positive ejection
Barrel length: 28″ **Weight:** 5 to 6 lbs.
Stock: American hardwood, black finish, full pistol grip stock with semi-beavertail forend; white line spacer; ventilated recoil pad
Finish: Satin nickel frame; blued barrel

SHOTGUNS

IGA SHOTGUNS

COACH GUN
Available in 12 and 20 Gauge or .410 Bore
$382.00

The **IGA CLASSIC SIDE-BY-SIDE COACH GUN** sports a 20-inch barrel. Lightning fast, it is the perfect shotgun for hunting upland game in dense brush or close quarters. This endurance-tested workhorse of a gun is designed from the ground up to give you years of trouble-free service. Two massive underlugs provide a super-safe, vise-tight locking system for lasting strength and durability. The mechanical extraction of spent shells and double-trigger mechanism assures reliability. The automatic safety is actuated whenever the action is opened, whether or not the gun has been fired. The polish and blue is deep and rich, and the solid sighting rib is matte-finished for glare-free sighting. Chrome-moly steel barrels with micro-polished bores give dense, consistent patterns. The classic stock and forend are of durable hardwood . . . oil finished, hand-rubbed and hand-checkered.

Improved Cylinder/Modified choking and its short barrel make the IGA coach gun the ideal choice for hunting in close quarters, security and police work. Three-inch chambers.

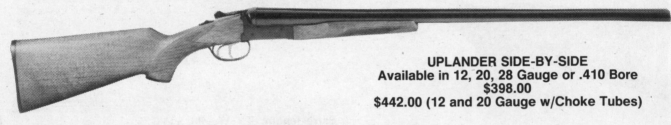

UPLANDER SIDE-BY-SIDE
Available in 12, 20, 28 Gauge or .410 Bore
$398.00
$442.00 (12 and 20 Gauge w/Choke Tubes)

The **IGA SIDE-BY-SIDE** is a rugged shotgun, endurance-tested and designed to give years of trouble-free service. A vise-tight, super-safe locking system is provided by two massive underlugs for lasting strength and durability. Two design features that make the IGA a standout for reliability are its positive mechanical extraction of spent shells and its traditional double-trigger mechanism. The safety is automatic in that every time the action is opened, whether or not the gun has been fired, the safety is actuated. The polish and bluing are deep and rich. The solid sighting rib carries a machined-in matte finish for glare-free sighting. Barrels are of chrome-moly steel with micro-polished bores to give dense, consistent patterns. The stock and forend are available with either traditional stock or the legendary English-style stock. Both are of durable Brazilian hardwood, oil-finished, hand-rubbed and hand-checkered.

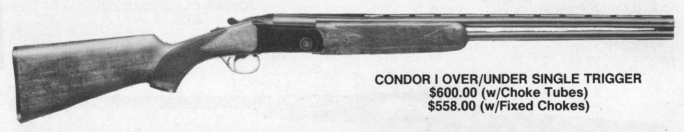

CONDOR I OVER/UNDER SINGLE TRIGGER
$600.00 (w/Choke Tubes)
$558.00 (w/Fixed Chokes)

The **IGA OVER/UNDER SINGLE TRIGGER** is a workhorse of a shotgun, designed for maximum dependability in heavy field use. The super-safe lock-up system makes use of a sliding underlug, the best system for over/under shotguns. A massive monobloc joins the barrel in a solid one-piece assembly at the breech end. Reliability is assured, thanks to the mechanical extraction system. Upon opening the breech, the spent shells are partially lifted from the chamber, allowing easy removal by hand. IGA barrels are of chrome-moly steel with micro-polished bores to give tight, consistent patterns. They are specifically formulated for use with steel shot where Federal migratory bird regulations require. Atop the barrel is a sighting rib with an anti-glare surface. The buttstock and forend are of durable hardwood, hand-checkered and finished with an oil-based formula that takes dents and scratches in stride.

The IGA **Condor I** over/under shotgun is available in 12 and 20 gauge with 26- and 28-inch barrels with choke tubes and 3-inch chambers; 12 and 20 gauge with 26- and 28-inch barrels choked IC/M and Mod./Full, 3-inch chambers.

Also available:
Condor II O/U in 12 gauge, double trigger, 26″ barrel IC/M or 28″ barrel M/F. **Price:** **$458.00**

IGA SHOTGUNS

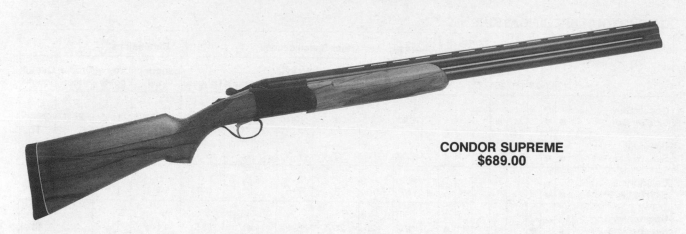

CONDOR SUPREME
$689.00

The IGA Condor Supreme truly complements its name. The stock is selected from upgraded Brazilian walnut, and the hand-finished checkering is sharp and crisp. A matte-laquered finish provides a soft warm glow, while maintaining a high resistance to dents and scratches.

A massive monoblock joins the barrel in a solid one-piece assembly at the breech end. Upon opening the breech, the automatic ejectors cause the spent shells to be thrown clear of the gun. The barrels are of moly-chrome steel with micro-polished bores to give tight, consistent patterns; they are specifically formulated for use with steel shot. Choke tubes are provided. Atop the barrel is a sighting rib with an anti-glare surface with both mid- and front bead. See table on the following page for additional specifications.

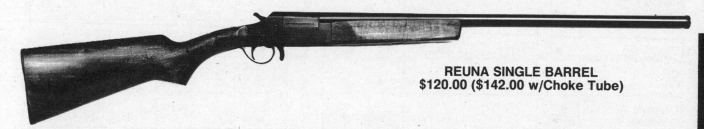

REUNA SINGLE BARREL
$120.00 ($142.00 w/Choke Tube)

IGA's entry-level single-barrel shotgun features a feeling of heft and quality not found in other shotguns similarly priced. Single mechanical extraction makes for convenient removal of spent shells. For ease of operation and maximum safety, the Reuna is equipped with an exposed hammer, which must be cocked manually before firing.

The Reuna single-barrel shotgun is available with a choke tube in 12 and 20 gauge and with fixed chokes in 12 and 20 gauge or .410 bore. Both the buttstock and semi-beavertail forearm are of durable Brazilian hardwood. The squared-off design of the firearm enhances stability and provides an additional gripping surface for greater comfort.

REUNA YOUTH MODEL SINGLE BARREL
$132.00

The Youth Model is designed especially for the young shooter. All the same features of the conventional-sized model are included in the youth version, complemented by an easy-handling shorter barrel (22″), shortened stock and ventilated recoil pad. In 20 gauge and .410 (Full choke).

IGA SHOTGUNS

IGA SHOTGUN SPECIFICATIONS

	Gauge				Barrel Length					Chokes		Other Specifications				Dimensions			
	12	20	28	410	20"	22"	24"	26"	28"	Fixed	Choke tubes	Chamber	Weight (lbs.)	Extractors	Triggers	Length of pull	Drop at comb	Drop at heel	Overall length
Coach Gun Side by Side	■	■		■	■					IC/M		3"	6¾	■	D.T.	14½"	1½"	2½"	36½"
Uplander Side by Side	■	■						■		IC/M	IC/M	3"	7½	■	D.T.	14½"	1½"	2½"	42"
Uplander Side by Side	■	■							■	M/F	M/F	3"	7½	■	D.T.	14½"	1½"	2½"	44"
Uplander Side by Side			■					■		IC/M		2¾"	6¾	■	D.T.	14½"	1½"	2½"	42"
Uplander Side by Side				■				■		F/F		3"	6¾	■	D.T.	14½"	1½"	2½"	42"
English Stock Side by Side		■					■	■		IC/M		3"	6½	■	D.T.	14½"	1⅜"	2⅜"	40"/42"
Condor Supreme*	■	■						■	■	IC/M M/F		3"	8		S.T.	14½"	1½"	2½"	43½"
Condor I Over/Under	■	■						■		IC/M	IC/M	3"	8	■	S.T.	14½"	1½"	2½"	43½"
Condor I Over/Under	■	■							■	M/F	M/F	3"	8	■	S.T.	14½"	1½"	2½"	45½"
Condor II Over/Under	■							■		IC/M		3"	8	■	D.T.	14½"	1½"	2½"	43½"
Condor II Over/Under	■								■	M/F		3"	8	■	D.T.	14½"	1½"	2½"	45½"
Reuna Single Barrel	■								■	F	F	3"	6¼	■		14½"	1½"	2½"	44½"
Reuna Single Barrel	■							■		M		3"	6¼	■		14½"	1½"	2½"	42½"
Reuna Single Barrel		■						■		F	F	3"	6¼	■		14½"	1½"	2½"	42½"
Reuna Single Barrel			■					■		F		3"	6	■		14½"	1½"	2½"	42½"
Reuna-Youth Model Single Barrel		■		■		■				F		3"	5	■		13	1½"	2½"	37

* Condor Supreme equipped with automatic ejectors.

ITHACA SHOTGUNS

For Specifications, see following page.

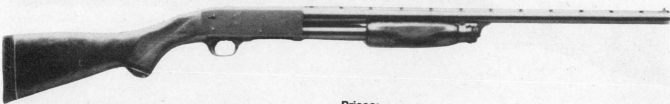

MODEL 87 FIELD GRADES

Made in much the same manner as 50 years ago, Ithaca's Model 37 pump (now designated as Model 87) features Roto-forged barrels hammered from 11″ round billets of steel, then triple-reamed, lapped and polished. The receivers are milled from a solid block of ordnance grade steel, and all internal parts—hammer, extractors, slides and carriers—are milled and individually fitted to each gun.

Prices:

Model 87 Basic	$477.00
Model 87 Supreme	808.50
Model 87 Deluxe Vent	533.25
Model 87 Camo Field	542.00
Model 87 English	545.50
Model 87 Turkey Field w/Tube	508.25
Model 87 Turkey Field w/Fixed Choke	465.75
Model 87 Turkey Field Camo w/Tube	550.75
Model 87 Turkey Field Camo w/Fixed Choke	508.25

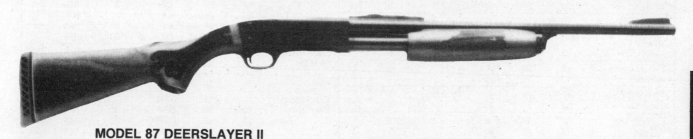

MODEL 87 DEERSLAYER II

The first shotgun developed to handle rifled slugs successfully, the Deerslayer's design results in an ''undersized'' cylinder bore—from the forcing cone all the way to the muzzle. This enables the slug to travel smoothly down the barrel with no gas leakage or slug rattle. The new Deerslayer II features the world's first production rifled barrel for shotguns; moreover, the Deerslayer's barrel is permanently screwed into the receiver for solid frame construction, which insures better accuracy to about 85 yards.

Prices:

Deerslayer Smoothbore Basic	$424.50
Deerslayer Smoothbore Deluxe	464.75
Deerslayer DSR Deluxe	498.25
Deerslayer II Basic	464.75
Deerslayer II Brenneke	566.50

SHOTGUNS

MODEL 87 FIELD TURKEY GUN
(Camo-seal Finish)

ITHACA SHOTGUNS

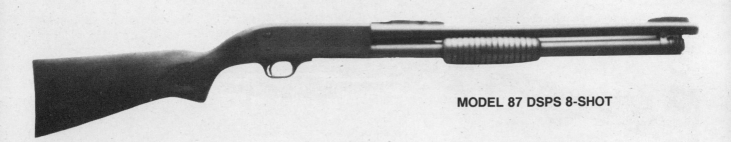

MODEL 87 DSPS 8-SHOT

SPECIFICATIONS ITHACA MODEL 87 SHOTGUNS

Type	Gauge	Bbl In.	Chokes	Chamb.	Cap.	Wt.	Type	Gauge	Bbl In.	Chokes	Chamb.	Cap.	Wt.
Supreme	12	30	CT	3″	5	7	Deerslayer II	12	25	DSR1	3″	5	7
	12	28	CT	3″	5	7		12	20	DSR1	3″	5	7
	12	26	CT	3″	5	7		20	25	DSR1	3″	5	6.8
	20	26	CT	3″	5	6.8		20	20	DSR1	3″	5	6.8
Deluxe Vent	12	30	CT	3″	5	7	Deerslayer II Brenneke	12	25	DSR2	3″	5	7
	12	28	CT	3″	5	7	Extra Barrel Vent Rib	12	30	CTt	3″	N/A	2.1
	12	26	CT	3″	5	7		12	28	CT	3″	N/A	2.1
	20	26	CT	3″	5	6.8		12	26	CT	3″	N/A	2.1
English	20	26	CT	3″	5	6.8		12	24	CT	3″	N/A	2.1
	20	24	CT	3″	5	6.8		20	28	CT	3″	N/A	2
Camo Field	12	28	CT	3″	5	7		20	26	CT	3″	N/A	2
	12	24	CT	3″	5	7		20	24	CT	3″	N/A	2
	12	28	CT	3″	5	7	Extra Barrel Deer SB	12	25	DS	3″	N/A	2.1
	12	24	CT	3″	5	7		12	20	DS	3″	N/A	2.1
Turkey Camo w/Tube	12	24	Full CT	3″	5	7		20	25	DS	3″	N/A	2
	12	22	Full CT	3″	5	7		20	20	DS	3″	N/A	2
Turkey Camo Fixed Choke	12	24	Full	3″	5	7	Extra Barrel Deer DSR	12	25	DSR	3″	N/A	2.1
	12	22	Full	3″	5	7		12	20	DSR	3″	N/A	2.1
Turkey Field w/Tube	12	24	Full CT	3″	5	7		20	25	DSR	3″	N/A	2
	12	22	Full CT	3″	5	7		20	20	DSR	3″	N/A	2
Turkey Field Fixed Choke	12	24	Full	3″	5	7	Hand Grip	12	19	CYL	3″	5	6.5
	12	22	Full	3″	5	7		20	19	CYL	3″	5	6
Deerslayer SB Deluxe	12	25	DS	3″	5	7		12	20	CYL	3″	8	5.5
	20	20	DS	3″	5	7	DSPS	12	20	DS	3″	5	7
	20	25	DS	3″	5	6.8		12	19	DS	3″	5	7
	20	20	DS	3″	5	6.8		12	19	DS	3″	8	7
Deerslayer SB Basic	12	25	DS	3″	5	7		12	20	DSR1	3″	5	7
	12	20	DS	3″	5	7		12	25	DSR1	3″	5	7
	20	25	DS	3″	5	6.8	M&P	12	20	CYL	3″	5	7
	20	20	DS	3″	5	6		20	20	CYL	3″	5	6.5
Deerslayer DSR Deluxe	12	25	DSR	3″	5	7		12	20	CYL	3″	8	7
	12	20	DSR	3″	5	7							
	20	25	DSR	3″	5	6.8							
	20	20	DSR	3″	5	6.8							

CT = Full Mod IC Tubes; DS = Smooth Bore Deer; DSR = Rifled Bore Deer; DSR1 = Fixed Rifled Bore Deer 1/34; DSR2 = Fixed Rifled Bore Deer 1/25

KBI SHOTGUNS

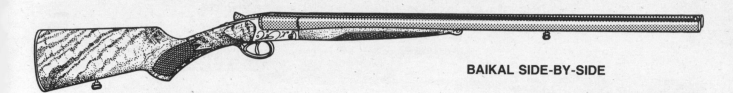

BAIKAL SIDE-BY-SIDE

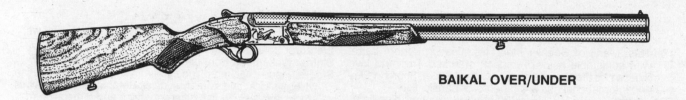

BAIKAL OVER/UNDER

BAIKAL SHOTGUNS

Baikal shotguns are manufactured by Russian arms producers. The **Single Barrel** model has a chrome-lined barrel bore and chamber, non-auto safety and automatic ejector with external disengaging lever. Available in 12 or 20 gauge. Additional specifications are listed in the table below.
Price: . . . . **$ 79.00**

Prices:
The **Side by Side** model has chrome-lined barrels and chambers and comes with double triggers and extractors.
28 gauge **$310.00**
20 gauge **369.00**

The **Over/Under** model features either double trigger and extractors or single selective trigger and auto ejectors.
12 ga. double triggers **$429.00**
12 ga. single trigger, Auto Ejector **469.00**

BAIKAL SHOTGUN SPECIFICATIONS

Model	Article Number	Gauge	Barrel Length in.	Chamber in.	Chokes	Weight lb.
Single	GD1000	12	28	2¾"	F	6
Single	GD1019	12	28	2¾"	M	6
Single	GD1027	12	26	2¾"	IC	6
Single	GD1035	20	28	3"	F	5½
Single	GD1043	20	28	3"	M	5½
Single	GD1051	20	26	3"	IC	5½
Single	GD1078	410	26	3"	F	5½
Side/Side—DT	GE1006	12	28	2¾"	M/F	6¾
Side/Side—DT	GE1014	12	26	2¾"	IC/F	6¾
Side/Side-DT	GE1022	12	20	2³/₄"	Cyl./Cyl.	6
O/U—DT	GP1006	12	28	2¾"	M/F	7
O/U—DT	GP1014	12	26	2¾"	IC/F	7
O/U—SST	GP1065	12	28	2¾"	M/F	7
O/U—SST	GP1073	12	26	2¾"	IC/F	7

SHOTGUNS

KRIEGHOFF SHOTGUNS

(See following page for additional Specifications and Prices)

MODEL K-80 SPORTING CLAY

MODEL K-80 TRAP, SKEET, SPORTING CLAY AND LIVE BIRD

Barrels: Made of Boehler steel; free-floating bottom barrel with adjustable point of impact; standard Trap and Live Pigeon ribs are tapered step; standard Skeet, Sporting Clay and International ribs are tapered or parallel flat.
Receivers: Hard satin-nickel finish; casehardened; blue finish available as special order
Triggers: Wide profile, single selective, position adjustable. Removable trigger option available (add'l **$1000.00**)
Weight: 8½ lbs. (Trap); 8 lbs. (Skeet)

Ejectors: Selective automatic
Sights: White pearl front bead and metal center bead
Stocks: Hand-checkered and epoxy-finished Select European walnut stock and forearm; quick-detachable palm swell stocks available in five different styles and dimensions
Safety: Push button safety located on top tang.
Also available:
SKEET SPECIAL (28″ and 30″ barrel; tapered flat or 8mm rib; 2 choke tubes). **Price: $7100.00** (Standard)

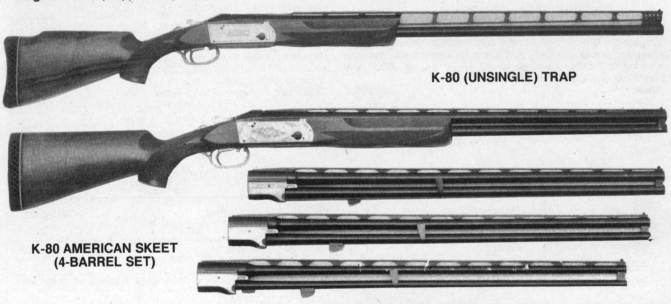

K-80 (UNSINGLE) TRAP

**K-80 AMERICAN SKEET
(4-BARREL SET)**

MODEL ULM-P
O/U SIDELOCK LIVE BIRD GUN
$15,400.00

SPECIFICATIONS
Gauge: 12
Chamber: 2³/₄″
Barrel: 28″ or 30″ long; tapered, ventilated rib
Choke: Top, Full; bottom, Imp. Mod.
Trigger action: Single trigger, non-selective bottom-top; hand-detachable sidelocks with coil springs; optional release trigger

Stock: Selected fancy English walnut, oil finish; length, 14³/₈″; drop at comb, 1³/₈″; optional custom-made stock
Forearm: Semi-beavertail
Engraving: Light scrollwork; optional engravings available
Weight: Approx. 8 lbs.
Also available in Skeet and Trap models

KRIEGHOFF SHOTGUNS

SPECIFICATIONS AND PRICES MODEL K-80 (see also preceding page)

Model	Description	Bbl. Length	Choke	Standard	Bavaria	Danube	Gold Target	Extra Barrels
Trap	Over & Under	30″/32″	IM/F	$ 6895.00	$11,545.00	$19,895.00	$24,895.00	$2450.00
		30″/32″	CT/CT	7485.00	12,135.00	20,485.00	25,485.00	2995.00
	Unsingle	32″/34″	Full	7595.00	12,245.00	20,595.00	25,595.00	3250.00
	Combo (Top Single)	30″ + 34″ 32″ + 34″	IM/F&F	9595.00	14,245.00	22,595.00	27,595.00	3250.00
	Combo (unsingle)	30″ + 32″ 30″ + 34″	IM/F+F	9595.00	14,245.00	22,595.00	27,595.00	
		32″ + 34″	CT/CT + CT	10,585.00	15,235.00	23,590.00	28,585.00	
Skeet	4-Barrel Set	28″/12 ga.	Tula	$14,625.00	$19,275.00	$27,625.00	$32,625.00	$2750.00
		28″/20 ga.	Skeet					2700.00
		28″/28 ga.	Skeet					2700.00
		28″/.410 ga.	Skeet					2700.00
	2-Barrel Set	28″/12 ga.	Tula	9,335.00	13,985.00	22,335.00	27,335.00	2450.00
	Lightweight	28″ + 30″/12 ga.	Skeet	6450.00	11,100.00	N/A	N/A	2450.00
	Standardweight	28″/12 ga.	Tula	6750.00	11,400.00	19,750.00	24,750.00	2750.00
		28″ + 30″/12 ga.	Skeet	6450.00	11,100.00	19,450.00	24,450.00	2450.00
	International	28″/12 ga.	Tula	7200.00	11,850.00	20,200.00	25,200.00	2800.00
	Skeet Special			7100.00	11,750.00	20,100.00	25,100.00	2995.00
Sporting Clays	Over/Under w/screw-in tubes (5)	28″ + 30″ + 32″/ 12 ga.	Tubes	$7550.00	$12,200.00	$20,550.00	$25,550.00	$2995.00

Optional Features:*
Screw-in chokes (O/U, Top or Unsingle) $400.00
Single factory release 390.00
Double factory release 675.00

Optional engravings: Super Scroll $1150.00

* Choke tubes in single barrel (w/tubes): add $400.00. In O/U barrel (5 tubes) add $590.00.

MODEL KS-5

The KS-5 is a single barrel trap gun with a ventilated, tapered and adjustable step rib, casehardened receiver in satin grey matte or blue, finished in electroless nickel. It features an adjustable point of impact by means of different optional fronthangers. Screw-in chokes and adjustable stock are optional. Trigger is adjustable externally for poundage.

SPECIFICATIONS
Gauge: 12
Chamber: 2³/₄″
Barrel length: 32″ or 34″
Choke: Full; optional screw-in chokes
Rib: Tapered step; ventilated
Trigger: Weight of pull adjustable; optional release

Receiver: Casehardened; satin grey finished in electroless nickel; now available in blue
Grade: Standard; engraved models on special order
Weight: Approximately 8.6-8.8 lbs.
Case: Aluminum
Prices:
With full choke and case $3575.00
With screw-in choke and case 4350.00
Screw-in choke barrels 2350.00
Regular barrels 1950.00
Special barrels (F) 2600.00

Also available:
KS-5 SPECIAL. Same as **KS-5** except barrel has fully adjustable rib and stock. **Price: $4580.00** (add **$3000.00** for special screw-in choke barrel)

SHOTGUNS

LANBER SHOTGUNS

MODEL 82 FIELD GRADE O/U
$584.95

SPECIFICATIONS
Gauge: 12 or 20   **Capacity:** 3″
Barrel lengths: 26″ and 28″
Weight: 7¹/₈ lbs. (12 ga.); 6¹/₄ lbs. (20 ga.)
Chokes: IC/M (26″ barrel) or M/F (28″ barrel)
Length of pull: 14³/₈″
Drop at comb: 1¹/₂″   **Drop at heel:** 2¹/₂″
Sights: Target bead front
Features: Silver engraved receiver; rubber recoil pad; European walnut stock with deep-cut checkering; chrome-lined barrels; raised vent. ribs; extractors; reinforced monobloc barrels

MODEL 87 DELUXE FIELD GRADE O/U

Same specifications as Model 82, except it features 5 choke tubes (F, M, IM, IC, CYL), single selective trigger, automatic ejectors and solid boxlock action.

MODEL 97 SPORTING CLAYS
$964.95

SPECIFICATIONS
Gauge: 12   **Chamber:** 2³/₄″
Barrel length: 28″   **Weight:** 7³/₈ lbs.
Chokes: 5 choke tubes (F, IM, IC, CYL, M)
Length of pull: 14³/₈″
Drop at comb: 1¹/₂″   **Drop at heel:** 2³/₈″
Sights: Target front; mid-rib bead
Features: Same as Model 87, but with extra engravings and vents

LAURONA SHOTGUNS

NEW CLASSIC X

MODEL 85X

MODEL 85X/NEW CLASSIC X
$1195.00 HUNTING
$1449.00 TRAP $1424.00 SPORTING

SPECIFICATIONS
Gauge: 12 (2³/₄" chamber; 3" avail. in Hunting model)
Barrel length: 28" (29" in Trap model)
Weight: 6 lbs. 14 oz. (Classic X Hunting); 7 lbs. (Model 85X Hunting); 8 lbs. (Classic Trap); 8 lbs. 2 oz. (Model 85X Trap); 7 lbs. 10 oz. (Classic X Sporting); 7 lbs. 12.oz. (Model 85X Sporting)
Chokes: Multichoke or screw-in choke

Trigger: Single
Stock: Full pistol grip with beavertail forend (tulip forend in Hunting model); hand-checkered
Length of pull: 14³/₄"   **Drop at comb:** 1⁷/₁₆"
Features: Automatic selective ejectors; select-grade walnut w/non-glare finish; 5-year warranty; high-alloy full-length hinge pin; anti-rust black chrome; under-bolt locking method; high-precision monoblock barrel construction; imitation leather carrying case; spare 20-gauge barrel optional

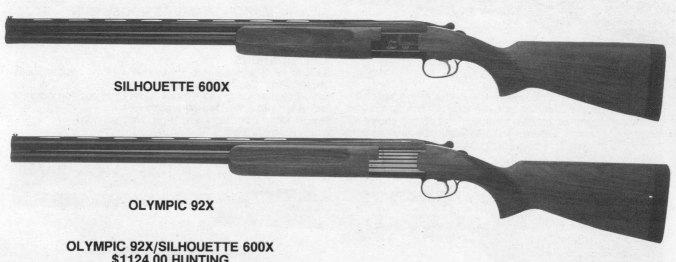

SILHOUETTE 600X

OLYMPIC 92X

OLYMPIC 92X/SILHOUETTE 600X
$1124.00 HUNTING
$1429.00 TRAP $1409.00 SPORTING

SPECIFICATIONS
Gauge: 12 (2³/₄" chamber; 3" avail. in Hunting model)
Barrel length: 28"
Weight: 6 lbs. 14 oz. (Silhouette 600X Hunting); 7 lbs. (Olympic 92X Hunting); 8 lbs. (Silhouette 800X Trap); 8 lbs. 2 oz. (Olympic 92X Trap); 7 lbs. 10 oz. (Silhouette 600X Sporting); 7 lbs. 12 oz. (Olympic 92X Sporting)

Choke: Multichoke
Trigger: Single
Forend: Beavertail (tulip in Hunting model)
Length of pull: 14³/₄"   **Drop at comb:** 1⁷/₁₆"
Features: Same as Models 85X and New Classic X

MAGTECH SHOTGUNS

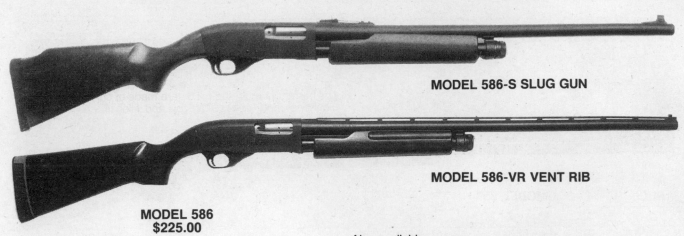

MODEL 586-S SLUG GUN

MODEL 586-VR VENT RIB

**MODEL 586
$225.00**

SPECIFICATIONS
Gauge: 12 (2¾″ or 3″ chamber)
Chokes: Full, Mod. or Imp.
Barrel length: 28″   **Overall length:** 48½″
Weight: 8 lbs.

Also available:
MODEL 586-P Pump Action w/19″ hammer-forged
  Cylinder choke barrel, 7.75 lbs. $219.00
MODEL 586-S Slug Gun w/Monte Carlo stock,
  24″ Cyl. bore barrel, rifle sights 235.00
MODEL 586-VR w/vent rib (26″ or 28″ bbl.) 255.00

MARLIN SHOTGUNS

**MARLIN MODEL 55 GOOSE GUN
$284.90**

High-flying ducks and geese are the Goose Gun's specialty. The Marlin Goose Gun has an extra-long 36-inch full-choked barrel and Magnum capability, making it the perfect choice for tough shots at wary waterfowl. It also features a quick-loading 2-shot clip magazine, a convenient leather carrying strap and a quality ventilated recoil pad.

SPECIFICATIONS
Gauge: 12; 2¾″ Magnum, 3″ Magnum or 2¾″ regular shells
Choke: Full   **Capacity:** 2-shot clip magazine
Action: Bolt action; positive thumb safety; red cocking indicator
Barrel length: 36″   **Overall length:** 56¾″
Sights: Bead front sight and U-groove rear sight
Weight: About 8 lbs.
Stock: Walnut-finish hardwood with pistol grip and ventilated recoil pad; swivel studs; tough Mar-Shield® finish

**MODEL 512 SLUGMASTER
$343.05**

Overall length: 44.75″
Weight: 8 lbs. (w/o scope and mount)
Sights: Adjustable folding semi-buckhorn rear; ramp front with brass bead and removable Wide-Scan® hood; receiver drilled and tapped for scope mount
Stock: Walnut finished, press-checkered Maine birch w/pistol grip and Mar-Shield® finish, swivel studs, vent. recoil pad

SPECIFICATIONS
Gauge: 12 (up to 3″ shells)
Capacity: 2-shot box magazine (+1 in chamber)
Action: Bolt action; thumb safety; red cocking indicator
Barrel length: 21″ rifled (1:28″ right-hand twist)

MAROCCHI AVANZA SHOTGUNS

AVANZA OVER/UNDER

Marocchi's ultralight, quick handling over and under is an ideal field gun that moves fast for sporting clays as well. It houses a strong boxlock action fitted with a single selective trigger. Barrel cycling is controlled mechanically and features automatic selective extractors and ejectors, plus unbreakable firing pins.

The barrels, chambered for 3″, are made of chrome steel with highly polished bores. Both top and filler ribs have modern wide ventilated styling.

Features: Single selective trigger (5½ lb. pull); cut-checkered stock of select walnut; top and middle ribs are ventilated; 3″ chambers; mechanism is lightweight all steel with Mono-Block boxlock, automechanical barrel cycling, selective automatic ejectors/extractors, and automatic safety.

SPECIFICATIONS MAROCCHI AVANZA O/U

Gauge	Barrel Length	Chokes	Overall Length	Weight	Prices
12	28″	M & F	46½″	6 lbs. 9 oz.	$769.00
12	28″	IC & M & F Interchokes	46½″	6 lbs. 13 oz.	829.00
12	26″	IC & M	44½″	6 lbs. 6 oz.	769.00
12	26″	IC & M & F Interchokes	44½″	6 lbs. 9 oz.	829.00

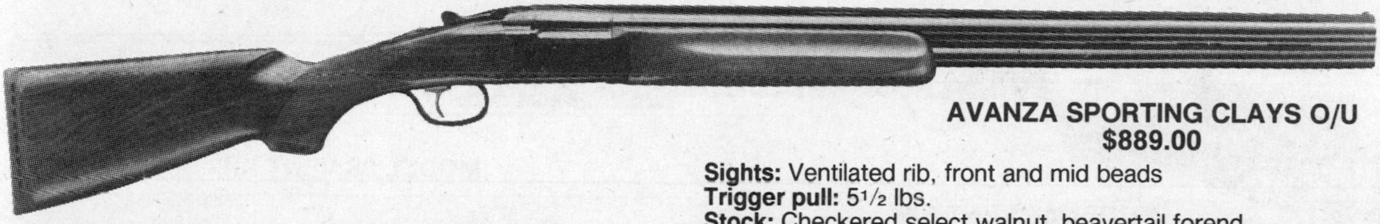

AVANZA SPORTING CLAYS O/U
$889.00

Sights: Ventilated rib, front and mid beads
Trigger pull: 5½ lbs.
Stock: Checkered select walnut, beavertail forend
Features: Chrome steel double barrel with highly polished bores; all steel mono block boxlock action; automatic safety; single selective trigger; automechanical barrel cycling; selective automatic ejectors/extractors

SPECIFICATIONS
Gauge: 12 (3″ chamber) **Barrel length:** 28″
Choke: Interchangeable tubes in IC & Modified
Weight: 7 lbs.

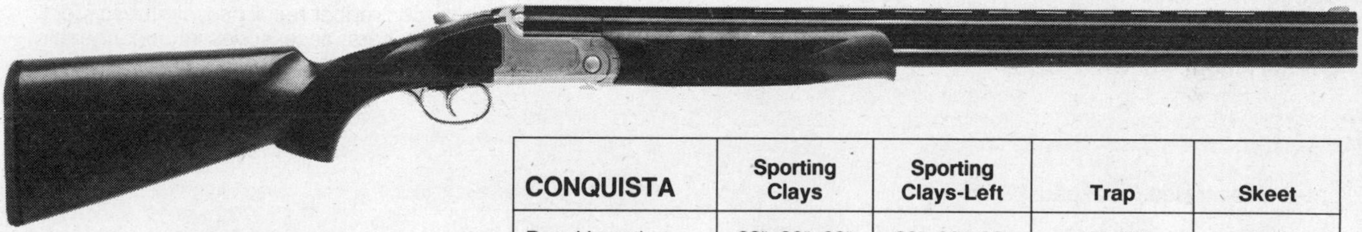

CONQUISTA
(12 Gauge, 2¾″ Chambers)

Prices:
Grade 1 (Sporting Clays, Sporting Clays Left, Trap, Skeet) **$1885.00**
Grade 2 (Sporting Clays, Trap, Skeet) 2250.00
Grade 3 (Sporting Clays, Trap, Skeet) 3550.00

CONQUISTA	Sporting Clays	Sporting Clays-Left	Trap	Skeet
Barrel Lengths	28″, 30″, 32″	28″, 30″, 32″	30″, 32″	28″
Chokes	Contrechokes	Contrechokes	Imp. Mod./Full	Skeet
Overall Length	45″–49″	45″–49″	47″–49″	45″
Weight	7⅞ lbs.	7⅞ lbs.	8¼ lbs.	7¾ lbs.
Front Sight	Competition White	Competition White	Competition White	Competition White
Extractors/Ejectors	Automatic	Automatic	Automatic	Automatic
Trigger Type	Instajust Selective	Instajust Selective	Instajust Selective	Instajust Selective
Trigger Pull (Weight)	3.5–4.0 lbs.	3.5–4.0 lbs.	2.9–3.9 lbs.	2.9–4.0 lbs.
Trigger Pull (Length)	14½″–14⅞″	14½″–14⅞″	14½″–14⅞″	14⅜″–14¾″
Stock	Select American Walnut			
Drop at comb	1⁷/₁₆″	1⁷/₁₆″	1⁹/₃₂″	1½″
Drop at heel	2³/₁₆″	2³/₁₆″	1¹¹/₁₆″	2³/₁₆″

SHOTGUNS

MAVERICK OF MOSSBERG

PUMP ACTION SHOTGUNS

MODEL 91 ULTI-MAG

MODEL 91 ULTI-MAG
$259.00

SPECIFICATIONS
Gauge: 12 w/3¹/₂″ chamber; also fires 2³/₄″ and 3″ shells
Capacity: 6 shots (2³/₄″ shells)
Barrel length: 28″ vent rib   **Overall length:** 48¹/₂″

Weight: 7.7 lbs.
Choke: Threaded for Accu-Mag tubes
Features: Crossbolt safety; rubber recoil pad; synthetic stock and forearm; dual slide bars; accessories interchangeable with Mossberg M835; cablelock included

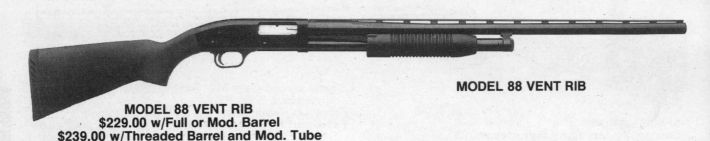

MODEL 88 VENT RIB

MODEL 88 VENT RIB
$229.00 w/Full or Mod. Barrel
$239.00 w/Threaded Barrel and Mod. Tube

SPECIFICATIONS
Gauge: 12; 3″ chamber
Capacity: 6 shots (2³/₄″ shells)
Barrel length: 30″ Full Fixed, all others 28″
Overall length: 48″ w/28″ barrel

Weight: 7.2 lbs. w/28″ barrel
Chokes: Fixed or threaded for Accu-Choke
Features: Crossbolt safety; rubber recoil pad; synthetic stock and forearm; dual slide bars; accessories interchangeable with Mossberg M500; cablelock included

MODEL 88
w/30″ Plain Barrel

MODEL 88
$219.00 w/30″ Full or 28″ Mod. Barrel

SPECIFICATIONS
Gauge: 12; 3″ chamber
Capacity: 6 shots (2³/₄″ shells)
Barrel length: 30″ Full or 28″ Mod.

Overall length: 48″ w/28″ barrel
Weight: 7.1 lbs. w/28″ barrel
Chokes: Fixed, Full or Mod.
Features: Crossbolt safety; rubber recoil pad; synthetic stock and forearm; dual slide bars; accessories interchangeable with Mossberg M500; cablelock included

MAVERICK OF MOSSBERG

PUMP ACTION SHOTGUNS

MODEL 88 DEER GUN

MODEL 88 DEER GUN
$236.00

SPECIFICATIONS
Gauge: 12; 3″ chamber
Capacity: 6 shots (2³/₄″ shells)
Choke: Cylinder bore
Barrel length: 24″   **Overall length:** 44″

Weight: 7 lbs.
Sights: Adjustable rear leaf and rifle-style front
Features: Crossbolt safety; rubber recoil pad; synthetic stock and forearm; dual slide bars; accessories interchangeable with Mossberg M500; cablelock included

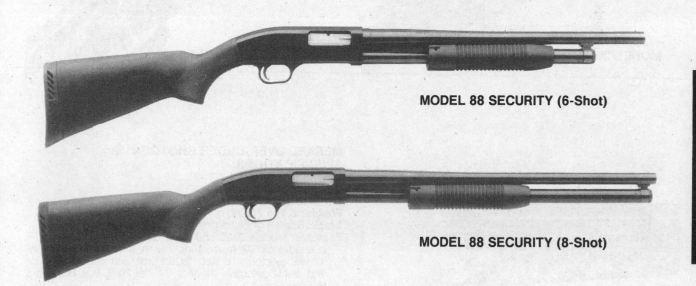

MODEL 88 SECURITY (6-Shot)

MODEL 88 SECURITY (8-Shot)

MODEL 88 SECURITY
$234.00 (8-Shot) $249.00 (8-Shot w/Heat Shield)
$219.00 (6-Shot) $234.00 (6-Shot w/Heat Shield)

SPECIFICATIONS
Gauge: 12; 3″ chamber   **Choke:** Cyl. bore
Barrel length: 18¹/₂″ (6-shot); 20″ (8-shot)
Overall length: 28″ w/18¹/₂″ barrel

Weight: 5.6 lbs. (6-shot)
Features: Synthetic pistol grip and forearm; crossbolt safety; dual slide bars; optional heat shield; optional capacities; accessories interchangeable with Mossberg M500; cable-lock included

MERKEL OVER/UNDER SHOTGUNS

Merkel over-and-unders are the first hunting guns with barrels arranged one above the other, and they have since proved to be able competitors of the side-by-side gun. Merkel superiority lies in the following details:
- Available in 12, 16 and 20 gauge.
- Lightweight (6.4 to 7.28 lbs.).
- The high, narrow forend protects the shooter's hand from the barrel in hot or cold climates.
- The forend is narrow and therefore lies snugly in the hand to permit easy and positive swinging.

- The slim barrel line provides an unobstructed field of view and thus permits rapid aiming and shooting.
- The over-and-under barrel arrangement reduces recoil error; the recoil merely pushes the muzzle up vertically.

All Merkel shotguns are manufactured by Jagd und Sport-waffen GmbH, Suhl, Thuringia, Germany; imported, distributed and retailed in the U.S. by GSI Inc. (see Directory of Manufacturers and Distributors).

MODEL 200E BOXLOCK

MODEL 201E BOXLOCK

MODEL 203E SIDELOCK

MERKEL OVER/UNDER SHOTGUN SPECIFICATIONS

Gauges: 12 and 20
Barrel lengths: 26³/₄″ and 28″
Weight: 6.4 to 7.28 lbs.
Stock: English or pistol grip in European walnut
Features: Models 200E and 201E are boxlocks; Model 203E is a sidelock. All models include three-piece forearm, automatic ejectors, Kersten double crossbolt lock, Blitz action and single selective triggers. Model 203E has Holland & Holland ejectors.

Prices:
MODEL 200E Boxlock (w/scroll engraved casehardened receiver) . . . . $ 3,395.00
MODEL 201ES Skeet or Trap (Skeet/Skeet) . . . 5,595.00
MODEL 200ES Skeet or Trap (Skeet/Skeet) 12 ga. w/26³/₄″ barrel 4,995.00
MODEL 200ET Skeet or Trap (Full/Full) 12 ga. w/30″ barrel 4,795.00
MODEL 201E Boxlock (w/hunting scenes) 4,195.00
MODEL 202E Sidelock (w/hunting scenes) 7,995.00
MODEL 203E Sidelock (w/English-style engraving) 9,695.00
MODEL 303E Sidelock (w/quick-detachable side-lock plates w/integral retracting hook 21,295.00

MERKEL SIDE-BY-SIDE SHOTGUNS

MODEL 47E BOXLOCK

MODEL 147E BOXLOCK

SIDE-BY-SIDE SHOTGUNS

SPECIFICATIONS
Gauges: 12 and 20 (28 ga. and .410 in Models 47S and 147S)
Barrel lengths: 20″ and 20″ (25½″ in Models 47S and 147S)
Weight: 6 to 7 lbs.
Stock: English or pistol grip in European walnut
Features: Models 47E and 147E are boxlocks; Models 47S and 147S are sidelocks. All guns have cold hammer-forged barrels, double triggers, double lugs and Greener crossbolt locking systems and automatic ejectors.

Prices:
MODEL 8	$1295.00
MODEL 47E (Holland & Holland ejectors)	1595.00
MODEL 122 (H&H ejectors, engraved hunting scenes)	3195.00
MODEL 147 (H&H ejectors)	1795.00
MODEL 147E (engraved hunting scenes)	1995.00
MODEL 47S Sidelock (H&H ejectors)	4195.00
MODEL 147S Sidelock	5195.00
MODEL 247S (English-style engraving)	6895.00
MODEL 347S (H&H ejectors)	7895.00
MODEL 447S	8995.00
MODEL 47LSC Sporting Clays (H&H ejectors) 12 ga. w/28″ barrel	2995.00

MODEL 47S SIDE-BY-SIDE

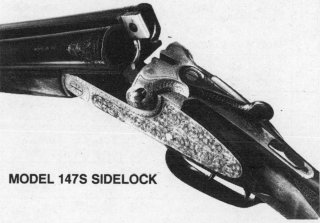

MODEL 147S SIDELOCK

SHOTGUNS

MOSSBERG PUMP SHOTGUNS

MODEL 500 SPORTING

All Mossberg Model 500 pump-action shotguns feature Milspec tough, lightweight alloy receivers with "top thumb safety." Standard models include 6-shot capacity with 2³/₄" chamber, cut-checkered stocks, Quiet Carry forearms, gold trigger, engraved receiver, blue metal finish and the largest selection of accessory barrels. All models include a Cablelock and 10-year limited warranty.

SPECIFICATIONS & PRICES MODEL 500

Ga.	Stock #	Bbl. Length	Bbl. Type	Sights	Chokes	Stock	Length O/A	Wt.	Q.D. Studs	Notes	Prices
12	54120	28"	Vent Rib	2 Beads	Accu-Choke	Walnut Finish	48"	7.2		I.C., Mod. & Full Tubes	$285.00
12	58126	24"	Plain	Iron	Accu-Choke	Walnut Finish	44"	7.0	Y	Mod. & X-Full Tubes	274.00
12	58117	28"	Vent Rib	2 Beads	Accu-Choke	Walnut Finish	48"	7.2		Mod. Tube Only	269.00
12	58116	26"	Vent Rib	2 Beads	Accu-Choke	Walnut Finish	46"	7.1		I.C. & Mod. Tubes	269.00
20	58132	22"	Vent Rib	2 Beads	Accu-Choke	Walnut Finish	42"	6.9		Mod. Tube Only, Bantam Stock	269.00
20	54136	26"	Vent Rib	2 Beads	Accu-Choke	Walnut Finish	46"	7.0		I.C., Mod. & Full Tubes	285.00
20	58137	26"	Vent Rib	2 Beads	Accu-Choke	Walnut Finish	46"	7.0		Mod. Tube Only	269.00
.410	50149	24"	Plain	2 Beads	Full	Synthetic	43"	6.8		Fixed Choke, Bantam Stock	274.00
.410	58104	24"	Vent Rib	2 Beads	Full	Walnut Finish	44"	6.8		Fixed Choke	275.00
12	54032	24"	Trophy Slugster™	None	Rifled Bore	Walnut Finish	44"	7.3	Y	Dual-Comb™ Stock	340.00
12	54044	24"	Slug	Iron	Rifled Bore	Walnut Finish	44"	7.0	Y		312.00
12	58045	24"	Slug	Iron	Cyl. Bore	Walnut Finish	44"	7.0	Y		276.00
20	54051	24"	Slug	Iron	Rifled Bore	Walnut Finish	44"	6.9	Y		312.00
20	58050	24"	Slug	Iron	Cyl. Bore	Walnut Finish	44"	6.9	Y		276.00

500 COMBOS

Ga.	Stock #	Bbl. Length	Bbl. Type	Sights	Chokes	Stock	Length O/A	Wt.	Q.D. Studs	Notes	Prices
12	54043	28" 24"	Vent Rib Trophy Slugster™	2 Beads None	Accu-Choke Rifled Bore	Walnut Finish	48"	7.2	Y	I.C., Mod. & Full Tubes Dual-Comb™ Stock	$388.00
12	54164	28" 24"	Vent Rib Slug	2 Beads Iron	Accu-Choke Rifled Bore	Walnut Finish	48"	7.2	Y	I.C., Mod. & Full Tubes	354.00
12	58483	28" 18.5"	Plain Plain	Bead Bead	Modified Cyl. Bore	Walnut Finish	48"	7.2		Fixed Choke, w/Pistol Grip	363.00
20	54182	26" 24"	Vent Rib Slug	2 Beads Iron	Accu-Choke Rifled Bore	Walnut Finish	46"	7.0	Y	I.C., Mod. & Full Tubes	354.00
12	58158	28" 24"	Vent Rib Slug	2 Beads Iron	Accu-Choke Cyl. Bore	Walnut Finish	48"	7.2		Mod. Tube Only	318.00
12	58169	28" 18.5"	Vent Rib Plain	2 Beads Bead	Modified Cyl. Bore	Walnut Finish	48"	7.2		Mod. Tube Only, w/Pistol Grip	318.00
.410	58456	24" 18.5"	Vent Rib Plain	2 Beads Bead	Full Cyl. Bore	Walnut Finish	44"	6.8		Fixed Choke, w/Pistol Grip	324.00
12	54153	24" 24"	Slug Muzzleloader	Iron Iron	Rifled Bore Rifled Bore	Walnut Finish	44"	7.0		.50 Cal., 1/26" Right Hand Twist	418.00

MOSSBERG PUMP SHOTGUNS

MODEL 500 SPORTING

MODEL 500 TROPHY SLUGSTER

MODEL 500 OFM MARSH CAMO

MODEL 500 OFM WOODLAND CAMO

SPECIFICATIONS
MODEL 500 CAMO

12	50192	28″	Vent Rib	2 Beads	Accu-Choke	Synthetic	48″	7.2	Y	Mod. & X-Full Tubes	**$282.00**
12	50190	28″	Vent Rib	2 Beads	Accu-Choke	Synthetic	48″	7.2	Y	I.C., Mod. & Full Tubes	**310.00**
12	50195	24″	Vent Rib	2 Beads	Accu-Choke	Synthetic	44″	7.1	Y	I.C., Mod., Full & X-Full Tubes	**310.00**
12	50196	24″	Ghost Ring™ Sight	Ghost Ring™	Accu-Choke	Synthetic	44″	7.1	Y	I.C., Mod. Full & X-Full Tubes	**368.00**
12	50217	28″ 24″	Vent Rib Slug	2 Beads Iron	Accu-Choke Rifled Bore	Synthetic	48″	7.2	Y	I.C., Mod. & Full Tubes	**385.00**

MODEL 500 AMERICAN FIELD (Pressed Checkered)

12	50117	28″	Vent Rib	2 Beads	Accu-Choke	Walnut Finish	48″	7.2		Mod. Tube Only	**$266.00**
20	50137	26″	Vent Rib	2 Beads	Accu-Choke	Walnut Finish	46″	7.1		Mod. Tube Only	**266.00**
.410	50104	24″	Vent Rib	2 Beads	Full	Walnut Finish	44″	7.0		Fixed Choke	**272.00**
12	50044	24″	Slug	Iron	Rifled Bore	Walnut Finish	44″	7.0	Y		**309.00**
12	50045	24″	Slug	Iron	Cyl. Bore	Walnut Finish	44″	7.0	Y		**274.00**
20	50050	24″	Slug	Iron	Cyl. Bore	Walnut Finish	44″	7.0	Y		**274.00**

MOSSBERG PUMP SHOTGUNS

MODEL 500/590 SPECIAL PURPOSE

Since 1979, Mossberg's Special Purpose Models 500 and 590 have been the only pump shotguns that meet or exceed U.S. Government MILSPEC 3443D and 3443E requirements. These shotguns feature lightweight alloy receivers with ambidextrous "top thumb safety" button, walnut-finished wood or durable synthetic with Quiet Carry™ forearms, rubber recoil pads, dual extractors, two slide bars and twin cartridge stops. All models include a Cablelock™ and a 10-year limited warranty.

SPECIFICATIONS AND PRICES

Gauge	Barrel Length	Sight	Choke	Finish	Stock	Capacity	Overall Length	Weight	Notes	Price
MODEL 500/590 MARINER™										
12	18.5″	Bead	Cyl. Bore	Marinecote™	Synthetic	6	38.5″	6.8	Includes Pistol Grip	$396.00
12	20″	Bead	Cyl. Bore	Marinecote™	Synthetic	9	40″	7.0	Includes Pistol Grip	406.00
MODEL 500 SPECIAL PURPOSE										
12	18.5″	Bead	Cyl. Bore	Blue	Walnut Finish	6	38.5″	6.8	Includes Pistol Grip	272.00
12	18.5″	Bead	Cyl. Bore	Blue	Synthetic	6	38.5″	6.8	Includes Pistol Grip	272.00
12	18.5″	Bead	Cyl. Bore	Parkerized	Synthetic	6	38.5″	6.8	Includes Pistol Grip	304.00
12	18.5″	Bead	Cyl. Bore	Blue	Pistol Grip	6	28″	5.6	Includes Heat Shield	265.00
20	18.5″	Bead	Cyl. Bore	Blue	Walnut Finish	6	38.5″	6.8	Includes Pistol Grip	272.00
20	18.5″	Bead	Cyl. Bore	Blue	Pistol Grip	6	28″	5.6		265.00
20	18.5″	Bead	Cyl. Bore	Blue	Pistol Grip	6	28″	5.6	Includes Camper Case	297.00
.410	18.5″	Bead	Cyl. Bore	Blue	Pistol Grip	6	28″	5.3		271.00
.410	18.5″	Bead	Cyl. Bore	Blue	Pistol Grip	6	28″	5.3	Includes Camper Case	303.00
12	20″	Iron	Cyl. Bore	Blue	Walnut Finish	8	40″	7.0		296.00
12	20″	Bead	Cyl. Bore	Blue	Walnut Finish	8	40″	7.0	Includes Pistol Grip	272.00
12	20″	Bead	Cyl. Bore	Blue	Synthetic	8	40″	7.0	Includes Pistol Grip	272.00
12	20″	Bead	Cyl. Bore	Blue	Pistol Grip	8	40″	7.0	Includes Heat Shield	265.00
MODEL 590 SPECIAL PURPOSE										
12	20″	Bead	Cyl. Bore	Blue	Synthetic	9	40″	7.2	w/Acc. Lug & Heat Shield	322.00
12	20″	Bead	Cyl. Bore	Blue	Speed Feed	9	40″	7.2	w/Acc. Lug & Heat Shield	351.00
12	20″	Bead	Cyl. Bore	Parkerized	Synthetic	9	40″	7.2	w/Acc. Lug & Heat Shield	371.00
12	20″	Bead	Cyl. Bore	Parkerized	Speed Feed	9	40″	7.2	w/Acc. Lug & Heat Shield	403.00
MODEL 500/590 GHOST RING™										
12	18.5″	Ghost Ring™	Cyl. Bore	Blue	Synthetic	6	38.5″	6.8		321.00
12	18.5″	Ghost Ring™	Cyl. Bore	Parkerized	Synthetic	6	38.5″	6.8		374.00
12	20″	Ghost Ring™	Cyl. Bore	Blue	Synthetic	9	40″	7.2	w/Acc. Lug	376.00
12	20″	Ghost Ring™	Cyl. Bore	Parkerized	Synthetic	9	40″	7.2	w/Acc. Lug	426.00
12	20″	Ghost Ring™	Cyl. Bore	Parkerized	Speed Feed	9	40″	7.2	w/Acc. Lug	457.00
HS 410 HOME SECURITY										
.410	18.5″	Bead	Spreader	Blue	Synthetic	6	39.5″	6.6	Includes Vertical Foregrip	284.00

MOSSBERG PUMP SHOTGUNS

MODEL 835 ULTI-MAG

Mossberg's Model 835 Ulti-Mag pump action shotgun has a 3½" 12 gauge chamber but can also handle standard 2¾" and 3" shells. Field barrels are backbored for optimum patterns and felt recoil reduction. Cut checkered walnut and walnut-finished stocks and Quiet Carry™ forearms are standard, as are gold triggers and engraved receivers. Camo models are drilled and tapped for scope and feature detachable swivels and sling. All models include a Cablelock™ and 10-year limited warranty.

MODEL 835 ULTI-MAG

MODEL 835 ULTI-MAG TROPHY SLUGSTER

MODEL 835 ULTI-MAG with Mossy Oak™ Camo Finish

SPECIFICATIONS AND PRICES MODEL 835 ULTI-MAG (12 Gauge, 6 Shot)

Barrel Length	Barrel Type	Sights	Chokes	Finish	Stock	Overall Length	Weight	Notes	Prices
28"	Vent Rib	2 Beads	Accu-Mag	Blue	Walnut	48.5"	7.7	4 Tubes* & Dual-Comb™ Stock	$400.00
24"	Trophy Slugster™	None	Rifled Bore	Blue	Walnut	44.5"	7.3	Dual-Comb™ Stock	421.00
28"	Vent Rib	2 Beads	Accu-Mag	Blue	Walnut	48.5"	7.7	4 Tubes*	393.00
28"	Vent Rib	2 Beads	Accu-Mag	Blue	Walnut Finish	48.5"	7.7	Mod. Tube Only	297.00
24"	Vent Rib	2 Beads	Accu-Mag	Blue	Walnut Finish	44.5"	7.3	Mod. & X-Full Tubes	297.00
28" 24"	Vent Rib Trophy Slugster™	2 Beads None	Accu-Mag Rifled Bore	Blue	Walnut	48.5"	7.7	4 Tubes* Dual-Comb™ Stock	472.00
28" 24"	Vent Rib Slug	2 Beads Iron	Accu-Mag Rifled Bore	Blue	Walnut	48.5"	7.7	4 Tubes* Dual-Comb™ Stock	463.00
24"	Vent Rib	2 Beads	Accu-Mag	Realtree®	Synthetic	44.5"	7.3	4 Tube Turkey Pack**	468.00
24"	Vent Rib	2 Beads	Accu-Mag	Mossy Oak®	Synthetic	44.5"	7.7	4 Tube Turkey Pack**	468.00
28"	Vent Rib	2 Beads	Accu-Mag	Realtree®	Synthetic	48.5"	7.7	4 Tubes*	468.00

MOSSBERG SHOTGUNS

MODEL 9200 AUTOLOADERS

All Mossberg Model 9200 award-winning autoloader shotguns handle light 2³⁄₄″ or heavy 3″ magnum loads. Features include cut-checkered walnut stock and forearm, gold trigger, engraved receiver, top thumb safety, light weight and easy shooting. All models include a Cablelock™ and a Lifetime Limited Warranty.

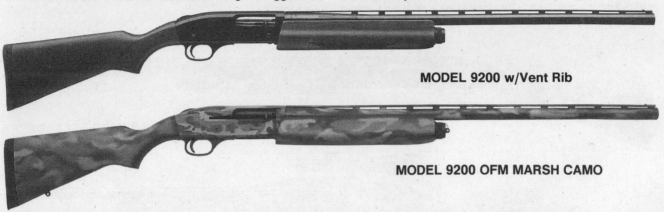

MODEL 9200 w/Vent Rib

MODEL 9200 OFM MARSH CAMO

SPECIFICATIONS AND PRICES MODEL 9200 (12 Gauge, 5 Shot)

Bbl. Length	Bbl. Type	Sights	Chokes	Finish	Stock	Length O/A	Wt.	Q.D. Studs	Notes	Prices
28″	Vent Rib	2 Beads	Accu-Choke	Blue	Walnut	48″	7.7		I.C., Mod. & Full Tubes	**$457.00**
24″	Vent Rib	2 Beads	Accu-Choke	Blue	Walnut	44″	7.3		I.C., Mod. & Full Tubes	457.00
24″	Trophy Slugster™	None	Rifled Bore	Blue	Walnut	44″	7.3	Y	Dual-Comb™ Stock	478.00
24″	Slug	Iron	Rifled Bore	Blue	Walnut	44″	7.3			457.00
26″	Vent Rib	2 Beads	Accu-Choke	Blue	Walnut	46″	7.5		USST, I.C., Mod., Full & Skeet	457.00
18.5″	Plain	Bead	Mod.	Matte Blue	Synthetic	39.5″	7.0	Y	Fixed Choke	443.00
24″	Vent Rib	2 Beads	Accu-Choke	Mossy Oak®	Synthetic	44″	7.3	Y	I.C., Mod., Full & X-Full Tubes	538.00
24″	Vent Rib	2 Beads	Accu-Choke	Realtree®/Matte Blue	Synthetic	44″	7.3	Y	I.C., Mod., Full & X-Full Tubes	538.00
28″	Vent Rib	2 Beads	Accu-Choke	OFM Marsh	Synthetic	48″	7.7	Y	I.C., Mod. & Full Tubes	443.00

MOSSBERG LINE LAUNCHER
$999.00   ($735.00 Launcher Kit)

The Line Launcher is the first shotgun devoted to rescue and personal safety. It provides an early self-contained rescue opportunity for boaters, police and fire departments, salvage operations or whenever an extra-long throw of line is the safest alternative. This shotgun uses a special 12-gauge blank cartridge to propel a convertible projectile with a line attached. With a floating head attached, the projectile will travel 250 to 275 feet. Removing the floating head increases the projectile range to approx. 700 feet. The braided 360-pound test floating line is highly visible and is supplied inside on an 800-foot coiled spool.

NEW ENGLAND FIREARMS

NWTF TURKEY SPECIAL
$199.95

SPECIFICATIONS
Gauge: 10 (3½ chamber)
Chokes: Screw-in chokes (Turkey Full Choke provided; Extra Full and steel shot choke available)
Barrel length: 24″   **Overall length:** 40″

Weight: 9¼ lbs.
Sights: Bead sights; drilled and tapped for scope mounting
Stock: American hardwood; Mossy Oak camo finish; full pistol grip; swivel and studs.   **Length of pull:** 14½″
Also available: **YOUTH TURKEY.** 20-gauge model with 22″ barrel and 3″ chamber, Mod. choke, Mossy Oak Bottomland Camo finish **$149.95**

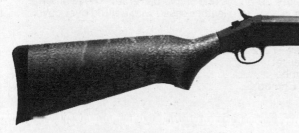

TURKEY & GOOSE GUN
$149.95
($159.00 w/Camo Paint, Swivels & Sling)

SPECIFICATIONS
Gauge: 10 (3½ chamber)   **Choke:** Full
Barrel length: 28″   **Overall length:** 44″

Weight: 9½ lbs.
Sights: Bead sights
Stock: American hardwood; walnut or camo finish; full pistol grip; ventilated recoil pad.   **Length of pull:** 14½″

TRACKER II RIFLED SLUG GUN
$129.95

SPECIFICATIONS
Gauges: 12 and 20 (3″ chamber)   **Choke:** Rifled Bore
Barrel length: 24″   **Overall length:** 40″
Weight: 6 lbs.

Sights: Adjustable rifle sights
Length of pull: 14½″
Stock: American hardwood; walnut or camo finish; full pistol grip; recoil pad; sling swivel studs
Also available:
TRACKER SLUG GUN w/Cylinder Bore: **$124.95**

PARDNER YOUTH (.410 Ga.)
$99.95

PARDNER SINGLE BARREL SHOTGUNS

SPECIFICATIONS
Gauges: 12, 16, 20, 28 and .410

Barrel lengths: 22″ (Youth); 26″ (20, 28, .410); 28″ (12 and 16 ga.)
Chokes: Full (all gauges, except 28); Modified (12, 20 and 28 ga.)
Chamber: 2¾″ (16 and 28 ga.); 3″ (all others)

PARKER REPRODUCTIONS

PARKER A-1 SPECIAL

Recognized by the shooting fraternity as the finest American shotgun ever produced, the Parker A-1 Special is again available. Exquisite engraving and rare presentation-grade French walnut distinguish the A-1 Special from any other shotguns in the world. Currently offered in 12 and 20 gauge, each gun is custom-fitted in its own oak and leather trunk case. Two models are offered: Hand Engraved and Custom Engraved. Also available in D Grade.

Standard features: Automatic safety, selective ejectors, skeleton steel butt plate, splinter forend, engraved snap caps, fitted leather trunk case, canvas and leather case cover, chrome barrel interiors, hand-checkering. The A-1 Special also features a 24k gold initial plate or pistol cap, 32 lines-per-inch checkering, selected wood and fine hand-engraving. Choose from single or double trigger, English or pistol grip stock (all models). Options include beavertail forend, additional barrels.

In addition to the A-1 Special, the D-Grade is available in 12, 20, 16/20 and 28 gauge. A 16-gauge, 28″ barrel can be ordered with a 20-gauge one or two-barrel set. The two-barrel sets come in a custom leather cased with a fitted over cover.

Prices: D-GRADE
One Barrel—12, 20, 28 gauge **$3370.00**
Two-barrel set **4200.00**
16/20 Combo **4870.00**
20/20/16 Combo **5630.00**

SPECIFICATIONS

Gauge	Barrel Length	Chokes	Chambers	Drop At Comb	Drop At Heel	Length of Pull	Nominal Weight (lbs.)	Overall Length
12	26″	Skeet I & II or IC/M	2³/₄″	1³/₈″	2³/₁₆″	14¹/₈″	6³/₄	42⁵/₈″
12	28″	IC/M or M/F	2³/₄ & 3″	1³/₈″	2³/₁₆″	14¹/₈″	6³/₄	44⁵/₈″
20	26″	Skeet I & II or IC/M	2³/₄″	1³/₈″	2³/₁₆″	14³/₈″	6¹/₂	42³/₈″
20	28″	M/F	3″	1³/₈″	2³/₁₆″	14³/₈″	6¹/₂	44⁵/₈″
16 on 20 frame	28″	Skeet I & II, IC/M or M/F	2³/₄″	1³/₈″	2³/₁₆″	14³/₈″	6¹/₄	44⁵/₈″
28	26″	Skeet I & II or IC/M	2³/₄″	1³/₈″	2³/₁₆″	14³/₈″	5¹/₃	42⁵/₈″
28	28″	M/F	2³/₄ & 3″	1³/₈″	2³/₁₆″	14³/₈″	5¹/₃	44⁵/₈″

* *Note:* The 16-gauge barrels are lighter than the 20-gauge barrels.

PERAZZI SHOTGUNS

For the past 20 years or so, Perazzi has concentrated solely on manufacturing competition shotguns for the world market. Today the name has become synonymous with excellence in competitive shooting. The heart of the Perazzi line is the classic over/under, whose barrels are soldered into a monobloc that holds the shell extractors. At the sides are the two locking lugs that link the barrels to the action, which is machined from a solid block of forged steel. Barrels come with flat, step or raised ventilated rib. The walnut forend, finely checkered, is available with schnabel, beavertail or English styling, and the walnut stock can be of standard, Monte Carlo, Skeet or English design. Double or single non-selective or selective triggers. Sideplates and receiver are masterfully engraved and transform these guns into veritable works of art.

GAME MODELS

GAME MODEL MX20C

GAME MODELS MX8, MX12, MX20, MX8/20, MX28 & MX410

SPECIFICATIONS
Gauges: 12, 20, 28 & .410
Chambers: 2³/₄"; also available in 3"
Barrel lengths: 26" and 27¹/₂"
Weight: 6 lbs. 6 oz. to 7 lbs. 4 oz.
Trigger group: Non-detachable with coil springs and selective trigger

Stock: Interchangeable and custom; schnabel forend
Prices:

Standard Grade	$ 7,550.00–$15,100.00
SC3 Grade	12,850.00– 20,400.00
SCO Grade	21,900.00– 29,450.00
SCO Gold Grades	24,650.00– 46,500.00

AMERICAN TRAP COMBO MODELS

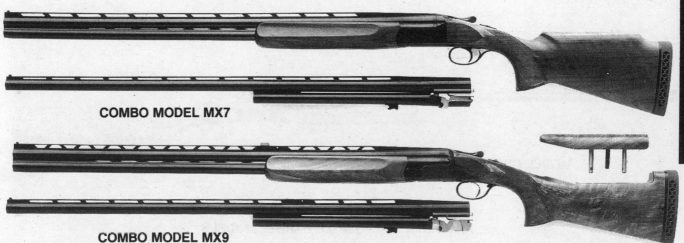

COMBO MODEL MX7

COMBO MODEL MX9

AMERICAN TRAP COMBO MODELS MX6, MX7, MX9, MX10, MX8 SPECIAL, DB81 SPECIAL & GRANDAMERICA 1

SPECIFICATIONS
Gauge: 12   **Chamber:** 2³/₄"
Barrel lengths: 29¹/₂" and 31¹/₂" (O/U); 32" and 34" (single barrel)   **Chokes:** Mod./Full (O/U); Full (single barrel)
Weight (avg.): 8 lbs. 6 oz.

Trigger group: Detachable and interchangeable with flat "V" springs
Stock: Interchangeable and custom; beavertail forend
Prices:

Standard Grade	$ 6,930.00–$13,700.00
SC3 Grade	16,650.00– 19,800.00
SCO Grade	26,600.00– 29,700.00
Gold Grade	29,500.00– 32,650.00

PERAZZI SHOTGUNS

AMERICAN TRAP SINGLE BARREL MODELS

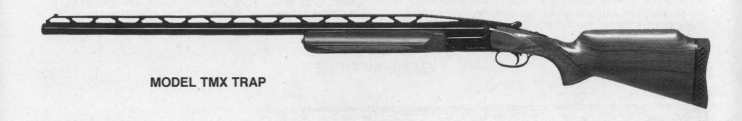

MODEL TMX TRAP

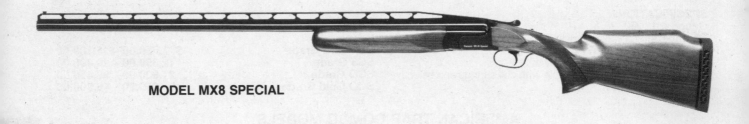

MODEL MX8 SPECIAL

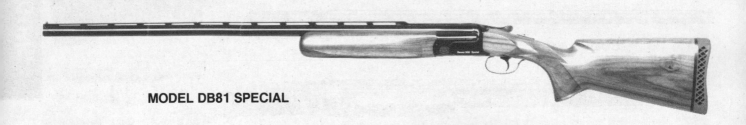

MODEL DB81 SPECIAL

AMERICAN TRAP SINGLE BARREL MODELS
MX6, MX7, TM1 SPECIAL, TMX SPECIAL

SPECIFICATIONS
Gauge: 12
Chamber: 2³/₄″
Barrel lengths: 32″ and 34″
Weight: 8 lbs. 6 oz.
Choke: Full
Trigger group: Detachable and interchangeable with coil springs

Stock: Interchangeable and custom made
Forend: Beavertail
Prices:
Standard Grade . . . . $ 5,950.00–$ 8,850.00
SCO Grade 17,350.00– 20,300.00
Gold Grade 19,350.00– 22,300.00

PERAZZI SHOTGUNS

COMPETITION OVER/UNDER SHOTGUNS
OLYMPIC, DOUBLE TRAP, SKEET, PIGEON & ELECTROCIBLES

MODEL MX10

MODEL MX7 SPORTING

MODEL DB81 TRAP

MX8 SKEET

MIRAGE SPORTING

STANDARD GRADE

SPECIFICATIONS
Gauges: 12 and 20; 4 (Mirage Special only)
Barrel lengths: 27¹/₂″, 28³/₈″, 29¹/₂″, 30³/₄″, 31¹/₂″
Prices:
MX6 12 ga. removable trigger group
   29¹/₂″, 30³/₄″ and 31¹/₂″ barrels . . . . **$ 5500.00**
MX7 12 ga., non-removable trigger group
   29¹/₂″, 30³/₄″ and 31¹/₂″ barrels **5900.00**
MX8 12 ga. removable trigger group
   29¹/₂″, 30³/₄″ and 31¹/₂″ barrels **7550.00**
MX9 12 ga. w/adj. stock and rib inserts
   29¹/₂″ and 31¹/₂″ barrels **9900.00**
MX10 12 & 20 ga. w/adj. stock and rib
   29¹/₂″, 30³/₄″ and 31¹/₂″ bbl. **10,300.00**
MX8/20 20 ga. w/removable trigger group
   27¹/₂″, 28³/₈″, 29¹/₂″, 30³/₄″ and 31¹/₂″
   barrels **7550.00**
MIRAGE SPECIAL 12 ga. w/adj. trigger, all
   barrel lengths **8000.00**
MIRAGE SPECIAL SPORTING 12 ga. w/external
   selector and 5 chokes; 27¹/₂″, 28³/₈″,
   29¹/₂″ and 31¹/₂″ barrels **8000.00**

MIRAGE SPORTING CLASSIC 12 ga. **9500.00**
MX8 SPECIAL 12 ga. w/adjustable trigger
   29¹/₂″ and 31¹/₂″ barrels . . . . **$ 8000.00**
DB81 SPECIAL w/adjustable trigger
   29¹/₂″ and 31¹/₂″ barrels **8250.00**

NOTE: **PIGEON & ELECTROCIBLE MODELS** available in MX1B, Mirage, Mirage Special & MX10 only.
Also available:
SC3 Grade (Models MX8, MX9, MX10,
   MX8/20, MX8 Special, Mirage,
   Mirage Spec., DB81 Spec.) . . . . **$12,850.00–$15,600.00**
SCO Grade (same models as SC3
   Grade) **21,850.00– 24,650.00**
SCO GOLD Grade (same models
   as above) **24,650.00– 27,450.00**
SCO Grade Sideplates (same
   models as above) **33,550.00– 36,300.00**
SCO GOLD Grade Sideplates (same
   models above) **38,950.00– 41,750.00**

SHOTGUNS

PIOTTI SHOTGUNS

One of Italy's top gunmakers, Piotti limits its production to a small number of hand-crafted, best-quality double-barreled shotguns whose shaping, checkering, stock, action and barrel work meets or exceeds the standards achieved in London before WWII. The Italian engravings are the finest ever and are becoming recognized as an art form in themselves.

All of the sidelock models exhibit the same overall design, materials and standards of workmanship; they differ only in the quality of the wood, shaping and sculpturing of the action, type of engraving and gold inlay work and other details. The Model Piuma differs from the other shotguns only in its Anson & Deeley boxlock design.

Also available: Bass-style sidelock over/under in 12 and 20 ga. **Price: $3400.00** (and up).

SPECIFICATIONS
Gauges: 10, 12, 16, 20, 28, .410
Chokes: As ordered
Barrels: 12 ga., 25" to 30"; other gauges, 25" to 28"; chopper lump (demi-bloc) barrels with soft-luster blued finish; level, file-cut rib or optional concave or ventilated rib
Action: Boxlock, Anson & Deeley; Sidelock, Holland & Holland pattern; both have automatic ejectors, double triggers with front trigger hinged (non-selective single trigger optional), coin finish or optional color case-hardening
Stock: Hand-rubbed oil finish (or optional satin luster) on straight grip stock with checkered butt (pistol grip optional)
Forend: Classic (splinter); optional beavertail
Weight: Ranges from 4 lbs. 15 oz. (.410 ga.) to 8 lbs. (12 ga.)

MODEL PIUMA BOXLOCK
$10,000.00

Anson & Deeley boxlock ejector double with chopper lump (demi-bloc) barrels, and scalloped frame. Very attractive scroll and rosette engraving is standard. A number of optional engraving patterns including game scene and gold inlays are available at additional cost.

MODEL KING NO. 1 SIDELOCK
$17,400.00

Best-quality Holland & Holland pattern sidelock ejector double with chopper lump barrels, level filecut rib, very fine, full coverage scroll engraving with small floral bouquets, gold crown in top lever, name in gold, and gold crest in forearm, finely figured wood.

MODEL LUNIK SIDELOCK
$18,800.00

Best-quality Holland & Holland pattern sidelock ejector double with chopper lump (demi-bloc) barrels, level, filecut rib, Renaissance-style, large scroll engraving in relief, gold crown in top lever, gold name, and gold crest in forearm, finely figured wood.

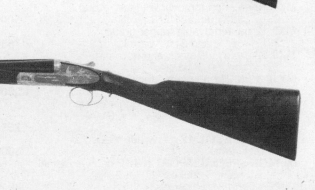

MODEL KING EXTRA (With Gold)
$27,500.00

Best-quality Holland & Holland pattern sidelock ejector double with chopper lump barrels, level filecut rib, choice of either bulino game scene engraving or game scene engraving with gold inlays, engraved and signed by a master engraver, exhibition grade wood.

REMINGTON SHOTGUNS

MODEL 90-T SINGLE BARREL TRAP GUN
$2995.00
($3595.00 w/High Post, Adj. Rib)

Remington's **Model 90-T Single Barrel Trap** features a top-lever release and internal, full-width, horizontal bolt lockup. Barrel is overbored, with elongated forcing cone, and is available in 34″ length. A medium-high, tapered, ventilated rib includes a white, Bradley-type front bead and stainless steel center bead. Choice of stocks includes Monte Carlo style with 1³/₈″, 1¹/₂″ or 1¹/₄″ drop at comb, or a conventional straight stock with 1¹/₂″ drop. Standard length of pull is 14³/₈″. Stocks and modified beavertail forends are made from semi-fancy American walnut. Wood finish is low-lustre satin with positive, deep-cut checkering 20 lines to the inch. All stocks come with black, vented-rubber recoil pads. **Overall length:** 51″. **Weight:** Approx. 8³/₄ lbs. **Choke:** Full.

MODEL 870 EXPRESS "YOUTH" GUN
20 Gauge Lightweight
$292.00

The **Model 870 Express "Youth" Gun** has been specially designed for youths and smaller-sized adults. It's a 20-gauge lightweight with a 1-inch shorter stock and 21-inch barrel. Yet it is still all 870, complete with REM Choke and ventilated rib barrel. Also available with a 20″ fully rifled, rifle-sighted deer barrel. **Barrel length:** 21″. **Stock Dimensions:** Length of pull 12¹/₂″ (including recoil pad); drop at heel; 2¹/₂″ drop at comb 1⁵/₈″. **Overall length:** 39″. **Average Weight:** 6 lbs. **Choke:** REM Choke-Mod. (vent-rib version).

MODEL 870 EXPRESS SMALL GAUGE
$307.00

This shotgun is designed for shooters who want the light weight (6¹/₂ lbs.) and maneuverability of a .410 with the concealment advantages of the non-reflective metal and wood finish of Remington's Express line. The .410 comes with a 25″ Full Choke barrel (45¹/₂″ overall), with Modified REM Choke tube. Now available in 28 gauge with 26″ or 28″ Rem. choke barrel with Mod. choke tube.

MODEL 870 EXPRESS SECURITY
$283.00 (not shown)

Intended for home defense, with an 18¹/₂″ Cylinder choke barrel and front bead sight. **Overall length:** 38¹/₂″. **Weight:** 7¹/₄ lbs.

SHOTGUNS

REMINGTON PUMP SHOTGUNS

MODEL 870 EXPRESS (20 GA.)

MODEL 870 EXPRESS
$292.00 (12 & 20 GA.)
($299.00 w/Black Synthetic Stock & Forend)

Model 870 Express features the same action as the Wingmaster and is available with 3″ chamber and 26″ or 28″ vent-rib barrel. It has a hardwood stock with low-luster finish and solid buttpad. Choke is Modified REM Choke tube and wrench. **Overall length:** 48¹/₂″ (28″ barrel). **Weight:** 7¹/₄ lbs (26″ barrel).

MODEL 870 EXPRESS TURKEY GUN
$305.00

The **Model 870 Express Turkey Gun** boasts all the same features as the Model 870 Express, except has 21″ vent rib barrel and Turkey Extra-Full REM Choke.

MODEL 870 EXPRESS DEER GUN
With Rifle Sights $287.00
($320.00 Fully Rifled)

This 12-gauge pump-action deer gun is for hunters who prefer open sights. Features a 20″ barrel, quick-reading iron sights, fixed Imp. Cyl. choke and Monte Carlo stock. Also available with fully rifled barrel.

MODEL 870 EXPRESS COMBO (not shown)
$395.00

The **Model 870 Express** in 12 and 20 gauge offers all the features of the standard Model 870, including twin-action bars, quick-changing 28″ barrels, REM Choke and vent rib plus low-luster, checkered hardwood stock and no-shine finish on barrel and receiver. The Model 870 Combo is packaged with an extra 20″ deer barrel, fitted with rifle sights and fixed, Improved Cylinder choke (additional REM chokes can be added for special applications). The 3-inch chamber handles all 2³/₄″ and 3″ shells without adjustment. **Weight:** 7¹/₂ lbs.

REMINGTON PUMP SHOTGUNS

SPECIAL PURPOSE

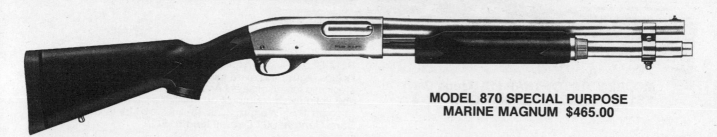

**MODEL 870 SPECIAL PURPOSE
MARINE MAGNUM $465.00**

Remington's **Model 870 Special Purpose Marine Magnum** is a versatile, multi-purpose security gun featuring a rugged synthetic stock and extensive, electroless nickel plating on all metal parts. This new shotgun utilizes a standard 12-gauge Model 870 receiver with a 7-round magazine extension tube and an 18″ cylinder barrel (38½″ overall) with bead front sight.

The receiver, magazine extension and barrel are protected (inside and out) with heavy-duty, corrosion-resistant nickel plating. The synthetic stock and forend reduce the effects of moisture. The gun is supplied with a black rubber recoil pad, sling swivel studs, and positive checkering on both pistol grip and forend. **Weight:** 7½ lbs.

**MODEL 870 SPS-CAMO
$440.00**

This Mossy Oak Bottomland™ Camo version of Model 11-87 and Model 870 Special Purpose Synthetic shotguns features a durable camo finish and synthetic stocks that are immune to the effects of ice, snow and mud. Available with 26″ or 28″ vent rib barrels with twin bead sights and Imp. Cyl., Modified, and Full REM Choke tubes.
Also available:
Big-Game 20″ Rem Choke model w/rifle sights. **Weight:** 7 lbs.
Price: $452.00

**MODEL 870 SPST ALL BLACK
TURKEY GUN $399.00**

Same as the Model 870 SPS above, except with a 21″ vent rib turkey barrel and Extra-Full REM Choke tube. Also available with Mossy Oak Greenleaf Camo finish **Price: $453.00**
Also available: 20″ fully rifled cantilever deer barrel (All Black).
Price: $452.00 ($419.00 with Rem Choke; **$395.00** with 20″ fully rifled deer barrel, rifle sights).

REMINGTON PUMP SHOTGUNS

MODEL 870 "TC" TRAP (12 Gauge Only)
$613.00 ($628.00 w/Monte Carlo Stock)

The **870 "TC"** is a trap version of Model 870 that features REM Choke and a high step-up ventilated rib. REM chokes include regular full, extra full and super full. **Stock:** Redesigned stock and forend of select American walnut with cut-checkering and satin finish; length of pull 14³/₈"; drop at heel 1⁷/₈"; drop at comb 1³/₈". **Weight:** 8 lbs. (8¼ lbs. w/Monte Carlo stock). **Barrel length:** 30". **Overall length:** 51".

MODEL 870 WINGMASTER
12 Gauge, Light Contour Barrel
$479.00 ($527.00 Left Hand, Satin Only)

This restyled **870 "Wingmaster"** pump has cut-checkering on its satin-finished American walnut stock and forend for confident handling, even in wet weather. Also available in Hi-Gloss finish. An ivory bead "Bradley"-type front sight is included. Rifle is available with 26", 28" and 30" barrel with REM Choke and handles 3" and 2³/₄" shells interchangeably. **Overall length:** 46½" (26" barrel), 48½" (28" barrel), 50½" (30" barrel). **Weight:** 7¼ lbs. (w/26" barrel).

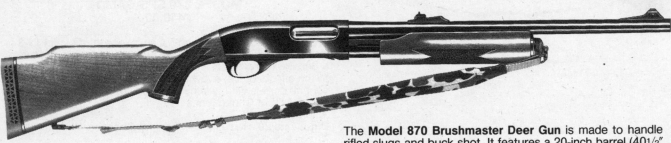

MODEL 870 BRUSHMASTER DEER GUN
$452.00 ($495.00 Left Hand)

The **Model 870 Brushmaster Deer Gun** is made to handle rifled slugs and buck shot. It features a 20-inch barrel (40½" overall) with 3-inch chamber and fully adjustable rifle-type sights. Stock fitted with rubber recoil pad and white-line spacer. Also available in standard model, but with lacquer finish, no checkering, recoil pad, grip cap; special handy short forend. Rem Choke with rifle sights. Imp. Cyl. **Weight:** 7¼ lbs.

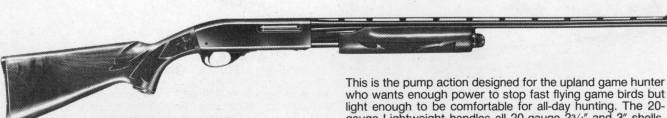

MODEL 870 WINGMASTER
20 Gauge Lightweight
$467.00

This is the pump action designed for the upland game hunter who wants enough power to stop fast flying game birds but light enough to be comfortable for all-day hunting. The 20-gauge Lightweight handles all 20-gauge 2³/₄" and 3" shells. REM choke and ventilated rib. **Stock:** American walnut stock and forend. Satin or Hi-Gloss finish. **Barrel lengths:** 26" and 28". **Average weight:** 6½ lbs.

REMINGTON PUMP SHOTGUNS

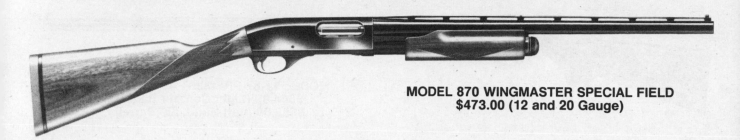

MODEL 870 WINGMASTER SPECIAL FIELD
$473.00 (12 and 20 Gauge)

The **Model 870 Wingmaster "Special Field"** shotgun combines the traditional, straight-stock styling of years past with features never before available on a Remington pump. Its 23-inch vent-rib barrel, slimmed and shortened forend, straight, cut-checkered stock offers upland hunters a quick, fast-pointing shotgun. The "Special Field" is chambered for 3-inch shells and will also handle all 2¾-inch shells interchangeably. Barrels will not interchange with standard 870 barrels. **Overall length:** 41½". **Weight:** 7 lbs. (12 ga.); 6¼ lbs. (20 ga.).

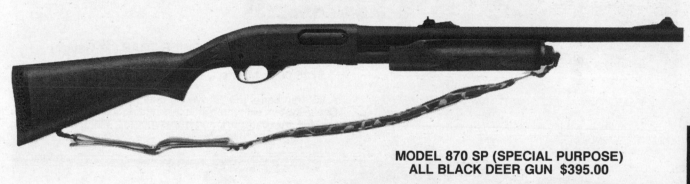

MODEL 870 SP (SPECIAL PURPOSE)
ALL BLACK DEER GUN $395.00

Gauge: 12. **Choke:** Fully rifled with rifle sights, recoil pad. **Barrel length:** 20". **Overall length:** 40½." **Average weight:** 7 lbs.

MODEL 870 WINGMASTER
CANTILEVER SCOPE MOUNT DEER GUN
(12 & 20 Ga.)
$532.00 (Fully Rifled; American Walnut Stock)
$452.00 (w/I.C. Deer Barrel and Rifle Sights)

Also available: 28 ga. and .410 **Model 870 Wingmaster** models w/25" barrels (45½" overall), Full or Modified. **Weight:** 6 lbs. (6½ lbs. in .410). **Price:** $504.00

REMINGTON AUTO SHOTGUNS

MODEL 11-87 PREMIER AUTOLOADER
$644.00 (Light Contour Barrel)
$692.00 (Left Hand, 28″ Barrel)

Remington's redesigned 12-gauge **Model 11-87 Premier Autoloader** features new, light-contour barrels that reduce both barrel weight and overall weight (more than 8 ounces). The shotgun has a standard 3-inch chamber and handles all 12-gauge shells interchangeably—from 2³/₄″ field loads to 3″ magnums. The gun's interchangeable REM choke system includes Improved Cylinder, Modified and Full chokes. Select American walnut stocks with fine-line, cut-checkering in satin or high-gloss finish are standard. Right-hand models are available in 26″, 28″ and 30″ barrels (left-hand models are 28″ only). A two-barrel gun case is supplied.
Also available: 21″ fully rifled cantilever deer barrel (41″ overall). **Weight:** 8¹/₂ lbs. **Price: $699.00**

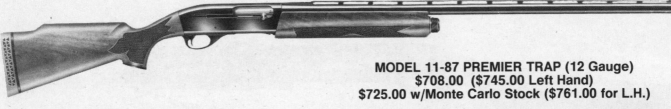

MODEL 11-87 PREMIER TRAP (12 Gauge)
$708.00 ($745.00 Left Hand)
$725.00 w/Monte Carlo Stock ($761.00 for L.H.)

A 30″ trap barrel (50¹/₂″ overall) offers trap shooters a REM Choke system with three interchangeable choke constrictions: trap full, trap extra full, and trap super full. **Weight:** 8³/₄ lbs.

MODEL 11-87 PREMIER SKEET (12 Gauge)
$700.00 ($735.00 Left Hand)

This model features American walnut wood and distinctive cut checkering with satin finish, plus new two-piece butt plate. REM Choke system includes option of two skeet chokes— skeet and improved skeet. Trap and skeet guns are designed for 12-gauge target loads and are set to handle 2³/₄″ shells only. **Barrel length:** 26″. **Overall length:** 46″. **Weight:** 8¹/₈ lbs.

MODEL 11-87 PREMIER DEER GUN
With Cantilever Scope Mount and
Fully Rifled 21″ Barrel
$699.00 (Satin Finish)

REMINGTON AUTO SHOTGUNS

MODEL 11-87 SPS (Special Purpose Synthetic)
12 Gauge Autoloader, 3″ Chamber w/Wood or
Synthetic Stock and REM Chokes
26″ or 28″ Vent Rib Barrels
$625.00

MODEL 11-87 SPST TURKEY GUN
12 Gauge Autoloader, 3″ Chamber with
21″ Barrel and Synthetic Stock
Extra-Full REM Choke Turkey Tube
$639.00
$707.00 w/Mossy Oak Greenleaf Camo Finish

MODEL 11-87 SPS-CAMO
12 Gauge Autoloader, 3″ Chamber
(26″ or 28″ barrel) with
REM Chokes and Synthetic Stock
In Mossy Oak Bottomland™ Camo
$693.00

MODEL 11-87 SPORTING CLAYS
$725.00

Remington's new **Model 11-87 Premier Sporting Clays** features a target-grade, American walnut competition stock with a length of pull that is ³/₁₆″ longer and ¹/₄″ higher at the heel. The tops of the receiver, barrel and rib have a non-reflective matte finish. The rib is medium high with a stainless mid-bead and ivory front bead. The barrel (26″ or 28″) has a lengthened forcing cone to generate greater pattern uniformity; and there are 5 REM choke tubes—Skeet, Improved Skeet, Improved Cylinder, Modified and Full. All sporting clays choke tubes have a knurled end extending .45″ beyond the muzzle for fast field changes. Both the toe and heel of the buttpad are rounded. **Weight:** 7¹/₂ lbs. (26″); 7⁵/₈ lbs. (28″)

SHOTGUNS

REMINGTON SHOTGUNS

MODEL 11-87 SPS BIG GAME
12 Gauge w/IC, Rifled & Turkey Super
Full Chokes, 21″ Rifle-Sighted Barrel and
Mossy Oak Bottomland Camo Pattern
$692.00

MODEL 11-87 SPS SPECIAL PURPOSE
SYNTHETIC ALL BLACK DEER GUN
(3″ Magnum) $665.00   ($699.00 Fully Rifled)

Features same finish as other SP models plus a padded, camostyle carrying sling of Cordura nylon with Q.D. sling swivels. Barrel is 21″ (41″ overall) with rifle sights and rifled and IC choke (handles all 2¾″ and 3″ rifled slug and buckshot loads as well as high-velocity field and magnum loads; does not function with light 2¾″ field loads). **Weight:** 8½ lbs.

PEERLESS OVER/UNDER
with Vent Rib and Engraved Sideplates

PEERLESS OVER/UNDER SHOTGUN
$1172.00

Practical, lightweight, well-balanced and affordable are the attributes of this new Remington shotgun. Features include an all-steel receiver, boxlock action and removable side plates (engraved with a pointer on one side and a setter on the other). The bottom of the receiver has the Remington logo, plus the words "Peerless, Field" and the serial number. Cut-checkering appears on both pistol grip and forend (shaped with finger grooves and tapered toward the front). The buttstock is fitted with a black, vented recoil pad. The stock is American walnut with an Imron finish.

SPECIFICATIONS
Gauge: 12 (3″ chamber)
Chokes: REM Choke System (1 Full, 1 Mod., 1 Imp. Cyl.)
Barrel lengths: 26″, 28″, 30″ with vent rib
Overall length: 43″ (26″ barrel); 45″ (28″ barrel); 47″ (30″ barrel)
Weight: 7¼ lbs. (26″); 7⅜ lbs. (28″); 7½ lbs. (30″)
Trigger: Single, selective, gold-plated
Safety: Automatic safety
Sights: Target gun style with mid-bead and Bradley-type front bead
Length of pull: 14³/₁₆″
Drop at comb: 1½″   **Drop at heel:** 2¼″
Features: Solid, horseshoe-shaped locking bar with two rectangular lug extensions on either side of the barrel's mid-bore; fast lock time (3.28 milliseconds)

REMINGTON AUTO SHOTGUNS

MODEL 1100 AUTOLOADING SHOTGUNS

The Remington Model 1100 is a 5-shot gas-operated auto-loading shotgun with a gas metering system designed to reduce recoil effect. This design enables the shooter to use all 2³/₄-inch standard velocity "Express" and 2³/₄-inch Magnum loads without any gun adjustments. Barrels, within gauge and versions, are interchangeable. The 1100 is made in gauges of 12, Lightweight 20, 28 and .410. All 12 and 20 gauge versions include REM Choke; interchangeable choke tubes in 26″ and 28″ (12 gauge only) barrels. The solid-steel receiver features decorative scroll work. Stocks come with fine-line checkering in a fleur-de-lis design combined with American walnut and a scratch-resistant finish. Features include white-diamond inlay in pistol-grip cap, white-line spacers, full beavertail forend, fluted-comb cuts, chrome-plated bolt and metal bead front sight. Made in U.S.A.

MODEL 1100 SPECIAL FIELD (12 & 20 GA.)
$605.00

The **Model 1100 "Special Field"** shotgun combines traditional, straight-stock styling with its 23-inch vent-rib barrel and slimmed and shortened forend, which offer upland hunters a quick, fast-pointing shotgun. Non-engraved receiver; non-Magnum extra barrels are interchangeable with standard Model 1100 barrels. **Overall length:** 41″. **Stock dimensions:** Length of pull 14¹/₈″; drop at comb 1¹/₂″; drop at heel 2¹/₂″. **Choke:** REM Choke system. **Weight:** 7¹/₄ lbs. (12 ga.); 6¹/₂ lbs. (20 ga.).

MODEL 1100 LT-20
$605.00

Designed for 2³/₄-inch Magnum shells; accepts and functions with any 1100 standard 2³/₄-inch chambered barrel. Available in 20 gauge, 26″ or 28″ ventilated rib barrels, Rem Choke. **Stock dimensions:** 14″ long including pad; 1¹/₂″ drop at comb; furnished with recoil pad. Satin or Hi-Gloss finish. **Weight:** About 7 lbs.
Also available: 3″ Magnum (28″ and satin finish only).

MODEL 1100 DEER GUN
Lightweight 20 Gauge
$545.00

Features 21-inch barrels, Improved Cylinder choke. Rifle sights adjustable for windage and elevation. Recoil pad. Choked for both rifled slugs and buck shot. **Weight:** 6¹/₂ lbs. **Overall length:** 41″. Also availble w/fully rifled cantilever 21″ deer barrel. **Price:** $652.00

SHOTGUNS

REMINGTON AUTO SHOTGUNS

MODEL 1100 TOURNAMENT SKEET
$692.00 (20, 28 and .410 Gauge)

The world's winningest skeet gun, with high-grade positive cut-checkering on selected American walnut stock and forend. The LT-20, 28 and .410 gauge Model 1100 Tournament Skeet guns have a higher vent rib to match the sight picture of the 12-gauge model. A true "matched set," with all the reliability, superb balance, and low recoil sensation that make it the choice of over 50% of the entrants in the world skeet shooting championships. **Barrel lengths:** 25″ (28 ga. & .410 bore) and 26″ (20 ga.). **Chokes:** REM Choke (20 ga.) and Skeet choke (28 and .410 ga.). **Weight:** 6¾ lbs. (20 ga.); 6½ lbs. (28 ga.); 7 lbs. (.410).

MODEL 1100 AUTOLOADER
$647.00 (28 and .410 Gauge)

The Remington Model 1100 Autoloading shotguns in 28 and .410 gauges are scaled-down models of the 12-gauge version. Built on their own receivers and frames, these small-gauge shotguns are available in full (.410 only) and modified chokes with ventilated rib barrels.

SPECIFICATIONS. **Type:** Gas-operated. **Capacity:** 5-shot with 28 ga. shells; 4-shot with 3″ .410 ga. shells; 3-shot plug furnished. **Barrel:** 25″ of special Remington ordnance steel; extra barrels interchangeable within gauge. **Chamber:** 3″ in .410, 2¾″ in 28 ga. **Overall length:** 45″. **Safety:** Convenient crossbolt type. **Receiver:** Made from solid steel, top matted, scroll work on bolt and both sides of receiver. **Stock dimensions:** Walnut; 14″ long; 2½″ drop at heel; 1½″ drop at comb. **Average weight:** 6½ lbs. (28 ga.); 7 lbs. (.410).

MODEL 1100 LT-20 YOUTH GUN
Lightweight, 20 Gauge Only
$605.00

The Model 1100 LT-20 Youth Gun autoloading shotgun features a shorter barrel (21″) and stock. **Overall length:** 39½″. **Weight:** 6½ lbs.

REMINGTON SHOTGUNS

SP-10 MAGNUM SHOTGUN
$993.00

Remington's **SP-10 Magnum** is the only gas-operated semi-automatic 10-gauge shotgun made today. Engineered to shoot steel shot, the SP-10 delivers up to 34 percent more pellets to the target than standard 12-gauge shotgun and steel shot combinations. This autoloader features a non-corrosive, stainless-steel gas system, in which the cylinder moves—not the piston. This reduces felt recoil energy by spreading the recoil over a longer time. The SP-10 has a 3/8″ vent rib with middle and front sights for a better sight plane. It is also designed to appear virtually invisible to the sharp eyes of waterfowl. The American walnut stock and forend have a protective, low-gloss satin finish that reduces glare, and positive deep-cut checkering

for a sure grip. The receiver and barrel have a matte finish, and the stainless-steel breech bolt features a non-reflective finish. Remington's new autoloader also has a brown-vented recoil pad and a padded camo sling of Cordura nylon for easy carrying. The receiver is machined from a solid billet of ordnance steel for total integral strength. The SP-10 vented gas system reduces powder residue buildup and makes cleaning easier.

Gauge: 10. **Barrel lengths & choke:** 26″ REM Choke and 30″ REM Choke. **Overall length:** 51 1/2″ (30″ barrel) and 47 1/2″ (26″ barrel). **Weight:** 11 lbs. (30″ barrel) and 10 3/4 lbs. (26″ barrel).

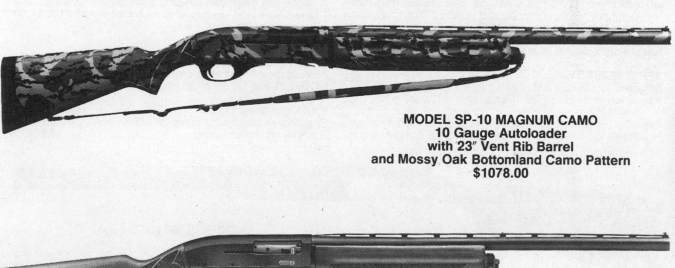

MODEL SP-10 MAGNUM CAMO
10 Gauge Autoloader
with 23″ Vent Rib Barrel
and Mossy Oak Bottomland Camo Pattern
$1078.00

MODEL SP-10 MAGNUM COMBO
10 Gauge Autoloader (26″ barrel) with REM Chokes
Extra 22″ Barrel with Rifle Sights and
Extra-Full Turkey Choke Tube
$1132.00

RIZZINI SHOTGUNS

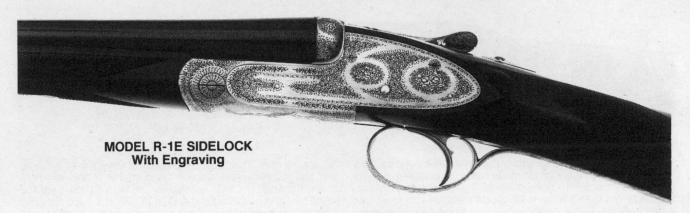

MODEL R-1E SIDELOCK
With Engraving

MODEL R-1 and R-2 SHOTGUNS

Rizzini offers the **R-1**, a sidelock ejector, and **R-2**, an Anson & Deeley boxlock with a removable inspection plate on the bottom of the action. The basic price of these Rizzini shotguns includes the choice of single or double triggers and a fitted leather trunk case (but not the cost of engraving). A wide variety of ornamental and game scene engravings are available.

Multi-gauge two-barrel sets (supplied with a single forearm) are available in .410/28 ga., 28/20 ga. or 20/16 ga. The actions and locks are carefully shaped and polished. The stock finish is hand-rubbed oil applied after the stock is filled and sealed. Options include stock ovals, hand-detachable locks, rib type, barrel length, weight of gun, and more.

SPECIFICATIONS
Gauges: 12, 16, 20, 28, .410
Barrel lengths: 25″ to 28″ (28 ga. and .410); 25″ to 30″ (12, 16 and 20 ga.)
Top ribs: Straight matted concave ribs are standard; swamped smooth concave and level file cut ribs are optional

Action: Anson & Deeley boxlock with inspection plate (Model R-2 only); Holland & Holland-type sidelock (Model R-1 only)
Weight: 4 lbs. 14 oz. to 6 lbs. (28 and .410); 5½ lbs. to 6 lbs. 12 oz. (20 ga.); 5 lbs. 12 oz. to 7 lbs. (16 ga.); 6 lbs. 4 oz. to 8 lbs. 4 oz. (12 ga.)
Stock & Forearm: High-luster hand-rubbed oil finish; straight-grip stock with checkered butt; classic forearm; pistol-grip stock and beavertail forearm optional
Trigger: Non-selective trigger (double trigger with front trigger hinged optional)
Prices:
MODEL R-1 in 12, 16 or 20 ga.
  (w/o engraving) **$38,800.00**
  In 28 ga. or .410 (w/o engraving) 43,800.00
MODEL R-2 in 12, 16 or 20 ga.
  (w/o engraving) 23,000.00
Engravings available:
Fine English scroll or ornamental with swans . . 8,500.00
Fracassi-style ornamental 18,800.00

MODEL S790 (20 Gauge)

MODEL 780 O/U SERIES

MODEL S780N. Double trigger, standard extractors. **Gauges:** 12 and 16. **Barrel length:** 27.3″. **Weight:** 6.4 lbs. **Features:** Schnabel forend. **Stock:** Standard walnut. **Price: $1650.00**
MODEL S780 EM. Single selective trigger, standard ejectors. **Gauges:** 12, 16, 20, 28, 36. **Barrel lengths:** 26″ and 27.3″. **Weight:** 6.25 lbs. (20, 28 and 36 ga.); 6.8 lbs. (12 and 16 ga.). **Price:** **$1300.00**
MODEL 780 EML. Same specifications as above, but in 12 and 16 ga. only. **Weight:** 6.8 lbs. **Price:** **$1450.00**
MODEL S780 EMEL. Same specifications as above, but with hand finishing and hand engraving. **Price:** **$7100.00**

MODEL S782 SERIES

MODEL S782. Single selective trigger, ejectors, sideplates. **Gauges:** 12 and 16. **Barrel lengths:** 26″ and 27.3″. **Weight:** 6.9 lbs. **Price:** **$1550.00**
MOODEL S782 EML. Same specifications as above, but with select walnut stock. **Price:** **$1900.00**
MODEL S782 EM SLUG. Same features as above except **Gauges:** 12 and 20. **Barrel lengths:** 24″ and 25.75″. **Weight:** 6.6 lbs. **Price:** **$3000.00**

RIZZINI SHOTGUNS

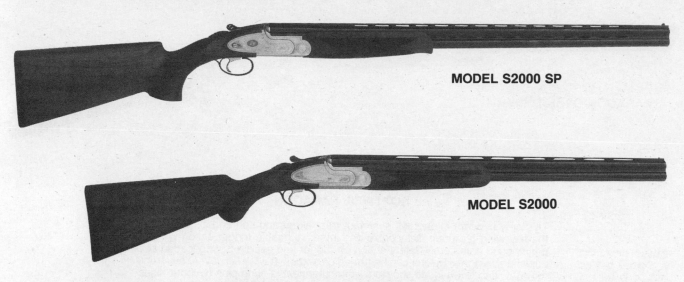

MODEL S2000 SP

MODEL S2000

TRAP, SKEET & SPORTING MODELS

MODEL S780 SKEET. Gauge: 12. **Barrel lengths:** 26.1″ and 27.3″. **Weight:** 7 lbs. Same features as Model S780 Series. **Price:** **$1450.00**

MODEL S780 TRAP. Stock is standard walnut. **Barrel length:** 29.25″. **Weight:** 7.8 lbs. **Price:** **$1450.00**

MODEL S780 SPORTING. Same features and specifications as Model S780N except **Gauge:** 12. **Weight:** 7.25 lbs. Features sporting-style forend. **Price:** **$1450.00**

MODEL S790 SKEET SL. Same specifications as above with select walnut stock. **Weight:** 7½ lbs. **Price:** . . **$2050.00**

MODEL S790 TRAP SL. Same features as above but with select walnut stock **Weight:** 7.9 lbs. **Price:** . . . **$2050.00**

MODEL S790 SL. Same as above, but with select walnut stock. **Weight:** 7.4 lbs. **Price:** **$2050.00**

MODEL S790 EL. Same as above but with hand finishing and hand engraving. **Weight:** 8.15 lbs. Select walnut stock. **Price:** **$5650.00**

MODEL S2000 TRAP. Same specifications as above, but with select walnut stock, sideplates, single trigger and automatic ejector. **Gauges:** 12 and 20. **Price:** **$2100.00**

MODEL S2000 SP. Same specifications as above, but with select walnut stock, sideplates, single trigger, automatic ejector and double vent ribs. **Gauges:** 12 and 20. **Price:** . . . . **$0000.00 ??**

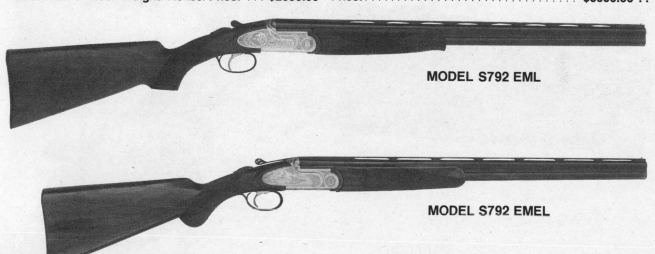

MODEL S792 EML

MODEL S792 EMEL

Also available:
MODEL S792 EML. Single selective trigger, ejectors, sideplates. **Gauges:** 20, 28, 36. Schnabel forend. **Weight:** 6 lbs. **Price:** **$1800.00**

MODEL S792 EMEL. Same as above, except in 20 gauge only. **Weight:** 6.1 lbs. Hand finishing and hand engraving. **Price:** **$7600.00**

MODEL S792 TRAP SL. Same as other trap models, but in 20 gauge only. **Price:** **$1900.00**

ROTTWEIL SHOTGUNS

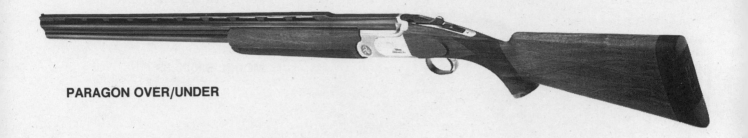

PARAGON OVER/UNDER

ROTTWEIL PARAGON

This new concept in shotgun systems, trap, skeet and sporting clays includes the following features: Detachable and interchangeable trigger action with superimposed hammers • Safety action on trigger and sears • Spring-loaded self-adjusting wedges • Ejector can be turned on and off at will • Top lever convertible for right- and left-handed shooters • Interchangeable firing pins (without disassembly) • Length and weight of barrels selected depending on application (see below) • Module system: Fully interchangeable receiver, barrels, stocks trigger action and forends • Select walnut stocks

Barrel lengths:

Field & Skeet	27$1/2$″	Sporting	28$1/2$″
American Skeet	28″	Trap	29″ & 30″
Parcours	28$3/8$″	American Trap Single	32″ & 34″

Price: . . . . **$5,200.00** to **$7,200.00**

PARAGON
(Close-up Open)

RUGER SHOTGUNS

RED LABEL OVER/UNDER SHOTGUN
$1157.50 (Incl. Screw-in Chokes)

These shotguns are made of hardened chrome molybdenum, other alloy steels and music wire coil springs. Features include boxlock action with single-selective mechanical trigger, selective automatic ejectors, automatic top safety, free-floating vent rib with serrated top surface to reduce glare, standard brass bead front sight, and stainless steel receiver. Stock and semi-beavertail forearm are shaped from American walnut with hand-cut checkering (20 lines per inch). Pistol-grip cap and rubber recoil pad are standard, and all wood surfaces are polished

and beautifully finished. Available in 12 or 20 gauge with 3″ chambers.

Screw-in choke inserts. Designed especially for the popular 12-gauge "Red Label" over/under shotgun. Easily installed with a key wrench packaged with each shotgun. Choke fits flush with the muzzle. Every shotgun is equipped with a Full, Modified, Improved Cylinder and two Skeet screw-in chokes. The muzzle edge of the chokes has been slotted for quick identification in or out of the barrels. Full choke has 3 slots; Modified, 2 slots; Improved Cylinder, 1 slot (Skeet has no slots). For additional specifications, please see table below.

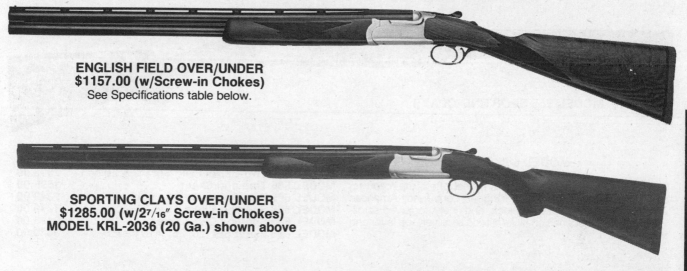

ENGLISH FIELD OVER/UNDER
$1157.00 (w/Screw-in Chokes)
See Specifications table below.

SPORTING CLAYS OVER/UNDER
$1285.00 (w/2⁷/₁₆″ Screw-in Chokes)
MODEL KRL-2036 (20 Ga.) shown above

Catalog Number	Gauge	Chamber	Choke*	Barrel Length	Overall Length	Length Pull	Drop Comb	Drop Heel	Sights**	Approx. Wt. (lbs.)	Type Stock
KRL-2826	28	2¾″	F, M, IC, S+	26″	43″	14⅛″	1½″	2½″	GBF	6	Pistol Grip
KRL-2827	28	2¾″	F, M, IC, S+	28″	45″	14⅛″	1½″	2½″	GBF	6¼	Pistol Grip
KRL-2029	20	3″	F, M, IC, S+	26″	43″	14⅛″	1½″	2½″	GBF	7	Pistol Grip
KRL-2030	20	3″	F, M, IC, S+	28″	45″	14⅛″	1½″	2½″	GBF	7¼	Pistol Grip
KRLS-2029	20	3″	F, M, IC, S+	26″	43″	14⅛″	1½″	2½″	GBF	6¾	Straight
KRLS-2030	20	3″	F, M, IC, S+	28″	45″	14⅛″	1½″	2½″	GBF	7	Straight
KRL-2036	20	3″	M, IC, S+	30″	47″	14⅛″	1½″	2½″	GBF/GBM	7	Pistol Grip
KRL-1226	12	3″	F, M, IC, S+	26″	43″	14⅛″	1½″	2½″	GBF	7¾	Pistol Grip
KRL-1227	12	3″	F, M, IC, S+	28″	45″	14⅛″	1½″	2½″	GBF	8	Pistol Grip
KRLS-1226	12	3″	F, M, IC, S+	26″	43″	14⅛″	1½″	2½″	GBF	7½	Straight
KRLS-1227	12	3″	F, M, IC, S+	28″	45″	14⅛″	1½″	2½″	GBF	7¾	Straight
KRL-1236	12	3″	M, IC, S+	30″	47″	14⅛″	1½″	2½″	GBF/GBM	7¾	Pistol Grip

*F-Full, M-Modified, IC-Improved Cylinder, S-Skeet. +Two skeet chokes standard with each shotgun.
**GBF-Gold-Bead Front Sight, GBM-Gold-Bead Middle.

SHOTGUNS

SKB SHOTGUNS

MODEL 585 SPORTING CLAYS

MODEL 585 SERIES

SPECIFICATIONS
Gauges: 12 (2³/₄″ or 3″), 20 (3″), 28 (2³/₄″), .410 (3″)
Barrel lengths:
  12 gauge—26″, 28″, 30″, 32″, 34″ w/Inter-Choke tube
  20 or 28 gauge—26″ and 28″ w/Inter-Choke tube
  .410 gauge—26″ and 28″ w/Imp. Cyl., Mod./Mod. & Full
Overall length: 43″ to 51³/₈″
Weight: 6.6 lbs. to 8¹/₂ lbs.
Sights: Metal front bead (Field Model); target sights on Trap, Skeet and Sporting Clay models
Stock: Hand-checkered walnut with high-gloss finish; target stocks available in standard or Monte Carlo

Features: Silver nitride receiver with game scene engraving; boxlock action; manual safety; automatic ejectors; single selective trigger; ventilated rib

Prices:
MODEL 585 Field and Youth (12 & 20 ga.) $1149.00
MODEL 585 Two-Barrel Field Set (12 & 20 ga.) . 1849.00
MODEL 585 Trap and Skeet (12 & 20 ga.) 1195.00
MODEL 585 Two-Barrel Trap Combo 1849.00
MODEL 585 Sporting Clays (12 & 20 ga.) 1249.00
MODEL 585 Skeet Set (20, 28, .410 ga.) 2849.00

MODEL 685 SPORTING CLAYS

MODEL 685 SERIES

Similar to Model 585 Deluxe, this model offers the following additional features: gold-plated trigger, semi-fancy American walnut stock, jeweled barrel block, and finely engraved scrollwork and game scenes on silver nitride receiver, top lever and trigger guard.

Prices:
MODEL 685 Field (12 & 20 ga.) $1649.00
MODEL 685 Two-Barrel Field Set (12 & 20 ga.) . 2395.00
MODEL 685 Trap and Skeet 1695.00
MODEL 685 Two-Barrel Trap Combo 2349.00
MODEL 685 Sporting Clays (12 & 20 ga.) 1749.00
MODEL 685 Sporting Clays Sets 2495.00
MODEL 685 Skeet Set (20, 28, .410 ga.) 3449.00

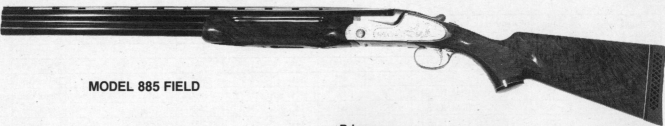

MODEL 885 FIELD

MODEL 885 SERIES

Similar to Model 685. Additional features include intricately engraved scrollwork and game scenes on silver nitride receiver, sideplates, top lever and trigger guard; select semi-fancy American walnut stock.

Prices:
MODEL 885 Field* (12 & 20 ga.) $2049.00
MODEL 885 Two-Barrel Field Set*
  (12 & 20 ga.) 2149.00
MODEL 885 Trap and Skeet* 2095.00
MODEL 885 Two-Barrel Trap Combo 2995.00
MODEL 885 Sporting Clays (12 & 20 ga.) 2149.00
MODEL 885 Skeet Set (20, 28, .410 ga.) 3895.00

* Also available in 28 ga. and .410 (approx. **$50** add'l.)

TIKKA SHOTGUNS

(Formerly Valmet)

MODEL 512S OVER/UNDER FIELD

TIKKA 512S OVER/UNDER
FIELD GRADE $1225.00
PREMIUM GRADE $1275.00

Designed for the experienced hunter, Tikka's 512S represents the pride and skill of "Old World" European craftsmanship. The barrels are polished to a mirror finish and deeply blued. Select semi-fancy American walnut stock and forearm highlight fine, deep-cut checkering. Other features include:

Time-proven action: Designed to handle large centerfire calibers for more durability and reliability.

Mechanical trigger: Fires two shots as fast as you can pull the trigger. Does not rely on the inertia from the recoil of the first shot to set the trigger for the second. In the event of a faulty primer or light hit, inertia trigger shotguns cannot function on the second round.

Single selective trigger: Selector button is located on the trigger for fast, easy selection.

Large trigger guard opening: Designed for cold weather shooting; permits easy finger movement when wearing gloves.

American walnut stock and forearm: Add greatly to overall appearance.

Superior stock design: A palm swell provides additional hand comfort. Length and angle (pitch) can be adjusted for a perfect fit with addition of factory spacers. Fine, deep-cut checkering.

Palm-filling forearm: Rounded and tapered for comfort and smooth, true swing, plus fine, deep-cut checkering.

Automatic ejectors: Select and eject fired rounds. Raise unfired shells for safe removal.

Chrome-lined barrels: For more consistent patterns. Eliminates pitting and corrosion, extends barrel life even with steel shot.

Stainless steel choke tubes: Added strength over regular carbon and alloy materials. Easily handles steel shot. Recessed so as not to detract from appearance. Tight tolerances enable truer patterns and enhance choke versatility.

Sliding locking bolt: Secure lockup between receiver and barrels. Wears in, not loose.

Polished blue receiver: Fully engraved with gold inlay.

Wide vent rib: Cross-file pattern reduces glare. Fluorescent front and middle beads.

Automatic safety: Goes to safe position automatically when gun is opened.

Cocking indicators: Allow shooter to determine (through sight or feel) which barrel has been fired.

Steel receiver: Forged and machined for durability.

Chamber: 3-inch on all models

Two-piece firing pin: For more durability

Versatility: Change from over/under shotgun to shotgun/rifle, trap, skeet or double rifle. Precision tolerances require only minor initial fitting.

SPECIFICATIONS
Gauge: 12
Chambers: 3″
Weight: 7¼ lbs. w/26″ barrels; 7½ lbs. w/28″ barrels
Barrel lengths/chokes:
26″, 5 chokes (F, M, IM, IC & Skeet)
28″, 5 chokes (F, M, IM, IC & Skeet)

SPORTING CLAYS SHOTGUN (not shown)
$1315.00

Designed to accommodate the specific requirements of the shooter in this, the fastest growing shooting sport in America today. The Sporting Clays shotgun features a specially designed American walnut stock with a double palm swell finished with a soft satin lacquer for maximum protection with minimum maintenance. Available in 12 gauge with a selection of 5 recessed choke tubes. Other features include a 3″ chamber, manual safety, customized sporting clay recoil pad, single selective trigger, blued receiver and 28″ and 30″ barrel with ventilated side and top rib with two iridescent beads. In addition, the shotgun is furnished with an attractive carrying case.

Manufactured in Italy, Tikka is designed and crafted by Sako of Finland, which has enjoyed international acclaim for the manufacture of precision sporting firearms since 1918.

TIKKA SHOTGUNS

(Formerly Valmet)

MODEL 512S SHOTGUN/RIFLE

TIKKA 512S SHOTGUN/RIFLE
$1350.00

Tikka's unique 512S Shotgun/Rifle combination continues to be the most popular gun of its type in the U.S. Its features are identical to the 512S Field Grade over/under shotguns, including strong steel receiver, superior sliding locking mechanism with automatic safety, cocking indicators, mechanical triggers and two-piece firing pin. In addition, note the other features of this model—

Barrel regulation: Adjusts for windage simply by turning the screw on the muzzle. Elevation is adjustable by regulating the sliding wedge located between the barrels.

Compact: 24-inch barrels mounted on the low-profile receiver limit the overall length to 40 inches (about 5″ less than most bolt-action rifles with similar 24-inch barrels).

Single selective trigger: A barrel selector is located on the trigger for quick, easy selection. Double triggers are also available.

Choice of calibers: Choose from 222 or 308. Both are under the 12 gauge, 3″ chamber with Improved Modified choke.

Sighting options: The vent rib is cross-filed to reduce glare. The rear sight is flush-folding and permits rapid alignment with the large blade front sight. The rib is milled to accommodate Tikka's one-piece scope mount with 1″ rings. Scope mount is of ''quick release'' design and can be removed without altering zero.

European walnut stock: Stocks are available with palm swell for greater control and comfort. Quick detachable sling swivel. Length or pitch adjustable with factory spacers. Semi-Monte Carlo design.

Interchangeability: Receiver will accommodate Tikka's over/under shotgun barrels and double-rifle barrels with minor initial fitting.

SPECIFICATIONS
Gauge/Caliber: 12/222 or 12/308
Chamber: 3″ with Improved Modified choke
Barrel length: 24″
Overall length: 40″
Weight: 8 lbs.
Stock: European walnut with semi-Monte Carlo design

Extra Barrel Sets:
Over/Under **$650.00**
Shotgun/Rifle. **725.00**
Double Rifle **825.00**

WEATHERBY SHOTGUNS

ATHENA GRADE V CLASSIC FIELD

ATHENA GRADE IV $2200.00
ATHENA GRADE V $2575.00

Receiver: The Athena receiver houses a strong, reliable box-lock action, yet it features side lock-type plates to carry through the fine floral engraving. The hinge pivots are made of a special high-strength steel alloy. The locking system employs the time-tested Greener cross-bolt design.

Single selective trigger: It is mechanically rather than recoil operated. This provides a fully automatic switchover, allowing the second barrel to be fired on a subsequent trigger pull, even during a misfire. A flick of the trigger finger and the selector lever, located just in front of the trigger, is all the way to the left, enabling you to fire the lower barrel first, or to the right for the upper barrel. The Athena trigger is selective as well.

Barrels: The breech block is hand-fitted to the receiver, providing closest possible tolerances. Every Athena is equipped with a matted, ventilated rib and bead front sight.

Selective automatic ejectors: The Athena contains ejectors that are fully automatic both in selection and action.

Slide safety: The safety is the traditional slide type located conveniently on the upper tang atop the pistol grip.

Stock: Each stock is carved from specially selected Claro walnut, with fine line hand-checkering and high-luster finish. Trap model has Monte Carlo stock only.

See the Athena and Orion table on the following page for additional information and specifications.

GRADE IV CHOKES
Fixed Choke
Field, .410 Gauge
Skeet, 12 or 20 Gauge
IMC Multi-Choke
Field, 12, 20 or 28 Gauge
Trap, 12 Gauge
Trap, single barrel, 12 Gauge
Trap Combo, 12 Gauge

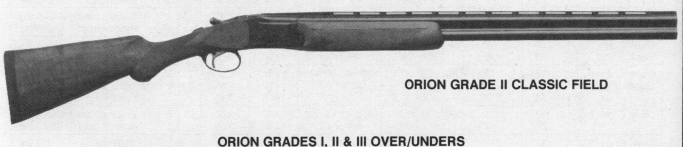

ORION GRADE II CLASSIC FIELD

ORION GRADES I, II & III OVER/UNDERS

For greater versatility, the Orion incorporates the integral multichoke (IMC) system. Available in Extra-full, Full, Modified, Improved Modified, Improved Cylinder and Skeet, the choke tubes fit flush with the muzzle without detracting from the beauty of the gun. Three tubes are furnished with each gun. The precision hand-fitted monobloc and receiver are machined from high-strength steel with a highly polished finish. The box-lock design uses the Greener cross-bolt locking system and special sears maintain hammer engagement. Pistol grip stock and forearm are carved of Claro walnut with hand-checkered diamond inlay pattern and high-gloss finish. Chrome-moly steel barrels, and the receiver, are deeply blued. The Orion also features selective automatic ejectors, single selective trigger, front bead sight and ventilated rib. The trap model boasts a curved trap-style recoil pad and is available with Monte Carlo stock only. **Weight:** 12 ga. Field, 7½ lbs.; 20 ga. Field, 7½ lbs.; Trap, 8 lbs.

See following page for prices and additional specifications.

ORION CHOKES
Grade I
IMC Multi-Choke, Field, 12 or 20 Gauge
Grade II
Fixed Choke, Field, .410 Gauge
Fixed Choke, Skeet, 12 or 20 Gauge
IMC Multi-Choke, Field, 12, 20 or 28 Gauge
IMC Multi-Choke, Trap, 12 Gauge
Grade II Sporting Clays
12 Gauge only
Grade III
IMC Multi-Choke, Field, 12 or 20 Gauge

WEATHERBY SHOTGUNS

ORION GRADE II CLASSIC SPORTING
12 Gauge Over/Under
$1335.00

ORION GRADES I, II, III
Prices:
Orion I $1165.00
Orion II Classic Field 1235.00
Orion II Sporting Clays 1335.00
Orion III Field & Classic Field 1470.00

ATHENA & ORION OVER/UNDER SHOTGUN SPECIFICATIONS

Model	Gauge	Chamber	Barrel Length	Overall Length	Length of Pull	Drop at Heel	Drop at Comb	Bead Sight	Approx. Weight
Athena Grade V Classic Field	12 20	3″ 3″	30″, 28″ or 26″ 28″ or 26″	47″, 45″ or 43″ 45″ or 43″	14¼″ 14¼″	2.5″ 2.5″	1.5″ 1.5″	Brilliant front Brilliant front	6½–8 lbs. 6½–8 lbs.
Athena Grade IV Field	12 20	3″ 3″	30″, 28″ or 26″ 28″ or 26″	47″, 45″ or 43″ 45″ or 43″	14¼″ 14¼″	2.5″ 2.5″	1.5″ 1.5″	Brilliant front Brilliant front	6½–8 lbs. 6½–8 lbs.
Orion Grade III Classic Field	12 20	3″ 3″	28″ 26″	45″ 43″	14¼″ 14¼″	2.5″ 2.5″	1.5″ 1.5″	Brilliant front Brilliant front	6½–8 lbs. 6½–8 lbs.
Orion Grade III Field	12 20	3″ 3″	30″, 28″ or 26″ 28″ or 26″	47″, 45″ or 43″ 45″ or 43″	14¼″ 14¼″	2.5″ 2.5″	1.5″ 1.5″	Brilliant front Brilliant front	6½–8 lbs. 6½–8 lbs.
Orion Grade II Classic Field	12 20 28	3″ 3″ 2¾″	30″, 28″ or 26″ 28″ or 26″ 26″	47″, 45″ or 43″ 45″ or 43″ 43″	14¼″ 14¼″ 14¼″	2.5″ 2.5″ 2.5″	1.5″ 1.5″ 1.5″	Brilliant front Brilliant front Brilliant front	6½–8 lbs. 6½–8 lbs. 6½–8 lbs.
Orion Grade I	12 20	3″ 3″	30″, 28″ or 26″ 28″ or 26″	47″, 45″ or 43″ 45″ or 43″	14¼″ 14¼″	2.5″ 2.5″	1.5″ 1.5″	Brilliant front Brilliant front	6½–8 lbs. 6½–8 lbs.
Orion Grade II Classic Sporting	12	2¾″	28″	45″	14¼″	2.25″	1.5″	Midpoint w/ white front	7½ lbs.
Orion Grade II Sporting	12	2¾″	30″ or 28″	47″ or 45″	14¼″	2.25″	1.5″	Midpoint w/ white front	7½–8 lbs.

WINCHESTER SECURITY SHOTGUNS

These tough 12-gauge shotguns provide backup strength for security and police work as well as all-around utility. The action is one of the fastest second-shot pumps made. It features a front-locking rotating bolt for strength and secure, single-unit lockup into the barrel. Twin-action slide bars prevent binding.

The shotguns are chambered for 3-inch shotshells. They handle 3-inch Magnum, 2³/₄-inch Magnum and standard 2³/₄-inch shotshells interchangeably. They have cross-bolt safety,

walnut-finished hardwood stock and forearm, black rubber butt pad and plain 18-inch barrel with Cylinder Bore choke. All are ultra-reliable and easy to handle.

Special chrome finishes on Police and Marine guns are actually triple-plated: first with copper for adherence, then with nickel for rust protection, and finally with chrome for a hard finish. This triple-plating assures durability and quality. Both guns have a forend cap with swivel to accommodate sling.

MODEL 1300 DEFENDER
(8-Shot Wood Model Shown) $277.00
$381.00 DEFENDER/FIELD COMBO

Security Defender™ is ideal for home security use. The compact 38⁵/₈″ overall length (28¹/₂″ w/pistol grip) handles and stores easily. The Defender has a deep blued finish on metal surfaces and features a traditional ribbed forearm for sure pumping grip. It has a metal bead front sight. The magazine holds eight 12-gauge, 3″ Magnum shells.

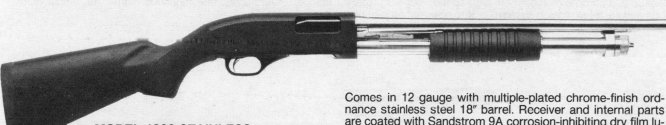

MODEL 1300 STAINLESS
MARINE DEFENDER
$440.00

Comes in 12 gauge with multiple-plated chrome-finish ordnance stainless steel 18″ barrel. Receiver and internal parts are coated with Sandstrom 9A corrosion-inhibiting dry film lubricant. Stock and forend are made of corrosion and moisture-resistant material. **Capacity:** 7 shells (2³/₄″). **Sights:** Bead front (sling swivels incl.). Additional specifications in chart below.

SPECIFICATIONS MODEL 1300 DEFENDER

Model	Gauge	Barrel Length	Chamber	Capacity*	Choke	Overall Length	Length of Pull	Drop At Comb/Heel		Weight (lbs.)	Sights
Combo, Hardwood Stock and Pistol Grip	12	18″	3″ Mag.	5	Cyl	38⁵/₈″	14″	1¹/₂″	2¹/₂″	6¹/₄″	MBF
	12	28 VR	3″ Mag.	5	W1M	48⁵/₈	14	1¹/₂	2¹/₂	7¹/₂	MBF
Hardwood Stock 5 Shot	12	18″	3″ Mag.	5	Cyl	38⁵/₈″	14″	1³/₈″	2³/₄″	6¹/₂	MBF
Hardwood Stock 8 Shot	12	18″	3″ Mag.	8¹	Cyl.	38⁵/₈″	14″	1³/₈″	2³/₄″	6³/₄	MBF
Synthetic Pistol Grip	12	18″	3″ Mag.	8¹	Cyl.	28⁵/₈″	—	—	—	5³/₄	MBF
Synthetic Stock 12 Gauge	12	18″	3″ Mag.	8¹**	Cyl.	38⁵/₈″	14″	1³/₈″	2³/₄″	6³/₄	MBF
Synthetic Stock 20 Gauge	20	18″	3″ Mag.	5	Cyl.	38⁵/₈″	14″	1³/₈″	2³/₄″	6	MBP
Stainless Marine	12	18″	3″ Mag.	7¹	Cyl.	38⁵/₈	14″	1³/₈″	2³/₄″	6³/₄	MBF
Pistol Grip (Stainless)	12	18	3″ Mag.	7¹	Cyl.	28⁵/₈	14	—	—	5³/₄	MBF

* Includes one shotshell in chamber when ready to fire. ** Also available in 5-shot. ¹ Subtract one shell capacity for 3″ shells. VR = Ventilated rib. Cyl. – Cylinder Bore. MBF = Metal bead front.

WINCHESTER SHOTGUNS

**MODEL 1300 BLACK SHADOW DEER
SMOOTHBORE
$287.00**

MODEL 1300 WALNUT PUMP DEER

Winchester's Model 1300 Walnut Deer pump-action shotgun features a rifled barrel with 8 lands and grooves, rifle-type sights, and a receiver that is factory-drilled and tapped for scope. Model 1300 Walnut Deer has a smooth-bore barrel that comes with an extra long Sabot-rifled choke tube. Also included is an improved Cylinder Winchoke tube for traditional slug or buckshot shooting.

The Walnut models feature sculptured, cut-checkered forends with honeycomb recoil pad and a crossbolt safety. The lockup is a chrome molybdenum high-speed, four-slug rotary bolt and barrel extension system (lockup does not require use of the receiver top as part of the locking system). The Model 1300 receiver is made of lightweight, corrosion-resistant, space age alloy. Because the rotary lockup is concentric with the bore of the barrel, recoil forces are used to unlock the bolt and drive both bolt and forend rearward to help the shooter set up the next shot.

Also available:
Model 1300 Black Shadow Deer with all-black, non-glare composite stock and forearm. Improved cylinder Winchoke is standard. Additional specifications same as Model 1300 Walnut.

**MODEL 1300 WALNUT DEER (fully rifled)
Sights Drilled and Tapped
$393.00**

SPECIFICATIONS (12 GAUGE):
Model: 1300 Walnut
Chamber: 3″ Mag.
Shotshell Capacity: 5 + 1 in chamber
Choke: Cylinder
Barrel length: 22″ Rifled
Overall length: 42⅝″

Weight: 7¼ lbs.
Rate of Twist: 1 Turn in 35″
Sights: Rifle
Length of pull: 14″
Drop at comb: 1½″
Drop at heel: 2½″

WINCHESTER SHOTGUNS

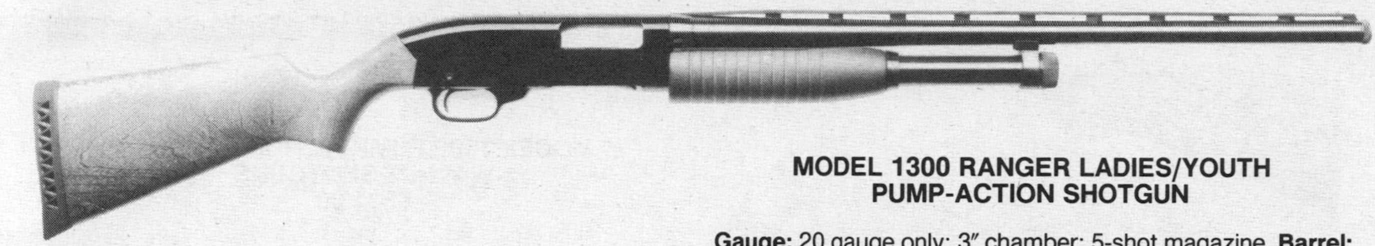

MODEL 1300 RANGER LADIES/YOUTH PUMP-ACTION SHOTGUN

Gauge: 20 gauge only; 3″ chamber; 5-shot magazine. **Barrel:** 22″ barrel w/vent. rib; Winchoke (Full, Modified, Improved Cylinder). **Weight:** 6½ lbs. **Length:** 41⅝″. **Stock:** Walnut or American hardwood with ribbed forend. **Sights:** Metal bead front. **Features:** Cross-bolt safety; black rubber butt pad; twin-action slide bars; front-locking rotating bolt; removable segmented magazine plug to limit shotshell capacity for training purposes.

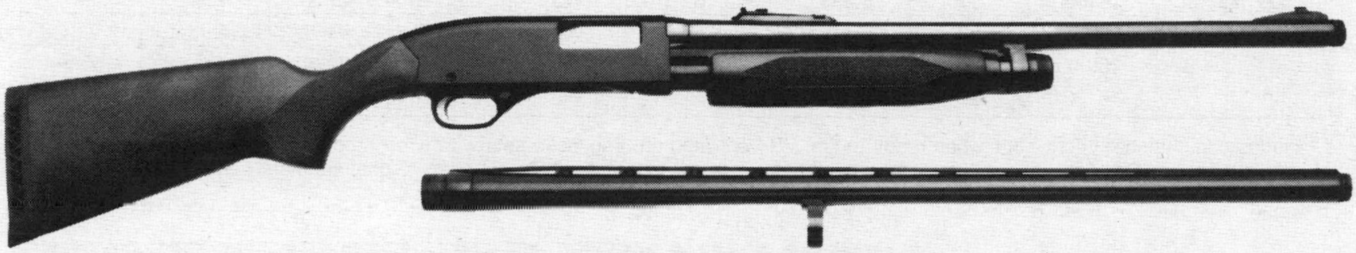

MODEL 1300 RANGER 12 GAUGE DEER COMBO
22″ Rifled w/Sights & 28″ Vent Rib Barrels

SHOTGUNS

SPECIFICATIONS MODEL 1300 RANGER, RANGER DEER & LADIES/YOUTH

Model	Gauge	Choke	Barrel Length & Type	Overall Length	Nominal Length of Pull	Nominal Weight (Lbs.)	Sights	Prices
1300 Ranger	12	W3	28″ VR	48⅝″	14″	7½	MBF	**$300.00**
	20	W3	28 VR	48⅝	14	7¼	MBF	
	12	W3	26 VR	46⅝	14	7¼	MBF	
	20	W3	26 VR	46⅝	14	7¼	MBF	
1300 Ranger Deer Combo 12 ga. Extra Vent Rib Barrel	12 12	Cyl WIM	22 Smooth 28 VR	42⅝ 48⅝	14 14	6¾ 7½	Rifle MBF	**368.00***
1300 Ranger Deer	12	W2	22 Smooth	42⅝	14	6¾	Rifle	**345.00**
	12	Rifled Barrel	22 Rifled	42⅝	14	6¾	Rifle	**333.00**
1300 Ranger Ladies/Youth	20	W3	22 VR	41⅝	13	6½	MBF	**300.00**

All models have 3″ Mag. chambers and 5-shot shell capacity, including one shotshell in chamber when ready to fire. VR-Ventilated rib. Cyl.-Cylinder Bore, R-Rifled Barrel. MBF-Metal bead front. RT-Rifle type front and rear sights. Model 1300 and Ranger pump action shotguns have factory-installed plug which limits capacity to three shells. Ladies/Youth has factory-installed plug which limits capacity to one, two or three shells as desired. Extra barrels for Model 1300 and Ranger shotguns are available in 12 gauge, plain or ventilated rib, in a variety of barrel lengths and chokes; interchangeable with gauge. Winchoke sets with wrench come with gun as follows: W3W-Extra Full, Full, Modified tubes. W3-Full, Modified, Improved Cylinder tubes. W1M-Modified tube. W1F-Full tube. Nominal drop at comb: 1½″; nominal drop at heel: 2½″ (2⅜″-Ladies' models).

WINCHESTER SHOTGUNS

**MODEL 1300 PUMP-ACTION DELUXE
WALNUT SHOTGUNS
$331.00**

SPECIFICATIONS MODEL 1300 DELUXE WALNUT

Model	Symbol Number	Gauge	Chamber	Shotshell Capacity*	Choke	Barrel Length & Type	Overall Length	Drop At Heel	Weight (Lbs.)
1300 Walnut	16105	12	3″ Mag.	5	W3	28″ VR	48⅝″	2½″	7½
	16023	20	3″ Mag.	5	W3	28″ VR	48⅝″	2½″	7¼
	16072	12	3″ Mag.	5	W3	26″ VR	46⅝″	1½″	7¼
	16122	20	3″ Mag.	5	W3	26″ VR	46⅝″	2½″	7¼

* Includes one shotshell in chamber when ready to fire. VR-Ventilated rib. Winchoke sets with wrench come with gun as follows: W3-Full, Modified, Improved Cylinder tubes. All models have 14″ nominal length of pull; 1½″ drop at comb. Sights are metal bead front.

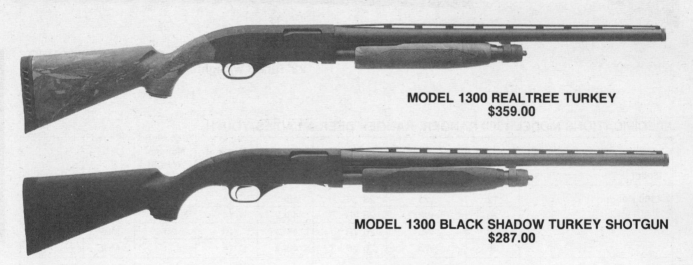

**MODEL 1300 REALTREE TURKEY
$359.00**

**MODEL 1300 BLACK SHADOW TURKEY SHOTGUN
$287.00**

SPECIFICATIONS MODEL 1300 TURKEY

Model	Gauge	Barrel Length & Type	Chamber	Shotshell Capacity*†	Choke	Overall Length	Length of Pull	Drop At Comb/Heel	Weight (Lbs.)	Sights
Realtree All Purpose Pattern	12	22″ VR	3″ Mag.	5	W3W	42⅝″	14″	1½″ 2½″	7	MBF
Black Shadow	12	22″ VR	3″ Mag.	5	W1F	42⅝″	14	1½″ 2½″	7¼	MBF

* Includes one shotshell in chamber when ready to fire. VR-Ventilated rib. MBF-Metal bead front. Winchoke sets with wrench come with gun as follows: W3W-Extra Full, Full, Modified tubes. W3-Full, Modified, Improved Cylinder tubes. † Includes one shotshell in chamber when ready to fire.

WINCHESTER SHOTGUNS

MODEL 1400 SEMIAUTO SHOTGUNS

MODEL 1400 SEMIAUTO WALNUT

SPECIFICATIONS MODEL 1400 SEMIAUTO

Model	Gauge	Barrel Length & Type	Chamber	Shotshell Capacity*	Choke	Overall Length	Length of Pull	Drop At Comb/Heel		Weight (lbs.)	Sights	Prices
1400	12	28" VR	2³/₄"	3	W3"	48⁵/₈"	14"	1¹/₂"	2¹/₂"	7³/₄	MBF	$419.00
	20	28" VR	2³/₄"	3	W3"	48¹/₈"	13¹/₂"	1¹/₂"	2¹/₂"	7¹/₂	MBF	
1400 Ranger	12	28" VR	2³/₄"	3	W3"	48¹/₈"	13¹/₂"	1¹/₂"	2¹/₂"	7³/₄	MBF	$377.00
	20	26" VR	2³/₄"	3	W3"	46¹/₈	13¹/₂"	1¹/₂"	2¹/₂"	7¹/₄	MBF	
1400 Ranger Deer Combo (With Extra 28" Barrel)	12	22" Smooth	2³/₄"	3	Cyl	42⁵/₈"	14"	1¹/₂"	2¹/₂"	7¹/₄	Rifle	$430.00
	12	28" VR	2³/₄"	3	WIM	48¹/₈"					MBF	

* Includes one shotshell in chamber when ready to fire. VR-Ventilated rib. Cyl.-Cylinder Bore. MBF-Metal bead front. Winchoke sets with wrench come with gun as follows: WIM-Modified tube. W2 Improved Cylinder and rifled Sabot tubes. W3-Full, Modified, Improved Cylinder tubes.

SHOTGUNS

WINCHESTER SHOTGUNS

MODEL 1001 O/U SHOTGUNS
$1099.00 Field Grade (shown)
$1253.00 Sporting Clays

SPECIFICATIONS MODEL 1001

Model	Gauge	Barrel Length & Type	Chamber	Choke*	Nominal Overall Length	Nominal Length of Pull	Nominal Drop at Comb	Heel	Weight (lbs.)	Sights
Field	12	28″	3″	IC, M, F	46″	14¹/₄″	1¹/₂″	2″	7	MBF
Sporting Clays I	12	28″	2³/₄″	SK, IC, M, I-MOD	46″	14³/₈″	1³/₈″	2¹/₈″	7³/₄	MBM-WBF
Sporting Clays II	12	30″	2³/₄″	SK, IC, M, I-MOD	48″	14³/₈″	1³/₈″	2¹/₈″	7³/₄	MBM-WBF

* WinPlus choke tubes supplied with your 1001. Choke Code: IC = Improved Cylinder, M = Modified, I-MOD = Improved Modified, F = Full, SK = Skeet, MBF = Metal bead front. MBM = Metal bead middle. WBF = White bead front.

MODEL 12
$879.00 Grade I
$1431.00 Grade IV

SPECIFICATIONS CLASSIC TRADITION—LIMITED EDITIONS

Model 12	Gauge	Barrel Length	Chamber	Shotshell Capacity**	Choke	Overall Length*	Length of Pull*	Drop at Comb*/Heel		Weight (lbs.)	Sights
Grade I	20	26″	2³/₄″	6	IC	45″	14″	1¹/₂″	2¹/₂″	7	MBF
Grade IV	20	26″	2³/₄″	6	IC	45″	14	1¹/₂″	2¹/₂″	7	WBF

* Nominal weight/measurement. **Includes one shotshell in chamber when ready to fire. IC = Improved Cylinder. F = Full Choke, MBF = Metal Bead Front, MBM = Metal Bead Middle, WBF = White Bead Front.

Black Powder

For addresses and phone numbers of manufacturers and distributors included in this section, turn to *DIRECTORY OF MANUFACTURERS AND SUPPLIERS* at the back of the book.

AMERICAN ARMS

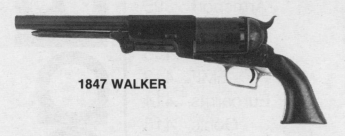

1847 WALKER

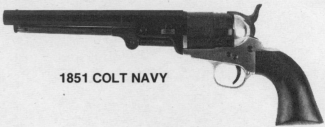

1851 COLT NAVY

1847 WALKER
$259.00 (Percussion)

This replica of the Texas Ranger's and Mexican War "Horse Pistol" was designed by Capt. Samuel Walker and built by Samuel Colt.

SPECIFICATIONS
Caliber: 44
Capacity: 6 shots
Barrel length: 9″ round w/hinged loading lever
Overall length: 15½″   **Weight:** 72 oz.
Features: Engraved blued steel cylinder; color casehardened steel frame and backstrap; solid brass trigger guard; one-piece walnut grip

1851 COLT NAVY
$134.00 (Percussion)

This replica of the most famous revolver of the percussion era was used extensively during the Civil War and on the Western frontier.

SPECIFICATIONS
Caliber: 36   **Capacity:** 6 shots
Barrel length: 7½″ octagonal w/hinged loading lever
Overall length: 13″   **Weight:** 44 oz.
Features: Solid brass frame, trigger guard and backstrap; one-piece walnut grip; engraved blued steel cylinder

1858 REMINGTON ARMY
$153.00 (Percussion)

This replica of the last of Remington's percussion revolvers saw extensive use in the Civil War.

SPECIFICATIONS
Caliber: 44   **Capacity:** 6 shots
Barrel length: 8″ octagonal w/creeping loading lever
Overall length: 13″   **Weight:** 38 oz.

1858 REMINGTON ARMY

Features: Two-piece walnut grips
Also available w/stainless-steel frame, barrel and cylinder, adj. rear target sight and ramp blade. **Price:** $305.00

1858 ARMY STAINLESS STEEL TARGET

1860 COLT ARMY
$137.00 (Percussion)

Union troops issued this sidearm during the Civil War and subsequent Indian Wars.

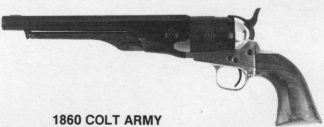

1860 COLT ARMY

SPECIFICATIONS
Caliber: 44   **Capacity:** 6 shots
Barrel length: 8″ round w/creeping loading lever
Overall length: 13½″   **Weight:** 44 oz.
Features: Solid brass frame, trigger guard and backstrap; one-piece walnut grip; engraved blued steel cylinder

ARMSPORT

REPLICA REVOLVERS

**MODEL 5145
COLT 1847 WALKER
$275.00**

The largest of all Colt revolvers, this true copy of the original weighs 4½ lbs., making it the most powerful (and only) revolver made at the time. **Caliber:** 44.

**MODEL 5152
ENGRAVED REMINGTON 44 CALIBER
Gold & Nickel-Plated
$290.00**

**MODEL 5153
ENGRAVED COLT ARMY 44 CALIBER
$290.00**

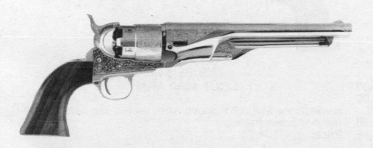

**MODEL 5154
ENGRAVED COLT NAVY 36 CALIBER
$290.00**

ARMSPORT

REPLICA REVOLVERS

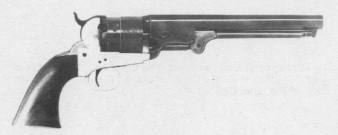

MODEL 5133
COLT 1851 NAVY "REB"

A modern replica of a Confederate percussion revolver in 36 or 44 caliber, this has a polished brass frame, rifled blued barrel and polished walnut grips.
Price: **$138.00**

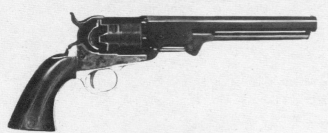

MODEL 5136
COLT 1851 NAVY STEEL

This authentic reproduction of the Colt Navy Revolver in 36 or 44 caliber, which helped shape the history of America, features a rifled barrel, casehardened steel frame, engraved cylinder, polished brass trigger guard and walnut grips.
Price: **$178.00**

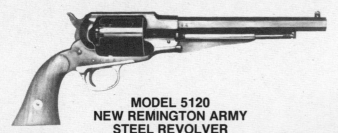

MODEL 5120
NEW REMINGTON ARMY
STEEL REVOLVER

One of the most accurate cap-and-ball revolvers of the 1860s. Its rugged steel frame and top strap made this 44 caliber the favorite of all percussion cap revolvers.
Price: **$210.00**
Model 5121 with brass frame: 160.00
Also available:
Steel Target Model 5122 . . . . 245.00
Stainless Target Model 5149 . . . . 375.00

MODEL 5138
REMINGTON ARMY STAINLESS STEEL

This stainless-steel version of the 44-caliber Remington New Army Revolver is made for the shooter who seeks the best. Its stainless-steel frame assures lasting good looks and durability.
Price: **$335.00**

MODEL 5139
COLT 1860 ARMY

This authentic 44-caliber reproduction offers the same balance and ease of handling for fast shooting as the original 1860 Army model.
Price: **$210.00**
Also available:
Model 5150 Stainless Steel 355.00

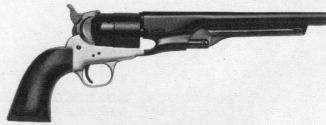

MODEL 5140
COLT 1860 ARMY

Same as the Model 5139 Colt Army replica, but with brightly polished brass frame.
Price: **$149.00**

CVA REVOLVERS

**1858 ARMY REVOLVER
STEEL FRAME**

Same specifications as **Target Model** (below), except rear sight is grooved in frame.
Prices:
Brass Frame: **$169.95**
Steel Frame: **189.95**
Kit—Brass only: **145.95**

1858 TARGET MODEL

Caliber: 44
Cylinder: 6-shot
Barrel length: 8″ octagonal
Overall length: 13″
Weight: 38 oz.
Sights: Blade front; adjustable target
Grip: Two-piece walnut
Price: **$204.95**

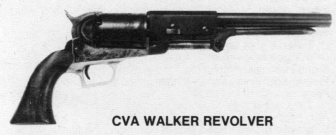

CVA WALKER REVOLVER

Caliber: 44
Barrel: 9″ rounded with hinged-style loading lever
Cylinder: 6-shot engraved
Overall length: 15$\frac{1}{2}$″
Weight: 72 oz.
Grip: One-piece walnut
Front sight: Blade
Finish: Solid brass trigger guard
Price: **$249.95**

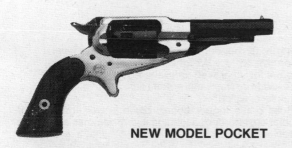

NEW MODEL POCKET

Caliber: 31 percussion
Barrel length: 4″ octagonal
Cylinder: 5 shots
Overall length: 7$\frac{1}{2}$″
Sights: Post in front; groove in frame in rear
Weight: 15 oz.
Finish: Solid brass frame
Price: **$129.95**

BLACK POWDER

CVA REVOLVERS

**1851 NAVY REVOLVER
BRASS FRAME**

Caliber: 36 and 44
Barrel length: 7½″ octagonal; hinged-style loading lever
Overall length: 13″
Weight: 44 oz.
Cylinder: 6-shot, engraved
Sights: Post front; hammer notch rear
Grip: One-piece walnut
Finish: Solid brass frame, trigger guard and backtrap; blued barrel and cylinder; color casehardened loading lever and hammer
Prices:
Finished **$138.95**
Kit **129.95**

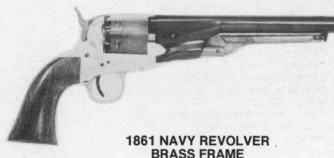

**1861 NAVY REVOLVER
BRASS FRAME**

Calibers: 44
Barrel length: 8″ rounded; creeping style
Overall length: 13″
Weight: 44 oz.
Cylinder: 6-shot, engraved
Sights: Blade front; hammer notch rear
Finish: Solid brass frame, trigger guard and backstrap; blued barrel and cylinder
Grip: One-piece walnut
Prices:
Finished **$139.95**
Kit **129.95**

**SHERIFF'S MODEL REVOLVER
BRASS FRAME**

Caliber: 36
Barrel length: 5½″ (octagonal w/creeping-style loading lever)
Overall length: 11½″
Weight: 40 oz.
Cylinder: 6-shot semi-fluted
Grip: One-piece walnut
Sight: Hammer notch in rear
Finish: Solid brass frame, trigger guard and backstrap
Price: **$149.95**

Also available:
Engraved Nickel Plated Model with matching flask
Price: **$194.95**

1860 ARMY REVOLVER

Caliber: 44
Barrel length: 8″ rounded; creeping-style loading lever
Overall length: 13½″   **Weight:** 44 oz.
Cylinder: 6-shot, engraved and rebated
Sights: Blade front; hammer notch rear
Grip: One-piece walnut
Finish: Solid brass trigger guard; blued barrel and cylinder with color casehardened loading lever, hammer and frame
Price: Steel Frame Only **$179.95**

CVA REVOLVERS

WELLS FARGO MODEL

Caliber: 31
Capacity: 5-shot cylinder (engraved)
Barrel length: 4″ octagonal
Overall length: 9″
Weight: 28 oz. (w/extra cylinder)
Sights: Post front; hammer notch rear
Grip: One-piece walnut
Price: w/Brass Frame $129.95

Caliber: 36
Capacity: 5-shot cylinder
Barrel length: 5½″ octagonal, with creeping-style loading lever
Overall length: 10½″
Weight: 26 oz.
Sights: Post front; hammer notch rear
Price: w/Brass Frame $139.95

POCKET POLICE

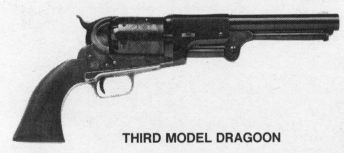

THIRD MODEL DRAGOON

Caliber: 44
Cylinder: 6-shot engraved
Barrel length: 7½″ rounded with hinged-style loading lever
Overall length: 14″
Weight: 66 oz.
Sights: Blade front; hammer notch rear
Grip: One-piece walnut
Price: w/Steel Frame $219.95

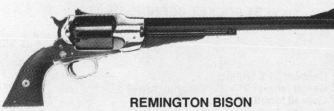

REMINGTON BISON

Caliber: 44
Cylinder: 6-shot
Barrel length: 10¼″ octagonal
Overall length: 18″
Weight: 48 oz.
Sights: Fixed blade front; screw adjustable target rear
Grip: Two-piece walnut
Finish: Solid brass frame
Price: $194.95

BLACK POWDER

CVA PISTOLS

KENTUCKY PISTOL

Caliber: 50 percussion
Barrel: 9³/₄″, rifled, octagonal
Overall length: 15¹/₂″
Weight: 40 oz.
Finish: Blued barrel, brass hardware
Sights: Brass blade front; fixed open rear
Stock: Select hardwood
Ignition: Engraved, color casehardened percussion lock, screw adjustable sear engagement
Accessories: Brass-tipped, hardwood ramrod; stainless steel nipple or flash hole liner
Prices:
Finished **$149.95**
Percussion Kit **99.95**

HAWKEN PISTOL

Caliber: 50 percussion
Barrel length: 9³/₄″, octagonal
Overall length: 16¹/₂″
Weight: 50 oz.
Trigger: Early-style brass
Sights: Beaded steel blade front; fully adjustable rear (click adj. screw settings lock into position)
Stock: Select hardwood
Finish: Solid brass wedge plate, nose cap, ramrod thimbles, trigger guard and grip cap
Prices:
Finished **$149.95**
Kit **109.95**

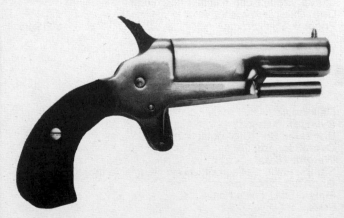

VEST POCKET DERRINGER

Caliber: 31 Derringer
Barrel length: 2¹/₂″ (single shot) brass
Overall length: 5″
Weight: 16 oz.
Grip: Two-piece walnut
Frame: Brass
Price: Finished **$69.95**

CVA RIFLES

VARMINT RIFLE

Caliber: 32 percussion
Barrel length: 24″ octagonal; 7/8″ across flats
Rifling: 1 in 56″
Overall length: 40″
Weight: 6 3/4 lbs.
Lock: Color casehardened with 45° offset hammer

Trigger: Single trigger
Sights: Steel blade front; Patridge-style click-adjustable rear
Stock: Select hardwood
Features: Brass trigger guard, nose cap, wedge plate, thimble and buttplate
Price: $219.95

BUSHWACKER RIFLE

Caliber: 50
Barrel length: 26″ blued octagonal
Overall length: 40″
Weight: 7 1/2 lbs.
Sights: Brass blade front; fixed open semi-buckhorn rear

Stock: Walnut-stained select hardwood with rounded nose
Trigger: Single trigger with oversized trigger guard
Features: Blued steel wedge plates; ramrod thimbles; blackened trigger guard and black plastic buttplate
Price: $159.95

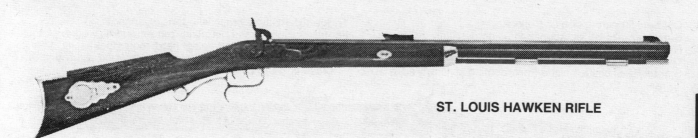

ST. LOUIS HAWKEN RIFLE

Calibers: 50 and 54 percussion or flintlock (50 cal. only)
Barrel: 28″ octagonal 15/16″ across flats; hooked breech; rifling one turn in 66″, 8 lands and deep grooves
Overall length: 44″
Weight: 8 lbs.
Sights: Dovetail, beaded blade (front); adjustable open hunting-style dovetail (rear)
Stock: Select hardwood with beavertail cheekpiece
Triggers: Double set; fully adjustable trigger pull

Finish: Solid brass wedge plates, nose cap, ramrod thimbles, trigger guard and patchbox
Prices:
50 Caliber Flintlock	$234.95
50 Caliber Flintlock Left Hand (finished)	249.95
50 Caliber Percussion Left Hand (finished)	234.95
50 Caliber Percussion	209.95
50 Caliber Percussion Kit	169.95
Percussion Combo Kit	229.95

BLACK POWDER

CVA RIFLES

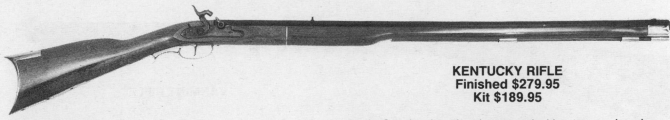

KENTUCKY RIFLE
Finished $279.95
Kit $189.95

Caliber: 50 percussion
Barrel length: 33½″ octagonal; ⅞″ across flats
Rifling: 1 in 66″
Overall length: 48″ **Weight:** 7½ lbs.

Lock: Casehardened and engraved with v-type mainspring
Trigger: Early brass-style single trigger
Sights: Brass blade front; fixed open rear
Features: Solid brass trigger guard, buttplate, toe plate, nose cap and thimble; select hardwood stock

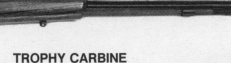

TROPHY CARBINE
$259.95

Calibers: 50 and 54 percussion
Barrel length: 24″ half-round, half-octagonal; 15/16″ across flats
Rifling: 1 in 32″
Overall length: 40″ **Weight:** 6¾ lbs.
Lock: Hawken-style with bridle and fly; engraved

Trigger: Single modern-style trigger integral with trigger guard
Sights: Ivory-colored beaded ramp mounted front; fully click adjustable rear
Stock: Walnut with Monte Carlo comb, pistol grip and formed cheekpiece
Features: Ventilated rubber recoil pad, molded black oversized trigger guard and blued thimble

FRONTIER CARBINE
$189.95 Percussion $199.95 Flintlock
$129.95 Kit (Percussion Only)

Caliber: 50 percussion or flintlock
Barrel length: 24″ octagonal; 15/16″ across flats
Rifling: 1 turn in 48″ (8 lands and deep grooves)
Overall length: 40″ **Weight:** 6 lbs. 12 oz.
Sights: Steel-beaded blade front; hunting-style fully adjustable rear

Trigger: Early-style brass with tension spring
Features: Solid brass buttplate, trigger guard, wedge plate, nose cap and thimble, select hardwood stock
Accessories: Stainless steel nipple, hardwood ramrod with brass tips and cleaning jag

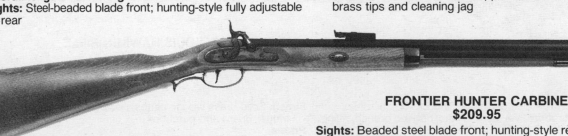

FRONTIER HUNTER CARBINE
$209.95

Calibers: 50 and 54 percussion
Barrel length: 24″, blued, octagonal
Overall length: 40″ **Weight:** 7½ lbs.

Sights: Beaded steel blade front; hunting-style rear, fully click adjustable for windage and elevation
Stock: Laminated hardwood with straight grip; solid rubber recoil pad
Triggers: Single
Features: Black-chromed nose cap, trigger guard and blued wedge plate

CVA RIFLES

PANTHER CARBINE
$189.95 ($204.95 Left Hand)

Tigger: Single modern style
Sights: Steel-beaded blued blade front; click adjustable rear
Stock: Textured black Dura Grip hardwood w/Monte Carlo comb, pistol grip and formed cheekpiece
Features: Vent. rubber recoil pad; molded black oversized trigger guard

SPECIFICATIONS
Calibers: 50 and 54 percussion
Barrel length: 24″ carbine blued octagonal
Overall length: 40″   **Weight:** 7¹/₂ lbs.

PLAINSMAN RIFLE
$139.95

Trigger: Single trigger with large trigger guard
Sights: Brass blade front; adjustable rear
Stock: Select hardwood
Finish: Black trigger guard, wedge plate and thimble
Accessories: Color casehardened nipple, hardwood ramrod with brass tip and cleaning jag

Caliber: 50 percussion
Barrel length: 26″ octagonal (¹⁵/₁₆″ across flats)
Overall length: 40″
Weight: 6¹/₂ lbs.

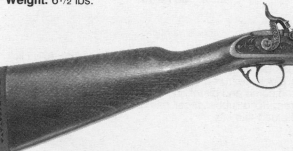

TRACKER CARBINE LS
$229.95

Trigger: Single trigger integral with trigger guard
Sights: Steel-beaded blued blade front; fully click adjustable rear
Stock: Matte finish laminated hardwood with straight grip
Features: Ventilated rubber recoil pad; molded black oversized trigger guard and blued wedge plates and thimbles; drilled and tapped for scope mount; non-glare black chrome furniture

Caliber: 50 percussion
Barrel length: 21″ blued, half-round, half-octagonal
Overall length: 37″   **Weight:** 6¹/₂ lbs.
Lock: Hawken-style with bridle and fly, color casehardened

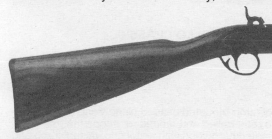

WOODSMAN RIFLE LS
$169.95

Overall length: 40″   **Weight:** 6¹/₂ lbs.
Trigger: Single trigger w/oversized trigger guard
Stock: Dark brown stained laminated hardwood
Features: Blued wedge plate and ramp-mounted thimble; black polymer buttplate

SPECIFICATIONS
Caliber: 50 percussion
Barrel length: 26″ blued octagon

CVA RIFLES

APOLLO SERIES

APOLLO CARBELITE RIFLE
$349.95

Caliber: 50 percussion
Barrel length: 27″ blued round, tapered w/octagonal one-piece receiver (drilled and tapped w/foul weather cover)
Rifling: 1 in 32″
Overall length: 43″ **Weight:** 7½ lbs.
Trigger: Box-type with hooking tumbler; auto safety system

Lock: In-line stainless-steel percussion bolt with push lock safety
Sights: Front, ramp-mounted beaded blade; fully click adjustable rear
Stock: Bell & Carlson Carbelite composite w/Monte Carlo fluted comb, fully formed cheekpiece and pistol grip
Features: Solid rubber recoil pad; molded black oversized trigger guard; bottom screw attachment and blued thimble; sling swivels; coated fiberglass ramrod with blackened tip

APOLLO 90 SHADOW SS RIFLE
$314.95

Calibers: 50 and 54 percussion
Barrel length: 24″ blued round taper with octagonal one-piece receiver
Overall length: 42″ **Weight:** 9 lbs.
Trigger: Box-style with hooking tumbler; auto safety system

Sights: Ramp mounted brass bead with hood front; fully click adjustable rear; drilled and tapped
Stock: Black, textured epoxicoat Dura Grip hardwood with raised comb and pistol grip
Features: Solid rubber recoil pad; bottom screw attachment and blued thimble

APOLLO CLASSIC
$266.95

Calibers: 50 and 54 percussion
Barrel length: 25″ blued round taper with octagonal one-piece receiver
Overall length: 42″ **Weight:** 8½ lbs.
Trigger: Box-style with hooking tumbler; auto safety system

Sights: Steel ramp-mounted beaded blade front; click adjustable-style rear; drilled and tapped for scope mounts
Stock: Laminated hardwood with pistol grip, raised comb and recoil pad; sling swivel stud on buttstock
Features: Molded black oversized trigger guard; bottom screw attachment and blued thimble

CVA RIFLES/SHOTGUNS

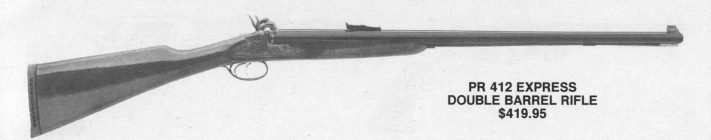

**PR 412 EXPRESS
DOUBLE BARREL RIFLE
$419.95**

Caliber: 50 percussion
Barrels: Two laser-aligned, tapered 28″ round; button-style breech
Rifling: 1 turn in 48″
Overall length: 44″   **Weight:** 10 lbs.
Locks: Engraved, color casehardened plate; includes bridle, fly, screw-adjustable sear engagement

Triggers: Double, gold tone
Sights: Hunting-style rear, fully adjustable for windage and elevation; blade front
Stock: Select hardwood
Features: Casehardened engraved locks, hammer, trigger guard and tang

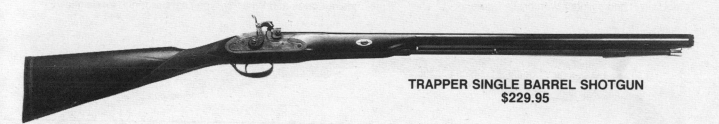

**TRAPPER SINGLE BARREL SHOTGUN
$229.95**

Gauge: 12 percussion
Barrel length: 28″ round, smoothbore; hooked breech; three interchangeable chokes (Imp., Mod. and Full)
Overall length: 46″   **Weight:** 6 lbs.
Trigger: Early-style steel
Lock: Color casehardened; engraved with v-type mainspring, bridle and fly

Sights: Brass bead front (no rear sight)
Stock: Select hardwood; English-style checkered straight grip
Features: German silver wedge plates; color casehardened engraved lock plate; hammer, trigger guard and tang

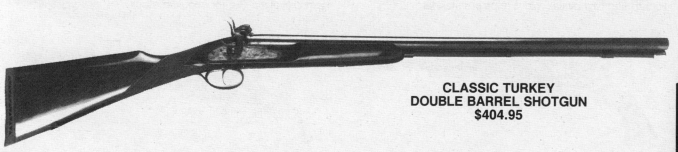

**CLASSIC TURKEY
DOUBLE BARREL SHOTGUN
$404.95**

Gauge: 12 percussion
Barrel length: 28″ round, smoothbore; double button-style breech, Modified choke
Overall length: 45″   **Weight:** 9 lbs.
Triggers: Hinged, gold-tone double triggers
Lock: Color casehardened; engraved with v-type mainspring, bridle and fly

Sights: Brass bead front (no rear sight)
Stock: European walnut; English-style checkered straight grip; wraparound forearm with bottom screw attachment
Features: Color casehardened engraved lock plates, trigger guard and tang

BLACK POWDER

DIXIE

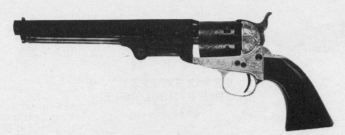

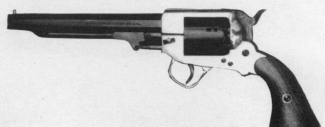

DIXIE NAVY REVOLVER
Plain Model $135.00
Engraved Model $139.95
Kit $114.00

This 36-caliber revolver was a favorite of the officers of the Civil War. Although called a Navy type, it is somewhat misnamed since many more of the Army personnel used it. Made in Italy; uses .376 mold or ball to fit and number 11 caps. Blued steel barrel and cylinder with brass frame.

SPILLER & BURR 36 CALIBER
BRASS FRAME REVOLVER
$125.00   Kit $115.00

The 36-caliber octagonal barrel on this revolver is 7 inches long. The six-shot cylinder chambers mike .378, and the hammer engages a slot between the nipples on the cylinder as an added safety device. It has a solid brass trigger guard and frame with backstrap cast integral with the frame, two-piece walnut grips and Whitney-type casehardened loading lever.

REMINGTON 44 ARMY REVOLVER
$169.95

All steel external surfaces finished bright blue, including 8″ octagonal barrel (hammer is casehardened). Polished brass guard and two-piece walnut grips are standard.

DIXIE 1860 ARMY REVOLVER
$169.95

The Dixie 1860 Army has a half-fluted cylinder and its chamber diameter is .447. Use .451 round ball mold to fit this 8-inch barrel revolver. Cut for shoulder stock.

"WYATT EARP" REVOLVER
$130.00

This 44-caliber revolver has a 12-inch octagon rifled barrel and rebated cylinder. Highly polished brass frame, backstrap and trigger guard. The barrel and cylinder have a deep blue luster finish. Hammer, trigger, and loading lever are case-hardened. Walnut grips. Recommended ball size is .451.

RHO200 WALKER REVOLVER
$225.95   Kit $184.95

This 4¹/₂-pound, 44-caliber pistol is the largest ever made. Steel backstrap; guard is brass with Walker-type rounded-to-frame walnut grips; all other parts are blued. Chambers measure .445 and take a .450 ball slightly smaller than the originals.

DIXIE

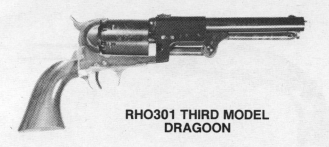

RHO301 THIRD MODEL DRAGOON

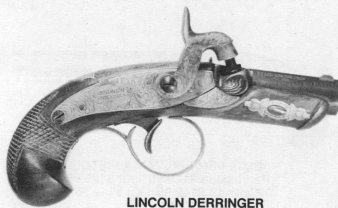

LINCOLN DERRINGER

SCREW BARREL PISTOL

RHO301 THIRD MODEL DRAGOON
$199.95

This engraved-cylinder, 4¹/₂-pounder is a reproduction of the last model of Colt's 44-caliber "horse" revolvers. Barrel measures 7³/₈ inches, ¹/₈ inch shorter than the original; color casehardened steel frame, one-piece walnut grips. Recommended ball size: .454.

LINCOLN DERRINGER
$285.00   Kit $89.95

This 41-caliber, 2-inch browned barrel gun has 8 lands and 8 grooves and will shoot a .400 patch ball.

FHO201 FRENCH CHARLEVILLE FLINT PISTOL
(Not shown)
$195.00

Reproduction of the Model 1777 Cavalry, Revolutionary War-era pistol. Has reversed frizzen spring; forend and lock housing are all in one; casehardened, round-faced, double-throated hammer; walnut stock; casehardened frizzen and trigger; shoots .680 round ball loaded with about 40 grains FFg black powder.

SCREW BARREL (FOLDING TRIGGER) PISTOL
$89.95 ($75.00 Kit)

This little gun, only 6¹/₂" overall, has a unique loading system that eliminates the need for a ramrod. The barrel is loosened with a barrel key, then unscrewed from the frame by hand. The recess is then filled with 10 grains of FFFg black powder, the .445 round ball is seated in the dished area, and the barrel is then screwed back into place. The .245×32 nipple uses #11 percussion caps. The pistol also features a sheath trigger that folds into the frame, then drops down for firing when the hammer is cocked. Comes with color casehardened frame, trigger and center-mounted hammer.

BLACK POWDER

DIXIE

LePAGE PERCUSSION DUELING PISTOL
$259.95

This 45-caliber percussion pistol features a blued 10″ octagonal barrel with 12 lands and grooves; a brass-bladed front sight with open rear sight dovetailed into the barrel; polished silver-plated trigger guard and butt cap. Right side of barrel is stamped "LePage á Paris." Double-set triggers are single screw adjustable. **Overall length:** 16″. **Weight:** 2¹/₂ lbs.

QUEEN ANNE PISTOL
$189.95
Kit $154.95

Named for the Queen of England (1702-1714), this flintlock pistol has a 7¹/₂″ barrel that tapers from rear to front with a cannon-shaped muzzle. The brass trigger guard is fluted and the brass butt on the walnut stock features a grotesque mask worked into it. **Overall length:** 13″. **Weight:** 2¹/₄ lbs.

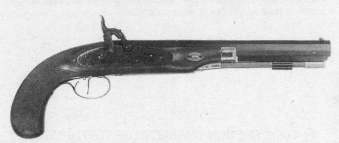

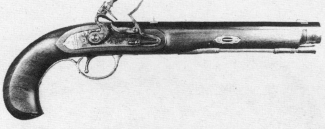

PEDERSOLI ENGLISH DUELING PISTOL
$265.00

This reproduction of an English percussion dueling pistol, created by Charles Moore of London, features a European walnut halfstock with oil finish and checkered grip. The 45-caliber octagonal barrel is 11″ with 12 grooves and a twist of 1 in 15″. Nose cap and thimble are silver. Barrel is blued; lock and trigger guard are color casehardened.

DIXIE PENNSYLVANIA PISTOL
Percussion $149.95   Kit $129.95
Flintlock $149.95   Kit $119.95

Available in 44-caliber percussion or flintlock. The bright luster blued barrel measures 10″ long; rifled, ⁷/₈-inch octagonal and takes .430 ball; barrel is held in place with a steel wedge and tang screw; brass front and rear sights. The brass trigger guard, thimbles, nose cap, wedge plates and side plates are highly polished. Locks are fine quality with early styling. Plates measure 4³/₄ inches × ⁷/₈ inch. Percussion hammer is engraved and both plates are left in the white. The flint is an excellent style lock with the gooseneck hammer having an early wide thumb piece. The stock is walnut stained and has a wide bird's-head-type grip.

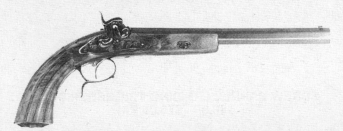

PEDERSOLI MANG TARGET PISTOL
$749.00

Designed specifically for the precision target shooter, this 38-caliber pistol has a 10⁷/₁₆″ octagonal barrel with 7 lands and grooves. Twist is 1 in 15″. Blade front sight is dovetailed into the barrel, and rear sight is mounted on the breech-plug tang, adjustable for windage. **Overall length:** 17¹/₄″. **Weight:** 2¹/₂ lbs.

DOUBLE BARREL MAGNUM
MUZZLELOADING SHOTGUN (not shown)

A full 12-gauge, high-quality, double-barreled percussion shotgun with 30-inch browned barrels. Will take the plastic shot cups for better patterns. Bores are choked, modified and full. Lock, barrel tang and trigger are casehardened in a light gray color and are nicely engraved.

12 Gauge	**$399.00**
12 Gauge Kit	350.00
10 Gauge Magnum (double barrel—right-hand = cyl. bore, left-hand = Mod.)	495.00
10 Gauge Magnum Kit	375.00

DIXIE

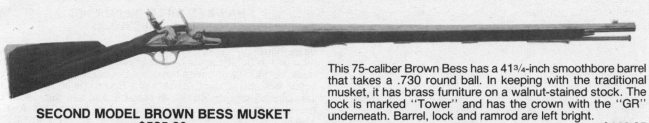

SECOND MODEL BROWN BESS MUSKET
$525.00

This 75-caliber Brown Bess has a 41¾-inch smoothbore barrel that takes a .730 round ball. In keeping with the traditional musket, it has brass furniture on a walnut-stained stock. The lock is marked "Tower" and has the crown with the "GR" underneath. Barrel, lock and ramrod are left bright.

Kit: **$446.25**

THE KENTUCKIAN RIFLE
Flintlock $269.95
Percussion $259.95

This 45-caliber rifle, in flintlock or percussion, has a 33½-inch blued octagonal barrel that is ¹³⁄₁₆ inch across the flats. The bore is rifled with 6 lands and grooves of equal width and about .006″ deep. Land-to-land diameter is .453 with groove-to-groove diameter of .465. Ball size ranges from .445 to .448.

The rifle has a brass blade front sight and a steel open rear sight. The Kentuckian is furnished with brass buttplate, trigger guard, patchbox, side plate, thimbles and nose cap plus casehardened and engraved lock plate. Highly polished and finely finished stock in European walnut. **Overall length:** 48″. **Weight:** Approx. 6¼ lbs.

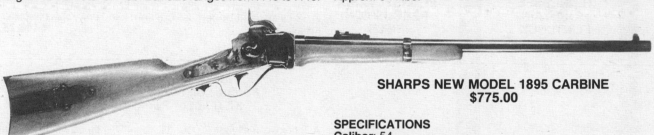

SHARPS NEW MODEL 1895 CARBINE
$775.00

SPECIFICATIONS
Caliber: 54
Barrel length: 22″ (1 in 48″ twist); blued, round barrel has 7-groove rifling
Overall length: 37½″   **Weight:** 7¾ lbs.
Sights: Blade front; adjustable rear
Stock: Oil-finished walnut
Features: Barrel band, hammer, receiver, saddle bar and ring all color casehardened

About 115,000 Sharps New Model 1859 carbines and its variants were made during the Civil War. Characterized by durability and accuracy, they became a favorite of cavalrymen on both sides. Made in Italy by David Pedersoli & Co.

SHARPS NEW MODEL 1859 MILITARY RIFLE
$895.00

Initially used by the First Connecticut Volunteers, this rifle is associated mostly with the 1st U.S. (Berdan's) Sharpshooters. There were 6,689 made with most going to the Sharpshooters (2,000) and the U.S. Navy (2,780). Made in Italy by David Pedersoli & Co.

SPECIFICATIONS
Caliber: 54
Barrel length: 30″ (1 in 48″ twist)

Overall length: 45½″
Weight: 9 lbs.
Sights: Blade front; rear sight adjustable for elevation and windage
Features: Buttstock and forend straight-grained and oil-finished walnut; three barrel bands, receiver, hammer, nose cap, lever, patchbox cover and butt are all color casehardened; sling swivels attached to middle band and butt

DIXIE RIFLES

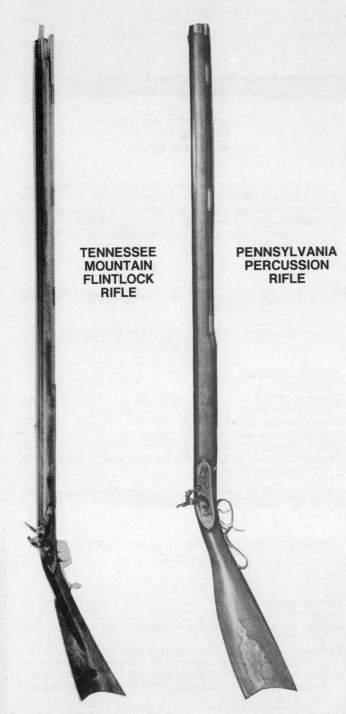

TENNESSEE MOUNTAIN FLINTLOCK RIFLE

PENNSYLVANIA PERCUSSION RIFLE

HAWKEN RIFLE (not shown)
$250.00   Kit $220.00

Blued barrel is 15/16″ across the flats and 30″ in length with a twist of 1 in 64″. Stock is of walnut with a steel crescent buttplate, halfstock with brass nosecap. Double set triggers, front-action lock and adjustable rear sight. Ramrod is equipped with jag. **Overall length:** 46½″. Average actual **weight:** about 8 lbs., depending on the caliber; shipping weight is 10 lbs. Available in either finished gun or kit. **Calibers:** 45, 50, and 54.

DIXIE TENNESSEE MOUNTAIN RIFLE
$525.00 Percussion or Flintlock

This 50-caliber rifle features double-set triggers with adjustable set screw, bore rifled with 6 lands and grooves, barrel of 15/16 inch across the flats, brown finish and cherry stock. **Overall length:** 41½ inches. Right- and left-hand versions in flint or percussion. **Kit:** **$446.25**

DIXIE TENNESSEE SQUIRREL RIFLE
$525.00   (not shown)

In 32-caliber flint or percussion, right hand only, cherry stock. Kit available: **$446.25**

PENNSYLVANIA RIFLE
Percussion or Flintlock $395.00
Kit (Flint or Perc.) $345.00

A lightweight at just 8 pounds, the 41½″ blued rifle barrel is fitted with an open buckhorn rear sight and front blade. The walnut one-piece stock is stained a medium darkness that contrasts with the polished brass buttplate, toe plate, patchbox, side plate, trigger guard, thimbles and nose cap. Featuring double-set triggers, the rifle can be fired by pulling only the front trigger, which has a normal trigger pull of 4 to 5 pounds; or the rear trigger can first be pulled to set a spring-loaded mechanism that greatly reduces the amount of pull needed for the front trigger to kick off the sear in the lock. The land-to-land measurement of the bore is an exact .450 and the recommended ball size is .445. **Overall length:** 51½″.

PEDERSOLI WAADTLANDER RIFLE (not shown)
$1295.00

This authentic re-creation of a Swiss muzzleloading target rifle features a heavy octagonal barrel (31″) that has 7 lands and grooves. **Caliber:** 45. Rate of twist is 1 turn in 48″. Double-set triggers are multi-lever type and are easily removable for adjustment. Sights are fitted post front and tang-mounted Swiss-type diopter rear. Walnut stock, color casehardened hardware, classic buttplate and curved trigger guard complete this reproduction. The original was made between 1839 and 1860 by Marc Bristlen, Morges, Switzerland.

DIXIE

MISSISSIPPI RIFLE
$430.00

Commonly called the U.S. Rifle Model 1841, this Italian-made replica is rifled in a 58 caliber to use a round ball or a Minie ball; 3 grooves and regulation sights; solid brass furniture; casehardened lock.

This 44-40 caliber gun can use modern or black powder cartridges. **Overall length:** 39″. **Barrel:** 20″ round. Its full tubular magazine will hold 11 shots. The walnut forearm and buttstock complement the high-luster bluing of the all steel parts such as the frame, barrel, magazine and buttplate. Comes with the trap door in the butt for the cleaning rod; leaf rear sight and blade front sight. This carbine is marked "Model 1873" on the tang and caliber "44-40" on the brass carrier block.

WINCHESTER '73 CARBINE
$895.00
ENGRAVED WINCHESTER '73 RIFLE
$1250.00

1863 SPRINGFIELD CIVIL WAR MUSKET
$525.00   Kit $446.25

This is an exact copy of the Model 1863 Springfield, which was the last of the regulation muzzleloading rifles. The barrel on this .58-caliber gun measures 40 inches. The action and all-metal furniture is finished bright. The oil-finished walnut-stained stock is 53 inches long. **Overall length:** 56″. **Weight:** 9½ lbs.

IN-LINE CARBINE
$349.95

Made in Italy by D. Pedersoli, this rifle in 50 or 54 caliber features a sliding "bolt" that completely encloses cap and nipple, making it the most weatherproof muzzleloader available. **Barrel length:** 24″. **Overall length:** 41″. **Weight:** 6½ lbs. **Sights:** Ramp front with red insert; rear sight adjustable for windage and elevation. **Stock:** Walnut-colored wood with Monte Carlo comb and black plastic buttplate. Features include fully adj. trigger, automatic slide safety, and chromed bolt and handle.

TRYON CREEDMOOR RIFLE (not shown)
$595.00

This Tryon rifle features a high-quality back-action lock, double-set triggers, steel buttplate, patchbox, toe plate and curved trigger guard. **Caliber:** 45. **Barrel:** 32¾″, octagonal, with 1 twist in 20.87″. **Sights:** Hooded post front fitted with replaceable inserts; rear is tang-mounted and adjustable for windage and elevation.

DIXIE

U.S. MODEL 1861 SPRINGFIELD PERCUSSION RIFLE-MUSKET
$525.00   Kit $446.25

An exact re-creation of an original rifle produced by Springfield National Armory, Dixie's Model 1861 Springfield .58-caliber rifle features a 40″ round, tapered barrel with three barrel bands. Sling swivels are attached to the trigger guard bow and middle barrel band. The ramrod has a trumpet-shaped head with swell; sights are standard military rear and bayonet-attachment lug front. The percussion lock is marked "1861" on the rear of the lockplate with an eagle motif and "U.S. Springfield" in front of the hammer. "U.S." is stamped on top of buttplate. All furniture is "National Armory Bright." **Overall length:** 55$^{13}/_{16}$″. **Weight:** 8 lbs.

1862 THREE-BAND ENFIELD RIFLED MUSKET
$485.00   Kit $425.00

One of the finest reproduction percussion guns available, the 1862 Enfield was widely used during the Civil War in its original version. This rifle follows the lines of the original almost exactly. The .58-caliber musket features a 39-inch barrel and walnut stock. Three steel barrel bands and the barrel itself are blued; the lockplate and hammer are case colored and the remainder of the furniture is highly polished brass. The lock is marked, "London Armory Co." **Weight:** 10^1/$_2$ lbs. **Overall length:** 55″.

U.S. MODEL 1816 FLINTLOCK MUSKET
$725.00

The U.S. Model 1816 Flintlock Musket was made by Harpers Ferry and Springfield Arsenals from 1816 until 1864. It had the highest production of any U.S. flintlock musket and after conversion to percussion saw service in the Civil War. It has a .69-caliber, 42″ smoothbore barrel held by three barrel bands with springs. All metal parts are finished in "National Armory Bright." The lockplate has a brass pan and is marked "Harpers Ferry" vertically behind the hammer, with an American eagle placed in front of the hammer. The bayonet lug is on top of the barrel and the steel ramrod has a button-shaped head. Sling swivels are mounted on the trigger guard and middle barrel band. **Overall length:** 56^1/$_2$″. **Weight:** 9^3/$_4$ lbs.

1858 TWO-BAND ENFIELD RIFLE
$450.00

This 33-inch barrel version of the British Enfield is an exact copy of similar rifles used during the Civil War. The .58-caliber rifle sports a European walnut stock, deep blue-black finish on the barrel, bands, breech-plug tang and bayonet mount. The percussion lock is color casehardened and the rest of the furniture is brightly polished brass.

EMF REVOLVERS

SHERIFF'S MODEL 1851

SHERIFF'S MODEL 1851 REVOLVER
$140.00 (Brass) $172.00 (Steel)

SPECIFICATIONS
Caliber: 36 Percussion
Ball diameter: .376 round or conical, pure lead
Barrel length: 5″
Overall length: 10½″
Weight: 39 oz.
Sights: V-notch groove in hammer (rear); truncated cone in front
Percussion cap size: #11

MODEL 1860 ARMY REVOLVER
$145.00 (Brass) $173.00 (Engraved, Brass)
$200.00 (Steel) $330.00 (Engraved, Steel)

SPECIFICATIONS
Caliber: 44 Percussion
Barrel length: 8″
Overall length: 13⅝″
Weight: 41 oz.
Frame: Casehardened
Finish: High-luster blue with walnut grips
Also available.
Cased set with steel frame, wood case,
   flask and mold $303.00
   Engraved cased set 438.00
Fluted cylinder model (steel frame only) 245.00

MODEL 1860 ARMY

SECOND MODEL 44 DRAGOON
$275.00

SPECIFICATIONS
Caliber: 44
Barrel length: 7½″ (round)
Overall length: 14″
Weight: 4 lbs.
Finish: Steel casehardened frame
Also available:
Third Model Dragoon $275.00
Texas Dragoon 290.00

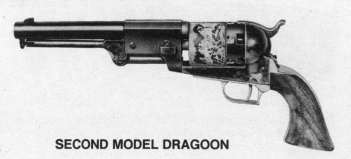

SECOND MODEL DRAGOON

MODEL 1862 POLICE REVOLVER
$200.00 (Steel) $150.00 (Brass)

SPECIFICATIONS
Caliber: 36 Percussion
Capacity: 5-shot
Barrel length: 6½″
Also available:
Cased set (steel only) $297.00

MODEL 1862 POLICE

BLACK POWDER

EUROARMS OF AMERICA

COOK & BROTHER CONFEDERATE CARBINE
Model 2300: $378.00

Classic re-creation of the rare 1861, New Orleans-made Artillery Carbine. The lockplate is marked "Cook & Brother N.O. 1861" and is stamped with a Confederate flag at the rear of the hammer.

SPECIFICATIONS
Caliber: 58 percussion
Barrel length: 24"   **Overall length:** 40⅓"
Weight: 7½ lbs.
Sights: Fixed blade front and adjustable dovetailed rear

Ramrod: Steel
Finish: Barrel is antique brown; buttplate, trigger guard, barrel bands, sling swivels and nose cap are polished brass; stock is walnut
Recommended ball sizes: .575 r.b., .577 Minie and .580 maxi; uses musket caps
Also available:
MODEL 2301 COOK & BROTHER FIELD with 33" barrel
Price: $549.60

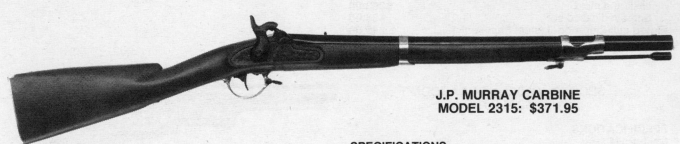

J.P. MURRAY CARBINE
MODEL 2315: $371.95

Replica of an extremely rare CSA Cavalry Carbine based on an 1841 design of parts and lock.

SPECIFICATIONS
Caliber: 58 percussion
Barrel length: 23"
Features: Brass barrel bands and buttplate; oversized trigger guard; sling swivels

C.S. RICHMOND MUSKET
MODEL 2370: $485.35

SPECIFICATIONS
Caliber: 58 percussion. **Barrel length:** 40" with three bands.

EUROARMS OF AMERICA

**LONDON ARMORY COMPANY
2-BAND RIFLE MUSKET
Model 2270: $398.80**

SPECIFICATIONS
Caliber: 58 percussion
Barrel length: 33″, blued and rifled
Overall length: 49″
Weight: 8½ to 8¾ lbs., depending on wood density
Stock: One-piece walnut; polished "bright" brass butt plate, trigger guard and nose cap; blued barrel bands
Sights: Inverted 'V' front sight; Enfield folding ladder rear
Ramrod: Steel

**LONDON ARMORY COMPANY
ENFIELD MUSKETOON
Model 2280: $371.95**

SPECIFICATIONS
Caliber: 58; Minie ball
Barrel length: 24″; round high-luster blued barrel
Overall length: 40½″
Weight: 7 to 7½ lbs., depending on density of wood
Stock: Seasoned walnut stock with sling swivels
Ramrod: Steel
Ignition: Heavy-duty percussion lock
Sights: Graduated military-leaf sight
Furniture: Brass trigger guard, nose cap and butt plate; blued barrel bands, lock plate, and swivels

**LONDON ARMORY COMPANY
3-BAND ENFIELD RIFLED MUSKET
Model 2260: $424.40**

SPECIFICATIONS
Caliber: 58 percussion
Barrel length: 39″, blued and rifled
Overall length: 54″
Weight: 9½ to 9¾ lbs., depending on wood density
Stock: One-piece walnut; polished "bright" brass butt plate, trigger guard and nose cap; blued barrel bands
Ramrod: Steel; threaded end for accessories
Sights: Traditional Enfield folding ladder rear sight; inverted 'V' front sight
Also available:
MODEL 2261 with white barrel $593.45

BLACK POWDER

EUROARMS OF AMERICA

1803 HARPERS FERRY FLINTLOCK RIFLE
Model 2305: $487.80

SPECIFICATIONS
Caliber: 54 Flintlock
Barrel length: 33″, octagonal
Features: Walnut half stock with cheekpiece; browned barrel

1841 MISSISSIPPI RIFLE
Model 2310: $426.80

SPECIFICATIONS
Caliber: 54 percussion
Barrel length: 33″, octagonal
Features: Walnut stock; brass barrel bands and buttplate; sling swivels

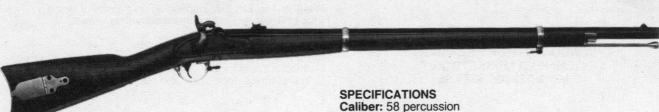

1863 ZOUAVE RIFLE (2-Barrel Bands)
Model 2255: $436.60 (Range Grade)
Model 2250: $317.00 (Field Grade)

SPECIFICATIONS
Caliber: 58 percussion
Barrel length: 33″, octagonal
Overall length: 48¹/₂″
Weight: 9¹/₂ to 9³/₄ lbs.
Sights: U.S. Military 3-leaf rear; blade front
Features: Two brass barrel bands; brass buttplate and nose cap; sling swivels

1861 SPRINGFIELD RIFLE
Model 2360: $485.35

SPECIFICATIONS
Caliber: 58 percussion
Barrel length: 40″
Features: 3 barrel bands

EUROARMS OF AMERICA

MODEL 1005

ROGERS & SPENCER ARMY REVOLVER
Model 1006 (Target): $218.30

SPECIFICATIONS
Caliber: 44; takes .451 round or conical lead balls; #11 percussion cap
Weight: 47 oz.
Barrel length: 7¹/₂″   **Overall length:** 13³/₄″
Finish: High gloss blue; flared walnut grip; solid frame design; precision-rifled barrel
Sights: Rear fully adjustable for windage and elevation; ramp front sight

ROGERS & SPENCER REVOLVER
LONDON GRAY (Not shown)
Model 1007: $224.40

Revolver is the same as Model 1005, except for London Gray finish, which is heat treated and buffed for rust resistance; same recommended ball size and percussion caps.

REMINGTON 1858
NEW MODEL ARMY REVOLVER
Model 1020: $195.10

This model is equipped with blued steel frame, brass trigger guard in 44 caliber.

SPECIFICATIONS
Weight: 40 oz.
Barrel length: 8″   **Overall length:** 14³/₄″
Finish: Deep luster blue rifled barrel; polished walnut stock; brass trigger guard.
Also available:
MODEL 1010. Same as Model 1020, except with 6¹/₂″ barrel and in 36 caliber: **$195.10**

ROGERS & SPENCER REVOLVER
Model 1005: $207.30

SPECIFICATIONS
Caliber: 44 Percussion; #11 percussion cap
Barrel length: 7¹/₂″   **Overall length:** 13³/₄″
Weight: 47 oz.
Sights: Integral rear sight notch groove in frame; brass truncated cone front sight
Finish: High gloss blue; flared walnut grip; solid frame design; precision-rifled barrel
Recommended ball diameter: .451 round or conical, pure lead

MODEL 1006

REMINGTON 1858
NEW MODEL ARMY ENGRAVED (Not shown)
Model 1040: $251.20

Classical 19th-century style scroll engraving on this 1858 Remington New Model revolver.

SPECIFICATIONS
Caliber: 44 Percussion; #11 cap
Barrel length: 8″   **Overall length:** 14³/₄″
Weight: 41 oz.
Sights: Integral rear sight notch groove in frame; blade front sight
Recommended ball diameter: .451 round or conical, pure lead

MODEL 1010
(36 Cal. w/6¹/₂″ barrel)

GONIC ARMS

MODEL GA-87 RIFLE
$547.00 Standard Walnut
$593.00 (Open Sights) $597.00 (Peep Sights)

SPECIFICATIONS
Calibers: 30, 38, 44, 45, 50 Mag., 54 and 20 ga. smoothbore
Barrel length: 26″   **Overall length:** 43″
Weight: 6½ lbs.
Sights: Bead front; open rear (adjustable for windage and elevation); drilled and tapped for scope bases
Stock: American walnut (grey or brown laminated stock optional).   **Length of pull:** 14″

Trigger: Single stage (4-lb. pull)
Mechanism type: Closed-breech muzzleloader
Features: Ambidextrous safety; non-glare satin finish; newly designed loading system; all-weather performance guaranteed; faster lock time
Also available:
Brown or Grey Laminated Stock Models (Std.) . . **$595.00**
  With open sights **640.00**
  With peep sights **644.00**

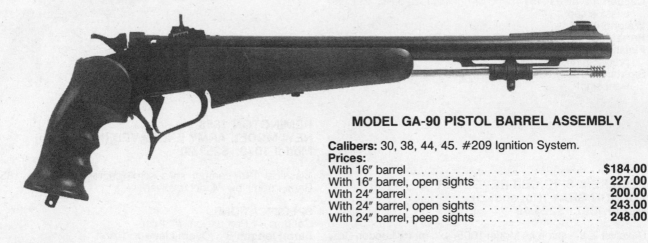

MODEL GA-90 PISTOL BARREL ASSEMBLY

Calibers: 30, 38, 44, 45. #209 Ignition System.
Prices:
With 16″ barrel **$184.00**
With 16″ barrel, open sights **227.00**
With 24″ barrel **200.00**
With 24″ barrel, open sights **243.00**
With 24″ barrel, peep sights **248.00**

MODEL 93 MAGNUM RIFLE
$442.00
$521.00 (Stainless w/Black Stock)

Features: Walnut-stained hardwood stock; adjustable trigger; nipple wrench; drilled and tapped for scope bases; ballistics and instruction manual
Also available:
Brown or Grey Laminated Thumbhole Stock
Blued Barrel **$754.00**
  With open sights **796.50**
  With peep sights **799.00**
Stainless Barrel **597.00**
With thumbhole stock **833.00**
  Same as above w/open sights **875.50**
  With peep sights **878.00**

Gonic Arms' new blackpowder rifle has a unique loading system that produces better consistency and utilizes the full powder charge of the specially designed penetrator bullet (ballistics = 2,650 foot-pounds at 1,600 fps w/465-grain .500 bullet).

SPECIFICATIONS
Caliber: 50 Magnum
Barrel length: 26″   **Overall length:** 43″
Weight: 6 to 6½ lbs.
Sights: Open hunting sights (adjustable)

LYMAN CARBINES/RIFLES

DEERSTALKER RIFLE
Percussion $304.95
Flintlock $324.95

Lyman's Deerstalker rifle incorporates many features most desired by muzzleloading hunters: higher comb for better sighting plane • non-glare hardware • 24" octagonal barrel • casehardened side plate • Q.D. sling swivels • Lyman sight package (37MA beaded front; fully adjustable fold-down 16A rear) • walnut stock with 1/2" black recoil pad • single trigger. Left-hand models available (same price). **Calibers:** 50 and 54, flintlock or percussion. **Weight:** 7 1/2 lbs.

DEERSTALKER CARBINE
Percussion $314.95
Flintlock $339.95

This carbine version of the famous Deerstalker Hunting Rifle is now available in 50 caliber percussion or flintlock and features a precision-rifled ''stepped octagon'' barrel with a 1 in 24" twist for optimum performance with conical projectiles.

The specially designed Lyman sight package features a fully adjustable Lyman 16A fold-down in the rear. The front sight is Lyman's 37MA white bead on an 18 ramp. Each rifle comes complete with a darkened nylon ramrod and modern sling and swivels set. Left-hand models available. **Weight:** 6 3/4 lbs.

GREAT PLAINS RIFLE
Percussion $409.95 (Kit $329.95)
Flintlock $439.95 (Kit $359.95)

The Great Plains Rifle has a 32-inch deep-grooved barrel and 1 in 66-inch twist to shoot patched round balls. Blued steel furniture including the thick steel wedge plates and steel toe plate; correct lock and hammer styling with coil spring dependability; and a walnut stock without a patchbox. A Hawken-style trigger guard protects double-set triggers. Steel front sight and authentic buckhorn styling in an adjustable rear sight. Fixed primitive rear sight also included. Left-hand models available (same price). **Calibers:** 50 and 54.

LYMAN TRADE RIFLE
Percussion $309.95 (Kit $289.95)
Flintlock $339.95 (Kit $319.95)

The Lyman Trade Rifle features a 28-inch octagonal barrel, rifled 1 turn at 48 inches, designed to fire both patched round balls and the popular maxistyle conical bullets. Polished brass furniture with blued finish on steel parts; walnut stock; hook breech; single spring-loaded trigger; coil-spring percussion lock; fixed steel sights; adjustable rear sight for elevation also included. Steel barrel rib and ramrod ferrrule. **Caliber:** 50 or 54 percussion and flint. **Overall length:** 45".

BLACK POWDER

MK-85 KNIGHT RIFLES

The MK-85 muzzleloading rifles (designed by William A. "Tony" Knight) are handcrafted, lightweight rifles capable of 1½-inch groups at 100 yards. They feature a one-piece, in-line bolt assembly, patented double safety system, Timney feather-weight deluxe trigger system, recoil pad, and Lothar Walther barrels (1 in 28" twist in 50 and 54 caliber).

SPECIFICATIONS
Calibers: 50 and 54
Barrel lengths: 20" and 24"   **Overall length:** 39" to 43"

MK-85 GRAND AMERICAN

Weight: 6 to 7¼ lbs.
Sights: Adjustable high-visibility open sights
Stock: Classic walnut, laminated or composite
Features: Swivel studs installed; hard anodized aluminum ramrod; combo tool; hex keys, and more.
Prices:
MK-85 HUNTER (Walnut) . . . . $ 529.95
MK-85 KNIGHT HAWK (24" Blued barrel) 689.95
  Stainless barrel 759.95
MK-85 PREDATOR (Stainless) 649.95
MK-85 STALKER (Laminated) 579.95
MK-85 GRAND AMERICAN (Blued barrel,
  Shadow brown or black) 995.95
  In Stainless Steel 1095.95

SPECIFICATIONS
Calibers: 50 and 54
Barrel length: 24" (tapered non-glare w/open breech system)
Sights: Adjustable; tapped and drilled for scope mount

MODEL BK-92 BLACK KNIGHT
$349.95

Stock: Black synthetic coated (see also below)
Features: Patented double safety system; in-line ignition system; 1 in 28" twist; stainless-steel breech plug; adjustable trigger; ½" recoil pad; Knight precision loading ramrod; Monte Carlo stock
Also available:
  With Hardwood stock **$379.95**
  With Composite stock **429.95**

MAGNUM ELITE
w/Black Composite Stock

MAGNUM ELITE
$849.95 (Realtree All-Purpose Camouflage)
$769.95 (Black Composite)

SPECIFICATIONS
Calibers: 50 and 54
Barrel length: 24"   **Overall length:** 41"   **Weight:** 6.75 lbs.
Sights: Adjustable high-visibility open sights
Features: "Posi-Fire" ignition system; Knight Double Safety System; aluminum ramrod; sling swivel studs

SPECIFICATIONS
Caliber: 50
Barrel length: 22"; blued rifle-grade steel (1:28" twist)

WOLVERINE
$249.95

Overall length: 41"   **Weight:** 6 lbs.
Sights: Adjustable high-visibility rear sight; Quick Detachable scope bases and rings
Stock: Lightweight Fiber-Lite molded stock
Features: Patented double safety system; adjustable Accu-Lite trigger; removable breech plug; stainless-steel hammer

NAVY ARMS REVOLVERS

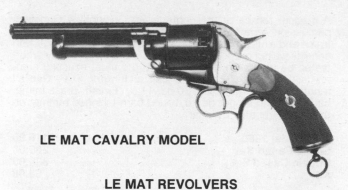

LE MAT CAVALRY MODEL

LE MAT NAVY MODEL

LE MAT ARMY MODEL

LE MAT REVOLVERS

Once the official sidearm of many Confederate cavalry officers, this 9-shot .44-caliber revolver with a central single-shot barrel of approx. 65 caliber gave the cavalry man 10 shots to use against the enemy. **Barrel length:** 7⅝″. **Overall length:** 14″. **Weight:** 3 lbs. 7 oz.

Cavalry Model	$ 595.00
Navy Model	595.00
Army Model	595.00
18th Georgia (engraving on cylinder, display case)	795.00
Beauregard (hand-engraved cylinder and frame; display case and mold)	1000.00

1862 NEW MODEL POLICE

COLT 1847 WALKER

This is the last gun manufactured by the Colt plant in the percussion era. It encompassed all the modifications of each gun, starting from the early Paterson to the 1861 Navy. It was favored by the New York Police Dept. for many years. One-half fluted and rebated cylinder, 36 cal., 5 shot, .375 dia. ball, 18 grains of black powder, brass trigger guard and backstrap. Casehardened frame, loading lever and hammer—balance blue. **Barrel length:** 5½″.

The 1847 Walker replica comes in 44 caliber with a 9-inch barrel. **Weight:** 4 lbs. 8 oz. Well suited for the collector as well as the blackpowder shooter. Features include: rolled cylinder scene; blued and casehardened finish; and brass guard. Proof tested.

1862 Police	$285.00
Law and Order Set	360.00

Colt 1847 Walker	$260.00
Single Cased Set	385.00
Deluxe	505.00

ROGERS & SPENCER REVOLVER

This revolver features a six-shot cylinder, octagonal barrel, hinged-type loading lever assembly, two-piece walnut grips, blued finish and casehardened hammer and lever. **Caliber:** 44. **Barrel length:** 7½″. **Overall length:** 13¾″. **Weight:** 3 lbs.

Rogers & Spencer	$240.00
London Gray	260.00
Target Model (w/adjustable sights)	260.00

ROGERS & SPENCER REVOLVER

NAVY ARMS REVOLVERS

REB MODEL 1860

A modern replica of the confederate Griswold & Gunnison percussion Army revolver. Rendered with a polished brass frame and a rifled steel barrel finished in a high-luster blue with genuine walnut grips. All Army Model 60s are completely proof-tested by the Italian government to the most exacting standards. **Calibers:** 36 and 44. **Barrel length:** 7¼". **Overall length:** 13". **Weight:** 2 lbs. 10 oz.-11 oz. **Finish:** Brass frame, backstrap and trigger guard, round barrel, hinged rammer on the 44 cal. rebated cylinder.

Reb Model 1860	$110.00
Single Cased Set	205.00
Double Cased Set	335.00
Kit	90.00

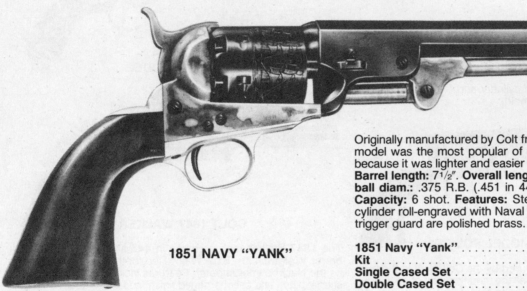

1851 NAVY "YANK"

Originally manufactured by Colt from 1850 through 1876, this model was the most popular of the Union revolvers, mostly because it was lighter and easier to handle than the Dragoon. **Barrel length:** 7½". **Overall length:** 14". **Weight:** 2 lbs. **Rec. ball diam.:** .375 R.B. (.451 in 44 cal) **Calibers:** 36 and 44. **Capacity:** 6 shot. **Features:** Steel frame, octagonal barrel, cylinder roll-engraved with Naval battle scene, backstrap and trigger guard are polished brass.

1851 Navy "Yank"	$145.00
Kit	120.00
Single Cased Set	245.00
Double Cased Set	405.00

The 1860 Army satisfied the Union Army's need for a more powerful .44-caliber revolver. The cylinder on this replica is roll engraved with a polished brass trigger guard and steel strap cut for shoulder stock. The frame, loading lever and hammer are finished in high-luster color case-hardening. Walnut grips. **Weight:** 2 lbs. 9 oz. **Barrel length:** 8". **Overall length:** 13⅝". **Caliber:** 44. **Finish:** Brass trigger guard, steel backstrap, round barrel, creeping lever, rebated cylinder, engraved Navy scene. Frame cut for s/stock (4 screws).

1860 Army	$165.00
Single Cased Set	265.00
Double Cased Set	430.00
Kit	145.00

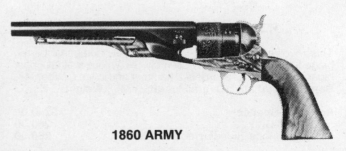

1860 ARMY

NAVY ARMS REVOLVERS

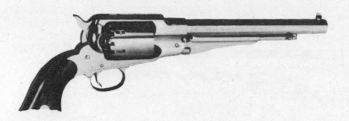

1858 NEW MODEL ARMY
REMINGTON-STYLE, STAINLESS STEEL

Exactly like the standard 1858 Remington (below) except that every part except for the grips and trigger guard is manufactured from corrosion-resistant stainless steel. This gun has all the style and feel of its ancestor with all of the conveniences of stainless steel. **Caliber: 44.**

1858 Remington Stainless $260.00
Single Cased Set 355.00
Double Cased Set 620.00

1858 TARGET MODEL

With its top strap solid frame, the Remington Percussion Revolver is considered the magnum of Civil War revolvers and is ideally suited to the heavy 44-caliber charges. Based on the Army Model, the target gun has target sights for controlled accuracy. Ruggedly built from modern steel and proof tested.

1858 Target Model $200.00

NEW MODEL ARMY
REMINGTON-STYLE

NEW MODEL ARMY
REMINGTON-STYLE REVOLVER

This rugged, dependable, battle-proven Civil War veteran with its top strap and rugged frame was considered the magnum of C.W. revolvers, ideally suited for the heavy 44 charges. Blued finish. **Caliber: 44. Barrel length: 8″. Overall length: 14¼″. Weight: 2 lbs. 8 oz.**

New Model Army Revolver $160.00
Single cased set 255.00
Double cased set 420.00
Kit 140.00
Also available:
Brass Frame 120.00
Brass Frame Kit 110.00
Single Cased Set 210.00
Double Cased Set 325.00

REB 60 SHERIFF'S MODEL

A shortened version of the Reb Model 60 Revolver. The Sheriff's model version became popular because the shortened barrel was fast out of the leather. This is actually the original snub nose, the predecessor of the detective specials or belly guns designed for quick-draw use. **Calibers: 36 and 44.**

Reb 60 Sheriff's Model $110.00
Kit 90.00
Single Cased Set 205.00
Double Cased Set 335.00

DELUXE NEW MODEL 1858
REMINGTON-STYLE 44 CALIBER (not shown)

Built to the exact dimensions and weight of the original Remington 44, this model features an 8″ barrel with progressive rifling, adjustable front sight for windage, all-steel construction with walnut stocks and silver-plated trigger guard. Steel is highly polished and finished in rich charcoal blue. **Barrel length: 8″. Overall length: 14¼″. Weight: 2 lbs. 14 oz.**

Deluxe New Model 1858 . . . . $365.00

BLACK POWDER

NAVY ARMS PISTOLS

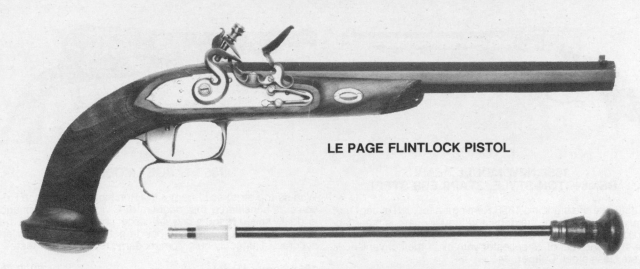

LE PAGE FLINTLOCK PISTOL

LE PAGE FLINTLOCK PISTOL
(44 Caliber)

The Le Page pistol is a beautifully hand-crafted reproduction featuring hand-checkered walnut stock with hinged buttcap and carved motif of a shell at the forward portion of the stock. Single-set trigger and highly polished steel lock and furniture together with a brown-finished rifled barrel make this a highly desirable target pistol. **Barrel length:** 10½″. **Overall length:** 17″. **Weight:** 2 lbs. 2 oz.

Le Page Flintlock (rifled or smoothbore) **$550.00**

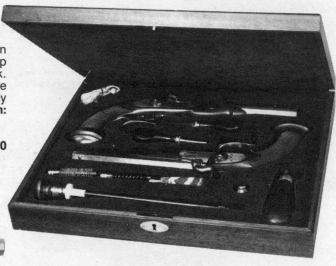

LE PAGE PERCUSSION PISTOL
(44 Caliber)

The tapered octagonal rifled barrel is in the traditional style with 7 lands and grooves. Fully adjustable single-set trigger. Engraved overall with traditional scrollwork. The European walnut stock is in the Boutet style. Spur-style trigger guard. Fully adjustable elevating rear sight. Dovetailed front sight adjustable for windage. **Barrel length:** 9″. **Overall length:** 15″. **Weight:** 2 lbs. 2 oz. **Rec. ball diameter:** 424 R.B.

Le Page Percussion **$475.00**

CASED LE PAGE PISTOL SETS

The case is French-fitted and the accessories are the finest quality to match.

Double Cased Sets
French-fitted double-cased set comprising two Le Page pistols, turn screw, nipple key, oil bottle, cleaning brushes, leather covered flask and loading rod. Rifled or smoothbore barrel.

Double Cased Flintlock Set **$1430.00**
Double Cased Percussion Set 1290.00

Single Cased Sets
French-fitted single-cased set comprising one Le Page pistol, turn screw, nipple key, oil bottle, cleaning brushes, leather covered flask and loading rod. Rifled or smoothbore barrel.

Single Cased Flintlock Set **$760.00**
Single Cased Percussion Set 685.00

NAVY ARMS

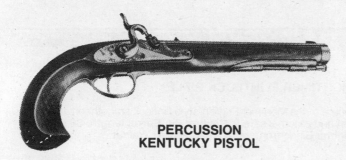

PERCUSSION KENTUCKY PISTOL

FLINTLOCK KENTUCKY PISTOL

KENTUCKY PISTOLS

The Kentucky Pistol is truly a historical American gun. It was carried during the Revolution by the Minutemen and was the sidearm of "Andy" Jackson in the Battle of New Orleans. Navy Arms Company has conducted extensive research to manufacture a pistol representative of its kind, with the balance and handle of the original for which it became famous.

Prices:

Flintlock	**$205.00**
Single Cased Flintlock Set	300.00
Double Cased Flintlock Set	515.00
Percussion	195.00
Single Cased Percussion Set	290.00
Double Cased Percussion Set	495.00

1806 HARPERS FERRY FLINTLOCK PISTOL

1806 HARPERS FERRY PISTOL

Of all the early American martial pistols, Harpers Ferry is one of the best known and was carried by both the Army and the Navy. Navy Arms Company has authentically reproduced the Harper's Ferry to the finest detail, providing a well-balanced and well-made pistol. **Weight:** 2 lbs. 9 oz. **Barrel length:** 10". **Overall length:** 16". **Caliber:** 58 smoothbore. **Finish:** Walnut stock; casehardened lock; brass-mounted browned barrel.

Harpers Ferry	**$265.00**
Single Cased Set	325.00

1816 M.T. WICKHAM MUSKET

This version of the French 1777 Charleville musket was chosen by the U.S. Army in 1816 to replace the 1808 Springfield. Manufactured in Philadelphia by M.T. Wickham, it was one of the last contract models. **Caliber:** 69. **Barrel length:** 44 1/2". **Overall length:** 56 1/4". **Weight:** 10 lbs. **Sights:** Brass blade front. **Stock:** European walnut. **Feature:** Brass flashpan.

1816 M.T. Wickham Musket	**$690.00**

BLACK POWDER

NAVY ARMS RIFLES

MORTIMER FLINTLOCK RIFLE

This big-bore flintlock rifle, a replica of the Mortimer English-style flintlock smoothbore, features a waterproof pan, roller frizzen and external safety. **Caliber:** 54. **Barrel length:** 36″. **Overall length:** 53″. **Weight:** 7 lbs. **Sights:** Blade front; notch rear. **Stock:** Walnut.

Mortimer Flintlock Rifle $690.00
   12-gauge drop-in barrel 285.00

MORTIMER FLINTLOCK MATCH RIFLE

This is the sleek match version of the large-bore rifle above. **Caliber:** .54. **Barrel length:** 36″. **Overall length:** 52¼″. **Weight:** 9 lbs. **Sights:** Precision aperture match rear; globe-style front. **Stock:** Walnut with cheekpiece, checkered wrist, sling swivels. **Features:** Waterproof pan; roller frizzen; external safety.

Mortimer Flintlock Match Rifle $850.00

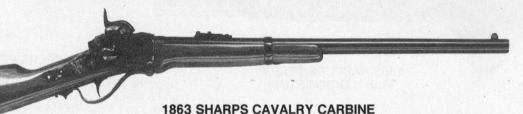

1863 SHARPS CAVALRY CARBINE

This percussion version of the Sharps is a copy of the popular breechloading Cavalry Carbine of the Civil War. It features a patchbox and bar and saddle ring on left side of the stock. **Caliber:** 54. **Barrel length:** 22″. **Overall length:** 39″. **Weight:** 7¾ lbs. **Sights:** Blade front; military ladder rear. **Stock:** Walnut.

Sharps Cavalry Carbine $750.00
Also available:
1859 Sharps Infantry Rifle (54 cal.) 850.00

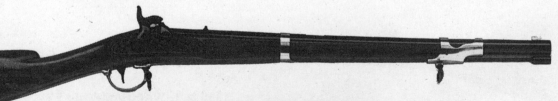

J.P. MURRAY CARBINE

Popular with the Confederate Cavalry, the J.P. Murray percussion carbine was originally manufactured in Columbus, Georgia, during the Civil War. **Caliber:** 58. **Barrel length:** 23½″. **Overall length:** 39¼″. **Weight:** 8 lbs. 5 oz. **Finish:** Walnut stock with polished brass.

J.P. Murray Carbine $380.00

NAVY ARMS RIFLES

1853 ENFIELD RIFLE MUSKET

The Enfield Rifle Musket marked the zenith in design and manufacture of the military percussion rifle and this perfection has been reproduced by Navy Arms Company. This and other Enfield muzzleloaders were the most coveted rifles of the Civil War, treasured by Union and Confederate troops alike for their fine quality and deadly accuracy. **Caliber:** 58. **Barrel length:** 39″. **Weight:** 10 lbs. 6 oz. **Overall length:** 55″. **Sights:** Fixed front; graduated rear. **Stock:** Seasoned walnut with solid brass furniture.

1853 Enfield Rifle Musket **$480.00**

1858 ENFIELD RIFLE

In the late 1850s the British Admiralty, after extensive experiments, settled on a pattern rifle with a 5-groove barrel of heavy construction, sighted to 1,100 yards, designated the Naval rifle, Pattern 1858. **Caliber:** 58. **Barrel length:** 33″. **Weight:** 9 lbs. 10 oz. **Overall length:** 48.5″. **Sights:** Fixed front; graduated rear. **Stock:** Seasoned walnut with solid brass furniture.

1858 Enfield Rifle **$450.00**

1861 MUSKETOON

The 1861 Enfield Musketoon was the favorite long arm of the Confederate Cavalry. **Caliber:** 58. **Barrel length:** 24″. **Weight:** 7 lbs. 8 oz. **Overall length:** 40.25″. **Sights:** Fixed front; graduated rear. **Stock:** Seasoned walnut with solid brass furniture.

1861 Musketoon **$375.00**
Kit 345.00

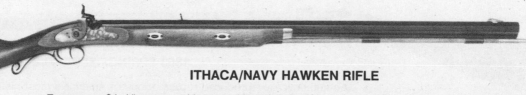

ITHACA/NAVY HAWKEN RIFLE

Features a 31½″ octagonal browned barrel crowned at the muzzle with buckhorn-style rear sight, blade front sight. Color casehardened percussion lock is fitted on walnut stock. **Calibers:** 50 and 54.

Ithaca/Navy Hawken Rifle **$400.00**
Kit 360.00

BLACK POWDER

NAVY ARMS RIFLES

MISSISSIPPI RIFLE MODEL 1841

This historic percussion weapon gained its name because of its performance in the hands of Jefferson Davis' Mississippi Regiment during the heroic stand at the Battle of Buena Vista. Also known as the "Yager" (a misspelling of the German Jaeger), this was one of the first percussion rifles adopted by Army Ordnance. The Mississippi is handsomely furnished in brass, including patchbox for tools and spare parts. **Weight:** 9½ lbs. **Barrel length:** 32½". **Overall length:** 48½". **Calibers:** 54 and 58. **Finish:** Walnut finish stock, brass mounted.

Mississippi Rifle Model 1841 .... **$460.00**

SMITH CARBINE

The Smith Carbine was considered one of the finest breechloading carbines of the Civil War period. The hinged breech action allowed fast reloading for cavalry units. Available in either the **Cavalry Model** (with saddle ring and bar) or **Artillery Model** (with sling swivels). **Caliber:** 50. **Barrel length:** 21½". **Overall length:** 39". **Weight:** 7¾ lbs. **Sights:** Brass blade front; folding ladder rear. **Stock:** American walnut.

Smith Carbine **$600.00**

1861 SPRINGFIELD RIFLE

One of the most popular Union rifles of the Civil War, the 1861 used the 1855-style hammer. The lockplate on this replica is marked "1861, U.S. Springfield." **Caliber:** 58. **Barrel length:** 40". **Overall length:** 56". **Weight:** 10 lbs. **Finish:** Walnut stock with polished metal lock and stock fitting.

1861 Springfield Rifle **$550.00**

1863 SPRINGFIELD RIFLE

An authentically reproduced replica of one of America's most historical firearms, the 1863 Springfield rifle features a full-size, three-band musket and precision-rifled barrel. **Caliber:** 58. **Barrel length:** 40". **Overall length:** 56". **Weight:** 9½ lbs. **Finish:** Walnut stock with polished metal lock and stock fittings. Casehardened lock available upon request.

1863 Springfield Rifle **$550.00**
Springfield Kit **450.00**

NAVY ARMS

PENNSYLVANIA LONG RIFLE

This new version of the Pennsylvania Rifle is an authentic reproduction of the original model. Its classic lines are accented by the long, browned octagon barrel and polished lockplate. **Caliber:** 32 or 45 (flint or percussion. **Barrel length:** 40½″. **Overall length:** 56½″. **Weight:** 7½ lbs. **Sights:** Blade front; adjustable Buckhorn rear. **Stock:** Walnut.

Pennsylvania Long Rifle Flintlock $410.00
    Percussion 395.00

BROWN BESS MUSKET

Used extensively in the French and Indian War, the Brown Bess Musket proved itself in the American Revolution as well. This fine replica of the "Second Model" is marked "Grice" on the lock plate. **Caliber:** 75. **Barrel length:** 42″. **Overall length:** 59″. **Weight:** 9½ lbs. **Sights:** Lug front. **Stock:** Walnut.

Brown Bess Musket $675.00
Kit 510.00

Also available:
Brown Bess Carbine
Caliber: 75. **Barrel length:** 30″. **Overall length:** 47″. **Weight:** 7¾ lbs.
Price $675.00

1803 HARPERS FERRY RIFLE

This 1803 Harpers Ferry rifle was carried by Lewis and Clark on their expedition to explore the Northwest territory. This replica of the first rifled U.S. Martial flintlock features a browned barrel, casehardened lock and a brass patchbox. **Caliber:** 54. **Barrel length:** 35″. **Overall length:** 50½″. **Weight:** 8½ lbs.

1803 Harpers Ferry Rifle $555.00

"BERDAN" 1859 SHARPS RIFLE

A replica of the Union sniper rifle used by Col. Hiram Berdan's First and Second U.S. Sharpshooters Regiments during the Civil War. **Caliber:** 54. **Barrel length:** 30″. **Overall length:** 46¾″. **Weight:** 8 lbs. 8 oz. **Sights:** Military-style ladder rear; blade front. **Stock:** Walnut. **Features:** Double set triggers, casehardened receiver; patchbox and furniture

"Berdan" 1859 Sharps Rifle $895.00
Also available:
Single Trigger Infantry Model 850.00

BLACK POWDER

NAVY ARMS RIFLES

1862 C.S. RICHMOND RIFLE

This model was manufactured by the Confederacy at the Richmond Armory utilizing 1855 Rifle Musket parts captured from the Harpers Ferry Arsenal. This replica features the unusual 1855 lockplate, stamped "1862 C.S. Richmond, V.A." **Caliber:** 58. **Barrel length:** 40″. **Overall length:** 56″. **Weight:** 10 lbs. **Finish:** Walnut stock with polished metal lock and stock fittings.

1862 C.S. Richmond Rifle **$550.00**

TRYON CREEDMOOR RIFLE

This replica of the Tryon Creedmoor match rifle won a Gold Medal at the 13th World Shoot in Germany. It features a blued octagonal heavy match barrel, hooded target front sight, adjustable Vernier tang sight, double-set triggers, sling swivels and a walnut stock. **Caliber:** 451. **Barrel length:** 33″. **Overall length:** 48¼″. **Weight:** 9½ lbs.

Tryon Creedmoor Rifle **$680.00**
   Brass telescopic sight **125.00**

1874 SHARPS SNIPER RIFLE

A replica of the 1874 three-band sharpshooters rifle. **Caliber:** .45-70. **Barrel length:** 30″. **Overall length:** 46¾″. **Weight:** 8 lbs. 8 oz. **Stock** Walnut. **Features:** Double set triggers; casehardened receiver; patchbox and furniture.

1874 Sharps Sniper Rifle **$920.00**
Also available:
Single Trigger Infantry Model **875.00**

KODIAK DOUBLE RIFLE

The powerful double-barreled Kodiak percussion rifle has fully adjustable sights mounted on blued steel barrels. The lockplates are engraved and highly polished. **Calibers:** 50, 54 and 58. **Barrel length:** 28½″. **Overall length:** 45″. **Weight:** 11 lbs. **Sights:** Folding notch rear; ramp bead front. **Stock:** Hand-checkered walnut.

Kodiak Double Rifle **$680.00**

NAVY ARMS SHOTGUNS

MORTIMER FLINTLOCK SHOTGUN

This replica of the Mortimer Shotgun features a browned barrel, casehardened furniture, sling swivels and checkered walnut stock. The lock contains waterproof pan, roller frizzen and external safety. **Gauge:** 12. **Barrel length:** 36″. **Overall length:** 53″. **Weight:** 7 lbs.

Mortimer Flintlock Shotgun $670.00

STEEL SHOT MAGNUM SHOTGUN

This shotgun, designed for the hunter who must use steel shot, features engraved polished lockplates, English-style checkered walnut stock (with cheekpiece) and chrome-lined barrels. **Gauge:** 10. **Barrel length:** 28″. **Overall length:** 45$\frac{1}{2}$″. **Weight:** 7 lbs. 9 oz. **Choke:** Cylinder/Cylinder.

Steel Shot Magnum Shotgun $510.00

FOWLER SHOTGUN

A traditional side-by-side percussion field gun, this fowler model features blued barrels and English-style straight stock design. It also sports a hooked breech, engraved and color casehardened locks, double triggers and checkered walnut stock. **Gauge:** 12. **Chokes:** Cylinder/Cylinder. **Barrel length:** 28″. **Overall length:** 44$\frac{1}{2}$″. **Weight:** 7$\frac{1}{2}$ lbs.

Fowler Shotgun $325.00
Kit 285.00

T & T SHOTGUN

This Turkey and Trap side-by-side percussion shotgun, choked full/full, features a genuine walnut stock with checkered wrist and oil finish, color casehardened locks and blued barrels. **Gauge:** 12. **Barrel length:** 28″. **Overall length:** 44″. **Weight:** 7$\frac{1}{2}$ lbs.

T & T Shotgun $500.00

BLACK POWDER

PARKER-HALE RIFLES

1853 THREE-BAND MUSKET

Commonly known as the 3-band Enfield, this is a high-quality replica of the 1853 Enfield rifle musket that was manufactured between 1853 and 1863 by various British contractors. Manufactured today by the Gibbs Rifle Company. **Caliber:** 577. **Barrel length:** 39″. **Overall length:** 55″. **Weight:** 9 lbs. **Rate of twist:** 1:48″; 3-groove barrel. **Price:** $585.00

1858 TWO-BAND MUSKET

One of the most accurate military rifles of the percussion era, the 1858 two-band Enfield was developed for the British Admiralty in the late 1850s. This replica of the Enfield Naval Pattern rifle has won many blackpowder national and world championships. Manufactured by the Gibbs Rifle Company. **Caliber:** 577. **Barrel length:** 33″. **Overall length:** 48½″. **Weight:** 8½ lbs. **Rate of twist:** 1:48″, 5-groove barrel. **Price:** $550.00

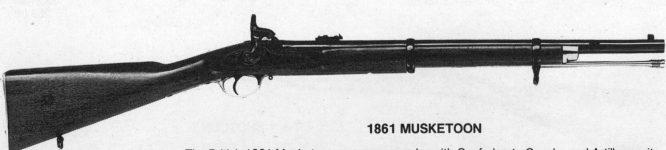

1861 MUSKETOON

The British 1861 Musketoon was very popular with Confederate Cavalry and Artillery units. This reproduction, like all the Parker-Hale replicas, is constructed using the original 130-year-old gauges for authentic reference. Manufactured by the Gibbs Rifle Company. **Caliber:** 577. **Barrel length:** 24″. **Overall length:** 40¼″. **Weight:** 7½ lbs. **Rate of twist:** 1:48″; 5-groove barrel. **Price:** $450.00

PARKER-HALE RIFLES

VOLUNTEER RIFLE

Originally designed by Irish gunmaker William John Rigby, this relatively small-caliber rifle was issued to volunteer regiments during the 1860s. Today it is rifled by the cold-forged method, making one turn in 20 inches. Sights are adjustable: globe front and ladder-type rear with interchangeable leaves; hand-checkered walnut stock. Manufactured by the Gibbs Rifle Company. **Caliber:** 451. **Barrel length:** 32″. **Weight:** 9¹/₂ lbs. **Price:** $750.00

WHITWORTH MILITARY TARGET RIFLE

Recreation of Sir Joseph Whitworth's deadly and successful sniper and target weapon of the mid-1800s. Devised with a hexagonal bore with a pitch of 1 turn in 20 inches. Barrel is cold-forged from ordnance steel, reducing the build-up of black powder fouling. Globe front sight; open military target rifle rear sight has interchangeable blades of different heights. Walnut stock is hand-checkered. Manufactured by the Gibbs Rifle Company. **Caliber:** 451. **Barrel length:** 36″. **Weight:** 9¹/₂ lbs. **Price:** $815.00

RUGER

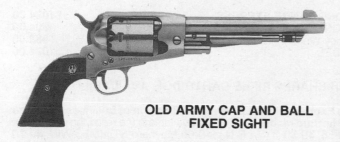

**OLD ARMY CAP AND BALL
FIXED SIGHT**

OLD ARMY CAP AND BALL
$378.50   ($428.50 Stainless Steel)

This Old Army cap-and-ball revolver with fixed sights is reminiscent of the Civil War era martial revolvers and those used by the early frontiersmen in the 1800s. This Ruger model comes in both blued and stainless-steel finishes and features modern materials, technology and design throughout, including steel music-wire coil springs.

SPECIFICATIONS
Caliber: 45 (.443″ bore; .45″ groove)
Barrel length: 7¹/₂″
Rifling: 6 grooves, R.H. twist (1:16″)
Weight: 2⁷/₈ lbs.
Sights: Fixed, ramp front; topstrap channel rear
Percussion cap nipples: Stainless steel (#11)

BLACK POWDER

SHILOH SHARPS

MODEL 1874 BUSINESS RIFLE
$910.00

Calibers: 45-70, 45-90, 45-120, 50-70 and 50-90. **Barrel:** 28-inch heavy-tapered round; dark blue. Double-set triggers adjustable set. **Sights:** Blade front, and sporting rear with leaf. Buttstock is straight grip rifle buttplate, forend sporting schnabel style. Receiver group and butt plate case-colored; wood is American walnut oil-finished. **Weight:** 9 lbs. 8 oz.
Also available:
MODEL 1874 SADDLE RIFLE w/26″ tapered octagonal barrel **$962.00**

MODEL 1874 SPORTING RIFLE NO. 1
$1008.00

Calibers: 45-70, 45-90, 45-120, 50-70 and 50-90. Features 30-inch tapered octagon barrel. Double-set triggers with adjustable set, blade front sight, sporting rear with elevation leaf and sporting tang sight adjustable for elevation and windage. Buttstock is pistol grip, shotgun butt, sporting forend style. Receiver group and buttplate case colored. Barrel is high-finish blue-black; wood is American walnut oil finish. **Weight:** 10 lbs.

MODEL 1874 SPORTING RIFLE NO. 3
$870.00

Calibers: 45-70, 45-90, 45-120, 50-70 and 50-90. **Barrel:** 30-inch tapered octagonal; with high finish blue-black. Double-set triggers with adjustable set, blade front sight, sporting rear with elevation leaf and sporting tang sight adjustable for elevation and windage. Buttstock is straight grip with rifle butt plate; trigger plate is curved and checkered to match pistol grip. Forend is sporting schnabel style. Receiver group and butt plate is case colored. Wood is American walnut oil-finished. **Weight:** 9 lbs. 8 oz.
Also available:
MODEL 1874 LONG RANGE EXPRESS $1034.00
MODEL 1874 MONTANA ROUGHRIDER 904.00
   With Semi-fancy Wood 988.00
HARTFORD MODEL 1074.00

SHILOH SHARPS RIFLE CARTRIDGE AVAILABILITY

The **Long Range Express, No. 1 Sporting, No. 3 Sporting, Business, Montana Roughrider** and **Hartford Model rifles** all are available in the following cartridges: 30-40, 38-55, 38-56, 40-50 ($1^{11}/_{16}$ B.N.), 40-60 Maynard ST, 40-65 WIN, 40-70 ($2^1/_{10}$ B.N.), 40-70 ($2^1/_4$ B.N.), 40-70 ($2^1/_2$ ST), 40-82 Shiloh, 40-90 ($3^1/_4$ ST), 40-90 ($2^5/_8$ B.N.), 44-77 B.N., 44-90 B.N., 45-70 ($2^1/_{10}$ ST), 45-90 ($2^4/_{10}$ ST), 45-100 ($2^6/_{10}$ ST), 45-110 ($2^7/_8$ ST), 45-120 ($3^1/_4$ ST), 50-70 ($1^3/_4$ ST) and 50-100 ($2^1/_2$ ST). The **1874 Saddle Rifle** is available in all of the above calibers, except the 40-82 Shiloh, 44-90, 45-100, 45-110 and 45-120. (B.N. = Bottleneck ST = Straight)

THOMPSON/CENTER

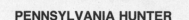

PENNSYLVANIA HUNTER

The 31″ barrel on this model is cut rifled (.010″ deep) with 1 turn in 66″ twist. Its outer contour is stepped from octagon to round. Sights are fully adjustable for both windage and elevation. Stocked with select American black walnut; metal hardware is blued steel. Features a hooked breech system and coil-spring lock. **Caliber:** 50. **Overall length:** 48″. **Weight:** Approx. 7.6 lbs.

Pennsylvania Hunter Caplock **$340.00**
Pennsylvania Hunter Flintlock 355.00

PENNSYLVANIA HUNTER CARBINE

Thompson/Center's new Pennsylvania Hunter Carbine is a 50-caliber carbine with 1:66″ twist and cut-rifling. It was designed specifically for the hunter who uses patched round balls only and hunts in thick cover or brush. The 21″ barrel is stepped from octagonal to round. **Overall length:** 38″. **Weight:** 6½ lbs. **Sights:** Fully adjustable open hunting-style rear with bead front. **Stock:** Select American walnut. **Trigger:** Single hunting-style trigger. **Lock:** Color cased, coil spring, with floral design.

Pennsylvania Hunter Carbine Caplock **$330.00**
Pennsylvania Hunter Carbine Flintlock 345.00

THE NEW ENGLANDER RIFLE

This percussion rifle features a 26″ round, 50- or 54-caliber rifled barrel (1 in 48″ twist). **Weight:** 7 lbs. 15 oz.

New Englander Rifle **$285.00**
  With Rynite stock (24″ barrel, right-hand only) 270.00
Left-Hand Model 305.00

THE NEW ENGLANDER SHOTGUN

This 12-gauge muzzleloading percussion shotgun weighs only 5 lbs. 2 oz. It features a 28-inch (screw-in full choke) round barrel and is stocked with select American black walnut.

New Englander Shotgun **$305.00**

THOMPSON/CENTER

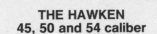

THE HAWKEN
45, 50 and 54 caliber

Similar to the famous Rocky Mountain rifles made during the early 1800s, the Hawken is intended for serious shooting. Button-rifled for ultimate precision, the Hawken is available in 45, 50 or 54 caliber, flintlock or percussion. It features a hooked breech, double-set triggers, first-grade American walnut stock, adjustable hunting sights, solid brass trim and color casehardened lock. Beautifully decorated. **Weight:** Approx. 8 1/2 lbs.

Hawken Caplock 45, 50 or 54 caliber $395.00
Hawken Flintlock 50 caliber 405.00
Kit: Caplock 290.00
Kit: Flintlock 310.00

HAWKEN CUSTOM RIFLE

T/C's Hawken Custom features a crescent buttplate inletted in a select American walnut stock. The barrel, lockplate, buttplate, trigger guard and forend cap are all polished and buffed to a high-luster sheen with a deep blue finish. The 50-caliber 28-inch octagonal barrel has a 1:48" twist and handles patched round balls and conical projectiles. **Overall length:** 45 1/4". **Weight:** 8 1/2 lbs. **Sights:** Fully adjustable open hunting-style rear; bead-style front. **Features:** Hooked-breech system; heavy-duty, coil-type internal springs; fully adjustable trigger (functions as double set or single stage).

Hawken Custom Rifle $475.00

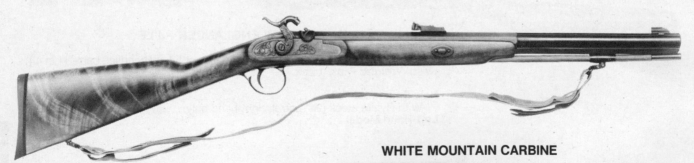

WHITE MOUNTAIN CARBINE

WHITE MOUNTAIN
CAPLOCK CARBINE

This hunter's rifle with single trigger features a wide trigger guard bow that allows the shooter to fire the rifle in cold weather without removing his gloves. Its stock is of select American black walnut with a rifle-type rubber recoil pad, and is equipped with swivel studs and quick detachable sling swivels. A soft leather hunting-style sling is included. The barrel is stepped from octagonal to round. **Calibers:** 50 and 54 (Hawken or Renegade loads). **Barrel length:** 21". **Overall length:** 38". **Weight:** 6 1/2 lbs. **Sights:** Open hunting (Patridge) style, fully adjustable. **Lock:** Heavy-duty coil springs; decorated with floral design and color-cased. **Breech:** Hooked-breech system.

White Mountain Carbine–Caplock (right-hand only) $350.00

THOMPSON/CENTER

THE RENEGADE

Available in 50- or 54-caliber percussion, the Renegade was designed to provide maximum accuracy and maximum shocking power. It is constructed of superior modern steel with investment cast parts fitted to an American walnut stock, featuring a precision-rifled (26-inch carbine-type) octagonal barrel, hooked-breech system, coil spring lock, double-set triggers, adjustable hunting sights and steel trim. **Weight:** Approx. 8 lbs.

Renegade Caplock 50 and 54 caliber $350.00
Renegade Caplock Left Hand 360.00
Renegade Caplock Kit Right Hand 268.00

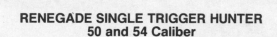

RENEGADE SINGLE TRIGGER HUNTER
50 and 54 Caliber

This single trigger hunter model, fashioned after the double-triggered Renegade, features a large bow in the shotgun-style trigger guard. This allows shooters to fire the rifle in cold weather without removing their gloves. The octagon barrel measures 26 inches and the stock is made of select American walnut. **Weight:** About 8 pounds.

Renegade Hunter $325.00

BIG BOAR CAPLOCK RIFLE

This large 58-caliber caplock rifle is designed for the muzzleloading hunter who prefers larger game. The rifle features a 26-inch octagonal barrel, rubber recoil pad, leather sling with QD sling swivels, and an adjustable open-style hunting rear sight with bead front sight. Stock is of American black walnut.

Big Boar Caplock $355.00

BLACK POWDER

THOMPSON/CENTER

SCOUT CARBINE with Rynite Stock

SCOUT CARBINE & PISTOL

Thompson/Center's Scout Carbine & Pistol uses the in-line ignition system with a special vented breech plug that produces constant pressures from shot to shot, thereby improving accuracy. The patented trigger mechanism consists of only two moving parts—the trigger and the hammer—thus providing ease of operation and low maintenance. Both the carbine and pistol are available in 50 and 54 caliber. The carbine's 21-inch barrel and the pistol's 12-inch barrel are easily removable and readily interchangeable in either caliber. Their lines are reminiscent of the saddle guns and pistols of the "Old West," combining modern-day engineering with the flavor of the past. Both are suitable for left-hand shooters.

Scout Carbine	$415.00
Scout Carbine with Rynite Stock	325.00
Scout Pistol	340.00

THUNDER HAWK

Thompson/Center's in-line caplock rifle, the Thunder Hawk, combines the features of an old-time caplock with the look and balance of a modern bolt-action rifle. The in-line ignition system ensures fast, positive ignition, plus an adjustable trigger for a crisp trigger pull. The 21-inch barrel has an adjustable rear sight and bead-style front sight (barrel is drilled and tapped to accept T/C's Thunder Hawk scope rings or Quick Release Mounting System). The stock is American black walnut with rubber recoil pad and sling swivel studs. Rifling is 1 in 38" twist, designed to fire patched round balls, conventional conical projectiles and sabot bullets. **Weight:** Approx. 6¾ lbs.

Thunder Hawk (50 or 54 caliber)	$290.00
Stainless Steel with Rynite Stock	290.00

GREY HAWK

T/C's Grey Hawk is a stainless steel caplock rifle in 50 and 54 caliber with a Rynite buttstock and a round 24-inch barrel. It also features a stainless-steel lockplate, hammer, thimble and trigger guard. Adjustable rear sight and bead-style front sight are blued. **Weight:** Approx. 7 lbs.

Grey Hawk	$310.00

TRADITIONS

PIONEER PISTOL

PIONEER PISTOL
$169.00   ($119.00 Kit)

SPECIFICATIONS
Caliber: 45 percussion
Barrel length: 9⅝" octagonal with tenon; ¹³/₁₆" across flats, rifled 1 in 16"; hooked breech
Overall length: 15"   **Weight:** 1 lb. 15 oz.
Sights: Blade front; fixed rear
Trigger: Single
Stock: Beech, rounded
Lock: V-type mainspring
Features: German silver furniture; blackened hardware

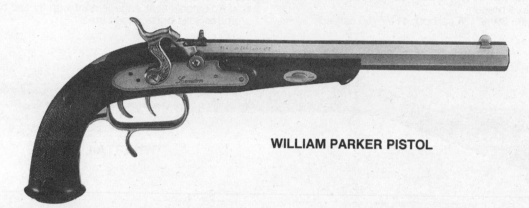

WILLIAM PARKER PISTOL

WILLIAM PARKER PISTOL
$265.00

SPECIFICATIONS
Caliber: 50 percussion
Barrel length: 10⅜" octagonal (¹⁵/₁₆" across flats)
Overall length: 17½"   **Weight:** 2 lbs. 5 oz.
Sights: Brass blade front; fixed rear
Stock: Walnut, checkered at wrist
Triggers: Double set; will fire set and unset
Lock: Adjustable sear engagement with fly and bridle; V-type mainspring
Features: Brass percussion cap guard; polished hardware, brass inlays and separate ramrod

BUCKSKINNER PISTOL (not shown)
$157.00   ($182.00 Laminated)

SPECIFICATIONS
Caliber: 50 percussion
Barrel length: 10" octagonal (¹⁵/₁₆" across flats)
Overall length: 15"   **Weight:** 2 lbs. 9 oz.
Sights: Fixed rear; blade front
Stock: Beech or laminated
Trigger: Single
Features: Blackened furniture; wood ramrod

TRAPPER PISTOL
$170.00 Percussion   ($130.00 Percussion Kit)
$190.00 Flintlock

SPECIFICATIONS
Caliber: 50 percussion or flintlock
Barrel length: 9¾"; octagonal (⅞" across flats) with tenon
Overall length: 16"   **Weight:** 2 lbs. 11 oz.
Stock: Beech
Lock: Adjustable sear engagement with fly and bridle
Triggers: Double set, will fire set and unset
Sights: Primitive-style adjustable rear; brass blade front
Furniture: Solid brass; blued steel on assembled pistol

TRAPPER PISTOL

BLACK POWDER

TRADITIONS

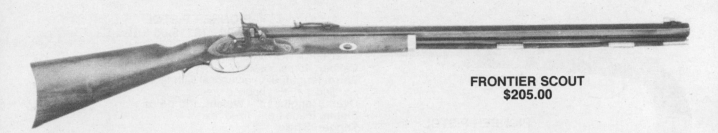

FRONTIER SCOUT
$205.00

SPECIFICATIONS
Calibers: 36 and 50 percussion
Barrel length: 24″ (36 caliber); 26″ (50 caliber); octagonal (⁷/₈″ across flats) with tenon; rifled 1:48″ (36 cal.) and 1:66″ (50 cal.); hooked breech
Overall length: 39¹/₈″ (36 caliber); 41¹/₈″ (50 caliber)

Weight: 5 lbs. 10 oz.
Sights: Primitive, adjustable rear, brass blade front
Stock: Beech   **Length of pull:** 12¹/₄″
Lock: Adjustable sear engagement with fly and bridle
Furniture: Solid brass, blued steel

WHITETAIL RIFLE

WHITETAIL RIFLE & CARBINE
$257.00 (Percussion)
$274.00 (Flintlock)

SPECIFICATIONS
Caliber: 50 (flintlock or percussion) and 54 (percussion only)
Barrel length: 24″ octagonal tapering to round

Overall length: 39¹/₄″   **Weight:** 5 lbs. 12 oz.
Sights: Hunting-style rear, click adjustable for windage and elevation; beaded blade front with fluorescent dot
Trigger: Single
Stock: Select hardwood w/walnut stain, rubber recoil pad
Features: Sling swivels; oversized trigger guard; inletted wedge plates; engraved and color casehardened lock

DEERHUNTER RIFLE
$165.00 (Percussion) $182.00 (Flintlock)
$152.75 (Percussion Kit, 50 and 54 caliber)

SPECIFICATIONS
Calibers: 32, 50, 54
Barrel length: 24″ octagonal
Rifling twist: 1:48″ (percussion only); 1:66″ (flint or percussion)
Overall length: 39¹/₄″
Weight: 6 lbs. (6.25 lbs. in All-Weather Model)

Trigger: Single
Sights: Fixed rear; blade front
Features: Wooden ramrod; blackened furniture; inletted wedge plates
Also available:
DEERHUNTER ALL-WEATHER MODEL. Beech stock is epoxy-covered; finish is C-Nickel. **Price: $206.00** (1:66″ twist flintlock); **$189.50** (1:48″ twist, percussion, 50 or 54 caliber, drilled and tapped)

TRADITIONS

HAWKEN WOODSMAN
$240.00 (Percussion) $260.00 (Left Hand)
$256.00 (50 Caliber Flintlock w/Blackened
Furniture)

SPECIFICATIONS
Calibers: 50 and 54 percussion or flintlock
Barrel length: 26″ (octagonal); hooked breech; rifled 1 turn in 66″ (1 turn in 48″ in 50 or 54 caliber also available)
Overall length: 45³/₄″ (Brass); 42″ (Blackened)
Weight: 6 lbs. 14 oz.
Triggers: Double set; will fire set or unset

Lock: Adjustable sear engagement with fly and bridle
Stock: Beech
Sights: Beaded blade front; hunting-style rear, fully screw adjustable for windage and elevation
Furniture: Solid brass, blued steel or blackened (50 cal. only); unbreakable ramrod

PENNSYLVANIA RIFLE
$495.00 (Flintlock)
$467.00 (Percussion)

SPECIFICATIONS
Caliber: 50
Barrel length: 40¹/₄″; octagonal (⁷/₈″ across flats) with 3 tenons; rifled 1 turn in 66″
Overall length: 57¹/₂″   **Weight:** 9 lbs.
Lock: Adjustable sear engagement with fly and bridle

Stock: Walnut, beavertail style
Triggers: Double set; will fire set and unset
Sights: Primitive-style adjustable rear; brass blade front
Furniture: Solid brass, blued steel

PIONEER CARBINE

PIONEER RIFLE & CARBINE
$198.00

SPECIFICATIONS
Calibers: 50 and 54 percussion (rifle only)
Barrel length: Carbine—24″ (1:32″); Rifle—28″ (1:48″ or 1:66″), octagonal w/tenon
Overall length: 40¹/₂″ (carbine); 44″ (rifle)

Weight: 6 lbs. 3 oz. (carbine); 6 lbs. 14 oz. (rifle)
Trigger: Single
Stock: Beech
Sights: Buckhorn rear with elevation ramp, ajustable for windage and elevation; German silver blade front
Lock: Adjustable sear engagement; V-type mainspring
Features: Blackened hardware; German silver furniture; unbreakable ramrod; inletted wedge plates

TRADITIONS

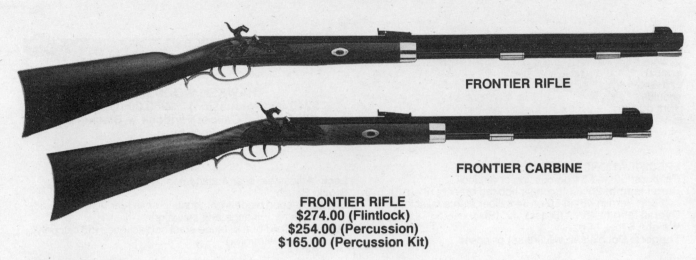

FRONTIER RIFLE

FRONTIER CARBINE

FRONTIER RIFLE
$274.00 (Flintlock)
$254.00 (Percussion)
$165.00 (Percussion Kit)

SPECIFICATIONS
Caliber: 50
Barrel length: 28″ octagonal (15/16″ across flats) with tenon; hooked breech, rifled 1 turn in 66″ (1 turn in 48″ optional)
Overall length: 44³/₄″ **Weight:** 7 lbs. 5 oz.
Lock: Adjustable sear engagement with fly and bridle
Triggers: Double set; will fire set and unset
Stock: Beech

Sights: Beaded blade front; hunting-style rear, adjustable for windage and elevation
Furniture: Solid brass, blued steel
Also available:
FRONTIER CARBINE with 24″ barrel (1:66″ or 1:32″ twist), 40¹/₂″ overall length; weight 6¹/₂ lbs.; percussion only in 50 caliber. **Price:** $254.00

BUCKSKINNER CARBINE
Laminated Stock

BUCKSKINNER CARBINE
$290.00 (Flintlock and Left Hand)
$274.00 (Percussion)
$320.00 (Laminated Stock, Percussion)
$337.00 (Laminated Stock, Flintlock)

SPECIFICATIONS
Caliber: 50 percussion and flintlock
Barrel length: 21″ octagonal-to-round with tenon; 15/16″ across flats; 1:66″ twist (1:20″ twist also available)
Overall length: 36¹/₄″ **Weight:** 5 lbs. 15 oz.
Sights: Hunting-style, click adjustable rear; beaded blade front with white dot

Trigger: Single
Features: Blackened furniture; German silver ornamentation; belting leather sling and sling swivels; unbreakable ramrod
Also available: **BUCKSKINNER CARBINE DELUXE** Percussion w/1:20″ twist: **$352.00**; or Flintlock w/1;66″ twist: **$337.00**

TRADITIONS

IN-LINE RIFLE SERIES

This new family of hunting rifles is designed for serious black-powder hunters as well as those who are just getting started in muzzleloading. These rifles deliver the balance, fast handling and long-range accuracy of a modern hunting gun. They shoulder naturally, pick up the target quickly, and hold the point of aim. A unique three-position safety (1) allows the bolt to be placed forward without engaging the nipple or cap; (2) allow the bolt to lock open in an elevated notch position, preventing the trigger from being engaged; or (3) locks the bolt with a push of the bolt handle to prevent any bolt movement during stalking. The breech design disperses ignition gases safely to the side. No special tools are needed for cleaning. The oversized trigger guard is ample for gloved trigger fingers. All rifles have rubber recoil pads, are drilled and tapped (with 8-40 plug screws) for easy scoping and come with click-adjustable hunting sights and PVC ramrod. Specifications and prices for all models in the series are listed in the table below.

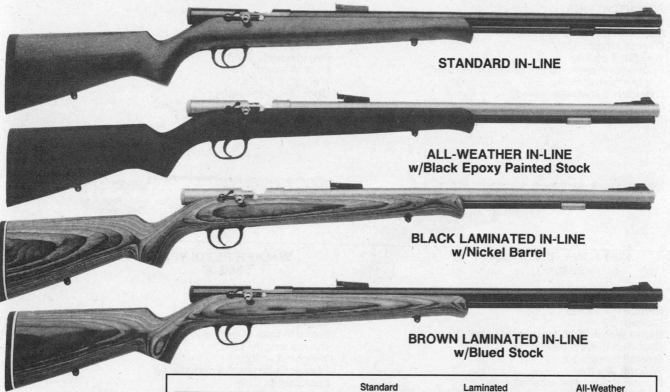

STANDARD IN-LINE

ALL-WEATHER IN-LINE
w/Black Epoxy Painted Stock

BLACK LAMINATED IN-LINE
w/Nickel Barrel

BROWN LAMINATED IN-LINE
w/Blued Stock

		Standard	Laminated	All-Weather
Lock:	ignition	in-line percussion	in-line percussion	in-line percussion
Stock:		beech	laminated	epoxy covered beech
Calibers:	rifling twist	.50p (1:32") .54p (1:48")	.50p (1:32") .54p (1:48")	.50p (1:32") .54p (1:48")
Barrel:	length/shape finish breech other	24" tapered round blued open drilled & tapped	24" tapered round blued or C-Nickel open drilled & tapped	24" tapered round C-Nickel open drilled & tapped
Trigger:		single	single	single
Sights:	rear front	click adjustable beaded blade	click adjustable beaded blade	click adjustable beaded blade
Features:	furniture ramrod other	blued PVC sling swivel studs	blued PVC sling swivel studs	blued PVC sling swivel studs
Overall length:		41"	41"	41"
Weight:		7 lbs. 8 oz.	8 lbs. 1 oz.	7 lbs. 9 oz.
Models/prices:		R40002 .50p $239 R40048 .54p $239	R40202 .50p blk $322 R40248 .54p blk $322 R40302 .50p brn $305 R40348 .54p brn $305	R40102 .50p $255 R40148 .54p $255

A. UBERTI

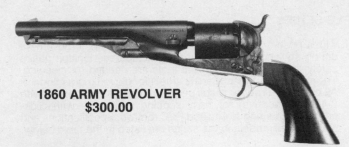

1860 ARMY REVOLVER
$300.00

SPECIFICATIONS
Caliber: 44
Barrel length: 8″ (round, tapered)
Overall length: 13³/₄″
Weight: 2.65 lbs.
Frame: One-piece, color casehardened steel
Trigger guard: Brass
Cylinder: 6 shots (engraved)
Grip: One-piece walnut

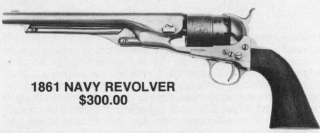

1861 NAVY REVOLVER
$300.00

SPECIFICATIONS
Caliber: 36
Capacity: 6 shots
Barrel length: 7¹/₂″
Overall length: 13″
Weight: 2.75 lbs.
Grip: One-piece walnut
Frame: Color casehardened steel

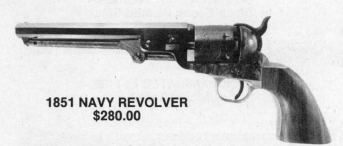

1851 NAVY REVOLVER
$280.00

SPECIFICATIONS
Caliber: 36
Barrel length: 7¹/₂″ (octagonal, tapered)
Cylinder: 6 shots (engraved)
Overall length: 13″
Weight: 2³/₄ lbs.
Frame: Color casehardened steel
Backstrap and trigger guard: Brass
Grip: One-piece walnut

WALKER REVOLVER
$360.00

SPECIFICATIONS
Caliber: 44
Barrel length: 9″ (round in front of lug)
Overall length: 15³/₄″
Weight: 4.41 lbs.
Frame: Color casehardened steel
Backsstrap: Steel
Cylinder: 6 shots (engraved with ''Fighting Dragoons'' scene)
Grip: One-piece walnut

1st MODEL DRAGOON REVOLVER

SPECIFICATIONS
Caliber: 44
Capacity: 6 shots
Barrel length: 7¹/₂″ round forward of lug
Overall length: 13¹/₂″
Weight: 4 lbs.
Frame: Color casehardened steel
Grip: One-piece walnut
Also available:
2nd Model Dragoon w/square cylinder bolt shot . . $325.00
3rd Model Dragoon w/loading lever latch, steel
    backstrap, cut for shoulder stock 330.00

1st MODEL DRAGOON REVOLVER
$325.00

A. UBERTI

1858 REMINGTON NEW ARMY 44 REVOLVER

Prices:
8" barrel, open sights	$280.00
With stainless steel and open sights	380.00
Target Model w/black finish	330.00
Target Model w/stainless steel	399.00

Also available:
1858 New Navy (36 cal.)	280.00
1858 New Army Revolving Carbine (18" barrel)	425.00

1858 REMINGTON NEW ARMY TARGET MODEL

PATERSON REVOLVER

Manufactured at Paterson, New Jersey, by the Patent Arms Manufacturing Company from 1836 to 1842, these were the first revolving pistols created by Samuel Colt. All early Patersons featured a five-shot cylinder, roll-engraved with one or two scenes, octagon barrel and folding trigger that extends when the hammer is cocked.

SPECIFICATIONS
Caliber: 36
Capacity: 5 shots (engraved cylinder)
Barrel length: 7¹/₂" octagonal
Overall length: 11¹/₂"
Weight: 2.552 lbs.
Frame: Color casehardened steel
Grip: One-piece walnut

PATERSON REVOLVER
$395.00
($450.00 w/Lever)

SANTA FE HAWKEN RIFLE
$495.00
$449.00 Kit

SANTA FE HAWKEN RIFLE

SPECIFICATIONS
Calibers: 50 and 54
Barrel length: 32" octagonal
Overall length: 50"
Weight: 9¹/₂ lbs.
Stock: Walnut with beavertail cheekpiece
Features: Brown finish; double-set trigger; color casehardened lockplate; German silver wedge plates

ULTRA LIGHT ARMS

**MODEL 90
$950.00**

This muzzleloader comes with a 28″ button-rifled barrel (1 in 48″ twist) and 13½″ length of pull. Fast ignition with in-line action and fully adjustable Timney trigger create consistent shots. Available in 45 or 50 caliber, each rifle has a Kevlar/ Graphite stock and Williams sights, plus integral side safety. Recoil pad, sling swivels and a hard case are all included. **Weight:** 6 lbs.

WHITE SYSTEMS

MODEL SUPER 91

WHITETAIL MUZZLELOADER

MODEL SUPER 91

This modern muzzleloading system features the following: Ordnance-grade stainless-steel construction • Fast twist, shallow groove rifling • Stainless-steel nipple and breech plug • Side swing safety (locks the striker, not just the trigger) • Classic stock configuration (fits right- or left-handed shooters • Fast second shot and easy access to nipple from either side for quick capping • Fully adjustable trigger

Calibers: 41, 45 and 50
Barrel length: 24″
Rifling: 1 in 20″ (45 cal.); 1 in 24″ (50 cal.)
Weight: 7¾ lbs.
Sights: Fully adjustable Williams sights
Prices:
Blued $599.00
Stainless Steel 699.00
Sporting 699.00
Also available:
SUPER SAFARI SS TAPERED
   41, 45, 50 caliber $799.00

WHITETAIL AND BISON MUZZLELOADING RIFLES

White's ''G Series'' rifles feature straight-line action with easy no-tool takedown in the field. A stainless-steel hammer system has an ambidextrous cocking handle that doubles as a sure-safe hammer-lock safety. Other features include the ''Insta-Fire'' one-piece nipple/breech-plug system (with standard #11 percussion caps); fully adjustable open hunting sights; 22″ barrel with integrated ramrod guide and swivel studs.

Calibers: 41, 45, 50 and 54
Prices:
Bison (50 and 54 cal.) Blued $299.00
Whitetail (45 and 50 cal.) Blued 399.00
   Stainless steel 499.00

sights & scopes

For addresses and phone numbers of manufacturers and distributors included in this section, turn to *DIRECTORY OF MANUFACTURERS AND SUPPLIERS* at the back of the book.

AIMPOINT SIGHTS

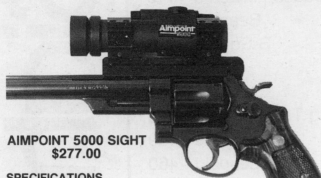

SERIES 3000 UNIVERSAL
$232.00 (Black or Stainless)

SPECIFICATIONS
Weight: 5.15 oz.
Length: 6⁷/₈″ and 6.15″
Magnification: 1X
Scope Attachment: 3X
Battery Choices: Lithium CR 1/3 N, 2L76, DL 1/3 N, Mercury (2) MP 675 or SP 675, Silver Oxide (2) D 375 H, Alkaline (2) LR 44
Material: Anodized Aluminum, Blue or Stainless Finish
Mounting: 1″ Rings (Medium or High)

AIMPOINT 5000 SIGHT
$277.00

SPECIFICATIONS
System: Parallax free
Optical: Anti-reflex coated lenses
Adjustment: 1 click = ¹/₄-inch at 100 yards
Length: 5¹/₂″
Weight: 5.8 oz.
Objective diameter: 36mm
Mounting system: 30mm rings
Magnification: 1X
Material: Anodized aluminum; blue or stainless finish
Diameter of dot: 3″ at 100 yds. or Mag Dot reticle, 10″ at 100 yards.

AIMPOINT COMP
$308.00

SPECIFICATIONS
System: 100% Parallax free
Optics: Anti-reflex coated lenses
Eye relief: Unlimited
Batteries: 3 × Mercury SP 675
Adjustment: 1 click = ¹/₄-inch at 100 yards
Length: 4³/₈″
Weight: 4.75 oz.
Objective diameter: 36mm
Dot diameter: 2″ at 30 yds. (7 MOA); 3″ at 30 yds. (10 MOA)
Mounting system: 30mm rings
Magnification: 1X
Material: Anodized aluminum, blue or stainless finish

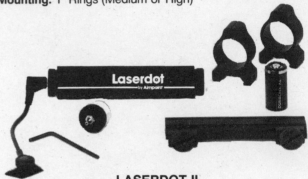

LASERDOT II
$319.00 (Black or Stainless)

SPECIFICATIONS
Length: 4″
Weight: 4.06 oz.
Diameter: 1″
Switch pad: 5.5″ long cable and pressure switch (optional toggle switch available)
Material: 6061T aluminum
Finish: Black or stainless
Output beam: Wavelength 670 nm; Class IIIa limit; output aperture approx. ¹/₄″ (6mm); beam divergence 0.5 Rad
Batteries: 1 X 3v Lithium
Battery life: Up to 15 hours continuous
Environmental: 0-30 C. operating 0-95% rh.; will withstand 2 meter drop; one meter immersion proof

AIMPOINT 5000 2-POWER
$367.00

SPECIFICATIONS
System: Parallax free
Optical: Anti-reflex coated lens
Adjustment: clock = ¹/₄″ at 100 yards
Length: 8¹/₂″
Weight: 9 oz.
Objective diameter: 46mm
Diameter of dot: 1¹/₂″ at 100 yards
Mounting system: 30mm rings
Magnification: 2X
Material: Anodized aluminum; blue finish

LASERDOT AUTO LASER (not shown)
$351.00 (w/Mounts)

SPECIFICATIONS
Length: 4″
Weight: 4.5 oz.
Dot diameter: 1¹/₂″ at 100 yards
Switch pad: Toggle
Material: 6061T aluminum
Finish: Black or stainless
Batteries: 1x6v Lithium
Mounting: Laser Module w/choice of mount

BAUSCH & LOMB RIFLESCOPES

SPECIFICATIONS MODEL ELITE™ 3000 RIFLESCOPES

Model	Special Features	Actual Magnifi-cation	Obj. Lens Aperture (mm)	Field of View at 100 yds. (ft.)	Weight (oz.)	Length (in.)	Eye Relief (in.)	Exit Pupil (mm)	Click Value at 100 yds. (in.)	Adjust Range at 100 yds.	Selections	Price
30-1545G		1.5X–4.5X	32	73–24	10	9.7	3.3	21–7	.25	100″	Low power variable ideal for brush, medium range sluggun hunting.	$394.95
30-1545M	Matte Finish	1.5X–4.5X	32	73–24	10	9.7	3.3	21–7	.25	100″	Low power variable ideal for brush, medium range sluggun hunting.	413.95
30-1642E	European Recticle 30mm Tube	1.5X–6X	42	5.7.7–15	21	14.4	3	17.6–7	.36	60″	Large exit pupil and tube for maximum brightness.	615.95
30-4124A	Adjustable Objective	4X–12X	40	26.9@4X 9@12X	15.2	13.2	3	10@4X 3.33@12X	.25	±25″	Medium to long-range variable makes superb choice for varmint or big game hunting.	401.95
30-3940G		3X–9X	40	33.8@3X 11.5@9X	13.1	12.6	3	13.3@3X 4.44@9X	.25	±25″	For the full range of hunting. From varmint to big game. Tops in versatility.	346.95
30-3940M	Matte Finish	3X–9X	40	33.8@3X 11.5@9X	13.1	12.6	3	13.3@3X 4.44@9X	.25	±25″	For the full range of hunting. From varmint to big game. Tops in versatility.	367.95
30-3940S	Silver Finish	3X–9X	40	33.8@3X 11.5@9X	13.1	12.6	3	13.3@3X 4.44@9X	.25	±25″	For the full range of hunting. From varmint to big game. Tops in versatility.	367.95
30-2732G		2X–7X	32	44.6@2X 12.7@7X	11.7	11.6	3	12.2@2X 4.6@7X	.25	±25″	Compact variable for close-in brush or medium range shooting.	322.95
30-2732M	Matte Finish	2X–7X	32	44.6@2X 12.7@7X	11.7	11.6	3	12.2@2X 4.6@7X	.25	±25″	Compact variable for close-in brush or medium range shooting.	341.95
30-3950G		3X–9X	50	31.5@3X 10.5@9X	19	15.7	3	12.5@3X 5.6@9X	.25	±25″	All purpose variable with extra brightness.	411.95
30-3950M		3X–9X	50	31.5@3X 10.5@9X	19	15.7	3	12.5@3X 5.6@9X	.25	±25″	All purpose variable with extra brightness.	424.95
30-3955E	European Recticle 30mm Tube	3X–9X	50	31.5–10.5	22	15.6	3	12.5–5.6	.36	70″	Large exit pupil and tube for maximum brightness.	604.95
30-4124A*	Adjustable Objective	4X–12X	40	26.9–9	15	13.2	3	10–3.33	.25	50″	Medium to long-range variable makes superb choice for varmint or big game hunting.	401.95
30-2028G	Handgun	2X	28	23	6.9	8.4	9–26	14	.25	30″	Excellent short to medium range hunting and target scope with maximum recoil resistance.	284.95
30-2028S	Handgun, Silver Finish	2X	28	23	6.9	8.4	9–26	14	.25	30″	Excellent short to medium range hunting and target scope with maximum recoil resistance.	303.95
30-2632F	Handgun	2X–6X	32	10–4	10	9	20	16–5.3	.25	30″	Constant eye relief at all powers with maximum recoil resistance.	392.95
30-2632S	Handgun, Silver Finish	2X–6X	32	10–4	10	9	20	16–5.3	.25	30″	Constant eye relief at all powers with maximum recoil resistance.	411.95

* Custom reticles available for adjustable objective Elite Riflescopes.

BAUSCH & LOMB RIFLESCOPES

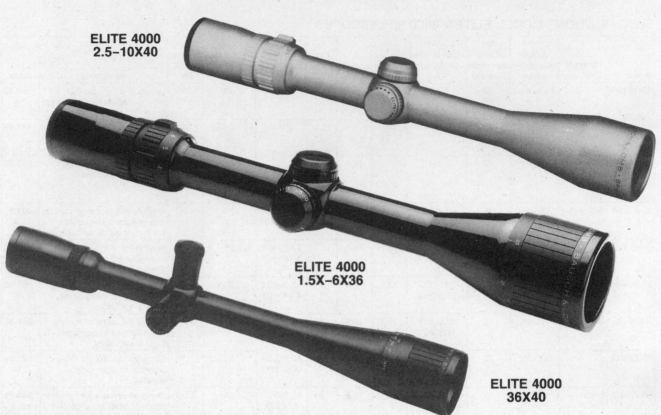

ELITE 4000
2.5–10X40

ELITE 4000
1.5X–6X36

ELITE 4000
36X40

SPECIFICATIONS MODEL ELITE™ 4000 RIFLESCOPES

Model	Special Features	Actual Magnification	Obj. Lens Aperture (mm)	Field of View at 100 yds. (ft.)	Weight (oz.)	Length (in.)	Eye Relief (in.)	Exit Pupil (mm)	Click Value at 100 yds. (in.)	Adjust Range at 100 yds.	Selections	Price
40-6244A*	Adjustable Objective Sunshade	6X–24X	40	18@6X 4.5@24X	20.2	16.9	3	6.66@6X 1.66@24X	.125	±13″	Varmint, target and silhouette long range shooting. Parallax focus adjustments for pinpoint accuracy.	$ 675.00
40-2104G		2.5X–10X	40	41.5@2.5X 10.8@10X	16	13.5	3	15.6@2.5 4@10X	.25	±25″	All purpose hunting scope with 4 times zoom range for close-in brush and long range shooting.	600.00
40-2104M	Matte Finish Silver Finish	2.5X–10X	40	41.5@2.5X 10.8@10X	16	13.5	3	15.6@2.5X 4@10X	.25	±25″	All purpose hunting scope with 4 times zoom range for close-in brush and long range shooting.	625.00
40-1636G		1.5X–6X	36	61.8@1.5X 16.1@6X	15.4	12.8	3	14.6@1.5X 6@6X	.25	±25″	Compact wide angle for close-in and brush hunting. Maximum brightness. Excellent for shot guns.	550.00
40-1636M	Matte Finish	1.5X–6X	36	61.8@1.5X 16.1@6X	15.4	12.8	3	14.6@1.5X 6@6X	.25	±30″	Compact wide angle for close-in and brush hunting. Maximum brightness. Excellent for shot guns.	575.00
40-1040	Ranging reticle, 30mm body tube	10X	40	10.5@10X	22.1	13.8	3.6	4@10X	.25	±60″	The ultimate for precise pinpoint accuracy with parallax focus and target adjustment.	1725.00
40-3640A*	Adj. Obj.	36X	40	3	17.6	15	3.2	1.1	.25	30″	Ideal benchrest scope	825.00

* Custom reticles available for adjustable objective Elite Riflescopes.

BEEMAN SCOPES

BEEMAN SS-2 W/FIREARM

SS-1 AND SS-2 SERIES

Beeman SS-1 and SS-2 short scopes are extra compact and rugged, largely due to breakthroughs in optical engineering and computer programming of lens formulas. Less than 7 inches long, both scopes pack 11 lenses that actually gather light for bigger, brighter targets than "projected spot" targets. Scope body and built-in mounts are milled as a single unit from a solid block of hi-tensile aircraft aluminum.

SS-1 Series **$235.00**
SS-2 Series . . . . **from $290.00–305.00**

BEEMAN SS-2

BEEMAN SS-2L "SKYLITE" RIFLESCOPE

BEEMAN SS-2L "SKYLITE" RIFLESCOPE

Features a brightly illuminated reticle powered by daylight and even moonlight (no batteries necessary). In addition to standard black reticle, supplementary color filters are available for different lighting and shooting situations. Filter options include: white (for silhouette or target); red (for twilight and general purpose); yellow (for haze, fog and low light); green (for bright light and snow). A small electrical illuminator is also available for use in total darkness.

Beeman SS-2L w/color reticle, 3x **$340.00**
Beeman SS-2L w/color reticle, 4x 370.00
Micro Lamp 35.00
Filter Kit (green or yellow) 19.00

BEEMAN SS-3

SS-3 SERIES

Offers 1.5-4x zoom power for greater flexibility. Glare-free black matte finish is anodized into metal for deep sheen and extra toughness. Instant action dial around front of scope dials away parallax error and dials in perfect focus from 10 feet to infinity. Scope measures only 5¾ inches in length and weighs only 8.5 ounces. **SS-3 Series** **$300.00**

BEEMAN SCOPES

BLUE RIBBON

BEEMAN MODEL 66R

Every feature of the Model 66R has been carefully developed to make it the ultimate scope for centerfire, .22-caliber rimfire and adult air rifles. It can zoom instantly from 2 to 7X—an excellent range of magnification that provides sufficient high power for long-range shots requiring pinpoint accuracy, and very low power for times when speed and a broad field of view are needed. Speed Dials (with full saddle) with 1/2-minute adjustments per click and Range Focus are two of the 66R's star features. The 32mm objective lens up front provides a bright, wide field view, extends daily shooting time, and affords better viewing into shadows.

Model 66R 2-7X32 with Range Focus **$310.00**

BEEMAN MODEL 66RL

The ultimate scope for centerfire, rimfire and adult airguns. Zoom instantly from 2X to 7X—range that provides sufficient high power for long shots as well as very low power when a broad field of view is demanded. Features speed dials, Range Focus and 32mm objective lens.

Model 66RL w/Color Reticle Scope **$350.00**

BEEMAN MODEL 68R

The Model 68R with 4-12 zoom power meets the demand for a high-power scope suitable for airguns, rimfire and centerfire rifles. Higher magnifications are a necessity for metallic silhouette shooters. Field shooters who must make precise head shots on small game and vermin for a humane kill find the higher magnifications helpful. The big 40mm objective lens provides a brilliant sight picture, even at high power. Other features include Speed Dials with 1/4-minute clicks, Range Focus, full saddle, and the essential extra lens bracing required of airgun scopes.

Model 68R, Blue Ribbon 4-12X40 .... **$440.00**

BEEMAN MODEL 25
(Shown on Beeman P1 Air Pistol)

Beeman has improved upon the modified domestic pistol scope that was formerly recommended. Starting with the same dedication to quality, special Beeman Blue Ribbon® Scope features have been added to produce one of the finest scopes for air pistols (and a variety of other pistols). Features include Speed Dial elevation and windage knobs and the brightness and performance that only a 1″ tube, top-quality lens system can offer. Full saddle. Especially recommended for Beeman P1/P2 air pistols.

Model 25 2X20 Blue Ribbon Pistol Scope **$155.00**

B-SQUARE

MINI-LASERS
$299.95 (Blue) $309.95 (Stainless)

SPECIFICATIONS
Power: 5mW max (class IIIA)
Size: 1.1″×1.1″×.6″
Batteries: Common A76 (lithium or alkaline)
Aiming method: Omnidirectional
Visibility: 1″ at 25 yds.
Min. Life: 60 min.
Features: Quick detachable laser allows holstering; moisture proof and shock resistant; ''Aim-Lock'' screw type; windage and elevation adjustable; no trigger finger interference; cord or integral switch; on-off switch optional; wide selection of mounts available

BSL-1 LASER SIGHT
$229.95 (Blue) $239.95 (Stainless)

SPECIFICATIONS
Power: 5mW max (class IIIA)
Size: .75″; dia.×2.65″
Aiming method: Omnidirectional (screw type, lockable)
Visibility: Pulsed dot 1″ at 100 yds.
Features: ''Aim-Lock'' no-slip mounts; T-Slot cap; cord or integral switch; wide selection of mounts

BURRIS SCOPES

6X HBR
$294.00 (Fine Plex)

Engineered and designed to meet Hunter Benchrest specifications, this scope has precise ⅛″ minute click adjustments with re-zeroable target-style knobs. Complete with parallax adjustment from 25 yards to infinity. Weighs only 13 ounces.

**1.5X GUNSITE
SCOUT SCOPE**

GUNSITE SCOUT SCOPES

Made for hunters who need a 7- to 14-inch eye relief to mount just in front of the ejection port opening, allowing hunters to shoot with both eyes open. The 15-foot field of view and 2¾X magnification are ideal for brush guns and handgunners.

1.5X Plex XER	**$208.95**
1.5X Plex XER Safari Finish	226.95
1.5X Heavy Plex	216.95
1.5X Heavy Plex Safari	235.95
2.75X Heavy Plex	222.95
2.75X Heavy Plex Safari	241.95
2.75X Plex XER	213.95

BURRIS SCOPES

SIGNATURE SERIES

All models in the Signature Series have **Hi-Lume** (multi-coated) lenses for maximum light transmission. Also features new **Posi-Lock** to prevent recoil and protect against rough hunting use and temperature change. Allows shooter to lock internal optics of scope in position after rifle has been sighted in.

2.5X10X SIGNATURE

3X-9X SIGNATURE

8X-32X SIGNATURE

Models	Prices
1.5X-6X Plex	$395.95
1.5X-6X Plex Safari	414.95
1.5X-6X Plex Silver Safari	422.95
1.5X-6X Heavy Plex	405.95
1.5X-6X Heavy Plex Safari	422.95
1.5X-6X Plex Posi-Lock	472.95
1.5X-6X Plex Posi-Lock Safari	491.95
2X-8X Plex	459.95
2X-8X Plex Safari	478.95
2X-8X Plex Silver Safari	487.95
2X-8X Plex Posi-Lock	535.95
2X-8X Plex Posi-Lock Safari	552.95
2X-8X Plex Posi-Lock Silver Safari	562.95
2.5X-10X Peep Plex	538.95
2.5X-10X Peep Plex Safari	557.95
2.5X-10X Plex	530.95
2.5X-10X Plex Safari	548.95
2.5X-10X Plex Silver Safari	557.95
2.5X-10X Plex Posi-Lock	602.95
2.5X-10X Plex Posi-Lock Safari	621.95
2.5X-10X Plex Posi-Lock Silver Safari	630.95
3X-9X Plex	469.95
3X-9X Plex Safari	487.95
3X-9X Plex Silver Safari	495.95
3X-9X Plex Posi-Lock	544.95
3X-9X Plex Posi-Lock Safari	562.95
3X-9X Plex Posi-Lock Silver Safari	571.95
3X-12X Plex	587.95
3X-12X Plex Safari	605.95
3X-12X Peep Plex	595.95
3X-12X Peep Plex Safari	614.95
3X-12X Plex Posi-Lock	660.95
3X-12X Plex Posi-Lock Safari	678.95
4X Plex	325.95
4X Plex Safari	341.95
6X Plex	340.95
6X Plex Safari	353.95
6X-24X Plex	611.95
6X-24X 2"-.5" Dot Silhouette	650.95
6X-24X Fine Flex Silhouette	635.95
6X-24X Plex Safari	630.95
6X-24X Fine Plex Silhouette Safari	652.95
6X-24X Peep Plex	621.95
6X-24X Peep Plex Safari	638.95
6X-24X Plex Posi-Lock	685.95
8X-32X Fine Plex Silhouette	671.95
8X-32X Peep Plex Silhouette	680.95
8X-32X 2"-5" Dot Silhouete	688.95
8X-32X Fine Plex Posi-Lock	760.95

3X-9X RAC SCOPE
w/Automatic Rangefinder & Hi-Lume Lenses

When the crosshair is zeroed at 200 yards (or 1.8″ high at 100 yards), it will remain zeroed at 200 yards regardless of the power ring setting. The Range Reticle automatically moves to zero at ranges up to 500 yards as power is increased to fit the target between the stadia range wires. No need to adjust the elevation knob; bullet drop compensation is automatic.

3X-9X RAC CHP Safari Finish	$384.95
3X-9X RAC Crosshair Dot	369.95
3X-9X RAC Crosshair Plex	369.95

FULLFIELD SCOPES
Fixed Power with Hi-Lume Lenses

12X FULLFIELD

1½X Plex	$236.95
1½X Heavy Plex	245.95
2½X Plex	248.95
2½X Heavy Plex	256.95
4X Plex	265.95
4X Plex Safari Finish	284.95
6X Plex	285.95
6X Plex Safari Finish	302.95
12X Plex	358.95
12X Fine Plex	358.95
12X Fine Plex Silhouette	385.95
12X ½″ Dot Silhouette	402.95

BURRIS FULLFIELD SCOPES

VARIABLE POWER W/HI-LUME LENSES

MINI 6X

1.75X-5X Plex	$312.95
1.75X-5X Plex Safari	329.95
1.75X-5X Heavy Plex	321.95
1.75X-5X Heavy Plex Safari	338.95
1.75x-5x Plex Posi-Lock	384.95
1.75X-5X Plex Safari Posi-Lock	401.95
1.75X-5X Plex Silver Posi-Lock	409.95
2X-7X Plex	334.95
2X-7X Plex Silver Safari	359.95
2X-7X Plex Safari Finish	351.95
2X-7X Post Crosshair	342.95
3X-9X Plex	331.95
3X-9X Plex Safari	356.95
3X-9X Plex Silver Finish	369.95
3X-9X Post Crosshair	355.95
3X-9X 3."-1." Dot	359.95
3X-9X Peep Plex	355.95
3X-9X Plex Posi-Lock	402.95
3X-9X Plex Safari Posi-Lock	427.95
3X-9X Plex Silver Safari Posi-Lock	441.95
3x-9X Crosshair Plex RAC	369.95
3X-9X Crosshair Plex Safari RAC	384.95
3.5X-10X-50mm Plex	422.95
3.5X-10X-50mm Plex Safari	441.95
3.5X-10X-50mm Plex Silver Safari	449.95
3.5X-10X-50mm Peep Plex	431.95
3.5X-10X-50mm Plex Posi-Lock	494.95
3.5X-10X-50mm Plex Safari Posi-Lock	513.95
3.5X-10X-50mm Silver Safari Posi-Lock	521.95
4X-12X Plex P.A.	419.95
4X-12X Fine Plex	419.95
4X-12X Peep Plex P.A.	428.95
6X-18X Plex	437.95
6X-18X Fine Plex P.A.	437.95
6X-18X Fine Plex Silhouette	459.95
6X-18X 2."-.7" Dot Silhouette	478.95
6X-18X Plex Safari	455.95
6X-18X Peep Plex P.A.	445.95
6X-18X Peep Plex Safari P.A.	464.95

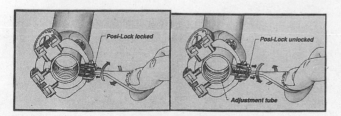

BURRIS POSI-LOCK

MINI SCOPES

4X Plex	$213.95
4X Plex P.A.	248.95
4X Plex Silver Safari	238.95
6X Plex	228.95
6X Plex P.A.	262.95
6X HBR Fine Plex Target	294.95
6X HBR .375 Dot Target	312.95
6X HBR .375 Dot P.A. Silver Target	338.95
6X HBR FCH Target	294.95
6X HBR FCH P.A. Silver Target	321.95
2X-7X Plex	292.95
2X-7X Plex Safari	310.95
2X-7X Plex Silver Safari	319.95
3X-9X Plex	299.95
3X-9X Plex Safari	316.95
3X-9X Plex Silver Safari	326.95
4X-12X Plex P.A.	395.95

3X LER

HANDGUN LONG EYE RELIEF SCOPE
with Plex Reticle:

1X LER Plex	$202.95
2X LER Plex	209.95
2X LER Plex Silver Safari	238.95
2X Plex Posi-Lock	281.95
2X-Plex Silver Posi-Lock	308.95
3X LER Plex	226.95
4X LER Plex	235.95
4X LER Plex P.A.	272.95
4X LER Plex P.A. Silver Safari	299.95
4X LER Plex Silver Safari	251.95
4X Plex Posi-Lock	306.95
4X Plex Silver Posi-Lock	334.95
7X Plex	258.98
7X Plex P.A	294.95
7X Plex Silver P.A.	319.95
10X Plex Silhouette	348.95
1.5X-4X LER Plex	328.95
1.5X-4X LER Plex Silver Safari	355.95
1.5X-4X Plex Posi-Lock	399.95
1.5X-4X Plex Silver Posi-Lock	427.95
2X-7X LER Plex	321.95
2X-7X LER Plex Safari	338.95
2X-7X LER Plex Silver Safari	348.95
2X-7X LER Plex P.A.	359.95
2X-7X Plex Posi-Lock	392.95
2X-7X Plex Silver Posi-Lock	419.95
3X-9X LER Plex	359.95
3X-9X LER Plex P.A.	392.95
3X-9X LER Plex Safari	378.95
3X-9X LER Plex Silver Safari	387.95
3X-9X Plex Posi-Lock	431.95
3X-9X Plex Silver Posi-Lock	458.95
3X-9X Peep Plex	369.95

BURRIS SCOPES

LONG EYE-RELIEF HANDGUN SCOPES

1X LONG EYE RELIEF (LER)

This model has been tested and proved in numerous competitions over the years. The 1X magnification eliminates any parallax. Has super crisp optics and the widest field of view of all our handgun scopes.

2X-7X LONG EYE-RELIEF (LER)

The perfect handgun scope for varmint or big game hunters. Versatile, compact and available with parallax adjustment.

1.5X-4X LONG EYE RELIEF (LER)

Designed especially for hunting with a handgun. At 1½X there is a big field of view for those fast, close shots. 4X magnification permits precision, long range shots. Eye relief is 11 inches minimum, 24 inches maximum. Weight is 11 ounces, overall length is 10¼ inches. Burris mounts recommended for this scope.

3X-9X LONG EYE RELIEF (LER)

The highest variable powered handgun scope made. Compact and versatile, this scope is the ultimate for hunting game or testing loads. Fog proof and magnum proof. Also available with parallax adjustment.

INTERMEDIATE EYE-RELIEF HANDGUN SCOPES

7X INTERMEDIATE EYE RELIEF (IER)
$258.95

Developed for the long range, accurate handgunner. Requires a two handed hold. Eye relief is 10 inches minimum, 16 inches maximum. Weight is 10 ounces. Overall length is 11¼ inches. Burris mounts recommended for mounting this scope. Also available with parallax adjustment.

10X IER P.A.
$348.95

Designed for the precision shooting handgunner to be shot from a rest or a two handed hold. Handgun should have a 14 inch barrel minimum. Overall length of scope is 13.6 inches. Weight is 14 ounces. Parallax adjustment is 50 yards to infinity.

BUSHNELL RIFLESCOPES

SPECIFICATIONS BUSHNELL TROPHY RIFLESCOPES

MODEL	Special Features	Actual Magni-fication	Obj. Lens Aperture (mm)	Field of View at 100 yds. (ft.)	Weight (oz.)	Length (in.)	Eye Relief (in.)	Exit Pupil (mm)	Click Value at 100 yds. (in.)	Adjust Range at 100 yds. (in.)	Selection
73-1500	Wide Angle	1.75x-5x	32	68@1.75x 23@5x	12.3	10.8	3.5	18.3@1.75x 6.4@5x	.25	120	Shotgun, black powder or centerfire. Close-in brush hunting.
73-2733	Wide Angle	2x-7x	32	63@2x 18@7x	11.3	10	3	16@2x 4.6@7x	.25	110	Low power variable ideal for brush, medium range sluggun hunting. Ideal for sluggun shooting.
73-2545	45mm objective for max. light transmission	2.5x-10x	45	39@2.5x 10@10x	14	13.75	3	18@2.5x 4.5@10x	.25	60	All purpose hunting with four times zoom for close-in and long range shooting.
73-3940	Wide Angle										All purpose variable, excellent for use from close to long range. Circular view provides a definite advantage over "TV screen" type scopes for running game-uphill or down
73-3940S	Wide Angle Silver Finish	3x-9x	40	42@3x 14@9x	13.2	11.7	3	13.3@3x 4.4@9x	.25	60	
73-3948	Wide Angle Matte Finish										
73-0440	Wide Angle	4x	40	36	12.5	12.5	3	10	.25	100	General purpose. Wide angle field of view.
73-4124	Wide Angle, adjustable objective	4x-12x	40	32@4x 11@12x	16.1	12.6	3	10@4x 3.3@12x	.25	60	Medium to long range variable for varmit & big game. Range focus adjustment. Excellent for air riflescope.
73-6184	Semi-turret target adjustments, adjustable objective	6x-18x	40	17.3@6x 6@18x	17.9	14.8	3	6.6@6x 2.2@18x	.125	40	Long range varmit centerfire or short range air rifle target precision accuracy.
TROPHY HANDGUN SCOPES											
73-0232		2x	32	20	7.7	8.7	9-26	16	.25	90	Designed for target and short to medium range hunting. Magnum recoil resistant.
73-0232S	Silver Finish										
73-2632		2x-6x	32	21@2x 7@6x	9.6	9.1	9-26	16@2x 5.3@6x	.25	50	Verstile, all purpose, four times zoom range of close-in brush, long range shooting and shotgun shooting.
73-2632S	Silver Finish										
SHOTGUN/HANDGUN SCOPES											
73-0130	Iluminated dot reticle	1x	25	61	5.5	5.25	Unltd.	18	1.0	80	30mm tube with rings, ext. tube polarization filter, amber coating. For black powder, shotgun and handgun shooting.
73-0150	Red dot w/ TV View	1x	50	26	8.3	4.8	Unltd.	38	1.0	140	Excellent handgun or shotgun scope. Includes rings and mounts for Weaver base.
73-1420	Turkey/Brush Scope w/ Circle x Reticle	1.75x-4x	32	73@1.75x 30@4x	10.9	10.8	3.5	18@1.75x 8@4x	.25	120	Ideal for slugguns, turkey hunting or black powder guns.
73-1421	Brush Scope	1.75x-4x	32	73@1.75x 30@4x	10.9	10.8	3.5	18@1.75x 8@4x	.25	120	Ideal for slugguns, turkey hunting or black powder guns.

BUSHNELL TROPHY RIFLESCOPE 1.75-4×32 TURKEY SCOPE MODEL NO. 1420

Model Trophy Rifle Scopes	$Price	Trophy Handgun Scopes	
73-1500	$257.95	73-0232	$189.95
73-2733	248.95	73-0232S	204.95
73-2545	301.95	73-2632	235.95
73-3940	180.95	73-2632S	248.95
73-3940S	191.95		
73-3948	191.95	**Shotgun/Handgun Scopes**	
73-0440	151.95	73-0130	$269.95
73-4124	273.95	73-0150	403.95
73-6184	322.95	73-1420	263.95
		73-1421	263.95

EMERGING TECHNOLOGIES

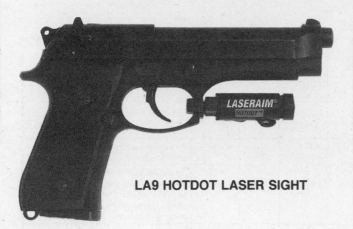

LA9 HOTDOT LASER SIGHT

LA10 HOTDOT LASER SIGHT

LA9 HOTDOT LASERAIM LASER SIGHT

Ten times brighter than other laser sights, Laseraim's LA9 Hotdot Lasersight includes a rechargeable NICad battery and in-field charger. Produces a 2″ dot at 100 yards with a 500-yard range. **Length:** 2″. **Diameter:** .75″. Features pressure switch and 4-way microlock windage and elevation adjustments. Can be used with handguns, rifles, shotguns and bows. Fits all Laseraim mounts. Available in black or satin.
Prices:

LA5 LASERAIM SIGHT

LA9 HOTDOT LASERSIGHT (Black or Satin) **$239.95**
    For 20-hour battery or 1½-mile range, add **20.00**
    For trigger guard mount (fits 8 S&W, 4 Glock,
    4 Ruger and 4 Beretta handguns) add **16.00**
LA10 HOTDOT LASER SIGHT: Same as Model LA9
    but with 20-hour battery and 1½-mile range . . . **249.95**
LA5 LASERAIM SIGHT (Black or Satin) **239.00**
LA5 MAGNUM LASER SIGHT (1½-mile range) . . . **265.00**
LA6 SHOTGUN LASER SIGHT MOUNT
    (Remington 870, 1100, 11-87; Mossberg 500;
    Winchester 1300, 1400; Browning A5, BPS) . . . **265.00**

LASERAIM ILLUSION ELECTRONIC RED DOT SIGHTS

LA9750B GRAND ILLUSION 2″
    6 MOA Medium Dot **$230.00**
    Same as above with 10 MOA Large Dot **230.00**

GRAND ILLUSION SIGHT

INTERAIMS SIGHTS

RED DOT ELECTRONIC SIGHTS

The following features are incorporated into each model, including the MONO TUBE:

—5 YEAR WARRANTY—

- Sharp Red Dot
- Lightweight
- Compact
- Wide Field of View
- Parallaxfree
- True 1X for Unlimited Eye Relief
- Nitrogen Filled Tube
- Waterproof, Moisture Proof, Shockproof

- Rugged Aluminum Body
- Easy 1″ and 30mm Ring Mounting
- Manually Adjustable Light Intensity
- Windage and Elevation Adjustments
- Dielectrical Coated Lenses
- Battery—Polarized Filter—Extension Tube—Protective Rubber Eye Piece—All included

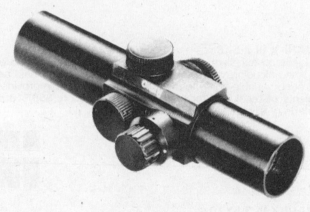

MONOTUBE CONSTRUCTIONS ONE V

Weight	Length	Battery	Finish
3.9 oz.	4½″	(1) 3 V Lithium	Black Satin Nickel

ONE V 1″ MODEL
$139.95

Also available:
1″ or 30mm rings in black or satin nickel **$11.95**

ONE V 30
$159.95

MONOTUBE CONSTRUCTIONS ONE V 30

Weight	Length	Battery	Finish
5.5 oz.	5.4″	(1) 3 V Lithium DL 2032	Black or Satin Nickel

LEUPOLD RIFLE SCOPES

VARI-X III LINE

The Vari-X III scopes feature a power-changing system that is similar to the sophisticated lens systems in today's finest cameras. Some of the improvements include an extremely accurate internal control system and a sharp, superb-contrast sight picture. All lenses are coated with **Multicoat 4**. Reticles are the same apparent size throughout power range, stay centered during elevation/windage adjustments. Eyepieces are adjustable and fog-free.

VARI-X III 1.5X5
Here's a fine selection of hunting powers for ranges varying from very short to those at which big game is normally taken. The exceptional field at 1.5X lets you get on a fast-moving animal quickly. With the generous magnification at 5X, you can hunt medium and big game around the world at all but the longest ranges. Duplex or Heavy Duplex **$516.10** In black matte finish: **$537.50**
Also available:
VARI-X III 1.75X32mm: $539.30. With matte finish: **$560.70**

VARI-X III 1.5X5

VARI-X III 2.5X8
This is an excellent range of powers for almost any kind of game, inlcuding varmints. In fact, it possibly is the best all-around variable going today. The top magnification provides plenty of resolution for practically any situation. **$557.10** In matte or silver finish: **$578.60.**

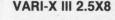

VARI-X III 2.5X8

VARI-X III 3.5X10
The extra power range makes these scopes the optimum choice for year-around big game and varmint hunting. The adjustable objective model, with its precise focusing at any range beyond 50 yards, also is an excellent choice for some forms of target shooting. **$576.80.** With matte finish: **$598.20** With silver: **$598.20**

VARI-X III 3.5X10

VARI-X III 3.5X10–50mm
Leupold announces its first hunting scope designed specifically for low-light situations. The 3.5X10–50mm scope, featuring lenses coated with Multicoat 4, is ideal for twilight hunting (especially whitetail deer) because of its efficient light transmission. The new scope delivers maximum available light through its large 50mm objective lens, which translates into an exit pupil that transmits all the light the human eye can handle in typical low-light circumstances, even at the highest magnification: **$680.40.** With matte or silver finish: **$701.80.**

Also available:
VARI-X III 3.5X10-50mm Adj. Objective: $735.70. With matte finish: **$757.00.**

VARI-X III 3.5X10–50mm

VARI-X III 4.5X14 (Adj. Objective)
This new model has enough range to double as a hunting scope and as a varmint scope. Duplex or Heavy Duplex: **$669.60**
Same as above with 50mm adj. obj., Duplex or Heavy Duplex: **$766.10**

**VARI-X III 6.5X20
(With Adjustable Objective)**

VARI-X III 6.5X20 (Adj. Objective)
This scope has the widest range of power settings in our variable line, with magnifications that are especially useful to hunters of all types of varmints. In addition, it can be used for any kind of big-game hunting where higher magnifications are an aid: **$675.00.** With matte finish: **$696.40**

LEUPOLD RIFLE SCOPES

VARIABLE POWER SCOPES

VARI-X II 1X4

VARI-X II 1X4 DUPLEX
This scope, the smallest of Leupold's VARI-X II line, is reintroduced in response to consumer demand for its large field of view: 70 feet at 100 yards. **$346.40**

VARI-X II 2X7

VARI-X II 2X7 DUPLEX
A compact scope, no larger than the Leupold M8-4X, offering a wide range of power. It can be set at 2X for close ranges in heavy cover or zoomed to maximum power for shooting or identifying game at longer ranges. **$387.50**

VARI-X II 3X9–50mm

VARI-X II 3X9–50mm
This LOV scope delivers a 5.5mm exit pupil, providing excellent low-light visibility. **$458.90**
With matte finish: 480.40

VARI-X II 3X9 DUPLEX
A wide selection of powers lets you choose the right combination of field of view and magnification to fit the particular conditions you are hunting at the time. Many hunters use the 3X or 4X setting most of the time, cranking up to 9X for positive identification of game or for extremely long shots. The adjustable objective eliminates parallax and permits precise focusing on any object from less than 50 yards to infinity for extra-sharp definition. **$391.10**
Also available in matte or silver: 412.50

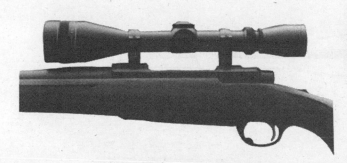

VARI-X II 4X12 MATTE FINISH

VARI-X II 4X12 (Adj. Objective)
The ideal answer for big game and varmint hunters alike. At 12.25 inches, the 4X12 is virtually the same length as Vari-X II 3X9. . . . . **$514.30**
With matte or silver finish: 535.70

ARMORED SPOTTING SCOPE

SPOTTING SCOPES
Leupold's Golden Ring Armored Spotting Scopes feature extraordinary eye relief and crisp, bright roof prism optics housed in a lightweight, sealed, waterproof body. The Spotting Scopes come complete with a self-storing screw-on sunshade, lens caps, and a green canvas case. Now available in 12X40-60mm variable power with 30.8mm eye relief at 20X.
Prices:
20X50mm Compact Armored **$646.40**
25X50mm Compact Armored 689.30
30X60mm 710.70
12X40-60mm Variable Power 1026.80

LEUPOLD SCOPES

THE COMPACT SCOPE LINE

The introduction of the Leupold Compacts has coincided with the increasing popularity of the new featherweight rifles. Leupold Compact scopes give a more balanced appearance atop these new scaled-down rifles and offer generous eye relief, magnification and field of view, yet are smaller inside and out. Fog-free.

4X COMPACT & 4X RF SPECIAL

M8-4X COMPACT RF SPECIAL
The 4X RF Special is focused to 75 yards and has a Duplex reticle with finer crosshairs. **$317.90**

2X7 COMPACT

2X7 COMPACT
Two ounces lighter and a whole inch shorter than its full-size counterpart, this 2X7 is one of the world's most compact variable power scopes. It's the perfect hunting scope for today's trend toward smaller and lighter rifles. **$405.40**
Also available: **RF Special** (Fine Duplex) **$405.40**

3X9 COMPACT

3X9 COMPACT
The 3X9 Compact is a full-blown variable that's 3½ ounces lighter and 1.3 inches shorter than a standard 3X9. **$419.60**
Also available in black matte finish or silver **$441.10**

3X9 COMPACT SILVER

SHOTGUN SCOPES (not shown)

Leupold shotgun scopes are parallax-adjusted to deliver precise focusing at 75 yards (as opposed to 150 yards usually prescribed for rifle scopes). Each scope features a special Heavy Duplex reticle that is more effective against heavy, brushy backgrounds.
Prices:

Vari-X II 1X4 Model	367.90
M8-4X Heavy Duplex	339.30
Vari-X III 2X7 Heavy Duplex	408.90

LEUPOLD SCOPES

HANDGUN SCOPES

M8-2X EER
With an optimum eye relief of 12–24 inches, the 2X EER is an excellent choice for most handguns. It is equally favorable for carbines and other rifles with top ejection that calls for forward mounting of the scope. Available in black anodized or silver finish to match stainless steel and nickel-plated handguns. **$257.10**. In silver finish: **$278.60**

LASERLIGHT HANDGUN SIGHT (not shown)
This advanced electronic device projects a laser dot, the result of a collimated coherent laser beam. Micro circuitry pulses the diode thousands of times per second, placing a dot of light where the bullet will impact the target: **$258.90**

2X EER

M8-4X EER
Only 8.4 inches long and 7.6 ounces. Optimum eye relief 12–24 inches. Available in black anodized or silver finish to match stainless steel and nickel-plated handguns. In matte or silver finish: **$346.40**

Also available:
VARI-X 2.5X8 EER w/Multicoat 4: $500.00. In silver: **$521.40**

FIXED-POWER SCOPES

4X

M8-4X
The all-time favorite is the 4X, which delivers a widely used magnification and a generous field of view. **$317.90**. In black matte finish: **$339.30**

6X

M8-6X
Gaining popularity fast among fixed power scopes is the 6X, which can extend the range for big-game hunting and double, in some cases, as a varmint scope: **$337.50**

6X42mm

M8-6X42mm W/Multicoat 4
Large 42mm objective lens features increased light gathering capability and a 7mm exit pupil. Great for varmint shooting at night. Duplex or Heavy Duplex: **$419.60**. In matte finish: **$441.10**

VARMINT SCOPES

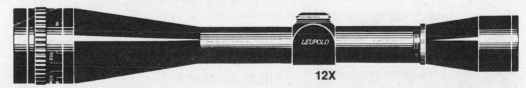

12X

8-12X STANDARD (Adj. Obj.)
Superlative optical qualities, outstanding resolution and magnification make the 12X a natural for the varmint shooter. Adjustable objective is standard for parallax-free focusing. **$544.60**. With Multicoat 4: **$464.30.** With CPC reticle or Dot: **$517.90**

Also available:
VARI-X III 6.5X20 VARMINT (Adj. Obj.) Target Dot w/Multicoat 4: **$755.40**

LYMAN SIGHTS

93 MATCH SIGHT

Lyman's globe front sight, the "93 Match," adapts to any rifle with a standard dovetail mounting block. The sight has a diameter of $7/8''$ and comes complete with 7 target inserts. The 93 Match has a special hooked locking bolt and nut to allow quick removal or installation. Bases are available in .860 (European) and .562 (American) hole spacing. The sight height is .550 from the top of the dovetail to the center of the aperture. **$48.00**

90 MJT UNIVERSAL TARGET RECEIVER SIGHT
(Not shown)

Designed to mount on a Marlin Model 2000 Target Rifle using standard Williams FP bases, the Target Sight features target knobs scribed with audible click detents in minute and quarter-minute graduations, plus elevation and windage direction arrows. Adjustable zero scales allow adjustments to be made without disturbing pre-set zero; quick release slide allows slide to be removed with a press of the release button. Large $7/8''$ diameter non-glare .040 target aperture disk is standard. Adjustable from 1.060 to 1.560 above centerline of bore. **$79.95**

20 MJT $7/8''$ DIAMETER GLOBE FRONT SIGHT

Machined from one solid piece of steel designed for use with dovetail slot mounting in the barrel or with Lyman's 25A dovetail base. Height is .700″ from bottom of dovetail to center of aperture. Supplied with 7 Anschutz-size steel apertures. **$36.00**

NO. 16 FOLDING LEAF SIGHT

Designed primarily as open rear sights with adjustable elevation, leaf sights make excellent auxiliary sights for scope-mounted rifles. They fold close to the barrel when not in use, and they can be installed and left on the rifle without interfering with scope or mount. Two lock screws hold the elevation blade adjustments firmly in place. Leaf sights are available in the following heights. **$13.50**

16A—.400″ high; elevates to .500″
16B—.345″ high; elevates to .445″
16C—.500″ high; elevates to .600″

LYMAN NO.2 TANG SIGHT

Recreated for the Special Edition Winchester 1894 100th Anniversary Edition, this version of the tang sight features high index marks on the aperture post with a maximum elevation of .800 for long-range shooting. Comes with both .093 quick-sighting aperture and .040 large disc aperture, plus replacement stock screw and front tang screw.
Price: **$69.00**

PEEP SIGHT

The 66-MK "peep" sight fits all versions of the Knight MK-85 rifle with flat-sided receiver. The 57 SME and 57 SMET are designed for White Systems Model 91 and Whitetail models with round receivers. Both sights feature quick release slides, adjustable windage and elevation scales.

66 MK **$66.00**
57 SME/57 SMET **$66.00**

SHOTGUN SIGHTS

No. 10 Front Sight (press fit) for use on double barrel, or ribbed single -barrel guns **$5.00**
No. 10D Front Sight (screw fit) for use on non-ribbed single-barrel guns; supplied with a wrench **6.50**
No. 11 Middle Sight (press fit). This small middle sight is intended for use on double-barrel and ribbed single-barrel guns **5.00**

MILLETT SIGHTS

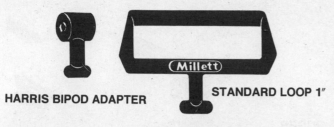

HARRIS BIPOD ADAPTER **STANDARD LOOP 1"**

FLUSH-MOUNT HARRIS BIPOD ADAPTER

Millett's flush-mount sling swivels have a simple-to-use adapter for the Harris bipod that detaches quickly so the loop can then be installed in the bipod loop receptacle. Will also fit Pachmayr flush-mount bases.

Harris Bipod Adapter	SS00004	**$9.15**

DUAL-CRIMP INSTALLATION TOOL KIT

The Dual-Crimp System is a new revolutionary way of installing front sights on autos. Now it is not necessary to heliarc or silver solder to get a good secure job. Dual-Crimp has a two-post, hollow rivet design that works very much like an aircraft rivet and withstands the heavy abuse of hardball ammo. Your choice of four styles and nine heights. Dual Crimp is the quick and easy system for professionals. Requires a drill press.

Dual-Crimp Tool Set, Complete	**$150.00**
Application Tool	**80.80**
Reverse counterbore (solid carbide)	**38.70**
3/16" Drill (solid carbide)	**17.90**
Drill Jig	**23.05**

FLUSH-MOUNT SLING SWIVELS

Millett's flush-mount redesigned Pachmayr sling swivels are quick detachable and beautifully styled in heat-treated nickel steel. The sling swivel loop has been redesigned to guide the sling into the loop, eliminating twisitng and fraying on edges of sling. Millett flush-mount bases are much easier to install than the old Pachmayr design, with no threading and an easy to use step drill.

Flush-Mount Swivels (pair)	SS00001	**$16.45**
Loops Only	SS00002	**9.15**
Installation Drill	SS00003	**17.58**

3-DOT SYSTEM SIGHTS

Millett announces 3-Dot System sights for a wide variety of popular handguns.

3-Dot System Front and Rear Sight Selection Chart (partial listing only)

DUAL-CRIMP® FRONTS (White Dot) **$16.00**

DC 18500	.185 Height
DC 20004	.200 Height
DC 22512	.225 Height
DC31216	.312 Height
DC34020	.340 Height
DC36024	.360 Height

WIDE STAKE-ON FRONTS (White Dot) **$16.00**
(for Colt pistols only, after June 1988)

WS18504	.185 Height
WS20008	.200 Height
WS31220	.312 Height
WS22504	.225 Height

SPECIAL-APPLICATION PISTOL FRONTS **$16.00**

GL00006	Glock 17, 17L & 19
RP85009	Ruger-P-85
RS22015	Ruger Std. Auto (Fixed Model)
SP22567	Sig Sauer P225/226, Dovetail
SW40513	S&W 3rd Generation, Dovetail
SW46913	S&W 3rd Generation, Dovetail
BE00010	Beretta Accurizer **$25.14**

AUTOPISTOL REAR SIGHTS **$56.40**

BE00003	Beretta	55.60
BA00008	Browning Hi-Power, Adjustable	55.60
BF00008	Browning Hi-Power, Fixed	55.60
CA00008	Colt-Hi-Profile	55.60
CC00008	Colt Custom Combat Lo-Profile	55.60
GC00008	Colt Gold Cup	49.30
RP85008	Ruger P-85	55.60
RS22003	Ruger Std. Auto	55.60
SP22005	Sig P220, 225, 226	55.60
SW40504	Smith & Wesson, _ALL Factory Adjustable_ (incl. 2nd & 3rd Generation)* . . .	55.60
SW46904	Smith & Wesson, _ALL Factory Fixed_ (incl. 2nd & 3rd Generation)*	55.60

* 2nd Generation use DC Fronts; 3rd Generation use Dovetail Fronts

MILLETT REVOLVER SIGHTS

COLT REVOLVER
The Series 100 Adjustable Sight System offers today's discriminating Colt owner the finest quality replacement sight available. 12 crisp click stops for each turn of adjustment, delivers 5/8" of adjustment per click at 100 yards with a 6" barrel. For Colt Python, Trooper, Diamond Back and new Frontier single action army.

Rear Only (White Outline)	CR00001	$49.30
Rear Only (Target Blade)	CR00002	49.30
Rear Only (Silhouette)	CR00003	49.30

Colt owners will really appreciate the high-visability feature of Colt front sights. Easy to install—just drill 2 holes in the new sight and pin on. All steel. Your choice of blaze orange or white bar. Fits 4", 6" & 8" barrels only.

Colt Python & Anaconda (White or Orange Bar)	FB00007-8	$13.60
Diamond Back, King Cobra, Peacemaker	FB00015-16	13.60

SMITH & WESSON
The Series 100 Adjustable Sight System for Smith & Wesson revolvers provides the sight picture and crisp click adjustments desired by the discriminating shooter. 1/2" of adjustment per click, at 100 yards on elevation, and 5/8" on windage, with a 6" barrel. Can be installed in a few minutes, using factory front sight.

Smith & Wesson N Frame:
N.312—Model 25-5, all bbl., 27-3 1/2" & 5", 28-4" & 6"
N.360—Model 25, 27, 29, 57, & 629-4, 6 & 6 1/2" bbl.
N.410—Model 27, 29, 57, 629 with 8 3/8" bbl.

Smith & Wesson K&L Frame:
K.312—Models 14, 15, 18, 48-4", & 53
K&L360—Models 16, 17, 19, 48-6", 8 3/8", 66, 686, 586

Smith & Wesson K&L-Frame $49.30	
Rear Only .312 (White Outline)	SK00001
Rear Only .312 (Target Blade)	SK00002
Rear Only .360 (White Outline)	SK00003
Rear Only .360 (Target Blade)	SK00004
Rear Only .410 (White Outline)	SK00005
Rear Only .410 (Target Blade)	SK00006

Smith & Wesson K&N Old Style $49.30	
Rear Only .312 (White Outline)	KN00001
Rear Only .312 (Target Blade)	KN00002
Rear Only .360 (White Outline)	KN00003
Rear Only .360 (Target Blade)	KN00004
Rear Only .410 (White Outline)	KN00005
Rear Only .410 (Target Blade)	KN00006

Smith & Wesson N-Frame $49.30	
Rear Only .312 (White Outline)	SN00001
Rear Only .312 (Target Blade)	SN00002
Rear Only .360 (White Outline)	SN00003
Rear Only .360 (Target Blade)	SN00001
Rear Only .410 (White Outline)	SN00005
Rear Only .410 (Target Blade)	KN00006

RUGER
The high-visibility white outline sight picture and precision click adjustments of the Series 100 Adjustable Sight System will greatly improve the accuracy and fast sighting capability of your Ruger. 3/4" per click at 100 yard for elevation, 5/8" per click for windage, with 6" barrel. Can be easily installed, using factory front sight or all-steel replacement front sight which is a major improvement over the factory front. Visibility is greatly increased for fast sighting. Easy to install by drilling one hole in the new front sight.

The Red Hawk all-steel replacement front sight is highly visible and easy to pickup under all lighting conditions. Very easy to install. Fits the factory replacement system.

SERIES 100 Ruger Double Action Revolver Sights

Rear Sight (fits all adjustable models)	$49.30
Front Sight (Security Six, Police Six, Speed Six)	13.60
Front Sight (Redhawk)	9.60

SERIES 100 Ruger Single Action Revolver Sights

Rear Sight (Black Hawk Standard & Super; Bisley Large Frame, Single-Six	$49.30

Front Sight (Millett Replacement sights not available for Ruger single action revolvers).

TAURUS

Rear, .360 White Outline	$49.30
Rear, .360 Target Blade	49.30

DAN WESSON
This sight is exactly what every Dan Wesson owner has been looking for. The Series 100 Adjustable Sight System provides 12 crisp click stops for each turn of adjustment, with 5/8" per click for windage, with a 6" barrel. Can be easily installed, using the factory front or new Millett high-visibility front sights.

Choice of white outline or target blade.

Rear Only (White Outline)	DW00001	$49.30
Rear Only (Target Blade)	DW00002	49.30
Rear Only (White Outline) 44 Mag.	DW00003	49.30
Rear Only (Target Blade) 44 Mag.	DW00004	49.30

If you want super-fast sighting capability for your Dan Wesson, the new Millett blaze orange or white bar front is the answer. Easy to install. Fits factory quick-change system. All steel, no plastic. Available in both heights.

Dan Wesson .44 Mag (White Bar) (high)	FB00009	$12.95
Dan Wesson .44 Mag (Orange Bar) (high)	FB00010	13.60
Dan Wesson 22 Caliber (White Bar) (low)	FB00011	13.60
Dan Wesson 22 Caliber (Orange Bar) (low)	FB00012	13.60

MILLETT AUTO PISTOL SIGHTS

RUGER STANDARD AUTO

The Ruger Standard Auto Combo provides a highly visible sight picture even under low-light conditions. The blaze orange or white bar front sight allows the shooter to get on target fast. Great for target use or plinking. Uses Factory Front Sight on adjustable model guns when using Millett target rear only. All other installations use Millett Front Sight. Easy to install.

Rear Only (White Outline)	$55.60
Rear Only (Silhouette Target Blade)	55.60
Rear Only (Target Blade)	55.60
Front Only (White), Fixed Model	16.00
Front Only (Orange), Fixed Model	16.00
Front Only (Serrated Ramp), Fixed Model	16.00
Front Only (Target-Adjustable Model/White Bar)	16.00
Front Only (Target-Adjustable Model/Orange Bar)	16.00
Front Only Bull Barrel (White or Orange Ramp)	17.60

RUGER P85

Rear (White Outline)	$55.60
Rear (Target Blade)	55.60
Front (White Ramp)	16.00
Front (Orange Ramp)	16.00
Front (Serrated Ramp)	16.00

TAURUS PT92

Rear (White Outline, use Beretta Front)	$55.60
Rear (Target Blade, use Beretta Front)	55.60
Front (White Bar)	25.14
Front (Orange Bar)	25.14
Front (Serrated Ramp)	25.14

GLOCK 17, 17L & 19

Rear (White Outline)	$55.60
Rear (Target Blade)	55.60
Rear (3-Dot)	55.60

GLOCK STAKE-ON FRONT SIGHTS

Front .340 White Bar	$16.00
Front .340 Orange Bar	16.00
Front .340 Serrated Ramp	16.00
Front .340 White Dot	16.00

COLT

Colt Gold Cup Marksman Speed Rear Only (Target .410 Blade)	$49.30
Custom Combat Low Profile Marksman Speed Rear Only (Target .410 Blade)	55.60
Colt Gold Cup Rear (use DC or WS 200 Frt)	49.30
Colt Mark I Fixed Rear Only (use .200 front)	19.75
Colt Mark II Fixed Rear Only (use .200 front)	34.60

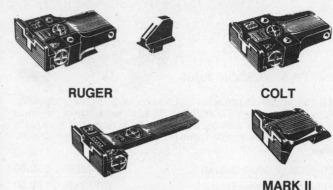

RUGER   **COLT**

COLT GOLD CUP   **MARK II HI-PROFILE**

COLT WIDE STAKE FRONT SIGHTS (POST 6/88) $16.00

.185 WS White Bar
.185 WS Orange Bar
.185 WS Serrated Ramp
.185 WS White Dot
.200 WS White Bar
.200 WS Orange Bare with Skirt
.200 WS Serrated Ramp with Skirt
.200 WS White Dot with Skirt
.312 WS White Bar with Skirt
.312 WS Orange Bar with Skirt
.312 WS Serrated Ramp with Skirt
.312 WS White Dot with Skirt

SIG/SAUER P-220, P-225, P-226

Now Sig Pistol owners can obtain a Series-100 adjustable sight system for their guns. Precision click adjustment for windage and elevation makes it easy to zero when using different loads. The high-visibility features assures fast sight acquisition when under the poorest light conditions. Made of high-quality heat-treated nickel steel and built to last. Extremely easy to install on P-225 and P-226. The P-220 and Browning BDA 45 require the Dual-Crimp front sight installation.

Sig P220-25-26 Rear Only (White)*	SP22003	$55.60
Sig P220-25-26 Rear Only (Target)*	SP22004	55.60
Sig P225-6 (White) Dovetail Front*	SP22565	16.00
Sig P225-6 (Orange) Dovetail Front*	SP22566	16.00

* The Sig P220 Uses .360 Dual-Crimp Front Sight. The Sig P225-6 Uses a Dovetail Mount Front Sight

MILLETT AUTO PISTOL SIGHTS

SMITH & WESSON 39/59

This sight system provides fast and accurate sighting capability even under low-light conditions. The unique white outline rear blade teamed up with the blaze orange or white bar front sight creates a highly visible sight picture, ideal for match or duty use.

Rear Only (White outline)	SW39595	**$59.30**
Rear Only (Target Blade)	SW39596	59.30

Requires .340 Dual-Crimp Front

SMITH & WESSON 2ND AND 3RD GENERATION
FITS FACTORY ADJUSTABLE $56.80
Rear (3-Dot)
Rear (White Outline)
Rear (Target Blade)
FITS FACTORY FIXED $55.60
Rear (3-Dot)
Rear (White Outline)
Rear (Target Blade)
THIRD GENERATION FRONT SIGHTS $16.00
Front Dovetail White Bar .260
Front Dovetail Orange Bar .260
Front Dovetail Serrated Ramp .260
Front Dovetail White Dot .260
Front Dovetail White Bar .385
Front Dovetail Orange Bar .385
Front Dovetail Serrated Ramp .385
Front Dovetail White Dot .385

BROWNING HI-POWER

The Series 100 Adjustable Sight System for Browning Hi-Power will provide accurate high-visibility sighting for both fixed and adjustable slides with no machine modifications required to the dovetail. Most adjustable slide model Hi-Powers can use the factory front sight as shown in the photo. The fixed slide model requires a new front sight installation. We highly recommend the Dual-Crimp front sight installation on this gun.

BROWNING HI-POWER (Adjustable Slide Model)

Rear Only (White Outline)	BA00009	**$55.60**
Rear Only (Target Blade)	BA00010	55.60

High-Power Requires .340 High Front Sight.

BROWNING HI-POWER (Fixed Slide Model)

Rear Only (White Outline)	BF00009	**$55.60**
Rear Only (Target Blade)	BF00010	55.60

High-Power Requires .340 High Front Sight.

MODELS CZ75/TZ75/TA90 AUTOPISTOL SIGHTS

Rear Sight (White Outline or Target Blade)		**$55.60**

COLT 45

This Series 100 High Profile Adjustable Sight is rugged, all steel, precision sight which fits the standard factory dovetail with no machine modifications required. This sight provides a highly visible sight picture even under low-light conditions. Blaze orange or white bar front sight, precision click adjustments for windage and elevation makes the Colt .45 Auto Combo the handgunner's choice.

Rear Only (White Outline)	CA00009	**$55.60**
Rear Only (Target Blade)	CA00010	55.60
Rear (Marksman, .410 Blade)	CA00018	55.60

Colt Gov. and Com. Require .312 High Front Sight.

BERETTA ACCURIZER

This amazing new sight system not only provides a highly visible sight picture but also tunes the barrel lockup to improve your accuracy and reduce your group size by as much as 50%. The Beretta Accurizer sight system fits the 92S, 92SB, 84 and 85 models. Easy to install. Requires the drilling of one hole for installation. Your choice of rear blade styles. Front sight comes in white bar, serrated ramp or blaze orange.

Rear Only (White Outline)	BE00005	**$56.40**
Rear Only (Target Blade)	BE00006	56.40
Front Only (White Bar)	BE00007	25.14
Front Only (Orange Bar)	BE00008	25.14
Front Only (Serrated Ramp)	BE00009	25.14

Fits Models 92S, 92SB, 85, 84

NEW BAR-DOT-BAR™ TRITIUM NIGHT SIGHT COMBOS

Ruger P-85, 89, 90 Combo	**$135.00**
Sig Sauer P225/226/228 & New P220 Combo	135.00
Sig Sauer P220 (Prior to 10-90) Combo	135.00
Browning Hi-Power (Fixed Model) Combo	135.00
Browning Hi-Power (Fixed Model Dovetail Front)	135.00
Colt Auto Combo	135.00
Colt Custom Combat Low Profile Combo	135.00
CZ-75/TZ-75, TA-90 Combo	135.00
Glock 17, 19, 20, 21, 22, 23 Combo	135.00
S & W 3rd Generation Fixed Combo	135.00
S & W 2nd Generation Fixed Combo	135.00
Beretta 92SB, 85, 84 Combo	143.50
Taurus PT-92 Combo	143.50

MILLET

BENCHMASTER PISTOL REST
$39.95

BENCHMASTER RIFLE REST
$39.95

BENCHMASTER WINDOW REST (not shown)
$19.95

ONE-PIECE COMBO PAKS $56.25

Model	Smooth Turn-In Med	Smooth Angle-Loc Weaver-Style Med
Browning A-Bolt	CP00041	CP00052
Interarms Mk X, FN Mauser 98	CP00043	CP00054
Remington 700 (LA-RH)	CP00040	CP00051
Remington 700 (SA-RH) 600, 7, XP-100	CP00039	CP00050
Remington 7400/7600/4/6	CP00044	CP00055
Winchester 70 (LA-RH)	CP00045	CP00056

MAGAZINES

Description	Price
Ruger 30 rounds, 10/22, M77, AMT Lighting	$19.95
Remington Short Action .308, .243, 10 rounds; Auto & Pumps 740, 742, 760, 7400, 7600, 4 & 6	19.95
Remington Long Action .30-06, .270, .280, 7mm, 10 rounds; Auto and Pumps 740, 742, 760, 7400, 7600, 4 and 6	19.95
45 ACP, 7 rounds, flush fit, blued steel	24.95
45 ACP, 7 rounds, flush fit, stainless steel	29.95
45 ACP, 8 rounds, flush fit, safety orange Follower, fixed base, blued steel	24.95
45 ACP, 8 rounds, flush fit, safety orange Follower, fixed base, stainless steel	29.95
Mini-14, 35 rounds	20.00
Mini-30, 17 rounds	20.00

SEE-THRU MOUNTS
SILVER FINISH $27.95

Browning A-Bolt
Knight Muzzle Loader MK-85
Remington 700, 721, 722, 725
Savage 110
Winchester 70

SHOTGUN SADDLE MOUNTS
$29.97

Mossberg 500, 5500
Remington 870/1100 Shotgun
Winchester 1300/1400 Shotgun

TWO-PIECE COMBO PACKS

Model		Low	Med	Price
Browning A-Bolt		CP00017 (Smooth)	CP00018 (Smooth)	$56.25
		CP00019 (Engraved)	CP00020 (Engraved)	71.30
	New	CP00037 (Nickel)	CP00038 (Nickel)	68.15
	New	CP00717 (Matte)	CP00718 (Matte)	56.25
Browning BAR/BLR		CP00013 (Smooth)	CP00014 (Smooth)	56.25
		CP00015 (Engraved)	CP00016 (Engraved)	71.30
Interarms Mk X, FN		CP00005 (Smooth)	CP00006 (Smooth)	56.25
		CP00007 (Engraved)	CP00008 (Engraved)	71.30
Marlin 336		CP00025 (Smooth)	CP00026 (Smooth)	56.25
		CP00027 (Engraved)	CP00028 (Engraved)	71.30
Remington 700		CP00001 (Smooth)	CP00002 (Smooth)	56.25
		CP00023 (Engraved)	CP00024 (Engraved)	71.30
		CP00901 (Nickel)	CP00902 (Nickel)	68.15
	New		CP00702 (Matte)	56.25
Remington 7400/7600/4/6		CP00021 (Smooth)	CP00022 (Smooth)	56.25
		CP00003 (Engraved)	CP00004 (Engraved)	71.30
Savage 110		CP00033 (Smooth)	CP00034 (Smooth)	56.25
		CP00035 (Engraved)	CP00036 (Engraved)	71.30
	New		CP00934 (Nickel)	68.15
	New		CP00734 (Matte)	56.25
Winchester 70		CP00009 (Smooth)	CP00010 (Smooth)	56.25
		CP00011 (Engraved)	CP00012 (Engraved)	71.30
		CP00909 (Nickel)	CP00910 (Nickel)	68.15
Winchester 94		CP00029 (Smooth)	CP00030 (Smooth)	56.25
		CP00031 (Engraved)	CP00032 (Engraved)	71.30

NIKON SCOPES

PISTOL SCOPES

Key Features:
- Edge to edge sharpness for precise detection of camouflaged game
- Super multicoating and blackened internal metal parts provide extreme reduction of flare and image ghost-out
- Aluminum alloy one-piece 1″ tube provides lightweight but rugged construction (fully tested on Magnum calibers)

- ¼ MOA windage/elevation adjustment
- Extended eye relief
- Nitrogen gas filled and 0-ring sealed for water and fogproofing
- Black Lustre or satin silver finish

Prices:

2X20 EER	$234.00
1.5-4.5X24 EER	387.00

2X20 EER PISTOL SCOPE

1.5-4.5X20 RIFLESCOPE

1.5-4.5X24 EER PISTOL SCOPE

RIFLESCOPES

Key Features: Essentially the same as the **Pistol Scopes** (above). Available in black lustre and black matte finishes (3-9 available and 4X40 in silver).

Prices:

4X40	$295.00
1.5-4.5X20	387.00
2-7X32	426.00
3-9X40	443.00
4-12X40 AO	563.00
6.5-20X44 AO	653.00
3.5-10X50	653.00
4-12X50	712.00

4-12X40 AO RIFLESCOPE

6.5-20X44 AO RIFLESCOPE
⅛ MOA

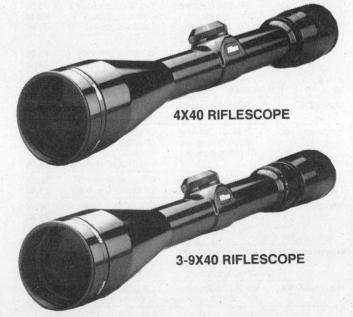

4X40 RIFLESCOPE

3.5-10X50 RIFLESCOPE

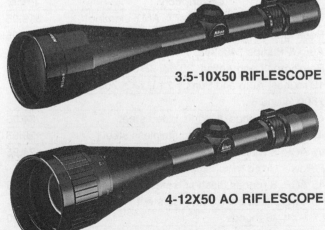

3-9X40 RIFLESCOPE

4-12X50 AO RIFLESCOPE

PENTAX SCOPES

FIXED POWER

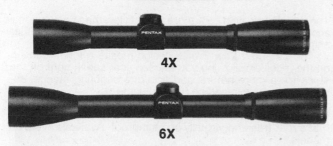

4X

6X

VARIABLE POWER

3X-9X

2X-7X

MINI 3X-9X

FIXED POWER RIFLESCOPES

Magnification: 4X
Field of view: 35'
Eye relief: 3.25"
Diameter: 1"
Weight: 12.2 oz.
Length: 11.6"
Prices: $285.00 (Glossy)
305.00 (ProFinish)

Magnification: 6X
Field of view: 20'
Eye relief: 3.25"
Diameter: 1"
Weight: 13½ oz.
Length: 13.4"
Prices: $310.00 (Glossy)
330.00 (ProFinish)

VARIABLE POWER RIFLESCOPES

Magnification: 1.5X-5X
Field of view: 66'-25'
Eye relief: 3"-3¼"
Diameter: 1"
Weight: 13 oz.
Length: 11"
Price: $340.00 (ProFinish)
320.00 (Glossy)
360.00 (Satin Chrome)

Magnification: 3X-9X
Field of view: 33'-13½'
Eye relief: 3"-3¼"
Diameter: 1"
Weight: 15 oz.
Length: 13"
Prices: $380.00 (Glossy)
400.00 (ProFinish)
420.00 (Satin Chrome)

Magnification: 2X-7X
Field of view: 42.5'-17'
Eye relief: 3"-3¼"
Diameter: 1"
Weight: 14 oz.
Length: 12"
Prices: $360.00 (Glossy)
380.00 (ProFinish)
400.00 (Satin Chrome)

Magnification: Mini 3X-9X
Field of view: 26½'-10½'
Eye relief: 3¼"
Diameter: 1"
Weight: 13 oz.
Length: 10.4"
Prices: $320.00 (Mini Glossy)
340.00 (Mini ProFinish)

Also available:
4X-12X (Mini Glossy): **$430.00** (**$450.00** ProFinish)
6X-12X (Glossy): **$480.00** (**$500.00** ProFinish)
6X-12X Silhouette: $500.00

LIGHTSEEKER 3X-11X

LIGHTSEEKER 3X-9X RIFLE SCOPE

Field of view: 36-14'
Eye relief: 3"
Diameter: 1"
Weight: 15 oz.
Length: 12.7"
Prices: $520.00 (Glossy)
540.00 (ProFinish)
560.00 (Satin)

Also available:
2X-8X (Glossy): **$490.00**
2X-8X (ProFinish): **$510.00**
2X-8X (Satin Chrome): **$530.00**
3.5X-10X (Glossy): **$540.00**
3.5X-10X (Pro Matte): **$560.00**
3.5X-10X (Chrome): **$580.00**
3X-11X (Glossy-XL): **$680.00**
3X-11X (Pro Matte): **$700.00**
3X-11X (Satin Chrome): **$720.00**
6X-24X Fine Plex (Glossy): **$760.00**
6X-24X Fine Plex (Pro Matte): **$780.00**
6X-24X Dot Reticle (Glossy): **$770.00**
6X-24X Dot Reticle (Pro Matte): **$790.00**
2.5X SG Plus (Glossy): **$280.00**
2.5X SG Plus (Pro Matte): **$300.00**
2.5 Mossy Oak SG: $320.00

PISTOL SCOPES

Magnification: 2X
Field of view: 21'
Eye relief: 10"-24"
Diameter: 1"
Weight: 6.8 oz.
Length: 8¼"
Prices: $230.00 (Glossy)
260.00 (Chrome-Matte)

Magnification: 1.5X-4X
Field of view: 16'-11'
Eye relief: 11"-25"/11"-18"
Diameter: 1"
Weight: 11 oz.
Length: 10"
Prices: $360.00 (Glossy)
390.00 (Chrome-Matte)

Also available:
1.5X-4X (Glossy): **$350.00**
1.5X-4X (Satin Chrome): **$380.00**
2.5X-7X (Glossy): **$370.00**
2.5X-7X (Chrome-Matte): **$390.00**

REDFIELD SCOPES

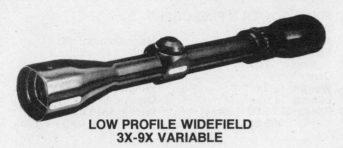

**LOW PROFILE WIDEFIELD
3X-9X VARIABLE**

**GOLDEN FIVE STAR TARGET
ADJUST KNOBS (6X-18X)**

GOLDEN FIVE STAR SCOPES

This series of seven scopes incorporates the latest variable and fixed power scope features, including multi-coated and magnum recoil-resistant optical system, plus maximum light-gathering ability. Positive quarter-minute click adjustments for ease of sighting and optimum accuracy. Anodized finish provides scratch-resistant surface.

Golden Five Star Scopes:
4X Fixed Power	$247.95
6X Fixed Power	272.95
1X-4X Variable Power	305.95
1X-4X Black Matte Variable Power	315.95
2X-7X Variable Power	319.95
3X-9X Variable Power	343.95
3X-9X Nickel Plated Variable Power	361.95
3X-9X Black Matte Variable Power	351.95
3X-9X Nickel Matte Variable Power	361.95
3X-9X Accu-Trac Variable Power	393.95
4X-12X Variable Power (Adj. Objective)	438.95
4X-12X w/Target Knob (AO)	455.95
4X-12X Accu-Trac (AO)	486.95
6X-18X Variable Power (Adj. Objective)	465.95
6X-18X Accu-Trac Variable Power (Adj. Obj.)	512.95
6X-18X Black Matte (AO)	473.95
6X-18X w/Target Knob (AO)	481.95
6X-18X Black Matte w/Targt Knob (AO)	496.95

50mm Golden Five Star Scopes:
3X-9X 50mm Five Star Variable	
116500 4 Plex	$416.95
3X-9X 50mm Five Star RealTree Camo	
116505 4 Plex	436.95
3X-9X 50mm Five Star Matte Finish	
116508 4 Plex	426.95
3X-9X 50mm Five Star Nickel Matte Finish	
116900 4 Plex	436.95

LOW PROFILE WIDEFIELD

The Widefield®, with 25% more field of view than conventional scopes, lets you spot game quicker, stay with it and see other animals that might be missed.

The patented Low Profile design means a low mounting on the receiver, allowing you to keep your cheek tight on the stock for a more natural and accurate shooting stance, especially when swinging on running game.

The one-piece, fog-proof tube is machined with high tensile strength aluminum alloy and is anodized to a lustrous finish that's rust-free and virtually scratch-proof. Available in seven models.

WIDEFIELD LOW PROFILE SCOPES

1³/₄X-5X Low Profile Black Matte Variable Power	
113807 1³/₄X-5X 4 Plex	$386.95
1³/₄X-5X Low Profile Variable Power	
113806 1³/₄X-5X 4 Plex	376.95
2X-7X Low Profile Variable Power	
111806 2X-7X 4 Plex	387.95
2X-7X Low Profile Nickel Matte Variable Power	
111808 2X-7X 4 Plex	354.95
2X-7X Low Profile Accu-Trac Variable Power	
111810 2X-7X 4 Plex AT	455.95
3X-9X Low Profile Variable Power	
112806 3X-9X 4 Plex	430.95
3X-9X Low Profile Accu-Trac Variable Power	
112810 3X-9X 4 Plex AT	498.95
2³/₄X Low Profile Fixed Power	
141807 2³/₄X 4 Plex	274.95
4X Low Profile Fixed Power	
143806 4X 4 Plex	307.95
6X Low Profile Fixed Power	
146806 6X 4 Plex	330.95
3X-9X Low Profile Nickel Matte Variable Power	
112814 4 Plex	452.95
3X-9X Low Profile Black Matte Variable Power	
112812 4 Plex	438.95

**3X-9X NICKEL-PLATED GOLDEN
FIVE STAR SCOPE**

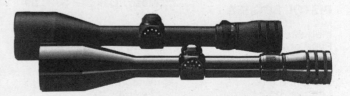

50mm GOLDEN FIVE STAR SCOPE

REDFIELD SCOPES

3X-9X WIDEFIELD® ILLUMINATOR
w/Nickel Matte Finish

THE ILLUMINATOR

With the Illuminator series, you can add precious minutes to morning and evening hunting. These scopes actually compensate for the low light, letting you "see" contrasts between field and game.

Optimum resolution, contrast, color correction, flatness of field, edge-to-edge sharpness and absolute fidelity are improved by the unique air-spaced, triplet objective, and the advanced 5-element erector lens system.

The Illuminators also feature a zero tolerance nylon cam follower and thrust washers to provide absolute point of impact hold through all power ranges. The one-piece tube construction is virtually indestructible, tested at 1200g acceleration forces, and fog-free through the elimination of potential leak paths.

Offered in both the Traditional and Widefield® variable power configurations, the Illuminator is also available with the Accu-Trac® feature.

Also offered in 30mm 3X-12X with a 56mm adj. obj.

ILLUMINATOR SCOPES
4X Widefield Fixed Power
112906 4 Plex $395.95
2X-7X Widefield Variable Power
112910 4 Plex 523.95
3X-9X Traditional Variable Power
123886 3X-9X 4 Plex 530.95
3X-9X Widefield Variable Power
112886 3X-9X 4 Plex 591.95
3X-9X Widefield Accu-Trac Variable Power
112880 3X-9X 4 Plex 645.95
3X9 Widefield Var. Power Black Matte Finish
112888 . . . . 601.95
3X-9X Widefield Nickel Matte Variable Power
112892 4 Plex AT 610.95
3X-9X Widefield Accu-Trac Black Matte Variable
112890 4 Plex AT 654.95

GOLDEN FIVE STAR EXTENDED EYE RELIEF HANDGUN SCOPES

2X Fixed
140002 4 Plex $216.95
2X Nickel Plated Fixed
10003 4 Plex 232.95
4X Fixed
140005 4 Plex 216.95
4X Nickel Plated Fixed
140006 4 Plex 232.95
2-1/2X-7X Variable
140008 4 Plex 295.95
2-1/2X-7X Nickel Plated Variable
140009 4 Plex 310.95
2-1/2X-7X Black Matte Variable
140010 4 Plex 310.95

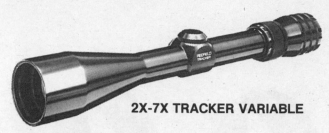

2X-7X TRACKER VARIABLE

THE TRACKER

The Tracker series brings you a superior combination of price and value. It provides the same superb quality, precision and strength of construction found in all Redfield scopes, but at an easily affordable price. Features include the tough, one-piece tube, machined and hand-fitted internal parts, excellent optical quality and traditional Redfield styling.

TRACKER SCOPES:
2X-7X Tracker Variable Power
122300 2X-7X 4 Plex $229.95
2X-7X Tracker Nickel Matte Variable Power
122310 4 Plex 246.95
3X-9X Tracker Variable Power
123300 3X-9X 4 Plex 259.95
3X-9X Tracker Nickel Matte Variable Power
123320 4 Plex 277.95
3X-9X Tracker Nickel-Plated
123303 4 Plex 277.95
3X-9X Tracker RealTree Camo
123305 4 Plex 277.95
4X Tracker Fixed Power
135300 4X 4 Plex 180.95
4X 40mm Tracker Nickel Matte Fixed Power
135312 4 Plex 207.95
4X 40mm Tracker Black Matte Fixed Power
135320 4 Plex 199.95
4X 40mm Tracker RealTree Camo
135305 4 Plex 207.95
6X Tracker Fixed Power
135600 6X 4 Plex 202.95
Matte Finish
122308 2X-7X 4 Plex 239.95
123308 3X-9X 4 Plex 267.95
135608 6X 4 Plex 210.95
135308 4X 32mm 189.95

VARIABLE GOLDEN FIVE STAR
(2-1/2X-7X) HANDGUN SCOPES
(Black, Nickel, Black Matte)

SAKO SCOPE MOUNTS

SCOPE MOUNTS

These new Sako scope mounts are lighter, yet stronger than ever. Tempered steel allows the paring of every last gram of unnecessary weight without sacrificing strength. Like the original mount, these rings clamp directly to the tapered dovetails on Sako rifles, thus eliminating the need for separate bases. Grooves inside the rings preclude scope slippage even under the recoil of the heaviest calibers. Nicely streamlined and finished in a rich blue-black to complement any Sako rifle.
Price:
Low, medium, or high (1″) **$70.00**
Medium or high (30mm) **85.00**

"ORIGINAL" SCOPE MOUNTS

Sako's "Original" scope mounts are designed and engineered to exacting specifications, which is traditional to all Sako products. The dovetail mounting system provides for a secure and stable system that is virtually immovable. Unique to this Sako mount is a synthetic insert that provides maximum protection against possible scope damage. It also affords additional rigidity by compressing itself around the scope. Manufactured in Finland.
Prices:
1″ Medium & High (Short, Medium
& Long Action) **$130.00**
30mm Medium & High (Short, Medium
& Long Action) **145.00**

SCHMIDT & BENDER RIFLE SCOPES

2½-10X56 VARIABLE POWER SCOPE
$1117.00 ($1137.00 w/Glass Reticle)

4-12X42 VARIABLE POWER SCOPE
$989.00

Also available:
1.5-6X42 SNIPER $1434.00

SCHMIDT & BENDER SCOPES

1¼-4X20 VARIABLE POWER SCOPE
$858.00 ($849.00 w/Glass Reticle)

1½-6X42 VARIABLE POWER SCOPE
$930.00 ($950.00 w/Glass Reticle)

1½X15 FIXED POWER SCOPE
(Steel Tube w/o Mounting Rail)
$631.00

3-12X50 VARIABLE POWER SCOPE
$1048.00 ($1068.00 w/Glass Reticle)

4X36 FIXED POWER SCOPE
(Steel Tube w/o Mounting Rail)
$648.00

6X42 FIXED POWER SCOPE
(Steel Tube w/o Mounting Rail)
$707.00

4-12X42 VARIABLE POWER RIFLE SCOPE
$1024.00

10X42 FIXED POWER SCOPE
(Steel Tube w/o Mounting Rail)
$773.00

8X56 FIXED POWER SCOPE
(Steel Tube w/o Mounting Rail)
$793.00

SIMMONS SCOPES

44 MAG RIFLESCOPES

MODEL 1045

MODEL 1043
2-7X44mm
Field of view: 56'-16'
Eye relief: 3.3"
Length: 11.8"
Weight: 13 oz.
Price: $256.95

MODEL 1044
3-10X44mm
Field of view: 38'-12'
Eye relief: 3"
Length: 12.8"
Weight: 16.9 oz.
Price: $268.95

MODEL 1045
4-12X44mm
Field of view: 27'-9'
Eye relief: 3"
Length: 12.8"
Weight: 19.5 oz.
Price: $280.95

MODEL 3044
3-10X44mm
Field of view: 38'-11'
Eye relief: 3"
Length: 13.1"
Weight: 16.4 oz.
Price: $269.95

PROHUNTER RIFLESCOPES

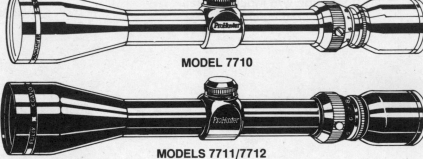

MODEL 7710
3-9X40mm Wide Angle Riflescope
Field of view: 40'-15' at 100 yards
Eye relief: 3"
Length: 12.6"
Weight: 11.6 oz.
Features: Triplex reticle; silver finish
Price: $169.95 (Same in black matte or
    black polish, Models 7711 and 7712)

MODEL 7710

MODELS 7711/7712

Also available:
Model 7700 2-7X32 Black Matte or Black Polish .. $159.95
Model 7715 4-12X40 Black Polish 179.95
Model 7720 6-18X40 (adj. obj. Black) 209.95
Model 7725 4.5X32 Silver 109.95
Model 7740 6X40 Black Matte 139.95

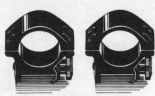

#1401

#1403

#1406

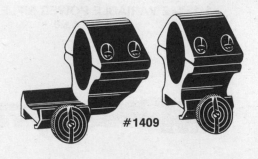

#1409

RIFLESCOPE ALLOY RINGS
Model 1401 Low 1" Set $11.95
Model 1403 Medium 1" Set 11.95
Model 1404 High 1" Set 11.95
Model 1405 1" See-Thru Set 13.95
Model 1406 1" Rings for 22 Grooved Receiver 11.95
Model 1409 1" Rings extention for
    Compact Scopes 20.95

SIMMONS SCOPES

WHITETAIL CLASSIC RIFLESCOPES

Simmons' Whitetail Classic Series features fully coated lenses and glare-proof BlackGranite finish. The Mono-Tube construction means that front bell and tube, saddle and rear tube are all turned from one piece of aircraft aluminum. This system eliminates 3 to 5 joints found in most other scopes in use today, making the Whitetail Classic up to 400 times stronger than comparably priced scopes.

MODEL WTC10

MODEL WTC10
4X32mm
Field of view: 35'
Eye relief: 4"
Length: 12"
Weight: 11.0 oz.
Price: $139.95

MODEL WTC11

MODEL WTC11
1.5-5X20mm
Field of view: 80'–23.5'
Eye relief: 3.5"
Length: 9½"
Weight: 9.9 oz.
Price: $174.95

MODEL WTC12

MODEL WTC12
2.5-8X36mm
Field of view: 48'–14.8'
Eye relief: 3"
Length: 12.8"
Weight: 12.9 oz.
Price: $189.95

MODEL WTC13

MODEL WTC13
3.5-10X40mm
Field of view: 35'–12'
Eye relief: 3"
Length: 12.8"
Weight: 16.9 oz.
Price: $209.95

MODEL WTC14

MODEL WTC14
2-10X44mm
Field of view: 50'–11'
Eye relief: 3"
Length: 12.8"
Weight: 16.9 oz.
Price: $256.95

MODEL WTC16
4X40 Black Granite
Field of view: 36.8'
Eye relief: 4"
Length: 9.9"
Weight: 12 oz.
Price: $149.95

MODEL WTC15/35
3.5-10X50 Black
or Silver Granite
Field of view: 30.3'–11.3'
Eye relief: 3.2"
Length: 12.25"
Weight: 13.6 oz.
Price: $329.95

MODEL WTC 17
4-12X44 WA/AO Black Granite
Field of view: 26'–7.9'
Eye relief: 3"
Length: 12.8"
Weight: 19.5 oz.
Price: $329.95

MODEL WTC23
3.5-10X40
Field of view: 34'–11.5'
Eye relief: 3.2"
Length: 12.4"
Weight: 12.8 oz.
Price: $209.95

MODEL WTC33
3.5-10X40 Silver
Same specifications as
Model WTC23
Price: $209.95

SIMMONS SCOPES

GOLD MEDAL SILHOUETTE SERIES

Simmons Gold Medal Silhouette Riflescopes are made of state-of-the-art drive train and erector tube design, a new windage and elevation indexing mechanism, camera-quality 100% multi-coated lenses, and a super smooth objective focusing device.

High silhouette-type windage and elevation turrets house 1/8 minute click adjustments. The scopes have a black matte finish and crosshair reticle and are fogproof, waterproof and shockproof.

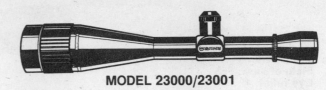

MODEL 23000/23001

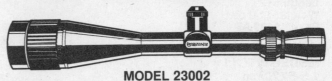

MODEL 23002

MODEL #23000
12X44mm
Field of view: 8.7'
Eye relief: 3.17"
Length: 14.5"
Weight: 18.3 oz.
Feature: Truplex Reticle, 100% Multi-Coat Lens system, black matte finish, obj. focus
Price: $449.95 (Crosshair)
        454.95 (Dot Reticle)

MODEL #23001
24X44mm
Field of view: 4.3'
Eye relief: 3"
Length: 14.5"
Weight: 18.3 oz.
Feature: Truplex reticle, 100% Multi-Coat Lens System, black matte finish, obj. focus
Price: $455.95 (Crosshair)
        460.95 (Dot Reticle)

MODEL #23002
6-20X44mm
Field of view: 17.4'-5.4'
Eye relief: 3"
Length: 14.5"
Weight: 18.3 oz.
Feature: Truplex reticle, 100% Multi-Coat Lens System, black matte finish, obj. focus
Price: $499.95 (Crosshair)
        504.95 (Dot Reticle)

GOLD MEDAL HANDGUN SERIES

Simmons gold medal handgun scopes offer long eye relief, no tunnel vision, light weight, high resolution, non-critical head alignment, compact size and durability to withstand the heavy

recoil of today's powerful handguns. In black and silver finishes, all have fully multicoated lenses and a Truplex reticle.

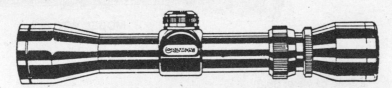

MODEL 22001

MODEL #22001
2.5-7X28mm
Field of view: 9.7'-4.0'
Eye relief: 8.9"-19.4"
Length: 9.2"
Weight: 9 oz.
Feature: Truplex reticle, 100% Multi-Coat Lens System, black polished finish.
Price: $319.95

MODEL #22002
2.5-7X28mm
Field of view: 9.7'-4.0'
Eye relief: 8.9"-19.4"
Length: 9.2"
Weight: 9 oz.
Feature: Truplex reticle, 100% Multi-Coat Lens System, black polished finish.
Price: $319.95

Also Available:
MODEL #22003
2X20 .... $219.95
MODEL #22004
2X20 .... 219.95
MODEL #22005
4X32 .... 249.95
MODEL #22006
4X32 .... 249.95

SWAROVSKI RIFLESCOPES

PH SERIES

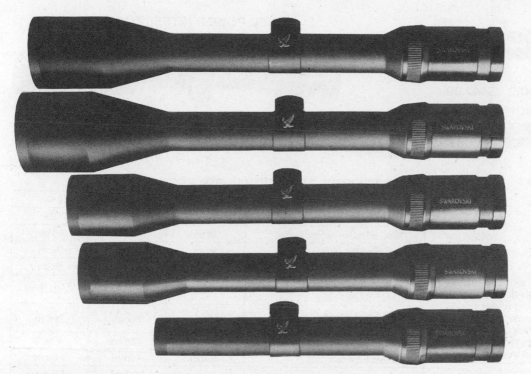

3-12X50 $1166.00

2.5-10X56 $1216.00

2.5-10X42 $1150.00

1.5-6X42 $972.00

SPECIFICATIONS PROFESSIONAL HUNTER PH SERIES

Type	Maintube	Magnification	Max. effective objective lens ø in/mm	Exit pupil ø in/mm	Exit pupil distance in/mm	Field of view ft/100yds m/100 m	Twilight performance acc. to DIN 58388	Middle tube ø standard in/mm a	Objective lens tube ø in/mm b	Total length in/mm c	1 click in/100 yds mm/100 m	Max. adjustment range in/100 yds m/100 m	Weight S/LS (approx.) oz/g	Licencenumber
1.25-4x24	S L LS	1.25-4	0.94 24	0.49-0.24 12.5-6	3.15 80	10.8-3.5 32.8-10.4	3.5-9.8	1.18 30	1.18 30	10.6 270	0.54 15	119 3.3	450 350 385	15.9 12.3 13.6
1.5-6x42	S L LS	1.5-6	1.65 42	0.52-0.28 13.1-7	3.15 80	7.3-2.3 21.8-7	4.2-15.9	1.18 30	1.89 48	13.0 330	0.36 10	79 2.2	580 450 485	20.5 15.9 17.1
2.5-10x42	S L LS	2.5-10	1.65 42	0.52-0.17 13.1-4.2	3.15 80	4.4-1.4 13.2-4.1	7.1-20.5	1.18 30	1.89 48	13.2 336	0.36 10	47 1.3	550 420 455	19.4 14.8 16.0
2.5-10x56	S L LS	2.5-10	2.20 56	0.52-0.22 13.1-5.6	3.15 80	4.4-1.4 13.2-4.1	7.1-23.7	1.18 30	2.44 62	14.7 374	0.36 10	47 1.3	690 520 560	24.3 18.3 19.8
3-12x50	S L LS	3-12	1.97 50	0.52-0.17 13.1-4.2	3.15 80	3.7-1.2 11-3.5	8.5-24.5	1.18 30	2.20 56	14.3 364	0.36 10	40 1.1	625 470 510	22.0 16.6 18.0

S = steel body, L = light alloy body, LS = light alloy body with mounting rail

SWAROVSKI RIFLESCOPES

TRADITIONAL RIFLESCOPES

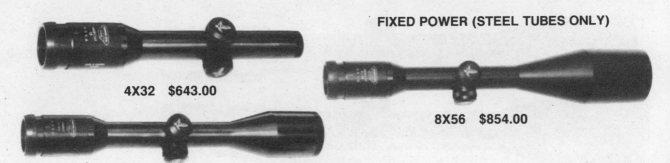

4X32 $643.00

6X42 $721.00

FIXED POWER (STEEL TUBES ONLY)

8X56 $854.00

These fine Austrian-made sights feature brilliant optics with high-quality lens coating and optimal sighting under poor light and weather conditions. The Nova ocular system with telescope recoil damping reduces the danger of injury, especially with shots aimed in an upward direction.

SPECIFICATIONS TRADITIONAL RIFLESCOPES

Telescopic Sights	4X32	6X42	8X56
Magnification	4X	6X	8X
Max. effective objective dia.	32mm	42mm	56mm
Exit pupil dia.	8mm	7mm	7mm
Field of view at 100m	10m	7m	5.2m
Twilight effective factor (DIN 58388)	11.3	15.9	21.1
Intermediary tube dia. Steel-Standard	26mm	26mm	26mm
Objective tube dia.	38mm	48mm	62mm
Ocular tube dia.	40mm	40mm	40mm
Scope length	290mm	322mm	370mm
Weight Steel	430g	500g	660g
(approx.) Light metal with rail	NA	NA	NA
A change of the impact point per click in mm/100m	7	6	4

AMERICAN LIGHTWEIGHT RIFLESCOPE

This model features precision ground, coated and aligned optics sealed in a special aluminum alloy tube to withstand heavy recoil. Eye relief is 85mm and the recoiling eyepiece protects the eye. Positive click adjustments for elevation and windage change the impact point (approx. ¼") per click at 100 yards, with parallax also set at 100 yards. Weight is only 13 ounces.

Prices:
1.5-4.5X20 with duplex reticle **$576.00**
4X32 with duplex reticle **494.00**
6X36 with duplex reticle **550.00**
3-9X36 with duplex reticle **610.00**

SWIFT RIFLESCOPES

RIFLESCOPE SPECIFICATIONS

MODEL#	DESCRIPTION	FIELD OF VIEW AT 100 YDS. (FT')	ZERO PARALLAX AT	EYE RELIEF (INCH)	TUBE DIAMETER (INCH)	CLICK ADJUST-MENT (INCH)	LENGTH (INCH)	WEIGHT (OZ.)	LENS ELEMENT (PC'E)
650	4x,32mm	26'	100 YDS.	4	1"	1/4"	12"	9.1	9
653	4x,40mm, W.A.	35'	100 YDS.	4	1"	1/4"	12.2"	12.6	11
660	4x,20mm	25'	35 YDS.	4	1"	1/4"	11.8"	9	9
666	1x,20mm	113'	-	3.2	1"	1/4"	7.5"	9.6	-
654	3-9x,32mm	35' @ 3x 12' @ 9x	100 YDS.	3.4 @ 3x 2.9 @ 9x	1"	1/4"	12"	9.8	11
656	3-9x,40mm, W.A.	40' @ 3x 14' @ 9x	100 YDS.	3.4 @ 3x 2.8 @ 9x	1"	1/4"	12.6"	12.3	11
664R	4-12x,40mm	27' / 9'	Adjust-able	3.0 / 2.8	1"	1/4"	13.3"	14.8	
665	1.5-4.5x,21mm	69' / 24.5'	100 YDS.	3.5 / 3.0	1"	1/4"	10.9"	9.6	
659	3.5-10x,44mm, W.A.	34' @ 3.5x 12' @ 10x	-	3.0 / 2.8	1"	1/4"	12.8"	13.5	
649	4-12x,50mm, W.A.	30' @ 4x 10' @ 12x	-	3.2 / 3.0	1"	1/4"	13.2"	14.6	
658	2-7x,40mm, W.A.	55' @ 2x 18' @ 7x	-	3.3 / 3.0	1"	1/4"	11.6"	12.5	
667	FIRE-FLY, 1x,30mm	40'	100YDS.	Unlimited	30mm	1/2"	5 3/8"	5	-

MODEL 649
$215.00

MODEL 650 $78.00

MODEL 653 $95.00

MODEL 654 $94.00

MODEL 656 $102.00

MODEL 658 $135.00

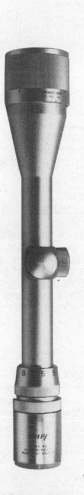

MODEL 664R
$144.00

MODEL 659
$212.00

TASCO SCOPES

MODEL WA1.35×20

WORLD-CLASS WIDE-ANGLE® RIFLESCOPES

Features:
- 25% larger field of view
- Exceptional optics

- Fully coated for maximum light transmission
- Waterproof, shockproof, fogproof
- Non-removable eye bell
- Free haze filter lens caps
- TASCO's unique World Class Lifetime Warranty

This member of Tasco's World Class Wide Angle line offers a wide field of view—115 feet at 1X and 31 feet at 3.5X—and quick sighting without depending on a critical view. The scope is ideal for hunting deer and dangerous game, especially in close quarters or in heavily wooded and poorly lit areas. Other features include 1/2-minute positive click stops, fully coated lenses (including Supercon process), nonremovable eyebell and windage/elevation screws. Length is 9³/₄″, with 1″ diameter tube. Weight is 10.5 ounces.

WORLD-CLASS, WIDE-ANGLE VARIABLE ZOOM RIFLESCOPES

Model No.	Description	Reticle	Price
WA13.5X20	1X-3.5X Zoom (20mm)	Wide Angle 30/30	$260.00
WA2.58X40ST	2.5X-8X (40mm)	Wide Angle 30/30	206.00
WA2.58X40	2.5X-8X (40mm)	Wide Angle 30/30	199.00
WA39X40TV	3X-9X (40mm)	Wide Angle 30/30	199.00
WA4X32ST	4X (32mm)	Wide Angle 30/30	218.00
WA4X40	4X (40mm)	Wide Angle 30/30	160.00
WA6X40	6X (40mm)	Wide Angle 30/30	227.00
WA1.755X20	1.75X-5X Zoom (20mm)	Wide Angle 30/30	266.00
WA27X32	2X-7X Zoom (32mm)	Wide Angle 30/30	191.00
WA39X40	3X-9X Zoom (40mm)	Wide Angle 30/30	199.00
WA39X40ST	3X-9X (40mm)	Wide Angle 30/30	206.00
CW28X32 COMPACT	2X-8X (32mm)	Wide Angle 30/30	255.00
CW4X32 LE COMPACT	4X (32mm)	Wide Angle 30/30	221.00
ER39X40WA	3X-9X 40m Electronic Reticle	Electronic Red	543.00

WORLD-CLASS 1″ PISTOL SCOPES

Built to withstand the most punishing recoil, these scopes feature a 1″ tube that provides long eye relief to accommodate all shooting styles safely, along with fully coated optics for a bright, clear image and shot-after-shot durability. The 2X22 model is recommended for target shooting, while the 4X28 model and 1.25X-4X28 are used for hunting as well. All are fully waterproof, fogproof, shockproof and include haze filter caps.

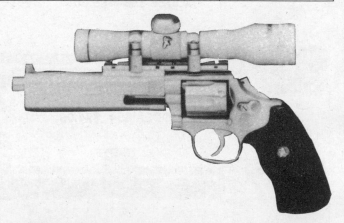

SPECIFICATIONS

Model	Power	Objective Diameter	Finish	Reticle	Field of View @ 100 Yds	Eye Relief	Tube Diam.	Scope Length	Scope Weight	Prices
PWC2X22	2X	22mm	Blk Gloss	30/30	25′	11″–20″	1″	8.75″	7.3 oz.	$206.00
PWC2X22MA	2X	22mm	Matte Alum.	30/30	25′	11″–20″	1″	8.75″	7.3 oz.	206.00
PWC4X28	4X	28mm	Blk Gloss	30/30	8′	12″–19″	1″	9.45″	7.9 oz.	252.00
PWC4X28MA	4X	28mm	Matte Alum.	30/30	8′	12″–19″	1″	9.45″	7.9 oz.	252.00
P1.254X28	1.25X-4X	28mm	Blk Gloss	30/30	23′-9′	15″-23″	1″	9.25″	8.2 oz.	289.00
P1.254X28MA	1.25X-4X	28mm	Matte Alum.	30/30	23′-9′	15″-23″	1″	9.25″	8.2 oz.	289.00

TASCO SCOPES

PROPOINT MULTI-PURPOSE SCOPES

Tasco's ProPoint is a true 1X-30mm scope with electronic red dot reticle that features unlimited eye relief, enabling shooters to shoot with both eyes open. It is available with a 3X booster, plus a special, open T-shaped electronic reticle with a dot in the center, making it ideal for fast-action pistol competition and bull's-eye marksmanship. It also has application for rifle, shotgun, bow and black powder. The compact version (PDP2) houses a lithium battery pack, making it 1¼ inches narrower than previous models and lighter as well (5.5 oz.). A mercury battery converter is provided for those who prefer standard batteries.

Tasco's 3X booster with crosshair reticle weighs 6.1 oz. and is 5½ inches long. Model PB2 fits the new PDP2/PDP2MA, and because both units include separate windage and elevation systems the electronic red dot is movable within the crosshair. That means it can be set for two different distances, making it the ultimate rangefinder. Another 3X booster—the PB1— has no crosshairs and fits all other Pro-Point models. Specifications and prices are listed below.

3X BOOSTER

SPECIFICATIONS PROPOINT SCOPES*

Model	Power	Objective Diameter	Finish	Reticle	Field of View @ 100 Yds.	Eye Relief	Tube Diam.	Scope Length	Scope Weight	Prices
PDP2	1X	25mm	Matte Black	Illum. Red Dot	40'	Unltd.	30mm	5″	5.5 oz.	$267.00
PDP2ST	1X	25mm	Stainless Steel	Illum. Red Dot	40'	Unltd.	30mm	5″	5.5 oz.	267.00
PDP2BD	1X	25mm	Matte Black	Illum. Red Dot	40'	Unltd.	30mm	5″	5.5 oz.	267.00
PDP2BDST	1X	25mm	Stainless Steel	Illum. Red Dot	40'	Unltd.	30mm	5″	5.5 oz.	267.00
PDP3	1X	25mm	Black Matte	Illumn. Red Dot	52'	Unltd.	30mm	5″	5.5 oz.	367.00
PDP3ST	1X	25mm	Stainless Steel	Illum. Red Dot	52'	Unltd.	30mm	5″	5.5 oz.	367.00
PDP3BD	1X	25mm	Black Matte	Illum. Red Dot	52'	Unltd.	30mm	5″	5.5 oz.	367.00
PDP3BDST	1X	25mm	Stainless Steel	Illum. Red Dot	52'	Unltd.	30mm	5″	5.5 oz.	367.00
PB2	3X	14mm	Black Matte	Crosshair	35'	3″	30mm	5.5″	6 oz.	183.00
PB3	2X	38mm	Black Matte	NA	30'	Unltd.	30mm	1.25″	2.6 oz.	214.00

* **Model PDP4-10** (1X40) available. Specifications to be announced. **$458.00**

TASCO RIFLE SCOPES

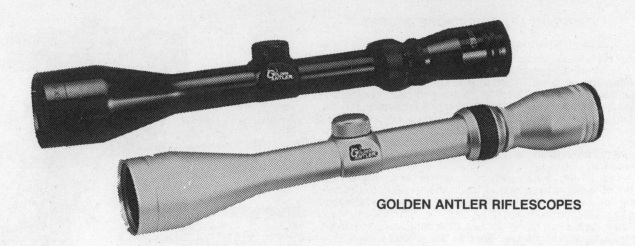

GOLDEN ANTLER RIFLESCOPES

GOLDEN ANTLER™ RIFLESCOPES

Model	Power	Objective Diameter	Finish	Reticle	Field of View @ 100 Yds.	Eye Relief	Tube Diam.	Scope Length	Scope Weight	Prices
GA4X32TV	4X	32mm	Blk. Gloss/ Matte Alum.	30/30TV	32′	3″	1″	13″	12.7 oz.	$102.00
GA4X40TV	4X	40mm	Black Gloss	30/30TV	32′	3″	1″	12″	12.5 oz.	99.00
GA2.510X44TV	2.5X-10X	44mm	Black Gloss	30/30TV	35′-9′	3.5″	1″	12.5″	14.4 oz.	255.00
GA39X32TV	3X-9X	32mm	Black Gloss	30/30TV	39′-13′	3″	1″	13.25″	12.2 oz.	102.00
GA39X32MA	3X-9X	32mm	Matte Aluminum	30/30TV	39′-13′	3″	1″	13.25″	12.2 oz.	102.00
GA39X40TV	3X-9X	40mm	Black Gloss	30/30TV	39′-13′	3″	1″	12.5″	13 oz.	135.00
GA39X40MA	3X-9X	40mm	Matte Aluminum	30/30TV	39′-13′	3″	1″	12.5″	13 oz.	135.00

SILVER ANTLER™ RIFLESCOPES

Model	Power	Objective Diameter	Finish	Reticle	Field of View @ 100 Yds.	Eye Relief	Tube Diam.	Scope Length	Scope Weight	Prices
SA2.5X32	2.5X	32mm	Black Gloss	30/30	42′	3.25″	1″	11″	10 oz.	$ 86.00
SA4X32	4X	32mm	Blk. Gloss/ Matte Alum.	30/30	32′	3″	1″	12″	12.5 oz.	79.00
SA4X40	4X	40mm	Black Gloss	30/30	32′	3″	1″	12″	12.5 oz.	99.00
SA39X32	3X-9X	32mm	Black Gloss	30/30	39′-13′	3″	1″	12″	11 oz.	102.00
SA39X32MA	3X-9X	32mm	Matte Aluminum	30/30	39′-13′	3″	1″	12″	11 oz.	102.00
SA39X40	3X-9X	40mm	Black Gloss	30/30	39′-13′	3″	1″	12.5″	13 oz.	135.00
SA39X40MA	3X-9X	40mm	Matte Aluminum	30/30	39′-13′	3″	1″	12.5″	13 oz.	135.00
SA2.51OX44	2.5X-10X	44mm	Black Gloss	30/30	35′-9′	3.5″	1″	12.5″	14.4 oz.	255.00

TASCO SCOPES

WORLD-CLASS PLUS RIFLESCOPES

WORLD-CLASS TARGET SCOPE 36X50mm

SPECIFICATIONS WORLD CLASS PLUS RIFLESCOPES

Model	Power	Objective Diameter	Finish	Reticle	Field of View @ 100 Yds.	Eye Relief	Tube Diam.	Scope Length	Scope Weight	Prices
WCP4X44	4X	44mm	Black Gloss	30/30	32'	3¼"	1"	12.75"	13.5 oz.	$260.00
WCP6X44	6X	44mm	Black Gloss	30/30	21'	3¼"	1"	12.75"	13.6 oz.	260.00
WCP39X44	3X-9X	44mm	Black Gloss*	30/30	39'-14'	3½"	1"	12.75"	15.8 oz.	306.00
DWCP39X44	3X-9X	44mm	Black Matte	30/30	39'-14'	3½"	1"	12.75"	15.8 oz.	306.00
WCP39X44ST	3X-9X	44 mm	Stainless	30/30	39'-14'	3½"	1"	12.75"	15.8 oz.	000.ST
WCP3.510X50	3.5X-10X	50mm	Black Gloss	30/30	30'-10.5'	3¾"	1"	13"	17.1 oz.	382.00
DWCP3.510X50	3.5X-10X	50mm	Black Matte	30/30	30'-10.5'	3¾"	1"	13"	17.1 oz.	382.00
WCP24X50	24X	50mm	Black Gloss	Crosshair	4.8'	3½"	1"	13.25"	15.9 oz.	645.00
WCP36X50	36X	50mm	Black Gloss	Crosshair	3'	3½"	1"	14"	15.9 oz.	679.00

RUBBER ARMORED SCOPES

Model	Power	Objective Diameter	Finish	Reticle	Field of View @ 100 Yards	Eye Relief	Tube Diam.	Scope Length	Scope Weight	Price
RC39X40 A,B	3-9	40mm	Green Rubber	30/30	35'-14'	3¼"	1"	12⅝"	14.3 oz.	$206.00

"A" fits standard dove tail base.   "B" fits ⅜" grooved receivers—most 22 cal. and airguns.

MAG IV RIFLESCOPES (not shown)

MAG IV scopes yield four times magnification range in a standard size riflescope and one-third more zooming range than most variable scopes. Features include: Fully coated optics and large objective lens to keep target in low light . . . Non-removable eye bell . . . ¼-minute positive click stops . . . Non-removable windage and elevation screws. . . Opticentered 30/30 rangefinding reticle . . . Waterproof, fogproof, shockproof.

SPECIFICATIONS

Model	Power	Objective Diameter	Finish	Reticle	Field of View @ 100 Yds.	Eye Relief	Tube Diam.	Scope Length	Scope Weight	Price
W312X40	3–12	40mm	Black	30/30	35'–9'	3⅛"	1"	12³⁄₁₆"	12 oz.	$183.00
W416X40†	4–16	40mm	Black	30/30	26'–6'	3⅛"	1"	14⅛"	15.6 oz.	229.00
W624X40†	6–24	40mm	Black	30/30	17'–4'	3"	1"	15⅜"	16.75 oz.	290.00

† Indicates focusing objective.

TASCO SCOPES

HIGH COUNTRY RIFLESCOPES

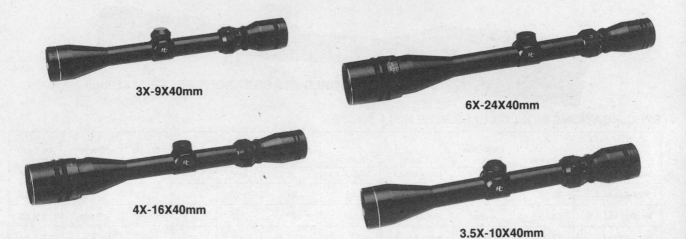

3X-9X40mm

6X-24X40mm

4X-16X40mm

3.5X-10X40mm

SPECIFICATIONS HIGH COUNTRY RIFLESCOPES $199.00–$290.00

Model	Power	Objective Diameter	Field of view @ 100 yds.	Reticle	Eye Relief	Tube Diam.	Finish	Length	Weight
HC416X40	4X-16X	40mm	26'-7'	30/30	3.25″	1″	Black gloss	14.25″	15.6 oz.
HC624X40	6X-24X	40mm	17'-4'	30/30	3″	1″	Black gloss	15.25″	16.8 oz.
HC39X40	3X-9X	40mm	41'-15'	30/30	3″	1″	Black gloss	12.75″	13 oz.
HC3.510X40	3.5X-10X	40mm	30'-10.5'	30/30	3″	1″	Black gloss	11.75″	14.25 oz.

LASER POINT

Tasco's LaserPoint model is the first compact to have a multi-mode red dot (one second continuous followed by one second pulsating), making it the fastest and easiest dot to locate on the target. An index-guided diode designed with minimum astigmatism, maximum efficiency and battery life results in a much improved laser dot. Additional features include adjustable windage and elevation system, waterproofing and several optional mounts that require no gunsmithing.

Price: **$435.00**

WEAVER SCOPES

V16 4X-16X

	Fixed				Variable							
Scope	K2.5	K4	K6	KT15	V3		V9		V10		V16	
Actual					1x	3x	3x	9x	2x	10x	4x	16x
Magnification	2.5x	3.7	5.7	14.6	1.1	2.8	2.8	8.7	2.2	9.6	3.8x	15.5x
Field of View @ 100 Yds. (ft.)	35	26.5	18.5	7.5	88	32	33	11	38.5	9.5	26.8	6.8
Eye Relief (in.)	3.7	3.3	3.3	3.2	3.9	3.7	3.5	3.4	3.4	3.3	3.1	3.1
Unrestricted Obj. Lens Dia. (in.)	.78	1.48	1.48	1.65	.78		1.48		1.48		1.65	
Objective Bell Dia. (in.)	1	1.79	1.79	2.08	1		1.79		1.79		2.08	
Eye Piece Dia. (in.)	1.49	1.49	1.49	1.49	1.49		1.49		1.49		1.49	
Tube Dia. (in.)	1	1	1	1	1		1		1		1	
Adj. per Click @ 100 Yds. (in.)	1/4	1/4	1/4	1/4	1/4		1/4		1/4		1/4	
Max. Adj. Range @ 100 Yds. (in.)	±55	±55	±50	±35	±85		±40		±40		±30	
Overall Length (in.)	9.5	11.3	11.4	12.9	9.2		12.1		12.2		13.9	
Weight (oz.)	7.3	10	10	14.7	8.5		11.1		11.2		16.5	
Reticle Size @ 100 Yds. A (in.)	1	.5	.35	.15	2	.75	.65	.25	1	.21	.52	.14
B (in.)	3	2	1.35	.55	8.1	2.8	2.5	.9	3.6	.8	2	.52
C (in.)	30	21	12.5	5.05	74.6	25.6	27	9	33	7.5	18.3	4.8

Note: V16 Fine reticle size at 100yds = .22" (4x), .06" (16x).
V16 Dot reticle size at 100yds = 1" (4x), .25" (16x).

V16 4X-16X VARIABLE POWER

Offers large magnification for extreme long shots like varmint or bench rest shooting. Or twist it down for close-range shots that require a large field of view. The adjustable objective allows a parallax-free view at any distance. Its single piece tube design offers strength and moisture resistance in the foulest of conditions. Multi-coated lenses provide clear, crisp images from dawn to dusk, while the matte finish eliminates glare.
Model V16 MDX (dual X reticle) **$342.00**
Model V16 MFC (fine crosshair reticle) 342.00
Model V16 MDT (1/4-minute dot reticle) 342.00

K2.5 2.5X FIXED POWER

Optimum power, field of view and size for close-range hunting with slug or brush guns.
Model K2.5 . . . . $130.00

K4 4X FIXED POWER

The K4's 4X magnification and wide field of view make it the ideal all-purpose scope. Especially good for medium-range small- and big-game hunting.
Model K4 . . . . $140.00
Model K4 Matte Finish 148.00

K6 6X FIXED POWER

Identical to the K4, except with more power. Use the K6 for longer distance shooting in wide-open country.
Model K6 . . . . $153.00

KT15 15X FIXED POWER

The KT15's range-focus mechanism allows parallax-free sighting at almost any distance. This is an ideal scope for precise target and varmint shooting.
Model KT15 . . . . $305.00

V3 1X-3X VARIABLE POWER

At 1X the V3 provides excellent field of view. At 3X you get the crisp magnification needed for critical shot placement. With extra eye relief for big recoils and fast targeting, giant exit pupil for dim light and moving game.
Model V3 . . . . $170.00

V9 3X-9X VARIABLE POWER

Easy adjustment ring lets you dial the magnification. Precision ground lenses offer crystal clear resolution. Long eye relief and rubber guards protect against facial impact from recoil.
Model V9 . . . . $183.00
Model V9 Matte Finish 192.50

V10 2X-10X VARIABLE POWER

The 5 to 1 magnification range of the V10 lets it handle a variety of situations, from close-in moving game to long-distaance varmint or target shooting.
Model V10 . . . . $195.00
Model V10 Matte Finish 205.00
Model V10 Stainless 205.00

WIDEVIEW SCOPE MOUNTS

PREMIUM SEE-THRU SCOPE MOUNTS

Rifle	Model
Browning A Bolt	NN
Browning Semi-Auto	AA
Browning Lever Action	AA
Browning F.N. Bolt Action	CB
BSA, Medium & Long Action	DB
Glenfield 30 by Marlin	GG
H & R Bolt Action F.N. 300, 301, 317, 330, 370	CB
Husquvarna F.N. Action	CB
Interarms Mark X	CB
Ithaca BSA (Bolt Action)	DB
Marlin 336, 62, 36, 444	GG
Marlin 1893, 1894, 1895, 9, 45	GG
Marlin 465 F.N. Action	CB
Mauser F.N. 98, 2000, 3000	CB
Mossberg 800, 500	AA
Parker Hale 1000, 1000C, 1100, 1200, 1200C	CB
Parker Hale 2100	JB
Remington 7	SB
Remington 700, 721, 722, 725	DB
Remington 740, 742, 760	EB
Remington 788	FH
Remington 4, 6, 7400, 7600	EB8
Remington XP-100 Pistol 600, 660	HB
Revelation 200 Lever Action	GG
Ruger M-77 round receiver	DB
Ruger 44 rifle	QO
Ruger 10/22	OP
Savage 99	KM
Savage 110, 111, 112	IB
Savage 170	EB
Smith & Wesson 1500 Bolt	DB
Weatherby Mark V & Vanguard	DB
Western Field 740	GG
Winchester 88, 100	BB
Winchester 70, 70A, 670, 770 and Ser. #700, (except 375 H & H)	FB
Winchester 94AE Angle Eject	94 AE
Winchester 94 Side Mount	94

SHOTGUNS (Must Be Drilled and Tapped)

Ithaca, Remington, Winchester, Mossberg, etc.
Note: Remington 1100 and Browning 5 Auto not recommended by Wideview for top mounts. Receivers should be .150 thousands or more in thickness. .... BB

FOR WEAVER-STYLE BASES

Savage 24V	U-20
Straight Cut	U-10
20 Degree Angle Cut	U-20
Straight Cut 30 Millimeter	U-1030
20 Degree Angle Cut 30 Millimeter	U-2030

RING-STYLE MOUNTS

Dove Tail Solid Lock Ring Fits Any Redfield, Tasco, or Weaver Style Base	SR
Dove Tail Solid Lock 30 Millimeter	SR30
True-Fit Lo Ring	L-Ring

Rifle	Model
True-Fit Hi Ring	H-Ring
True-Fit Grooved Receiver	GR-Ring
Ruger High Rings M77R, M77RS, M77V	TU
Ruger High Rings No. 1-A, RSI, B, S, H, 3. 72/22	TU
Ruger Redhawk Hunter	UU
Ruger Mini-14/5R, K-Mini Thirty	TU
Ruger Redhawk, Hunter 30 millimeter	UU30

PREMIUM FLASHLIGHT MOUNT

Site Lite Mount	SL

22 RIMFIRE SEE-THRU MOUNT

For All 22 Caliber Rimfire Rifles With Grooved Receiver. Designed For 3/4 and 1" Diameter Scopes. Also Used On Air Rifles 22R

BLACK POWDER MOUNTS (See Thru)
Barrels Must Be Drilled and Tapped

CVA Frontier Carbine, Plainsman, Pennsylvania Long Rifle	GG
CVA Hawken, Hunter Hawken, Mountain Rifle, Blazer	GG
CVA Squirrel Rifle, Kentuckey, Kentuckey Hunter & Kit Rifles	GG
CVA St Louis Hawken (Except 12 Gage and 58 Caliber)	GG
Thompson Center, Renegade, Hawken, White Mountain, New Englander	GG
Traditions Frontier Carbine, Pioneer Rifle	GG
Traditions Pennsylvania Rifle, Hawken Woodsman, Frontier Rifle	GG
Traditions Trapper, Frontier Scout Kits for Hawken Woodsman, Frontier Rifle, Frontier Carbine	GG

BLACK POWDER SEE-THRU MOUNTS NO DRILLING OR TAPPING

Thompson Center Hawken (Ultra Precision See Thru Mounts Included)	TCH
Thompson Center Renegade (Ultra Precision See Thru Mounts Included)	TCR
CVA Stalker Rifle / Carbine	ii
CVA Apollo 90 Rifle/Carbine	ii
CVA Apollo Sporter	ii
CVA Shadow Rifle	ii
CVA Tracker Carbine	ii
CVA Hawken Deerslayer Rifle / Carbine	ii
CVA Trophy Carbine	ii
CVA Frontier Hunter Carbine	ii
Knight MK - 85, BK - 90	ii

SHOTGUN MOUNTS (No Drilling or Tapping)

Mossberg 500	500
Remington 870	870
Remington 1100	1100

WILLIAMS TWILIGHT SCOPES

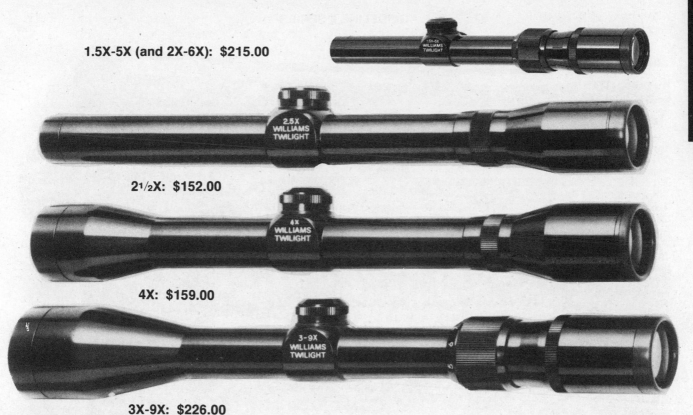

1.5X-5X (and 2X-6X): $215.00

2½X: $152.00

4X: $159.00

3X-9X: $226.00

The "Twilight" series of scopes was introduced to accommodate those shooters who want a high-quality scope in the medium-priced field. The "Twilight" scopes are waterproof and shockproof, have coated lenses and are nitrogen-filled. Resolution is sharp and clear. All "Twilight" scopes have a highly polished, rich, black, hard anodized finish.

There are five models available: the 2½x, the 4x, the 1.5x-5x, the 2x-6x, and the 3x-9x. They are available in T-N-T reticle only (which stands for "thick and thin"). Pistol scopes are also available in 1.5X and 2X (black only). **Price: $158.00**

			1.5X-5X		2X-6X		3X-9X	
OPTICAL SPECIFICATIONS	**2.5X**	**4X**	**At 1.5X**	**At 5X**	**At 2X**	**At 6X**	**At 3X**	**At 9X**
Clear aperture of objective lens	20mm	32mm	20mm	Same	32mm	Same	40mm	Same
Clear aperture of ocular lens	32mm	32mm	32mm	Same	32mm	Same	32mm	Same
Exit Pupil	8mm	8mm	13.3mm	4mm	16mm	5.3mm	13.3mm	44.4mm
Relative Brightness	64	64	177	16	256	28	161.2	17.6
Field of view (degree of angle)	6°10'	5°30'	11°	4°	8°30'	3°10'	7°	2°20'
Field of view at 100 yards	32'	29'	57¾'	21'	45½'	16¾'	36½'	12¾'
Eye Relief	3.7"	3.6"	3.5"	3.5"	3"	3"	3.1"	2.9"
Parallax Correction (at)	50 yds.	100 yds.	100 yds.	Same	100 yds.	Same	100 yds.	Same
Lens Construction	9	9	10	Same	11	Same	11	Same
MECHANICAL SPECIFICATIONS								
Outside diameter of objective end	1.00"	1.525"	1.00"	Same	1.525"	Same	1.850"	1.850"
Outside diameter of ocular end	1.455"	1.455"	1.455"	Same	1.455"	Same	1.455"	Same
Ouside diameter of tube	1"	1"	1"	Same	1"	Same	1"	Same
Internal adjustment graduation	¼ min.	¼ min.	¼ min.	Same	¼ min.	Same	¼ min.	
Minimum internal adjustment	75 min.	75 min.	75 min.	Same	75 min.	Same	60 min.	Same
Finish			Glossy Hard Black Anodized					
Length	10"	11¾"	10¾"	Same	11½"	11½"	12¾"	12¾"
Weight	8½ oz.	9½ oz.	10 oz.	Same	11½ oz.	Same	13½ oz.	Same

WILLIAMS

GUIDELINE II SERIES

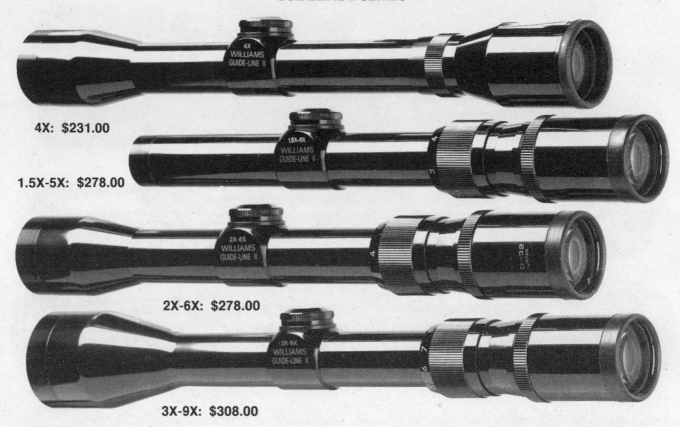

4X: $231.00

1.5X-5X: $278.00

2X-6X: $278.00

3X-9X: $308.00

Patterned after the popular Twilight Series, Williams' new Guideline II Series features silent adjustment screws, streamlined adjustment caps and power adjustment rings. Fully multi-coated lenses ensure superior light gathering and brightness. Comes equipped with T-N-T reticle and a choice of matte or glossy black finish.

		1.5X–5X		2X–6X		3X–9X	
OPTICAL SPECIFICATIONS	**4X**	**At 1.5X**	**At 5X**	**At 2X**	**At 6X**	**At 3X**	**At 9X**
Clear aperture of objective lens	32mm	20mm	Same	32mm	Same	40mm	Same
Clear aperture of ocular lens	32mm	32mm	Same	32mm	Same	32mm	Same
Exit Pupil	8mm	13.3mm	4mm	16mm	5.3mm	13.3mm	44.4mm
Relative Brightness	64	177	16	256	28	161.2	17.6
Field of view (degree of angle)	5°30'	11°	4°	8°30'	3°10'	7°	2°20'
Field of view at 100 yards	29'	57³/₄'	21'	45¹/₂'	16³/₄'	36¹/₂'	12³/₄'
Eye Relief	3.6"	3.5"	3.5"	3"	3"	3.1"	2.9"
Parallax Correction (at)	100 yds.	100 yds.	Same	100 yds.	Same	100 yds.	Same
Lens Construction	9	10	Same	11	Same	11	Same
MECHANICAL SPECIFICATIONS							
Outside diameter of objective end	1.525"	1.00"	Same	1.525"	Same	1.850"	1.850"
Outside diameter of ocular end	1.455"	1.455"	Same	1.455"	Same	1.455"	Same
Outside diameter of tube	1"	1"	Same	1"	Same	1"	Same
Internal adjustment graduation	1/4 min.	1/4 min.	Same	1/4 min.	Same	1/4 min.	Same
Minimum internal adjustment	75 min.	75 min.	Same	75 min.	Same	60 min.	Same
Finish				Glossy Hard Black Anodized			
Length	11³/₄"	10³/₄"	Same	11¹/₂"	11¹/₂"	12³/₄"	12³/₄"
Weight	9¹/₂ oz.	10 oz.	Same	11¹/₂ oz.	Same	13¹/₂ oz.	Same

ZEISS RIFLESCOPES

THE C-SERIES

The C-Series was designed by Zeiss specifically for the American hunter. It is based on space-age alloy tubes with integral objective and ocular bells, and an integral adjustment turret. This strong, rigid one-piece construction allows perfect lens alignment, micro-precise adjustments and structural integrity. Other features include quick focusing, a generous 3¹/₂″ of eye relief, rubber armoring, T-Star multi-layer coating, and parallax setting (free at 100 yards).

DIATAL-C 10X36T
$610.00

DIAVARI-C 3-9X36T
$760.00

DIATAL-C 6X32T
$540.00

DIATAL-C 4X32T
$520.00

DIAVARI-C 1.5-4.5X18T
$695.00

SPECIFICATIONS

	4X32	6X32	10X36	3-9X36		C1.5-4.5X18	
Magnification	4X	6X	10X	3X	9X	1.5X	4.5X
Objective Diameter (mm)/(inch)	1.26″	1.26″	1.42″	1.42″		15.0/0.6	18.0/0.7
Exit Pupil	0.32″	0.21″	0.14″	0.39″	0.16″	10.0	4.0
Twilight Performance	11.3	13.9	19.0	8.5	18.0	4.2	9.0
Field of View at 100 yds.	30′	20′	12′	36′	13′	72′	27′
Eye Relief	3.5″	3.5″	3.5″	3.5″	3.5″	3.5″	
Maximum Interval Adjustment (elevation and windage (MOA)	80	80	50	50		10.5′ @ 100 yds.	
Click-Stop Adjustment 1 click = 1 interval (MOA)	¹/₄	¹/₄	¹/₄	¹/₄		.36″ @ 100 yds.	
Length	10.6″	10.6″	12.7″	11.2″		11.8″	
Weight approx. (ounces)	11.3	11.3	14.1	15.2		13.4	
Tube Diameter	1″	1″	1″	1″		1″	
Objective Tube Diameter	1.65″	1.65″	1.89″	1.73″		1″	
Eyepiece O.D.	1.67″	1.67″	1.67″	1.67″		1.8″	

ZEISS RIFLESCOPES

THE "Z" SERIES

These new Zeiss riflescopes feature a surface that is harder and more resistant to abrasion and mechanical damage than the multiple coatings or bluings. The black, silken matte finish suppresses reflections and prevents finger prints. All optical elements are provided with the Zeiss T Star multi-coating. The size of the reticles in the Diavari Z-types changes with the power set; therefore, the ratio of the reticle size to the target size always remains the same.

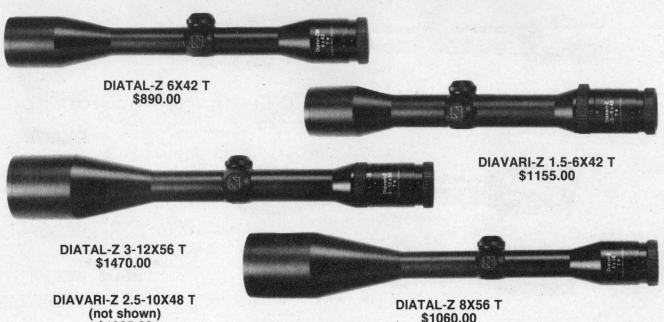

DIATAL-Z 6X42 T
$890.00

DIAVARI-Z 1.5-6X42 T
$1155.00

DIATAL-Z 3-12X56 T
$1470.00

DIAVARI-Z 2.5-10X48 T
(not shown)
$1365.00

DIATAL-Z 8X56 T
$1060.00

ZM/Z SERIES RIFLESCOPE SPECIFICATIONS

Model	Diatal-ZM/Z 6X42 T	Diavari-ZM/Z 1.5-6X42 T	Diavari-ZM/Z 3-12X56 T	Diatal-ZM/Z 8X56 T	Diavari-ZM/Z 2.5-10X48 T	Diavera-ZM/Z 1.25-4X24
Magnification	6X	1.5X 6X	3X 12X	8X	2.5X-10X	1.25-4X
Effective obj. diam.	42mm/1.7″	19.5/0.8″ 42/1.7″	38/1.5″ 56/2.2″	56mm/2.2″	33/1.30″ 48/1.89″	NA
Diameter of exit pupil	7mm	13mm 7mm	12.7mm 4.7mm	7mm	13.2mm 4.8mm	12.6mm 6.3mm
Twilight factor	15.9	4.2 15.9	8.5 25.9	21.2	7.1 21.9	3.54 9.6
Field of view at 100 m/ ft. at 100 yds.	6.7m/20.1′	18/54.0′ 6.5/19.5′	9.2/27.6′ 3.3/9.9′	5m/15.0′	11.0/33.0 3.9/11.7	32 10
Approx. eye relief	8cm/3.2″	8cm/3.2″	8cm/3.2″	8cm/3.2″	8cm/3.2″	8cm/3.2″
Click-stop adjustment 1 click = (cm at 100 m)/ (inch at 100 yds.)	1cm/0.36″	1cm/0.36″	1cm/0.36″	1cm/0.36″	1cm/0.36″	1cm/0.36″
Max. adj. (elev./wind.) at 100 m (cm)/at 100 yds.	187	190	95	138	110/39.6	300
Center tube dia.	25.4mm/1″	30mm/1.18″	30mm/1.18″	25.4mm/1″	30mm/1.18″	30mm/1.18″
Objective bell dia.	48mm/1.9″	48mm/1.9″	62mm/2.44″	62mm/2.44″	54mm/2.13″	NA
Ocular bell dia.	40mm/1.57″	40mm/1.57″	40mm/1.57″	40mm/1.57″	40mm/1.57″	NA
Length	324mm/12.8″	320mm/12.6″	388mm/15.3″	369mm/14.5″	370mm/14.57″	290mm/11.46″
Approx. weight: ZM	350g/15.3 oz.	586g/20.7 oz.	765g/27.0 oz.	550g/19.4 oz.	715g/25.2 oz.	490g/17.3 oz.
Z	400g/14.1 oz.	562g/19.8	731g/25.8 oz.	520g/18.3 oz.	680g/24 oz.	NA

Ammunition

For addresses and phone numbers of manufacturers and distributors included in this section, turn to *DIRECTORY OF MANUFACTURERS AND SUPPLIERS* at the back of the book.

FEDERAL AMMUNITION

The following pages include Federal's new or recent lines of cartridges and shotshells for 1994—1995. For a complete listing of all Federal ammunition, call or write the Federal Cartridge Company (see Directory of Manufacturers in the Reference section for address and phone number).

PREMIUM CARTRIDGES

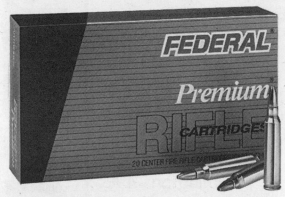

A Sierra 40-grain hollow-point bullet is featured in Federal's 223 Rem. Varmint load.

Federal's 30-06 Springfield cartridge (Premium Safari) is loaded with a Trophy Bonded 165-grain bullet.

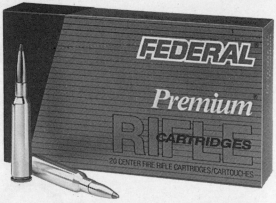

A 140-grain Nosler Partition bullet is a recent addition to Federal's 6.5×55 Swedish cartridge.

Federal's new Premium Trophy Bonded bullet retains up to 95 percent of its weight to ensure optimum expansion and penetration on big-game animals.

This varmint cartridge for the 22-250 Rem. delivers a muzzle velocity of 4000 fps.

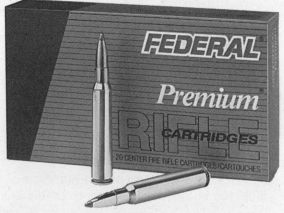

Federal's 7×64 Brenneke with a 160-grain Nosler Partition bullet is part of the Premium line.

FEDERAL AMMUNITION

PREMIUM CARTRIDGES/SHOTSHELL PRIMERS

A Sierra 60-grain hollow-point bullet is available to all 243 Win. shooters.

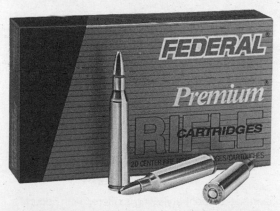

This 25-06 Rem. Varmint load features a Sierra 90-grain hollow-point bullet.

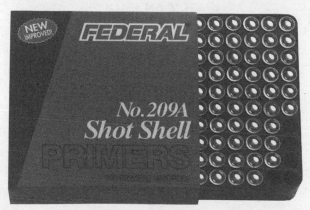

The new Federal 209A primer replaces the 209. It provides instant, full and dependable ignition—and uniform ballistics.

CLASSIC CARTRIDGES

Federal offers a 30-06 Springfield cartridge with a 150-grain Hi-Shok soft-point bullet.

A 230-grain Hi-Shok jacketed hollow-point bullet complements Federal's 45 Auto cartridge.

Federal's Classic 22 LR High Velocity cartridges use 40-grain solid copper-plated bullets with a muzzle velocity of 1255 fps.

FEDERAL AMMUNITION

GOLD MEDAL LINE

Federal's Auto Match cartridge with its 185-grain Semi-Wadcutter bullet joins the Gold Medal line.

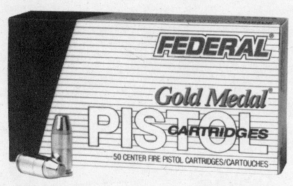

Federal's Gold Medal line, which includes this 9mm Luger Match cartridge, now has a 356 TSW with a 147-grain FMJ-SWC Match bullet.

Two new paper trap loads—an Extra-Lite Skeet load and two new 24-gram International loads—have been added to the Gold Medal line.

PREMIUM SHOTSHELLS

This 20-gauge 3″ Premium shotshell contains 1¼ oz. of copper-plated No. 6 shot.

A copper-plated Sabot-style slug is offered with Federal's 12-gauge 3″ Premium slug load.

AMERICAN EAGLE

The American Eagle line includes this 22 LR cartridge with a 40-grain copper-plated hollow-point bullet.

REMINGTON AMMUNITION

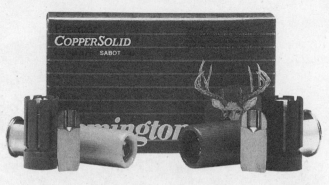

PREMIER STEEL SHOT TARGET LOADS

Remington's new Premier Steel Shot Target Loads are designed to equal the performance of high-grade lead target loads, utilizing a one-piece, uni-body hull with brass head for easy resizing and reloadability. A high density polyethylene shot cup provides maximum protection for shotgun bore and chokes. Remington's Premier steel shotshell design delivers equal or better pellet counts with similar recoil sensation when compared to Premier lead target loads.

Four different specifications of 12 gauge are available, all in 2³/₄" shells, featuring 1¹/₈ ounce of 6¹/₂, 7 and 8 shot. The 20-gauge skeet load features 8 shot and weighs a full ⁷/₈ ounce. All five specifications are safe for use in modern competition grade shotguns.

20 GAUGE COPPER SOLID SABOT SLUG

Remington's Copper Solid Sabot Slug is the industry's first 100% solid copper projectile designed for shotgun use. Originally introduced in a 12-gauge 2³/₄" shell, the Copper Solid gained a reputation as the most accurate slug on the market, delivering 2¹/₂" groups at 100 yards consistently. This accuracy comes from the precision machining of the slug from solid copper rather than swaging from soft lead—to produce greater dimensional uniformity and optimum concentricity.

The original 12-gauge slug weighs approximately 1 oz. and offers a muzzle velocity of 1450 feet per second. The result is better down-range energy and flatter trajectories than other sabot slugs, regardless of shell length. The new slug measures a full .45 caliber and at 317 grains is the heaviest 20-gauge sabot slug available. It provides 1480 foot-pounds of energy with a muzzle velocity of 1450 fps. It delivers sub-3" groups at 100 yards consistently when used with Remington's new cantilever fully rifled 20- gauge barrels.

Also available:

GOLDEN SABER™ HPJ
SPIRAL NOSE CUT DESIGN

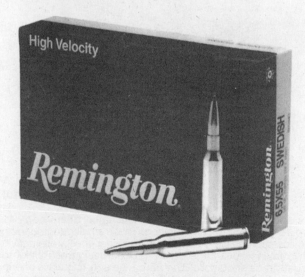

6.5 X 55 SWEDISH CENTERFIRE AMMO
**High-Velocity 140-grain, Pointed
Soft-Point, Core-Lokt™ Loading**

**PREMIER MAGNUM
TURKEY LOADS**

Remington has expanded its line of Premier Turkey shotgun ammunition with a new 12-gauge Magnum #5 specification, plus a new 20-gauge offering. This load utilizes the hardest grade copper-plated shot available. A granulated polyethylene buffer cushions the shot charge and protects the pellets from deformation even through the tightest turkey choke tubes, including Remington's Super Extra Full Rem Choke tube.

The 12-gauge 2³/₄″ Magnum loading with #5 shot features

a 1¹/₂ oz. payload with a velocity of 1260 fps. This #5 loading provides greater pattern energy than #6 and better pattern density over #4 shot.

Also available is a 20-gauge 3″ loading with 1¹/₄ oz. of #6 shot for a dense, small-gauge turkey hunting option. Reaching a velocity of 1185 fps, these Premier Turkey loads offer a lower recoil alternative to current 10- or 12-gauge magnum loading, particularly for younger turkey hunters.

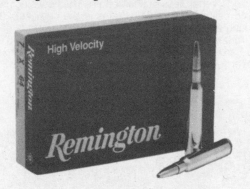

**7 X 64 CENTERFIRE AMMO
175- or 140-Grain, Pointed
Soft-Point, Core-Lokt™ Loading**

**22 LONG RIFLE SUBSONIC AMMO
39-Grain, Lead Hollow Point,
1050 fps Velocity**

**SPORTING CLAYS TARGET LOADS
(not shown)**

For sporting clays enthusiasts, Remington offers new .410 and 28-gauge target ammunition. These small-gauge loadings provide sporting clays shooters with higher energy alternatives, while taking advantage of the proven design of Remington target-grade shotshells. The 28-gauge sporting clays offering is loaded with target-grade #8 shot for added long-range pellet energy and enhanced pattern performance. The 28-gauge 2³/₄″ shells feature a new one-piece unibody hull with an all-brass head for longer reloading life, plus the proven Remington Power

Piston wad, which cushions and protects the shot column. This sporting clays load is built to standard skeet specifications (2 dram, ³/₄ ounce, 1200 fps).

The new Remington .410 Sporting Clays offering is loaded with #8¹/₂ shot for extra pellet energy and minimal loss in pattern density. Loaded to standard skeet specifications (Max. dram, ¹/₂ oz. shot, 1225 fps), these loads share all components with Remington's current .410 skeet loading.

Super-X® Game Loads—High Brass

Gauge	Symbol	Length of Shell In.	Powder Dram Equivalent	Velocity fps @ 3 ft.	Oz. Shot	Standard Shot Sizes
12	X12	2-3/4	3-3/4	1330	1-1/4	2,4,5,6,7-1/2,8
16	X16H	2-3/4	3-1/4	1295	1-1/8	4,6,7-1/2
20	X20	2-3/4	2-3/4	1220	1	4,5,6,7-1/2,8
28	X28	2-3/4	2-1/4	1295	3/4	6,7-1/2
28	X28H	2-3/4	Max	1125	1	6,7-1/2,8
410	X413	3	Max	1135	11/16	4,6,7-1/2
410	X41	2-1/2	Max	1225	1/2	4,6,7-1/2

Super-X Small Game Hunter™ Loads

Gauge	Symbol	Length of Shell In.	Powder Dram Equivalent	Velocity fps @ 3 ft.	Oz. Shot	Standard Shot Sizes
New 12	XHS12	2-3/4	3-1/4	1220	1-1/4	6,7-1/2,8
12	XH12	2-3/4	3-1/4	1255	1-1/8	6,7-1/2,8
20	XH20	2-3/4	2-1/2	1165	1	6,7-1/2,8

Double X® Magnum Turkey Loads—Copperplated, Buffered Shot

Gauge	Symbol	Length of Shell In.	Powder Dram Equivalent	Velocity fps @ 3 ft.	Oz. Shot	Standard Shot Sizes
10	X103XCT	3-1/2	4-1/2	1210	2-1/4	6
12	XXT12L	3-1/2	Mag	1150	2-1/4	4,6
12	X123MXCT	3	Mag	1125	2	4,5,6
12	X12HXCT	2-3/4	Mag	1250	1-5/8	4,5,6

Double X Magnum Buckshot Loads—Copperplated, Buffered Shot

Gauge	Symbol	Length of Shell In.	Powder Dram Equivalent	Velocity fps @ 3 ft.	Total Pellets	Standard Shot Sizes
10	X10C00B	3-1/2	N/A	1150	18	00 Buck
12	X123C000B	3	N/A	1225	10	000 Buck
12	X12XC3B5	3	N/A	1210	15	00 Buck
12	X12XC0B5	2-3/4	N/A	1290	12	00 Buck
12	X12C1B	2-3/4	N/A	1075	20	1 Buck
12	X12XCMB5	3	N/A	1210	41	4 Buck
20	X203C3B	3	N/A	1150	24	3 Buck

Double X Magnum Game Loads—Copperplated, Buffered Shot

Gauge	Symbol	Length of Shell In.	Powder Dram Equivalent	Velocity fps @ 3 ft.	Oz. Shot	Standard Shot Sizes
10	X103XC	3-1/2	4-1/2	1210	2-1/4	4
12	X123XC	3	4	1210	1-7/8	BB,2,4,6
12	X12MXC	3	4	1280	1-5/8	4,6
12	X12XC	2-3/4	3-3/4	1260	1-1/2	BB,2,4,5,6
16	X16XC	2-3/4	3-1/4	1260	1-1/4	4,6
20	X203XC	3	3	1185	1-1/4	4,6
20	X20XC	2-3/4	2-3/4	1175	1-1/8	6,7-1/2

Super-X Drylok™* Steel Non-Toxic Waterfowl Loads

Gauge	Symbol	Length of Shell In.	Powder Dram Equivalent	Velocity fps @ 3 ft.	Oz. Shot	Standard Shot Sizes
12	XSD12	2-3/4	Max	1375	1	2,4,6
12	XS12	2-3/4	Max	1365	1-1/8	2,4,6
16	XS16	2-3/4	Max	1300	7/8	2,4
20	XS20	2-3/4	Max	1425	3/4	4,6

*except 16 gauge.

Super-X Drylok Steel Non-Toxic Magnum Waterfowl Loads

Gauge	Symbol	Length of Shell In.	Powder Dram Equivalent	Velocity fps @ 3 ft.	Oz. Shot	Standard Shot Sizes
10	XSM10	3-1/2	Mag	1260	1-3/4	BB,1,2
12	XSM12L	3-1/2	Mag	1300	1-9/16	1,2
12	XSM123	3	Mag	1265	1-3/8	BB,1,2,3,4
12	XSV123	3	Mag	1375	1-1/4	BB,1,2,3,4,5
12	XSM12	2-3/4	Mag	1275	1-1/4	BB,1,2,3,4,6
20	XSM203	3	Mag	1330	1	2,3,4

Super-X Drylok Steel Non-Toxic Copperplated Magnum Goose Loads

Gauge	Symbol	Length of Shell In.	Powder Dram Equivalent	Velocity fps @ 3 ft.	Oz. Shot	Standard Shot Sizes
10	XSC10	3-1/2	Mag	1350	1-5/8	T,BBB
12	XSC12L	3-1/2	Mag	1300	1-9/16	T,BBB
12	XSC123	3	Mag	1375	1-1/4	T,BBB
12	XSC12	2-3/4	Mag	1300	1-1/8	T,BBB

Super-X Buckshot Loads With Buffered Shot

Gauge	Symbol	Length of Shell In.	Powder Dram Equivalent	Velocity fps @ 3 ft.	Total Pellets	Standard Shot Sizes
12	XB12L00	3-1/2 Magnum	N/A	1200	18	00 Buck
12	XB1231	3 Magnum	N/A	1040	24	1 Buck
12	XB12300	3 Magnum	N/A	1210	15	00 Buck
12	XB12000	2-3/4	N/A	1325	8	000 Buck
New12	XB120012	2-3/4 Magnum	N/A	1290	12	00 Buck
12	XB120025	2-3/4	N/A	1325	9	00 Buck
12	XB1200	2-3/4	N/A	1325	9	00 Buck
12	XB120	2-3/4	N/A	1275	12	0 Buck
12	XB121	2-3/4	N/A	1250	16	1 Buck
12	XB124	2-3/4	N/A	1325	27	4 Buck
16	XB161	2-3/4	N/A	1225	12	1 Buck
20	XB203	2-3/4	N/A	1200	20	3 Buck

Supreme® HI-IMPACT® Sabot Slug Loads

Gauge	Symbol	Length of Shell In.	Powder Dram Equivalent	Velocity fps @ 3 ft.	Slug Wt. (Oz.)	Slug Type
12	SRSH123	3	N/A	1550	1	Sabot
12	SRSH12	2-3/4	N/A	1450	1	Sabot

Super-X BRI® Sabot Slug Loads

Gauge	Symbol	Length of Shell In.	Powder Dram Equivalent	Velocity fps @ 3 ft.	Slug Wt. (Oz.)	Slug Type
12	XRS123	3	N/A	1300	1	Sabot
12	XRS12	2-3/4	N/A	1200	1	Sabot
20	XRS20	2-3/4	N/A	1400	5/8	Sabot

Super-X Rifled Slug Loads

Gauge	Symbol	Length of Shell In.	Powder Dram Equivalent	Velocity fps @ 3 ft.	Slug Wt. (Oz.)	Slug Type
12	X123RS15	3 Magnum	N/A	1760	1	Rifled
12	X12RS15	2-3/4	N/A	1600	1	Rifled HP
16	X16RS5	2-3/4	N/A	1600	4/5	Rifled HP
20	X20RSM5	2-3/4	N/A	1600	3/4	Rifled HP
410	X41RS5	2-1/2	N/A	1830	1/5	Rifled HP

Double A® Target Loads

Gauge	Symbol	Length of Shell In.	Powder Dram Equivalent	Velocity fps @ 3 ft.	Oz. Shot	Standard Shot Sizes
12	AAST12 (Steel Target Load)	2-3/4	3	1235	1	7 Steel
New12	AANL12 (International Target Load)	2-3/4	3-1/4	1325	24 gm.	7-1/2,8,9
12	AAP12 (Super Pigeon®)	2-3/4	3-1/4	1220	1-1/4	7-1/2,8
12	AAH12 (Super-Handicap)	2-3/4	3	1200	1-1/8	7-1/2,8,9
12	AAM12 (Heavy Target Load)	2-3/4	3	1200	1-1/8	7-1/2,8,9
12	AASL12 (Super-Lite)	2-3/4	2-3/4	1125	1-1/8	7-1/2,8,8-1/2,9
12	AA12 (Target Load)	2-3/4	2-3/4	1145	1-1/8	7-1/2,8,9
12	AAL12 (Xtra-Lite™)	2-3/4	2-3/4	1180	1	7-1/2,8,9
20	AAH20 (Heavy Target Load)	2-3/4	2-1/2	1165	1	8,9
20	AA20 (Target Load)	2-3/4	2-1/2	1200	7/8	8,9
28	AA28 (Target Load)	2-3/4	2	1200	3/4	9
410	AA410 (Target Load)	2-1/2	Max	1200	1/2	9

WINCHESTER AMMUNITION

SUPER-X RIMFIRE

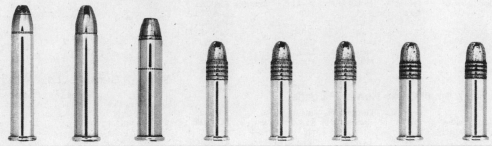

Cartridge Bullet Type		22 Win. Mag. JHP	22 Win. Mag. FMJ	**NEW** 22 WRF LFN*	22 Long Rifle CP – LRN*	22 Long Rifle CP – LRN*	**NEW** 22 Long Rifle CP – LRN*	22 Long Rifle CP – LHP*	22 Long Rifle CP – LHP*
Symbol		X22MH	X22M	22WRF	X22LR	X22LR1	X22LRB	X22LRH	X22LRH1
Bullet Wt. (grs.)		40	40	45	40	40	40	37	37
Rounds per box		50	50	50	50	100	325	50	100
RIFLE	Velocity (fps) — Muzzle	1910	1910	1320	1255	1255	1255	1280	1280
	Velocity (fps) — 100 yds.	1326	1326	1055	1017	1017	1017	1015	1015
	Energy (ft.-lbs.) — Muzzle	324	324	174	140	140	140	135	135
	Energy (ft.-lbs.) — 100 yds.	156	156	111	92	92	92	85	85
	Mid-Range Traj. 100 yds.	1.7	1.7	3.3	3.6	3.6	3.6	3.5	3.5
HANDGUN	Barrel Length (in.)	6.5	6.5	–	6	6	6	–	–
	Muzzle Velocity (fps)	1480	1480	–	1060	1060	1060	–	–
	Muzzle Energy (ft.-lbs.)	195	195	–	100	100	100	–	–
Game Guide		P,S,H	P,S,H	P,S,T	P,S,T,H	P,S,T,H	P,S,T,H	P,S	P,S

Cartridge Bullet Type		Super Silhouette® 22 Long Rifle LTC	T22® 22 Long Rifle LRN – SV	22 Long Rifle #12 Shot	22 Short Blank	22 Short CP – LRN*	22 Short C.B. LRN
Symbol		XS22LR	XT22LR	X22LRS	22BL	X22S	X22SCB
Bullet Wt. (grs.)		42	40	–	Blk. Pwd.	29	29
Rounds per box		50	50	50	50	50	50
RIFLE	Velocity (fps) — Muzzle	1220	1150	–	–	1095	725
	Velocity (fps) — 100 yds.	1003	976	–	–	903	–
	Energy (ft.-lbs.) — Muzzle	139	117	–	–	77	–
	Energy (ft.-lbs.) — 100 yds.	94	85	–	–	52	–
	Mid-Range Traj. 100 yds.	3.6	4.0	–	–	4.5	–
HANDGUN	Barrel Length (in.)	6/10	6	–	–	6	–
	Muzzle Velocity (fps)	1025/1105	950	–	–	1010	–
	Muzzle Energy (ft.-lbs.)	98/114	80	–	–	66	–
Game Guide		P,S,T,H	P,T	–	–	P,S	P,T

*Lubaloy® Coated

Game Guide
P – Plinking   S – Small Game   T – Target   H – Silhouette

Abbreviations:
Blk. Pwd.	Black Powder	JHP	Jacketed Hollow Point	LRN	Lead Round Nose
CP	Copperplated	LFN	Lead Flat Nose	LTC	Lead Truncated Cone
FMJ	Full Metal Jacket	LHP	Lead Hollow Point	SV	Standard Velocity

Ballistics

FEDERAL BALLISTICS

CENTERFIRE RIFLE

CLASSIC® CENTERFIRE RIFLE BALLISTICS (Approximate)

Usage Key: [1] = Varmints, predators, small game   [2] = Medium game   [3] = Large, heavy game

USAGE	FEDERAL LOAD NO.	CALIBER	BULLET WGT. GRAINS	BULLET WGT. GRAMS	BULLET STYLE**	FACTORY PRIMER NO.	VELOCITY MUZZLE	100 YDS.	200 YDS.	300 YDS.	400 YDS.	500 YDS.	ENERGY MUZZLE	100 YDS.	200 YDS.	300 YDS.	400 YDS.	500 YDS.
[1]	222A	222 Rem. (5.56x43mm)	50	3.24	Hi-Shok Soft Point	205	3140	2600	2120	1700	1350	1110	1095	750	500	320	200	135
[5]	222B		55	3.56	Hi-Shok FMJ Boat-tail	205	3020	2740	2480	2230	1990	1780	1115	915	750	610	485	385
[1]	223A	223 Rem. (5.56x45mm)	55	3.56	Hi-Shok Soft Point	205	3240	2750	2300	1910	1550	1270	1280	920	650	445	295	195
[5]	223B		55	3.56	Hi-Shok FMJ Boat-tail	205	3240	2950	2670	2410	2170	1940	1280	1060	875	710	575	460
[1]	22250A	22-250 Rem.	55	3.56	Hi-Shok Soft Point	210	3680	3140	2660	2220	1830	1490	1655	1200	860	605	410	270
[1]	243A	243 Win. (6.16x51mm)	80	5.18	Hi-Shok Soft Point	210	3350	2960	2590	2260	1950	1670	1995	1550	1195	905	675	495
[2]	243B		100	6.48	Hi-Shok Soft Point	210	2960	2700	2450	2220	1990	1790	1945	1615	1330	1090	880	710
[1]	6A	6mm Rem.	80	5.18	Hi-Shok Soft Point	210	3470	3060	2690	2350	2040	1750	2140	1665	1290	980	735	540
[2]	6B		100	6.48	Hi-Shok Soft Point	210	3100	2830	2570	2330	2100	1890	2135	1775	1470	1205	985	790
[2]	2506B	25-06 Rem.	117	7.58	Hi-Shok Soft Point	210	2990	2730	2480	2250	2030	1830	2320	1985	1645	1350	1100	885
[2]	270A	270 Win.	130	8.42	Hi-Shok Soft Point	210	3060	2800	2560	2330	2110	1900	2700	2265	1890	1565	1285	1045
[2]	270B		150	9.72	Hi-Shok Soft Point RN	210	2850	2500	2180	1890	1620	1390	2705	2085	1585	1185	870	640
[2]	7A	7mm Mauser	175	11.34	Hi-Shok Soft Point RN	210	2440	2140	1860	1600	1380	1200	2315	1775	1340	1000	740	565
[2]	7B	(7x57mm Mauser)	140	9.07	Hi-Shok Soft Point	210	2660	2450	2260	2070	1890	1730	2200	1865	1585	1330	1110	930
[2]	280B	280 Rem.	150	9.72	Hi-Shok Soft Point	210	2890	2670	2460	2260	2060	1880	2780	2370	2015	1695	1420	1180
[2]	7RA	7mm Rem. Magnum	150	9.72	Hi-Shok Soft Point	215	3110	2830	2570	2320	2090	1870	3220	2670	2200	1790	1450	1160
[3]	7RB		175	11.34	Hi-Shok Soft Point	215	2860	2650	2440	2240	2060	1880	3180	2720	2310	1960	1640	1370
[1]	30CA	30 Carbine (7.62x33mm)	110	7.13	Hi-Shok Soft Point RN	205	1990	1570	1240	1040	920	840	965	600	375	260	210	175
[2]	76239B	7.62x39mm Soviet	123	7.97	Hi-Shok Soft Point	210	2300	2030	1780	1550	1350	1200	1445	1125	860	655	500	395
[2]	3030A	30-30 Win.	150	9.72	Hi-Shok Soft Point FN	210	2390	2020	1680	1400	1180	1040	1900	1355	945	650	460	355
[2]	3030B		170	11.01	Hi-Shok Soft Point FN	210	2200	1900	1620	1380	1190	1060	1830	1355	990	720	535	425
[1]	3030C		125	8.10	Hi-Shok Hollow Point	210	2570	2090	1660	1320	1080	960	1830	1210	770	480	320	260
[2]	300A	300 Savage	150	9.72	Hi-Shok Soft Point	210	2630	2350	2100	1850	1630	1430	2305	1845	1460	1145	885	685
[2]	300B		180	11.66	Hi-Shok Soft Point	210	2350	2140	1940	1750	1570	1410	2205	1825	1495	1215	985	800
[2]	308A	308 Win. (7.62x51mm)	150	9.72	Hi-Shok Soft Point	210	2820	2530	2260	2010	1770	1560	2650	2140	1705	1345	1050	810
[2]	308B		180	11.66	Hi-Shok Soft Point	210	2620	2390	2180	1970	1780	1600	2745	2290	1895	1555	1270	1030
[2]	3006A	30-06 Springfield	150	9.72	Hi-Shok Soft Point	210	2910	2620	2340	2080	1840	1620	2820	2280	1825	1445	1130	875
[3]	3006B	(7.62x63mm)	180	11.66	Hi-Shok Soft Point	210	2700	2470	2250	2040	1850	1660	2915	2435	2025	1665	1360	1105
[1]	3006C		125	8.10	Hi-Shok Soft Point	210	3140	2780	2450	2140	1850	1600	2735	2145	1660	1270	955	705
[3]	3006H		220	14.25	Hi-Shok Soft Point RN	210	2410	2130	1870	1630	1420	1250	2835	2215	1705	1300	985	760
[3]	3006J		180	11.66	Hi-Shok Soft Point RN	210	2700	2350	2020	1730	1470	1250	2915	2200	1630	1190	860	620
[3]	300WB	300 Win. Magnum	180	11.66	Hi-Shok Soft Point	215	2960	2750	2540	2340	2160	1980	3500	3010	2580	2195	1860	1565
[2]	303A	303 British	180	11.66	Hi-Shok Soft Point	210	2460	2230	2020	1820	1630	1460	2420	1995	1625	1315	1060	850
[2]	303B		150	9.72	Hi-Shok Soft Point	210	2690	2440	2210	1980	1780	1590	2400	1980	1620	1310	1055	840
[2]	32A	32 Win. Special	170	11.01	Hi-Shok Soft Point	210	2250	1920	1630	1370	1180	1040	1910	1395	1000	710	520	410
[2]	*8A	8mm Mauser (8x57mm JS Mauser)	170	11.01	Hi-Shok Soft Point	210	2360	1970	1620	1330	1120	1000	2100	1465	995	670	475	375
[3]	338C	338 Win. Magnum	225	14.58	Hi-Shok Soft Point	215	2780	2570	2370	2180	2000	1830	3860	3305	2815	2380	2000	1670
[2]	357G	357 Magnum	180	11.66	Hi-Shok Hollow Point	100	1550	1160	980	860	770	680	960	535	385	295	235	185
[2]	35A	35 Rem.	200	12.96	Hi-Shok Soft Point	210	2080	1700	1380	1140	1000	910	1920	1280	840	575	445	370
[3]	375A	375 H&H Magnum	270	17.50	Hi-Shok Soft Point	215	2690	2420	2170	1920	1700	1500	4340	3510	2810	2220	1740	1355
[4]	375B		300	19.44	Hi-Shok Soft Point	215	2530	2270	2020	1790	1580	1400	4265	3425	2720	2135	1665	1295
[2]	44A	44 Rem. Magnum	240	15.55	Hi-Shok Hollow Point	150	1760	1380	1090	950	860	790	1650	1015	640	485	395	330
[2]	4570A	45-70 Government	300	19.44	Hi-Shok Hollow Point	210	1880	1650	1430	1240	1110	1010	2355	1815	1355	1015	810	680

*Only for use in barrels intended for .323 inch diameter bullets. Do not use in 8x57mm J Commission Rifles (M1888) or in sporting or other military arms of .318 inch bore diameter.
**RN = Round Nose   FN = Flat Nose   FMJ = Full Metal Jacket   HP = Hollow Point

GOLD MEDAL® MATCH RIFLE BALLISTICS (Approximate)

USAGE	FEDERAL LOAD NO.	CALIBER	BULLET WGT. GRAINS	BULLET WGT. GRAMS	BULLET STYLE*	FACTORY PRIMER NO.	VELOCITY MUZZLE	100 YDS.	200 YDS.	300 YDS.	400 YDS.	500 YDS.	600 YDS.	700 YDS.	800 YDS.	900 YDS.	1000 YDS.
[5]	GM223M	223 Rem. (5.56x45mm)	69	4.47	Boat-tail HP Match	205M	3000	2720	2460	2210	1980	1760	1560	1390	1240	1130	1060
[5]	GM308M	308 Win. (7.62x51mm)	168	10.88	Boat-tail HP Match	210M	2600	2420	2240	2070	1910	1760	1610	1480	1360	1260	1170
[5]	GM3006M	30-06 Springfield (7.62x63mm)	168	10.88	Boat-tail HP Match	210M	2700	2510	2330	2150	1990	1830	1680	1540	1410	1300	1210

*HP = Hollow Point

FEDERAL BALLISTICS

CENTERFIRE RIFLE

④ = Dangerous game   ⑤ = Target shooting, training, practice

| WIND DRIFT IN INCHES 10 MPH CROSSWIND | | | | | HEIGHT OF BULLET TRAJECTORY IN INCHES ABOVE OR BELOW LINE OF SIGHT IF ZEROED AT ⊕ YARDS. SIGHTS 1.5 INCHES ABOVE BORE LINE. | | | | | | | | | | TEST BARREL LENGTH INCHES | FEDERAL LOAD NO. |
| | | | | | AVERAGE RANGE | | | | LONG RANGE | | | | | | | |
100 YDS.	200 YDS.	300 YDS.	400 YDS.	500 YDS.	50 YDS.	100 YDS.	200 YDS.	300 YDS.	50 YDS.	100 YDS.	200 YDS.	300 YDS.	400 YDS.	500 YDS.		
1.7	7.3	18.3	36.4	63.1	-0.2	⊕	-3.7	-15.3	+0.7	+1.9	⊕	-9.7	-31.6	-71.3	24	222A
0.9	3.4	8.5	16.8	26.3	-0.2	⊕	-3.1	-12.0	+0.6	+1.6	⊕	-7.3	-21.5	-44.6	24	222B
1.4	6.1	15.0	29.4	50.8	-0.3	⊕	-3.2	-12.9	+0.5	+1.6	⊕	-8.2	-26.1	-58.3	24	223A
0.8	3.3	7.8	14.5	24.0	-0.3	⊕	-2.5	-9.9	+0.3	+1.3	⊕	-6.1	-18.3	-37.8	24	223B
1.2	5.2	12.5	24.4	42.0	-0.4	⊕	-2.1	-9.1	+0.1	+1.0	⊕	-6.0	-19.1	-42.6	24	22250A
1.0	4.3	10.4	19.8	33.3	-0.3	⊕	-2.5	-10.2	+0.3	+1.3	⊕	-6.4	-19.7	-42.2	24	243A
0.9	3.6	8.4	15.7	25.8	-0.2	⊕	-3.3	-12.4	+0.4	+1.6	⊕	-7.5	-22.0	-45.4	24	243B
1.0	4.1	9.9	18.8	31.6	-0.3	⊕	-2.2	-9.3	+0.2	+1.1	⊕	-5.9	-18.2	-39.0	24	6A
0.8	3.3	7.9	14.7	24.1	-0.3	⊕	-2.9	-11.0	+0.5	+1.4	⊕	-6.7	-19.8	-40.6	24	6B
0.8	3.4	8.1	15.1	24.9	-0.2	⊕	-3.2	-12.0	+0.6	+1.6	⊕	-7.2	-21.4	-44.0	24	2506B
0.8	3.2	7.6	14.2	23.3	-0.2	⊕	-2.9	-11.2	+0.5	+1.5	⊕	-6.8	-20.0	-41.1	24	270A
1.2	5.3	12.8	24.5	41.3	-0.1	⊕	-4.1	-15.5	+0.9	+2.0	⊕	-9.4	-28.6	-61.0	24	270B
1.5	6.2	15.0	28.7	47.8	-0.1	⊕	-6.2	-22.6	+1.6	+3.1	⊕	-13.3	-40.1	-84.6	24	7A
1.3	3.2	8.2	15.4	23.4	-0.1	⊕	-4.3	-15.4	+1.0	+2.1	⊕	-9.0	-26.1	-52.9	24	7B
0.7	3.1	7.2	13.4	21.9	-0.2	⊕	-3.4	-12.6	+0.7	+1.7	⊕	-7.5	-21.8	-44.3	24	280B
0.8	3.4	8.1	15.1	24.9	-0.3	⊕	-2.9	-11.0	+0.5	+1.4	⊕	-6.7	-19.9	-41.0	24	7RA
0.7	3.1	7.2	13.3	21.7	-0.2	⊕	-3.5	-12.8	+0.7	+1.7	⊕	-7.6	-22.1	-44.9	24	7RB
3.4	15.0	35.5	63.2	96.7	+0.6	⊕	-12.8	-46.9	+3.9	+6.4	⊕	-27.7	-81.8	-167.8	18	30CA
1.5	6.4	15.2	28.7	47.3	+0.2	⊕	-7.0	-25.1	+1.9	+3.5	⊕	-14.5	-43.4	-90.6	20	76239B
2.0	8.5	20.9	40.1	66.1	+0.2	⊕	-7.2	-26.7	+1.9	+3.6	⊕	-15.9	-49.1	-104.5	24	3030A
1.9	8.0	19.4	36.7	59.8	+0.3	⊕	-8.3	-29.8	+2.4	+4.1	⊕	-17.4	-52.4	-109.4	24	3030B
2.2	10.1	25.4	49.4	81.6	+0.1	⊕	-6.6	-26.0	+1.7	+3.3	⊕	-16.0	-50.9	-109.5	24	3030C
1.1	4.8	11.6	21.9	36.3	0	⊕	-4.8	-17.6	+1.2	+2.4	⊕	-10.4	-30.9	-64.4	24	300A
1.1	4.6	10.9	20.3	33.3	+0.1	⊕	-6.1	-21.6	+1.7	+3.1	⊕	-12.4	-36.1	-73.8	24	300B
1.0	4.4	10.4	19.7	32.7	-0.1	⊕	-3.9	-14.7	+0.8	+2.0	⊕	-8.8	-26.3	-54.8	24	308A
0.9	3.9	9.2	17.2	28.3	-0.1	⊕	-4.0	-16.5	+1.1	+2.3	⊕	-9.7	-28.3	-57.8	24	308B
1.0	4.2	9.9	18.7	31.2	-0.2	⊕	-3.6	-13.6	+0.7	+1.8	⊕	-9.0	-24.4	-50.9	24	3006A
0.9	3.7	8.8	16.5	27.1	-0.1	⊕	-4.2	-15.3	+1.0	+2.1	⊕	-9.0	-26.4	-54.0	24	3006B
1.1	4.5	10.8	20.5	34.4	-0.3	⊕	-3.0	-11.9	+0.5	+1.5	⊕	-7.3	-22.3	-47.5	24	3006C
1.4	6.0	14.3	27.2	45.0	-0.1	⊕	-6.2	-22.4	+1.7	+3.1	⊕	-13.1	-39.3	-82.2	24	3006H
1.5	6.4	15.7	30.4	51.2	-0.1	⊕	-4.9	-18.3	+1.1	+2.4	⊕	-11.0	-33.6	-71.9	24	3006J
0.7	2.8	6.6	12.3	20.0	-0.2	⊕	-3.1	-11.7	+0.6	+1.6	⊕	-7.0	-20.3	-41.1	24	300WB
1.1	4.5	10.6	19.9	32.7	0	⊕	-5.5	-19.6	+1.4	+2.8	⊕	-11.3	-33.2	-68.1	24	303A
1.0	4.1	9.6	18.1	29.9	-0.1	⊕	-4.4	-15.9	+1.0	+2.2	⊕	-9.4	-27.6	-56.8	24	303B
1.9	8.4	20.3	38.6	63.0	+0.3	⊕	-8.0	-29.2	+2.3	+4.0	⊕	-17.2	-52.3	-109.8	24	32A
2.1	9.3	22.9	43.9	71.7	+0.2	⊕	-7.6	-28.5	+2.1	+3.8	⊕	-17.1	-52.9	-111.9	24	8A
0.8	3.1	7.3	13.6	22.2	-0.1	⊕	-3.8	-13.7	+0.8	+1.9	⊕	-8.1	-23.5	-47.5	24	338C
5.8	21.7	45.2	76.1	NA	⊕	-3.4	-29.7	-88.2	+1.7	⊕	-22.8	-77.9	-173.8	-321.4	18	357G
2.7	12.0	29.0	53.3	83.3	+0.5	⊕	-10.7	-39.3	+3.2	+5.4	⊕	-23.3	-70.0	-144.0	24	35A
1.1	4.5	10.8	20.3	33.7	-0.4	⊕	-5.5	-18.4	+1.0	+2.4	⊕	-10.9	-33.3	-71.2	24	375A
1.2	5.0	11.9	22.4	37.1	+0.5	⊕	-6.3	-21.2	+1.3	+2.6	⊕	-11.2	-33.3	-69.1	24	375B
4.2	17.8	39.8	68.3	102.5	⊕	-2.2	-21.7	-67.2	+1.1	⊕	-17.4	-60.7	-136.0	-250.2	20	44A
1.7	7.6	18.6	35.7	NA	⊕	-1.3	-14.1	-43.7	+0.7	⊕	-11.5	-39.7	-89.1	-163.1	24	4570A

These trajectory tables were calculated by computer using the best available data for each load. Trajectories are representative of the nominal behavior of each load at standard conditions (59°F temperature; barometric pressure of 29.53 inches; altitude at sea level). Shooters are cautioned that actual trajectories may differ due to variations in altitude, atmospheric conditions, guns, sights, and ammunition.

| ENERGY IN FOOT-POUNDS (TO NEAREST 5 FOOT-POUNDS) | | | | | | | | | | | WIND DRIFT IN INCHES 10 MPH CROSSWIND | | | | | | | | | | HEIGHT OF BULLET TRAJECTORY IN INCHES ABOVE OR BELOW LINE OF SIGHT IF ZEROED AT ⊕ YARDS. SIGHTS 1.5 INCHES ABOVE BORE LINE. | | | | | | | | | | FEDERAL LOAD NO. |
MUZZLE	100 YDS.	200 YDS.	300 YDS.	400 YDS.	500 YDS.	600 YDS.	700 YDS.	800 YDS.	900 YDS.	1000 YDS.	100 YDS.	200 YDS.	300 YDS.	400 YDS.	500 YDS.	600 YDS.	700 YDS.	800 YDS.	900 YDS.	1000 YDS.	100 YDS.	200 YDS.	300 YDS.	400 YDS.	500 YDS.	600 YDS.	700 YDS.	800 YDS.	900 YDS.	1000 YDS.	
1380	1135	925	750	600	475	375	295	235	195	170	0.9	3.7	8.7	16.3	27.0	41.3	59.5	82.2	109.2	140.0	+1.6	⊕	-7.4	-21.9	-45.3	-79.8	-128.7	-194.1	-280.2	-388.7	GM223M
2520	2180	1870	1600	1355	1150	970	815	690	590	510	0.8	3.1	7.4	13.6	22.2	33.3	47.1	64.1	84.2	107.5	+17.5	+30.5	+36.6	+34.5	+22.9	⊕	-36.1	-87.8	-157.5	-247.4	GM308M
2720	2350	2025	1730	1470	1245	1050	880	740	630	540	0.7	3.0	7.0	13.0	21.2	31.8	45.1	61.5	81.0	103.6	+16.1	+28.1	+33.8	+31.9	+21.1	⊕	-33.4	-81.3	-146.0	-230.1	GM3006M

CENTERFIRE RIFLE

PREMIUM® HUNTING RIFLE BALLISTICS (Approximate)

Usage Key: [1] = Varmints, predators, small game   [2] = Medium game   [3] = Large, heavy game

USAGE	FEDERAL LOAD NO.	CALIBER	BULLET WGT. GRAINS	BULLET WGT. GRAMS	BULLET STYLE*	FACTORY PRIMER NO.	VELOCITY IN FEET PER SECOND (TO NEAREST 10 FEET) MUZZLE	100 YDS.	200 YDS.	300 YDS.	400 YDS.	500 YDS.	ENERGY IN FOOT-POUNDS (TO NEAREST 5 FOOT-POUNDS) MUZZLE	100 YDS.	200 YDS.	300 YDS.	400 YDS.	500 YDS.
[1]	P223E	223 Rem. (5.56x45mm)	55	3.56	Boat-tail HP	205	3240	2770	2340	1950	1610	1330	1280	935	670	465	315	215
[1]	P22250B	22-250 Rem.	55	3.56	Boat-tail HP	210	3680	3280	2920	2590	2280	1990	1655	1315	1040	815	630	480
[2]	P243C	243 Win. (6.16x51mm)	100	6.48	Boat-tail SP	210	2960	2760	2570	2380	2210	2040	1950	1690	1460	1260	1080	925
[1]	P243D	243 Win. (6.16x51mm)	85	5.50	Boat-tail HP	210	3320	3070	2830	2600	2380	2180	2080	1770	1510	1280	1070	890
[2]	P243E	243 Win. (6.16x51mm)	100	6.48	Nosler Partition**	210	2960	2730	2510	2300	2100	1910	1945	1650	1395	1170	975	805
[2]	P6C	6mm Rem.	100	6.48	Nosler Partition	210	3100	2830	2570	2330	2100	1890	2135	1775	1470	1205	985	790
[2]	P257B	257 Roberts (High-Velocity + P)	120	7.77	Nosler Partition	210	2780	2560	2360	2160	1970	1790	2060	1750	1480	1240	1030	855
[2]	P2506C	25-06 Rem.	117	7.58	Boat-tail SP	210	2990	2770	2570	2370	2190	2000	2320	2000	1715	1465	1240	1045
[2]	P6555A	6.5x55 Swedish	140	9.07	Nosler Partition	210	2550	2350	2170	1990	1820	1660	2020	1725	1460	1230	1030	860
[2]	P270C	270 Win.	150	9.72	Boat-tail SP	210	2850	2660	2480	2300	2130	1970	2705	2355	2040	1760	1510	1290
[2]	P270D	270 Win.	130	8.42	Boat-tail SP	210	3060	2830	2620	2410	2220	2030	2700	2320	1980	1680	1420	1190
[2]	P270E	270 Win.	150	9.72	Nosler Partition	210	2850	2590	2340	2100	1880	1670	2705	2225	1815	1470	1175	930
[2]	P270T1	270 Win.	140	9.07	Trophy Bonded	210	2940	2700	2480	2260	2060	1860	2685	2270	1905	1590	1315	1080
[2]	P730A	7-30 Waters	120	7.77	Boat-tail SP	210	2700	2300	1930	1600	1330	1140	1940	1405	990	685	470	345
[2]	P7C	7mm Mauser (7x57mm Mauser)	140	9.07	Nosler Partition	210	2660	2450	2260	2070	1890	1730	2200	1865	1585	1330	1110	930
[2]	P764A	7x64 Brenneke	160	10.37	Nosler Partition	210	2650	2480	2310	2150	2000	1850	2495	2180	1895	1640	1415	1215
[2]	P280A	280 Rem.	150	9.72	Nosler Partition	210	2890	2620	2370	2140	1910	1710	2780	2295	1875	1520	1215	970
[2]	P708A	7mm-08	140	9.07	Nosler Partition	210	2800	2590	2390	2200	2020	1840	2435	2085	1775	1500	1265	1060
[2]	P7RD	7mm Rem. Magnum	150	9.72	Boat-tail SP	215	3110	2920	2750	2580	2410	2250	3220	2850	2510	2210	1930	1690
[3]	P7RE	7mm Rem. Magnum	165	10.69	Boat-tail SP	215	2950	2800	2650	2510	2370	2230	3190	2865	2570	2300	2050	1825
[3]	P7RF	7mm Rem. Magnum	160	10.37	Nosler Partition	215	2950	2770	2590	2420	2250	2090	3090	2715	2375	2075	1800	1555
[2]	P7RG	7mm Rem. Magnum	140	9.07	Nosler Partition	215	3150	2930	2710	2510	2320	2130	3085	2660	2290	1960	1670	1415
[3]NEW	P7RT1	7mm Rem. Magnum	175	11.34	Trophy Bonded	215	2860	2660	2470	2290	2120	1950	3180	2750	2375	2040	1740	1475
[2]	P3030D	30-30 Win.	170	11.01	Nosler Partition	210	2200	1900	1620	1380	1190	1060	1830	1355	990	720	535	425
[2]	P308C	308 Win. (7.62x51mm)	165	10.69	Boat-tail SP	210	2700	2520	2330	2160	1990	1830	2670	2310	1990	1700	1450	1230
[3]	P308E	308 Win. (7.62x51mm)	180	11.66	Nosler Partition	210	2620	2430	2240	2060	1890	1730	2745	2355	2005	1700	1430	1200
[3]NEW	P308T1	308 Win.	165	10.69	Trophy Bonded	210	2700	2480	2280	2080	1900	1720	2670	2280	1900	1590	1315	1085
[2]	P3006D	30-06 Spring (7.62x63mm)	165	10.69	Boat-tail SP	210	2800	2610	2420	2240	2070	1910	2870	2490	2150	1840	1580	1340
[3]	P3006F	30-06 Spring (7.62x63mm)	180	11.66	Nosler Partition	210	2700	2500	2320	2140	1970	1810	2910	2510	2150	1830	1550	1350
[2]	P3006G	30-06 Spring (7.62x63mm)	150	9.72	Boat-tail SP	210	2910	2690	2480	2270	2070	1880	2820	2420	2040	1710	1430	1180
[3]	P3006L	30-06 Spring (7.62x63mm)	180	11.66	Nosler Partition	210	2700	2540	2380	2220	2080	1930	2915	2570	2260	1975	1720	1495
[2]	P3006T1	30-06 Spring (7.62x63mm)	165	10.69	Trophy Bonded	210	2800	2540	2290	2050	1830	1630	2870	2360	1915	1545	1230	975
[3]	P300WC	300 Win. Magnum	200	12.96	Boat-tail SP	215	2830	2680	2530	2380	2240	2110	3560	3180	2830	2520	2230	1970

PREMIUM® SAFARI™ RIFLE BALLISTICS (Approximate)

USAGE	FEDERAL LOAD NO.	CALIBER	BULLET WGT. GRAINS	BULLET WGT. GRAMS	BULLET STYLE*	FACTORY PRIMER NO.	VELOCITY IN FEET PER SECOND (TO NEAREST 10 FEET) MUZZLE	100 YDS.	200 YDS.	300 YDS.	400 YDS.	500 YDS.	ENERGY IN FOOT-POUNDS (TO NEAREST 5 FOOT-POUNDS) MUZZLE	100 YDS.	200 YDS.	300 YDS.	400 YDS.	500 YDS.
[3]	P300HA	300 H&H Magnum	180	11.66	Nosler Partition	215	2880	2620	2380	2150	1930	1730	3315	2750	2260	1840	1480	1190
[3]	P300WD2	300 Win. Magnum	180	11.66	Nosler Partition	215	2960	2700	2450	2210	1990	1780	3500	2905	2395	1955	1585	1270
[3]	P300WT1	300 Win. Magnum	200	12.96	Trophy Bonded	215	2800	2570	2350	2150	1950	1770	3480	2935	2460	2050	1690	1385
[3]	P338A2	338 Win. Magnum	210	13.60	Nosler Partition	215	2830	2590	2370	2160	1960	1770	3735	3140	2620	2170	1785	1455
[3]	P338B2	338 Win. Magnum	250	16.20	Nosler Partition	215	2660	2400	2150	1910	1690	1500	3925	3185	2555	2055	1590	1245
[3]	P338T1	338 Win. Magnum	225	14.58	Trophy Bonded	215	2800	2560	2330	2110	1900	1710	3915	3265	2700	2220	1800	1455
[4]	P375D	375 H&H Magnum	300	19.44	Solid	215	2530	2170	1840	1550	1310	1140	4265	3140	2260	1605	1140	860
[4]	P375F	375 H&H Magnum	300	19.44	Nosler Partition	215	2530	2320	2120	1930	1750	1590	4265	3585	2995	2475	2040	1675
[4]	P375T1	375 H&H Magnum	300	19.44	Trophy Bonded	215	2530	2280	2040	1810	1610	1425	4265	3450	2765	2190	1725	1350
[3]	P458A	458 Win. Magnum	350	22.68	Soft Point	215	2470	1990	1570	1250	1060	950	4740	3065	1915	1205	870	705
[4]	P458B	458 Win. Magnum	510	33.04	Soft Point	215	2090	1820	1570	1360	1190	1080	4945	3730	2790	2080	1605	1320
[4]	P458C	458 Win. Magnum	500	32.40	Solid	215	2090	1870	1670	1480	1320	1190	4850	3880	3085	2440	1945	1585
[4]NEW	P458T1	458 Win. Magnum	400	25.92	Trophy Bonded	215	2380	2170	1960	1770	1590	1430	5030	4165	3415	2785	2255	1825
[4]NEW	P458T2	458 Win. Magnum	500	32.40	Trophy Bonded	215	2090	1870	1660	1480	1310	1180	4850	3870	3065	2420	1915	1550
[4]NEW	P458T3	458 Win. Magnum	500	32.40	Trophy Bonded Solid	215	2090	1860	1650	1460	1300	1170	4850	3845	3025	2365	1865	1505
[4]	P416A	416 Rigby	410	26.57	Weldcore SP	216	2370	2110	1870	1640	1440	1280	5115	4050	3165	2455	1895	1485
[4]	P416B	416 Rigby	410	26.57	Solid	216	2370	2110	1870	1640	1440	1280	5115	4050	3165	2455	1895	1485
[4]NEW	P416T1	416 Rigby	400	25.92	Trophy Bonded	216	2370	2210	2050	1900	1750	1620	4990	4315	3720	3185	2720	2315
[4]NEW	P416T2	416 Rigby	400	25.92	Trophy Bonded Solid	216	2370	2120	1890	1660	1460	1290	4990	3975	3130	2440	1895	1480
[4]	P470A	470 Nitro Express	500	32.40	Woodleigh SP	216	2150	1890	1650	1440	1270	1140	5130	3965	3040	2310	1790	1435
[4]	P470B	470 Nitro Express	500	32.40	Woodleigh Solid	216	2150	1890	1650	1440	1270	1140	5130	3965	3040	2310	1790	1435
[4]NEW	P470T1	470 Nitro Express	500	32.40	Trophy Bonded	216	2150	1940	1740	1560	1400	1260	5130	4170	3360	2695	2160	1750
[4]NEW	P470T2	470 Nitro Express	500	32.40	Trophy Bonded Solid	216	2150	1940	1740	1560	1400	1260	5130	4170	3360	2695	2160	1750

PREMIUM® VARMINT RIFLE BALLISTICS (Approximate)

USAGE	FEDERAL LOAD NO.	CALIBER	BULLET WGT. GRAINS	BULLET WGT. GRAMS	BULLET STYLE*	FACTORY PRIMER NO.	VELOCITY IN FEET PER SECOND (TO NEAREST 10 FEET) MUZZLE	100 YDS.	200 YDS.	300 YDS.	400 YDS.	500 YDS.	ENERGY IN FOOT-POUNDS (TO NEAREST 5 FOOT-POUNDS) MUZZLE	100 YDS.	200 YDS.	300 YDS.	400 YDS.	500 YDS.
[1]	P223V	223 Rem. (5.56x45mm)	40	2.59	Hollow Point Varmint	205	3650	3010	2450	1950	1530	1210	1185	805	535	340	205	130
[1]	P22250V	22-250 Rem.	40	2.59	Hollow Point Varmint	210	4000	3320	2720	2200	1740	1360	1420	980	660	430	265	165
[1]	P243V	243 Win. (6.16x51mm)	60	3.89	Hollow Point Varmint	210	3600	3110	2660	2260	1890	1560	1725	1285	945	680	475	325
[1]	P2506V	25-06 Rem.	90	5.83	Hollow Point Varmint	210	3440	3040	2680	2340	2030	1750	2365	1850	1435	1100	825	610

*HP = Hollow Point   SP = Soft Point   **"Nosler" and "Partition" are registered trademarks of Nosler Bullets, Inc.
+P ammunition is loaded to a higher pressure. Use only in firearms so recommended by the gun manufacturer.

CENTERFIRE RIFLE

4 = Dangerous game   5 = Target shooting, training, practice

100 YDS.	200 YDS.	WIND DRIFT IN INCHES 10 MPH CROSSWIND 300 YDS.	400 YDS.	500 YDS.	50 YDS.	AVERAGE RANGE 100 YDS.	200 YDS.	300 YDS.	50 YDS.	100 YDS.	LONG RANGE 200 YDS.	300 YDS.	400 YDS.	500 YDS.	TEST BARREL LENGTH INCHES	FEDERAL LOAD NO.
1.3	5.8	14.2	27.7	47.6	-0.3	⊕	-2.7	-10.8	+0.4	+1.4	⊕	-6.7	-20.5	-43.4	24	P223E
0.8	3.6	8.4	15.8	26.3	-0.4	⊕	-1.7	-7.6	0	+0.9	⊕	-5.0	-15.1	-32.0	24	P22250B
0.6	2.6	6.1	11.3	18.4	-0.2	⊕	-3.1	-11.4	+0.6	+1.5	⊕	-6.8	-19.8	-39.9	24	P243C
0.7	2.7	6.3	11.6	18.8	-0.3	⊕	-2.2	-8.8	+0.2	+1.1	⊕	-5.5	-16.1	-32.8	24	P243D
0.7	3.1	7.3	13.5	22.1	-0.2	⊕	-3.2	-11.9	+0.6	+1.6	⊕	-7.1	-20.9	-42.5	24	P243E
0.8	3.3	7.9	14.7	24.1	-0.3	⊕	-2.9	-11.0	+0.5	+1.4	⊕	-6.7	-19.8	-39.0	24	P6C
0.8	3.3	7.7	14.3	23.5	-0.1	⊕	-3.8	-14.0	+0.8	+1.9	⊕	-8.2	-24.0	-48.9	24	P257B
0.7	2.8	6.5	12.0	19.6	-0.2	⊕	-3.0	-11.4	+0.5	+1.5	⊕	-6.8	-19.9	-40.4	24	P2506C
0.8	3.5	8.3	15.1	25.1	0	⊕	-4.8	-17.1	+1.2	+2.4	⊕	-9.8	-28.2	-57.7	24	P6555A
0.7	2.7	6.3	11.6	18.9	-0.2	⊕	-3.4	-12.5	+0.7	+1.7	⊕	-7.4	-21.4	-43.0	24	P270C
0.7	2.8	6.6	12.1	19.7	-0.2	⊕	-2.8	-10.7	+0.5	+1.4	⊕	-6.5	-19.0	-38.5	24	P270D
0.9	3.9	9.2	17.3	28.5	-0.2	⊕	-3.7	-13.8	+0.8	+1.9	⊕	-8.3	-24.4	-50.5	24	P270E
0.8	3.2	7.6	14.2	23.0	-0.2	⊕	-3.3	-12.2	+0.6	+1.6	⊕	-7.3	-21.5	-43.7	24	P270T1
1.6	7.2	17.7	34.5	58.1	0	⊕	-5.2	-19.8	+1.2	+2.6	⊕	-12.0	-37.6	-81.7	24	P730A
1.3	3.2	8.2	15.4	23.4	-0.1	⊕	-4.3	-15.4	+1.0	+2.1	⊕	-9.0	-26.1	-52.9	24	P7C
0.7	2.8	6.6	12.3	19.5	-0.1	⊕	-4.2	-14.9	+0.9	+2.1	⊕	-8.7	-24.9	-49.4	24	P764A
0.9	3.8	9.0	16.8	27.8	-0.2	⊕	-3.6	-13.4	+0.7	+1.8	⊕	-8.0	-23.8	-49.2	24	P280A
0.8	3.1	7.3	13.5	21.8	-0.2	⊕	-3.7	-13.5	+0.8	+1.8	⊕	-8.0	-23.1	-46.6	24	P708A
0.5	2.2	5.1	9.3	15.0	-0.3	⊕	-2.6	-9.8	+0.4	+1.3	⊕	-5.9	-17.0	-34.2	24	P7RD
0.5	2.0	4.6	8.4	13.5	-0.2	⊕	-3.0	-10.9	+0.5	+1.5	⊕	-6.4	-18.4	-36.6	24	P7RE
0.6	2.5	5.6	10.4	16.9	-0.2	⊕	-3.1	-11.3	+0.6	+1.5	⊕	-6.7	-19.4	-39.0	24	P7RF
0.6	2.6	6.0	11.1	18.2	-0.3	⊕	-2.6	-9.9	+0.4	+1.3	⊕	-6.0	-17.5	-35.6	24	P7RG
0.7	2.8	6.5	12.1	19.6	-0.2	⊕	-3.4	-12.5	+0.7	+1.7	⊕	-7.4	-21.5	-43.3	24	P7RT1
0.9	8.0	19.4	36.7	59.8	-0.3	⊕	-8.3	-29.8	+2.4	+4.1	⊕	-17.4	-52.4	-109.4	24	P3030D
0.7	3.0	7.0	13.0	21.1	-0.1	⊕	-4.0	-14.4	+0.9	+2.0	⊕	-8.4	-24.3	-49.0	24	P308C
0.8	3.3	7.7	14.3	23.3	-0.1	⊕	-4.4	-15.8	+1.0	+2.2	⊕	-9.2	-26.5	-53.6	24	P308E
0.8	3.5	8.2	15.2	24.9	-0.1	⊕	-4.1	-15.0	+0.9	+2.1	⊕	-8.8	-25.6	-52.2	24	P308T1
0.7	2.8	6.6	12.3	19.9	-0.2	⊕	-3.6	-13.2	+0.8	+1.8	⊕	-7.8	-22.4	-45.2	24	P3006D
0.7	3.0	7.3	13.4	27.7	-0.1	⊕	-4.0	-14.6	+0.9	+2.0	⊕	-8.6	-24.6	-49.6	24	P3006F
0.7	3.0	7.1	13.4	22.0	-0.2	⊕	-3.3	-12.4	+0.6	+1.7	⊕	-7.4	-21.5	-43.7	24	P3006G
0.6	2.6	6.0	11.0	17.8	-0.1	⊕	-3.9	-13.9	+0.9	+1.9	⊕	-8.1	-23.1	-46.1	24	P3006L
1.0	4.0	9.6	17.8	29.7	-0.1	⊕	-3.9	-14.5	+0.8	+2.0	⊕	-8.7	-25.4	-53.1	24	P3006T1
0.5	2.2	5.0	9.2	14.9	-0.2	⊕	-3.4	-12.2	+0.7	+1.7	⊕	-7.1	-20.4	-40.5	24	P300WC

100 YDS.	200 YDS.	WIND DRIFT IN INCHES 10 MPH CROSSWIND 300 YDS.	400 YDS.	500 YDS.	50 YDS.	AVERAGE RANGE 100 YDS.	200 YDS.	300 YDS.	50 YDS.	100 YDS.	LONG RANGE 200 YDS.	300 YDS.	400 YDS.	500 YDS.	TEST BARREL LENGTH INCHES	FEDERAL LOAD NO.
0.9	3.7	8.8	16.3	27.1	-0.3	⊕	-3.5	-13.3	+0.7	+1.8	⊕	-8.0	-23.4	-48.6	24	P300HA
0.9	3.5	8.4	15.8	25.9	-0.2	⊕	-3.3	-12.4	+0.6	+1.6	⊕	-7.5	-22.1	-45.4	24	P300WD2
0.9	3.4	8.1	14.9	24.5	-0.1	⊕	-3.7	-13.8	+0.8	+1.9	⊕	-8.2	-23.9	-48.8	24	P300WT1
0.9	3.4	8.2	15.2	24.9	-0.2	⊕	-3.6	-13.6	+0.8	+1.8	⊕	-8.1	-23.6	-48.3	24	P33BA2
1.1	4.5	10.8	20.3	33.6	-0.1	⊕	-4.6	-16.7	+1.1	+2.3	⊕	-9.8	-29.1	-60.2	24	P338B2
0.9	3.7	8.7	16.1	26.7	-0.2	⊕	-3.8	-14.1	+0.8	+1.9	⊕	-8.4	-24.5	-50.6	24	P338T1
1.7	7.2	17.6	33.9	56.5	⊕	-1.1	-9.1	-27.5	+0.5	⊕	-7.0	-24.2	-55.8	-106.5	24	P375D
0.9	3.9	9.1	17.0	27.8	0	⊕	-5.0	-17.7	+1.2	+2.5	⊕	-10.3	-29.9	-60.8	24	P375F
1.1	4.8	11.3	21.5	35.4	-0.1	⊕	-5.3	-18.8	+1.3	+2.6	⊕	-10.9	-32.8	-67.8	24	P375T1
2.5	11.0	27.6	52.6	83.9	⊕	-1.5	-11.0	-34.9	+0.1	⊕	-7.5	-29.1	-71.1	-138.0	24	P458A
1.9	7.9	18.9	35.3	56.8	⊕	-1.8	-13.7	-39.7	+0.4	⊕	-9.1	-32.3	-73.9	-138.0	24	P458B
1.5	6.1	14.5	26.9	43.7	⊕	-1.7	-12.9	-36.7	+0.4	⊕	-8.5	-29.5	-66.2	-122.0	24	P458C
1.1	4.5	10.7	19.9	32.7	⊕	-0.2	-6.3	-21.5	+0.1	⊕	-5.9	-20.9	-47.1	-87.0	24	P458T1
1.5	6.2	14.6	27.3	44.5	⊕	-0.7	-10.0	-31.8	+0.4	⊕	-8.5	-29.7	-66.8	-124.2	24	P458T2
1.5	6.4	15.2	28.3	46.1	⊕	-0.7	-10.1	-32.2	+0.4	⊕	-8.6	-30.0	-67.8	-126.3	24	P458T3
1.3	5.7	13.6	25.6	42.3	⊕	-0.8	-9.8	-28.5	+0.6	⊕	-7.4	-24.8	-55.0	-101.6	24	P416A
1.3	5.7	13.6	25.6	42.3	⊕	-1.2	-9.8	-28.5	+0.6	⊕	-7.4	-24.8	-55.0	-101.6	24	P416B
0.8	3.4	7.9	14.7	23.8	⊕	-0.2	-6.0	-20.1	+0.1	⊕	-5.7	-19.6	-43.3	-78.4	24	P416T1
1.3	5.5	13.2	24.8	41.0	⊕	-0.2	-6.8	-23.2	+0.1	⊕	-6.3	-22.5	-51.5	-96.7	24	P416T2
1.6	7.0	16.6	31.1	50.6	⊕	-1.6	-12.6	-36.2	+0.8	⊕	-9.3	-31.3	-69.7	-128.6	24	P470A
1.6	7.0	16.6	31.1	50.6	⊕	-1.6	-12.6	-36.2	+0.8	⊕	-9.3	-31.3	-69.7	-128.6	24	P470B
1.3	5.5	13.0	24.3	39.7	⊕	-0.6	-9.0	-28.9	+0.3	⊕	-7.8	-27.1	-60.8	-112.4	24	P470T1
1.3	5.5	13.0	24.3	39.7	⊕	-0.6	-9.0	-28.9	+0.3	⊕	-7.8	-27.1	-60.8	-112.4	24	P470T2

100 YDS.	200 YDS.	WIND DRIFT IN INCHES 10 MPH CROSSWIND 300 YDS.	400 YDS.	500 YDS.	50 YDS.	AVERAGE RANGE 100 YDS.	200 YDS.	300 YDS.	50 YDS.	100 YDS.	LONG RANGE 200 YDS.	300 YDS.	400 YDS.	500 YDS.	TEST BARREL LENGTH INCHES	FEDERAL LOAD NO.
1.5	6.5	16.1	32.3	56.9	-0.4	⊕	-2.4	-10.7	+0.2	+1.2	⊕	-7.1	-23.4	-54.2	24	P223V
1.3	5.7	14.0	27.9	49.2	-0.4	⊕	-1.7	-8.1	0	+0.8	⊕	-5.6	-18.4	-42.8	24	P22250V
1.1	4.8	11.7	22.6	38.7	-0.4	⊕	-2.1	-9.2	+0.2	+1.1	⊕	-6.0	-18.9	-41.6	24	P243V
1.0	4.1	9.8	18.7	31.3	-0.3	⊕	-2.3	-9.4	+0.2	+1.1	⊕	-6.0	-18.3	-39.2	24	P2506V

These trajectory tables were calculated by computer using the best available data for each load. Trajectories are representative of the nominal behavior of each load at standard conditions (59°F temperature; barometric pressure of 29.53 inches; altitude at sea level). Shooters are cautioned that actual trajectories may differ due to variations in altitude, atmospheric conditions, guns, sights, and ammunition.

FEDERAL BALLISTICS

PISTOL/REVOLVER

CLASSIC® AUTOMATIC PISTOL BALLISTICS (Approximate)

Usage Key: [1] = Varmints, predators, small game   [2] = Medium game   [3] = Self defense   [4] = Target shooting, training, practice

USAGE	FEDERAL LOAD NO.	CALIBER	BULLET WGT. IN GRAINS	GRAMS	BULLET STYLE**	FACTORY PRIMER NO.	VELOCITY IN FEET PER SECOND MUZZLE	25 YDS.	50 YDS.	75 YDS.	100 YDS.	ENERGY IN FOOT-POUNDS MUZZLE	25 YDS.	50 YDS.	75 YDS.	100 YDS.	MID-RANGE TRAJECTORY 25 YDS.	50 YDS.	75 YDS.	100 YDS.	TEST BARREL LENGTH INCHES
[3],[4]	25AP	25 Auto (6.35mm Browning)	50	3.24	Full Metal Jacket	100	760	750	730	720	700	65	60	59	55	55	0.5	1.9	4.5	8.1	2
[3],[4]	32AP	32 Auto (7.65mm Browning)	71	4.60	Full Metal Jacket	100	905	880	855	830	810	129	20	115	110	105	0.3	1.4	3.2	5.9	4
[3],[4]	380AP	380 Auto (9x17mm Short)	95	6.15	Full Metal Jacket	100	955	910	865	830	790	190	175	160	145	130	0.3	1.3	3.1	5.8	3¾
[3]	380BP	380 Auto (9x17mm Short)	90	5.83	Hi-Shok JHP	100	1000	940	890	840	800	200	175	160	140	130	0.3	1.2	2.9	5.5	3¾
[3],[4]	9AP	9mm Luger (9x19mm Parabellum)	124	8.03	Full Metal Jacket	100	1120	1070	1030	990	960	345	315	290	270	255	0.2	0.9	2.2	4.1	4
[3]	9BP	9mm Luger (9x19mm Parabellum)	115	7.45	Hi-Shok JHP	100	1160	1100	1060	1020	990	345	310	285	270	250	0.2	0.9	2.1	3.8	4
[3]	9MS	9mm Luger (9x19mm Parabellum)	147	9.52	Hi-Shok JHP	100	975	950	930	900	880	310	295	285	265	255	0.3	1.2	2.8	5.1	4
[3]	40SWA	40 S&W	180	11.06	Hi-Shok JHP	100	985	955	930	905	885	390	365	345	330	315	0.3	1.2	2.8	5.0	4
[3]	40SWB	40 S&W	155	10.04	Hi-Shok JHP	100	1140	1080	1030	990	950	445	400	365	335	315	0.2	0.9	2.2	4.1	4
[3]	10C	10mm Auto	180	11.06	Hi-Shok JHP	150	1030	995	970	945	920	425	400	375	355	340	0.3	1.1	2.5	4.7	5
[3]	10E	10mm Auto	155	10.04	Hi-Shok JHP	150	1325	1225	1140	1075	1025	605	515	450	400	360	0.2	0.7	1.8	3.3	5
[3]	45A	45 Auto	230	14.90	Full Metal Jacket	150	850	830	810	790	770	370	350	335	320	305	0.4	1.6	3.6	6.6	5
[4]	45C	45 Auto	185	11.99	Hi-Shok JHP	150	950	920	900	880	860	370	350	335	315	300	0.3	1.3	2.9	5.3	5
[3]	45D	45 Auto	230	14.90	Hi-Shok JHP	150	850	830	810	790	770	370	350	335	320	300	0.4	1.6	3.7	6.7	5

CLASSIC® REVOLVER BALLISTICS (Approximate)

Usage Key: [1] = Varmints, predators, small game   [2] = Medium game   [3] = Self defense   [4] = Target shooting, training, practice

USAGE	FEDERAL LOAD NO.	CALIBER	BULLET WGT. IN GRAINS	GRAMS	BULLET STYLE**	FACTORY PRIMER NO.	VELOCITY IN FEET PER SECOND MUZZLE	25 YDS.	50 YDS.	75 YDS.	100 YDS.	ENERGY IN FOOT-POUNDS MUZZLE	25 YDS.	50 YDS.	75 YDS.	100 YDS.	MID-RANGE TRAJECTORY 25 YDS.	50 YDS.	75 YDS.	100 YDS.	TEST BARREL LENGTH INCHES
[4]	32LA	32 S&W Long	98	6.35	Lead Wadcutter	100	780	700	630	560	500	130	105	85	70	55	0.5	2.2	5.6	11.1	4
[4]	32LB	32 S&W Long	98	6.35	Lead Round Nose	100	705	690	670	650	640	115	105	98	95	90	0.6	2.3	5.3	9.6	4
[3]	32HRA	32 H&R Magnum	95	6.15	Lead Semi-Wadcutter	100	1030	1000	940	930	900	225	210	190	185	170	0.3	1.1	2.5	4.7	4½
[3]	32HRB	32 H&R Magnum	85	5.50	Hi-Shok JHP	100	1100	1050	1020	970	930	230	210	195	175	165	0.2	1.0	2.3	4.3	4½
[4]	38BB	38 Special	158	10.23	Lead Round Nose	100	755	740	723	710	690	200	190	183	175	170	0.5	2.0	4.6	8.3	4-V
[3],[4]	38C	38 Special	158	10.23	Lead Semi-Wadcutter	100	755	740	.723	710	690	200	190	183	175	170	0.5	2.0	4.6	8.3	4-V
[1],[3]	38E	38 Special (High-Velocity+P)	125	8.10	Hi-Shok JHP	100	945	920	898	880	860	248	235	224	215	205	0.3	1.3	2.9	5.4	4-V
[1],[3]	38F	38 Special (High-Velocity+P)	110	7.13	Hi-Shok JHP	100	995	960	926	900	870	242	225	210	195	185	0.3	1.2	2.7	5.0	4-V
[1],[3]	38G	38 Special (High-Velocity+P)	158	10.23	Semi-Wadcutter HP	100	890	870	855	840	820	278	265	257	245	235	0.3	1.4	3.3	5.9	4-V
[3],[4]	38H	38 Special (High-Velocity+P)	158	10.23	Lead Semi-Wadcutter	100	890	870	855	840	820	270	265	257	245	235	0.3	1.4	3.3	5.9	4-V
[1],[3]	38J	38 Special (High-Velocity+P)	125	8.10	Hi-Shok JSP	100	945	920	898	880	860	248	235	224	215	205	0.3	1.3	2.9	5.4	4-V
[2],[3]	357A	357 Magnum	158	10.23	Hi-Shok JSP	100	1235	1160	1104	1060	1020	535	475	428	395	365	0.2	0.8	1.9	3.5	4-V
[1],[3]	357B	357 Magnum	125	8.10	Hi-Shok JHP	100	1450	1350	1240	1160	1100	583	495	427	370	335	0.1	0.6	1.5	2.8	4-V
[4]	357C	357 Magnum	158	10.23	Lead Semi-Wadcutter	100	1235	1160	1104	1060	1020	535	475	428	395	365	0.2	0.8	1.9	3.5	4-V
[1],[3]	357D	357 Magnum	110	7.13	Hi-Shok JHP	100	1295	1180	1094	1040	990	410	340	292	260	235	0.2	0.8	1.9	3.5	4-V
[2],[3]	357E	357 Magnum	158	10.23	Hi-Shok JHP	100	1235	1160	1104	1060	1020	535	475	428	395	365	0.2	0.8	1.9	3.5	4-V
[2]	357G	357 Magnum	180	11.66	Hi-Shok JHP	100	1090	1030	980	930	890	475	425	385	350	320	0.2	1.0	2.4	4.5	4-V
[2],[3]	357H	357 Magnum	140	9.07	Hi-Shok JHP	100	1360	1270	1200	1130	1080	575	500	445	395	360	0.2	0.7	1.6	3.0	4-V
[1],[3]	41A	41 Rem. Magnum	210	13.60	Hi-Shok JHP	150	1300	1210	1130	1070	1030	790	680	595	540	495	0.2	0.7	1.8	3.3	4-V
[1],[3]	44SA	44 S&W Special	200	12.96	Semi-Wadcutter HP	150	900	860	830	800	770	360	330	305	285	260	0.3	1.4	3.4	6.3	6½-V
[2],[3]	44A	44 Rem. Magnum	240	15.55	Hi-Shok JHP	150	1180	1130	1081	1050	1010	741	675	623	580	550	0.2	0.9	2.0	3.7	6½-V
[1],[3]	44B*	44 Rem. Magnum	180	11.66	Hi-Shok JHP	150	1610	1480	1365	1270	1180	1035	875	750	640	555	0.1	0.5	1.2	2.3	6½-V
[1],[3]	45LCA	45 Colt	225	14.58	Semi-Wadcutter HP	150	900	880	860	840	820	405	385	369	355	340	0.3	1.4	3.2	5.8	5½

+P ammunition is loaded to a higher pressure. Use only in firearms so recommended by the gun manufacturer.   "V" indicates vented barrel to simulate service conditions.

*Also available in 20-round box (A44B20).

**JHP = Jacketed Hollow Point   HP = Hollow Point   JSP = Jacketed Soft Point

FEDERAL BALLISTICS

PISTOL/REVOLVER

HYDRA-SHOK® PISTOL AND REVOLVER BALLISTICS (Approximate)

Usage Key: ①=Varmints, predators, small game   ②=Medium game   ③=Self defense   ④=Target shooting, training, practice

USAGE	FEDERAL LOAD NO.	CALIBER	BULLET WGT. IN GRAINS	GRAMS	BULLET STYLE*	FACTORY PRIMER NO.	VELOCITY IN FEET PER SECOND MUZZLE	25 YDS.	50 YDS.	75 YDS.	100 YDS.	ENERGY IN FOOT-POUNDS MUZZLE	25 YDS.	50 YDS.	75 YDS.	100 YDS.	MID-RANGE TRAJECTORY 25 YDS.	50 YDS.	75 YDS.	100 YDS.	TEST BARREL LENGTH INCHES
③	P380HS1	380 Auto (9x17mm Short)	90	5.83	Hydra-Shok HP	100	1000	940	890	840	800	200	175	160	140	130	0.3	1.2	2.9	5.5	3³/₄
③	P9HS1	9mm Luger (9x19mm Parabellum)	124	8.03	Hydra-Shok HP	100	1120	1070	1030	990	960	345	315	290	270	255	0.2	0.9	2.2	4.1	4
③	P9HS2	9mm Luger (9x19mm Parabellum)	147	9.52	Hydra-Shok HP	100	1000	960	920	890	860	325	300	275	260	240	0.3	1.2	2.8	5.1	4
③	P40HS1	40 S&W	180	11.06	Hydra-Shok HP	100	985	955	930	910	890	390	365	345	330	315	0.3	1.2	2.8	5.0	4
③	P40HS2	40 S&W	155	10.04	Hydra-Shok HP	100	1140	1080	1030	990	950	445	400	365	335	315	0.2	0.9	2.2	4.1	4
③ NEW	P40HS3	40 S&W	165	10.66	Hydra-Shok HP	100	975	850	927	908	887	348	331	315	301	288	0.3	1.2	2.7	5.1	4
③	P10HS1	10mm Auto	180	11.06	Hydra-Shok HP	150	1030	995	970	945	920	425	400	375	355	340	0.3	1.1	2.5	4.7	5
③	P45HS1	45 Auto	230	14.90	Hydra-Shok HP	150	850	830	810	790	770	370	350	335	320	305	0.4	1.6	3.6	6.6	5
③	P38HS1	38 Special (High Velocity +P)	129	8.36	Hydra-Shok HP	100	945	930	910	890	870	255	245	235	225	215	0.3	1.3	2.9	5.3	4-V
③	P357HS1	357 Magnum	158	10.23	Hydra-Shok HP	100	1235	1160	1104	1060	1020	535	475	428	395	365	0.2	0.8	1.9	3.5	4-V
③	P44HS1	44 Rem. Magnum	240	15.55	Hydra-Shok HP	150	1180	1130	1081	1050	1010	741	675	623	580	550	0.2	0.9	2.0	3.7	6¹/₂-V

+P ammunition is loaded to a higher pressure. Use only in firearms so recommended by the gun manufacturer. "V" indicates vented barrel to simulate service conditions.

*HP = Hollow Point SWC = Semi-Wadcutter

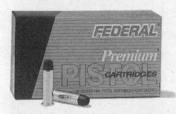

Nyclad Round Nose   Nyclad Semi-Wadcutter Hollow Point   Nyclad Hollow Point

NYCLAD® PISTOL AND REVOLVER BALLISTICS (Approximate)

Usage Key: ①=Varmints, predators, small game   ②=Medium game   ③=Self defense   ④=Target shooting, training, practice

USAGE	FEDERAL LOAD NO.	CALIBER	BULLET WGT. IN GRAINS	GRAMS	BULLET STYLE*	FACTORY PRIMER NO.	VELOCITY IN FEET PER SECOND MUZZLE	25 YDS.	50 YDS.	75 YDS.	100 YDS.	ENERGY IN FOOT-POUNDS MUZZLE	25 YDS.	50 YDS.	75 YDS.	100 YDS.	MID-RANGE TRAJECTORY 25 YDS.	50 YDS.	75 YDS.	100 YDS.	TEST BARREL LENGTH INCHES
③	P9BP	9mm Luger (9x19mm Parabellum)	124	8.03	Nyclad Hollow Point	100	1120	1070	1030	990	960	345	315	290	270	255	0.2	0.9	2.2	4.1	4
④	P38B	38 Special	158	10.23	Nyclad Round Nose	100	755	740	723	710	690	200	190	183	175	170	0.5	2.0	4.6	8.3	4-V
①, ③	P38G	38 Special (High Velocity +P)	158	10.23	Nyclad SWC-HP	100	890	870	855	840	820	270	265	257	245	235	0.3	1.4	3.3	5.9	4-V
③	P38M	38 Special	125	8.10	Nyclad Hollow Point	100	825	780	730	690	650	190	170	150	130	115	0.4	1.8	4.3	8.1	2-V
①, ③	P38N	38 Special (High Velocity +P)	125	8.10	Nyclad Hollow Point	100	945	920	898	880	860	248	235	224	215	205	0.3	1.3	2.9	5.4	4-V
②, ③	P357E	357 Magnum	158	10.23	Nyclad SWC-HP	100	1235	1160	1104	1060	1020	535	475	428	395	365	0.2	0.8	1.9	3.5	4-V

*HP = Hollow Point

Full Metal Jacket Semi-Wadcutter Match   Lead Wadcutter Match   Metal Case Profile Match

GOLD MEDAL® MATCH PISTOL AND REVOLVER BALLISTICS (Approximate)

Usage Key: ①=Varmints, predators, small game   ②=Medium game   ③=Self defense   ④=Target shooting, training, practice

USAGE	FEDERAL LOAD NO.	CALIBER	BULLET WGT. IN GRAINS	GRAMS	BULLET STYLE*	FACTORY PRIMER NO.	VELOCITY IN FEET PER SECOND MUZZLE	25 YDS.	50 YDS.	75 YDS.	100 YDS.	ENERGY IN FOOT-POUNDS MUZZLE	25 YDS.	50 YDS.	75 YDS.	100 YDS.	MID-RANGE TRAJECTORY 25 YDS.	50 YDS.	75 YDS.	100 YDS.	TEST BARREL LENGTH INCHES
④	GM9MP	9mm Luger (9x19mm Parabellum)	124	8.03	FMJ-SWC Match	100	1120	1070	1030	990	960	345	315	290	270	255	0.2	0.9	2.2	4.1	4
④	GM38A	38 Special	148	9.59	Lead Wadcutter Match	100	710	670	634	600	560	166	150	132	115	105	0.6	2.4	5.7	10.8	4-V
④	GM44D	44 Rem. Magnum	250	16.20	MC Profile Match	150	1180	1140	1100	1070	1040	775	715	670	630	600	0.2	0.8	1.9	3.6	6¹/₂-V
④	GM45B	45 Auto	185	11.99	FMJ-SWC Match	150	775	730	695	660	620	247	220	200	175	160	0.5	2.0	4.8	9.0	5

*MC Profile Match = Metal Case Profile Match   FMJ = Full Metal Jacket   SWC = Semi-Wadcutter

HORNADY BALLISTICS

RIFLE	MUZZLE VELOCITY	VELOCITY FEET PER SECOND					ENERGY FOOT - POUNDS						TRAJECTORY TABLES				
Caliber	Muzzle	100 yds.	200 yds.	300 yds.	400 yds.	500 yds.	Muzzle	100 yds.	200 yds.	300 yds.	400 yds.	500 yds.	100 yds.	200 yds.	300 yds.	400 yds.	500 yds.
222 Rem., 50 gr. SX	3140	2602	2123	1700	1350	1107	1094	752	500	321	202	136	+2.2	-0-	-10.0	-32.3	-73.8
222 Rem., 55 gr. SX	3020	2562	2147	1773	1451	1201	1114	801	563	384	257	176	+2.2	-0-	-9.9	-31.0	-68.7
223 Rem., 53 gr. HP	3330	2882	2477	2106	1710	1475	1305	978	722	522	369	356	+1.7	-0-	-7.4	-22.7	-49.1
223 Rem., 55 gr. SP	3240	2747	2304	1905	1554	1270	1282	921	648	443	295	197	+1.9	-0-	-8.5	-26.7	-59.6
223 Rem., 55 gr. FMJ	3240	2759	2326	1933	1587	1301	1282	929	660	456	307	207	+1.9	-0-	-8.4	-26.2	-57.9
223 Rem., 60 gr. SP	3150	2782	2442	2127	1837	1575	1322	1031	795	603	450	331	+1.6	-0-	-7.5	-22.5	-48.1
22-250 Rem., 53 gr. HP	3680	3185	2743	2341	1974	1646	1594	1194	886	645	459	319	+1.0	-0-	-5.7	-17.8	-38.8
22-250 Rem., 55 gr. SP	3680	3137	2656	2222	1832	1439	1654	1201	861	603	410	272	+1.1	-0-	-6.0	-19.2	-42.6
22-250 Rem., 55 gr. FMJ	3680	3137	2656	2222	1836	1439	1654	1201	861	603	410	273	+1.1	-0-	-6.0	-19.2	-42.6
22-250 Rem., 60 gr. SP	3600	3195	2826	2485	2169	1878	1727	1360	1064	823	627	470	+1.0	-0-	-5.4	-16.3	-34.8
220 Swift, 50 gr. SP	3850	3327	2862	2442	2060	1716	1645	1228	909	662	471	327	+0.8	-0-	-5.1	-16.1	-35.3
220 Swift, 55 gr. SP	3650	3194	2772	2384	2035	1724	1627	1246	939	694	506	363	+1.0	-0-	-5.6	-17.4	-37.5
220 Swift, 60 gr. HP	3600	3199	2824	2475	2156	1868	1727	1364	1063	816	619	465	+1.0	-0-	-5.4	-16.3	-34.8
243 Win., 75 gr. HP	3400	2970	2578	2219	1890	1595	1926	1469	1107	820	595	425	+1.2	-0-	-6.5	-20.3	-43.8
243 Win., 80 gr. FMJ	3350	2955	2593	2259	1951	1670	1993	1551	1194	906	676	495	+1.2	-0-	-6.5	-19.9	-42.5
243 Win., 100 gr. BTSP	2960	2728	2508	2299	2099	1910	1945	1653	1397	1174	979	810	+1.6	-0-	-7.2	-21.0	-42.8
6MM Rem., 100 gr. BTSP	3100	2861	2634	2419	2231	2018	2134	1818	1541	1300	1088	904	+1.3	-0-	-6.5	-18.9	-38.5
257 Roberts, 100 gr. SP	3000	2736	2486	2251	2028	1818	1998	1662	1373	1124	913	734	+1.6	-0-	-7.2	-21.3	-43.9
257 Roberts, 117 gr. BTSP	2780	2550	2331	2122	1925	1740	2007	1689	1411	1170	963	787	+1.9	-0-	-8.3	-24.4	-49.9
25-06 100 gr. SP	3230	2952	2690	2443	2210	1989	2316	1934	1607	1325	1084	879	+1.2	-0-	-6.1	-18.0	-37.0
25-06 117 gr. BTSP	2990	2749	2520	2302	2096	1900	2322	1962	1649	1377	1141	938	+1.6	-0-	-7.0	-20.7	-42.2
270 Win., 130 gr. SP	3060	2800	2560	2330	2110	1900	2700	2265	1890	1565	1285	1045	+1.8	-0-	-7.1	-20.6	-42.0
270 Win., 140 gr. BTSP	2940	2747	2562	2385	2214	2050	2688	2346	2041	1769	1524	1307	+1.6	-0-	-7.0	-20.2	-40.3
270 Win., 150 gr. SP	2800	2684	2478	2284	2100	1927	2802	2400	2046	1737	1469	1237	+1.7	-0-	-7.4	-21.6	-43.9
7 x 57, 139 gr. BTSP	2700	2504	2316	2137	1965	1802	2251	1936	1656	1410	1192	1002	+2.0	-0-	-8.5	-24.9	-50.3
7 x 57 Mau., 154 gr. SP	2600	2400	2208	2025	1852	1689	2312	1970	1668	1403	1173	976	+2.2	-0-	-9.5	-27.7	-55.8
7MM Rem. Mag., 139 gr. BTSP	3150	2933	2727	2530	2341	2160	3063	2656	2296	1976	1692	1440	+1.2	-0-	-6.1	-17.7	-35.5
7MM Rem. Mag., 154 gr. SP	3035	2814	2604	2404	2212	2029	3151	2708	2319	1977	1674	1408	+1.3	-0-	-6.7	-19.3	-39.3
7MM Rem. Mag., 162 gr. BTSP	2940	2757	2582	2413	2251	2094	3110	2735	2399	2095	1823	1578	+1.6	-0-	-6.7	-19.7	-39.3
7MM Rem. Mag., 175 gr. SP	2860	2650	2440	2240	2060	1880	3180	2720	2310	1960	1640	1370	+2.0	-0-	-7.9	-22.7	-45.8
7MM Wby. Mag., 154 gr. SP	3200	2971	2753	2546	2348	2159	3501	3017	2592	2216	1885	1593	+1.2	-0-	-5.8	-17.0	-34.5
7MM Wby. Mag., 175 gr. SP	2910	2709	2516	2331	2154	1985	3290	2850	2459	2111	1803	1531	+1.6	-0-	-7.1	-20.6	-41.7
30-30 Win., 150 gr. RN	2390	1973	1605	1303	1095	974	1902	1296	858	565	399	316	-0-	-8.2	-30.0		
30-30 Win., 170 gr. FP	2200	1895	1619	1381	1191	1064	1827	1355	989	720	535	425	-0-	-8.9	-31.1		
308 Win., 150 gr. SP	2820	2533	2263	2009	1774	1560	2648	2137	1705	1344	1048	810	+2.3	-0-	-9.1	-26.9	-55.7
308 Win., 150 gr. BTSP	2820	2560	2315	2084	1866	1644	2648	2183	1785	1447	1160	922	+2.0	-0-	-8.5	-25.2	-51.8
308 Win., 165 gr. SP	2700	2440	2194	1963	1748	1551	2670	2180	1763	1411	1119	881	+2.5	-0-	-9.7	-28.5	-58.8
308 Win., 165 gr. BTSP	2700	2496	2301	2115	1937	1770	2670	2283	1940	1639	1375	1148	+2.0	-0-	-8.7	-25.2	-51.0
308 Win., 168 gr. BTHP MATCH	2700	2524	2354	2191	2035	1885	2720	2377	2068	1791	1545	1326	+2.0	-0-	-8.4	-23.9	-48.0
30-06 150 gr. SP	2910	2617	2342	2083	1843	1622	2820	2281	1827	1445	1131	876	+2.1	-0-	-8.5	-25.0	-51.8
30-06 150 gr. BTSP	2910	2683	2467	2262	2066	1880	2820	2397	2027	1706	1421	1177	+2.0	-0-	-7.7	-22.2	-44.9
30-06 165 gr. BTSP	2800	2591	2392	2202	2020	1848	2873	2460	2097	1777	1495	1252	+1.8	-0-	-8.0	-23.3	-47.0
30-06 168 gr. BTHP MATCH	2790	2620	2447	2280	2120	1966	2925	2561	2234	1940	1677	1442	+1.7	-0-	-7.7	-22.2	-44.3
30-06 180 gr. SP	2700	2469	2258	2042	1846	1663	2913	2436	2023	1666	1362	1105	+2.4	-0-	-9.3	-27.0	-54.9
300 Wby. Mag., 180 gr. SP	3120	2891	2673	2466	2268	2079	3890	3340	2856	2430	2055	1727	+1.3	-0-	-6.2	-18.1	-36.8
300 Win Mag., 150 gr. BTSP	3275	2988	2718	2464	2224	1998	3573	2974	2461	2023	1648	1330	+1.2	-0-	-6.0	-17.8	-36.5
300 Win. Mag., 165 gr. BTSP	3100	2877	2665	2462	2269	2084	3522	3033	2603	2221	1887	1592	+1.3	-0-	-6.5	-18.5	-37.3
300 Win. Mag., 180 gr. SP	2960	2745	2540	2344	2157	1979	3501	3011	2578	2196	1859	1565	+1.9	-0-	-7.3	-20.9	-41.9
300 Win. Mag., 190 gr. BTSP	2900	2711	2529	2355	2187	2026	3549	3101	2699	2340	2018	1732	+1.6	-0-	-7.1	-20.4	-41.0
303 British, 150 gr. SP	2685	2441	2210	1992	1787	1598	2401	1984	1627	1321	1064	500	+2.2	-0-	-9.3	-27.4	-56.5
303 British, 174 gr. RN	2500	2181	1886	1669	1387	1201	2414	1837	1374	1012	743	557	+2.9	-0-	-12.8	-39.0	-83.4

BARREL LENGTH	PISTOL	MUZZLE VELOCITY	VELOCITY FT. PER SECOND		ENERGY		
	Caliber	Muzzle	50 yds.	100 yds.	Muzzle	50 yds.	100 yds.
2"	25 Auto, 35 gr. JHP/XTP	900	813	742	63	51	43
2"	25 Auto, 50 gr. FMJ/RN	760	707	659	64	56	48
4"	32 Auto, 71 gr. FMJ	900	845	797	128	112	100
3¾"	380 Auto, 90 gr. JHP/XTP	1000	902	823	200	163	135
3¾"	380 Auto, 100 gr. FMJ/RN	950	875	810	200	170	146
4"	9MM Luger, 90 gr. JHP/XTP	1360	1112	978	370	247	191
4"	9MM Luger, 100 gr. FMJ/RN	1220	1059	959	331	249	204
4"	9MM Luger, 115 gr. JHP/XTP	1155	1047	971	341	280	241
4"	9MM Luger, 115 gr. FMJ	1155	1047	971	341	280	241
4"	9MM Luger, 124 gr. FMJ/RN	1110	1030	971	339	292	259
4"	9MM Luger, 124 gr. FMJ/FP	1110	1030	971	339	292	259
4"	9MM Luger, 124 gr. JHP/XTP	1110	1030	971	339	292	259
4"	9MM Luger, 147 gr. JHP/XTP	975	935	899	310	285	264
4"	9 x 18 Makarov, 95 gr. JHP/XTP	1000	930	874	211	182	161
4"V	38 Special, 125 gr. JHP/XTP	900	856	817	225	203	185
4"V	38 Special, 140 gr. JHP/XTP	900	850	806	252	225	202
4"V	38 Special, 148 gr. HBWC	800	697	610	210	160	122
4"V	38 Special, 158 gr. JHP/XTP	800	765	731	225	205	188
4"V	38 Special, 158 gr. LRN	775	744	715	211	194	179
4"V	38 Special, 158 gr. SWC	775	739	705	211	191	174
4"V	38 Special, 158 gr. SWC/HP	775	740	706	211	192	175
8"V	357 Mag., 125 gr. JHP/XTP	1500	1314	1166	624	479	377
8"V	357 Mag., 125 gr. JFP/XTP	1500	1311	1161	624	477	374
8"V	357 Mag., 140 gr. JHP/XTP	1400	1249	1130	609	485	397
8"V	357 Mag., 158 gr. JHP/XTP	1250	1150	1073	548	464	404
8"V	357 Mag., 158 gr. JFP	1250	1147	1068	548	461	400

BARREL LENGTH	PISTOL	MUZZLE VELOCITY	VELOCITY FT. PER SECOND		ENERGY		
	Caliber	Muzzle	50 yds.	100 yds.	Muzzle	50 yds.	100 yds.
8"V	357 Mag., 158 gr. SWC	1050	971	911	387	331	291
8"V	357 Mag., 158 gr. SWC/HP	1050	973	914	387	332	293
5"	10MM Auto, 155 gr. JHP/XTP	1265	1119	1020	551	431	358
5"	10MM Auto, 180 gr. JHP/XTP Full	1180	1077	1004	556	464	403
5"	10MM Auto, 180 gr. JHP/XTP Sub	915	873	835	335	304	278
5"	10MM Auto, 200 gr. FMJ/FP	1050	989	940	490	434	393
5"	10MM Auto, 200 gr. JHP/XTP	1050	994	948	490	439	399
4"	40 S&W, 155 gr. JHP/XTP	1180	1061	980	479	388	331
4"	40 S&W, 180 gr. JHP/XTP	950	903	862	361	326	297
4"	40 S&W, 180 gr. FMJ/FP	950	903	862	361	326	297
7½"V	44 Special, 180 gr. JHP/XTP	1000	935	882	400	350	311
7½"V	44 Rem. Mag., 180 gr. JHP/XTP	1550	1340	1173	960	717	550
7½"V	44 Rem. Mag., 200 gr. JHP/XTP	1500	1284	1128	999	732	565
7½"V	44 Rem. Mag., 240 gr. JHP/XTP	1350	1188	1078	971	753	619
7½"V	44 Rem. Mag., 240 gr. SWC	975	928	888	507	459	421
7½"V	44 Rem. Mag., 240 gr. SWC/HP	975	933	897	507	464	428
7½"V	44 Rem. Mag., 300 gr. JHP/XTP	1150	1084	1031	881	782	708
5"	45 ACP, 185 gr. JHP/XTP	950	880	819	371	318	276
5"	45 ACP, 200 gr. JHP/XTP	900	938	885	444	391	348
5"	45 ACP, 200 gr. FMJ C/T (M)	950	885	830	401	348	306
5"	45 ACP, 200 gr. SWC	900	811	738	360	292	242
5"	45 ACP+P, 200 gr. HP/XTP	1055	982	925	494	428	380
5"	45 ACP+P, 230 gr. HP/XTP	950	904	865	462	418	382
5"	45 ACP, 230 gr. FMJ/RN	850	809	771	369	334	304
5"	45 ACP, 230 gr. FMJ/FP	850	809	771	369	334	304

CENTERFIRE PISTOL/REVOLVER

Caliber	Order No.	Primer No.	Wt. Grs.	Bullet Style	Velocity (FPS) Muzzle	50 Yds.	100 Yds.	Energy (FT-LB) Muzzle	50 Yds.	100 Yds.	Mid-Range Trajectory 50 Yds.	100 Yds.	B.L.
(1) 221 REM. FIREBALL	R221F	7 1/2	50°	Pointed Soft Point	2650	2380	2130	780	630	505	0.2"	0.8"	10"
(2) 25 (6.35MM) AUTO. PISTOL	R25AP	1 1/2	50°	Metal Case	760	707	659	64	56	48	2.0"	8.7"	2"
(3) 6MM BR REM.	R6MMBR	7 1/2	100°	Pointed Soft Point	Refer to page 30 for ballistics.								
(4) 7MM BR REM.	R7MMBR	7 1/2	140°	Pointed Soft Point	Refer to page 30 for ballistics.								
(5) 32 S. & W.	R32SW	1 1/2	88°	Lead	680	645	610	90	81	73	2.5"	0.5"	3"
(6) 32 S. & W. LONG	R32SWL	1 1/2	98°	Lead	705	670	635	115	98	88	2.3"	10.5"	4"
(7) 32 (7.65MM) AUTO. PISTOL	R32AP	1 1/2	71°	Metal Case	905	855	810	129	115	97	1.4"	5.8"	4"
(8) 357 MAG.	R357M7	5 1/2	110	Semi-Jacketed H.P.	1295	1094	975	410	292	232	0.8"	3.5"	4"
Vented Barrel Ballistics	R357M1	5 1/2	125	Semi-Jacketed H.P.	1450	1240	1090	583	427	330	0.6"	2.8"	4"
	GS357MA	5 1/2	125	Brass-Jacketed Hollow Point	1220	1095	1009	413	333	283	0.8"	3.5"	4"
(Refer to page 90 for test details)	R357M2	5 1/2	158	Semi-Jacketed H.P.	1235	1104	1015	535	428	361	0.8"	3.5"	4"
	R357M3	5 1/2	158	Soft Point	1235	1104	1015	535	428	361	0.8"	3.5"	4"
	R357M5	5 1/2	158	Semi-Wadcutter	1235	1104	1015	535	428	361	0.8"	3.5"	4"
	R357M9	5 1/2	140	Semi-Jacketed H.P.	1360	1195	1076	575	444	360	0.7"	3.0"	4"
	R357M10	5 1/2	180	Semi-Jacketed H.P.	1145	1053	985	524	443	388	0.9"	3.9"	8"
	R357M11	5 1/2	125°	Semi-Jacketed H.P. (Med. Vel.)	1220	1077	984	413	322	269	0.8"	3.7"	4"
	R357MB	5 1/2	140	"Multi-Ball"	1155	829	663	418	214	136	1.2"	6.4"	4"
(9) 357 REM. MAXIMUM°°	357MX1	7 1/2	158°	Semi-Jacketed H.P.	1825	1588	1381	1168	885	669	0.4"	1.7"	10"
(10) 9MM LUGER	R9MM1	1 1/2	115	Jacketed H.P.	1155	1047	971	341	280	241	0.9"	3.9"	4"
	R9MM10	1 1/2	124	Jacketed H.P.	1120	1028	971	346	291	254	1.0"	4.1"	4"
AUTO. PISTOL	R9MM2	1 1/2	124	Metal Case	1110	1030	971	339	292	259	1.0"	4.1"	4"
	R9MM3	1 1/2	115°	Metal Case	1135	1041	973	329	277	242	0.9"	4.0"	4"
	R9MM5	1 1/2	88	Jacketed H.P.	1500	1191	1012	440	277	200	0.6"	3.1"	4"
	R9MM6	1 1/2	115	Jacketed H.P. (+P)‡	1250	1113	1019	399	316	265	0.8"	3.5"	4"
	R9MM8	1 1/2	147	Jacketed H.P. (Subsonic)	990	941	900	320	289	264	1.1"	4.9"	4"
	R9MM9	1 1/2	147	Metal Case (Match)	990	941	900	320	289	264	1.1"	4.9"	4"
	GS9MMB	1 1/2	124	Brass-Jacketed Hollow Point	1125	1031	963	349	293	255	1.0"	4.0"	4"
	GS9MMC	1 1/2	147	Brass-Jacketed Hollow Point	990	941	900	320	289	264	1.1"	4.9"	4"
(11) 380 AUTO. PISTOL	R380AP	1 1/2	95°	Metal Case	955	865	785	190	160	130	1.4"	5.9"	4"
	R380A1	1 1/2	88	Jacketed H.P.	990	920	868	191	165	146	1.2"	5.1"	4"
(12) 38 SUPER AUTO. COLT PISTOL (A)	R38SU1	1 1/2	115°	Jacketed H.P. (+P)‡	1300	1147	1041	431	336	277	0.7"	3.3"	5"
(13) 38 S. & W.	R38SW	1 1/2	146°	Lead	685	650	620	150	135	125	2.4"	10.0"	4"
(14) 38 SPECIAL	R38S1	1 1/2	95	Semi-Jacketed H.P. (+P)‡	1175	1044	959	291	230	194	0.9"	3.9"	4"
Vented Barrel Ballistics	R38S10	1 1/2	110	Semi-Jacketed H.P. (+P)‡	995	926	871	242	210	185	1.2"	5.1"	4"
	R38S16	1 1/2	110	Semi-Jacketed H.P.	950	890	840	220	194	172	1.4"	5.4"	4"
	R38S2	1 1/2	125°	Semi-Jacketed H.P. (+P)‡	945	898	858	248	224	204	1.3"	5.4"	4"
	GS38SB	1 1/2	125	Brass-Jacketed Hollow Point (+P)	975	929	885	264	238	218	1.0"	5.2"	4"
	R38S3	1 1/2	148	Targetmaster Lead W.C. Match	710	634	566	166	132	105	2.4"	10.8"	4"
	R38S4	1 1/2	158	Targetmaster Lead	755	723	692	200	183	168	2.0"	8.3"	4"
	R38S5	1 1/2	158	Lead (Round Nose)	755	723	692	200	183	168	2.0"	8.3"	4"
	R38S14	1 1/2	158	Semi-Wadcutter (+P)‡	890	855	823	278	257	238	1.4"	6.0"	4"
	R38S6	1 1/2	158	Semi-Wadcutter	755	723	692	200	183	168	2.0"	8.3"	4"
	R38S12	1 1/2	158	Lead H.P. (+P)‡	890	855	823	278	257	238	1.4"	6.0"	4"
	R38SMB	1 1/2	140°	"Multi-Ball"	830	731	506	216	130	80	2.0"	10.6"	4"
(15) 38 SHORT COLT	R38SC	1 1/2	125°	Lead	730	685	645	150	130	115	2.2"	9.4"	6"
(16) 40 S. & W.	R40SW1	5 1/2	155°	Jacketed H.P.	1205	1095	1017	499	413	356	0.8"	3.6"	4"
	R40SW2	5 1/2	180	Jacketed H.P.	1015	960	914	412	368	334	1.3"	4.5"	4"
	GS40SWA	5 1/2	165	Brass-Jacketed Hollow Point	1150	1040	964	485	396	340	1.0"	4.0"	4"
	GS40SWB	5 1/2	180	Brass-Jacketed Hollow Point	1015	960	914	412	368	334	1.3"	4.5"	4"
(17) 10MM AUTO.	R10MM2	2 1/2	200	Metal Case	1050	994	948	490	439	399	1.0"	4.2"	5"
	R10MM3	2 1/2	180°	Jacketed H. P. (Subsonic)	1055	997	951	445	397	361	1.0"	4.6"	5"
	R10MM4	2 1/2	180	Jacketed H. P. (High Vel.)	1160	1079	1017	538	465	413	0.9"	3.8"	5"
(18) 41 REM. MAG.	R41MG1	2 1/2	210	Soft Point	1300	1162	1062	788	630	526	0.7"	3.2"	4"
Vented Barrel Ballistics	R41MG2	2 1/2	210	Lead	965	898	842	434	376	331	1.3"	5.4"	4"
	R41MG3	2 1/2	170°	Semi-Jacketed H.P.	1420	1166	1014	761	513	388	0.7"	3.2"	4"
(19) 44 REM. MAG.	R44MG5	2 1/2	180	Semi-Jacketed H.P.	1610	1365	1175	1036	745	551	0.5"	2.3"	4"
Vented Barrel Ballistics	R44MG1	2 1/2	240	Lead Gas Check	1350	1186	1069	971	749	608	0.7"	3.1"	4"
	R44MG2	2 1/2	240°	Soft Point	1180	1081	1010	741	623	543	0.9"	3.7"	4"
	R44MG3	2 1/2	240	Semi-Jacketed H.P.	1180	1081	1010	741	623	543	0.9"	3.7"	4"
	R44MG4	2 1/2	240	Lead (Med. Vel.)	1000	947	902	533	477	433	1.1"	4.8"	6"
	R44MG6	2 1/2	210	Semi-Jacketed H.P.	1495	1312	1167	1042	803	634	0.6"	2.5"	4"
(20) 44 S. & W. SPECIAL	R44SW	2 1/2	246	Lead	755	725	695	310	285	265	2.0"	8.3"	6"
	R44SW1	2 1/2	200°	Semi-Wadcutter	1035	938	866	476	391	333	1.1"	4.9"	6"
(21) 45 COLT	R45C	2 1/2	250	Lead	860	820	780	410	375	340	1.6"	6.6"	5"
	R45C1	2 1/2	225	Semi-Wadcutter (Keith)	960	890	832	460	395	346	1.3"	5.5"	5"
(22) 45 AUTO.	R45AP1	2 1/2	185	Targetmaster Lead W.C. Match	770	707	650	244	205	174	2.0"	8.7"	5"
	R45AP2	2 1/2	185°	Jacketed H.P.	1000	939	889	411	362	324	1.1"	4.9"	5"
	R45AP4	2 1/2	230	Metal Case	835	800	767	356	326	300	1.6"	6.8"	5"
	R45AP6	2 1/2	185	Jacketed H.P. (+P)‡	1140	1040	971	534	445	387	0.9"	4.0"	5"
	GS45APA	2 1/2	185	Brass-Jacketed Hollow Point	1015	951	899	423	372	332	1.1"	4.5"	5"
	GS45APB	2 1/2	230	Brass-Jacketed Hollow Point	875	833	795	391	355	323	1.5"	6.1"	5"
BLANK CARTRIDGES													
38 S. & W.	R38SWBL§	1 1/2	-	Blank	-	-	-	-	-	-			
32 S. & W.	R32BLNK	5 1/2	-	Blank	-	-	-	-	-	-			
38 SPECIAL	R38BLNK	1 1/2	-	Blank	-	-	-	-	-	-			

°Illustrated (not shown in actual size). °°Will not chamber in 357 Mag. or 38 Special handguns. ‡Ammunition with (+P) on the case headstamp is loaded to higher pressure. Use only in firearms designated for this cartridge and so recommended by the gun manufacturer. §Subject to stock on hand. (A)Adapted only for 38 Colt Super and Colt Commander automatic pistols. Not for use in sporting, military and pocket models.

REMINGTON BALLISTICS

CENTERFIRE RIFLE

Caliber	Remington Order No.	Wt.-Grs.	Bullet Style	Primer No.	Muzzle	100 Yds.	Velocity – Feet Per Second 200 Yds.	300 Yds.	400 Yds.	500 Yds.
17 Rem.	R17REM	25	Hollow Point Power-Lokt®	7 ½	4040	3284	2644	2086	1606	1235
22 Hornet	R22HN1	45	Pointed Soft Point	6 ½	2690	2042	1502	1128	948	840
	R22HN2	45	Hollow Point	6 ½	2690	2042	1502	1128	948	840
220 Swift	R220S1	50	Pointed Soft Point	9 ½	3780	3158	2617	2135	1710	1357
222 Rem.	R222R1	50	Pointed Soft Point	7 ½	3140	2602	2123	1700	1350	1107
	R222R3	50	Hollow Point Power-Lokt®	7 ½	3140	2635	2182	1777	1432	1172
222 Rem. Mag.	R222M1	55	Pointed Soft Point	7 ½	3240	2748	2305	1906	1556	1272
223 Rem.	R223R1	55	Pointed Soft Point	7 ½	3240	2747	2304	1905	1554	1270
	R223R2	55	Hollow Point Power-Lokt®	7½	3240	2773	2352	1969	1627	1341
	R223R3	55	Metal Case	7 ½	3240	2759	2326	1933	1587	1301
	R223R4	60	Hollow Point Match	7 ½	3100	2712	2355	2026	1726	1463
22-250 Rem.	R22501	55	Pointed Soft Point	9 ½	3680	3137	2656	2222	1832	1493
	R22502	55	Hollow Point Power-Lokt®	9 ½	3680	3209	2785	2400	2046	1725
243 Win.	R243W1	80	Pointed Soft Point	9 ½	3350	2955	2593	2259	1951	1670
	R243W2	80	Hollow Point Power-Lokt®	9 ½	3350	2955	2593	2259	1951	1670
	R243W3	100	Pointed Soft Point Core-Lokt®	9 ½	2960	2697	2449	2215	1993	1786
	ER243WA	105	Extended Range	9 ½	2920	2689	2470	2261	2062	1874
6MM Rem.	R6MM1	80	Pointed Soft Point	9 ½	3470	3064	2694	2352	2036	1747
	R6MM4	100	Pointed Soft Point Core-Lokt®	9 ½	3100	2829	2573	2332	2104	1889
	ER6MMRA	105	Extended Range	9 ½	3060	2822	2596	2381	2177	1982
6MM BR Rem.	R6MMBR	100	Pointed Soft Point	7 ½	2550	2310	2083	1870	1671	1491
25-20 Win.	R25202	86	Soft Point	6 ½	1460	1194	1030	931	858	797
250 Sav.	R250SV	100	Pointed Soft Point	9 ½	2820	2504	2210	1936	1684	1461
257 Roberts	R257	117	Soft Point Core-Lokt®	9 ½	2650	2291	1961	1663	1404	1199
	ER257A	122	Extended Range	9 ½	2600	2331	2078	1842	1625	1431
25-06 Rem.	R25062	100	Pointed Soft Point Core-Lokt®	9 ½	3230	2893	2580	2287	2014	1762
	R25063	120	Pointed Soft Point Core-Lokt®	9 ½	2990	2730	2484	2252	2032	1825
	ER2506A	122	Extended Range	9 ½	2930	2706	2492	2289	2095	1911
6.5x55 Swedish	R65SWEI	140	Pointed Soft Point Core-Lokt®	9 ½	2550	2353	2164	1984	1814	1654
264 Win. Mag.	R264W2	140	Pointed Soft Point Core-Lokt®	9 ½M	3030	2782	2548	2326	2114	1914
270 Win.	R270W1	100	Pointed Soft Point	9 ½	3320	2924	2561	2225	1916	1636
	R270W2	130	Pointed Soft Point Core-Lokt®	9 ½	3060	2776	2510	2259	2022	1801
	R270W3	130	Bronze Point	9 ½	3060	2802	2559	2329	2110	1904
	R270W4	150	Soft Point Core-Lokt®	9 ½	2850	2504	2183	1886	1618	1385
	ER270WB	135	Extended Range	9 ½	3000	2780	2570	2369	2178	1995
	ER270WA	140	Extended Range Boat Tail	9 ½	2960	2749	2548	2355	2171	1995
7MM Br Rem.	R7MMBR	140	Pointed Soft Point	7 ½	2215	2012	1821	1643	1481	1336
7MM Mauser (7 x 57)	R7MSR1	140	Pointed Soft Point	9 ½	2660	2435	2221	2018	1827	1648
7 x 64	R7X641	140	Pointed Soft Point	9 ½	2950	2714	2489	2276	2073	1881
	R7X642	175	Pointed Soft Point Core-Lokt®	9 ½	2650	2445	2248	2061	1883	1716
7MM-08 Rem.	R7M081	140	Pointed Soft Point	9 ½	2860	2625	2402	2189	1988	1798
	R7M083	120	Hollow Point	9 ½	3000	2725	2467	2223	1992	1778
	ER7M08A	154	Extended Range	9 ½	2715	2510	2315	2128	1950	1781
280 Rem.	R280R3	140	Pointed Soft Point	9 ½	3000	2758	2528	2309	2102	1905
	R280R1	150	Pointed Soft Point Core-Lokt®	9 ½	2890	2624	2373	2135	1912	1705
	R280R2	165	Soft Point Core-Lokt®	9 ½	2820	2510	2220	1950	1701	1479
	R280R4§	120	Hollow Point	9 ½	3150	2866	2599	2348	2110	1887
	ER280RA	165	Extended Range	9 ½	2820	2623	2434	2253	2080	1915
7MM Rem. Mag.	R7MM2	150	Pointed Soft Point Core-Lokt®	9 ½M	3110	2830	2568	2320	2085	1866
	R7MM3	175	Pointed Soft Point Core-Lokt®	9 ½M	2860	2645	2440	2244	2057	1879
	R7MM4	140	Pointed Soft Point	9 ½M	3175	2923	2684	2458	2243	2039
	RS7MMA	160	Swift A-Frame™ PSP	9 ½M	2900	2659	2430	2212	2006	1812
	ER7MMA	165	Extended Range	9 ½M	2900	2699	2507	2324	2147	1979
7MM Wby Mag.	R7MWB1	140	Pointed Soft Point	9 ½M	3225	2970	2729	2501	2283	2077
	R7MWB2	175	Pointed Soft Point Core-Lokt®	9 ½M	2910	2693	2486	2288	2098	1918
	ER7MWB4	165	Extended Range	9 ½M	2950	2747	2553	2367	2189	2019
30 Carbine	R30CAR	110	Soft Point	6 ½	1990	1567	1236	1035	923	842
30 Rem.	R30REM	170	Soft Point Core-Lokt®	9'½	2120	1822	1555	1328	1153	1036
30-30 Win. Accelerator®	R3030A	55	Soft Point	9 ½	3400	2693	2085	1570	1187	986
30-30 Win.	R30301	150	Soft Point Core-Lokt®	9 ½	2390	1973	1605	1303	1095	974
	R30302	170	Soft Point Core-Lokt®	9 ½	2200	1895	1619	1381	1191	1061
	R30303	170	Hollow Point Core-Lokt®	9 ½	2200	1895	1619	1381	1191	1061
	ER3030A	160	Extended Range	9 ½	2300	1997	1719	1473	1268	1116
300 Savage	R30SV3	180	Soft Point Core-Lokt®	9 ½	2350	2025	1728	1467	1252	1098
	R30SV2	150	Pointed Soft Point Core-Lokt	9 ½	2630	2354	2095	1853	1631	1432

REMINGTON BALLISTICS

CENTERFIRE RIFLE

Energy – Foot-Pounds						Short Range Trajectory[1]						Long Range Trajectory[2]							Barrel
Muzzle	100 Yds.	200 Yds.	300 Yds.	400 Yds.	500 Yds.	50 Yds.	100 Yds.	150 Yds.	200 Yds.	250 Yds.	300 Yds.	100 Yds.	150 Yds.	200 Yds.	250 Yds.	300 Yds.	400 Yds.	500 Yds	Length
906	599	388	242	143	85	0.1	0.5	0.0	-1.5	-4.2	-8.5	2.1	2.5	1.9	0.0	-3.4	-17.0	-44.3	24"
723	417	225	127	90	70	0.3	0.0	-2.4	-7.7	-16.9	-31.3	1.6	0.0	-4.5	-12.8	-26.4	-75.6	-163.4	24"
723	417	225	127	90	70	0.3	0.0	-2.4	-7.7	-16.9	-31.3	1.6	0.0	-4.5	-12.8	-26.4	-75.6	-163.4	
1586	1107	760	506	325	204	0.2	0.5	0.0	-1.6	-4.4	-8.8	1.3	1.2	0.0	-2.5	-6.5	-20.7	-47.0	24"
1094	752	500	321	202	136	0.5	0.9	0.0	-2.5	-6.9	-13.7	2.2	1.9	0.0	-3.8	-10.0	-32.3	-73.8	24"
1094	771	529	351	228	152	0.5	0.9	0.0	-2.4	-6.6	-13.1	2.1	1.8	0.0	-3.6	-9.5	-30.2	-68.1	
1282	922	649	444	296	198	0.4	0.8	0.0	-2.2	-6.0	-11.8	1.9	1.6	0.0	-3.3	-8.5	-26.7	-59.5	24"
1282	921	648	443	295	197	0.4	0.8	0.0	-2.2	-6.0	-11.8	1.9	1.6	0.0	-3.3	-8.5	-26.7	-59.6	
1282	939	675	473	323	220	0.4	0.8	0.0	-2.1	-5.8	-11.4	1.8	1.6	0.0	-3.2	-8.2	-25.5	-56.0	24"
1282	929	660	456	307	207	0.4	0.8	0.0	-2.1	-5.9	-11.6	1.9	1.6	0.0	-3.2	-8.4	-26.2	-57.9	
1280	979	739	547	397	285	0.5	0.8	0.0	-2.2	-6.0	-11.5	1.9	1.6	0.0	-3.2	-8.3	-25.1	-53.6	
1654	1201	861	603	410	272	0.2	0.5	0.0	-1.6	-4.4	-8.7	2.3	2.6	1.9	0.0	-3.4	-15.9	-38.9	24"
1654	1257	947	703	511	363	0.2	0.5	0.0	-1.5	-4.1	-8.0	2.1	2.5	1.8	0.0	-3.1	-14.1	-33.4	
1993	1551	1194	906	676	495	0.3	0.7	0.0	-1.8	-4.9	-9.4	2.6	2.9	2.1	0.0	-3.6	-16.2	-37.9	
1993	1551	1194	906	676	495	0.3	0.7	0.0	-1.8	-4.9	-9.4	2.6	2.9	2.1	0.0	-3.6	-16.2	-37.9	24"
1945	1615	1332	1089	882	708	0.5	0.9	0.0	-2.2	-5.8	-11.0	1.9	1.6	0.0	-3.1	-7.8	-22.6	-46.3	
1988	1686	1422	1192	992	819	0.5	0.9	0.0	-2.2	-5.8	-11.0	2.0	1.6	0.0	-3.1	-7.7	-22.2	-44.8	
2139	1667	1289	982	736	542	0.3	0.6	0.0	-1.6	-4.5	-8.7	2.4	2.7	1.9	0.0	-3.3	-14.9	-35.0	
2133	1777	1470	1207	983	792	0.4	0.8	0.0	-1.9	-5.2	-9.9	1.7	1.5	0.0	-2.8	-7.0	-20.4	-41.7	
2183	1856	1571	1322	1105	916	0.4	0.8	0.0	-2.0	-5.2	-9.8	1.7	1.5	0.0	-2.7	-6.9	-20.0	-40.4	
1444	1185	963	776	620	494	0.3	0.0	-1.9	-5.6	-11.4	-19.3	2.8	2.3	0.0	-4.3	-10.9	-31.7	-65.1	15"
407	272	203	165	141	121	0.0	-4.1	-14.4	-31.8	-57.3	-92.0	0.0	-8.2	-23.5	-47.0	-79.6	-175.9	-319.4	24"
1765	1392	1084	832	630	474	0.2	0.0	-1.6	-4.7	-9.6	-16.5	2.3	2.0	0.0	-3.7	-9.5	-28.3	-59.5	24"
1824	1363	999	718	512	373	0.3	0.0	-1.9	-5.8	-11.9	-20.7	2.9	2.4	0.0	-4.7	-12.0	-36.7	-79.2	24"
1831	1472	1170	919	715	555	0.3	0.0	-1.9	-5.5	-11.2	-19.1	2.8	2.3	0.0	-4.3	-10.9	-32.0	-66.4	
2316	1858	1478	1161	901	689	0.4	0.7	0.0	-1.9	-5.0	-9.7	1.6	1.4	0.0	-2.7	-6.9	-20.5	-42.7	
2382	1985	1644	1351	1100	887	0.5	0.8	0.0	-2.1	-5.6	-10.7	1.9	1.6	0.0	-3.0	-7.5	-22.0	-44.8	24"
2325	1983	1683	1419	1189	989	0.5	0.9	0.0	-2.2	-5.7	-10.8	1.9	1.6	0.0	-3.0	-7.5	-21.7	-43.9	
2021	1720	1456	1224	1023	850	0.3	0.0	-1.8	-5.4	-10.8	-18.2	2.7	2.2	0.0	-4.1	-10.1	-29.1	-58.7	24"
2854	2406	2018	1682	1389	1139	0.5	0.8	0.0	-2.0	-5.4	-10.2	1.8	1.5	0.0	-2.9	-7.2	-20.8	-42.2	24"
2448	1898	1456	1099	815	594	0.3	0.7	0.0	-1.8	-5.0	-9.7	2.7	3.0	2.2	0.0	-3.7	-16.6	-39.1	
2702	2225	1818	1472	1180	936	0.5	0.8	0.0	-2.0	-5.5	-10.4	1.8	1.5	0.0	-2.9	-7.4	-21.6	-44.3	
2702	2267	1890	1565	1285	1046	0.4	0.8	0.0	-2.0	-5.3	-10.1	1.8	1.5	0.0	-2.8	-7.1	-20.6	-42.0	24"
2705	2087	1587	1185	872	639	0.7	1.0	0.0	-2.6	-7.1	-13.6	2.3	2.0	0.0	-3.8	-9.7	-29.2	-62.2	
2697	2315	1979	1682	1421	1193	0.5	0.8	0.0	-2.0	-5.3	-10.1	1.8	1.5	0.0	-2.8	-7.1	-20.4	-41.0	
2723	2349	2018	1724	1465	1237	0.5	0.8	0.0	-2.1	-5.5	-10.3	1.9	1.5	0.0	-2.9	-7.2	-20.7	-41.6	
1525	1259	1031	839	681	555	0.5	0.0	-2.7	-7.7	-15.4	-25.9	1.8	0.0	-4.1	-10.9	-20.6	-50.0	-95.2	15"
2199	1843	1533	1266	1037	844	0.2	0.0	-1.7	-5.0	-10.0	-17.0	2.5	2.0	0.0	-3.8	-9.6	-27.7	-56.3	24"
2705	2289	1926	1610	1336	1100	0.5	0.9	0.0	-2.1	-5.7	-10.7	1.9	1.6	0.0	-3.0	-7.6	-21.8	-44.3	24"
2728	2322	1964	1650	1378	1144	0.2	0.0	-1.7	-4.9	-9.9	-16.8	2.5	2.0	0.0	-3.9	-9.4	-26.9	-54.3	
2542	2142	1793	1490	1228	1005	0.6	0.9	0.0	-2.3	-6.1	-11.6	2.1	1.7	0.0	-3.2	-8.1	-23.5	-47.7	
2398	1979	1621	1316	1058	842	0.5	0.8	0.0	-2.1	-5.7	-10.8	1.9	1.6	0.0	-3.0	-7.6	-22.3	-45.8	24"
2520	2155	1832	1548	1300	1085	0.7	1.0	0.0	-2.5	-6.7	-12.6	2.3	1.9	0.0	-3.5	-8.8	-25.3	-51.0	
2797	2363	1986	1657	1373	1128	0.5	0.8	0.0	-2.1	-5.5	-10.4	1.8	1.5	0.0	-2.9	-7.3	-21.1	-42.9	
2781	2293	1875	1518	1217	968	0.6	0.9	0.0	-2.3	-6.2	-11.8	2.1	1.7	0.0	-3.3	-8.3	-24.2	-49.7	
2913	2308	1805	1393	1060	801	0.2	0.0	-1.5	-4.6	-9.5	-16.4	2.3	1.9	0.0	-3.7	-9.4	-28.1	-58.8	24"
2643	2188	1800	1468	1186	949	0.4	0.7	0.0	-1.9	-5.1	-9.7	2.8	3.0	2.2	0.0	-3.6	-15.7	-35.6	
2913	2520	2171	1860	1585	1343	0.6	0.9	0.0	-2.3	-6.1	-11.4	2.1	1.7	0.0	-3.2	-8.0	-22.8	-45.6	
3221	2667	2196	1792	1448	1160	0.4	0.8	0.0	-1.9	-5.2	-9.9	1.7	1.5	0.0	-2.8	-7.0	-20.5	-42.1	
3178	2718	2313	1956	1644	1372	0.6	0.9	0.0	-2.3	-6.0	-11.3	2.0	1.7	0.0	-3.2	-7.9	-22.7	-45.8	24"
3133	2655	2240	1878	1564	1292	0.4	0.7	0.0	-1.8	-4.8	-9.1	2.6	2.9	2.0	0.0	-3.4	-14.5	-32.6	
2987	2511	2097	1739	1430	1166	0.6	0.9	0.0	-2.2	-5.9	-11.3	2.0	1.7	0.0	-3.2	-7.9	-23.0	-46.7	
3081	2669	2303	1978	1689	1434	0.5	0.9	0.0	-2.1	-5.7	-10.7	1.9	1.6	0.0	-3.0	-7.5	-21.4	-42.9	
3233	2741	2315	1943	1621	1341	0.3	0.7	0.0	-1.7	-4.6	-8.8	2.5	2.8	2.0	0.0	-3.2	-14.0	31.5	
3293	2818	2401	2033	1711	1430	0.5	0.9	0.0	-2.2	-5.7	-10.8	1.9	1.6	0.0	-3.0	-7.6	-21.8	-44.0	24"
3188	2765	2388	2053	1756	1493	0.5	0.8	0.0	-2.1	-5.5	-10.3	1.9	1.6	0.0	-2.9	-7.2	-20.6	-41.3	
967	600	373	262	208	173	0.9	0.0	-4.5	-13.5	-28.3	-49.9	0.0	-4.5	-13.5	-28.3	-49.9	-118.6	-228.2	20"
1696	1253	913	666	502	405	0.7	0.0	-3.3	-9.7	-19.6	-33.8	2.2	0.0	-5.3	-14.1	-27.2	-69.0	-136.9	24"
1412	886	521	301	172	119	0.4	0.8	0.0	-2.4	-6.7	-13.8	2.0	1.8	0.0	-3.8	-10.2	-35.0	-84.4	24"
1902	1296	858	565	399	316	0.5	0.0	-2.7	-8.2	-17.0	-30.0	1.8	0.0	-4.6	-12.5	-24.6	-65.3	-134.9	
1827	1355	989	720	535	425	0.6	0.0	-3.0	-8.9	-18.0	-31.1	2.0	0.0	-4.8	-13.0	-25.1	-63.6	-126.7	24"
1827	1355	989	720	535	425	0.6	0.0	-3.0	-8.9	-18.0	-31.1	2.0	0.0	-4.8	-13.0	-25.1	-63.6	-126.7	
1879	1416	1050	771	571	442	0.5	0.0	-2.7	-7.9	-16.1	-27.6	1.8	0.0	-4.3	-11.6	-22.3	-56.3	-111.9	
2207	1639	1193	860	626	482	0.5	0.0	-2.6	-7.7	-15.6	-27.1	1.7	0.0	-4.2	-11.3	-21.9	-55.8	-112.0	24"
2303	1845	1462	1143	806	685	0.3	0.0	-1.8	-5.4	11.0	18.8	2.7	2.2	0.0	-4.2	-10.7	-31.5	-65.6	

REMINGTON BALLISTICS

CENTERFIRE RIFLE

Caliber	Remington Order No.	Wt.-Grs.	Bullet Style	Primer No.	Muzzle	Velocity – Feet Per Second				
						100 Yds.	200 Yds.	300 Yds.	400 Yds.	500 Yds.
30-40 Krag	R30402	180°	Pointed Soft Point Core-Lokt®	9 ½	2430	2213	2007	1813	1632	1468
308 Win. Accelerator®	R308W5	55	Pointed Soft Point	9 ½	3770	3215	2726	2286	1888	1541
308 Win.	R308W1	150	Pointed Soft Point Core-Lokt®	9 ½	2820	2533	2263	2009	1774	1560
	R308W2	180	Soft Point Core-Lokt®	9 ½	2620	2274	1955	1666	1414	1212
	R308W3	180	Pointed Soft Point Core-Lokt®	9 ½	2620	2393	2178	1974	1782	1604
	R308W7	168	Boat Tail H.P. Match	9 ½	2680	2493	2314	2143	1979	1823
	ER308WA	165	Extended Range Boat Tail	9 ½	2700	2497	2303	2117	1941	1773
	ER308WB	178	Extended Range	9 ½	2620	2415	2220	2034	1857	1691
30-06 Accelerator®	R30069	55	Pointed Soft Point	9 ½	4080	3485	2965	2502	2083	1709
30-06 Springfield	R30061	125	Pointed Soft Point	9 ½	3140	2780	2447	2138	1853	1595
	R30062	150	Pointed Soft Point Core-Lokt®	9 ½	2910	2617	2342	2083	1843	1622
	R30063	150	Bronze Point	9 ½	2910	2656	2416	2189	1974	1773
	R3006B	165	Pointed Soft Point Core-Lokt®	9 ½	2800	2534	2283	2047	1825	1621
	R30064	180	Soft Point Core-Lokt®	9 ½	2700	2348	2023	1727	1466	1251
	R30065	180	Pointed Soft Point Core-Lokt®	9 ½	2700	2469	2250	2042	1846	1663
	R30066	180	Bronze Point	9 ½	2700	2485	2280	2084	1899	1725
	R30067	220	Soft Point Core-Lokt®	9 ½	2410	2130	1870	1632	1422	1246
	R3006C	168	Boat Tail H.P. Match	9 ½	2710	2522	2346	2169	2003	1845
	ER3006A	152	Extended Range	9 ½	2910	2654	2413	2184	1968	1765
	ER3006B	165	Extended Range Boat Tail	9 ½	2800	2592	2394	2204	2023	1852
	ER3006C	178	Extended Range	9 ½	2720	2511	2311	2121	1939	1768
300 H&H Mag.	R300HH	180	Pointed Soft Point Core-Lokt®	9 ½ M	2880	2640	2412	2196	1990	1798
300 Win. Mag.	R300W1	150	Pointed Soft Point Core-Lokt®	9 ½ M	3290	2951	2636	2342	2068	1813
	R300W2	180	Pointed Soft Point Core-Lokt®	9 ½ M	2960	2745	2540	2344	2157	1979
	RS300WA	200	Swift A-Frame™ PSP	9 ½ M	2825	2595	2376	2167	1970	1783
	ER300WA	178	Extended Range	9 ½ M	2980	2769	2568	2375	2191	2015
	ER300WB	190	Extended Range Boat Tail	9 ½ M	2885	2691	2506	2327	2156	1993
300 Wby Mag.	R300WB1	180	Pointed Soft Point Core-Lokt®	9 ½ M	3120	2866	2627	2400	2184	1979
	R300WB2	220	Soft Point Core-Lokt®	9 ½ M	2850	2541	2283	1984	1736	1512
	ER30WBA	178	Extended Range	9 ½ M	3120	2902	2695	2497	2308	2126
	ER30WBB	190	Extended Range Boat Tail	9 ½ M	3030	2830	2638	2455	2279	2110
303 British	R303B1	180	Soft Point Core-Lokt®	9 ½	2460	2124	1817	1542	1311	1137
7.62 x 39MM	R762391	125	Pointed Soft Point	7 ½	2365	2062	1783	1533	1320	1154
32-20 Win.	R32201	100	Lead	6 ½	1210	1021	913	834	769	712
	R32202	100	Soft Point	6 ½	1210	1021	913	834	769	712
32 Win. Special	R32WS2	170	Soft Point Core-Lokt®	9 ½	2250	1921	1626	1372	1175	1044
8MM Mauser	R8MSR	170	Soft Point Core-Lokt®	9 ½	2360	1969	1622	1333	1123	997
8MM Rem. Mag.	R8MM1	185	Pointed Soft Point Core-Lokt®	9 ½ M	3080	2761	2464	2186	1927	1688
338 Win. Mag.	R338W1	225	Pointed Soft Point	9 ½ M	2780	2572	2374	2184	2003	1832
	R338W2	250	Pointed Soft Point	9 ½ M	2660	2456	2261	2075	1898	1731
	RS338WA	225	Swift A-Frame PSP	9 ½ M	2785	2517	2266	2029	1808	1605
35 Rem.	R35R1	150	Pointed Soft Point Core-Lokt®	9 ½	2300	1874	1506	1218	1039	934
	R35R2	200	Soft Point Core-Lokt®	9 ½	2080	1698	1376	1140	1001	911
350 Rem. Mag.	R350M1	200	Pointed Soft Point Core-Lokt®	9 ½ M	2710	2410	2130	1870	1631	1421
35 Whelen	R35WH1	200	Pointed Soft Point	9 ½ M	2675	2378	2100	1842	1606	1399
	R35WH2	250	Soft Point	9 ½ M	2400	2066	1761	1492	1269	1107
	R35WH3	250	Pointed Soft Point	9 ½ M	2400	2197	2005	1823	1652	1496
375 H&H Mag.	R375M1	270	Soft Point	9 ½ M	2690	2420	2166	1928	1707	1507
	RS375MA	300	Swift A-Frame PSP	9 ½ M	2530	2245	1979	1733	1512	1321
416 Rem. Mag.	R416R1	400	Solid	9 ½ M	2400	2042	1718	1436	1212	1062
	R416R2	400	Swift A-Frame PSP	9 ½ M	2400	2175	1962	1763	1579	1414
	R416R3	350	Swift A-Frame PSP	9 ½ M	2520	2270	2034	1814	1611	1429
44-40 Win.	R4440W	200	Soft Point	2 ½	1190	1006	900	822	756	699
44 Rem. Mag.	R44MG2	240	Soft Point	2 ½	1760	1380	1114	970	878	806
	R44MG3	240	Semi-Jacketed Hollow Point	2 ½	1760	1380	1114	970	878	806
	R44MG6	210	Semi-Jacketed Hollow Point	2 ½	1920	1477	1155	982	880	802
444 Mar.	R444M	240	Soft Point	9 ½	2350	1815	1377	1087	941	846
45-70 Government	R4570G	405	Soft Point	9 ½	1330	1168	1055	977	918	869
	R4570L	300	Jacketed Hollow Point	9 ½	1810	1497	1244	1073	969	895
458 Win. Mag.	R458W1	500	Metal Case	9 ½ M	2040	1823	1623	1442	1237	1161
	R458W2	510	Soft Point	9 ½ M	2040	1770	1527	1319	1157	1046

REMINGTON BALLISTICS

CENTERFIRE RIFLE

Energy – Foot-Pounds						Short Range Trajectory[1]						Long Range Trajectory[2]							Barrel
Muzzle	100 Yds.	200 Yds.	300 Yds.	400 Yds.	500 Yds.	50 Yds.	100 Yds.	150 Yds.	200 Yds.	250 Yds.	300 Yds.	100 Yds.	150 Yds.	200 Yds.	250 Yds.	300 Yds.	400 Yds.	500 Yds.	Length
2360	1957	1610	1314	1064	861	0.4	0.0	-2.1	-6.2	-12.5	-21.1	1.4	0.0	-3.4	-8.9	-16.8	-40.9	-78.1	24"
1735	1262	907	638	435	290	0.2	0.5	0.0	-1.5	-4.2	-8.2	2.2	2.5	1.8	0.0	-3.2	-15.0	-36.7	24"
2648	2137	1705	1344	1048	810	0.2	0.0	-1.5	-4.5	-9.3	-15.9	2.3	1.9	0.0	-3.6	-9.1	-26.9	-55.7	
2743	2066	1527	1109	799	587	0.3	0.0	-2.0	-5.9	-12.1	-20.9	2.9	2.4	0.0	-4.7	-12.1	-36.9	-79.1	
2743	2288	1896	1557	1269	1028	0.2	0.0	-1.8	-5.2	-10.4	-17.7	2.6	2.1	0.0	-4.0	-9.9	-28.9	-58.8	24"
2678	2318	1998	1713	1460	1239	0.2	0.0	-1.6	-4.7	-9.4	-15.9	2.4	1.9	0.0	-3.5	-8.9	-25.3	-50.6	
2670	2284	1942	1642	1379	1152	0.2	0.0	-1.6	-4.7	-9.4	-16.0	2.3	1.9	0.0	-3.5	-8.9	-25.6	-51.5	
2713	2306	1948	1635	1363	1130	0.2	0.0	-1.7	-5.1	-10.2	-17.2	2.5	2.1	0.0	-3.8	-9.6	-27.6	-55.8	
2033	1483	1074	764	530	356	0.4	1.0	0.9	0.0	-1.9	-5.0	1.8	2.1	1.5	0.0	-2.7	-12.5	-30.5	24"
2736	2145	1662	1269	953	706	0.4	0.8	0.0	-2.1	-5.6	-10.7	1.8	1.5	0.0	-3.0	-7.7	-23.0	-48.5	
2820	2281	1827	1445	1131	876	0.6	0.9	0.0	-2.3	-6.3	-12.0	2.1	1.8	0.0	-3.3	-8.5	-25.0	-51.8	
2820	2349	1944	1596	1298	1047	0.6	0.9	0.0	-2.2	-6.0	-11.4	2.0	1.7	0.0	-3.2	-8.0	-23.3	-47.5	
2872	2352	1909	1534	1220	963	0.7	1.0	0.0	-2.5	-6.7	-12.7	2.3	1.9	0.0	-3.6	-9.0	-26.3	-54.1	
2913	2203	1635	1192	859	625	0.2	0.0	-1.8	-5.5	-11.2	-19.5	2.7	2.3	0.0	-4.4	-11.3	-34.4	-73.7	
2913	2436	2023	1666	1362	1105	0.2	0.0	-1.6	-4.8	-9.7	-16.5	2.4	2.0	0.0	-3.7	-9.3	-27.0	-54.9	24"
2913	2468	2077	1736	1441	1189	0.2	0.0	-1.6	-4.7	-9.6	-16.2	2.4	2.0	0.0	-3.6	-9.1	-26.2	-53.0	
2837	2216	1708	1301	988	758	0.4	0.0	-2.3	-6.8	-13.8	-23.6	1.5	0.0	-3.7	-9.9	-19.0	-47.4	-93.1	
2739	2372	2045	1754	1497	1270	0.7	1.0	-0.0	-2.5	-6.6	-12.4	2.3	1.9	-0.0	-3.5	-8.6	-24.7	-49.4	
2858	2378	1965	1610	1307	1052	0.6	0.9	0.0	-2.3	-6.0	-11.4	2.0	1.7	0.0	-3.2	-8.0	-23.3	-47.7	
2872	2462	2100	1780	1500	1256	0.6	1.0	0.0	-2.4	-6.2	-11.8	2.1	1.8	0.0	-3.3	-8.2	-23.6	-47.5	
2924	2491	2111	1777	1486	1235	0.7	1.0	0.0	-2.6	-6.7	-12.7	2.3	1.9	0.0	-3.5	-8.8	-25.4	-51.2	
3315	2785	2325	1927	1583	1292	0.6	0.9	0.0	-2.3	-6.0	-11.5	2.1	1.7	0.0	-3.2	-8.0	-23.3	-47.4	24"
3605	2900	2314	1827	1424	1095	0.3	0.7	0.0	-1.8	-4.8	-9.3	2.6	2.9	2.1	0.0	-3.5	-15.4	-35.5	
3501	3011	2578	2196	1859	1565	0.5	0.8	0.0	-2.1	-5.5	-10.4	1.9	1.6	0.0	-2.9	-7.3	-20.9	-41.9	24"
3544	2989	2506	2086	1722	1412	0.6	1.0	0.0	-2.4	-6.3	-11.9	2.1	1.8	0.0	-3.3	-8.3	-24.0	-48.8	
3509	3030	2606	2230	1897	1605	0.5	0.8	0.0	-2.0	-5.4	-10.2	1.8	1.5	0.0	-2.9	-7.1	-20.4	-40.9	
3511	3055	2648	2285	1961	1675	0.5	0.9	0.0	-2.2	-5.7	-10.7	1.9	1.6	0.0	-3.0	-7.5	-21.4	-42.9	
3890	3284	2758	2301	1905	1565	0.4	0.7	0.0	-1.9	-5.0	-9.5	2.7	3.0	2.1	0.0	-3.5	-15.2	-34.2	
3967	3155	2480	1922	1471	1117	0.6	1.0	0.0	-2.5	-6.7	-12.9	2.3	1.9	0.0	-3.6	-9.1	-27.2	-56.8	24"
3847	3329	2870	2464	2104	1787	0.4	0.7	0.0	-1.8	-4.8	-9.1	2.6	2.9	2.0	0.0	-3.3	-14.3	-31.8	
3873	3378	2936	2542	2190	1878	0.4	0.8	0.0	-1.9	-5.1	-9.6	1.7	1.4	0.0	-2.7	6.7	-19.2	-38.4	
2418	1803	1319	950	687	517	0.4	0.0	-2.3	-6.9	-14.1	-24.4	1.5	0.0	-3.8	-10.2	-19.8	-50.5	-101.5	24"
1552	1180	882	652	483	370	0.4	0.0	-2.5	-7.3	-14.3	-25.7	1.7	0.0	-4.8	-10.8	-20.7	-52.3	-104.0	24"
325	231	185	154	131	113	0.0	-6.3	-20.9	-44.9	-79.3	-125.1	0.0	-11.5	-32.3	-63.8	-106.3	-230.3	-413.3	24"
325	231	185	154	131	113	0.0	-6.3	-20.9	-44.9	-79.3	-125.1	0.0	-11.5	-32.3	-63.6	-106.3	-230.3	-413.3	
1911	1393	998	710	521	411	0.6	0.0	-2.9	-8.6	-17.6	-30.5	1.9	0.0	-4.7	-12.7	-24.7	-63.2	-126.9	24"
2102	1463	993	671	476	375	0.5	0.0	-2.7	-8.2	-17.0	-29.8	1.8	0.0	-4.5	-12.4	-24.3	-63.8	-130.7	24"
3896	3131	2494	1963	1525	1170	0.5	0.8	0.0	-2.1	-5.6	-10.7	1.8	1.6	0.0	-3.0	-7.6	-22.5	-46.8	24"
3860	3305	2815	2383	2004	1676	0.6	1.0	0.0	-2.4	-6.3	-12.0	2.2	1.8	0.0	-3.3	-8.4	-24.0	-48.4	
3927	3348	2837	2389	1999	1663	0.2	0.0	-1.7	-4.9	-9.8	-16.6	2.4	2.0	0.0	-3.7	-9.3	-26.6	-53.6	24"
3871	3165	2565	2057	1633	1286	0.2	0.0	-1.5	-4.6	-9.4	-16.0	2.3	1.9	0.0	-3.6	-9.1	-26.7	-54.9	
1762	1169	755	494	359	291	0.6	0.0	-3.0	-9.2	-19.1	-33.9	2.0	0.0	-5.1	-14.1	-27.8	-74.0	-152.3	24"
1921	1280	841	577	445	369	0.8	0.0	-3.8	-11.3	-23.5	-41.2	2.5	0.0	-6.3	-17.1	-33.6	-87.5	-176.4	
3261	2579	2014	1553	1181	897	0.2	0.0	-1.7	-5.1	-10.4	-17.9	2.6	2.1	0.0	-4.0	-10.3	-30.5	-64.0	20"
3177	2510	1958	1506	1145	869	0.2	0.0	-1.8	-5.3	-10.8	-18.5	2.6	2.2	0.0	-4.2	-10.6	-31.5	-65.9	
3197	2369	1722	1235	893	680	0.4	0.0	-2.5	-7.3	-15.0	-26.0	1.6	0.0	-4.0	-10.9	-21.0	-53.8	-108.2	24"
3197	2680	2230	1844	1515	1242	0.4	0.0	-2.2	-6.3	-12.6	-21.3	1.4	0.0	-3.4	-9.0	-17.0	-41.0	-77.8	
4337	3510	2812	2228	1747	1361	0.2	0.0	-1.7	-5.1	-10.3	-17.6	2.5	2.1	0.0	-3.9	-10.0	-29.4	-60.7	24"
4262	3357	2608	2001	1523	1163	0.3	0.0	-2.0	-6.0	-12.3	-21.0	3.0	2.5	0.0	-4.7	-12.0	-35.6	-74.5	
5115	3702	2620	1832	1305	1001	0.4	0.0	-2.5	-7.5	-15.5	-27.0	1.7	0.0	-4.2	-11.3	-21.9	-56.7	-115.1	
5115	4201	3419	2760	2214	1775	0.4	0.0	-2.2	-6.5	-13.0	-22.0	1.5	0.0	-3.5	-9.3	-17.6	-42.9	-82.2	24"
4935	4004	3216	2557	2017	1587	0.3	0.0	-2.0	-5.9	-11.9	-20.2	2.9	2.4	0.0	-4.5	-11.4	-33.4	-68.7	
629	449	360	300	254	217	0.0	-6.5	-21.6	-46.3	-81.8	-129.1	0.0	-11.8	-33.3	-65.5	-109.5	-237.4	-426.2	24"
1650	1015	661	501	411	346	0.0	-2.7	-10.0	-23.0	-43.0	-71.2	0.0	-5.9	-17.6	-36.3	-63.1	-145.5	-273.0	
1650	1015	661	501	411	346	0.0	-2.7	-10.0	-23.0	-43.0	-71.2	0.0	-5.9	-17.6	-36.3	-63.1	-145.5	-273.0	20"
1719	1017	622	450	361	300	0.0	-2.2	-8.3	-19.7	-37.6	-63.2	0.0	-5.1	-15.4	-32.1	-56.7	-134.0	-256.2	
2942	1755	1010	630	472	381	0.6	0.0	-3.2	-9.9	-21.3	-38.5	2.1	0.0	-5.6	-15.9	-32.1	-87.8	-182.7	24"
1590	1227	1001	858	758	679	0.0	-4.7	-15.8	-34.0	-60.0	-94.5	0.0	-8.7	-24.6	-48.2	-80.3	-172.4	-305.9	24"
2182	1492	1031	767	625	533	0.0	-2.3	-8.5	-19.4	-35.9	-59.0	0.0	-5.0	-14.8	-30.1	-52.1	-119.5		
4620	3689	2924	2308	1839	1469	0.7	0.0	-3.3	-9.6	-19.2	-32.5	2.2	0.0	-5.2	-13.6	-25.8	-63.2	-121.7	24"
4712	3547	2640	1970	1516	1239	0.8	0.0	-3.5	-10.3	-20.8	-35.6	2.4	0.0	-5.6	-14.9	-28.5	-71.5	-140.4	

SUGGESTED USAGE	Cartridge	Weight Grains	Bullet Type	VELOCITY in Feet per Second						ENERGY in Foot-Pounds						PATH OF BULLET Above or below line-of-sight of riflescopes mounted 1.5" above bore				
				Muzzle	100 Yards	200 Yards	300 Yards	400 Yards	500 Yards	Muzzle	100 Yards	200 Yards	300 Yards	400 Yards	500 Yards	100 Yards	200 Yards	300 Yards	400 Yards	500 Yards
V	.224 WM	55	Pt-Ex	3650	3192	2780	2403	2057	1742	1627	1244	944	705	516	370	2.8	3.7	0.0	-9.7	-27.7
V M	.240 WM	87	Pt-Ex	3523	3233	2943	2680	2432	2197	2398	2007	1673	1388	1143	933	2.5	3.3	0.0	-8.0	-22.0
		100	Pt-Ex	3406	3116	2844	2588	2346	2117	2577	2156	1796	1488	1222	996	2.8	3.5	0.0	-8.6	-23.6
		100	Partition	3406	3136	2881	2641	2413	2196	2577	2184	1843	1549	1293	1071	2.7	3.5	0.0	-8.3	-22.7
V M	.257 WM	87	Pt-Ex	3825	3456	3118	2803	2511	2236	2827	2308	1878	1518	1218	966	2.1	2.8	0.0	-7.2	-20.0
		100	Pt-Ex	3602	3280	2980	2701	2438	2190	2882	2389	1973	1620	1320	1065	2.4	3.2	0.0	-7.8	-21.6
		100	*Partition*	3555	3270	3004	2753	2516	2290	2806	2374	2053	1683	1405	1165	2.5	3.2	0.0	-7.8	-21.0
		120	Partition	3305	3045	2800	2568	2348	2139	2911	2472	2090	1758	1469	1219	3.0	3.7	0.0	-8.8	-24.0
V M	.270 WM	100	Pt-Ex	3760	3380	3033	2712	2412	2133	3139	2537	2042	1633	1292	1010	2.3	3.0	0.0	-7.8	-21.6
		130	Pt-Ex	3375	3100	2842	2598	2367	2147	3287	2773	2330	1948	1616	1331	2.9	3.6	0.0	-8.7	-23.7
		130	Partition	3375	3127	2893	2670	2458	2257	3287	2822	2415	2058	1714	1470	2.8	3.5	0.0	-8.3	-22.4
		150	Pt-Ex	3245	3019	2803	2598	2402	2215	3507	3034	2617	2248	1922	1634	3.0	3.7	0.0	-8.9	-23.8
		150	Partition	3245	3029	2823	2627	2439	2259	3507	3055	2655	2298	1981	1699	3.0	3.7	0.0	-8.7	-23.3
M	7MM WM	139	Pt-Ex	3340	3082	2838	2608	2389	2180	3443	2931	2486	2099	1761	1467	2.9	3.6	0.0	-8.7	-23.6
		140	Partition	3303	3069	2847	2636	2434	2241	3391	2927	2519	2159	1842	1562	2.9	3.6	0.0	-8.6	-23.1
		154	Pt-Ex	3260	3022	2797	2583	2379	2184	3633	3123	2675	2281	1934	1630	3.0	3.7	0.0	-9.0	-24.1
		160	Partition	3200	2991	2791	2600	2417	2241	3637	3177	2767	2401	2075	1784	3.1	3.8	0.0	-8.9	-23.8
B		175	Pt-Ex	3070	2879	2696	2520	2351	2188	3662	3220	2824	2467	2147	1861	3.4	4.1	0.0	-9.6	-25.4
M	.300 WM	150	Pt-Ex	3600	3297	3016	2751	2502	2266	4316	3621	3028	2520	2084	1709	2.4	3.1	0.0	-7.7	-21.0
		150	Partition	3600	3319	3057	2809	2575	2353	4316	3669	3111	2628	2208	1843	2.4	3.0	0.0	-7.5	-20.1
		165	Boat Tail	3450	3220	3003	2796	2598	2409	4360	3799	3303	2863	2473	2146	2.5	3.2	0.0	-7.6	-20.4
B		180	Pt-Ex	3300	3064	2841	2629	2426	2233	4352	3753	3226	2762	2352	1992	2.9	3.6	0.0	-8.6	-23.2
		180	Partition	3300	3085	2881	2686	2499	2319	4352	3804	3317	2882	2495	2150	2.8	3.5	0.0	-8.3	-22.3
		220	Rn-Ex	2905	2498	2125	1787	1491	1250	4122	3047	2206	1560	1085	763	5.3	6.6	0.0	-17.6	-51.3
B	.340 WM	200	Pt-Ex	3260	3011	2775	2552	2339	2137	4719	4025	3420	2892	2430	2027	3.1	3.8	0.0	-9.1	-24.7
		210	Partition	3250	3000	2763	2539	2325	2122	4924	4195	3559	3004	2520	2098	3.1	3.8	0.0	-9.2	-24.9
		250	Rn-Ex	3002	2672	2365	2079	1814	1574	5002	3963	3105	2399	1827	1375	1.7	0.0	-7.9	-24.0	-50.7
		250	Partition	2980	2780	2588	2404	2228	2059	4931	4290	3719	3209	2756	2354	3.7	4.4	0.0	-10.3	-27.5
B	.378 WM	270	Pt-Ex	3180	2976	2781	2594	2415	2243	6062	5308	4635	4034	3496	3015	1.2	0.0	-5.7	-16.6	-33.5
																3.1	3.8	0.0	-9.0	-23.9
		300	Rn-Ex	2925	2603	2303	2024	1764	1531	5701	4516	3535	2729	2074	1563	1.8	0.0	-8.3	-25.2	-53.5
A		300	FMJ	2925	2580	2262	1972	1710	1482	5701	4434	3408	2592	1949	1463	1.84	0.0	-8.6	-26.1	-55.9
A	.416 WM	400	Swift P	2650	2411	2185	1971	1770	1585	6239	5165	4242	3450	2783	2233	5.4	6.3	0.0	-15.1	-41.7
																2.2	0.0	-9.5	-27.7	-57.5
		400	Rn-Ex	2700	2390	2101	1834	1591	1379	6474	5073	3921	2986	2247	1688	5.7	6.8	0.0	-17.2	-48.4
																2.3	0.0	-10.2	-30.9	-65.4
		400	**Mono Solid®	2700	2397	2115	1852	1613	1402	6474	5104	3971	3047	2310	1747	5.7	6.7	0.0	-17.0	-47.4
																2.3	0.0	-10.1	-30.4	-64.3
A	.460 WM	500	RNSP	2600	2310	2039	1787	1559	1359	7507	5926	4618	3545	2701	2050	2.5	0.0	-10.8	-32.7	-69.2
		500	FMJ	2600	2330	2077	1839	1623	1426	7507	6030	4791	3755	2924	2258					

LEGEND: Pt-Ex = Pointed-Expanding   Rn-Ex = Round nose-Expanding   FMJ = Full Metal Jacket   P = Divided Lead Cavity or "H" Type

Note: These tables were calculated by computer using a standard modern scientific technique to predict trajectories and recoil energies from the best available cartridge data. Figures shown are expected to be reasonably accurate; however, the shooter is cautioned that performance will vary because of variations in rifles, ammunition, atmospheric conditions and altitude. Velocities were determined using 26-inch barrels; shorter barrels will reduce velocity by 30 to 65 fps per inch of barrel removed. Trajectories were computed with the line-of-sight 1.5 inches above the bore centerline. *B.C.*: Ballistic Coefficients supplied by the bullet manufacturers. * Partition is a registered trademark of Nosler, Inc. ** Monolithic Solid is a registered trademark of A-Square, Inc.

WINCHESTER BALLISTICS

CENTERFIRE PISTOL/REVOLVER

Cartridge	Symbol	Bullet Wt. Grs.	Type	Velocity (fps) Muzzle	50 Yds.	100 Yds.	Energy (ft-lbs.) Muzzle	50 Yds.	100 Yds.	Mid Range Traj. (In.) 50 Yds.	100 Yds.	Barrel Length Inches
25 Automatic	X25AXP	45	Expanding Point**	815	729	655	66	53	42	1.8	7.7	2
25 Automatic	X25AP	50	Full Metal Jacket	760	707	659	64	56	48	2.0	8.7	2
30 Luger (7.65mm)	X30LP	93	Full Metal Jacket	1220	1110	1040	305	255	225	0.9	3.5	4-1/2
30 Carbine #	X30M1	110	Hollow Soft Point	1790	1601	1430	783	626	500	0.4	1.7	10
32 Smith & Wesson	X32SWP	85	Lead-Round Nose	680	645	610	90	81	73	2.5	10.5	3
32 Smith & Wesson Long	X32SWLP	98	Lead-Round Nose	705	670	635	115	98	88	2.3	10.5	4
32 Short Colt	X32SCP	80	Lead-Round Nose	745	665	590	100	79	62	2.2	9.9	4
32 Automatic	X32ASHP	60	Silvertip® Hollow Point	970	895	835	125	107	93	1.3	5.4	4
32 Automatic	X32AP	71	Full Metal Jacket	905	855	810	129	115	97	1.4	5.8	4
38 Smith & Wesson	X38SWP	145	Lead-Round Nose	685	650	620	150	135	125	2.4	10.0	4
380 Automatic	X380ASHP	85	Silvertip Hollow Point	1000	921	860	189	160	140	1.2	5.1	3-3/4
380 Automatic	X380AP	95	Full Metal Jacket	955	865	785	190	160	130	1.4	5.9	3-3/4
38 Special	X38S9HP	110	Silvertip Hollow Point	945	894	850	218	195	176	1.3	5.4	4V
38 Special Super Unleaded™	X38SU	130	Full Metal Jacket-Encapsulated	775	743	712	173	159	146	1.9	7.9	4V
38 Special Super Match®	X38SMRP	148	Lead-Wad Cutter	710	634	566	166	132	105	2.4	10.8	4V
38 Special	X38S1P	158	Lead-Round Nose	755	723	693	200	183	168	2.0	8.3	4V
38 Special	X38WCPSV	158	Lead-Semi Wad Cutter	755	721	689	200	182	167	2.0	8.4	4V
38 Special + P	X38SSHP	95	Silvertip Hollow Point	1100	1002	932	255	212	183	1.0	4.3	4V
38 Special + P#	X38S6PH	110	Jacketed Hollow Point	995	926	871	242	210	185	1.2	5.1	4V
38 Special + P#	X38S7PH	125	Jacketed Hollow Point	945	898	858	248	224	204	1.3	5.4	4V
38 Special + P#	X38S8HP	125	Silvertip Hollow Point	945	898	858	248	224	204	1.3	5.4	4V
38 Special Subsonic® + P	XSUB38S	147	Jacketed Hollow Point	860	830	802	241	225	210	1.5	6.3	4V
38 Special + P	X38SPD	158	Lead-Semi Wad Cutter Hollow Point	890	855	823	278	257	238	1.4	6.0	4V
38 Special + P	X38WCP	158	Lead-Semi Wad Cutter	890	855	823	278	257	236	1.4	6.0	4V
9mm Luger Super Unleaded	Y9MMSU	115	Full Metal Jacket-Encapsulated	1155	1047	971	341	280	241	0.9	3.9	4
9mm Luger	X9LP	115	Full Metal Jacket	1155	1047	971	341	280	241	0.9	3.9	4
9mm Luger	X9MMSHP	115	Silvertip Hollow Point	1225	1095	1007	383	306	259	0.8	3.6	4
9mm Luger Subsonic	XSUM9MM	147	Jacketed Hollow Point	990	945	907	320	292	268	1.2	4.8	4
9mm Luger	X9MMST147	147	Silvertip Hollow Point	1010	962	921	333	302	277	1.1	4.7	4
9mm Luger Super Match	X9MMTCM	147	Full Metal Jacket-Truncated Cone-Match	990	945	907	320	292	268	1.2	4.8	4
38 Super Automatic + P*	X38ASHP	125	Silvertip Hollow Point	1240	1130	1050	427	354	306	0.8	3.4	5
38 Super Automatic + P*	X38A1P	130	Full Metal Jacket	1215	1099	1017	426	348	298	0.8	3.6	5
357 Magnum #	X3573P	110	Jacketed Hollow Point	1295	1095	975	410	292	232	0.8	3.5	4V
357 Magnum #	X3576P	125	Jacketed Hollow Point	1450	1240	1090	583	427	330	0.6	2.8	4V
357 Magnum #	X357SHP	145	Silvertip Hollow Point	1290	1155	1060	535	428	361	0.8	3.5	4V
357 Magnum	X3571P	158	Lead-Semi Wad Cutter**	1235	1104	1015	535	428	361	0.8	3.5	4V
357 Magnum	X3574P	158	Jacketed Hollow Point	1235	1104	1015	535	428	361	0.8	3.5	4V
357 Magnum	X3575P	158	Jacketed Soft Point	1235	1104	1015	535	428	361	0.8	3.5	4V
40 Smith & Wesson	X40SWSTHP	155	Silvertip Hollow Point	1205	1096	1018	500	414	357	0.8	3.6	4
40 Smith & Wesson Super Match	X40SWTCM	155	Full Metal Jacket-Truncated Cone-Match	1125	1046	986	436	377	335	0.9	3.9	4
40 Smith & Wesson Super Unleaded	X40SWSU	180	Full Metal Jacket-Encapsulated	990	933	886	392	348	314	1.2	5.0	4
40 Smith & Wesson Subsonic	XSUB40SW	180	Jacketed Hollow Point	1010	954	909	408	364	330	1.1	4.8	4
10mm Automatic Super Match	X10MMTCM	155	Full Metal Jacket - Truncated Cone-Match	1125	1046	986	436	377	335	0.9	3.9	5
10mm Automatic	X10MMSTHP	175	Silvertip Hollow Point	1290	1141	1037	649	506	418	0.7	3.3	5-1/2
10mm Automatic Subsonic	XSUB10MM	180	Jacketed Hollow Point	990	936	891	390	350	317	1.2	4.9	5
41 Remington Magnum #	X41MSTHP2	175	Silvertip Hollow Point	1250	1120	1029	607	488	412	0.8	3.4	4V
41 Remington Magnum #	X41MHP2	210	Jacketed Hollow Point	1300	1162	1062	788	630	526	0.7	3.2	4V
44 Smith & Wesson Special #	X44STHPS2	200	Silvertip Hollow Point	900	860	822	360	328	300	1.4	5.9	6-1/2
44 Smith & Wesson Special	X44SP	246	Lead-Round Nose	755	725	695	310	285	265	2.0	8.3	6-1/2
44 Remington Magnum	X44MSTHP2	210	Silvertip Hollow Point	1250	1106	1010	729	570	475	0.8	3.5	4V
44 Remington Magnum	X44MHSP2	240	Hollow Soft Point	1180	1081	1010	741	623	543	0.9	3.7	4V
45 Automatic	X45ASHP2	185	Silvertip Hollow Point	1000	938	888	411	362	324	1.2	4.9	5
45 Automatic Super Match	X45AWCP	185	Full Metal Jacket - Semi Wad Cutter	770	707	650	244	205	174	2.0	8.7	5
45 Automatic Subsonic	XSUB45A	230	Jacketed Hollow Point	925	880	840	437	396	361	1.3	5.6	5
45 Automatic	X45A1P2	230	Full Metal Jacket	835	800	767	356	326	300	1.6	6.8	5
45 Colt #	X45CSHP2	225	Silvertip Hollow Point	920	877	839	423	384	352	1.4	5.6	5-1/2
45 Colt	X45CP2	255	Lead-Round Nose	860	820	780	420	380	345	1.5	6.1	5-1/2
45 Winchester Magnum #	X45WM2	230	Full Metal Jacket	1400	1232	1107	1001	775	636	0.6	2.8	5
(Not for Arms Chambered for Standard 45 Automatic)												
45 Winchester Magnum #	X45WMA	260	Jacketed Soft Point	–	–	–	Under Development			–	–	–
(Not for Arms Chambered for Standard 45 Automatic)												

+P Ammunition with (+P) on the case head stamp is loaded to higher pressure. Use only in firearms designated for this cartridge and so recommended by the gun manufacturer.

V-Data is based on velocity obtained from 4" vented test barrels for revolver cartridges (38 Special, 357 Magnum, 41 Rem. Mag. and 44 Rem. Mag.)

Specifications are nominal. Test barrels are used to determine ballistics figures. Individual firearms may differ from test barrel statistics.

Specifications subject to change without notice.

**Lubaloy® Coated
*For use only in 38 Super Automatic Pistols.
#Acceptable for use in rifles also.

WINCHESTER BALLISTICS

WINCHESTER® CENTERFIRE RIFLE AMMUNITION

Game Selector	CXP Class	Examples
V-Varmint	1	Prairie dog, coyote, woodchuck
D-Deer	2	Antelope, deer, black bear
O/P-Open or Plains	3	Elk, moose
M-Medium Game	3D	All game in category 3 plus large dangerous game (i.e. Kodiak bear)
L-Large Game	4	Cape Buffalo, elephant
XL-Extra Large Game	M	Match

Acceptable for use in pistols and revolvers also.
Bold type indicates Supreme® product line

*Intended for use in fast twist barrels (e.g., 1 in 7 to 1 in 9). Slower twist barrels may not sufficiently stabilize bullet.

Cartridge	Symbol	Game Selector Guide	CXP Guide Number	Bullet Wt. Grs.	Bullet Type	Barrel Length (In.)	Velocity in Feet Per Second (fps) Muzzle	100	200	300	400	500	Energy in Foot Pounds (ft-lbs.) Muzzle	100	200	300	400	500	Trajectory, Short Range Yards 50	100	150	200	250	300	Trajectory, Long Range Yards 100	150	200	250	300	400	500
218 Bee	X218B	V	1	46	Hollow Point	24	2760	2102	1550	1155	961	850	778	451	245	136	94	74	0.3	0	-2.3	-7.2	-15.8	-29.4	1.5	0	-4.2	-12.0	-24.8	-71.4	-155.6
22 Hornet	X22H1	V	1	45	Soft Point	24	2690	2042	1502	1128	948	840	723	417	225	127	90	70	0.3	0	-2.4	-7.7	-16.9	-31.3	1.6	0	-4.5	-12.8	-26.4	-75.6	-163.4
22 Hornet	X22H2	V	1	46	Hollow Point	24	2690	2042	1502	1128	948	841	739	426	230	130	92	72	0.3	0	-2.4	-7.7	-16.9	-31.3	1.6	0	-4.5	-12.8	-26.4	-75.5	-163.3
22-250 Remington	**S22250**	**V**	**1**	**52**	**Hollow Point Boattail**	**24**	**3750**	**3268**	**2835**	**2442**	**2082**	**1755**	**1624**	**1233**	**928**	**689**	**501**	**356**	**0.1**	**0**	**-0.7**	**-2.4**	**-5.1**	**-9.1**	**1.2**	**1.1**	**0**	**-2.1**	**-5.5**	**-16.9**	**-36.3**
22-250 Remington	X222501	V	1	55	Pointed Soft Point	24	3680	3137	2656	2222	1832	1493	1654	1201	861	603	410	272	0.2	0.5	0	-1.6	-4.4	-8.7	2.3	2.6	1.9	0	-3.4	-15.9	-38.9
222 Remington	X222R	V	1	50	Pointed Soft Point	24	3140	2602	2123	1700	1350	1107	1094	752	500	321	202	136	0.5	0.9	0	-2.5	-6.9	-13.7	2.2	1.9	0	-3.8	-10.0	-32.3	-73.8
222 Remington	X222R1	V	1	55	Full Metal Jacket	24	3020	2675	2355	2057	1783	1537	1114	874	677	517	388	288	0.5	0.9	0	-2.2	-6.1	-11.7	2.0	1.7	0	-3.3	-8.3	-24.9	-52.5
223 Remington	X223RH	V	1	53	Hollow Point	24	3330	2882	2477	2106	1770	1475	1305	978	722	522	369	256	0.3	0.7	0	-1.9	-5.3	-10.3	1.7	1.4	0	-2.9	-7.4	-22.7	-49.1
223 Remington	X223R	V	1	55	Pointed Soft Point	24	3240	2747	2304	1905	1554	1270	1282	921	648	443	295	197	0.4	0.8	0	-2.2	-6.0	-11.8	1.9	1.6	0	-3.3	-8.5	-26.7	-59.6
223 Remington	X223R1	V	1	55	Full Metal Jacket	24	3240	2877	2543	2232	1943	1679	1282	1011	790	608	461	344	0.4	0.7	0	-1.9	-5.1	-9.9	1.7	1.4	0	-2.8	-7.1	-21.2	-44.6
223 Remington	X223R2	D	2	64	Power-Point®	24	3020	2621	2256	1920	1619	1362	1296	977	723	524	373	264	0.6	0.9	0	-2.4	-6.5	-12.5	2.1	1.8	0	-3.5	-9.0	-27.4	-59.6
223 Remington Match	**S223M***	**–**	**M**	**69**	**Hollow Point Boattail**	**24**	**3060**	**2740**	**2442**	**2164**	**1904**	**1665**	**1435**	**1151**	**914**	**717**	**555**	**425**	**-0.2**	**0**	**-0.9**	**-3.1**	**-6.8**	**-12.1**	**1.6**	**1.4**	**0**	**-2.9**	**-7.4**	**-22.3**	**-46.7**
225 Winchester	X2251	V	1	55	Pointed Soft Point	24	3570	3066	2616	2208	1838	1514	1556	1148	836	595	412	280	0.2	0.6	0	-1.7	-4.6	-9.0	2.4	2.8	2.0	0	-3.5	-16.3	-39.5
243 Winchester	X2431	V	1	80	Pointed Soft Point	24	3350	2955	2593	2259	1951	1670	1993	1551	1194	906	676	495	0.3	0.7	0	-1.8	-4.9	-9.4	2.6	2.9	2.1	0	-3.6	-16.2	-37.9
243 Winchester	X2432	D,O/P	2	100	Power-Point	24	2960	2697	2449	2215	1993	1786	1945	1615	1332	1089	882	708	0.5	0.9	0	-2.2	-5.8	-11.0	1.9	1.6	0	-3.1	-7.8	-22.6	-46.3
243 Winchester	**S243**	**D,O/P**	**2**	**100**	**Soft Point Boattail**	**24**	**2960**	**2712**	**2477**	**2254**	**2042**	**1843**	**1946**	**1633**	**1363**	**1128**	**926**	**754**	**0.1**	**0**	**-1.3**	**-3.8**	**-7.8**	**-13.3**	**1.9**	**1.6**	**0**	**-3.0**	**-7.6**	**-22.0**	**-44.8**
6mm Remington	X6MMR1	V	1	80	Pointed Soft Point	24	3470	3064	2694	2352	2036	1747	2139	1667	1289	982	736	542	0.3	0.6	0	-1.6	-4.5	-8.7	2.4	2.7	1.9	0	-3.3	-14.9	-35.0
6mm Remington	X6MMR2	D,O/P	2	100	Power-Point	24	3100	2829	2573	2332	2104	1889	2133	1777	1470	1207	983	792	0.4	0.8	0	-1.9	-5.2	-9.9	1.7	1.5	0	-2.8	-7.0	-20.4	-41.7
25-06 Remington	X25061	V	1	90	Positive Expanding Point	24	3440	3043	2680	2344	2034	1749	2364	1850	1435	1098	827	611	0.3	0.6	0	-1.7	-4.5	-8.8	2.4	2.7	2.0	0	-3.4	-15.0	-35.2
25-06 Remington	X25062	D,O/P	2	120	Positive Expanding Point	24	2990	2730	2484	2252	2032	1825	2382	1985	1644	1351	1100	887	0.5	0.8	0	-2.1	-5.6	-10.7	1.9	1.6	0	-3.0	-7.5	-22.0	-44.8
25-20 Winchester #	X25202	V	1	86	Soft Point	24	1460	1194	1030	931	858	798	407	272	203	165	141	122	0	-4.1	-14.4	-31.8	-57.3	-92.0	0	-8.2	-23.5	-47.0	-79.6	-175.9	-319.4
25-35 Winchester	X2535	D	2	117	Soft Point	24	2230	1866	1545	1282	1097	984	1292	904	620	427	313	252	0.6	0	-3.1	-9.2	-19.0	-33.1	2.1	0	-5.1	-13.8	-27.0	-70.1	-142.0
250 Savage	X2503	D,O/P	2	100	Silvertip®	24	2820	2467	2140	1839	1569	1339	1765	1351	1017	751	547	398	0.2	0	-1.6	-4.9	-10.0	-17.4	2.4	2.0	0	-3.9	-10.1	-30.5	-65.2
257 Roberts + P	X257P3	D,O/P	2	117	Power-Point	24	2780	2411	2071	1761	1488	1263	2009	1511	1115	806	576	415	0.8	1.1	0	-2.9	-7.8	-15.1	2.6	2.2	0	-4.2	-10.8	-33.0	-70.0
264 Winchester Mag.	X2642	D,O/P	2	140	Power-Point	24	3030	2782	2548	2326	2114	1914	2854	2406	2018	1682	1389	1139	0.5	0.8	0	-2.0	-5.4	-10.2	1.8	1.5	0	-2.9	-7.2	-20.8	-42.2
New 6.5 x 55 Swedish	X6555	D,O/P	2	140	Soft Point	–	–	–	–	–	–	–	Under Development						–	–	–	–	–	–	–	–	–	–	–	–	–
270 Winchester	X2701	V	1	100	Pointed Soft Point	24	3430	3021	2649	2305	1988	1699	2612	2027	1557	1179	877	641	0.3	0.6	0	-1.7	-4.6	-9.0	2.5	2.8	2.0	0	-3.4	-15.5	-36.4
270 Winchester	X2705	D,O/P	2	130	Power-Point	24	3060	2802	2559	2329	2110	1904	2702	2267	1890	1565	1285	1046	0.4	0.8	0	-2.0	-5.3	-10.1	1.8	1.5	0	-2.8	-7.1	-20.6	-42.0
270 Winchester	X2703	D,O/P	2	130	Silvertip	24	3060	2776	2510	2259	2022	1801	2702	2225	1818	1472	1180	936	0.5	0.8	0	-2.0	-5.5	-10.4	1.8	1.5	0	-2.9	-7.4	-21.6	-44.3
270 Winchester	**S270**	**D,O/P**	**2**	**140**	**Silvertip Boattail**	**24**	**2960**	**2753**	**2554**	**2365**	**2183**	**2009**	**2724**	**2356**	**2029**	**1739**	**1482**	**1256**	**0.1**	**0**	**-1.2**	**-3.7**	**-7.5**	**-12.7**	**1.8**	**1.5**	**0**	**-2.9**	**-7.2**	**-20.6**	**-41.3**
New 270 Winchester Fail Safe®	**S270X**	–	–	–	–	–	–	–	–	–	–	–	**Under Development**						–	–	–	–	–	–	–	–	–	–	–	–	–
270 Winchester	X2704	D,M	3	150	Power-Point	24	2850	2585	2336	2100	1879	1673	2705	2226	1817	1468	1175	932	0.6	1.0	0	-2.4	-6.4	-12.2	2.2	1.8	0	-3.4	-8.6	-25.0	-51.4
280 Remington	X280R	D,O/P	2	140	Power-Point	24	3050	2705	2428	2167	1924	1698	2799	2274	1833	1461	1151	897	0.5	0.8	0	-2.2	-5.8	-11.1	1.9	1.6	0	-3.1	-7.8	-23.1	-47.8
280 Remington	**S280R**	**D,O/P**	**3**	**160**	**Silvertip Boattail**	**24**	**2840**	**2637**	**2442**	**2256**	**2078**	**1909**	**2866**	**2471**	**2120**	**1809**	**1535**	**1295**	**0.1**	**0**	**-1.4**	**-4.1**	**-8.3**	**-14.0**	**2.1**	**1.7**	**0**	**-3.2**	**-7.9**	**-22.6**	**-45.4**
284 Winchester	X2842	D,O/P,M	3	150	Power-Point	24	2860	2595	2344	2108	1886	1680	2724	2243	1830	1480	1185	940	0.6	1.0	0	-2.4	-6.3	-12.1	2.1	1.8	0	-3.4	-8.5	-24.8	-51.0
7mm Mauser (7x57)	X7MM1	D	2	145	Power-Point	24	2660	2413	2180	1959	1754	1564	2279	1875	1530	1236	990	788	0.2	0	-1.7	-5.1	-10.3	-17.5	1.1	0	-2.8	-7.4	-14.1	-34.4	-66.1
7mm Remington Mag.	**S7MAG**	**D,O/P**	**2**	**139**	**Soft Point Boattail**	**24**	**3165**	**2935**	**2717**	**2509**	**2311**	**2121**	**3093**	**2660**	**2279**	**1944**	**1648**	**1389**	**0.1**	**0**	**-1.0**	**-3.2**	**-6.5**	**-11.1**	**1.6**	**1.3**	**0**	**-2.5**	**-6.3**	**-18.3**	**-36.6**
7mm Remington Mag.	X7MMR1	D,O/P,M	2	150	Power-Point	24	3110	2830	2568	2320	2085	1866	3221	2667	2196	1792	1448	1160	0.4	0.8	0	-1.9	-5.2	-9.9	1.7	1.5	0	-2.8	-7.0	-20.5	-42.1
7mm Remington Mag.	**S7MAGA**	**D,O/P,M,L**	**3**	**160**	**Silvertip Boattail**	**24**	**2950**	**2745**	**2550**	**2363**	**2184**	**2012**	**3093**	**2679**	**2311**	**1984**	**1694**	**1439**	**0.1**	**0**	**-1.2**	**-3.7**	**-7.5**	**-12.8**	**1.9**	**1.5**	**0**	**-2.9**	**-7.2**	**-20.6**	**-41.4**
New 7mm Remington Mag. Fail Safe®	**S7MAGX**	–	–	–	–	–	–	–	–	–	–	–	**Under Development**						–	–	–	–	–	–	–	–	–	–	–	–	–
7mm Remington Mag.	X7MMR2	D,O/P,M	3	175	Power-Point	24	2860	2645	2440	2244	2057	1879	3178	2718	2313	1956	1644	1372	0.6	0.9	0	-2.3	-6.0	-11.3	2.0	1.7	0	-3.2	-7.9	-22.7	-45.8
7.62 x 39mm Russian	X76239	D,V	2	123	Soft Point	20	2365	2033	1731	1465	1248	1093	1527	1129	818	586	425	327	0.5	0	-2.6	-7.6	-15.4	-26.7	3.8	3.1	0	-6.0	-15.4	-46.3	-98.4
30 Carbine #	X30M1	V	1	110	Hollow Soft Point	20	1990	1567	1236	1035	923	842	967	600	373	262	208	173	0.9	0	-4.5	-13.5	-28.3	-49.9	0	-4.5	-13.5	-28.3	-49.9	-118.6	-228.2
30-30 Winchester	**S3030W150**	**D**	**2**	**150**	**Silvertip**	**24**	**2390**	**2018**	**1684**	**1398**	**1177**	**1036**	**1902**	**1356**	**944**	**651**	**461**	**357**	**0.5**	**0**	**-2.6**	**-7.7**	**-16.0**	**-27.9**	**3.9**	**3.2**	**0**	**-6.2**	**-16.1**	**-49.4**	**-105.2**
30-30 Winchester	X30301	D	2	150	Hollow Point	24	2390	2018	1684	1398	1177	1036	1902	1356	944	651	461	357	0.5	0	-2.6	-7.7	-16.0	-27.9	1.7	0	-4.3	-11.6	-22.7	-59.1	-120.5
30-30 Winchester	X30306	D	2	150	Power-Point	24	2390	2018	1684	1398	1177	1036	1902	1356	944	651	461	357	0.5	0	-2.6	-7.7	-16.0	-27.9	1.7	0	-4.3	-11.6	-22.7	-59.1	-120.5
30-30 Winchester	X30302	D	2	150	Silvertip	24	2390	2018	1684	1398	1177	1036	1902	1356	944	651	461	357	0.5	0	-2.6	-7.7	-16.0	-27.9	1.7	0	-4.3	-11.6	-22.7	-59.1	-120.5

WINCHESTER BALLISTICS

Cartridge	Symbol	Game Selector Guide	CXP Guide Number	Bullet Wt. Grs.	Bullet Type	Barrel Length (in.)	Velocity in Feet Per Second (fps) Muzzle	100	200	300	400	500	Energy in Foot Pounds (ft-lbs.) Muzzle	100	200	300	400	500	Trajectory, Short Range Yards 50	100	150	200	250	300	Trajectory, Long Range Yards 100	150	200	250	300	400	500
30-30 Winchester	X30303	D	2	170	Power-Point	24	2200	1895	1619	1381	1191	1061	1827	1355	989	720	535	425	0.6	0	-3.0	-8.9	-18.0	-31.1	2.0	0	-4.8	-13.0	-25.1	-63.6	-126.7
30-30 Winchester	X30304	D	2	170	Silvertip	24	2200	1895	1619	1381	1191	1061	1827	1355	989	720	535	425	0.6	0	-3.0	-8.9	-18.0	-31.1	2.0	0	-4.8	-13.0	-25.1	-63.6	-126.7
30-06 Springfield	X30062	V	1	125	Pointed Soft Point	24	3140	2780	2447	2138	1853	1595	2736	2145	1662	1269	953	706	0.4	0.8	0	-2.1	-5.6	-10.7	1.8	1.5	0	-3.0	-7.7	-23.0	-48.5
30-06 Springfield	X30061	D,O/P	2	150	Power-Point	24	2920	2580	2265	1972	1704	1466	2839	2217	1704	1295	967	716	0.6	1.0	0	-2.4	-6.6	-12.7	2.2	1.8	0	-3.5	-9.0	-27.0	-57.1
30-06 Springfield	X30063	D,O/P	2	150	Silvertip	24	2910	2617	2342	2083	1843	1622	2820	2281	1827	1445	1131	876	0.6	0.9	0	-2.3	-6.3	-12.0	2.1	1.8	0	-3.3	-8.5	-25.0	-51.8
30-06 Springfield	**S3006**	**D,O/P,M**	**2**	**165**	**Silvertip Boattail**	**24**	**2800**	**2597**	**2402**	**2216**	**2038**	**1869**	**2873**	**2421**	**2111**	**1799**	**1522**	**1280**	**0.1**	**0**	**-1.4**	**-4.3**	**-8.6**	**-14.6**	**2.1**	**1.8**	**0**	**-3.3**	**-8.2**	**-23.4**	**-47.0**
30-06 Springfield	X30065	D,O/P,M	2	165	Soft Point	24	2800	2573	2357	2151	1956	1772	2873	2426	2036	1696	1402	1151	0.7	1.0	0	-2.5	-6.5	-12.2	2.2	1.9	0	-3.6	-8.4	-24.4	-49.6
30-06 Springfield	**S3006A**	**D,O/P,M,L**	**3**	**180**	**Silvertip Boattail**	**24**	**2700**	**2503**	**2314**	**2133**	**1960**	**1797**	**2914**	**2504**	**2140**	**1819**	**1536**	**1290**	**0.2**	**0**	**-1.6**	**-4.7**	**-9.4**	**-15.8**	**2.3**	**1.9**	**0**	**-3.5**	**-8.8**	**-25.3**	**-50.8**
30-06 Springfield Fail Safe	**S3006X**	**D,O/P,M,L**	**3**	**180**	**Fail Safe**	**24**	**2700**	**2486**	**2283**	**2089**	**1904**	**1731**	**2914**	**2472**	**2083**	**1744**	**1450**	**1198**	**-0.1**	**0**	**-1.3**	**-4.1**	**-8.6**	**-14.9**	**2.1**	**1.8**	**0**	**-3.5**	**-8.7**	**-25.5**	**-51.8**
30-06 Springfield	X30064	D,O/P,M	2	180	Power-Point	24	2700	2348	2023	1727	1466	1251	2913	2203	1635	1192	859	625	0.2	0	-1.8	-5.5	-11.2	-19.5	2.7	2.3	0	-4.4	-11.3	-34.4	-73.7
30-06 Springfield	X30066	D,O/P,M,L	3	180	Silvertip	24	2700	2469	2250	2042	1846	1663	2913	2436	2023	1666	1362	1105	0.2	0	-1.6	-4.8	-9.7	-16.5	2.4	2.0	0	-3.7	-9.3	-27.0	-54.9
30-06 Springfield	X30069	M,L	3	220	Silvertip	24	2410	2192	1985	1791	1611	1448	2837	2347	1924	1567	1268	1024	0.4	0	-2.2	-6.4	-12.7	-21.6	1.5	0	-3.5	-9.1	-17.2	-41.8	-79.9
30-40 Krag	X30401	D	2	180	Power-Point	24	2430	2099	1795	1525	1298	1128	2360	1761	1288	929	673	508	0.4	0	-2.4	-7.1	-14.5	-25.0	1.6	0	-3.9	-10.5	-20.3	-51.7	-103.9
300 Winchester Mag.	X30WM1	D,O/P	2	150	Power-Point	24	3290	2951	2636	2342	2068	1813	3605	2900	2314	1827	1424	1095	0.3	0.7	0	-1.8	-4.8	-9.3	2.6	2.9	2.1	0	-3.5	-15.4	-35.5
300 Winchester Mag. Fail Safe	**S300WX**	**M,L**	**3D**	**180**	**Fail Safe**	**24**	**2960**	**2732**	**2514**	**2307**	**2110**	**1923**	**3503**	**2983**	**2528**	**2129**	**1780**	**1478**	**-0.2**	**0**	**-1.0**	**-3.2**	**-6.8**	**-11.8**	**1.6**	**1.4**	**0**	**-2.8**	**-7.1**	**-20.7**	**-42.1**
300 Winchester Mag.	X30WM2	O/P,M,L	3	180	Power-Point	24	2960	2745	2540	2344	2157	1979	3501	3011	2578	2196	1859	1565	0.5	0.8	0	-2.1	-5.5	-10.4	1.9	1.6	0	-2.9	-7.3	-20.9	-41.9
300 Winchester Mag.	**S300W**	**O/P,M,L**	**3D**	**190**	**Silvertip Boattail**	**24**	**2885**	**2698**	**2519**	**2347**	**2181**	**2023**	**3512**	**3073**	**2679**	**2325**	**2009**	**1728**	**0.1**	**0**	**-1.3**	**-3.9**	**-7.8**	**-13.2**	**1.9**	**1.6**	**0**	**-3.0**	**-7.4**	**-21.1**	**-42.2**
300 Winchester Mag.	X30WM3	M,L,XL	3D	220	Silvertip	24	2680	2448	2228	2020	1823	1640	3508	2927	2424	1993	1623	1314	0.2	0	-1.7	-4.9	-9.9	-16.9	2.5	2.0	0	-3.8	-9.5	-27.5	-56.1
300 H&H Magnum	X300H2	O/P,M,L	3	180	Silvertip	24	2880	2640	2412	2196	1991	1798	3315	2785	2325	1927	1584	1292	0.6	0.9	0	-2.3	-6.0	-11.5	2.1	1.7	0	-3.2	-8.0	-23.3	-47.4
300 Savage	X3001	D,O/P	2	150	Power-Point	24	2630	2311	2015	1743	1500	1295	2303	1779	1352	1012	749	558	0.3	0	-1.9	-5.7	-11.6	-19.9	2.8	2.3	0	-4.5	-11.5	-34.4	-73.0
300 Savage	X3003	D,O/P	2	150	Silvertip	24	2630	2354	2095	1853	1631	1434	2303	1845	1462	1143	886	685	0.3	0	-1.8	-5.4	-11.0	-18.8	2.7	2.2	0	-4.2	-10.7	-31.5	-65.5
300 Savage	X3004	D	2	180	Power-Point	24	2350	2025	1728	1467	1252	1098	2207	1639	1193	860	626	482	0.5	0	-2.6	-7.7	-15.6	-27.1	1.7	0	-4.2	-11.3	-21.9	-55.8	-112.0
303 Savage	X3032	D	2	190	Silvertip	24	1890	1612	1372	1183	1055	970	1507	1096	794	591	469	397	1.0	0	-4.3	-12.6	-25.5	-43.7	2.9	0	-6.8	-18.3	-35.1	-88.2	-172.5
303 British	X303B1	D	2	180	Power-Point	24	2460	2233	2018	1816	1629	1459	2418	1993	1627	1318	1060	851	0.3	0	-2.1	-6.1	-12.2	-20.8	1.4	0	-3.3	-8.8	-16.6	-40.4	-77.4
307 Winchester	X3076	D,M	2	180	Power-Point	24	2510	2179	1874	1599	1362	1177	2519	1898	1404	1022	742	554	0.3	0	-2.2	-6.5	-13.3	-22.9	1.5	0	-3.6	-9.6	-18.6	-47.1	-93.7
308 Winchester	**S308W150**	**D,O/P**	**2**	**150**	**Silvertip Boattail**	**24**	**2820**	**2559**	**2312**	**2080**	**1861**	**1659**	**2649**	**2182**	**1782**	**1441**	**1154**	**917**	**0.2**	**0**	**-1.5**	**-4.4**	**-9.0**	**-15.4**	**2.2**	**1.8**	**0**	**-3.5**	**-8.7**	**-25.5**	**-52.3**
308 Winchester	X3085	D,O/P	2	150	Power-Point	24	2820	2488	2179	1893	1633	1405	2648	2061	1581	1193	888	657	0.2	0	-1.6	-4.8	-9.8	-16.9	2.4	2.0	0	-3.8	-9.8	-29.3	-62.0
308 Winchester	X3082	D,O/P	2	150	Silvertip	24	2820	2533	2263	2009	1774	1560	2648	2137	1705	1344	1048	810	0.2	0	-1.5	-4.5	-9.3	-15.9	2.3	1.9	0	-3.6	-9.1	-26.9	-55.7
308 Winchester Match	**S308M**	**–**	**M**	**168**	**Hollow Point Boattail**	**24**	**2680**	**2485**	**2297**	**2118**	**1948**	**1786**	**2680**	**2303**	**1970**	**1674**	**1415**	**1190**	**-0.1**	**0**	**-1.3**	**-4.1**	**-8.6**	**-14.9**	**2.1**	**1.8**	**0**	**-3.4**	**-8.7**	**-25.1**	**-50.7**
308 Winchester	**S308W180**	**D,O/P,M**	**3**	**180**	**Silvertip Boattail**	**24**	**2610**	**2424**	**2245**	**2074**	**1911**	**1756**	**2723**	**2348**	**2015**	**1719**	**1459**	**1232**	**0.2**	**0**	**-1.7**	**-5.0**	**-10.1**	**-17.0**	**2.5**	**2.1**	**0**	**-3.8**	**-9.4**	**-26.9**	**-54.0**
308 Winchester Fail Safe	**S308X**	**D,O/P,M,L**	**3**	**180**	**Fail Safe**	**24**	**2620**	**2409**	**2207**	**2015**	**1834**	**1664**	**2744**	**2319**	**1947**	**1624**	**1344**	**1107**	**-0.1**	**0**	**-1.4**	**-4.5**	**-9.4**	**-16.2**	**2.3**	**1.9**	**0**	**-3.7**	**-9.4**	**-27.4**	**-55.7**
308 Winchester	X3086	D,O/P,M	2	180	Power-Point	24	2620	2274	1955	1666	1414	1212	2743	2066	1527	1109	799	587	0.3	0	-2.0	-5.9	-12.1	-20.9	2.9	2.4	0	-4.7	-12.1	-36.9	-79.1
308 Winchester	X3083	M,L	3	180	Silvertip	24	2620	2393	2178	1974	1782	1604	2743	2288	1896	1557	1269	1028	0.2	0	-1.8	-5.2	-10.4	-17.7	2.6	2.1	0	-4.0	-9.9	-28.9	-58.8
32 Win Special	X32WS2	D	2	170	Power-Point	24	2250	1870	1537	1267	1082	971	1911	1320	892	606	442	356	0.6	0	-3.1	-9.2	-19.0	-33.2	2.0	0	-5.1	-13.8	-27.1	-70.9	-144.3
32 Win Special	X32WS3	D	2	170	Silvertip	24	2250	1870	1537	1267	1082	971	1911	1320	892	606	442	356	0.6	0	-3.1	-9.2	-19.0	-33.2	2.0	0	-5.1	-13.8	-27.1	-70.9	-144.3
32-20 Winchester #	X32201	V	1	100	Lead	24	1210	1021	913	834	769	712	325	231	185	154	131	113	0	-6.3	-20.9	-44.9	-79.3	-125.1	0	-11.5	-32.3	-63.6	-106.3	-230.3	-413.3
8mm Mauser (8 x 57)	X8MM	D	2	170	Power-Point	24	2360	1969	1622	1333	1123	997	2102	1463	993	671	476	375	0.5	0	-2.7	-8.2	-17.0	-29.8	1.8	0	-4.5	-12.4	-24.3	-63.8	-130.7
338 Winchester Mag.	X3381	D,O/P,M	3	200	Power-Point	24	2960	2658	2375	2110	1862	1635	3890	3137	2505	1977	1539	1187	0.5	0.9	0	-2.3	-6.1	-11.6	2.0	1.7	0	-3.2	-8.2	-24.3	-50.4
338 Winchester Mag.	X3383	M,L,XL	3D	225	Soft Point	24	2780	2572	2374	2184	2003	1832	3862	3306	2816	2384	2005	1677	1.2	1.3	0	-2.7	-7.1	-12.9	2.7	2.1	0	-3.6	-9.4	-25.0	-49.9
338 Winchester Mag. Fail Safe	**S338XA**	**M,L,XL**	**3D**	**230**	**Fail Safe**	**24**	**2780**	**2573**	**2375**	**2186**	**2005**	**1834**	**3948**	**3382**	**2881**	**2441**	**2054**	**1719**	**-0.1**	**0**	**-1.2**	**-3.8**	**-7.9**	**-13.7**	**1.9**	**1.7**	**0**	**-3.2**	**-8.1**	**-23.4**	**-47.4**
348 Winchester	Q3167	D,M	3	200	Silvertip	24	2520	2215	1931	1672	1443	1253	2820	2178	1656	1241	925	697	0.3	0	-2.1	-6.2	-12.7	-21.9	1.4	0	-3.4	-9.2	-17.7	-44.4	-87.9
35 Remington	X35R1	D	2	200	Power-Point	24	2020	1646	1335	1114	985	901	1812	1203	791	551	431	360	0.9	0	-4.1	-12.1	-25.1	-43.9	2.7	0	-6.7	-18.3	-35.8	-92.8	-185.5
356 Winchester	X3561	D,M	2	200	Power-Point	24	2460	2114	1797	1517	1284	1113	2688	1985	1434	1022	732	550	0.4	0	-2.3	-7.0	-14.3	-24.7	1.6	0	-3.8	-10.4	-20.1	-51.2	-102.3
356 Winchester	X3563	M,L	3	250	Power-Point	24	2160	1911	1682	1476	1299	1158	2591	2028	1571	1210	937	745	0.6	0	-3.0	-8.7	-17.4	-30.0	2.0	0	-4.7	-12.4	-23.7	-58.4	-112.9
357 Magnum #	X3575P	V,D	2	158	Jacketed Soft Point	20	1830	1427	1138	980	883	809	1175	715	454	337	274	229	0	-2.4	-9.1	-21.0	-39.2	-64.3	0	-5.5	-16.2	-33.1	-57.0	-128.3	-235.8
358 Winchester	X3581	D,M	3	200	Silvertip	24	2490	2171	1876	1610	1379	1194	2753	2093	1563	1151	844	633	0.4	0	-2.2	-6.5	-13.3	-23.0	1.5	0	-3.6	-9.7	-18.6	-47.2	-94.1
375 Winchester	X375W	D,M	2	200	Power-Point	24	2200	1841	1526	1268	1089	980	2150	1506	1034	714	527	427	0.6	0	-3.2	-9.5	-19.5	-33.8	2.1	0	-5.2	-14.1	-27.4	-70.1	-138.1
375 H&H Magnum	X375H1	M,L,XL	3D	270	Power-Point	24	2690	2420	2166	1928	1707	1507	4337	3510	2812	2228	1747	1361	0.2	0	-1.7	-5.1	-10.3	-17.6	2.5	2.1	0	-3.9	-10.0	-29.4	-60.7
375 H&H Magnum	X375H2	M,L,XL	3D	300	Silvertip	24	2530	2268	2022	1793	1583	1397	4263	3426	2723	2141	1669	1300	0.3	0	-2.0	-5.9	-11.9	-20.3	2.9	2.4	0	-4.5	-11.5	-33.8	-70.1
375 H&H Magnum	X375H3	XL	4	300	Full Metal Jacket	24	2530	2171	1843	1551	1307	1126	4263	3139	2262	1602	1138	844	0.3	0	-2.2	-6.5	-13.5	-23.4	1.5	0	-3.6	-9.8	-19.1	-49.1	-99.5
38-40 Winchester #	X3840	D	2	180	Soft Point	24	1160	999	901	827	764	710	538	399	324	273	233	201	0	-6.7	-22.2	-47.3	-83.2	-130.8	0	-12.1	-33.9	-66.4	-110.6	-238.3	-425.6
38-55 Winchester	X3855	D	2	255	Soft Point	24	1320	1190	1091	1018	963	917	987	802	674	587	525	476	0	-4.7	-15.4	-32.7	-57.2	-89.3	0	-8.4	-23.4	-45.6	-75.2	-158.8	-277.4
44 Remington Magnum #	X44MSTHP2	V,D	2	210	Silvertip Hollow Point	20	1580	1198	993	879	795	725	1164	670	460	361	295	245	0	-3.7	-13.3	-29.8	-54.2	-87.3	0	-7.7	-22.4	-44.9	-76.1	-168.0	-305.8
44 Remington Magnum #	X44MHSP2	D	2	240	Hollow Soft Point	20	1760	1362	1094	953	861	789	1650	988	638	484	395	332	0	-2.7	-10.2	-23.6	-44.2	-73.3	0	-6.1	-18.1	-37.4	-65.1	-150.3	-282.5
44-40 Winchester #	X4440	D	2	200	Soft Point	24	1190	1006	900	822	756	699	629	449	360	300	254	217	0	-6.5	-21.6	-46.3	-81.8	-129.1	0	-11.8	-33.3	-65.5	-109.5	-237.4	-426.2
45-70 Government	X4570H	D,M	2	300	Jacketed Hollow Point	24	1880	1650	1425	1235	1105	1010	2355	1815	1355	1015	810	680	0	-2.4	-8.2	-17.6	-31.4	-51.5	0	-4.6	-12.8	-25.4	-44.3	-95.5	–
458 Winchester Magnum	X4580	XL	4	500	Full Metal Jacket	24	2040	1823	1623	1442	1287	1161	4620	3689	2924	2308	1839	1496	0.7	0	-3.3	-9.6	-19.2	-32.5	2.2	0	-5.2	-13.6	-25.8	-63.2	-121.7
458 Winchester Magnum	X4581	L,XL	3D	510	Soft Point	24	2040	1770	1527	1319	1157	1046	4712	3547	2640	1970	1516	1239	0.8	0	-3.5	-10.3	-20.8	-35.6	2.4	0	-5.6	-14.9	-28.5	-71.5	-140.4

Reloading

For addresses and phone numbers of manufacturers and distributors included in this section, turn to *DIRECTORY OF MANUFACTURERS AND SUPPLIERS* at the back of the book.

HORNADY BULLETS

RIFLE BULLETS

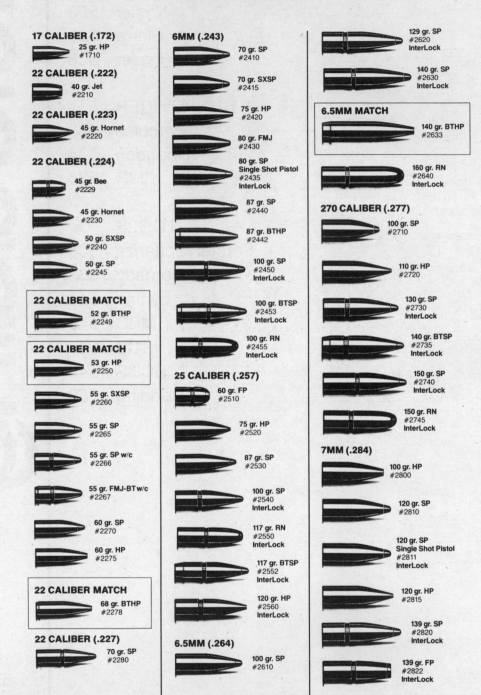

17 CALIBER (.172)
25 gr. HP
#1710

22 CALIBER (.222)
40 gr. Jet
#2210

22 CALIBER (.223)
45 gr. Hornet
#2220

22 CALIBER (.224)
45 gr. Bee
#2229

45 gr. Hornet
#2230

50 gr. SXSP
#2240

50 gr. SP
#2245

22 CALIBER MATCH
52 gr. BTHP
#2249

22 CALIBER MATCH
53 gr. HP
#2250

55 gr. SXSP
#2260

55 gr. SP
#2265

55 gr. SP w/c
#2266

55 gr. FMJ-BT w/c
#2267

60 gr. SP
#2270

60 gr. HP
#2275

22 CALIBER MATCH
68 gr. BTHP
#2278

22 CALIBER (.227)
70 gr. SP
#2280

6MM (.243)
70 gr. SP
#2410

70 gr. SXSP
#2415

75 gr. HP
#2420

80 gr. FMJ
#2430

80 gr. SP
Single Shot Pistol
#2435
InterLock

87 gr. SP
#2440

87 gr. BTHP
#2442

100 gr. SP
#2450
InterLock

100 gr. BTSP
#2453
InterLock

100 gr. RN
#2455
InterLock

25 CALIBER (.257)
60 gr. FP
#2510

75 gr. HP
#2520

87 gr. SP
#2530

100 gr. SP
#2540
InterLock

117 gr. RN
#2550
InterLock

117 gr. BTSP
#2552
InterLock

120 gr. HP
#2560
InterLock

6.5MM (.264)
100 gr. SP
#2610

129 gr. SP
#2620
InterLock

140 gr. SP
#2630
InterLock

6.5MM MATCH
140 gr. BTHP
#2633

160 gr. RN
#2640
InterLock

270 CALIBER (.277)
100 gr. SP
#2710

110 gr. HP
#2720

130 gr. SP
#2730
InterLock

140 gr. BTSP
#2735
InterLock

150 gr. SP
#2740
InterLock

150 gr. RN
#2745
InterLock

7MM (.284)
100 gr. HP
#2800

120 gr. SP
#2810

120 gr. SP
Single Shot Pistol
#2811
InterLock

120 gr. HP
#2815

139 gr. SP
#2820
InterLock

139 gr. FP
#2822
InterLock

HORNADY BULLETS

RIFLE BULLETS

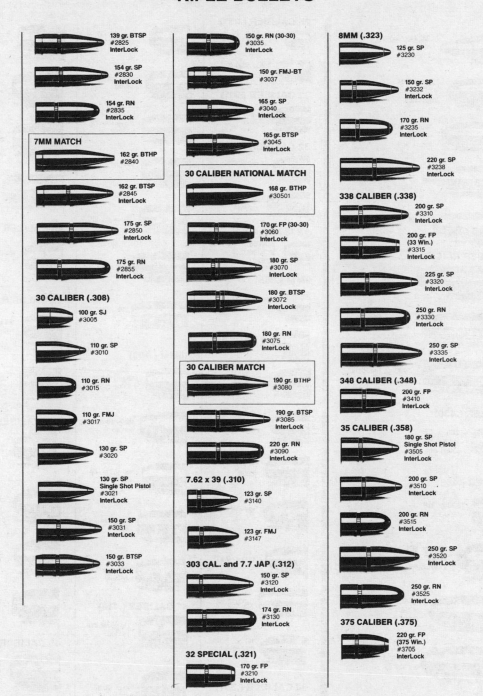

139 gr. BTSP
#2825
InterLock

154 gr. SP
#2830
InterLock

154 gr. RN
#2835
InterLock

7MM MATCH

162 gr. BTHP
#2840

162 gr. BTSP
#2845
InterLock

175 gr. SP
#2850
InterLock

175 gr. RN
#2855
InterLock

30 CALIBER (.308)

100 gr. SJ
#3005

110 gr. SP
#3010

110 gr. RN
#3015

110 gr. FMJ
#3017

130 gr. SP
#3020

130 gr. SP
Single Shot Pistol
#3021
InterLock

150 gr. SP
#3031
InterLock

150 gr. BTSP
#3033
InterLock

150 gr. RN (30-30)
#3035
InterLock

150 gr. FMJ-BT
#3037

165 gr. SP
#3040
InterLock

165 gr. BTSP
#3045
InterLock

30 CALIBER NATIONAL MATCH

168 gr. BTHP
#30501

170 gr. FP (30-30)
#3060
InterLock

180 gr. SP
#3070
InterLock

180 gr. BTSP
#3072
InterLock

180 gr. RN
#3075
InterLock

30 CALIBER MATCH

190 gr. BTHP
#3080

190 gr. BTSP
#3085
InterLock

220 gr. RN
#3090
InterLock

7.62 x 39 (.310)

123 gr. SP
#3140

123 gr. FMJ
#3147

303 CAL. and 7.7 JAP (.312)

150 gr. SP
#3120
InterLock

174 gr. RN
#3130
InterLock

32 SPECIAL (.321)

170 gr. FP
#3210
InterLock

8MM (.323)

125 gr. SP
#3230

150 gr. SP
#3232
InterLock

170 gr. RN
#3235
InterLock

220 gr. SP
#3238
InterLock

338 CALIBER (.338)

200 gr. SP
#3310
InterLock

200 gr. FP
(33 Win.)
#3315
InterLock

225 gr. SP
#3320
InterLock

250 gr. RN
#3330
InterLock

250 gr. SP
#3335
InterLock

348 CALIBER (.348)

200 gr. FP
#3410
InterLock

35 CALIBER (.358)

180 gr. SP
Single Shot Pistol
#3505
InterLock

200 gr. SP
#3510
InterLock

200 gr. RN
#3515
InterLock

250 gr. SP
#3520
InterLock

250 gr. RN
#3525
InterLock

375 CALIBER (.375)

220 gr. FP
(375 Win.)
#3705
InterLock

HORNADY BULLETS

RIFLE/PISTOL BULLETS

*270 gr. SP
#3710
InterLock

*270 gr. RN
#3715
InterLock

*300 gr. RN
#3720
InterLock

*300 gr. BTSP
#3725
InterLock

*300 gr. FMJ-RN
#3727

416 CALIBER (.416)

*340 gr. BTSP
#4163

*400 gr. RN
#4165
InterLock

*400 gr. FMJ-RN
#4167

44 CALIBER (.430)

265 gr. FP
#4300
InterLock

45 CALIBER (.458)

*300 gr. HP
#4500

*350 gr. RN
#4502
InterLock

*500 gr. RN
#4504
InterLock

*500 gr. FMJ-RN
#4507

50 CALIBER (.510)

**750 gr. A-MAX UHC
#5165

25 CALIBER (.251)

35 gr. HP/XTP
#35450

50 gr. FMJ-RN
#3545

32 CALIBER (.311)

71 gr. FMJ-RN
#3200

32 CALIBER (.312)

85 gr. HP/XTP
#32050

100 gr. HP/XTP
#32070

9 x 18 MAKAROV
NEW!
95 gr. HP/XTP
#36500

9MM (.355)

90 gr. HP/XTP
#35500

100 gr. FMJ-RN
#3552

115 gr. HP/XTP
#35540

115 gr. FMJ-RN
#3555

124 gr. FMJ-FP
#3556

124 gr. FMJ-RN
#3557

124 gr. HP/XTP
#35571

147 gr. HP-BT/XTP
35580

147 gr. FMJ-RN-BT
3559

38 CALIBER (.357)

110 gr. HP/XTP
#35700

125 gr. HP/XTP
#35710

125 gr. FP/XTP
#35730

140 gr. HP/XTP
#35740

158 gr. HP/XTP
#35750

158 gr. FP/XTP
#35780

160 gr. CL-SIL
#3572

180 gr. CL-SIL
#3577

180 gr. HP/XTP
35771

10MM (.400)

155 gr. HP/XTP
#40000

180 gr. HP/XTP
#40040

180 gr. FMJ-FP
#40041

200 gr. FMJ-FP
#4007

200 gr. HP/XTP
#40060

41 CALIBER (.410)

210 gr. HP/XTP
#41000

210 gr. CL-SIL
#4105

44 CALIBER (.430)

180 gr. HP/XTP
#44050

200 gr. HP/XTP
#44100

240 gr. HP/XTP
#44200

240 gr. CL-SIL
#4425

* 300 gr. HP/XTP
#44280

45 CALIBER (.451)

185 gr. HP/XTP
#45100

45 CALIBER MATCH

185 gr. SWC
#4513

200 gr. HP/XTP
#45140

45 CALIBER MATCH

200 gr. FMJ-C/T
#4515

230 gr. FMJ-RN
#4517

230 gr. FMJ-FP
#4518

230 gr. HP/XTP
45160

45 CALIBER (.452)

250 gr. Long Colt HP/XTP
#45200

*300 gr. HP/XTP
45230

*Packed 50 per box.
All others packed
100 per box.

NOSLER BULLETS

Caliber/Diameter	HANDGUN	Bullet Weight and Style	Sectional Density	Ballistic Coefficient	Part Number
9mm/.355"		90 Gr. Hollow Point	.102	.086	42050
		115 Gr. Full Metal Jacket	.130	.103	42059
		115 Gr. Hollow Point 250 Quantity Bulk Pack	.130	.110	43009 44848
38/.357"		125 Gr. Hollow Point 250 Quantity Bulk Pack	.140	.143	42055 44840
		NEW 135 Gr. IPSC 250 Quantity Bulk Pack	.151	.149	44836
		150 Gr. Soft Point	.168	.153	42056
		150 Gr. IPSC 250 Quantity Bulk Pack	.168	.157	44839
		158 Gr. Hollow Point 250 Quantity Bulk Pack	.177	.182	42057 44841
		180 Gr. Silhouette 250 Quantity Bulk Pack	.202	.210	42058 44851
10mm/.400"		135 Gr. Hollow Point 250 Quantity Bulk Pack	.121	.093	44838 44852
		150 Gr. Hollow Point	.134	.106	44849
		170 Gr. Hollow Point	.152	.137	44844
		180 Gr. Hollow Point	.161	.147	44837
41/.410"		210 Gr. Hollow Point	.178	.170	43012
44/.429"		200 Gr. Hollow Point 250 Quantity Bulk Pack	.155	.151	42060 44846
		240 Gr. Soft Point	.186	.177	42068
		240 Gr. Hollow Point 250 Quantity Bulk Pack	.186	.173	42061 44842
		300 Gr. Hollow Point	.233	.206	42069
45/.451"		185 Gr. Hollow Point 250 Quantity Bulk Pack	.130	.142	42062 44847
		230 Gr. Full Metal Jacket	.162	.183	42064

Caliber/Diameter	BALLISTIC TIP®	Bullet Weight and Style	Sectional Density	Ballistic Coefficient	Part Number
22/.224"		50 Gr. Spitzer (Orange Tip)	.142	.238	39522
		55 Gr. Spitzer (Orange Tip)	.157	.267	39526
6mm/.243"		70 Gr. Spitzer (Purple Tip)	.169	.310	39532
		95 Gr. Spitzer (Purple Tip)	.230	.379	39534
25/.257"		85 Gr. Spitzer (Blue Tip)	.183	.331	43004
		100 Gr. Spitzer (Blue Tip)	.216	.393	43005
6.5mm/.264"		100 Gr. Spitzer (Brown Tip)	.205	.350	43008
		120 Gr. Spitzer (Brown Tip)	.246	.458	43007
270/.277"		130 Gr. Spitzer (Yellow Tip)	.242	.433	39589
		140 Gr. Spitzer (Yellow Tip)	.261	.456	43983
		150 Gr. Spitzer (Yellow Tip)	.279	.496	39588
7mm/.284"		120 Gr. Spitzer (Red Tip)	.213	.417	39550
		140 Gr. Spitzer (Red Tip)	.248	.485	39587
		150 Gr. Spitzer (Red Tip)	.266	.493	39586
30/.308"		125 Gr. Spitzer (Green Tip)	.188	.366	43980
		150 Gr. Spitzer (Green Tip)	.226	.435	39585
		165 Gr. Spitzer (Green Tip)	.248	.475	39584
		180 Gr. Spitzer (Green Tip)	.271	.507	39583
338/.338"		200 Gr. Spitzer (Maroon Tip)	.250	.414	39595

NOSLER BULLETS

Caliber/Diameter	SOLID BASE® BOATTAIL	Bullet Weight and Style	Sectional Density	Ballistic Coefficient	Part Number
22/.224"		45 Gr. Hornet	.128	.144	35487
		52 Gr. Hollow Point Match	.148	.224	25857
		55 Gr. Spitzer w/cannelure	.157	.261	16339
		60 Gr. Spitzer	.171	.266	30323
6mm/.243"		100 Gr. Spitzer	.242	.388	30390
25/.257"		120 Gr. Spitzer	.260	.446	30404
270/.277"		130 Gr. Spitzer	.242	.420	30394
7mm/.284"		NEW 120 Gr. Flat Point	.213	.195	41722
		140 Gr. Spitzer	.248	.461	29599
30/.308"		150 Gr. Spitzer	.226	.393	27583
		165 Gr. Spitzer	.248	.428	27585
		180 Gr. Spitzer	.271	.491	27587

Caliber/Diameter	PARTITION®	Bullet Weight and Style	Sectional Density	Ballistic Coefficient	Part Number
6mm/.243"		85 Gr. Spitzer	.206	.315	16314
		95 Gr. Spitzer	.230	.365	16315
		100 Gr. Spitzer	.242	.384	35642
25/.257"		100 Gr. Spitzer	.216	.377	16317
		115 Gr. Spitzer	.249	.389	16318
		120 Gr. Spitzer	.260	.391	35643
6.5mm/.264"		125 Gr. Spitzer	.256	.449	16320
		140 Gr. Spitzer	.287	.490	16321

Caliber/Diameter	PARTITION®	Bullet Weight and Style	Sectional Density	Ballistic Coefficient	Part Number
270/.277"		130 Gr. Spitzer	.242	.416	16322
		150 Gr. Spitzer	.279	.465	16323
		160 Gr. Semi Spitzer	.298	.434	16324
7mm/.284"		140 Gr. Spitzer	.248	.434	16325
		150 Gr. Spitzer	.266	.456	16326
		160 Gr. Spitzer	.283	.475	16327
		175 Gr. Spitzer	.310	.519	35645
30/.308"		150 Gr. Spitzer	.226	.387	16329
		165 Gr. Spitzer	.248	.410	16330
		170 Gr. Round Nose	.256	.252	16333
		180 Gr. Spitzer	.271	.474	16331
		180 Gr. Protected Point	.271	.361	25396
		200 Gr. Spitzer	.301	.481	35626
		220 Gr. Semi Spitzer	.331	.351	16332
8mm/.323"		200 Gr. Spitzer	.274	.426	35277
338/.338"		210 Gr. Spitzer	.263	.400	16337
		225 Gr. Spitzer	.281	.454	16336
		250 Gr. Spitzer	.313	.473	35644
35/.358"		225 Gr. Spitzer	.251	.430	44800
		250 Gr. Spitzer	.279	.446	44801
375/.375"		260 Gr. Spitzer	.264	.314	44850
		300 Gr. Spitzer	.305	.398	44845

SIERRA BULLETS

RIFLE BULLETS

.22 Caliber Hornet
(.223/5.66MM Diameter)

40 gr. Hornet
Varminter #1100

45 gr. Hornet
Varminter #1110

.22 Caliber Hornet
(.224/5.69MM Diameter)

40 gr. Hornet
Varminter #1200

45 gr. Hornet
Varminter #1210

.22 Caliber
(.224/5.69MM Diameter)

40 gr. HP
Varminter #1385

45 gr. SMP
Varminter #1300

45 gr. SPT
Varminter #1310

50 gr. SMP
Varminter #1320

50 gr. SPT
Varminter #1330

50 gr. Blitz
Varminter #1340

52 gr. HPBT
MatchKing #1410

53 gr. HP
MatchKing #1400

55 gr. Blitz
Varminter #1345

55 gr. SMP
Varminter #1350

55 gr. FMJBT
GameKing #1355

55 gr. SPT
Varminter #1360

55 gr. SBT
GameKing #1365

55 gr. HPBT
GameKing #1390

60 gr. HP
Varminter #1375

63 gr. SMP
Varminter #1370

69 gr. HPBT
MatchKing #1380

7"-10" TWST BBLS

6MM .243 Caliber
(.243/6.17MM Diameter)

60 gr. HP
Varminter #1500

70 gr. HPBT
MatchKing #1505

75 gr. HP
Varminter #1510

80 gr. Blitz
Varminter #1515

85 gr. SPT
Varminter #1520

85 gr. HPBT
GameKing #1530

90 gr. FMJBT
GameKing #1535

100 gr. SPT
Pro-Hunter #1540

100 gr. SMP
Pro-Hunter #1550

100 gr. SBT
GameKing #1560

107 gr. HPBT
MatchKing #1570

7"-8" TWST BBLS

.25 Caliber
(.257/6.53MM Diameter)

75 gr. HP
Varminter #1600

87 gr. SPT
Varminter #1610

90 gr. HPBT
GameKing #1615

100 gr. SPT
Pro-Hunter #1620

100 gr. SBT
GameKing #1625

117 gr. SBT
GameKing #1630

117 gr. SPT
Pro-Hunter #1640

120 gr. HPBT
GameKing #1650

6.5MM .264 Caliber
(.264/6.71MM Diameter)

85 gr. HP
Varminter #1700

100 gr. HP
Varminter #1710

120 gr. SPT
Pro-Hunter #1720

120 gr. HPBT
MatchKing #1725

6.5MM .264 Caliber (cont.)
(.264/6.71MM Diameter)

140 gr. SBT
GameKing #1730

140 gr. HPBT
MatchKing #1740

160 gr. SMP
Pro-Hunter #1750

.270 Caliber
(.277/7.04MM Diameter)

90 gr. HP
Varminter #1800

110 gr. SPT
Pro-Hunter #1810

130 gr. SBT
GameKing #1820

130 gr. SPT
Pro-Hunter #1830

135 gr. HPBT
MatchKing #1833

140 gr. HPBT
GameKing #1835

140 gr. SBT
GameKing #1845

150 gr. SBT
GameKing #1840

150 gr. RN
Pro-Hunter #1850

7MM .284 Caliber
(.284/7.21MM Diameter)

100 gr. HP
Varminter #1895

120 gr. SPT
Pro-Hunter #1900

140 gr. SBT
GameKing #1905

140 gr. SPT
Pro-Hunter #1910

150 gr. SBT
GameKing #1913

150 gr. HPBT
MatchKing #1915

160 gr. SBT
GameKing #1920

160 gr. HPBT
GameKing #1925

168 gr. HPBT
MatchKing #1930

170 gr. RN
Pro-Hunter #1950

175 gr. SBT
GameKing #1940

SIERRA BULLETS

RIFLE BULLETS

.30 (30-30) Caliber (.308/7.82MM Diameter)

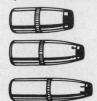

125 gr. HP
Pro-Hunter #2020

150 gr. FN
Pro-Hunter #2000
POWER JACKET

170 gr. FN
Pro-Hunter #2010
POWER JACKET

30 Caliber 7.62MM (.308/7.82MM Diameter)

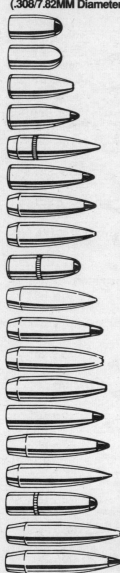

110 gr. RN
Pro-Hunter #2100

110 gr. FMJ
Pro-Hunter #2105

110 gr. HP
Varminter #2110

125 gr. SPT
Pro-Hunter #2120

150 gr. FMJBT
GameKing #2115

150 gr. SPT
Pro-Hunter #2130

150 gr. SBT
GameKing #2125

150 gr. HPBT
MatchKing #2190

150 gr. RN
Pro-Hunter #2135

155 gr. HPBT
1992 PALMA
MatchKing #2155

165 gr. SBT
GameKing #2145

165 gr. HPBT
GameKing #2140

168 gr. HPBT
MatchKing #2200

180 gr. SPT
Pro-Hunter #2150

180 gr. SBT
GameKing #2160

180 gr. HPBT
MatchKing #2220

180 gr. RN
Pro-Hunter #2170

190 gr. HPBT
MatchKing #2210

200 gr. SBT
GameKing #2165

30 Caliber 7.62MM (cont.) (.308/7.82MM Diameter)

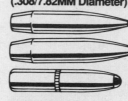

200 gr. HPBT
MatchKing #2230

220 gr. HPBT
MatchKing #2240

220 gr. RN
Pro-Hunter #2180

.303 Caliber 7.7MM (.311/7.90MM Diameter)

150 gr. SPT
Pro-Hunter #2300

180 gr. SPT
Pro-Hunter #2310

8MM (.323/8.20MM Diameter)

150 gr. SPT
Pro-Hunter #2400

175 gr. SPT
Pro-Hunter #2410

220 gr. SBT
GameKing #2420

.338 Caliber (.338/8.59MM Diameter)

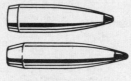

**215 gr. SBT
GameKing
#2610**

250 gr. SBT
GameKing #2600

.35 Caliber (.358/9.09MM Diameter)

200 gr. RN
Pro-Hunter #2800

225 gr. SBT
GameKing #2850

.375 Caliber (.375/9.53MM Diameter)

200 gr. FN
Pro-Hunter #2900
POWER JACKET

250 gr. SBT
GameKing #2950

300 gr. SBT
GameKing #3000

.45 Caliber (45.70) (.458/11.63MM Diameter)

300 gr. HP
Pro-Hunter #8900

Long Range & Specialty Bullets

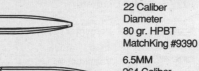

22 Caliber
Diameter
80 gr. HPBT
MatchKing #9390

6.5MM
264 Caliber
155 gr. HPBT
MatchKing #9570

.30 Caliber
7.62MM
240 gr. HPBT
MatchKing #9245

.338 Caliber
300 gr. HPBT
MatchKing #9300

SIERRA BULLETS

HANDGUN BULLETS

Single Shot Pistol Bullets

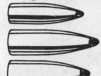

6MM .243 Dia. 80 gr. SPT
Pro-Hunter #7150

7MM .284 Dia. 130 gr. SPT
Pro-Hunter #7250

.30 cal. .308 Dia. 135 gr. SPT
Pro-Hunter #7350

.25 Caliber
(.251/6.38MM Diameter)

50 gr. FMJ
Tournament Master #8000

.32 Caliber 7.65MM
(.312/7.92MM Diameter)

71 gr. FMJ
Tournament Master #8010

.32 Mag.
(.312/7.92MM Diameter)

90 gr. JHC
Sports Master #8030
POWER JACKET

9MM .355 Caliber
(.355/9.02MM Diameter)

90 gr. JHP
Sports Master #8100
POWER JACKET

95 gr. FMJ
Tournament Master #8105

115 gr. JHP
Sports Master #8110
POWER JACKET

115 gr. FMJ
Tournament Master #8115

125 gr. FMJ
Tournament Master #8120

130 gr. FMJ
Tournament Master #8345

9MM Makarov
(.363 Diameter)

95 gr. JHP
Makarov #8200

100 gr. FPJ
Makarov #8210

.38 Super
(.356 Diameter)

.38 SUPER 150 gr. FPJ
Tournament Master #8250

.38 Caliber
(.357/9.07MM Diameter)

110 gr. JHC Blitz
Sports Master #8300
POWER JACKET

125 gr. JSP
Sports Master #8310

25 gr. JHC
Sports Master #8320
POWER JACKET

140 gr. JHC
Sports Master #8325
POWER JACKET

158 gr. JSP
Sports Master #8340

158 gr. JHC
Sports Master #8360
POWER JACKET

170 gr. JHC
Sports Master #8365
POWER JACKET

170 gr. FMJ Match
Tournament Master #8350

180 gr. FPJ Match
Tournament Master #8370

10MM .400 Caliber
(.400/10.16MM Diameter)

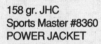

150 gr. JHP
Sports Master #8430
POWER JACKET

165 gr. JHP
Sports Master #8445
POWER JACKET

180 gr. JHP
Sports Master #8460
POWER JACKET

190 gr. FPJ
Tournament Master #8480

.41 Caliber
(.410/10.41MM Diameter)

170 gr. JHC
Sports Master #8500
POWER JACKET

210 gr. JHC
Sports Master #8520
POWER JACKET

220 gr. FPJ Match
Tournament Master #8530

.44 Magnum
(.4295/10.91MM Diameter)

180 gr. JHC
Sports Master #8600
POWER JACKET

210 gr. JHC
Sports Master #8620
POWER JACKET

220 gr. FPJ Match
Tournament Master #8605

240 gr. JHC
Sports Master #8610
POWER JACKET

250 gr. FPJ Match
Tournament Master #8615

300 gr. JSP
Sports Master #8630

.45 Caliber
(.4515/11.47MM Diameter)

185 gr. JHP
Sports Master #8800
POWER JACKET

185 gr. FPJ Match
Tournament Master #8810

200 gr. FPJ Match
Tournament Master #8825

230 gr. FMJ Match
Tournament Master #8815

240 gr. JHC
Sports Master #8820
POWER JACKET

300 gr. JSP
Sports Master #8830

SPEER RIFLE BULLETS

Bullet Caliber & Type	22 Spire Soft Point	22 Spitzer Soft Point	22 Spire Soft Point	22 Spitzer Soft Point	22 218 Bee Flat Soft Point w/Cann.	22 Spitzer Soft Point	22 "TNT" Hollow Point	22 Hollow Point	22 Hollow Point B.T. Match
Diameter	.223"	.223"	.224"	.224"	.224"	.224"	.224"	.224"	.224"
Weight (grs.)	40	45	40	45	46	50	50	52	52
Ballist. Coef.	0.145	0.166	0.144	0.167	0.094	0.231	0.223	0.225	0.253
Part Number	1005	1011	1017	1023	1024	1029	1030	1035	1036
Box Count	100	100	100	100	100	100	100	100	100

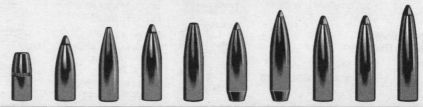

25-20 Win. Flat Soft Point w/Cann.	25 Spitzer Soft Point	25 "TNT" Hollow Point	25 Spitzer Soft Point	25 Hollow Point	25 Spitzer Soft Point B.T.	25 Spitzer Soft Point B.T.	25 Spitzer Soft Point	6.5mm Spitzer Soft Point	6.5mm Spitzer Soft Point
.257"	.257"	.257"	.257"	.257"	.257"	.257"	.257"	.263"	.263"
75	87	87	100	100	100	120	120	120	140
0.133	0.300	0.310	0.369	0.255	0.393	0.435	0.41	0.433	0.496
1237	1241	1246	1405	1407	1408	1410	1411	1435	1441
100	100	100	100	100	100	100	100	100	100

NEW!

7mm Spitzer Soft Point	7mm Match Hollow Point B.T.	7mm Spitzer Soft Point B.T.	7mm Spitzer Soft Point	7mm Mag-Tip™ Soft Point	7mm Mag-Tip™ Soft Point	30 Round Soft Point Plinker™	30 Hollow Point	30 Round Soft Point	30 Carbine Round FMJ	30 Spire Soft Point	30 "TNT" Hollow Point	30 Hollow Point
.284"	.284"	.284"	.284"	.284"	.284"	.308"	.308"	.308"	.308"	.308"	.308"	.308"
145	145	160	160	160	175	100	110	110	110	110	125	130
0.457	0.465	0.556	0.502	0.354	0.385	0.124	0.136	0.144	0.179	0.273	0.326	0.263
1629	1631	1634	1635	1637	1641	1805	1835	1845	1846	1855	1986	2005
100	100	100	100	100	100	100	100	100	100	100	100	100

30 Spitzer Soft Point B.T.	30 Spitzer Soft Point	30 Mag-Tip™ Soft Point	30 Match Hollow Point B.T.	30 Spitzer Soft Point	303 Spitzer Soft Point w/Cann.	303 (7.62x39) FMJ w/Cann.	303 Spitzer Soft Point	303 Round Soft Point	32 Flat Soft Point	8mm Spitzer Soft Point	8mm Semi-Spitzer Soft Point	8mm Spitzer Soft Point
.308"	.308"	.308"	.308"	.308"	.311"	.311"	.311"	.311"	.321"	.323"	.323"	.323"
180	180	180	190	200	125	123	150	180	170	150	170	200
0.540	0.483	0.352	0.540	0.556	0.292	0.256	0.411	0.328	0.297	0.369	0.354	0.411
2052	2053	2059	2080	2211	2213	2214	2217	2223	2259	2277	2283	2285
100	100	100	50	50	100	100	100	100	100	100	100	50

SPEER RIFLE BULLETS

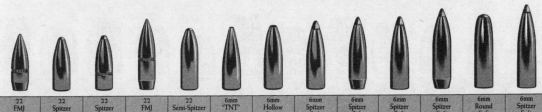

22 FMJ B.T. w/Cann.	22 Spitzer Soft Point	22 Spitzer S.P. w/Cann.	22 FMJ B.T. w/Cann.	22 Semi-Spitzer Soft Point	6mm "TNT" Hollow Point	6mm Hollow Point	6mm Spitzer Soft Point	6mm Spitzer Soft Point B.T.	6mm Spitzer Soft Point	6mm Spitzer Soft Point B.T.	6mm Round Soft Point	6mm Spitzer Soft Point
.224"	.224"	.224"	.224"	.224"	.243"	.243"	.243"	.243"	.243"	.243"	.243"	.243"
55	55	55	62	70	70	75	80	85	90	100	105	105
0.269	0.255	0.241	0.307	0.214	0.282	0.234	0.365	0.404	0.385	0.430	0.207	0.443
1044	1047	1049	1050	1053	1206	1205	1211	1213	1217	1220	1223	1229
100	100	100	100	100	100	100	100	100	100	100	100	100

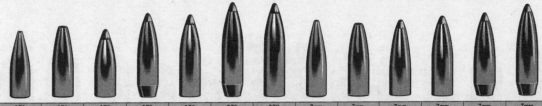

270 "TNT" Hollow Point	270 Hollow Point	270 Spitzer Soft Point	270 Spitzer Soft Point B.T.	270 Spitzer Soft Point	270 Spitzer Soft Point B.T.	270 Spitzer Soft Point	7mm "TNT" Hollow Point	7mm Hollow Point	7mm Spitzer Soft Point	7mm Spitzer Soft Point	7mm Spitzer Soft Point B.T.	7mm Spitzer Soft Point B.T.
.277"	.277"	.277"	.277"	.277"	.277"	.277"	.284"	.284"	.284"	.284"	.284"	.284"
90	100	100	130	130	150	150	110	115	120	130	130	145
0.275	0.225	0.319	0.449	0.408	0.496	0.481	0.338	0.257	0.386	0.394	0.411	0.502
1446	1447	1453	1458	1459	1604	1605	1616	1617	1620	1623	1624	1628
100	100	100	100	100	100	100	100	100	100	100	100	100

NEW! (270) **NEW!** (7mm)

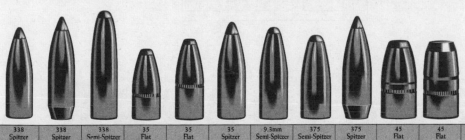

30 Flat Soft Point	30 Flat Soft Point	30 Round Soft Point	30 FMJ B.T. w/Cann.	30 Spitzer Soft Point B.T.	30 Spitzer Soft Point	30 Mag-Tip™ Soft Point	30 Round Soft Point	30 Spitzer Soft Point B.T.	30 Spitzer Soft Point	30 Match Hollow Point B.T.	30 Flat Soft Point	30 Round Soft Point
.308"	.308"	.308"	.308"	.308"	.308"	.308"	.308"	.308"	.308"	.308"	.308"	.308"
130	150	150	150	150	150	150	165	165	165	168	170	180
0.248	0.268	0.266	0.425	0.423	0.389	0.301	0.274	0.477	0.433	0.480	0.304	0.304
2007	2011	2017	2018	2022	2023	2025	2029	2034	2035	2040	2041	2047
100	100	100	100	100	100	100	100	100	100	100	100	100

338 Spitzer Soft Point B.T.	338 Spitzer Soft Point	338 Semi-Spitzer Soft Point	35 Flat Soft Point	35 Flat Soft Point	35 Spitzer Soft Point	9.3mm Semi-Spitzer Soft Point	375 Semi-Spitzer Soft Point	375 Spitzer Soft Point B.T.	45 Flat Soft Point	45 Flat Soft Point	50 BMG FMJ
.338"	.338"	.338"	.358"	.358"	.358"	.366"	.375"	.375"	.458"	.458"	.510"
200	225	275	180	220	250	270	235	270	350	400	647
0.448	0.484	0.456	0.245	0.316	0.446	0.361	0.317	0.429	0.232	0.214	0.701
2405	2406	2411	2435	2439	2453	2459	2471	2472	2478***	2479	2491
50	50	50	100	50	50	50	50	50	50	50	20

SPEER HANDGUN BULLETS

Gold Dot Hollow Point Bullets

Caliber & Type	9mm Gold Dot Hollow Point	9mm Gold Dot Hollow Point	9mm Gold Dot Hollow Point	9mm Gold Dot Hollow Point	38 Gold Dot Hollow Point	38 Gold Dot Hollow Point	9mm Makarov Gold Dot Hollow Point	40/10mm Gold Dot Hollow Point	40/10mm Gold Dot Hollow Point	45 Gold Dot Hollow Point	45 Gold Dot Hollow Point
Diameter	.355"	.355"	.355"	.355"	.357"	.357"	.364"	.400"	.400"	.451"	.451"
Weight (grs.)	90	115	124	147	125	158	90	155	180	185	230
Ballist. Coef.	0.101	0.125	0.134	0.164	0.140	0.168	0.107	0.123	0.143	0.109	0.143
Part Number	3992	3994	3998	4002	4012	4215	3999	4400	4406	4470	4483
Box Count	100	100	100	100	100	100	100	100	100	100	100
	NEW!				**NEW!**	**NEW!**	**NEW!**				

	38 JHP	38 JHP-SWC	38 TMJ	38 JHP	38 JSP	38 JSP-SWC	38 TMJ-Sil.	38 TMJ-Sil.	9mm Makarov TMJ	40/10mm TMJ	40/10mm TMJ	40/10mm TMJ	41 AE HP	41 JHP-SWC
	.357"	.357"	.357"	.357"	.357"	.357"	.357"	.357"	.364"	.400"	.400"	.400"	.410"	.410"
	140	146	158	158	158	160	180	200	95	155	180	200	180	200
	0.152	0.159	0.173	0.158	0.15	0.17	0.23	0.236	0.127	0.125	0.143	0.208	0.138	0.113
	4203	4205	4207	4211	4217	4223	4229	4231	4375	4399	4402	4403	4404	4405
	100	100	100	100	100	100	100	100	100	100	100	100	100	100
									NEW!	**NEW!**				

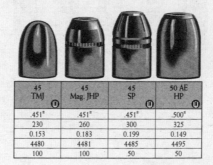

	45 TMJ	45 Mag. JHP	45 SP	50 AE HP
	.451"	.451"	.451"	.500"
	230	260	300	325
	0.153	0.183	0.199	0.149
	4480	4481	4485	4495
	100	100	50	50

Handgun Bullets Lead

Caliber & Type	32 HB-WC	32 HB-WC	38 BB-WC	38 HB-WC	38 SWC	38 HP-SWC	38 RN	44 SWC
Diameter	.314"	.314"	.358"	.358"	.358"	.358"	.358"	.430"
Weight (grs.)	98	98	148	148	158	158	158	240
Part Number	4600*	4600*	4605*	4617*	4623*	4627*	4647	4660*

SPEER BULLETS

25 TMJ	32 JHP	9mm TMJ	9mm JHP	9mm TMJ	9mm JHP	9mm SP	9mm TMJ	9mm TMJ	38 JHP	38 JSP	38 JHP	38 TMJ
.251"	.312"	.355"	.355"	.355"	.355"	.355"	.355"	.355"	.357"	.357"	.357"	.357"
50	100	95	100	115	115	124	124	147	110	125	125	125
0.110	0.167	0.131	0.111	0.177	0.118	0.115	0.114	0.208	0.122	0.14	0.135	0.146
3982	3981	4001	3983	3995*	3996	3997	4004	4006	4007	4011	4013	4015
100	100	100	100	100	100	100	100	100	100	100	100	100

41 JSP-SWC	41 TMJ-Sil.	44 Mag. JHP	44 JHP-SWC	44 JSP-SWC	44 Mag.JHP	44 Mag.JSP	44 TMJ-Sil.	44 Mag. SP	45 TMJ-Match	45 TMJ-Match	45 JHP	45 Mag. JHP
.410"	.410"	.429"	.429"	.429"	.429"	.429"	.429"	.429"	.451"	.451"	.451"	.451"
220	210	200	225	240	240	240	240	300	185	200	200	225
0.137	0.216	0.122	0.146	0.157	0.165	0.164	0.206	0.213	0.090	0.129	0.138	0.169
4417	4420	4425	4435	4447	4453	4457	4459	4463	4473	4475	4477	4479
100	100	100	100	100	100	100	50	100	100	100	100	100

45 SWC	45 RN	45 SWC
.452"	.452"	.452"
200	230	250
4677*	4690*	4683*

PLASTIC INDOOR AMMO

		Bullets	Cases
	No. Per Box	50	50
Part No.	38 Cal.	8510	8515
	44 Cal.	8520	8525
	45 Cal.	8530	See Note

Note: Shown are 38 bullet and 38 case. 45 bullet is used with regular brass case.

SHOT SHELL CAPSULES

Empty Capsules with Base Plugs		
	No. Per Box	50
Part No.	38/357	8780
	44	8782

SPEER BIG-GAME BULLETS

GRAND SLAM

Bullet Caliber & Type	6mm GS Soft Point	25 GS Soft Point	270 GS Soft Point	270 GS Soft Point	7mm GS Soft Point	7mm GS Soft Point	7mm GS Soft Point	30 GS Soft Point	30 GS Soft Point
Diameter	.243"	.257"	.277"	.277"	.284"	.284"	.284"	.308"	.308"
Weight (grs.)	100	120	130	150	145	160	175	150	165
Ballist. Coef.	0.351	0.328	0.345	0.385	0.327	0.387	0.465	0.305	0.393
Part Number	1222	1415	1465	1608	1632	1638	1643	2026	2038
Box Count	50	50	50	50	50	50	50	50	50

Bullet Caliber & Type	30 GS Soft Point	338 GS Soft Point	35 GS Soft Point	375 GS Soft Point
Diameter	.308"	.338"	..358"	.375"
Weight (grs.)	180	250	250	285
Ballist. Coef.	0.416	0.431	0.335	0.354
Part Number	2063	2408	2455	2473
Box Count	50	50	50	50

AFRICAN GRAND SLAM

Bullet Caliber & Type	338 AGS Tungsten Solid	375 AGS Tungsten Solid	416 AGS Soft Point	416 AGS Tungsten Solid	45 AGS Soft Point	45 AGS Tungsten Solid
Diameter	338"	.375"	.416"	.416"	.458"	.458"
Weight (grs.)	275	300	400	400	500	500
Ballist. Coef.	0.291	0.258	0.381	0.262	0.285	0.277
Part Number	2414	2474	2475	2476	2485	2486
Box Count	25	25	25	25	25	25

HERCULES SMOKELESS POWDERS

Twelve types of Hercules smokeless sporting powders are available to the handloader. These have been selected from the wide range of powders produced for factory loading to provide at least one type that can be used efficiently and economically for each type of ammunition. These include:

BULLSEYE® A high-energy, quick-burning powder especially designed for pistol and revolver. The most popular powder for .38 special target loads. Can also be used for 12 gauge-1 oz. shotshell target loads.

RED DOT® The preferred powder for light-to-medium shotshells; specifically designed for 12-gauge target loads. Can also be used for handgun loads.

GREEN DOT® Designed for 12-gauge medium shotshell loads. Outstanding in 20-gauge skeet loads.

UNIQUE® Has an unusually broad application from light to heavy shotshell loads. As a handgun powder, it is our most versatile, giving excellent performance in many light to medium-heavy loads.

HERCO® A long-established powder for high velocity shotshell loads. Designed for heavy and magnum 10-, 12-, 16- and 20-gauge loads. Can also be used in high-performance handgun loads.

BLUE DOT® Designed for use in magnum shotshell loads, 10-, 12-, 16-, 20- and 28-gauge. Also provides top performance with clean burning in many magnum handgun loads.

HERCULES 2400® For use in small-capacity rifle cartridges and .410-Bore shotshell loads. Can also be used for large-caliber magnum handgun cartridges.

RELODER® SERIES Designed for use in center-fire rifle cartridges. Reloder 7, 12, 15, 19 and 22 provide the right powder for the right use. From small capacity to magnum loads. All of them deliver high velocity, clean burn, round-to-round consistency, and economy.

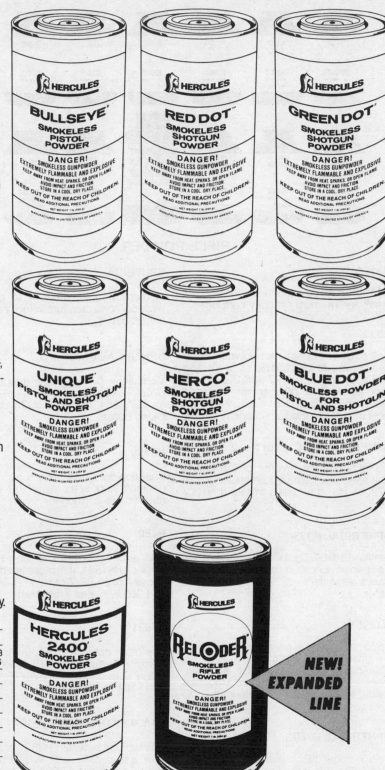

PACKAGING

POWDER	1-LB CANISTERS	4-LB CANISTERS	5-LB CANISTERS	8-LB KEG
Bullseye	●	●		●
Red Dot	●	●		●
Green Dot	●	●		●
Unique	●	●		●
Herco	●	●		●
Blue Dot	●		●	
Hercules 2400	●	●		●
Reloder Series	●		●	

HODGDON SMOKELESS POWDER

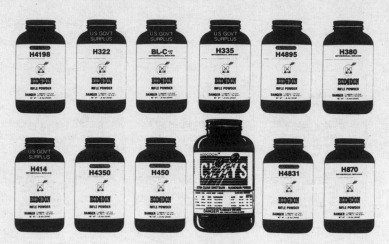

SHOTGUN AND PISTOL POWDER

RIFLE POWDER

H4198

H4198 was developed especially for small and medium capacity cartridges.

H322

Any extruded bench rest powder which has proved to be capable of producing fine accuracy in the 22 and 308 bench rest guns. This powder fills the gap betweenH4198 and BL-C(2). Performs best in small to medium capacity cases.

SPHERICAL BL-C®, Lot No. 2

A highly popular favorite of the bench rest shooters. Best performance is in the 222, and in other cases smaller than 30/06.

SPHERICAL H335®

Similar to BL-C(2), H335 is popular for its performance in medium capacity cases, especially in 222 and 308 Winchester.

H4895®

4895 may well be considered the most versatile of all propellants. It gives desirable performance in almost all cases from 222 Rem. to 458 Win. Reduced loads, to as low as 3/5 of maximum, still give target accuracy.

SPHERICAL H380®

This number fills a gap between 4320 and 4350. It is excellent in 22/250, 220 Swift, the 6mm's, 257 and 30/06.

SPHERICAL H414®

In many popular medium to medium-large calibers, pressure velocity relationship is better.

SPHERICAL H870®

Very slow burning rate adaptable to overbore capacity Magnum cases such as 257, 264, 270 and 300 Mags with heavy bullets.

H4350

This powder gives superb accuracy at optimum velocity for many large capacity metallic rifle cartridges.

H450

This slow-burning spherical powder is similar to H4831. It is recommended especially for 25-06, 7mm Mag., 30-06, 270 and 300 Win. and Wby. Mag.

H4831®

The most popular of all powders. Outstanding performance with medium and heavy bullets in the 6mm's, 25/06, 270 and Magnum calibers.

H1000 EXTRUDED POWDER

Fills the gap between H4831 and H870. Works especially well in overbore capacity cartridges (1,000-yard shooters take note).

HP38

A fast pistol powder for most pistol loading. Especially recommended for mid-range 38 specials.

CLAYS

A powder developed for 12-gauge clay target shooters. Also performs well in many handgun applications, including .38 Special, .40 S&W and 45 ACP. Perfect for 1 1/8 and 1 oz. loads.
Now available:
Universal Clays. Loads nearly all of the straight-wall pistol cartridges as well as 12 ga. 1 1/4 oz. thru 28 ga. 3/4 oz. target loads.
International Clays. Perfect for 12 and 20 ga. autoloaders who want reduced recoil.

HS-6 and HS-7

HS-6 and HS-7 for Magnum field loads are unsurpassed, since they do not pack in the measure. They deliver uniform charges and are dense to allow sufficient wad column for best patterns.

H110

A spherical powder made especially for the 30 M1 carbine. H110 also does very well in 357, 44 Spec., 44 Mag. or .410 ga. shotshell. Magnum primers are recommended for consistent ignition.

H4227

An extruded powder similar to H110, it is the fastest burning in Hodgdon's line. Recommended for the 22 Hornet and some specialized loading in the 45-70 caliber. Also excellent in magnum pistol and .410 shotgun.

IMR SMOKELESS POWDERS

SHOTSHELL POWDER

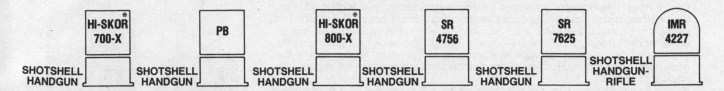

Hi-Skor 700-X Double-Base Shotshell Powder. Specifically designed for today's 12-gauge components. Developed to give optimum ballistics at minimum charge weight (which means more reloads per pounds of powder). 700-X is dense, easy to load, clean to handle and loads uniformly.
PB Shotshell Powder. Produces exceptional 20 and 28-gauge skeet reloads; preferred by many in 12-gauge target loads, it gives 3-dram equivalent velocity at relatively low chamber pressures.

Hi-Skor 800-X Shotshell Powder. An excellent powder for 12-gauge field loads and 20 and 28-gauge loads.
SR-4756 Powder. Great all-around powder for target and field loads.
SR-7625. A fast-growing favorite for reloading target as well as light and heavy field loads in 4 gauges. Excellent velocity-chamber pressure.
IMR-4227 Powder. Can be used effectively for reloading .410-gauge shotshell ammunition.

RIFLE POWDER

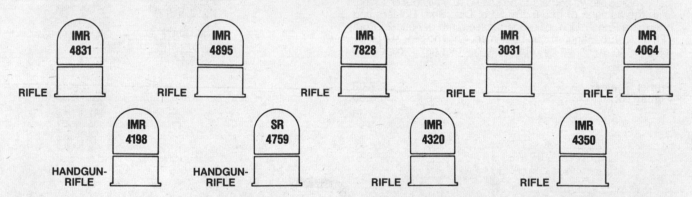

IMR-3031 Rifle Powder. Specifically recommended for medium-capacity cartridges.
IMR-4064 Rifle Powder. Has exceptionally uniform burning qualities when used in medium and large-capacity cartridges.
IMR-4198. Made the Remington 222 cartridge famous. Developed for small and medium-capacity cartridges.
IMR-4227 Rifle Powder. Fastest burning of the IMR Series. Specifically designed for the 22 Hornet class of cartridges.
SR-4759. Brought back by shooter demand. Available for cast bullet loads.
IMR-4320. Recommended for high-velocity cartridges.

IMR-4350 Rifle Powder. Gives unusually uniform results when loaded in Magnum cartridges.
IMR-4831. Produced as a canister-grade handloading powder. Packaged in 1 lb. canister, 8 lb. caddy and 20 lb. kegs.
IMR-4895 Rifle Powder. The time-tested standard for caliber 30 military ammunition; slightly faster than IMR-4320. Loads uniformly in all powder measures. One of the country's favorite powder.
IMR-7828 Rifle Powder. The slowest burning DuPont IMR canister powder, intended for large-capacity and magnum-type cases with heavy bullets.

PISTOL POWDER

PB Powder. Another powder for reloading a wide variety of centerfire handgun ammunition.
IMR-4227 Powder. Can be used effectively for reloading Magnum handgun ammunition.
Hi-Skor 700-X Powder. The same qualities that make it a superior powder contribute to its excellent performance in all the popular handguns.

Hi-Skor 800-X Powder. Good powder for heavier bullet handgun calibers.
SR-7625 Powder. For reloading a wide variety of centerfire handgun ammunition.
SR-4756. Clean burning with uniform performance. Can be used in a variety of handgun calibers.

FORSTER/BONANZA RELOADING

CO-AX® BENCH REST® RIFLE DIES

Bench Rest Rifle Dies are glass hard for long wear and minimum friction. Interiors are polished mirror smooth. Special attention is given to headspace, tapers and diameters so that brass will not be overworked when resized. Sizing die has an elevated expander button which is drawn through the neck of the case at the moment of the greatest mechanical advantage of the press. Since most of the case neck is still in the die when expanding begins, better alignment of case and neck is obtained.

Bench Rest® Seating Die is of the chamber type. The bullet is held in alignment in a close-fitting channel. The case is held in a tight-fitting chamber. Both bullet and case are held in alignment while the bullet is being seated. Cross-bolt lock ring included at no charge.

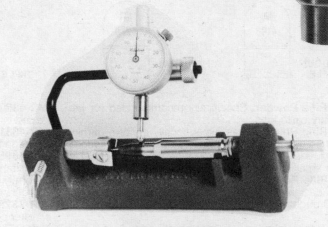

Bench Rest® Die Set $62.00
Full Length Sizer 28.30
Bench Rest Seating Die 34.00

PRIMER SEATER
With "E-Z-Just" Shellholder

The Bonanza Primer Seater is designed so that primers are seated Co-Axially (primer in line with primer pocket). Mechanical leverage allows primers to be seated fully without crushing. With the addition of one extra set of Disc Shell Holders and one extra Primer Unit, all modern cases, rim or rimless, from 222 up to 458 Magnum, can be primed. Shell holders are easily adjusted to any case by rotating to contact rim or cannelure of the case.

Primer Seater $58.00
Primer Tube 5.00

PRIMER SEATER

CO-AX® INDICATOR

Bullets will not leave a rifle barrel at a uniform angle unless they are started uniformly. The Co-Ax Indicator provides a reading of how closely the axis of the bullet corresponds to the axis of the cartridge case. The Indicator features a spring-loaded plunger to hold cartridges against a recessed, adjustable rod while the cartridge is supported in a "V" block. To operate, simply rotate the cartridge with the fingers; the degree of misalignment is transferred to an indicator which measures in one-thousandths.

Price: Without dial $48.00
Indicator Dial 56.60

FORSTER/BONANZA RELOADING

ULTRA BULLET SEATER DIE

Forster's new Ultra Die is available in 51 calibers, more than any other brand of micrometer-style seater. Adjustment is identical to that of a precision micrometer—the head is graduated to .001" increments with .025" bullet movement per revolution. The cartridge case, bullet and seating stem are completely supported and perfectly aligned in a close-fitting chamber before and during the bullet seating operation.
Price: **$49.50**

CO-AX LOADING PRESS B-2

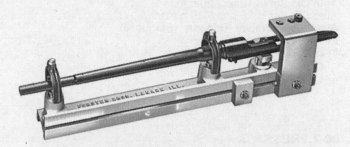

UNIVERSAL SIGHT MOUNTING FIXTURE

This product fills the exacting requirements needed for drilling and tapping holes for the mounting of scopes, receiver sights, shotgun beads, etc. The fixture handles any single-barrel gun—bolt-action, lever-action or pump-action—as long as the barrel can be laid into the "V" blocks of the fixture. Tubular guns are drilled in the same manner by removing the magazine tube. The fixture's main body is made of aluminum casting. The two "V" blocks are adjustable for height and are made of hardened steel ground accurately on the "V" as well as the shaft.
Price: **$324.00**

CO-AX® LOADING PRESS MODEL B-2

Designed to make reloading easier and more accurate, this press offers the following features: Snap-in and snap-out die change • Positive spent primer catcher • Automatic self-acting shell holder • Floating guide rods • Working room for right- or left-hand operators • Top priming device seats primers to factory specifications • Uses any standard $7/8" \times 14$ dies • No torque on the head • Perfect alignment of die and case • Three times the mechanical advantage of a "C" press
Price: **$258.00**

BENCH REST POWDER MEASURE

BENCH REST POWDER MEASURE

When operated uniformly, this measure will throw uniform charges from 2½ grains Bullseye to 95 grains #4320. No extra drums are needed. Powder is metered from the charge arm, allowing a flow of powder without extremes in variation while minimizing powder shearing. Powder flows through its own built-in baffle so that powder enters the charge arm uniformly.
Price: **$94.50**

HORNADY

APEX SHOTSHELL RELOADER

This new and versatile shotshell reloader has all the features of a progressive press along with the control, accuracy, easy operation and low price tag of a single-stage loader. You can load one shell at a time or seven shells at once, turning out a fully loaded shell with every pull of the handle. Other features include: extra-large shot hopper, short linkage arm, automatic dual-action crimp die, swing-out wad guide, and extra-long shot and powder feed tubes.

Apex Shotshell Reloader (Automatic)
In 12 and 20 gauge $385.00
In 28 and .410 gauge 425.00
Auto Die Set in 12 and 20 gauge 143.50
Auto Die Set in 28 and .410 gauge 184.00
Apex-91 Shotshell Reloader (Standard)
In 12 and 20 gauge 137.50
In 28 and .410 gauge 154.25

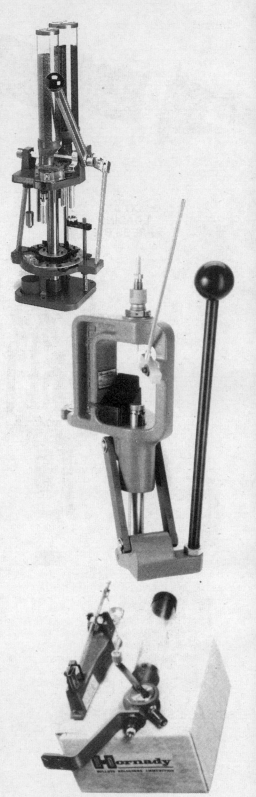

00-7 PRESS

- "Power-Pac" linkage multiplies lever-to-arm power.
- Frame of press angled 30° to one side, making the "O" area of press totally accessible.
- More mounting area for rock-solid attachment to bench.
- Special strontium-alloy frame provides greater stress resistance. Won't spring under high pressures needed for full-length resizing.

00-7 Press (does not include dies or shell holder) $63.25
00-7 Automatic Primer Feed (complete with large and small
  primer tubes) 19.25

THE 00-7 PRESS PACKAGE
A reloading press complete with dies and shell holder

Expanded and improved to include Automatic Primer Feed. It sets you up to load many calibers and includes choice of a basic 00-7 complete with: • Set of New Dimension Dies • Primer catcher • Removable head shell holder • Positive Priming System • Automatic Primer Feed.

00-7 Package with Series I & II Dies (14 lbs.) $158.00
00-7 Package Series II Titanium Nitride (15 lbs.) 170.00
00-7 Kit with Series I & II Dies 325.00
00-7 Kit with Titanium Nitride Dies 339.00

THE HANDLOADER'S ACCESSORY PACK I

Here's everything you need in one money-saving pack. It includes: • Deluxe powder measure • Powder scale • Two non-static powder funnels • Universal loading block • Primer turning plate • Case lube • Chamfering and deburring tool • 3 case neck brushes • Large and small primer pocket cleaners • Accessory handle. Plus one copy of the *Hornady Handbook of Cartridge Reloading.*

Handloader's Accessory Pack I No. 030300 . . . . $175.00

HORNADY

TRIMMER PACKAGE

Combines Hornady's Case Trimmer with the new Metric Caliper and Steel Dial Caliper, which measures case and bullet lengths plus inside/outside diameters. Made from machined steel, the caliper provides extremely accurate measurements with an easy-to-read large dial gauge.

Trimmer Package **$99.95**

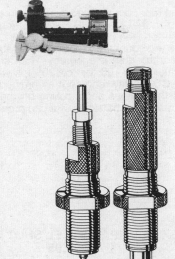

NEW DIMENSION RELOADING DIES

Features an Elliptical Expander that minimizes friction and reduces case neck stretch, plus the need for a tapered expander for "necking up" to the next larger caliber. Other recent design changes include a hardened steel decap pin that will not break, bend or crack even when depriming stubborn military cases. A bullet seater alignment sleeve guides the bullet and case neck into the die for in-line benchhrest alignment. All New Dimension Reloading Dies include: collar and collar lock to center expander precisely; one-piece expander spindle with tapered bottom for easy cartridge insertion; wrench flats on die body, Sure-Loc™ lock rings and collar lock for easy tightening; and built-in crimper.

New Dimension Reloading Dies:
Series I Two-die Rifle Set **$22.90**
Series I Three-die Rifle Set 25.50
Series II Three-die Pistol Set (w/Titanium Nitride) 34.35
Series III Two-die Rifle Set 29.85
Series IV Custom Die Set 52.65

PRO-JECTOR PRESS PACKAGE

- Includes Pro-Jector Press, set of New Dimension dies, automatic primer feed, brass kicker, primer catcher, shell plate, and automatic primer shut-off
- Just place case in shell plate, start bullet, pull lever and drop powder. Automatic rotation of shell plate prepares next round.
- Fast inexpensive changeover requires only shell plate and set of standard ⅞ × 14 threaded dies.
- Primes automatically.
- Power-Pac Linkage assures high-volume production even when full-length sizing.
- Uses standard powder measures and dies.

Series I & II **$380.00**
Series II Titanium Nitride Dies 395.00
Extra Shell Plates 21.95
Pro-Jector Kit with Series I & II Dies 549.95
Kit Series III with Titanium Nitride Dies 560.00

MODEL 366 AUTO SHOTSHELL RELOADER

The 366 Auto features full-length resizing with each stroke, automatic primer feed, swing-out wad guide, three-stage crimping featuring Taper-Loc for factory tapered crimp, automatic advance to the next station and automatic ejection. The turntable holds 8 shells for 8 operations with each stroke. The primer tube filler is fast. The automatic charge bar loads shot and powder. Right- or left-hand operation; interchangeable charge bushings, die sets and Magnum dies and crimp starters for 6 point, 8 point and paper crimps.

Model 366 Auto Shotshell Reloader:
12, 20, 28 gauge or .410 bore **$440.00**
Model 366 Auto Die Set 88.95
Auto Advance 50.00
Swing-out Wad Guide & Shell Drop Combo 125.00

HORNADY

CHRONOMAX
$144.95

The Chronomax pocket-sized box has an easy-to-read liquid crystal display that reads FIRE and a string of up to 5 shots. Separate buttons read ON/OFF, RESET and CONTRAST. The skyscreens are connected with a single cable and are spaced two feet apart for greater accuracy. The Chronomax can measure velocities from 100 fps to 4999 fps with an accuracy of 1 fps. Its LCD display reads out a continuous string of up to 5 shots, with previous shots rolling off the display area. It's powered by four AA batteries that can provide up to 72 hours of use.

APEX MULTIPAX STEEL SHOT COUNTER
$39.95

Loads steel and large-sized shot automatically. Features a cylindrical hopper of its own, along with a handle that rotates a series of internal disks to count out the proper number of shot pellets. The pellets then drop through the APEX die head and into the shell. Comes with multiple-sized disks and instructions.

M-3 CASE TUMBLER
$125.00

Twice as large as the M-2 Tumbler, this product will hold up to 1,000 cases (38 Special) for cleaning. Available in 110V and 220C models. A Case Tumbler Sifter (**$11.50**) conveniently separates tumbler media from cleaned shells.

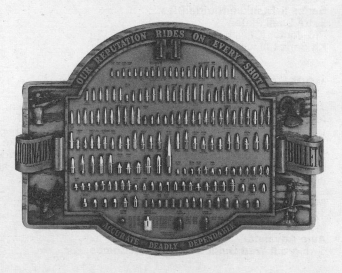

HORNADY BULLET DISPLAY
$165.00

This display features every one of Hornady's jacketed and lead bullets currently in production, along with muzzleloading bullets. Contains over 175 bullets in all, including the new A/MAX 50-caliber target bullet. Moulded from handcrafted oak, highlighting target shooters, a muzzleloader, whitetail deer, bighorn sheep and Hornady's slogan: "Our reputation rides on every shot."

LYMAN BULLET SIZING EQUIPMENT

MAG 20 ELECTRIC FURNACE

The MAG 20 is a new furnace offering several advantages to cast bullet enthusiasts. It features a steel crucible of 20-pound capacity and incorporates a proven bottom-pour valve system and a fully adjustable mould guide. The improved design of the MAG 20 makes it equally convenient to use the bottom-pour valve, or a ladle. A new heating coil design reduces the likelihood of pour spout "freeze." Heat is controlled from "Off" to nominally 825° F by a calibrated thermostat which automatically increases temperature output when alloy is added to the crucible. A pre-heat shelf for moulds is attached to the back of the crucible. Availalbe for 100 V and 200 V systems.

Price: 110 V **$229.95**
220 V **230.00**

BULLET MAKING EQUIPMENT

Deburring Tool
Lyman's deburring tool can be used for chamfering or deburring of cases up to 45 caliber. For precise bullet seating, use the pointed end of the tool to bevel the inside of new or trimmed cases. To remove burrs left by trimming, place the other end of the deburring tool over the mouth of the case and twist. The tool's centering pin will keep the case aligned . . **$14.00**

Mould Handles
These large hardwood handles are available in three sizes single-, double- and four-cavity.
Single-cavity handles (for small block, black powder and specialty moulds; 12 oz.) **$22.50**
Double-cavity handles (for two-cavity and large-block single-cavity moulds; 12 oz.) **22.00**
Four-cavity handles (1 lb.) **25.00**

Rifle Moulds
All Lyman rifle moulds are available in double cavity only, except those moulds where the size of the bullet necessitates a single cavity (12 oz.) **$44.95**

Hollow-Point Bullet Moulds
Hollow-point moulds are cut in single-cavity blocks only and require single-cavity handles (9 oz.) **$44.95**

Shotgun Slug Moulds
Available in 12 or 20 gauge; do not require rifling. Moulds are single cavity only, cut on the larger double-cavity block and require double-cavity handles (14 oz.) **$44.95**

Pistols Moulds
Cover all popular calibers and bullet designs in double-cavity blocks and, on a limited basis, four-cavity blocks.
Double-cavity mould block **$44.95**
Four-cavity mould block **69.95**

Lead Casting Dipper
Dipper with cast-iron head. The spout is shaped for easy, accurate pouring that prevents air pockets in the finished bullet **$15.00**

Gas Checks
Gas checks are gilding metal caps which fit to the base of cast bullets. These caps protect the bullet base from the burning effect of hot powder gases and permit higher velocities. Easily seated during the bullet sizing operation. Only Lyman gas checks should be used with Lyman cast bullets.

22 through 35 caliber (per 1000) **$22.50**
375 through 45 caliber (per 1000) **27.50**
Gas check seater **7.95**

Lead Pot
The cast-iron pot allows the bullet caster to use any source of heat. Pot capacity is about 8 pounds of alloy and the flat bottom prevents tipping **$15.00**

Universal Decapping Die
Covers all calibers .22 through .45 (except .378 and .460 Weatherby). Can be used before cases are cleaned or lubricated. Requires no adjustment when changing calibers; fits all popular makes of $7/8 \times 14$ presses, single station or progressive, and is packaged with 10 replacement pins **$11.50**

UNIVERSAL CARBIDE FOUR-DIE SET

Lyman's new 4-die carbide sets allow simultaneous neck expanding and powder charging. They feature specially designed hollow expanding plugs that utilize Lyman's 2-step neck-expansion system, while allowing powder to flow through the die into the cartridge case after expanding. Includes taper crimp die. All popular pistol calibers. **$44.95**

LYMAN RELOADING TOOLS

MAG TUMBLER

This new Mag Tumbler features an industrial strength motor and large 14-inch bowl design. With a working capacity of 2¾ gallons, it cleans more than 1,500 pistol cases in each cycle. The Mag Tumbler is also suitable for light industrial use in deburring and polishing metal parts. Available in 110V or 220V, with standard on/off switch.

Mag Tumbler	$229.95
Super Mag AutoFlo	249.95

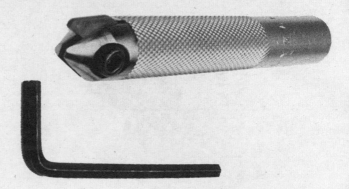

"INSIDE/OUTSIDE" DEBURRING TOOL

This unique new tool features an adjustable cutting blade that adapts easily to any rifle or pistol case from 22 caliber to 45 caliber with a simple hex wrench adjustment. Inside deburring is completed by a conical internal section with slotted cutting edges, thus providing uniform inside and outside deburring in one simple operation. The deburring tool is mounted on an anodized aluminum handle that is machine-knurled for a sure grip.

Deburring Tool $12.75

TUBBY TUMBLER

This popular tumbler now features a clear plastic "see thru" lid that fits on the outside of the vibrating tub. The Tubby has a polishing action that cleans more than 100 pistol cases in less than two hours. The built-in handle allows easy dumping of cases and media. An adjustable tab also allows the user to change the tumbling speed for standard or fast action.

Tubby Tumbler $58.50

MASTER CASTING KIT

Designed especially to meet the needs of blackpowder shooters, this new kit features Lyman's combination round ball and maxi ball mould blocks. It also contains a combination double cavity mould, mould handle, mini-mag furnace, lead dipper, bullet lube, a user's manual and a cast bullet guide. Kits are available in 45, 50 and 54 caliber.

Master Casting Kit $131.50

LYMAN RELOADING TOOLS

FOR RIFLE OR PISTOL CARTRIDGES

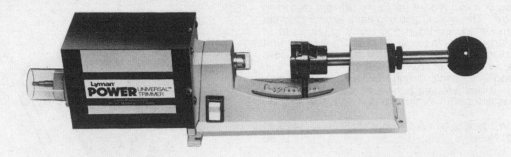

POWER CASE TRIMMER

The new Lyman Power Trimmer is powered by a fan-cooled electric motor designed to withstand the severe demands of case trimming. The unit, which features the Universal™ Chuckhead, allows cases to be positioned for trimming or removed with fingertip ease. The Power Trimmer package includes Nine Pilot Multi-Pack. In addition to two cutter heads, a pair of wire end brushes for cleaning primer pockets are included. Other features include safety guards, on-off rocker switch, heavy cast base with receptacles for nine pilots, and bolt holes for mounting on a work bench. Available for 110 V or 220 V systems.

Prices: 110 V Model $169.95
220 V Model 170.00

ACCULINE OUTSIDE NECK TURNER
(not shown)

To obtain perfectly concentric case necks, Lyman's Outside Neck Turner assures reloaders of uniform neck wall thickness and outside neck diameter. The unit fits Lyman's Universal Trimmer and AccuTrimmer. In use, each case is run over a mandrel, which centers the case for the turning operation. The cutter is carefully adjusted to remove a minimum amount of brass. Rate of feed is adjustable and a mechanical stop controls length of cut. Mandrels are available for calibers from .17 to .375; cutter blade can be adjusted for any diameter from .195" to .405".

Outside Neck Turner w/extra blade, 6 mandrels . . . $26.00
Individual Mandrels 4.00

STARTER KIT

Includes "Orange Crusher" Press, loading block, case lube kit, primer tray, Model 500 Pro scale, powder funnel and Lyman Reloading Handbook.

Starter Kit $149.95

LYMAN "ORANGE CRUSHER" RELOADING PRESS

The only press for rifle or pistol cartridges that offers the advantage of powerful compound leverage combined with a true magnum press opening. A unique handle design transfers power easily where you want it to the center of the ram. A 4½-inch press opening accommodates even the largest cartridges.

"Orange Crusher" Press:
With Priming Arm and Catcher $94.95

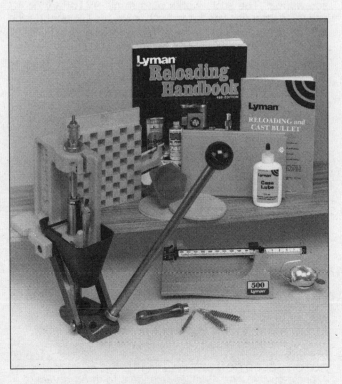

STARTER KIT

LYMAN RELOADING TOOLS

T-MAG TURRET RELOADING PRESS

With the T-Mag you can mount up to six different reloading dies on our turret. This means you can have all your dies set up, precisely mounted, locked in and ready to reload at all times. The T-Mag works with all $7/8 \times 14$ dies. The T-Mag turret with its quick-disconnect release system is held in rock-solid alignment by a 3/4-inch steel stud.

Also featured is Lyman's Orange Crusher compound leverage system. It has a longer handle with a ball-type knob that mounts easily for right- or left-handed operation.

T-Mag Press w/Priming Arm & Catcher **$129.95**
   Extra Turret Head **19.95**

Also available:
EXPERT KIT that includes T-MAG Press, Universal Case Trimmer and pilot Multi-Pak, Model 500 powder scale and Model 50 powder measure, plus accessories.
   Available in calibers 9mm Luger, 38/357, 44 Mag., 45 ACP and 30-06 . . . . **$320.00**

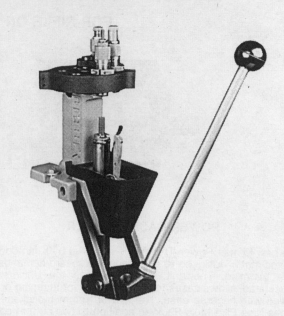

ELECTRONIC SCALE MODEL LE: 1000

Accurate to 1/10 grain, Lyman's new LE: 1000 measures up to 1000 grains of powder and easily converts to the gram mode for metric measurements. The push-botton automatic calibration feature eliminates the need for calibrating with a screwdriver. The scale works off a single 9V battery or AC power adaptor (included with each scale). Its compact design allows the LE: 1000 to be carried to the field easily. A sculpted carrying case is optional. 110 Volt or 220 Volt.

Model LE: 1000 Electronic Scale **$333.25**

PISTOL ACCUMEASURE

Lyman's Pistol AccuMeasure uses changeable brass rotors pre-drilled to drop precise charges of ball and flake pistol propellants (the tool is not intended for use with long grain IMR-type powders). Most of the rotors are drilled with two cavities for maximum accuracy and consistency. The brass operating handle, which can be shifted for left or right hand operation, can be removed. The Pistol AccuMeasure can be mounted on all turret and single station presses; it can also be hand held with no loss of accuracy.

Pistol AccuMeasure w/3-rotor starter kit **$29.95**

Also available:
AMMO HANDLER KIT that includes every tool (except reloading press) needed to produce high-quality ammunition **$150.00**
ROTOR SELECTION SET including 12 dual-cavity rotors and 4 single-cavity units. Enables reloaders to throw a variety of charges for all pistol calibers through 45 . . . . **$44.00**

LYMAN RELOADING TOOLS

DRILL PRESS CASE TRIMMER

Intended for competitive shooters, varmint hunters, and other sportsmen who use large amounts of reloaded ammunition, this new drill press case trimmer consists of the Universal™ Chuckhead, a cutter shaft adapted for use in a drill press, and two quick-change cutter heads. Its two major advantages are speed and accuracy. An experienced operator can trim several hundred cases in a hour, and each will be trimmed to a precise length.

Drill Press Case Trimmer **$44.00**

AUTO TRICKLER (not shown)

This unique device allows reloaders to trickle the last few grains of powder automatically into their scale powder pans. The Auto-Trickler features vertical and horizontal height adjustments, enabling its use with both mechanical and the new electronic scales. It also offers a simple push-button operation. The powder reservoir is easily removed for cleaning. Handles all conventional ball, stick or flare powder types.

Auto-Trickler **$39.95**

ACCU TRIMMER

Lyman's new Accu Trimmer can be used for all rifle and pistol cases from 22 to 458 Winchester Magnum. Standard shellholders are used to position the case, and the trimmer incorporates standard Lyman cutter heads and pilots. Mounting options include bolting to a bench, C-clamp or vise.

Accu Trimmer w/9-pilot multi-pak **$37.00**

UNIVERSAL TRIMMER WITH NINE PILOT MULTI-PACK

This trimmer with patented chuckhead accepts all metallic rifle or pistol cases, regardless of rim thickness. To change calibers, simply change the case head pilot. Other features include coarse and fine cutter adjustments, an oil-impregnated bronze bearing, and a rugged cast base to assure precision alignment and years of service. Optional carbide cutter available. Trimmer Stop Ring includes 20 indicators as reference marks.

Replacement carbide cutter	**$39.95**
Trimmer Multi-Pack (incl. 9 pilots: 22, 24, 27, 28/7mm, 30, 9mm, 35, 44 and 45A)	62.50
Nine Pilot Multi-Pack	9.95
Power Pack Trimmer	74.95
Universal Trimmer Power Adapter	14.95

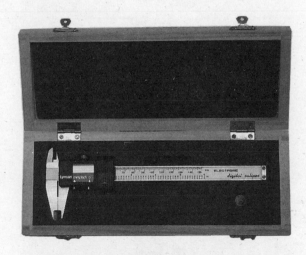

ELECTRONIC DIGITAL CALIPER

Lyman's new 6″ electronic caliper gives a direct digital readout for both inches and millimeters and can perform both inside and outside depth measurements. Its zeroing function allows the user to select zeroing dimensions and sort parts or cases by their plus or minus variation. The caliper works on a single, standard 1.5 volt silver oxide battery and comes with a fitted wooden storage case.

Electronic Caliper **$86.00**

MEC SHOTSHELL RELOADERS

MODEL 600 JR. MARK 5
$157.90

This single-stage reloader features a cam-action crimp die to ensure that each shell returns to its original condition. MEC's 600 Jr. Mark 5 can load 8 to 10 boxes per hour and can be updated with the 285 CA primer feed. Press is adjustable for 3″ shells. Die sets are available in all gauges at: **$59.50**

MODEL 8567 GRABBER
$445.50

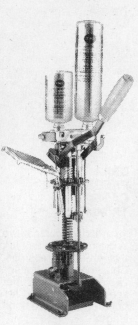

This reloader features 12 different operations at all 6 stations, producing finished shells with each stroke of the handle. It includes a fully automatic primer feed and Auto-Cycle charging, plus MEC's exclusive 3-stage crimp. The "Power Ring" resizer ensures consistent, accurately sized shells without interrupting the reloading sequence. Simply put in the wads and shell casings, then remove the loaded shells with each pull of the handle. Optional kits to load 3″ shells and steel shot make this reloader tops in its field. Resizes high and low base shells. Available in 12, 16, 20, 28 gauge and .410 bore. No die sets are available.

MODEL 650
$310.50

This reloader works on 6 shells at once. A reloaded shell is completed with every stroke. The MEC 650 does not resize except as a separate operation. Automatic Primer feed is standard. Simply fill it with a full box of primers and it will do the rest. Reloader has 3 crimping stations: the first one starts the crimp, the second closes the crimp, and the third places a taper on the shell. Available in 12, 16, 20 and 28 gauge and .410 bore. No die sets are available.

MODEL 8120
SIZEMASTER
$237.90

Sizemaster's "Power Ring" collet resizer returns each base to factory specifications. This new generation resizing station handles brass or steel heads, both high and low base. An 8-fingered collet squeezes the base back to original dimensions, then opens up to release the shell easily. The E-Z Prime auto primer feed is standard equipment. Press is adjustable for 3″ shells and is available in 12, 16, 20, 28 gauge and .410 bore. Die sets are available at: **$88.65 ($104.00** in 10 ga.).

MEC RELOADING

STEELMASTER SINGLE STAGE

The only shotshell relaoder equipped to load steel shotshells as well as lead ones. Every base is resized to factory specs by a precision "power ring" collet. Handles brass or steel heads in high or low base. The E-Z prime auto primer feed dispenses primers automatically and is standard equipment. Separate presses are available for 12 gauge 2³/₄", 3", 12 gauge 3¹/₂" and 10 gauge.

Steelmaster **$247.50**
In 12 ga. 3¹/₂" only 272.50

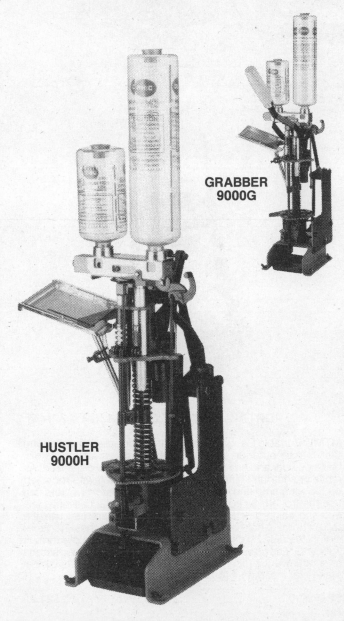

GRABBER 9000G

HUSTLER 9000H

E-Z PRIME "S" AND "V" AUTOMATIC PRIMER FEEDS

From carton to shell with security, these primer feeds provide safe, convenient primer positioning and increase rate of production. Reduce bench clutter, allowing more free area for wads and shells.

- Primers transfer directly from carton to reloader, tubes and tube fillers
- Positive mechanical feed (not dependent upon agitation of press)
- Visible supply
- Automatic. Eliminate hand motion
- Less susceptible to damage
- Adapt to all domestic and most foreign primers with adjustment of the cover
- May be purchased separately to replace tube-type primer feed or to update your present reloader

E-Z Prime "S" (for Super 600, 650) or
E-Z Prime "V" (for 600 Jr. Mark V & VersaMEC) . . . **$38.50**

MEC 9000 SERIES SHOTSHELL RELOADER

MEC's 9000 Series features automatic indexing and finished shell ejection for quicker and easier reloading. The factory set speed provides uniform movement through every reloading stage. Dropping the primer into the reprime station no longer requires operator "feel." The reloader requires only a minimal adjustment from low to high brass domestic shells, any one of which can be removed for inspection from any station.

MEC 9000H Hustler **$1306.50**
MEC 9000G Grabber 540.80

MTM RELOADING

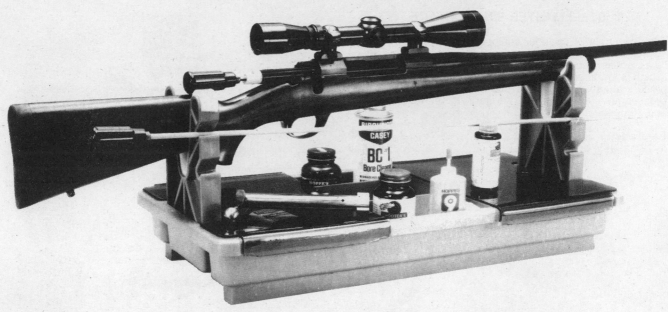

PORTABLE RIFLE MAINTENANCE CENTER

Holds rifles and shotguns for easy cleaning and maintenance (can also be used as a shooting rest). Features gun forks with built-in cleaning rod holders; sliding see-through dust covers; tough polypropylene material; fits conveniently on top of Case-Gard A-760.

Price: $29.15

MTM HANDLOADER'S LOG (not shown)

Space is provided for 1,000 entries covering date, range, group size or score, components, and conditions. Book is heavy-duty vinyl, reinforced 3-ring binder.

HL-74 . . . . $9.35
HL-50 (incl. 50 extra log sheets) 4.00

FUNNELS (not shown)

MTM Benchrest Funnel Set is designed specifically for the bench-rest shooter. One fits 222 and 243 cases only; the other 7mm and 308 cases. Both can be used with pharmaceutical vials popular with bench-rest competitors for storage of pre-weighed charges. Funnel design prevents their rolling off the bench.

MTM Universal Funnel fits all calibers from 222 to 45.
UF-1 . . . . $2.87

Patented MTM Adapt 5-in-1 Funnel Kit includes funnel, adapters for 17 Rem., 222 Rem. and 30 through 45. Long drop tube facilitates loading of maximum charges: 222 to 45.
AF-5 . . . . $4.14

PORTABLE RANGE ORGANIZER

MTM's Portable Range Organizer is a portable utility platform that makes outdoor shooting with shotguns, rifles, pistols, black powder and archery more convenient and organized. As a rifle or shotgun stand, it holds four long guns with room left on top for ammo and extra shooting year. Pistol shooters can use the troughs to hold an extra handgun or accessories like eye and ear protection. Blackpowder shooters can use the top platform to hold powder flask, caps and patches. A special slot in the side holds cleaning or slave rods. The platform is 13″ by 13″ and stands 32″ tall, allowing the shooter easy access to shooting equipment. Collapsible for storing. Uses two steel stakes for insertion into ground.

PRO-1-30 . . . . $22.90

RCBS RELOADING TOOLS

ROCK CHUCKER PRESS

The Rock Chucker Press, with patented RCBS compound leverage system, delivers up to 200% more leverage than most presses for heavy-duty reloading of even the largest rifle and pistol cases. Rugged, Block "O" Frame prevents press from springing out of alignment even under the most strenuous operations. It case-forms as easily as most presses full-length size; it full-length sizes and makes bullets with equal ease. Shell holders snap into sturdy, all-purpose shell holder ram. Non-slip handle with convenient grip. Operates on downstroke for increased leverage. Standard 7/8-inch×14 thread.

Rock Chucker Press
(Less dies) **$118.35**

PRIMER POCKET SWAGER COMBO

For fast, precision removal of primer pocket crimp from military cases. Leaves primer pocket perfectly rounded and with correct dimensions for seating of American Boxer-type primers. Will not leave oval-shaped primer pocket that reaming produces. Swager Head Assemblies furnished for large and small primer pockets no need to buy a complete unit for each primer size. For use with all presses with standard 7/8-inch×14 top thread, except RCBS "A-3" Press. The RCBS "A-2" Press requires the optional Case Stripper Washer.

**Primer Pocket
Swager Combo** **$21.88**

ROCK CHUCKER MASTER RELOADING KIT

For reloaders who want the best equipment, the Rock Chucker Master Reloading Kit includes all the tools and accessories needed. Included are the following: • Rock Chucker Press • RCBS 505 Reloading Scale • Speer Reloading Manual #11 • Uniflow Powder Measure • RCBS Rotary Case Trimmer-2 • deburring tool • case loading block • Primer Tray-2 • Automatic Primer Feed Combo • powder funnel • case lube pad • case neck brushes • fold-up hex ket set.

Rock Chucker Master Reloading Kit **$314.75**

PRIMER POCKET BRUSH COMBO

A slight twist of this tool thoroughly cleans residue out of primer pockets. Interchangeable stainless steel brushes for large and small primer pockets attach easily to accessory handle.

Primer Pocket Brush Combo **$10.90**

RCBS RELOADING TOOLS

RELOADER SPECIAL-5

This RCBS Reloader Special-5 Press is the ideal setup to get started reloading your own rifle and pistol ammo from 12 gauge shotshells and the largest Magnums down to 22 Hornets. This press develops ample leverage and pressure to perform all reloading tasks including: (1) resizing cases their full length; (2) forming cases from one caliber into another; (3) making bullets. Rugged Block "O" Frame, designed by RCBS, prevents press from springing out of alignment even under tons of pressure. Frame is offset 30° for unobstructed front access, and is made of 48,000 psi aluminum alloy. Compound leverage system allows you to swage bullets, full-length resize cases, form 30-06 cases into other calibers. Counter-balanced handle prevents accidental drop. Extra-long ram-bearing surface minimizes wobble and side play. Standard 7/8-inch-14 thread accepts all popular dies and reloading accessories.

Reloader Special-5
   (Less dies) **$92.25**

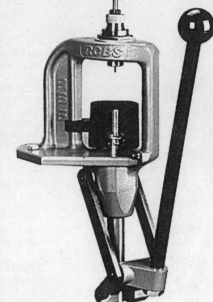

**RELOADER
SPECIAL-5**

RELOADING SCALE
MODEL 5-0-5

This 511-grain capacity scale has a three-poise system with widely spaced, deep beam notches to keep them in place. Two smaller poises on right side adjust from 0.1 to 10 grains, larger one on left side adjusts in full 10-grain steps. The first scale to use magnetic dampening to eliminate beam oscillation, the 5-0-5 also has a sturdy die-cast base with large leveling legs for stability. Self-aligning agate bearings support the hardened steel beam pivots for a guaranteed sensitivity to 0.1 grains.

Model 5-0-5 . . . . **$66.75**

**AMMOMASTER
SINGLE STAGE**

**AMMOMASTER
AUTO**

AMMOMASTER RELOADING
SYSTEM

The AmmoMaster offers the handloader the freedom to configure a press to his particular needs and preferences. It covers the complete spectrum of reloading, from single stage through fully automatic progressive reloading, from .32 Auto to .50 caliber. The **AmmoMaster Auto** has all the features of a five-station press.

AmmoMaster Single stage . . **$164.25**
AmmoMaster Auto **363.25**

RELOADING SCALE
MODEL 10-10

Up to 1010 Grain Capacity
Normal capacity is 510 grains, which can be increased, without loss in sensitivity, by attaching the included extra weight.
   Features include micrometer poise for quick, precise weighing, special approach-to-weight indicator, easy-to-read graduations, magnetic dampener, agate bearings, anti-tip pan, and dustproof lid snaps on to cover scale for storage. Sensitivity is guaranteed to 0.1 grains.

Model 10-10 Scale **$105.99**

RCBS RELOADING TOOLS

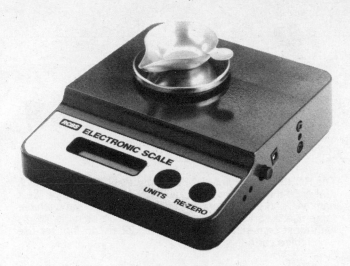

ELECTRONIC SCALE

This new RCBS Electronic Scale brings solid state electronic accuracy and convenience to handloaders. The LCD digital readings are ideal for weighing bullets and cases. The balance gives readings in grains, from zero to 500. The tare feature allows direct reading of the sample's weight with or without using the scale pan. The scale can be used on the range, operating on 8 AA batteries (approx. 50 hours).

Electronic Scale **$395.00**

POWDER CHECKER

Operates on a free-moving rod for simple, mechanical operation with nothing to break. Standard $7/8 \times 14$ die body can be used in any progressive loader that takes standard dies. Black oxide finish provides corrosion resistance with good color contrast for visibility.

Powder Checker **$21.50**

UPM MICROMETER ADJUSTMENT SCREW

Handloaders who want the convenience of a micrometer adjustment on their Uniflow Powder Measure can now add that feature to their powder measure. The RCBS Micrometer Adjustment Screw fits any Uniflow Powder Measure equipped with a large or small cylinder. It is easily installed by removing the standard metering screw, lock ring and bushing, which are replaced by the micrometer unit. Handloaders may then record the micrometer reading for a specific charge of a given powder and return to that setting at a later date when the same charge is used again.

UPM Micrometer Adjustment Screw
  (large or small) **$30.40**

PRECISION MIC

This ''Precisioneered Cartridge Micrometer'' provides micrometer readings of case heads to shoulder lengths, improving accuracy by allowing the best possible fit of cartridge to chamber. By allowing comparison of the chamber to SAAMI specifications, it alerts the handloader to a long chamber or excess headspace situation. It also ensures accurate adjustment of seater die to provide optimum seating depth. Available in 19 popular calibers.

Precision MIC **$34.60**

RCBS RELOADING TOOLS

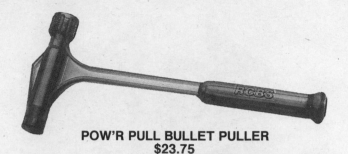

POW'R PULL BULLET PULLER
$23.75

The RCBS Pow'r Pull bullet puller features a three-jaw chuck that grips the case rim—just rap it on any solid surface like a hammer, and powder and bullet drop into the main chamber for re-use. A soft cushion protects bullets from damage. Works with most centerfire cartridges from .22 to .45 (not for use with rimfire cartridges).

HAND-PRIMING TOOL
$23.99

This hand-priming tool features a patented shielding mechanism that separates the seating operation from the primer supply, virtually ending the possibility of primer tray detonation. This tool fits comfortably in the palm of the hand for portable primer seating with a simple squeeze. The primer tray is easily removed, installed and filled, requiring no hand contact with the primers. Uses standard RCBS shell holders. Made of durable cast material and comes with large and small primer feed set ups.

PRO POWER CASE TRIMMER

CARTRIDGE COUNTER
$16.75

The RCBS Cartridge Counter enables reloaders to compare the number of cartridges actually loaded. It attaches to the RCBS Uniflow powder measure and registers each time the drum dispenses a charge. The cartridge counter also features a reset knob for quick return to zero.

PRO MANUAL CASE TRIMMER

TRIM PRO CASE TRIMMER
$198.50 (Power) $66.00 (Manual)
$83.25 (Power Kit)

Cartridge cases are trimmed quickly and easily with a few turns of the RCBS Trim Pro case trimmer. The lever-type handle is more accurate to use than draw collet systems. A flat plate shell holder keeps cases locked in place and aligned. A micrometer fine adjustment bushing offers trimming accuracy to within .001″. Made of die-cast metal with hardened cutting blades. The power model is like having a personal lathe, delivering plenty of torque. Positive locking handle and in-line power switch make it simple and safe.
Also available:
TRIM PRO CASE TRIMMER STAND $13.99
ACCESSORY BASE PLATE 29.75

REDDING RELOADING TOOLS

MODEL 721
"THE BOSS" PRESS

This "O" type reloading press features a rigid cast iron frame whose 36° offset provides the best visibility and access of comparable presses. Its "Smart" primer arm moves in and out of position automatically with ram travel. The priming arm is positioned at the bottom of ram travel for lowest leverage and best feel. Model 721 accepts all standard 7/8-14 threaded dies and universal shell holders.

Model 721 "The Boss" **$120.00**
    With Shellholder and 10A Dies **150.00**

Also available:
Boss Pro-Pak Deluxe Reloading Kit. Includes Boss Reloading Press, #2 Powder and Bullet Scale, Powder Trickler, Reloading Dies, and more **$315.00**

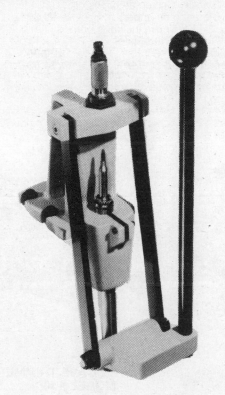

ULTRAMAG MODEL 700

Unlike other reloading presses that connect the linkage to the lower half of the press, the Ultramag's compound leverage system is connected at the top of the press frame. This allows the reloader to develop tons of pressure without the usual concern about press frame deflection. Huge frame opening will handle 50 × 3 1/4-inch Sharps with ease.

No. 700 Press, complete **$267.00**
No. 700K Kit, includes shell holder and one set of dies 298.50

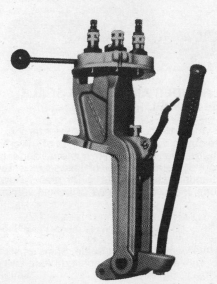

METALLIC TURRET RELOADING PRESS
MODEL 25

Extremely rugged, ideal for production reloading. No need to move shell, just rotate turret head to positive alignment. Ram accepts any standard snap-in shell holder. Includes primer arm for seating both small and large primers.

No. 25 Press, complete **$267.00**
No. 25K Kit, includes press, shell holder, and one set of dies 298.50

REDDING RELOADING TOOLS

MASTER POWDER MEASURE MODEL 3

Universal- or pistol-metering chambers interchange in seconds. Measures charges from 1/2 to 100 grains. Unit is fitted with lock ring for fast dump with large "clear" plastic reservoir. "See-thru" drop tube accepts all calibers from 22 to 600. Precision-fitted rotating drum is critically honed to prevent powder escape. Knife-edged powder chamber shears coarse-grained powders with ease, ensuring accurate charges.

No. 3 Master Powder Measure
(specify Universal- or Pistol-
Metering chamber) . . . . . . $ 99.00
No. 3K Kit Form, includes both
Universal and Pistol
chambers 120.00
Bench Stand 24.00

MATCH GRADE POWDER MEASURE MODEL 3BR

Designed for the most demanding reloaders—bench rest, silhouette and varmint shooters. The Model 3BR is unmatched for its precision and repeatability. Its special features include a powder baffle and zero backlash micrometer.

No. 3BR with Universal or Pistol
Metering Chamber **$135.00**
No. 3 BRK includes both
metering chambers 165.00

COMPETITION MODEL BR-30 POWDER MEASURE
(not shown)

This powder measure features a new drum and micrometer that limit the overall charging range from a low of 10 grains (depending on powder density) to a maximum of approx. 50 grains. For serious competitive shooters whose loading requirements are between 10 and 50 grains, this is the measure to choose. The diameter of Model 3BR's metering cavity has been reduced, and the metering plunger on the new model has a unique hemispherical or cup shape, creating a powder cavity that resembles the bottom of a test tube. The result: irregular powder setting is alleviated and charge-to-charge uniformity is enhanced.

Competition Model BR-30 Powder Measure **$159.00**

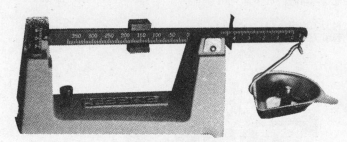

MASTER CASE TRIMMER MODEL 1400

This unit features a universal collet that accepts all rifle and pistol cases. The frame is solid cast iron with storage holes in the base for extra pilots. Both coarse and fine adjustments are provided for case length.

The case-neck cleaning brush and primer pocket cleaners attached to the frame of this tool make it a very handy addition to the reloading bench. Trimmer comes complete with:
• New speed cutter shaft
• Six pilots (22, 6mm, 25, 270, 7mm and 30 cal.)
• Universal collet
• Two neck cleaning brushes (22 thru 30 cal.)
• Two primer pocket cleaners (large and small)

No. 1400 Master Case Trimmer complete **$87.00**
No. 1500 Pilots 3.60

STANDARD POWDER AND BULLET SCALE MODEL RS-1

For the beginner or veteran reloader. Only two counterpoises need to be moved to obtain the full capacity range of 1/10 grain to 380 grains. Clearly graduated with white numerals and lines on a black background. Total capacity of this scale is 380 grains. An over-and-under plate graduate in 10th grains allows checking of variations in powder charges or bullets without further adjustments.

Model No. RS-1 . . . . **$48.50**

Also available: **Master Powder & Bullet Scale.** Same as standard model, but includes a magnetic dampened beam swing for extra fast readings. 505-grain capacity **$69.00**

Reference

THE SHOOTER'S BOOKSHELF

The following is a current listing of recently published titles of special interest to shooters and gun enthusiasts. Most of these books can be found at your local library, bookstore or gunshop. If not available, contact the publisher directly. For a complete listing of in-print titles covering all subjects of interest to shooters, use the *Subject Guide to Books in Print,* which is updated annually and is available at most public libraries. The following entries are listed alphabetically by author and include year of publication, number of pages, publisher and retail price.

ARMS AND ARMOR
Foss, C. **Jane's Armour & Artillery.** 1993. Jane's Info Group. $245.00

Gander, T. **Jane's Military Training Systems.** 1933. Jane's Info Group. $245.00

Hogg, I. **Jane's Infantry Weapons.** 1993. Jane's Info Group. $245.00

Randall, J. **Personal Defense Weapons.** 102 p. 1992. Loompanics. pap. $10.00

Saxon, Kurt. **The New Improved Poor Man's James Bond.** Vol. 1. 6th rev. ed. (Illus.). 477 p. 1988. Atlan Formularies. $18.00

DECOYS
Goldberger, Russ J. & Haid, Alan G. **Mason Decoys: A Pictorial Guide.** (Illus.). 144 p. 1993. Decoy Mag. $49.95

DEER HUNTING
Deer & Deer Hunting Magazine Staff. **Deer Hunter's Almanac.** (Illus.). 204 p. 1992. Krause Pubn. pap. $6.95

Miller, John. **Deer Camp: Last Light in the Northeast Kingdom.** (Illus.). 148 p. 1992. MIT Press. $29.95

Wagner, Robert. **Deer & Deer Hunting: The Serious Hunter's Guide.** 1992. Stackpole. pap. $16.95

Weiss, John. **The Advanced Deer Hunter's Bible.** 1993. Doubleday. pap. $12.00

ELK HUNTING
van Zwoll, Wayne. **The Elk Hunter's Handbook.** (Illus.). 416 p. 1992. Larsens Outdoor. $24.95

FIREARMS
Akehurst, Richard. **Game Guns & Rifles: Percussion to Hammerless.** (Illus.). 192 p. 1993. Sportsman's Pr. (UK) Trafalgar. $34.95

Gangarosa, Gene, Jr. **Modern Beretta Firearms.** (Illus.). 288 p. 1994. Stoeger Pub. Co. pap. $16.95

Long, Duncan. **Super Shotguns.** (Illus.). 96 p. 1992. Paladin Press. $18.00

Medlin, Eugene & Huon, Jean. **French Military Handguns.** (Illus.). 180 p. 1993. Excalibur NY. pap. $23.95

Murtz, Harold, ed. **Gun Digest Book of Exploded Long Gun Drawings.** (Illus.). 512 p. 1993. DBI. $29.95

Olsen, Robert. **Handgun Muzzle Flash Tests.** (Illus.). 144 p. 1993. Paladin Press. pap. $20.00

Preiser, J. **How to Shoot a Gun.** 96 p. 1993. Berkley Pub. pap. $3.99

Stephens, Charles. **How to Become a Master Handgunner.** (Illus.). 64 p. 1993. Paladin Pr. pap. $10.00

Wooters, John. **Complete Handloader.** 1988. Times Mirror Mag. Bk. Div. $29.95

FIREARMS—CATALOGS
Blue Book of Gun Values, 14th ed. (Illus.). 1152 p. 1993. Blue Bk. Pub. pap. $24.95

Gun Trader's Guide, 17th ed. (Illus.). 560 p. 1994. Stoeger Pub. $19.95

Jarrett, William, ed. **Shooter's Bible 1994,** No. 85. 576 p. 1993. Stoeger Pub. Co. $19.95

Traister, John. **Antique Guns: The Collector's Guide,** 2d ed. 1994 Stoeger Pub. Co. $19.95

Wilson, R.L. **Book of Colt Firearms.** (Illus.). 612 p. 1993. Blue Bk. Pub. $150.00

FIREARMS—HISTORY (*see also* RIFLES)
Henderson, Halton. **Artistry in Single Action: The Dallas Six Gun.** Ltd. ed. 125 p. 1989. Chamba Press. leather ed. $289.00

Munson, H. Lee. **Mortimer Gunmakers.** (Illus.). 320 p. A. Mowbray. $65.00

Phelps, Arthur J. **The Story of Merwin, Hulbert & Co. Firearms.** 226 p. 1993. Graphic Pubns. $56.00

Whiskey, Arthur J. **Arms Makers of Colonial America.** (Illus.). 224 p. 1992. Susquehanna U. Press. $55.00

FIREARMS—JUVENILE
Schleifer, Jay. **Everything You Had to Know about Weapons in School & at Home.** Rosen, Ruth, ed. 1993. Rosen Group. Price not set

FIREARMS—LAW & LEGISLATION
Davidson, Osha G. **Under Fire: The NRA and the Battle for Gun Control.** 304 p. 1993. H. Holt & Co. $25.00

Furnish, Brendan F. & Small, Dwight. **The Mounting Threat of Home Intruders: Weighing the Moral Option of Armed Self-Defense.** 274 p. 1993. C.C. Thomas. $54.75

Kruschke, Earl. **Gun Control.** 175 p. 1994. ABC-CLIO. lib. bdg. $39.50

Laughlin, Mark. **The Philosophy of Firearms: Reality & The Right to Bear Arms.** 190 p. 1992. Broad Reach. pap. $17.95

Otfinoski, Steve. **Gun Control: Is it a Right or a Danger to Bear Arms?** (Illus.). 64 p. 1993. TFC Books NY. $14.95

Reynolds, Morgan & Caruth, W.W. **Myths About Gun Control.** 1992. Nat'l Center Pol. pap. $10.00

GAME COOKERY
Chittenden, Russ. **Good Ole Boys Wild Game Cookbook.** 1990. Collector Books. $14.95

Deer & Deer Hunting Staff. **Three Hundred One Venison Recipes,** rev. ed. 128 p. 1992. Krause Pubns. $10.95

Hayes, James. **How to Cook a Deer & Other Critters: A Game Cookbook for Men.** (Illus.). 160 p. 1991. Country Pub. Inc. pap. $14.95

Phillips, John. **Deer & Fixings.** 188 p. 1991. Atlantic Pub. Co. pap. $8.95

Van Roden, Joanne. **Favorite Game Recipes.** 36 p. 1992. Wellspring. pap. $2.75

HANDGUNS (see also FIREARMS, FIREARMS—CATALOGS, PISTOLS, REVOLVERS*)*

Pistol & Revolver Handbook, 2d ed. 1994. Lyman. $17.95

HUNTING (see also DEER HUNTING, ELK HUNTING*)*

Bean, L. L., Co. **Hunting, Fishing & Camping.** 1993. Applewood. $19.95

Phillips, John E. **The Turkey Hunter's Bible.** 1992. Doubleday. pap. $12.00

Shuffett, Dave. **Back Trails & Fishing Tales: Dave Shuffett's Outdoor Adventures.** (Illus.). 144 p. 1993. Antex Corp. pap. $12.95

Zern, Ed. **Hunting & Fishing: A to Zern.** (Illus.). 312 p. 1993. Lyons & Burford. pap. $16.95

HUNTING DOGS

Jenkins, Len. **Gun Dog Training: Do It Yourself & Do It Right!** (Illus.). 250 p. 1991. C.J. Pubns. pap. $18.95

HUNTING WITH BOW & ARROW

Hardy, Robert. **Longbow.** 1993. Lyons & Burford. $24.95

Lawrence, H. Lea. **The Archers & Bowhunters Bible.** 1993. Doubleday. $12.00

KNIVES

Levine, Bernard. **Levine's Guide to Knives & Their Values,** 3d ed. 1993. Doublday. $12.00

PISTOLS (see also FIREARMS, FIREARMS—CATALOGS*)*

Gangarosa, Gene, Jr. **P.38 Automatic Pistol.** (Illus.). 288 p. 1993. Stoeger Pub. Co. pap. $16.95

RELOADING (see FIREARMS*)*

REVOLVERS (see FIREARMS—CATALOGS, FIREARMS—HISTORY*)*

RIFLES (see also FIREARMS—CATALOGS, FIREARMS—HISTORY*)*

Houze, Herbert. **To the Dreams of Youth: Winchester .22 Caliber Single Shot Rifle.** (Illus.). 208 p. 1993. Krause Pubns. $34.95

Lyman's Guide to Big Game Cartridges & Rifles. Ed. Matunas, ed. 1994. Lyman. $17.95

Murphy, John & Madaus, Howard. **Confederate Rifles & Muskets.** 300 p. 1993. Graphic Pubns. $46.95

Petrillo, Alan. **The Lee Enfield Number One Rifles.** (Illus.). 64 p. 1992. Excalibur NY. pap. $10.95

Schwing, Ned. **Winchester's Slide Action Rifles,** Vol. 2: Model 61 & Model 62. (Illus.). 1993. Krause Pubns. $34.95

Walter, John. **Rifles of the World.** (Illus.). 320 p. 1993. DBI. pap. $19.95

SHOOTING

Davies, K. **The Better Shot.** (Illus.). 136 p. 1992. Safari Pr. $37.50

Jackson, T. **Classic Game Shooting: The English Art.** (Illus.). 150 p. 1990. Safari Pr. $40.00

Reynolds, Mike & Barnes, Mike. **Shooting Made Easy.** (Illus.). 144 p. 1993. Crowood (UK) Trafalgar. pap. $29.95

Ruffer, J. **The Big Shots,** rev. ed. (Illus.). 160 p. 1989. Safari Pr. $30.00

SHOTGUNS (see also FIREARMS*)*

Lewis, Jack, ed. **Shotgun Digest,** 4th ed. (Illus.). 256 p. 1993. DBI. pap. $16.95

WILDLIFE CONSERVATION & PRESERVATION

Bonner, Raymond. **At the Hand of Man: Peril & Hope for Africa's Wildlife.** 1993. Knopf. $24.00

Mears, Raymond. **The Outdoor Survival Handbook: A Guide to the Resources & Material Available in the Wild & How to Use Them for Food, Shelter, Warmth & Navigation.** (Illus.). 240 p. 1993. St. Martin. pap. $13.95

Sanger, David. **North America's Endangered Species.** (Illus.). 97 p. Knopf. $24.00

Turbak, Gary. **Survivors in the Shadows: Threatened & Endangered Mammals of the American West.** (Illus.). 156 p. 1993. Northland AZ. pap. $19.95

DIRECTORY OF MANUFACTURERS AND SUPPLIERS

The following manufacturers, suppliers and distributors of firearms, ammunition, reloading equipment, sights, scopes and accessories all appear with their products in the catalog and/or "Manufacturers' Showcase" sections of this edition of SHOOTER'S BIBLE.

Action Arms, Ltd. (Brno handguns, rifles, Timber Wolf rifles, sights, scopes)
P.O. Box 9573
Philadelphia, Pennsylvania 19124
Ph: 215-744-0100   Fax: 215-533-2188

Aimpoint (sights, scopes, mounts)
580 Herndon Parkway, Suite 500
Herndon, Virginia 22070
Ph: 703-471-6828   Fax: 703-689-0575

Lou Alessandri & Son, Ltd. (Rizzini shotguns, rifles, extension tubes)
24 French Street
Rehoboth, Massachusetts 02769
PH: 508-252-5590   Fax: 508-252-3436

American Arms (handguns, rifles, blackpowder, Franchi shotguns)
715 E. Armour Road
N. Kansas City, Missouri 64116
Ph: 816-474-3161   Fax: 816-474-1225

American Derringer Corp. (handguns)
127 North Lacy Drive
Waco, Texas 76705
Ph: 817-799-9111   Fax: 817-799-7935

Anschutz (handguns, rifles)
Available through Precision Sales International

Arcadia Machine & Tool Inc. (AMT handguns, rifles)
6226 Santos Diaz Street
Irwindale, California 91702
Ph: 818-334-6629   Fax: 818-969-5247

Armes de Chasse (Francotte rifles and shotguns; AYA shotguns)
P.O. Box 827
Chadds Ford, Pennsylvania 19317
PH: 215-388-1146

Armsport, Inc. (blackpowder arms, Bernardelli handguns, shotguns)
3590 NW 49th Street, P.O. Box 523066
Miami, Florida 33142
Ph: 305-635-7850   Fax: 305-633-2877

A-Square Company Inc. (rifles)
One Industrial Park
Bedford, Kentucky 40006
Ph: 502-255-7456   Fax: 502-255-7657

Astra (handguns)
Available through European American Armory

Auto-Ordnance Corp. (handguns, rifles)
Williams Lane
West Hurley, New York 12491
Ph: 914-679-7225   Fax: 914-679-2698

AYA (shotguns)
Available through Armes de Chasse

Bausch & Lomb
Sports Optics Division
9200 Cody
Overland Park, Kansas 66214
Ph: 913-752-3400   Fax: 913-752-3550

Beeman Precision Airguns (scopes, mounts)
5454 Argosy Avenue
Huntington Beach, California 92649
Ph: 714-890-4800   Fax: 714-890-4808

Bell & Carlson, Inc. (rifle stocks)
509 North 5th
Atwood, Kansas 67730
Ph: 913-626-3204   Fax: 913-626-9602

Benelli (shotguns)
Available through Heckler & Koch;
(handguns) Available through European
American Armory

Beretta U.S.A. Corp. (handguns, shotguns)
17601 Beretta Drive
Accokeek, Maryland 20607
Ph: 301-283-2191   Fax: 301-283-0435

Bernardelli (handguns, shotguns)
Available through Armsport

Bersa (handguns)
Available through Eagle Imports Inc.

Blaser USA, Inc. (rifles)
c/o Autumn Sales, Inc.
1320 Lake Street
Fort Worth, Texas 76102
Ph: 817-335-1634   Fax: 817-338-0119

Blount, Inc. (RCBS reloading equipment, Speer and CCI bullets, Weaver sights)
P.O. Box 856
Lewiston, Idaho 83501
Ph: 208-746-2351   Fax: 208-746-2915

Bonanza (reloading tools)
See Forster Products

Brno (handguns, rifles)
Available through Action Arms

Brown Precision, Inc. (custom rifles, handguns)
7786 Molinas Avenue; P.O. Box 270 W.
Los Molinas, California 96055
Ph: 916-384-2506   Fax: 916-384-1638

Browning (handguns, rifles, shotguns)
Route One
Morgan, Utah 84050
Ph: 801-876-2711   Fax: 801-876-3331

B-Square Co. (sights, mounts)
P.O. Box 11281
Fort Worth, Texas 76110
Ph: 817-923-0964   Fax: 817-926-7012

Burris Company, Inc. (scopes, mounts)
331 East Eighth Street, P.O. Box 1747
Greeley, Colorado 80631
Ph: 303-356-1670   Fax: 303-356-8702

Butler Creek Corp. (stocks, scope covers, slings)
290 Arden Drive
Belgrade, Montana 59714
Ph: 406-388-1356   Fax: 406-388-7204

Charco Inc. (Charter Arms handguns)
26 Beaver Street
Ansonia, Connecticut 06401
Ph: 203-735-4686   Fax: 203-378-2846

Charter Arms Corp. (handguns)
Available through Charco Inc.

Clifton Arms (custom rifles)
P.O. Box 1471
Medina, Texas 78055
Ph: 210-589-2666

Colt's Manufacturing Co., Inc. (handguns, rifles)
P.O. Box 1868
Hartford, Connecticut 06144-1868
PH: 203-236-6311   Fax: 203-244-1449

Connecticut Shotgun Manufacturing Co. (A. H. Fox shotguns)
35 Woodland Street, P.O. Box 1692
New Britain, Connecticut 06051-1692
Ph: 203-225-6581   Fax: 203-832-8707

Connecticut Valley Classics (shotguns)
12 Taylor Lane
P.O. Box 2068
Westport, Connecticut 06880
Ph: 203-254-7864   Fax: 203-254-7866

Coonan Arms (handguns)
830 Hampden Ave.
St. Paul, Minnesota 55114
Ph: 612-646-0902

Cooper Firearms (rifles)
P.O. Box 114
Stevensville, Montana 59870
Ph: 406-777-5534

CVA (blackpowder guns)
5988 Peachtree Corners East
Norcross, Georgia 30071
Ph: 404-449-4687   Fax: 404-242-8546

Daewoo Precision Industries, Ltd. (handguns)
70 James Way
Southampton, Pennsylvania 18966
Ph: 800-726-3006   Fax: 215-322-5972

Dakota (handguns, rifles)
Available through E.M.F. Co., Inc.

Dakota Arms, Inc. (rifles)
HC 55, Box 326
Sturgis, South Dakota 57785
Ph: 605-347-4686   Fax: 605-347-4459

Davis Industries (handguns)
11186 Venture Drive
Mira Loma, California 91752
Ph: 909-360-5598   Fax: 909-360-1749

Dixie Gun Works (blackpowder guns)
P.O. Box 130, Highway 51 S.
Union City, Tennessee 38261
Ph: 901-885-0561   Fax: 901-885-0440

Dynamit Nobel/RWS (Rottweil shotguns)
81 Ruckman Road
Closter, New Jersey 07624
Ph: 201-767-1995   Fax: 201-767-1589

Eagle Arms Inc. (rifles)
131 East 22nd. Avenue, P.O. Box 457
Coal Valley, Illinois 61240
Ph: 309-799-5619   Fax: 309-799-5150

Eagle Imports, Inc. (Bersa handguns, Lanber
shotguns)
1750 Brielle Avenue, Unit B1
Wanamassa, New Jersey 07712
Ph: 908-493-0333   Fax: 908-493-0301

Emerging Technologies Inc. (handguns,
sights)
P.O. Box 3548
Little Rock, Arkansas 72203
Ph: 501-375-2227   Fax: 501-372-1445

E.M.F. Company, Inc. (Dakota handguns,
blackpowder arms)
1900 East Warner Avenue 1-D
Santa Ana, California 92705
Ph: 714-261-6611   Fax: 714-756-0133

Erma (handguns)
Available through Precision Sales

Euroarms of America Inc. (blackpowder guns)
1501 Lenoir Drive, P.O. Box 3277
Winchester, Virginia 22601
Ph: 703-662-1863

European American Armory (Astra and
Benelli handguns, E.A.A. handguns, rifles,
Sabatti shotguns)
P.O. Box 3498, Bright Station
Hialeah, Florida 33013
Ph: 305-688-4442   Fax: 305-688-5656

Feather Industries, Inc. (rifles)
2300 Central Avenue, Unit K
Boulder, Colorado 80301
Ph: 303-442-7021   Fax: 303-447-0944

Federal Cartridge Co. (ammunition)
900 Ehlen Drive
Anoka, Minnesota 55303-7503
Ph: 612-323-2506   Fax: 612-323-2506

Flintlocks, Etc. (Pedersoli replica rifles)
160 Rossiter Road
Richmond, Massachusetts 01254
Ph: 413-698-3822   Fax: 413-698-3866

Forster Products (Bonanza and Forster
reloading)
82 East Lanark Avenue
Lanark, Illinois 61046
Ph: 815-493-6360   Fax: 815-493-2371

A. H. Fox (shotguns)
Available thru Connecticut Shotgun
Mfg. Co.

Franchi (shotguns)
Available through American Arms

Francotte (rifles and shotguns)
Available through Armes de Chasse

Freedom Arms (handguns)
One Freedom Lane, P.O. Box 1776
Freedom, Wyoming 83120
Ph: 307-883-2468

Galaxy Imports, Ltd. (Laurona shotguns)
P.O. Box 3361
Victoria, Texas 77903
Ph: 512-573-4867

Garbi (shotguns)
Available through W. L. Moore & Co.

Gibbs Rifle Co. (Parker-Hale rifles and
blackpowder rifles)
Cannon Hill Industrial Park
Hoffman Road
Martinsburg, West Virginia 25401
Ph: 304-274-0458

Glaser Safety Slug, Inc. (ammunition and gun
accessories)
P.O. Box 8223
Foster City, California 94404
Ph: 415-345-7677   Fax: 415-345-8217

Glock, Inc. (handguns)
6000 Highlands Parkway
Smyrna, Georgia 30082
Ph: 404-432-1202   Fax: 404-433-8719

Gonic Arms (blackpowder rifles)
134 Flagg Road
Gonic, New Hampshire 03839
603-332-8456   Fax: 603-332-8457

Grendel, Inc. (handguns)
550 St. Johns Street
Cocoa, Florida 32922
Ph: 407-636-1211

Gun South Inc. (Merkel shotguns, Steyr-
Mannlicher rifles)
108 Morrow Ave., P.O. Box 129
Trussville, Alabama 35173
Ph: 205-655-8299   Fax: 205-655-7078

Carl Gustaf (rifles)
Available through Precision Sales Int'l

Hämmerli U.S.A. (handguns)
19296 Oak Grove Circle
Groveland, California 95321
Ph: 209-962-5311   Fax: 209-962-5931

Harrington & Richardson (handguns, rifles,
shotguns)
60 Industrial Rowe
Gardner, Massachusetts 01440
Ph: 508-632-9393   Fax: 508-632-2300

Heckler & Koch (handguns, rifles, Benelli
shotguns)
21480 Pacific Boulevard
Sterling, Virginia 20166
Ph: 703-450-1900   Fax: 703-450-8160

Helwan (handguns)
Available through Interarms

Hercules Inc. (powder)
Hercules Plaza
Wilmington, Delaware 19894
Ph: 302-594-5000   Fax: 302-594-5305

Heritage Manufacturing (handguns)
4600 NW 135 St.
Opa Locka, Florida 33054
Ph: 305-685-5966   Fax: 305-687-6721

Heym (rifles)
Available through JägerSport

Hi - point (handguns)
174 South Mulberry
Mansfield, Ohio 44902
Ph: 419-522-8330

Hodgdon Powder Co., Inc. (gunpowder)
6231 Robinson, P.O. Box 2932
Shawnee Mission, Kansas 66202
Ph: 913- 362-9455   Fax: 913-362-1307

Hornady Manufacturing Company (reloading,
ammunition)
P.O. Box 1848
Grand Island, Nebraska 68802-1848
Ph: 308-382-1390   Fax: 308-382-5761

IGA Shotguns
Available through Stoeger Industries

Impact Case Co. (gun cases)
Division of Knouff & Knouff
P.O. Box 99209
Ph: 509-467-3303   Fax: 509-326-5436

IMR Powder Company (gunpowder)
R.D. 5, Box 247E
Plattsburgh, New York 12901
Ph: 518-561-9530

InterAims (sights)
Available through Stoeger Industries

Interarms (handguns, shotguns and rifles,
including Helwan, Howa, Mark X, Rossi,
Star, Walther, Norinco)
10 Prince Street
Alexandria, Virginia 22314
Ph: 703-548-1400   Fax: 703-549-7826

Ithaca Gun (shotguns)
891 Route 34B
King Ferry, New York 13081
Ph: 315-364-7171   Fax: 315-364-5134

JägerSport (Heym and Voere rifles)
1 Wholesale Way
Cranston, Rhode Island 02920
Ph: 401-942-3380   Fax: 401-946-2587

Jarrett Rifles Inc. (custom rifles and
accessories)
383 Brown Road
Jackson, South Carolina 29831
Ph: 803-471-3616

Kahr Arms (handguns)
P.O. Box 220
Blauvelt, New York 10913
Ph: 914-353-5996

K.B.I., Inc. (handguns, shotguns)
P.O. Box 6346
Harrisburg, Pennsylvania 17112
Ph: 717-540-8518   Fax: 717-540-8567

K.D.F. Inc. (rifles)
2485 Highway 46 North
Seguin, Texas 78155
Ph: 210-379-8141   Fax: 210-379-5420

Kimber of America, Inc. (rifles)
9039 Southeast Jannsen Road
Clackamas, Oregon 97015
Ph: 503-656-1704   Fax: 503-657-5695

Kleen-Bore, Inc. (gun care products)
20 Ladd Avenue
Northampton, Massachusetts 01060
Ph: 413-586-7240   Fax: 413 586-0236

Kowa Optimed, Inc. (scopes)
20001 South Vermont Avenue
Torrance, California 90502
Ph: 310-327-1913   Fax: 310-327-4177

Krieghoff International Inc. (rifles, shotguns)
337A Route 611, P.O. Box 549
Ottsville, Pennsylvania 18942
Ph: 610-847-5173   Fax: 610-847-8691

LakeField Arms Ltd. (rifles)
P.O. Box 129
Lakefield, Ontario K0L 2H0 Canada
Ph: 705-652-8000   Fax: 705-652-8431

Lanber (shotguns)
Available through Eagle Imports, Inc.

L.A.R. Manufacturing, Inc. (Grizzly handguns)
4133 West Farm Road
West Jordan, Utah 84084
Ph: 801-255-7106   Fax: 801-569-1972

Laurona (shotguns)
Available through Galaxy Imports

Leupold & Stevens, Inc. (scopes, mounts)
P.O. Box 688
Beaverton, Oregon 97075
Ph: 503-646-9171   Fax: 503-526-1455

Llama (handguns)
Available through SGS Importers International

Luger, American Eagle (pistols)
Available through Stoeger Industries

Lyman Products Corp. (blackpowder guns,
sights, reloading tools)
Route 147
Middlefield, Connecticut 06455
PH: 203-349-3421   Fax: 203-349-3586

Magnum Research Inc. (Desert Eagle
handguns)
7110 University Avenue N.E.
Minneapolis, Minnesota 55432
Ph: 612-574-1868   Fax: 612-574-0109

Magtech Recreational Products (shotguns)
5030 Paradise Rd., Ste C-211
Las Vegas, Nevada 89119
Ph: 702-795-7191   Fax: 702-795-2769

Mail Order Video (videotapes)
7888 Ostrow Road #H
San Diego, California 92111
Ph: 800-942-8273   Fax: 619-569-0505

Mark X (rifles)
Available through Interarms

Marlin Firearms Company (rifles)
100 Kenna Drive
North Haven, Connecticut 06473
Ph: 203-239-5621   Fax: 203-234-2991

Marocchi (Avanza shotguns)
Available through Precision Sales

Maverick of Mossberg (shotguns)
Available through O. F. Mossberg

McMillan Gun Works (handguns, rifles)
302 W. Melinda Lane
Phoenix, Arizona 85027
Ph: 602-582-9627   Fax: 602-582-5178

MEC Inc. (reloading tools)
c/o Mayville Engineering Co.
715 South Street
Mayville, Wisconsin 53050
Ph: 414-387-4500   Fax: 414-387-2682

Merkel (shotguns)
Available through Gun South Inc.

Midway Arms, Inc. (reloading tools, shooting
accessories)
5875 West Van Horn Tavern Road
Columbia, Missouri 65203
Ph: 314-455-6363   Fax: 314-446-1018

Millett Sights (sights and mounts)
16131 Gothard Street
Huntington Beach, California 92647
Ph: 714-842-5575   Fax: 714-843-5707

Mitchell Arms (handguns, rifles, blackpowder
arms)
3400-1 West MacArthur Blvd.
Santa Ana, California 92704
Ph: 714-957-5711   Fax: 714-957-5732

M.O.A. Corp. (handguns)
2451 Old Camden Pike
Eaton, Ohio 45302
Ph: 513-456-3669

Modern Muzzle Loading Inc. (blackpowder
guns)
P.O. Box 130, 234 Airport Rd.,
Centerville, Iowa 52544
Ph: 515-856-2626   Fax :515-856-2628

William L. Moore & Co. (Garbi and Piotti
shotguns)
31360 Via Colinas, No. 109
Westlake Village, California 91361
Ph: 818-889-4160

O. F. Mossberg & Sons, Inc. (shotguns)
7 Grasso Avenue; P.O. Box 497
North Haven, Connecticut 06473
Ph: 203-230-5361   Fax: 203-230-5420

MTM Molded Products (reloading tools)
P.O. Box 14117
Dayton, Ohio 45413
Ph: 513-890-7461   Fax: 513-890-1747

Navy Arms Company, Inc. (handguns,
blackpowder guns, replicas)
689 Bergen Boulevard
Ridgefield, New Jersey 07657
Ph: 201-945-2500

New England Firearms Co., Inc. (handguns,
rifles, shotguns; Harrington & Richardson
handguns, shotguns)
Industrial Rowe
Gardner, Massachusetts 01440
Ph: 508-632-9393   Fax: 508-632-2300

Nikon Inc. (scopes)
1300 Walt Whitman Road
Melville, New York 11747
Ph: 516-547-4381   Fax: 516-547-0309

North American Arms (handguns)
1800 North 300 West
P.O. Box 707
Spanish Fork, Utah 84660
Ph: 800-821-5783   Fax: 801-798-9418

Nosler Bullets, Inc. (bullets)
P.O. Box 671
Bend, Oregon 97709
Ph: 503-382-3921   Fax: 503-388-4667

Nygord Precision Products (Unique
handguns, rifles)
P.O. Box 8394
La Crescenta, California 91224
Ph: 818-352-3027

Olin/Winchester (ammunition, primers, cases)
427 North Shamrock
East Alton, Illinois 62024
Ph: 618-258-2000

Palsa Outdoor Products (recoil pads)
P.O. Box 81336
Lincoln, Nebraska 68501-1336
Ph: 402-488-5288   Fax: 402-488-2321

Para-Ordnance (handguns)
3411 McNicoll Avenue #14
Scarborough, Ontario, Canada M1V 2V6
Ph: 416-297-7855   Fax: 416-297-1289

Parker-Hale (rifles)
Available through Navy Arms

Parker Reproduction (shotguns)
124 River Road
Middlesex, New Jersey 08846
Ph: 908-469-0100   Fax: 908-469-9692

Pedersoli, Davide (replica firearms)
Available through Flintlocks Etc.

Peltor Inc. (hearing protection products)
41 Commercial Way
East Providence, Rhode Island 02914
Ph: 401-438-4800   Fax: 401-434-1708

Pentax (scopes)
35 Inverness Drive East
Englewood, Colorado 80112
Ph: 303-799-8000   Fax: 303-790-1131

Perazzi U.S.A. (shotguns)
1207 S. Shamrock Ave.
Monrovia, California 91016
Ph: 818-303-0068   Fax: 818-303-2081

Piotti (shotguns)
Available through W. L. Moore & Co.

Precision Sales International (Anschutz
pistols and rifles, Erma pistols, Carl Gustaf
rifles, Marocchi shotguns)
P.O. Box 1776
Westfield, Massachusetts 01086
Ph: 413-562-5055

RCBS, Inc. (reloading tools)
See Blount, Inc.

Redding Reloading Equipment (reloading
tools)
1089 Starr Road
Cortland, New York 13045
Ph: 607-753-3331   Fax: 607-756-8445

Redfield (scopes)
5800 East Jewell Avenue
Denver, Colorado 80227
Ph: 303-757-6411   Fax: 303-756-2338

Remington Arms Company, Inc. (handguns,
rifles, shotguns, ammunition, primers)
1007 Market Street
Wilmington, Delaware 19898
Ph: 302-773-5291   Fax: 302-774-5776

Rizzini (shotguns)
Available through Lou Alessandri & Sons, Ltd.

Rossi (handguns, rifles, shotguns)
Available through Interarms

Rottweil (shotguns)
Available through Dynamit Nobel/RWS

Ruger (handguns, rifles, shotguns,
blackpowder guns)
See Sturm, Ruger & Company, Inc.

Ruko Products, Inc. (rifles, shotguns)
2245 Kenmore Avenue, Suite 102
Buffalo, New York 14207
Ph: 716-874-2707   Fax: 416-826-1353

Sako (rifles, actions, scope mounts)
Available through Stoeger Industries

Sauer (rifles)
c/o Paul Company, Inc.
27385 Pressonville Road
Wellsville, Kansas 66092
Ph: 913-883-4444   Fax: 913-883-2525

Savage Arms (rifles, shotguns)
Springdale Road
Westfield, Massachusetts 01085
Ph: 413-568-7001   Fax: 413-562-7764

Schmidt and Bender (scopes)
Schmidt & Bender U.S.A.
P.O. Box 134
Meriden, New Hampshire 03770
Ph: 800-468-3450   Fax: 603-469-3471

SGS Importers International Inc. (Bersa and
Llama handguns)
1750 Brielle Avenue
Wanamassa, New Jersey 07712
Ph: 908-493-0302   Fax: 908-493-0301

Shiloh Rifle Mfg. Co., Inc. (Shiloh Sharps
blackpowder rifles)
P.O. Box 279, Industrial Park
Big Timber, Montana 59011
Ph: 406-932-4454   Fax: 406-932-5627

Shooting Systems Group, Inc. (gun holsters,
cases)
1075 Headquarters Park
Fenton, Missouri 63026
Ph: 314-343-3575   Fax: 314-349-3311

Shotgun News (gun trading periodical)
Box 669
Hastings, Nebraska 68902

Sierra Bullets (bullets)
P.O. Box 818
1400 West Henry St.
Sedalia, Missouri 65301
Ph: 816-827-6300   Fax: 816-827-6300

Sigarms Inc. (Sig-Sauer handguns)
Corporate Park
Exeter, New Hampshire 03862
Ph: 603-772-2302   Fax: 603-772-9082

Simmons Outdoor Corp. (scopes)
2571 Executive Center Circle East, Suite 100
Tallahassee, Florida 32301
Ph: 904-878-5100   Fax: 904-878-0300

SKB Shotguns (shotguns, Nichols scopes)
4325 South 120th Street
P.O. Box 37669
Omaha, Nebraska 68137
Ph: 800-752-2767   Fax: 402-330-8029

Smith & Wesson (handguns)
2100 Roosevelt Avenue
Springfield, Massachusetts 01102-2208
Ph: 413-781-8300   Fax: 413-731-8980

Speer (bullets)
See Blount, Inc.

Springfield Inc. (handguns, rifles)
420 West Main Street
Geneseo, Illinois 61254
Ph: 309-944-5631   Fax: 309-944-3676

Star (handguns)
Available through Interarms

Steyr-Mannlicher (rifles)
Available through Gun South Inc.

Stoeger Industries (American Eagle Luger®,
IGA shotguns, InterAims sights, Sako
actions, mounts & rifles, Tikka rifles &
shotguns)
55 Ruta Court
South Hackensack, New Jersey 07606
Ph: 201-440-2700   Fax: 201-440-2707

Sturm, Ruger and Company, Inc. (Ruger
handguns, rifles, shotguns)
Lacey Place
Southport, Connecticut 06490
Ph: 203-259-4537   Fax: 203-259-2167

Swarovski Optik North America (scopes)
One Wholesale Way
Cranston, Rhode Island 02920
Ph: 401-942-3380   Fax: 401-946-2587

Swift Instruments, Inc. (scopes and mounts)
952 Dorchester Avenue
Boston, Massachusetts 02125
Ph; 800-446-1116   Fax: 617-436-3232

Tasco (scopes and mounts)
7600 N.W. 26th Street
Miami, Florida 33122
Ph: 305-591-3670   Fax: 305-592-5895

Taurus International, Inc. (handguns)
16175 N.W. 49th Avenue
Miami, Florida 33014
Ph: 305-624-1115   Fax: 305-623-7506

Texas Arms (handguns)
P.O. Box 154906
Waco, Texas 76715
Ph: 817-867-6972

Thompson/Center Arms (handguns, rifles,
blackpowder guns)
Farmington Road, P.O. Box 5002
Rochester, New Hampshire 03867
Ph: 603-332-2394   Fax: 603-332-5133

Tikka (rifles, shotguns)
Available through Stoeger Industries

Traditions, Inc. (blackpowder guns)
P.O. Box 235
Deep River, Connecticut 06417
Ph: 203-526-9555   Fax: 203-526-4564

Trius Products, Inc. (traps, clay targets)
221 South Miami Avenue, P.O. Box 25
Cleves, Ohio 45002
Ph: 513-941-5682   Fax: 513-941-7970

Uberti USA, Inc. (handguns, rifles,
blackpowder rifles and revolvers)
362 Limerock Rd., P.O. Box 469
Lakeville, Connecticut 06039
Ph: 203-435-8068

Ultra Light Arms Company (rifles,
blackpowder rifles)
214 Price Street, P.O. Box 1270
Granville, West Virginia 26505
Ph: 304-599-5687   Fax: 304-599-5687

Unique (handguns, rifles)
Available through Nygord Precision
Products

U.S. Repeating Arms Co. (Winchester rifles,
shotguns)
275 Winchester Avenue
New Haven, Connecticut 06511
Ph: 203-789-5000   Fax: 203-789-5071

Voere (rifles)
Available through JägerSport

Walther (handguns, rifles)
Available through Interarms

Weatherby, Inc. (rifles, shotguns, scopes,
ammunition)
2781 Firestone Boulevard
South Gate, California 90280
Ph: 213-569-7186   Fax: 213-569-5025

Weaver (scopes, mount rings)
See Blount, Inc.

Wesson Firearms Co., Inc. (handguns)
Maple Tree Industrial Center, Route 20
Wilbraham Road
Palmer, Massachusetts 01069
Ph: 413-267-4081   Fax: 413-267-3601

White Muzzleloading Systems (blackpowder
rifles)
P.O. Box 277
Roosevelt, Utah 84066
Ph: 801-277-3085   Fax: 802-722-3054

Wichita Arms (handguns)
P.O. Box 11371
Wichita, Kansas 67211
Ph: 316-265-0661

Wideview Scope Mount Corp. (mounts, rings)
26110 Michigan Avenue
Inkster, Michigan 48141
Ph: 313-274-1238   Fax: 313-274-2814

Wildey Inc. (handguns)
P.O. Box 475
Brookfield, Connecticut 06804
Ph: 203-355-9000

Williams Gun Sight Co. (sights, scopes,
mounts)
7389 Lapeer Road, P.O. Box 329
Davison, Michigan 48423
Ph: 313-653-2131   Fax: 313-658-2140

Winchester (ammunition, primers, cases)
See Olin/Winchester

Winchester (domestic rifles, shotguns)
See U.S. Repeating Arms Co.

Winslow Arms Co. (rifles)
P.O. Box 783
Camden, South Carolina 29020
PH: 803-432-2938

Zeiss Optical, Inc. (scopes)
1015 Commerce Street
Petersburg, Virginia 23803
Ph: 804-861-0033   Fax: 804-733-4024

CALIBERFINDER

How to use this guide: To find a 22 LR handgun, look under that heading in the **HANDGUNS** section below. You'll find several models of that description, including the Beretta Model 89 Sport Gold. Next, turn to the **GUNFINDER** section and locate the heading for **Beretta** (pistols, in this case). Beretta's Model 89 appears on page 108.

HANDGUNS

22 LONG RIFLE

American Arms Models CX, PK & PX, Model P-98
American Derringer Models 1, 2 and 7; High Standard DA
Anschutz Exemplar and Exemplar XIV
Auto Ordnance Model 1927 A3
Benelli Model MP90S Target
Beretta Models 21 Bobcat and 89 Sport Gold
Bernardelli Model PO10 Target Pistol
Bersa Model 23
Browning Buck Mark 22 Series, Cadet Semiauto DA Pistol
Charter Arms Off-Duty
Daewoo Model DP 52
Davis D-Series, Long-Bore Series
EMF/Dakota Dakota Target, Hartford Scroll Engraved Single Action Revolver
Erma ESP 85A Golden Target, Match, ESP 85A Sporting Pistol, Model 772 Match
European American Armory Windicator DA, Benelli Model MP90S Target
Feather Guardian Angel
Freedom Arms Model 252
Hämmerli Models 160 Free Pistol, 162 Electronic, 208S Standard Pistol, 212 Hunter, 280 Target Pistol
Harrington & Richardson Sportsman 999, Model 949 Classic Western
Heckler & Koch Model P7K3
High Standard Double Action
Heritage Rough Rider SA
Llama Automatic (Small Frame)
Magnum Research Mountain Eagle
Mitchell Arms High Standard Pistols, Sport King II
Navy Arms Model TT-Olympia
New England Firearms Standard Revolver, Ultra Revolver
North American Arms Mini-Revolvers and Mini-Master Series
Rossi Model 515
Ruger Bearcat SA, Bisley SA Target, Mark II Pistols, New Model Single-Six, Model SP101 Revolver
Smith & Wesson Models 41, 63, 422, 617 (K-22 Masterpiece), 2206, 2206 Target, 2213/2214
Taurus Models 94, 96
Thompson/Center Contender Bull Barrel, Octagon Barrel and "Super 14"
A. Uberti 1871 Rolling Block Target Pistol
Unique Models DES 69U, Int'l Silhouette & Sport
Walther DA Models PP and TPH
Wichita Arms International Pistol

22 RIMFIRE MAGNUM

American Derringer Models 1 and 7, High Standard Cop & Mini-Cop DA Derringers
AMT 22 Automag II
Davis D-Series Derringer
Grendel Model P-30
High Standard Cop and Mini-Cop DA
Mitchell Arms Sharpshooter II
Taurus Model 941
Dan Wesson 22 Rimfire Magnum, Silhouette
Wichita Arms International

22 SHORT

Harrington & Richardson Sportsman 999, Model 949 Classic Western
New England Firearms Standard, Ultra/Ultra Mag.
Unique Model DES 2000U

22 HORNET

Anschutz Exemplar Hornet
Magnum Research Lone Eagle SS
MOA Maximum
Thompson/Center Contender Bull Barrel and Super "14"

22 WIN. MAG.

Davis Long-Bore D-Series
Feather Guardian Angel
Heritage Rough Rider SA
Magnum Research Lone Eagle SS
New England Firearms Ultra Revolver
North American Mini-Revolvers and Mini-Master Series
Rossi Model 515
Smith & Wesson Models 648 and 651
Taurus Model 941
Thompson/Center Contender Bull Barrel
Uberti 1871 Rolling Block Target Pistol
Unique International Silhouette & Sport
Dan Wesson 22 Win. Mag. Revolvers

22-250

Magnum Research Lone Eagle SS
Remington Model XP-100 LR

223 REMINGTON

Magnum Research Lone Eagle SS
Remington XP Long-Range Custom, XP-100 Hunter, XP-100 Silhouette, XP-100 R
Thompson/Center Contender Bull Barrel and "Super 14"

223 REM. COMM. AUTO

American Derringer Model 1
Thompson/Center Contender Models and "Super 14"

6mm BR

Remington XP-100 Long-Range Custom

243

Magnum Research Lone Eagle SS

25 AUTO

American Derringer Model 2
Beretta Model 21, Model 950BS Jetfire
Davis D-Series Derringers
Heritage Model H25B/H25G Semiauto
KBI Model PSP-25
Walther DA Model TPH

250 SAVAGE

Remington XP-100 LR Custom

7mm BR

Magnum Research Lone Eagle SS
Remington Model XP-100 and XP-100 Hunter Target Pistols
Wichita Silhouette

7mm T.C.U.

Thompson/Center Contender Bull Barrel
Unique International Silhouette

7mm SUPER MAG.

Wichita Arms International Pistols

7mm-08

Magnum Research Model Lone Eagle SS
Remington Model XP-100 Silhouette, Long-Range Custom and Hunter

7-30 WATERS

Thompson/Center Contender Hunter, Super "14"
Wichita Arms International

30 CARBINE

American Derringer Model 1
AMT Automag III
Ruger Model Blackhawk SA

30 LUGER

Ruger P-Series Pistols

30-06

Magnum Research Lone Eagle SS

30-30 WIN.

American Derringer Model 1
Magnum Research Lone Eagle SS
Thompson/Center Contender Bull Barrel and Hunter, Super "14"
Wichita Arms International

300 WHISPER

Thompson/Center Contender Super "14"

308 WINCHESTER

Magnum Research Lone Eagle SS
Remington XP-100 Long-Range Custom
Wichita Arms Silhouette

32 MAGNUM

American Derringer Models 1, 3 & 7, Lady Derringer
Ruger Model SP101, Bisley SA Target, New Model Single Six
Dan Wesson Six-Shot

32 AUTO

American Derringer Model 2
Brno Model CZ-83
Davis D-Series Derringers, Long-Bore D-Series
European American Armory SA/DA Compacts
Walther Model PP Double Action

32 H&H MAGNUM

Wichita International

32 H&R

Charter Arms Police Undercover
Davis Big-Bore D-Series and Long-Bore D-Series
New England Firearms Lady Ultra Revolver

32 S&W LONG

American Derringer Model 1, Lady Derringer
Hämmerli Model 280 Target Pistol
Smith & Wesson Model 52

32 S&W WADCUTTER

Erma ESP 85A Competition Pistol, Model 773
European Amer. Armory Benelli Model MP90S
Ruger New Model Single-Six

32-20

American Derringer Model 1
EMF/Dakota Hartford Models
Thompson/Center Contender Bull Barrel

35 REMINGTON

Magnum Research Lone Eagle SS
Remington Model XP-100 Silhouette, XP-100R Hunter, Long-Range
Thompson/Center Contender Hunter and Super "14"

357 MAGNUM

American Derringer Models 1, 4 and 38 DA, Cop and Mini-Cop Double Derringers
Colt King Cobra, Python
Coonan Arms 357 Magnum, Cadet Compact
EMF/Dakota Target Model 1873, Model 1873 Dakota SA, 1875 Outlaw, 1890 Remington Police, Hartford Scroll Engraved SA
Erma Model 777 Sporting Revolver
European American Armory Big Bore Bounty Hunter, Windicator DA
Freedom Arms Model 353, Competition Models
L.A.R. Grizzly Mark I
Magnum Research Desert Eagle
Mitchell Arms SA Army Revolvers
Rossi Model 971 & 971 Compact
Ruger Blackhawk, Model GP-100, Model SP101, SA Target
Smith & Wesson Models 13, 19, 27, 65, 65 LadySmith, 66, 586
Taurus Models 65 and 66
Texas Arms Defender Derringer
Thompson/Center Contender Bull Barrel
Uberti 1871 Rolling Block Target Pistol, 1873 Cattleman Quick Draw, 1875 Remington Army Outlaw, Buckhorn SA
Unique Model Int'l Silhouette
Dan Wesson 357 Mag., 357 Super Mag., Fixed Barrel revolvers
Wichita Arms International

357 MAXIMUM

American Derringer Models 1 and 4
Magnum Research Lone Eagle SS
Thompson/Center Contender Bull Barrel
Dan Wesson 357 Super Mag., Hunter Series (Varmint)

358 WINCHESTER

Magnum Research Lone Eagle SS
MOA Maximum

36

EMF Colt 1851 Navy

375 WINCHESTER

Thompson/Center Contender Hunter and Super "14"

38 SPECIAL

American Derringer 1, 3, 7, 11, Lady Derringer, DA 38, Cop & 125th Anniversary Derringers
Charter Arms Off-Duty and Police Undercover
Colt King Cobra DA, Python Premium DA, Detective Special
Davis Long-Bore D-Series and Big-Bore D-Series
European American Armory Windicator DA
Heritage Sentry DA
Rossi Models 68, M88, 851, 971
Smith & Wesson Models 442, 649 Bodyguard, LadySmith

Taurus Models 66, 80, 82, 83, 85, 86
Dan Wesson 38 Special, Model 738P, Fixed Barrel revolvers

380 AUTO

American Derringer Models 1 and 7
AMT Model 380 Backup II
Beretta Cheetah Models 84, 85 and 86
Bernardelli PO18 Compact Target Pistol
Bersa Model 83 DA, Model 85
Browning Model BDA-380
Colt Government Model, Mustang, Mustang Plus II, Mustang Pocket Lite 380
Davis P-380 and Long-Bore D-Series
European American Armory SA/DA Compacts
Grendel Model P-12
Heckler & Koch Model P7K3
KBI FEG Model SMC-380
Llama Automatic (Small Frame)
Sig-Sauer Model 230
Taurus Model PT 58
Texas Arms Defender Derringer
Walther Model PPK & PPK/S, Model PP DA

38 SUPER

American Derringer Model 1
Colt Combat Commander MK Series 80, Gold Cup Combat Elite, Gov't. Model Series 80
European American Armory Witness Gold Team, Subcompact
McMillan Wolverine
Sig-Sauer Model 220 "American"
Springfield Model 1911-A1 Standard, Super, High-Capacity Full-House Racegun, Trophy Master Distinguished/Expert

38 S&W

Smith & Wesson Models 10, 13, 14, 15, 36, 38, 49, 60, 64, 65 and 640 Centennial

38-40

Colt Single Action Army
EMF/Dakota Hartford Model
Uberti 1873 Cattleman Quick Draw

9mm WIN. MAG.

AMT Automag III

9mm PARABELLUM (LUGER)

American Derringer Model 1, Model DA 38 Double Derringer, Semmerling LM-4
AMT On Duty DA Pistol
Astra Models A-70, A-75 and A-100
Bernardelli Model PO18 Target & Compact Target Pistols
Bersa Thunder 9 DA
Browning 9mm Hi-Power, Model BDM Double Action
Daewoo Model DP51
Davis Long-Bore D-Series
European American Armory Astra Models A-70, A-75, A-100, Witness, Witness Fab, Subcompact/Gold Team
Glock Models 17, 17L Competition, Model 19 Compact
Heckler & Koch Models P7M8, HK USP
Helwan Brigadier
Heritage Senty DA
Hi-Point Firearms JS-Series
Kahr Arms Model K9
KBI FEG Model PJK-9HP
Luger, American Eagle P-08
Magnum Research Baby Eagle
McMillan Wolverine
Ruger Model SP101, P-Series
Sig-Sauer Models 225, 226, 228
Smith & Wesson Model 915, Model 940 Centennial, Model 3900 Compact Series, 5900 and 6900 Compact Series

Springfield Model 1911-A1 Standard, Model XM4 High Capacity
Star Models Firestar, Ultrastar
Taurus Models PT92, PT-92AFC, PT99, PT-908
Texas Arms Defender Derringer
Walther Models P-38, P-88 DA, P-5 DA

9mm×19

Beretta Cougar 8000, G, F, D, Models 92 Brigadier, 92D, 92FS, 92 G, 92 Stock

10mm

American Derringer Model 1
Colt Delta Elite, Delta Gold Cup
Emerging Technologies Laseraim Series I, II, III
European American Armory Witness Gold Team
Glock Model 20
L.A.R. Grizzly Mark I
McMillan Wolverine
Star Model Megastar

40 S&W

American Derringer Model 1, 38 DA Derringer
AMT On Duty DA Pistol
Astra Models A-70, A-75, A-100
Beretta Cougar 8040G, D Inox, F Inox, Models 96, 96 Brigadier, 96 Centurion, 96D, 96DS, 96G, 96 Stock
Browning Hi-Power
Brno CZ-75 Standard
Daewoo Model DH40
European American Armory Astra Models A-70, A-75, A-100, Witness, Witness Fab, Subcompact & Gold Team
Glock Models 22, 23, 24
Heckler & Koch Model P7M10, Model HK USP
Hi-Point Firearms JS-Series
Magnum Research Baby Eagle
Sig-Sauer Model P229
Smith & Wesson Model 4000 Series Compact & Full Size, Model 411
Star Model Firestar
Taurus Model 100/101

40 AUTO

Ruger P-Series Pistols

41 MAGNUM

American Derringer Model 1
Magnum Research Desert Eagle
Ruger Bisley SA Target, Blackhawk
Smith & Wesson Model 657
Dan Wesson 41 Mag. Revolvers, Hunter Series (Boar)

41 ACTION EXPRESS

Beretta Cougar D Combo, F Inox/Combo, G Inox/Combo
Magnum Research Baby Eagle

.410

American Derringer Models 4 and 6
Thompson/Center Contender Super "16"

44 MAGNUM

American Derringer Model 1, Alaskan Survival
Colt Anaconda Revolver
European American Armory Big Boar Bounty Hunter
Freedom Arms Casull Premier & Field Grades, Competition Models
Magnum Research Desert Eagle, Lone Eagle SS
Mitchell Arms SA Army Model Revolvers
Ruger Redhawk, Super Blackhawk SA, Vaquero SA Revolvers
Smith & Wesson Model 29, 629, 629 Classic
Taurus Model 44 Revolver
Texas Arms Defender Derringer

Thompson/Center Contender Bull Barrel, Hunter and Super "14"
Uberti Buckhorn SA
Unique Model Int'l Silhouette
Dan Wesson 44 Mag. Revolvers, Hunter Series (Buck)

44 SPECIAL

American Derringer Models 1, 7
Charter Arms Bulldog Pug 44 Special
Colt Anaconda Revolver
EMF/Dakota Hartford Models
Rossi Model 720, 720 Covert Special DA
Smith & Wesson Model 629, 629 Classic
Taurus Model 431 & 441
Uberti 1873 Cattleman Quick Draw

44-40

American Derringer Model 1, 125th Anniversary Commemorative
Colt Single Action Army
EMF/Dakota Models 1873, 1873 Dakota SA, 1875 Outlaw, 1890 Remington Police, Hartford Scroll Engraved SA, Hartford Model
Navy Arms 1873 SA, 1875 Schofield
Ruger Vaquero SA Revolver
Uberti Buckhorn SA, 1873 Cattleman Quick Draw, 1875 Remington Army Outlaw

444 MARLIN

Magnum Research Lone Eagle SS

445 SUPERMAG

Dan Wesson 445 Supermag, Hunter Series (Grizzly)

45 AUTO

American Derringer Models 1, 4, 6, 10 Semmerling LM-4
AMT Longslide & Government Models, On Duty DA Pistol
Astra Models A-70, A-75 and A-100
Auto-Ordnance Model 1927A-1 Thompson, Model 1927A-1C, M1 Carbine
Colt Combat Commander, Combat Elite MKIV Series 80, Double Eagle, Gold Cup National Match, Government Model, Model 1991A1, Officer's ACP
Daewoo Model DH45 High Capacity
Emerging Technologies Laseraim Series I, II, III
European American Armory Astra Models A-75, A-100, Witness, Witness Fab/Gold Team/ Subcompact, Big Boar Bounty Hunter
Glock Model 21 Pistol
Hi-Point Firearms JS-Series
L.A.R. Grizzly Mark I
Laseraim Series I, II & III
Llama Automatics (Large and Compact Frames)
McMillan Wolverine
Para-Ordnance Models P12 & P14
Ruger P-Series, Model P220
Sig Sauer Model 220 "American"
Smith & Wesson Model 4500 Series Compact & Full Size, Model 625
Springfield Basic Competition, Bull's-eye Wadcutter, Champion, Compact, Defender, Factory Comp, Hi-Capacity Full-House Racegun, Model 1911-A1 Standard, Model XM4 High Capacity, Nat'l. Match Hardball, Trophy Master Distinguished Carry Gun, Trophy Match
Star Firestar, Megastar
Texas Arms Defender Derringer
Uberti 1873 Cattleman Quick Draw, 1875 Army SA Outlaw
Wesson Pin Gun

45 COLT

American Arms Regulator SA
American Derringer Models 1, 4, 6, 10, 125th Anniversary Commemorative
Colt Single Action Army Revolver, Anaconda
EMF/Dakota Hartford Models (Artillery, Cavalry Colt), Models 1873, 1873 Dakota SA, 1875 "Outlaw," Model 1890 Remington Police, Pinkerton Detective SA, Scroll Engraved SA, Target
Mitchell Arms SA Army Model Revolvers
Navy Arms 1873 SA, Artillery & Calvary, 1875 Schofield
Ruger Model Bisley SA Target, Blackhawk, Vaquero
Thompson/Center Contender Bull Barrel, Super "16"
Uberti 1871 Rolling Block Target Pistol, 1873 Cattleman Quick Draw
Wesson 45 Colt Revolver

45 WIN. MAG.

AMT Automag IV
American Derringer Model 1
L.A.R. Grizzly Mark I
Wildey Pistols

454 CASULL

Freedom Arms Casull Field & Premier Grades

45-70 GOV'T.

American Derringer Models 1 and 4, Alaskan Survival
Thompson/Center Contender Hunter

475 WILDEY MAG.

Wildey Pistols

50 MAG. AE

L.A.R. Grizzly 50 Mark 5
Magnum Reserach Desert Eagle 50

RIFLES

CENTERFIRE BOLT ACTION

STANDARD CALIBERS

17 BEE

Francotte Bolt Action

17 REMINGTON

Remington Model 700 ADL, BDL, Model Seven
Sako Deluxe, Hunter Lightweight, Varmint
Ultra Light Model 20 Series
Winslow Varmint

22 HORNET

Browning A-Bolt II Short Action
New England Firearms Handi-Rifle
Ruger Model 77/22RH
Ultra Light Model 20 Series

22 LONG RIFLE

Sako Finnfire
Unique Model T Dioptera Sporter, Model T UIT, Model T/SM Silhouette

22 MAGNUM

Unique Model T Dioptera Sporter, Model T/SM Silhouette

22 PPC

Ruger M-77 HB

22-250

Blaser Model R84
Browning Short Action A-Bolt II, Stalker Varmint
Dakota 76 Classic
McMillan Classic and Stainless Sporters, Talon Sporter, Varminter
New England Firearms Handi-Rifle
Parker-Hale Model M81, 1000 Clip & Standard, 1100 LWT, 1200 Super, 2100 Midland
Remington Model 700 ADL, BDL, Varmint Special & Synthetic
Ruger Model M-77 Mark II HB
Sako Deluxe, Hunter Lightweight, LS, Left-Handed Models, Varmint
Savage Model 111G Classic Hunter, Model 112FV
Steyr-Mannlicher Sporter Model L, Varmint
Tikka Continental, New Generation, Premium Grade
Ultra Light Model 20 Series
Voere Model 2150, 2165
Weatherby Mark V Deluxe, Varmintmaster
Winchester Model 70 Featherweight, FWT Classic, Classic Stainless, DBM Classic, Varmint
Winslow Basic

220 SWIFT

McMillan Varminter
Remington Model 700 VS Varmint
Ruger Model M-77R, M-77 Mark II HB
Winchester Model 70 Varmint HB

222 REM.

Francotte Bolt Action
Remington Model 700 ADL, BDL, Varmint Special (wood)
Sako Deluxe, Hunter Lightweight, Varmint
Steyr-Mannlicher Model SL, Varmint
Ultra Light Model 20 Series
Winslow Varmint

223 REM.

Browning A-Bolt II Short Action, Stalker Varmint
Brown Precision High Country Youth & Tactical Elite
Jarrett Lightweight Varmint
Mark X Barreled Actions, Mini-Mark X
McMillan Varminter
New England Firearms Handi-Rifle
Remington Models 700 ADL, BDL, BDL Stainless, Varmint Special & Synthetic, Model Seven
Ruger M-77 RL Ultra Light, M-77 Mark II HB, All-Weather, M-77 R
Sako Deluxe, Hunter Lightweight, Varmint
Savage Model 110 Series, Model 111G Classic Hunter, Model 112 BT Competition & 112FV, 116FCS & 116FSS
Steyr-Mannlicher Model SL, Varmint
Tikka Continental, New Generation, Premium Grade
Ultra Light Model 20 Series
Winchester Models 70 Featherweight, FWT Classic, Lightweight, Ranger, Youth, Varmint, Classic Stainless

223 UCC CASELESS

Voere Model VEC 91 Lightning Bolt
Winslow Varmint

243 WINCHESTER

Blaser Model R84
Brno Model 601
Brown Precision High Country Youth
Browning A-Bolt II Short Action
Clifton Arms Scout
Francotte Bolt Action
Carl Gustaf Model 2000
McMillan Benchrest, Classic Sporter, Standard & Stainless Sporters, Talon, Varminter

New England Firearms Handi-Rifle
Parker-Hale Models M81, 1000 Clip, 1100 LWT, 1200 Super, 2100 Midland, 1300C Scout
Remington Model Seven, Models 700 ADL, BDL, BDL European, BDL SS, CS, LS, Mountain, Varmint Special, 7400, 7600
Ruger M-77 Mark II HB & All-Weather, M-77R, M-77RS, M-77RL Ultra Light, M-77RSI
Sako Classic, Deluxe, Hunter Lightweight, Left-Handed Models, LS, Mannlicher-Style Carbine, Varmint
Savage Models 110 Series, Model 111G Classic Hunter, Model 116FCS & 116FSS
Steyr-Mannlicher Models Luxus, Sporter L, SSG, Varmint
Tikka Continental, New Generation, Premium Grade
Ultra Light Model 20 Series
Unique Model TGC
Voere Models 2150, 2155, 2165
Winchester Models 70 Featherweight, FWT Classic, Classic Stainless, DBM, Lightweight, Ranger and Ranger Youth, Varmint
Winslow Basic

244 REMINGTON

Winslow Basic

6mm BR

McMillan Benchrest Rifle, Classic & Talon Sporters

6mm REMINGTON

Blaser Model R84
Brown Precision High Country Youth & Tactical Elite
McMillan Benchrest, Classic Sporter, Stainless & Talon Sporters, Varminter
Parker-Hale Models 81, 1000 Clip, 1100 LWT, 1200 Super, Model 2100 Midland
Remington Model Seven, Models 700 ADL, BDL, BDL SS, Varmint Special
Steyr-Mannlicher Sporter Model L
Ultra Light Model 20 Series

6mm PPC

McMillan Benchrest
Ruger M-77 Mark II HB

6.5×55mm SWEDISH

Winchester Model 70 Walnut Featherweight

250-3000 SAVAGE

Savage Model 110
Ultra Light Model 20 Series

25-06

A-Square Hamilcar
Blaser Model R84
Browning A-Bolt II, Euro-Bolt
K.D.F. Model K15
McMillan LA Classic, Stainless Sporters, Talon Sporter, Varminter
Remington Models 700 ADL, BDL, BDL SS, Mountain, Sendero
Ruger Models 77RS, M-77V1 HB
Sako Deluxe, Fiberclass, Hunter Lightweight, LS, Left-Handed Models, TRG-S Magnum
Sauer Model 90
Ultra Light Model 24 Series
Winchester Model 70 Sporter & Classic Sporter
Winslow Basic

257 ACKLEY

Ruger Model M-77RL Ultra Light
Ultra Light Model 20 Series

257 ROBERTS

Browning A-Bolt II Short Action
Dakota Arms Model 76 Classic
Remington Model 700 Mountain
Ruger Model M-77R
Ultra Light Model 20 Series
Weatherby Mark V Sporter
Winslow Basic

264 WINCHESTER

A-Square Hamilcar
Ultra Light Arms Model 28 Series

270 WINCHESTER

A-Square Hamilcar
Blaser Models R84
Brno Models 537 and 600
Browning A-Bolt II, Euro-Bolt
Dakota Arms Model 76 Classic
Francotte Bolt Action
Carl Gustaf Model 2000
KDF Model K15
Mark X Whitworth, Barreled Actions, Mauser System Actions
McMillan Alaskan LA, Classic LA & Stainless Sporters, Talon Sporter, Titanium Mountain
Parker-Hale Models M81, 1000 Clip, 1100 LWT, 1200 Super, 2100 Midland
Remington Models 700 ADL & BDL, BDL SS, BDL European, Mountain, 7400, 7600, Sendero
Ruger Model M-77EXP Express, M77R & 77RS, RL Ultralight, 77RSI, International, 77R, M77 All-Weather, M77 Deluxe
Sako Classic, Deluxe, Fiberclass, Hunter Lightweight, LS, Mannlicher-Style Carbine, Left-Handed Models, Model TRG-S
Sauer Model 90
Savage Model 110 Series, Model 111GC & 111FC Classic Hunter, Model 114CU, 116FCS, 116FSK & 116FSS
Steyr-Mannlicher Luxus Model M (Sporter & Professional)
Tikka New Generation, Premium Grade, Whitetail/ Battue
Ultra Light Model 24 Series
Unique Model TGC
Voere Models 2150, 2155, 2165
Weatherby Models Mark V Alaskan, Lazermark, Sporter, Weathermark
Winchester Models 70 Featherweight, FWT Classic, Classic SM, Classic Super Grade, Classic Stainless, DBM, Lightweight, Ranger, Youth Sporter
Winslow Basic

280 REMINGTON

A-Square Hamilcar
Blaser Model R 84
Browning A-Bolt II, Euro-Bolt
Dakota Arms Model 76 Classic
KDF Model K15
McMillan Alaskan LA, Classic LA & Stainless Sporters, Talon Sporter, Titanium Mountain
Remington Models 700 ADL, 700 BDL, BDL European, BDL SS, CS, Mountain, 7400, 7600
Ruger Model M-77R, M-77 All-Weather
Sako Deluxe, Fiberclass, Hunter Lightweight, LS, Left-Handed Models
Winchester Models 70 Featherweight, FWT Classic Lightweight, **Winslow** Basic

280 IMP

Jarrett Standard Hunting

284 WINCHESTER

Browning A-Bolt II Short Action
McMillan Classic, Stainless & Talon Sporters
Ultra Light Model 20 Series
Winchester Model 70 Classic DBM
Winslow Basic

7mm ACKLEY

Ultra Light Model 20 Series

7mm BR

McMillan Classic, Stainless and Talon Sporters

7mm EXPRESS

Ultra Light Model 24 Series

7mm MAUSER

Remington Model 700 Mountain
Ultra Light Model 20 Series
Winslow Basic

7mm REM./WBY.

A-Square Hamilcar
Ultra Light Arms Model 28 Series

7mm STW

Jarrett Model 3

7mm-08

Brown Precision High Country Youth
Browning A-Bolt II Short Action
Clifton Arms Scout
McMillan Classic & Stainless Sporters, National Match, Talon, Varminter
Remington Model Seven, Models 700 ADL, BDL, BDL European, BDL SS, CS, Mountain and Varmint Special
Sako Deluxe, Hunter Lightweight, LS, Left-Handed Models, Varmint
Savage Model 111G Classic Hunter
Ultra Light Model 20 Series
Unique Model TGC
Winchester Model 70 Featherweight, FWT Classic

30-06

A-Square Caesar, Hamilcar & Hannibal
Blaser Model R84
Brno Models 537 and 600
Browning A-Bolt II, Euro-Bolt
Clifton Arms Scout
Dakota Arms Model 76 Classic
Francotte Bolt Action
Carl Gustaf Model 2000
K.D.F. Model K15
Mark X Viscount Sporter, Whitworth
McMillan Alaskan LA, Classic & Stainless Sporters LA, Talon Sporter, Titanium Mountain
Parker-Hale Models M81 Classic, 1000 Clip, 1100 Lightweight, 1200 Super, 2100 Midland
Remington Models 700 ADL, BDL, BDL European, CS, Mountain, 7400, 7400 Carbine, 7600
Ruger M-77 All-Weather, M-77EXP Express, Model 77R, 77RL Ultra Light, M-77RS, 77RSI International
Sako Classic, Deluxe, Mannlicher-Style Carbine, Fiberclass, Hunter Lightweight, LS, Left-Handed Models, Model TRG-S
Sauer Model 90
Savage Model 111FC & 111GC Classic Hunter, Model 114CU, 116FSS, 116FCS, 116FSK
Steyr-Mannlicher Models M (Sporter & Professional), Luxus M
Tikka New Generation, Premium Grade, Whitetail/ Battue
Ultra Light Model 24 Series
Unique Model TGC
Voere Models 2150, 2155, 2165
Weatherby Mark V Alaskan, Deluxe, Lazermark, Sporter, Weathermark
Winchester Models 70 Featherweight, Classic SM, DBM, FWT Classic, Lightweight, Ranger, Super Grade, Youth Sporter
Winslow Basic

30-06 CARBINE

Remington Models 7400 and 7600 Carbines

300 SAVAGE

Savage 110 Series, Model 111G Classic Hunter
Ultra Light Model 20 Series

300 WINCHESTER

Ultra Light Arms Model 28 Series

308 WINCHESTER

Brno Models 537 and 601
Browning A-Bolt II Short Action
Clifton Arms Scout
Francotte Bolt Action
Carl Gustaf Model 2000
Mark X Viscount Sporter
McMillan Benchrest, Classic & Stainless Sporters, National Match, Talon Sporter, Varminter
Parker-Hale Models M81, M-85 Sniper, 1000 Clip, 1100 LWT, 1200 Super, Model 1300C, 2100 Midland
Remington Models Seven, 700 ADL, BDL, BDL Stainless, BDL European, Mountain, Varmint Special & Synthetic, 7400, 7600
Ruger Models M-77 All Weather, M-77R, M-77RL Ultra Light, M-77RS, 77RSI Int'l. Mannlicher, M-77VI Heavy Barrel
Sako Deluxe, Hunter Lightweight, LS, Left-Handed Models, Mannlicher-Style Carbine, TRG-21, Varmint
Savage Model 110 Series, Model 111G Classic Hunter, Model 112BT Competition
Steyr-Mannlicher Luxus L, Match UIT, Sporter L, SSG & SSG Match UIT, Varmint
Tikka Continental, New Generation, Premium Grade, Whitetail/Battue
Ultra Light Model 20 Series
Unique Model TGC
Voere Models 2150 and 2165
Winchester Model 70 Featherweight, FWT Classic Stainless, Classic DBM, Lightweight, Ranger Youth, Varmint
Winslow Basic

35 WHELEN

Remington Models 7400, 7600

358 WINCHESTER

Ultra Light Model 20 Series
Winslow Basic

MAGNUM CALIBERS (RIFLE)

222 REM. MAG.

Steyr-Mannlicher Model Sporter (SL)
Ultra Light Model 20 Series

240 WBY. MAG.

Weatherby Mark V Deluxe, Lazermark

257 WBY./ROBERTS

A-Square Hamilcar
Blaser Model R84
Weatherby Mark V Alaskan, Deluxe, Lazermark, Sporter, Weathermark
Winslow Basic

264 WIN. MAG.

Blaser Model R84
Winchester Model 70 Sporter
Winslow Basic

270 WBY. MAG.

Weatherby Mark V Alaskan, Deluxe, Lazermark, Sporter, Weathermark
Winchester Model 70 Sporter
Winslow Basic

7mm REM./WBY. MAG.

A-Square Caesar and Hannibal
Blaser Model R84
Browning A-Bolt II, Euro-Bolt
Dakota Arms Model 76 Classic
Carl Gustaf Model 2000
K.D.F. Model K15
Mark X Viscount Sporter
McMillan Alaskan MA, Classic MA & Stainless MA Sporters, Long Range Rifle, Talon Sporter, Titanium Mountain
Remington African Plains, Alaskan Wilderness, Models 700 ADL, BDL, BDL European, BDL SS, CS, Sendero
Ruger Models M-77 All-Weather, M-77EXP Express, 77R, M-77RS
Sako Classic, Deluxe, Fiberclass, Hunter Lightweight, LS, Left-Handed Models, TRG-S
Sauer Model 90
Savage Model 114 CU, 116 FCS, 116FSK, 116FSS
Steyr-Mannlicher Luxus S, Sporter S & ST
Tikka New Generation, Premium Grade, Whitetail/Battue
Unique Model TGC
Voere Models 2155M & 2165M
Weatherby Mark V Alaskan, Deluxe, Lazermark, Sporter, Weathermark
Winchester Model 70 Classic SM, Classic Stainless, Classic Super Grade, DBM, Sporter, Super Grade, Winlite
Winslow Basic

8mm REM. MAG.

A-Square Caesar and Hannibal
Remington Model 700 Safari Grade

300 PHOENIX

McMillan Long Range, Safari, 300 Phoenix

300 WBY. MAG.

A-Square Caesar and Hannibal, Hamilcar
Blaser Model R 84
KDF Model K15
McMillan Alaskan MA, Classic MA & Stainless MA Sporters, Safari and Talon Sporter
Remington Model 700 BDL SS, African Plains, Alaskan Wilderness
Sako Deluxe, Hunter Lightweight, Left-Handed Models, Model TRG-S
Sauer Model 90
Ultra Light Arms Model 40 Series
Weatherby Mark V Alaskan, Deluxe, Lazermark, Sporter, Weathermark
Winchester Model 70 Classic Stainless, Sporter
Winslow Basic

300 WIN. MAG.

A-Square Caesar and Hannibal
Blaser Model R 84
Brno Model 602
Brown Precision Tactical Elite
Browning A-Bolt II, Euro-Bolt
Dakota Arms Model 76 Classic & Safari
Carl Gustaf Model 2000
KDF Model K15
Mark X Viscount Sporter, Whitworth
McMillan Alaskan, Classic MA & Stainless MA Sporters, Long Range Rifle, Safari, Talon Sporter, Titanium Mountain
Remington African Plains, Alaskan Wilderness, Models 700 ADL, BDL, BDL SS, Sendero
Ruger M-77 All-Weather, M-77EXP Express, M-77R, M-77RS
Sako Deluxe, Fiberclass, Left-Handed Models, Hunter Lightweight, LS, Model TRG-S

Sauer Model 90

Savage Model 110 Series, 114CU, 116FCS & 116FSS
Steyr-Mannlicher Luxus S, Sporter S & ST
Tikka New Generation, Premium Grade, Whitetall/Battue
Unique Model TGC
Voere Models 2155M & 2165M
Weatherby Mark V Alaskan, Sporter, Weathermark
Winchester Model 70, Classic SM, Classic Stainless, Classic Super Grade, DBM, Sporter
Winslow Basic

300 H&H

A-Square Caesar and Hannibal
McMillan Alaskan MA, Classic MA & Stainless MA Sporters, Safari, Talon Sporter
Winslow Basic

308 NORMA

Winslow Basic

338 LAPUA

Heym Model Express
McMillan Long Range, Safari
Sako Model TRG-S

338 WIN. MAG.

A-Square Caesar, Hamilcar & Hannibal
Blaser Model R 84
Browning A-Bolt II, Euro-Bolt, Stalker Series
Dakota Arms Model 76 Classic, Safari
KDF Model K15
McMillan Classic MA & Stainless MA Sporters, Talon Sporter & Safari
Remington African Plains, Alaskan Wilderness, Models 700 ADL, BDL, BDL SS
Ruger Models M-77 All-Weather, M-77EXP Express, 77R, M-77RS
Sako Carbine, Deluxe, Hunter Lightweight, Fiberclass, Left-Handed Models, LS, Mannlicher-Style Carbine, Safari Grade, TRG-S
Sauer Model 90
Savage Model 116FCS & 116FSK & 116FSS
Tikka New Generation, Premium Grade, Whitetail/Battue
Ultra Light Arms Model 28 Series
Weatherby Mark V Alaskan, Sporter, Weathermark
Winchester Model 70 Classic SM, Classic Stainless Sporter, Super Grade, Winlite
Winslow Basic

340 WBY. MAG.

A-Square Caesar and Hannibal
KDF Model K15
McMillan Alaskan MA, Classic MA & Stainless MA Sporters, Safari and Talon Sporter
Weatherby Mark V Alaskan, Deluxe, Lazermark, Sporter, Weathermark

350 REM. MAG.

Clifton Arms Scout
McMillan Varminter, Classic Stainless & Talon Sporters

35 WHELEN

Remington Model 700 ADL, BDL, Models 7400 & 7600

358 WIN./NORMA

McMillan Alaskan MA
Winslow Basic

375 H&H

A-Square Caesar and Hannibal
Blaser Model R84
Brno Model 602
Browning A-Bolt II & Composite Stalker
  (stainless), Euro-Bolt
Dakota Arms Model 76 Safari, Classic
Francotte Bolt Action
Heym Model Express
KDF Model K15
McMillan Alaskan MA, Classic MA & Stainless MA
  Sporters, Safari, Talon Sporter
Parker-Hale Model M81 African, 1100M African
Remington African Plains, Alaskan Wilderness,
  Model 700 Safari Grade
Ruger 77 Magnum
Sako Deluxe, Fiberclass, Hunter Lightweight, LS,
  Left-Handed Models, Mannlicher-Style Carbine,
  Safari Grade, Model TRG-S
Sauer Model 90
Steyr-Mannlicher Sporter Models S & S/T
Weatherby Mark V Alaskan, Sporter,
  Weathermark
Winchester Model 70 Classic SM
Winslow Basic

375 WEATHERBY

A-Square Caesar and Hannibal

378 WIN./WBY. MAG.

A-Square Hannibal
Heym Express
McMillan Safari
Weatherby Mark V, Deluxe, Lazermark

404 JEFFERY

A-Square Caesar and Hannibal
Dakota Arms 76 African
McMillan Safari
Ruger 77 Magnum

411 KDF

KDF Model K15

416 REM./WBY. MAG.

A-Square Caesar and Hannibal
KDF Model K15
McMillan Alaskan MA, Classic MA & Stainless MA
  Sporters, Safari, Talon Sporter
Remington Model 700 Safari Grade
Sako Deluxe, Fiberclass, Hunter Lightweight,
  Laminated Stock Models, Left-Handed Models,
  Safari Grade
Weatherby Mark V, Deluxe, Lazermark

416 RIGBY/DAKOTA

A-Square Hannibal
Dakota Arms 76 African
Francotte Bolt Action
Heym Express
McMillan Safari
Ruger 77 Magnum
Ultra Light Arms Model 40 Series

416 TAYLOR & HOFFMAN

A-Square Caesar and Hannibal

425 EXPRESS

A-Square Caesar and Hannibal

450 ACKLEY/DAKOTA

A-Square Caesar and Hannibal
Dakota Arms 76 African
Heym Express

45-70

Pedersoli Replica Rolling Block Target Rifle

458 WIN. MAG.

Brno Model 602
Dakota Arms Model 76 Classic, Safari
KDF Model K15
Mark X Barreled Action, Whitworth Express
McMillan Safari
Parker-Hale Model 1100M African
Remington Model 700 Safari Grade
Ruger Model-M-77RS MKII
Sauer Safari, Model 90
Steyr-Mannlicher Sporter Model S & S/T
Winslow Basic

460 WIN./WBY. MAG.

A-Square Hannibal
Francotte Bolt Action
Heym Express
McMillan Safari
Weatherby Mark V, Deluxe, Lazermark

500 N.E.

Heym Express

505 GIBBS

Francotte Bolt Action

600 N.E.

Heym Express 600

CENTERFIRE LEVER ACTION

22 LONG RIFLE

Uberti Model 1866 Sporting, Model 1866
  Yellowboy Carbine, Model 1871 Rolling Block
  Baby Carbine

22 MAGNUM

Uberti Model 1866 Sporting & Yellowboy Carbine,
  Model 1871 Rolling Block Baby Carbine

222 REM./223 REM.

Browning Model 81 BLR High Power

22-250

Browning Model 81 BLR High Power

243 WINCHESTER

Browning Model 81 BLR High Power
Savage Model 99C

270

Browning Model 81 BLR LA

284 WINCHESTER

Browning Model 81 BLR High Power

7mm-08

Browning Model 81 BLR High Power

7mm MAGNUM

Browning Model 81 BLR LA

7-30 WATERS

Winchester Model 94 Standard Walnut

30-06

Browning Model 81 BLR LA

30-30 WINCHESTER

Marlin Models 30AS, 336CS
Winchester Models 94 Ranger, Standard, Trapper
  Carbine, Walnut, Win-Tuff, Wrangler

307 WINCHESTER

Winchester Model 94 Big Bore Walnut

308 WINCHESTER

Browning Model 81 BLR High Power
Savage Model 99C

32 WINCHESTER

Winchester Model 94 Standard Walnut

35 REMINGTON

Marlin Model 336CS

356 WINCHESTER

Winchester Model 94 Big Bore Walnut

357 MAGNUM

EMF Model 1873 Sporting
Marlin Model 1894CS
Mitchell Arms 1873 Winchester
Rossi Models M92 SRS & SRC
Uberti Models 1871 Rolling Block Baby Carbine,
  1873 Sporting & Carbine
Winchester Model 94 Trapper Walnut Carbine

38 SPECIAL

EMF Model 1866 Yellow Boy Rifle/Carbine
Marlin Model 1894CS
Rossi Models M92 SRC & M92 SRS
Uberti Models 1866 Sporting, 1866 Yellowboy
  Carbine

44 REM. MAG.

Marlin Model 1894S
Winchester Model 94 Trapper Carbine, Wrangler

44 SPECIAL

Marlin Model 1894S
Winchester Model 94 Trapper Carbine, Wrangler

44-40

EMF Model 1860 Henry, 1866 Yellow Boy Rifle/
  Carbine, Model 1873 Sporting
Marlin Model 1894 Century Ltd.
Mitchell Arms 1858 Henry, 1866 and 1873
  Winchesters
Navy Arms 1866 Yellowboy (rifle & carbine),
  Henry Military, Iron Frame & Trapper Models,
  1873 Winchester-Style and Sporting Rifles
Uberti Model 1866 Sporting, Yellowboy Carbine,
  1873 Sporting & Carbine, Henry Rifle & Carbine

444 MARLIN

Marlin Model 444SS

45 COLT

EMF Model 1860 Henry, 1866 Yellow Boy Rifle/
  Carbine, 1873 Sporting, Carbine
Marlin Model 1894S
Mitchell Arms 1873 Winchester
Navy Arms 1873 Winchester Sporting Rifles,
  Sharps Cavalry Carbine
Uberti Model 1866, Model 1866 Yellowboy
  Carbine & Sporting, 1873 Sporting & Carbine,
  Henry
Winchester Model 94 Trapper Carbine

45-70 GOV'T.

Marlin Model 1895SS

SINGLE SHOT

17 REMINGTON

Thompson/Center Contender

218 BEE

Ruger No. 1B Standard, No. 1S Medium Sporter

22 S, L, LR

Dakota Arms Model 10
European American Armory HW 60 Target, HW 660 Match
Lakefield Mark I
Marlin 15YN "Little Buckaroo," Model 2000A Target
Thompson/Center Contender
Ultra Light Arms Model 20RF

22 BR REM.

Remington Model 40-XBBR Bench Rest

22 HORNET

Ruger No. 1B Standard
Thompson/Center Contender

22 PPC

Ruger No. 1V Special Varminter

22-250 REM.

Browning Model 1885
Harrington & Richardson Ultra Single Shot Varmint
Remington Model 40-XB Rangemaster
Ruger No. 1B Standard, Special Varminter

220 SWIFT

Remington Model 40XB Rangemaster
Ruger No. 1B Standard No. 1V Special Varminter

222 REMINGTON

Remington Models 40-XB Bench Rest, Rangemaster

222 REM. MAG.

Remington Model 40-XBBR

223

Browning Model 1885
Harrington & Richardson Ultra Single Shot Varmint
Ruger No. 1B Standard, No. 1V Special Varminter
Thompson/Center Contender

243 WINCHESTER

Ruger No. 1A Light Sporter, No. 1B, No. 1 RSI International Standard

25-06

Ruger No. 1B Standard, No. 1V Special Varminter

6mm PPC

Ruger No. 1V Special Varminter

6mm REMINGTON

Ruger No. 1B Standard, No. 1V Special Varminter

257 ROBERTS

Ruger No. 1B Standard

270 WEATHERBY

Ruger No. 1B Standard

270 WINCHESTER

Browning Model 1885
Ruger No. 1A Light Sporter, No. 1B Standard, RSI International

280 REMINGTON

Ruger No. 1B Standard

30-06

Browning Model 1885
Remington Model 40XB Rangemaster
Ruger No. 1A Light Sporter, No. 1B Standard, RSI International

300 WBY. MAG.

Ruger No. 1B Standard

300 WIN. MAG.

Remington Model 40-XB Rangemaster
Ruger No. 1B Standard, No. 1S Medium Sporter

30-30 WIN.

Thompson/Center Contender

308 WINCHESTER

Remington Model 40-XBBR

35 REMINGTON

Thompson/Center Contender

7mm BR REMINGTON

Remington Model 40-XB Rangemaster

7mm REM. MAG.

Browning Model 1885
Remington Model 40-XB Rangemaster
Ruger No. 1S Medium Sporter, No. 1B Standard

338 WIN. MAG.

Ruger No. 1S Medium Sporter, No. 1B Standard

375 H&H/WIN.

Ruger No. 1H Tropical

404 JEFFERY

Ruger No. 1H Tropical

416 REM./RIGBY

Ruger No. 1H Tropical

45-70 GOV'T.

Browning Model 1885
Navy Arms No. 2 Creedmoor Target, Remington Style Rolling Block Buffalo, Sharps Cavalry Carbine
Ruger No. 1S Medium Sporter

458 WIN. MAG.

Ruger No. 1H Tropical

50 BMG

L.A.R. Grizzly Big Boar Competitor

AUTOLOADING

22 LONG RIFLE

AMT Lightning and Small Game Hunting Rifle
Auto-Ordnance Thompson Model 1927-A3
Lakefield Model 64B
Mitchell Arms Models M-16A1 & CAR-15
Norinco Model 22ADT
Ruger 10/22 Carbine, Sporter & International Stainless Models

223 REMINGTON

Colt Sporter Lightweight and Sporter Competition, Match H-Bar, Target Models
Eagle Arms EA-15 Match
Ruger Mini-14, Mini-14 Ranch
Steyr-Mannlicher Aug S.A.

243 WINCHESTER

Browning BAR Mark II Safari
Springfield M1A National Match

270 WIN./WBY.

Browning BAR Mark II Safari

30-06

Browning BAR Mark II Safari
Laurona Model 2000X Express
Voere Model 2185MR Match, 2185/SR, 2185/ 1FH, 2185/2FS

300 WIN. MAG.

Browning BAR Mark II Safari

308 WINCHESTER

Browning BAR Mark II Safari
Heckler & Koch PSG-1 Marksman's
Springfield M1A Standard, National Match; M1A-A1 Bush
Voere Model 2185MR Match, 2185SR, 2185/ 1FH, 2185/2FS

338 WIN. MAG.

Browning BAR Mark II Safari

45 AUTO

Auto-Ordnance Thompson Models M1, 1927A1
Marlin Model 45

7mm REM. MAG.

Browning BAR Mark II Safari

7mm-08

Springfield M1A National Match

7.62×39

Ruger Mini-30

7×64

Voere Models 2185 Match, 2185SR, 2185/2FS

375 H&H MAG.

Laurona Model 2000X Express

9mm

Feather Model AT-9
Marlin Model 9 Camp

10mm

Auto-Ordnance Model 1927-A1

RIMFIRE BOLT ACTION

22 S, L, LR

Anschutz Achiever; BR-50 Bench Rest; Match 64 Sporters: Models 1416D, 1418D, 1516D, 1518D; DRT-Super Match, 64MS, 1907, 1910, 1911, 1913; Models 1700 Custom, Classic, Bavarian, Graphite Custom, Mannlicher 1733D, 1827 B/BT Biathlon, 2007 Super Match 54, 2013 Super Match 54
Brno Model 452
Browning Model A-Bolt 22
Dakota Arms Sporter
Kimber Model 82C LR Classic, SuperAmerica & Custom Shop SuperAm
Lakefield Sporting Models 90B, 91T & 91TR, 92S, Mark I, Mark II
Magtech Model MT-1222
Marlin Models 880, 881, Model 2000 Target
Mitchell Arms High Standard Model 9301-4
Navy Arms Model TU-KKW Training Rifles, TU-33/40 Sniper Trainer
Norinco Model JW-15 "Buckhorn"
Remington Models 40-XR, 541-T & 581-S
Ruger Model 77/22RS & K77/22VBZ Varmint
Ruko Model M14P
Ultra Light Arms Model 20 RF
Unique Model T Dioptra Sporter, UIT Standard, T/ SM Silhouette
Winchester Model 52B

22 HORNET

Anschutz Model 1700 Bavarian and 1700D Custom & Classic, Mannlicher 1733D
Brno Model 527 Mini-Mauser

22 MAGNUM

Anschutz Match 64 Sporter Models
Mitchell Arms High Standard Model 9301-4
Ruger Model K77 Varmint & 77/22RS

22 WMR

Marlin Models 25MN, 25N, 882L, 883, 883SS
Ruko Model M1500

222/223 REM.

Anschutz Model 1700 Bavarian, Custom and 1700D Classic, Mannlicher 1733D
Brno Model 527

7.62mm NATO

Navy Arms Model EM-331 Sporting Rifle
Remington Model 40-XC KS & 40-XR BR

RIMFIRE AUTOLOADING

22 S, L, LR

Anschutz Model 525
Browning Model 22 (Grades I & VI)
Feather Models AT-22 & F2 AT-22
Lakefield Model 64B
Marlin Models 60, 60SS, 70HC, 70P, 990L, 995
Mitchell Arms High Standard Model 15/22
Remington Model 522 Viper, Model 552 BDL Deluxe Speedmaster
Ruger Model 10/22 Carbine, Deluxe Sporter, K10/22RBI Int'l Stainless
Ruko Model M20P
Voere Model 2115

22 WIN. MAG.

Marlin Models 882L, 922 Mag.

RIMFIRE LEVER ACTION

22 S, L, LR

Browning Model BL-22 (Grades I & II)
Marlin Models 39TDS, Golden 39AS
Winchester Model 9422 (Walnut, Win-Tuff)

22 WMR MAG.

Winchester 9422 (Walnut, Win-Cam, Win-Tuff)

44

Navy Arms Henry Trapper Military and Iron Frame

RIMFIRE PUMP ACTION

22 S, L, LR

Remington Model 572 BDL Deluxe Fieldmaster
Rossi Models M62 SAC & SA

DOUBLE RIFLES

308

Krieghoff Models Ulm & Teck
Tikka Model 512S

30-06

Heym Model 88B Safari and Side-by-Side
Krieghoff Models Ulm & Teck
Tikka Model 512S

300 WIN. MAG.

Krieghoff Models Ulm & Teck

9.3×74R

Heym Model 88B
Tikka Model 512S

375 H&H

Heym Model 88B Safari
Krieghoff Model Ulm

458 WINCHESTER

Heym Model 88B Safari
Kreighoff Models Ulm & Teck

470 N.E.   500 N.E.

Heym Model 88B Safari

RIFLE/SHOTGUN COMBOS

22 LR, 22 HORNET, 223 REM.

Savage Model 24F (12 or 20 ga.)

30-30/12 or 20 ga.

Savage Model 24F

BLACK POWDER

HANDGUNS

30

Gonic Arms Model GA-90

31

CVA New Model Pocket, Vest Pocket Derringer, Wells Fargo Model

36

American Arms 1851 Colt Navy
Armsport Models 5133, 5136, 5154
CVA Model 1851, Navy Brass-Framed Revolver, Pocket Police, Sheriff's Model
Dixie Navy Revolver, Spiller & Burr Revolver
EMF 1851 Sheriff's Model, Model 1862 Police
Euroarms 1851 Navy & Navy Confederate Revolver, Model 1010 1858 Remington New Model Army
Navy Arms 1851 Navy "Yank" Revolver, Reb Sheriff's Model 1860, Army 1860 Sheriff's Model, 1862 Police Revolver
Uberti 1851 & 1861 Navy, 1858 Navy, Paterson

38

Dixie Pedersoli Mang Target Pistol
Gonic Arms Model GA-90

41

Dixie Abilene & Lincoln Derringers

44

American Arms 1847 Walker, 1858 Remington Army and Army SS Target, 1860 Colt Army
Armsport Models 5120, 5133, 5136, 5138, 5139, 5140, 5145, 5152, 5153
CVA 1851 & 1861 Navy Brass-Framed Revolvers, Remington Bison, 1858 Remington Target Model and Army Steel Frame Revolver (also Brass Frame), 1860 Army Revolvers, Third Model Dragoon, Walker

Dixie Pennsylvania Pistol, Remington Army Revolver, Third Model Dragoon, Walker Revolver, Wyatt Earp Revolver
EMF Model 1860 Army, Second Model 44 Dragoon
Euroarms Rogers & Spencer Models 1005, 1006 & 1007, 1851 Navy, Remington 1858 New Model Army
Gonic Arms Model GA-90
Navy Arms Colt Walker 1847, 1858 Target Model, Reb Model 1860 Revolver, Reb 60 Sheriff's Model, 1860 Army Revolver, Army 60 Sheriff's Model, Rogers & Spencer Revolver, 1858 Target Remington & Deluxe New Model Revolvers, Stainless Steel 1858 Remington New Army, Remington 1858 New Model Army, LeMat Revolvers, Le Page Flintlock/Percussion Pistols & Cased Sets
Uberti 1st, 2nd & 3rd Model Dragoon, Walker, 1858 Remington New Army, 1860 Army

45

Dixie LePage Dueling Pistol, Pedersoli English Dueling Pistol
Gonic Arms Model GA-90
Ruger Old Army Cap & Ball
Traditions Pioneer Pistol, Trapper Pistol

50

CVA Kentucky Pistol, Hawken Pistol
Thompson/Center Scout Pistol
Traditions Buckskinner, William Parker Pistol, Trapper Pistol, Whitetail (rifle & carbine)

54

Thompson/Center Scout Pistol

58

Navy Arms Harpers Ferry Pistol

RIFLES & CARBINES

30   38 & 44

Gonic Arms Model GA-87

32

CVA Varmint
Dixie Tennessee Squirrel
Navy Arms Pennsylvania Long
Traditions Deerhunter

36

Traditions Frontier Scout

41

White Systems Model Super 91, Whitetail/Bison

44-40

Dixie Winchester '73 Carbine

45

Dixie Hawken, Kentuckian, Pedersoli Waadtlander, Pennsylvania, Tryon Creedmoor
Gonic Arms Model GA-87
Navy Arms Pennsylvania Long
Thompson/Center Hawken
Ultra Light Arms Model 90
White Systems Model Super 91 and "G" Series

451

Parker-Hale Volunteer Rifle, Whitworth Military Target & Sniper Rifles

45-70   45-90   45-120

Shiloh Sharps Model 1874 Business, 1874 Sporting Rifles #1 and #3

CVA Apollo Carbelite, Classic & Sporter, Bushwacker, Express Double Rifle & Carbine, Frontier & Hunter Carbine, Kentucky, Panther Carbine, Plainsman, St. Louis Hawken, Tracker Carbine, Trophy Carbine, Woodsman
Dixie Hawken, In-Line Carbine, Tennessee Mountain
Gonic Arms Model GA-87 Rifle, GA-93 Carbine
Lyman Deerstalker Rifle and Carbine, Great Plains, Trade Rifle
Modern Muzzleloading Knight MK-85 Series, Model BK-92 Black Knight, Magnum Elite, Wolverine
Navy Arms Ithaca-Navy Hawken, Japanese Matchlock, Smith Carbine
Thompson/Center Grey Hawk, Hawken, Hawken Custom, New Englander, Pennsylvania Hunter (rifle & carbine), Renegade, Renegade Single Trigger Hunter, Scout Carbine, Thunder Hawk, White Mountain Carbine
Traditions Buckskinner Carbine, Deerhunter, Frontier Scout, Frontier (rifle & carbine), Hawken Woodsman, In-Line Rifle Series, Pennsylvania, Pioneer, Whitetail (rifle & carbine)
Uberti Santa Fe Hawken
Ultra Light Arms Model 90
White Systems Model Super 91 and ''G'' Series

Shiloh Sharps Model 1874 Business, Sporting #1 & #3, Military

CVA Apollo Classic & 90 Shadow, Frontier Hunter Carbine, Panther Carbine, St. Louis Hawken, Trophy Carbine
Dixie Hawken, In-Line Carbine, Sharps New Model 1859 Carbine & 1859 Military Rifle
Euroarms 1803 Harpers Ferry, 1841 Mississippi
Gonic Arms Model GA-87
Lyman Deerstalker Rifle, Great Plains, Trade Rifle
Modern Muzzleloading Knight MK-85 Series, BK-92 Black Knight, Magnum Elite
Navy Arms 1803 Harpers Ferry, 1841 Mississippi, 1859 Berdan Sharps, 1863 Sharps Cavalry Carbine, Ithaca/Navy Hawken, Mortimer Flintlock & Match Flintlock
Thompson/Center Grey Hawk, Hawken, New Englander, Renegade, Renegade Single Trigger Hunter, Scout Carbine, Thunder Hawk, White Mountain Carbine
Traditions Deerhunter, Hawken Woodsman, Hunter, In-Line Rifle Series, Pioneer, Whitetail (rifle & carbine)
Uberti Sante Fe Hawken

Parker-Hale 1853 & 1858 Enfields, 1861 Enfield Musketoon

Dixie 1858 Two-Band Enfield Rifle, U.S. Model 1861 Springfield, Mississippi, 1862 Three-Band

Enfield Rifle Musket, 1863 Springfield Civil War Musket
Euroarms Model 2260 London Armory Company Enfield 3-Band Rifle Musket, Models 2270 and 2280 London Armory Company Enfield Rifled Muskets, Model 2300 Cook & Brother Confederate Carbine, J. P. Murray Carbine, 1861 Springfield, 1863 Zouave, C. S. Richmond Musket
Navy Arms Mississippi Model 1841, 1853 Enfield Rifle & Musket, 1861 Enfield Musketoon, 1861 & 1863 Springfield, J.P. Murray Carbine
Thompson/Center Big Boar Caplock

Dixie U.S. Model 1816 Flintlock Musket
Navy Arms 1816 M.T. Wickham Musket

Dixie Second Model Brown Bess Musket
Navy Arms Brown Bess Musket & Carbine

SHOTGUNS (Black Powder)

CVA Trapper Single Barrel & Classic Turkey Double Barrel (12 ga.)
Dixie Double Barrel Magnum (12 ga.)
Navy Arms Model T&T, Fowler (12 ga.), Mortimer Flintlock 12 ga., Steel Shot Magnum 10 ga.
Thompson/Center New Englander (12 ga.)

DISCONTINUED MODELS

The following models, all of which appeared in the 1994 edition of SHOOTER'S BIBLE, have been discontinued by their manufacturers or are no longer imported by U.S. distributors or are now listed under a different manufacturer/distributor.

HANDGUNS

CHARTER ARMS Undercover 38 Special, Bonnie & Clyde revolvers
COLT All-American Model 2000
FREEDOM ARMS Model FA 454 ASM U.S. Deputy Marshal, Hunter Paks
HECKLER & KOCH Model SP89
HERITAGE Eagle Semiauto
LLAMA Comanche/Super Comanche Model M-82DA
MAUSER High Power 9mm Models 80SA/90DA/ 90 DAC Compact
MITCHELL ARMS American Eagle Parabellum '08
RUGER Model KP93DC
SMITH & WESSON Model 52 SA Pistol
   Model 57 Revolver
   Models 4006 and 4026 pistols
   Model 1000 Series pistols
SPRINGFIELD Model P9 DA
WALTHER Models GSP Match and OSP Match

RIFLES

AMT Lightning
ANSCHUTZ Biathlon Model 1450B
BEEMAN/WEIHRAUCH Model FWB 2600
   Models HW 60M Smallbore, HW 60J-ST, HW 660 Match
BROWNING Model 1886 Lever-action Carbine
HECKLER & KOCH Model SR-9(T) Target
JARRETT Custom Sniper
MAUSER Model 66 Model 86 Precision
   Model 99 Monte Carlo, Classic, Model 107, Models 201 Standard, Rimfire, Luxus
NAVY ARMS Cowboy's Companion
   Model 1873 Winchester-style Sporting Rifle
   Model TU 33/40
REMINGTON Model 700 SS Synthetic

Target Models 40 XB Rangemaster, 40-XB KS, 40-XC KS, 40-XBBR KS
RUGER Model No. 1H Tropical
   Model No. IRSI International
SAKO PPC Benchrest and Varmint rifles; actions
SAVAGE Models 112 BV, FVS, FVSS
STEYR-MANNLICHER Model SSG P-II Sniper/ SSG P-III
WEATHERBY Classicmark, Weathermark Ltd. Ed. Mark V Deluxe, Vanguard (all variations)
WINCHESTER Model 70 Classic SM (cal. 223, 22-250, 243, 308 only)
   Model 70 DBM (cal. 223 REM., 22-250 Rem., 308 Win. only)
   Model 70 Lightweight (280 Rem. only)
   Model 70 Sporter Wintuff
   Model 70 Walnut Featherweight (7mm Rem. Mag., 300 Win. Mag. only)
   Model 70 Wintuff Featherweight
   Model 94 Standard (32 Win., 7-30 Waters)

SHOTGUNS

AMERICAN ARMS Silver Trap & Skeet
   Specialty Model WS/SS
   Sporting & Falconet 2000
BERETTA Model 627 SS Field Grade
BERNARDELLI Model Brescia S/S
BROWNING GTI Citori O/U Sporting Clay
   Model A-500G Gas-Operated, Model A-500R Semiauto, Model BSA 10 Semiauto, Model 42 Grades I, V & Ltd. Ed., Model 325
CHURCHILL Sporting Clays
CONNECTICUT VALLEY CLASSICS Sporter (Stainless only)
FERLIB Model F. VII Boxlock S/S
KBI Sabatti Grade I
KRIEGHOFF Pigeon Model
IGA Model ERA 2000
MAVERICK (Note: All Maverick shotguns now distributed by O.F. Mossberg)

Bull Pup Model 88, Model 60, Model 91, Security Arms Auto
MOSSBERG Model 500 Persuader/Cruiser
   Model 590 Intimidator
   Model 835 Regal Ulti-Mag (now Crown Grade & Camo)
   Model 9200 Regal & Turkey
PRECISION SPORTS "600" Series
   Bill Hanus Birdguns
REMINGTON Model 870 SPS Magnum
   Express Deer Gun, SP
WINCHESTER Model 12 Ducks Unlimited
   Model 42
   Model 1300 NWTF Turkey Gun (Wincam Series III & NWTF Series III)
   Model 1300 Ranger Deer Combo
   Model 1300 Whitetails Unlimited

BLACK POWDER

AMERICAN ARMS Model Hawkeye rifle
CVA Philadelphia Derringer
   Siber pistol, Stalker Rifle/Carbine
DIXIE Abilene Derringer
   Model DSB-58 Screw Barrel Derringer
MITCHELL ARMS 1851 Colt Navy ''Wild Bill Hickok'' revolver, 1860 Colt Army
   General Custer Remington New Model Army, Remington ''Texas''
NAVY ARMS Japanese Matchlock
PARKER-HALE Model 1853 3-Band Musket
   Model 1858 2-Band Musket
   Model 1861 Musketoon
   Whitworth Sniper rifle, Ltd. Edition
THOMPSON/CENTER High Plains Sporter, New Englander shotgun (Rynite stock & Left Hand models), Renegade Flintlock
SHILOH SHARP Model 1863 Sporting Rifle
   Model 1874 Civilian Carbine, Mil. Rifle

GUNFINDER

To help you find the model of your choice, the following list includes each gun found in the catalog section of SHOOTER'S BIBLE 1995. The **Caliberfinder** and a supplemental listing of **Discontinued Models** precede this section.

HANDGUNS

PISTOLS

Action Arms/Brno
Models CZ-75 Standard & Compact ... 98
Model CZ-83, Model CZ-85 ... 98

American Arms
Models CX-22, P-98 Classic, PK-22 ... 99

American Derringer
Cop & Mini-Cop Derringers ... 101
Double Action 38 Derringer ... 100
High Standard Double Derringer ... 100
Lady Derringer ... 100
Models 1, 2, 3, 7, 10, 11 ... 100
Models 4, 6 ... 101
Semmerling LM-4 ... 101
125th Ann. Double Derringer (1866–1991) ... 101

AMT
AMT 380 Backup II ... 102
Automag II, III & IV ... 103
45 ACP Longslide ... 102
1911 Government ... 102
On Duty Double Action ... 102

Anschutz
Exemplar, Exemplar XIV, Hornet ... 104

Astra. See European American Armory

Auto-Ordnance
Duo-Tone ... 105
Model 1911 ''The General'' ... 105
Model 1911A-1 Thompson ... 105
Model 1927 A-5 ... 105
Pit Bull ... 105

Baikal. See KBI

Benelli. See European American Armory

Beretta
Model 21 Bobcat ... 107
Model 84 Cheetah ... 107
Model 85F Cheetah ... 108
Model 86 Cheetah ... 108
Model 87 Cheetah ... 108
Model 89 Sport Gold Standard ... 108
Models 92D, 92F, 92FS ... 109
Model 96, 98 ... 109
Model 950 BS Jetfire ... 107
Model 8000 Cougar ... 106

Bernardelli
Model PO18 Target & Compact Target ... 110
Model PO10 Target ... 110

Bersa
Model 23 DA ... 111
Model 83 DA, Models 85 & 86 ... 111
Thunder 9 DA ... 111

Brno. See Action Arms

Browning
Buck Mark 22 Series, Micro Buck Mark ... 113
Model BDM 9mm ... 112
Model BDA-380 ... 113
9mm Hi-Power ... 112

Colt
Cadet Semiauto DA ... 118
Combat Commander MK Series 80 ... 115
Delta Elite & Delta Gold Cup ... 117

Colt (cont.)
Double Eagle MK Series 90 ... 115
Gold Cup National Match ... 116
Government Model Series 80 ... 116
Government Model 380 Auto/Pocketlite ... 116
Model M1991-A1 ... 115
Mustang 380 ... 117
Officer's 45 ACP ... 117

Coonan Arms
''Cadet'' Compact Model ... 120
Coonan 357 Magnum ... 120

Daewoo
Model DP51/DH40 ... 121
Model DH45 High Capacity DA,
   Model DP52 ... 121

Davis
Long-Bore D-Series ... 122
Model D-Series Derringers ... 122
Model P-32, Model P-380 ... 122

Emerging Technologies
Laseraim Series I, II, III ... 123

Erma
Model ESP 85A Competition, Golden Target ... 126

European American Armory
Astra Models A-70, A-75, A-100 ... 127
Benelli Model MP90S Target ... 130
European SA/DA Compacts ... 129
Witness, Witness Fab ... 128
Witness Subcompact ... 129
Witness Gold Team ... 128

Feather Guardian Angel ... 130

FEG Handguns. See KBI

Glock
Models 17 & 17L ... 132
Model 19 Compact ... 132
Models 20, 21 and 22 ... 132
Model 23 Compact, 24 Competition ... 132

Grendel
Model P-12 ... 133
Models P-30, P-30M/31 ... 133

Hämmerli
Model 160 Free ... 134
Model 162 Electronic ... 134
Model 208S Target ... 134
Model 212 Hunter's Pistol ... 134
Model 280 Target ... 133

Heckler & Koch
Model HK USP ... 135
Model P7K3 ... 136
Models P7M8 ... 135
Model P7M10 ... 136
P7M13 Self-Loading ... 135

Helwan The Brigadier ... 136

Heritage Mfg.
Model H25B/H25C ... 137

High Standard. See Amer. Derringer, Mitchell

Hi-Point Firearms J-S Series ... 138

Kahr Arms Model K9 ... 138

KBI
Baikal Model IJ-70 ... 140
FEG Models GKK-45/SMC-22 ... 140
FEG Models PJK-9HP & SMC 380 ... 139
Model PSP-25 ... 139

Laseraim. See Emerging Technologies

L.A.R.
Grizzly Win. Mag. ... 141
Grizzly 50 Mark 5 ... 141
Mark I & Mark 4 Grizzly ... 141

Llama
Compact Models ... 142
Large-Frame Auto ... 142
Small-Frames ... 142

Luger American Eagle P-08 ... 144

Magnum Research
Desert Eagle Magnums ... 145
Desert Eagle Mark I ... 145
Lone Eagle Single Shot ... 146
Mountain Eagle & Baby Eagle ... 146

McMillan Wolverine ... 147

Mitchell Arms
High Standard Olympic I.S.U. ... 148
High Standard Signature Series ... 147
High Standard Sharpshooter II,
   Trophy II, Victor II ... 148

M.O.A. Maximum ... 149

Navy Arms TT-Olympia ... 149

Para-Ordnance
Model P12 · 45 Compact ... 152

Remington
Model XP-100 Bolt Action Silhouette
   & Hunter Models ... 152
Model XP-100 Long-Range Customs ... 152

Ruger
Mark II Bull Barrel & 22/45 ... 160
Mark II Models ... 159
Model P-Series ... 158

Sig-Sauer
Model P220 ''American'' ... 161
Model P225 ... 161
Model P226 ... 161
Model P228 ... 162
Model P229 ... 162
Model P230 ... 162

Smith & Wesson
Model 41 ... 166
Model 411 ... 165
Model 422 ... 166
Model 915 ... 165
Model 2206 ... 166
Model 2213/2214 Sportsman ... 166
Model 3900 Compact Series ... 163
Model 4000 Series ... 164
Model 4000 Compact Series ... 163
Model 4006 ... 164
Model 4500 Series ... 164
Model 4500 Compact Series ... 163
Model 5900 Series ... 165
Model 6900 Compact Series ... 163

Springfield
High-Capacity Full-House Racegun ... 176
Model 1911-A1 Models ... 175
Model XM4 High-Capacity 1911-A1 ... 176